W9-AWA-059

The sun is just a star.

Nuclear reactions make energy.

Earth Moon

(not to scale)

The La...

First hominids

One-inch line

Goal line

Ten thousand years ago, on the 0.0026 inch line, humans begin building cities and modern civilization begins.

50 40 30 20 10

Formation of the sun and planets from a cloud of interstellar gas and dust

Life begins in Earth's oceans.

Cambrian explosion 540 million years ago: Life in Earth's oceans becomes complex.

Life first emerges onto the land.

Age of Dinosaurs

40 30 20 10

TODAY

Over billions of years, generation after generation of stars have lived and died, cooking the hydrogen and helium of the big bang into the atoms of which you are made. Study the last inch of the time line to see the rise of human ancestors and the origin of civilization. Only in the last flicker of a moment on the time line have astronomers begun to understand the story.

Flash Reference: H–R Diagram

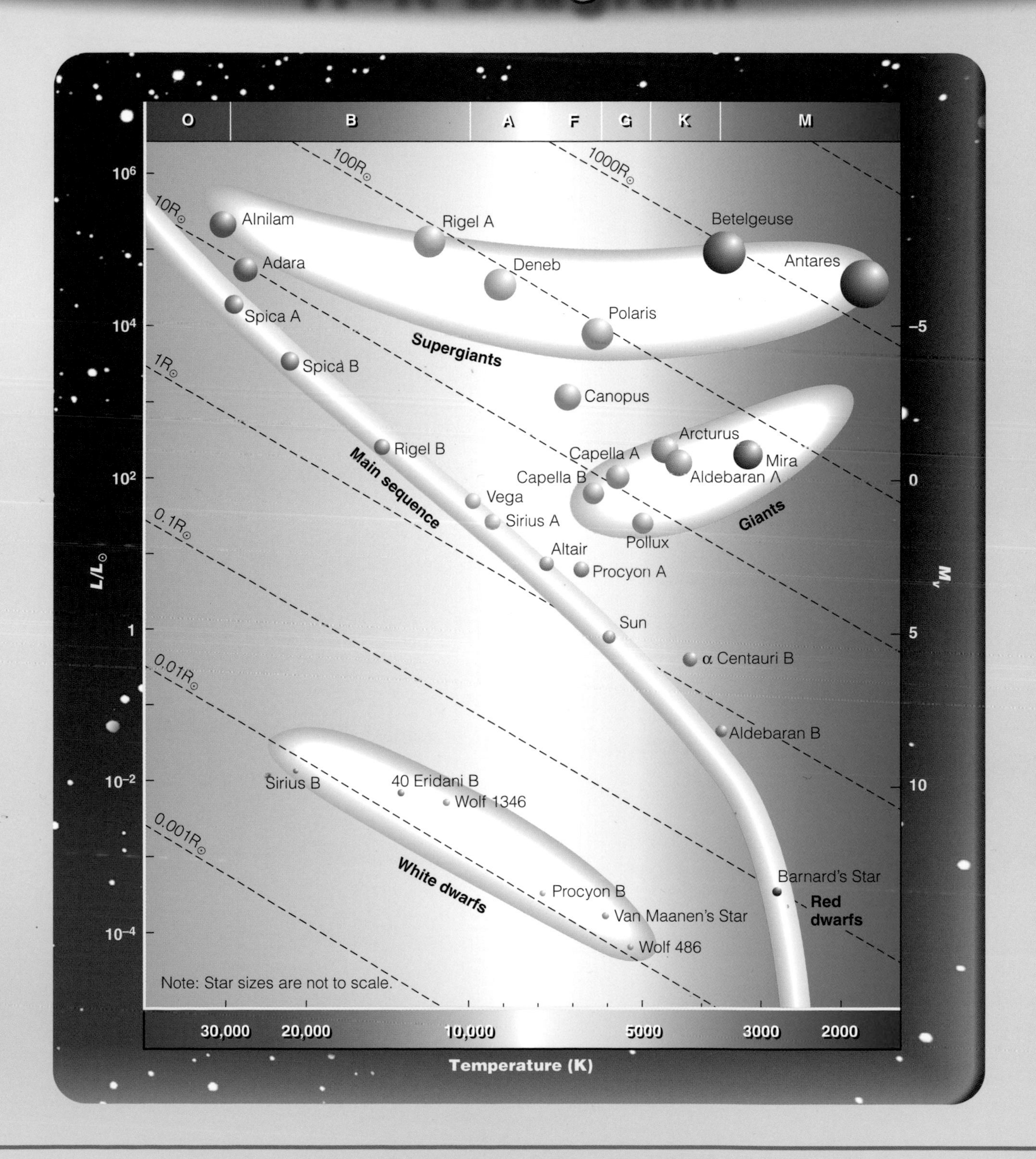

The H–R diagram is the key to understanding stars, their birth, their long lives, and their eventual deaths. Luminosity ($L/L_\odot$) refers to the total amount of energy that a star emits in terms of the sun's luminosity, and the temperature refers to the temperature of its surface. Together, the temperature and luminosity of a star locate it on the H–R diagram and tell astronomers its radius, its family relationships with other stars, and a great deal about its history and fate.

Flash Reference: Comparative Planetology

The terrestrial or Earthlike planets lie very close to the sun, and their orbits are hardly visible in a diagram that includes the outer planets.

Mercury, Venus, Earth and its moon, and Mars are small worlds made of rock and metal with little or no atmospheric gases.

The outer worlds of our solar system orbit far from the sun. Jupiter, Saturn, Uranus, and Neptune are Jovian or Jupiter-like planets much bigger than Earth. They contain large amounts of low-density gases.

Pluto is a tiny, icy world orbiting far from the sun, and some astronomers argue that it isn't really a planet at all.

The Terrestrial Worlds

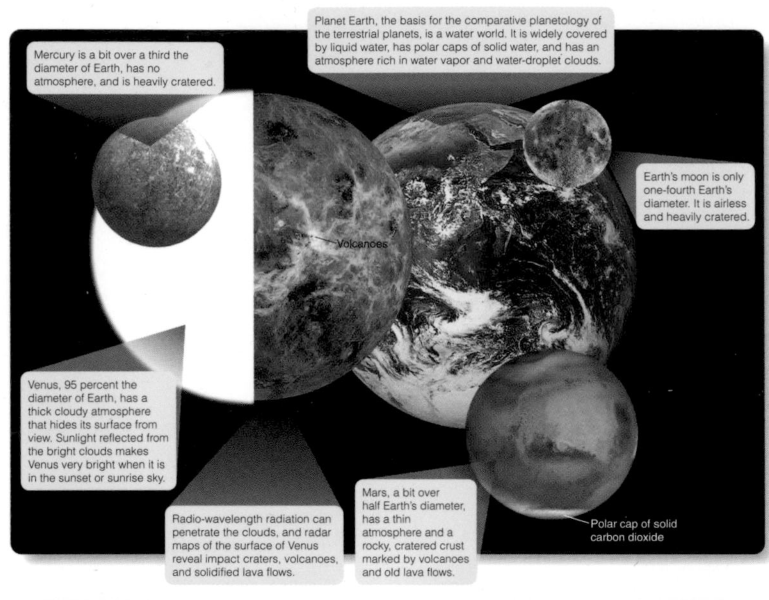
Planet Earth, the basis for the comparative planetology of the terrestrial planets, is a water world. It is widely covered by liquid water, has polar caps of solid water, and has an atmosphere rich in water vapor and water-droplet clouds.
Mercury is a bit over a third the diameter of Earth, has no atmosphere, and is heavily cratered.
Earth's moon is only one-fourth Earth's diameter. It is airless and heavily cratered.
Volcanoes
Venus, 95 percent the diameter of Earth, has a thick cloudy atmosphere that hides its surface from view. Sunlight reflected from the bright clouds makes Venus very bright when it is in the sunset or sunrise sky.
Radio-wavelength radiation can penetrate the clouds, and radar maps of the surface of Venus reveal impact craters, volcanoes, and solidified lava flows.
Mars, a bit over half Earth's diameter, has a thin atmosphere and a rocky, cratered crust marked by volcanoes and old lava flows.
Polar cap of solid carbon dioxide

Planetary Orbits

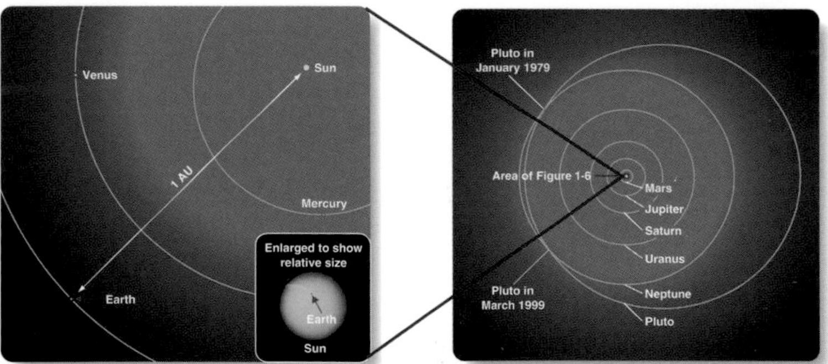
Venus
Sun
1 AU
Mercury
Earth
Enlarged to show relative size
Earth
Sun
Pluto in January 1979
Area of Figure 1-6
Mars
Jupiter
Saturn
Uranus
Pluto in March 1999
Neptune
Pluto

The Outer Worlds

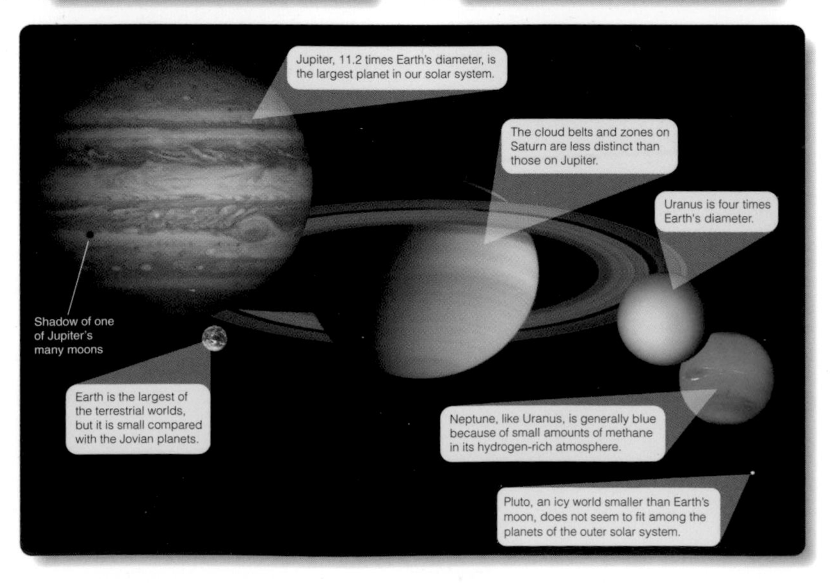
Jupiter, 11.2 times Earth's diameter, is the largest planet in our solar system.
The cloud belts and zones on Saturn are less distinct than those on Jupiter.
Uranus is four times Earth's diameter.
Shadow of one of Jupiter's many moons
Earth is the largest of the terrestrial worlds, but it is small compared with the Jovian planets.
Neptune, like Uranus, is generally blue because of small amounts of methane in its hydrogen-rich atmosphere.
Pluto, an icy world smaller than Earth's moon, does not seem to fit among the planets of the outer solar system.

• See page 478 for the terrestrial planets. See pages 6 and 7 for the two orbital diagrams. See page 555 for the outer worlds.

Flash Reference: Arrows

This book is designed to use arrows to alert you to important concepts in diagrams and graphs. Some arrows point things out, but others represent motion, force, or even the flow of light. Look at arrows in the book carefully and use this Flash Reference card to catch all of the arrow clues.

Point at things:

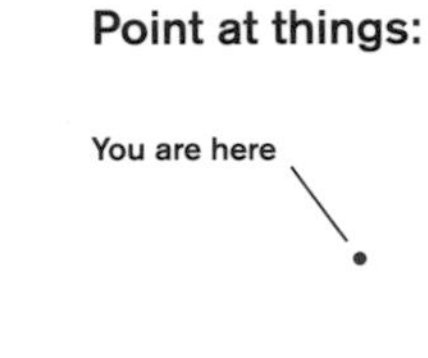
You are here

Force:

Process flow:

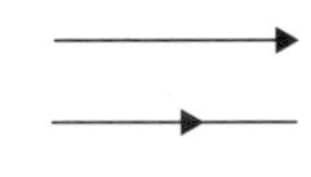

Measurement:

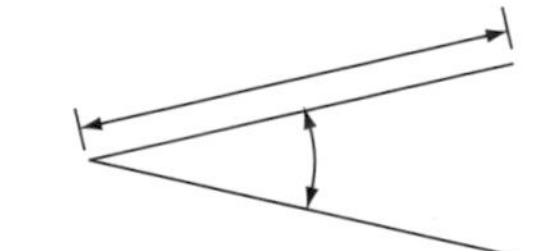

Direction:

Radio waves, infrared, photons:

Motion:

Rotation 2-D

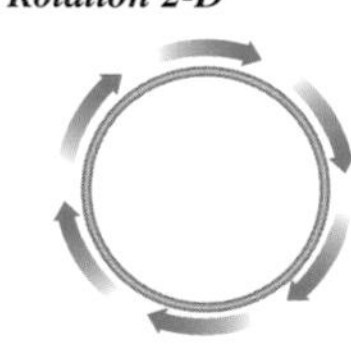

Rotation 3-D

Linear

Light flow:
Updated arrow style

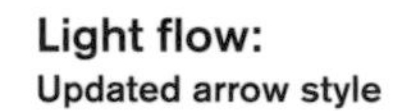

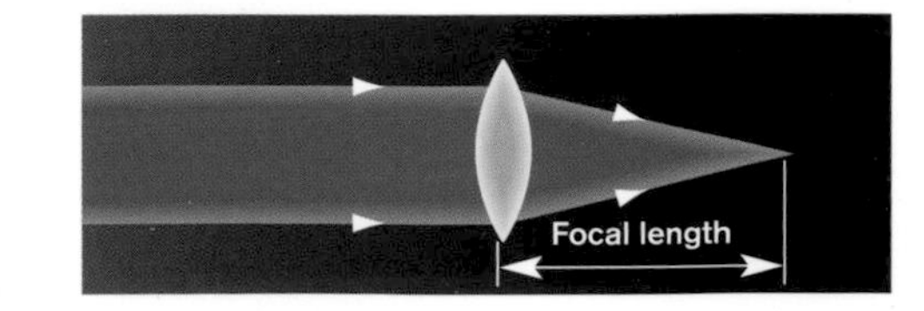
Focal length

Foundations of Astronomy

About the Author

Mike Seeds is Professor of Astronomy at Franklin and Marshall College, where he has taught astronomy since 1970. His research interests have focused on peculiar variable stars and the automation of astronomical telescopes. He has been serving since 1987 as Principal Astronomer in charge of the Phoenix 10, the first fully robotic telescope, located in southern Arizona. In 1989, he received the Christian R. and Mary F. Lindback Award for Distinguished Teaching. In addition to teaching, writing, and research, Mike has published educational tools for use in computer-smart classrooms. His interest in the history of astronomy led him to offer upper-level courses "Archaeoastronomy" and "Changing Concepts of the Universe." He has also published educational software for preliterate toddlers. Mike is Senior Consultant in the creation of the telecourse *Astronomy: Observations and Theories.*

He is the author of *Astronomy: The Solar System and Beyond,* Fourth Edition (2005), and *Horizons: Exploring the Universe,* Ninth Edition (2005), published by Brooks/Cole.

About the Cover

Located atop a 13,800-foot-high volcano in Hawaii, the Gemini North telescope is 8.1 m (26.6 feet) in diameter. Here it is shown in front of the Milky Way in Sagittarius. The center of our galaxy is visible as glowing clouds of billions of stars partially obscured by clouds of dust that lie in the plane of our disk-shaped galaxy. (Photo: Kris Koenig, 2005)

9 NINTH EDITION

Foundations of Astronomy

Michael A. Seeds

Joseph R. Grundy Observatory
Franklin and Marshall College

Australia • Canada • Mexico • Singapore • Spain
United Kingdom • United States

For Emery and Helen Seeds

THOMSON
BROOKS/COLE

Foundations of Astronomy, **Ninth Edition**
Michael A. Seeds

Astronomy Acquisitions Editor: *Chris Hall*
Development Editor: *Alyssa White*
Assistant Editor: *Brandi Kirksey*
Editorial Assistant: *Jessica Jacobs*
Editorial Intern: *David Marks*
Technology Project Manager: *Samuel Subity*
Publisher/Executive Editor: *David Harris*
Marketing Manager: *Mark Santee*
Marketing Assistant: *Michele Colella*
Managing Marketing Communications Manager: *Bryan Vann*
Project Manager, Editorial Production: *Teri Hyde*
Executive Art Director: *Vernon Boes*
Print and Media Buyer: *Karen Hunt*
Permissions Editor: *Stephanie Lee, Bob Kauser*
Production Service: *Heckman & Pinette*
Text Designer: *Linda Beaupré/Stone House Art*
Art Editor: *Lisa Torri*
Photo Researcher: *Mike Seeds and Kathleen Olson*
Copy Editor: *Margaret Pinette*
Illustrator: *Precision Graphics*
Cover Designer: *Irene Morris*
Cover Image: *Milky Way Over Gemini, North Observatory: Kris Koenig; Left: Helix Nebula, Hubble Images: NASA, ESA, C. R. O'Dell (Vanderbilt University), M. Melxner and P. I. McCullough (STScI); Middle: Large Magellanic Cloud N11, Hubble Images: NASA, C. Aguilera, S. Points, and C. Smith (CTIO), and Z. Levay (STScI); Right: Supernova Remnant Menagerie: NASA, ESA, HEIC, and The Hubble Heritage Team (STScI/AURA) with acknowledgment to Y.-H. Chu and R. M. Williams.*
Cover Printer: *Transcontinental-Interglobe*
Compositor: *Thompson Type*
Printer: *Transcontinental-Interglobe*

© 2007 Thomson Brooks/Cole, a part of The Thomson Corporation. Thomson, the Star logo, and Brooks/Cole are trademarks used herein under license.

ALL RIGHTS RESERVED. No part of this work covered by the copyright hereon may be reproduced or used in any form or by any means—graphic, electronic, or mechanical, including photocopying, recording, taping, Web distribution, information storage and retrieval systems, or in any other manner—without the written permission of the publisher.

Printed in Canada
1 2 3 4 5 6 7 10 09 08 07 06

Library of Congress Control Number: 2005936361

ISBN 0-495-01578-4

Thomson Higher Education
10 Davis Drive
Belmont, CA 94002-3098
USA

For more information about our products, contact us at:
Thomson Learning Academic Resource Center
1-800-423-0563

For permission to use material from this text or product, submit a request online at **http://www.thomsonrights.com.**
Any additional questions about permissions can be submitted by e-mail to **thomsonrights@thomson.com.**

Brief Contents

Part 1: Exploring the Sky

Part 2: The Stars

Part 3: The Universe

Part 4: The Solar System

Part 5: Life

Contents

Part 1: Exploring the Sky

Part 2: The Stars

Windows on Science

Concept Art Portfolios

Part 3: The Universe

Windows on Science

Concept Art Portfolios

Part 4: The Solar System

Windows on Science

Concept Art Portfolios

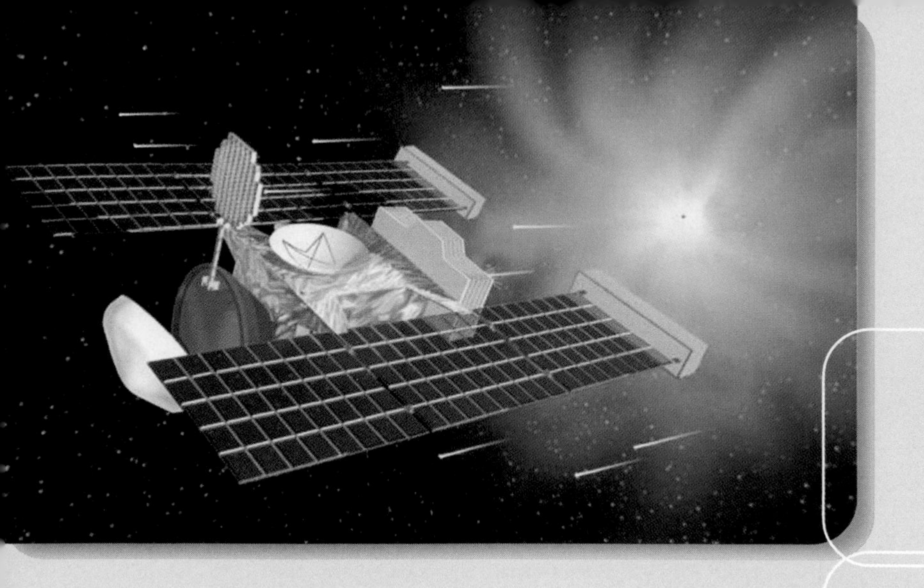

Part 5: Life

Windows on Science

Concept Art Portfolios

A Note to the Student

From Mike Seeds

Hi,
I'm really glad you are taking an astronomy course. You are going to see some amazing things, from the icy rings of Saturn to monster black holes. Our universe is so beautiful, it is sad to think that not everyone gets to take an astronomy course.

Two Goals

You will learn a lot of new stuff in this course, but there are two things I hope you find especially satisfying. This astronomy course will help you answer two important questions:

- What are we?
- How do we know?

By "What are we?" I mean, "Where do we fit into the history of the universe?" The atoms you are made of had their first birthday in the big bang when the universe began, but those atoms have been cooked and remade inside stars, and now they are inside you. Where will they be in a billion years? Astronomy is the only course on campus than can tell you that story, and it is a story that everyone should know.

By "How do we know?" I mean, "How does science work?" How can anyone know there was a big bang? In today's world, you need to think carefully about the things so-called experts say. You should demand explanations. Scientists have a special way of knowing based on evidence. Scientific knowledge isn't just opinion or policy or marketing or public relations. It is humanity's best understanding of nature. To understand the world around you, you should understand how science works.

These two questions are the focus of astronomy, and they are just for you. You need to know the answers to these questions so you can appreciate how wonderful the universe is and how special you are.

Favorite Stars

Along the way, I want to introduce you to eight of my favorite stars. Maybe you will add a few favorites of your own, but these are eight stars I enjoy looking at in the sky. When I see Betelgeuse in the winter sky, for example, I see more than just a bright red star. In my imagination, I see a star over 800 times larger than the sun—an aging star that will soon die, perhaps in a violent explosion. Here's my list of favorite stars:

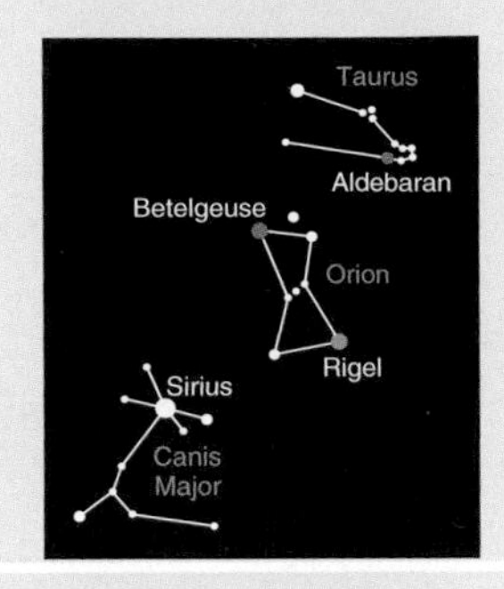

Sirius	**Brightest star in the sky**	**Winter**
Betelgeuse	**Bright red star in Orion**	**Winter**
Rigel	**Bright blue star in Orion**	**Winter**
Aldebaran	**Red eye of Taurus the Bull**	**Winter**
Polaris	**The North Star**	**Year round**
Vega	**Bright star overhead**	**Summer**
Spica	**Bright southern star**	**Summer**
Alpha Centauri	**Nearest star to the sun**	**Spring, far south**

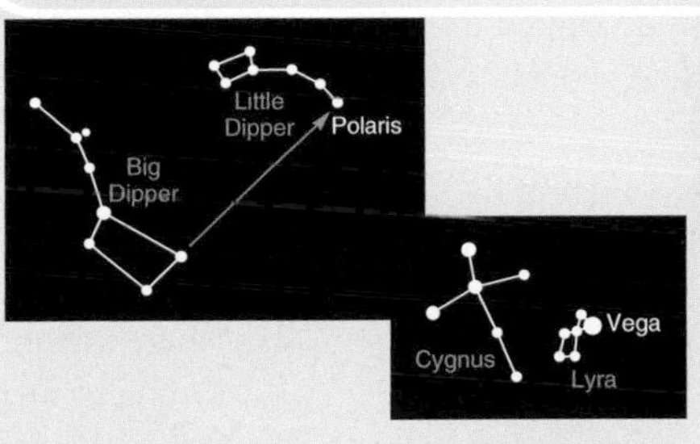

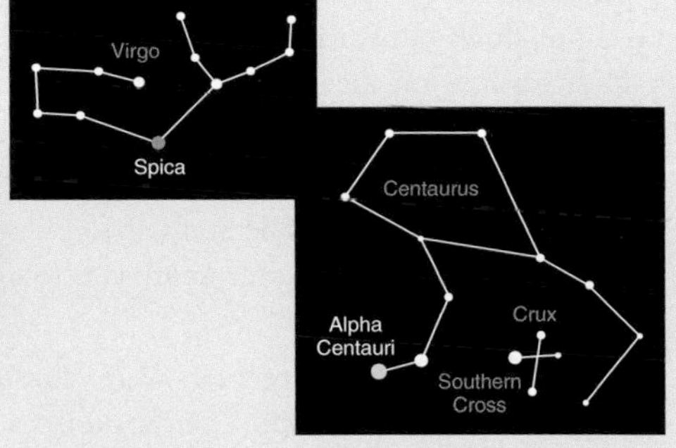

You can see Polaris year round, but Sirius, Betelgeuse, Rigel, and Aldebaran are in the winter sky. Spica is in the summer sky, and Vega is visible evenings in the late summer or fall. Alpha Centauri is a special star, but you will have to travel at least as far south as southern Florida to get a glimpse of it above the southern horizon. Use the charts here along with the star charts in the back of this book to locate these stars.

Every one of these stars has its own secret, and when you find out why they are my favorites, you'll enjoy finding them in the sky just as I do.

Expect to Be Astonished

One reason astronomy is exciting is that astronomers discover new things every day. Astronomers expect to be astonished. You can share in the excitement because I've worked hard to include the newest images, the newest discoveries, and the newest insights that will take you, in an introductory course, to the frontier of human knowledge. You'll see new evidence of ancient oceans and lakes on Mars and a vast ocean on Jupiter's moon Europa. You'll visit the moon Titan, where it rains liquid methane, and Pluto, where most gases freeze solid. You'll see stars die, and you'll share the struggle to understand new evidence that the expansion of the universe is speeding up. Huge telescopes in space and on remote mountaintops provide a daily dose of excitement that goes far beyond sensationalism. These new discoveries in astronomy are exciting because they are about us. They tell us more and more about what we are.

As you read this book, notice that it is not organized as lists of facts for you to memorize. That could make even astronomy boring. Rather this book is organized to show you how scientists use evidence and theory to create logical arguments that show how nature works. Look at the list of special features that follows this note. Those features were carefully designed to help you understand astronomy as evidence and theory. Once you see science as logical arguments, you hold the key to the universe.

Do Not Be Humble

As a teacher, my quest is simple. I want you to understand your place in the universe—not just your location in space but your location in the unfolding history of the physical universe. Not only do I want you to know where you are and what you are in the universe, but I want you to understand how scientists know. By the end of this book, I want you to know that the universe is very big, but that it is described by a small set of rules and that we humans have found a way to figure out the rules—a method called *science.*

Do not be humble. Astronomy tells us that the universe is vast and powerful, but it also tells us that we are astonishing creatures. We humans are the parts of the universe that think. You are a small creature, but remember that it is your human brain that is capable of understanding the depth and beauty of the cosmos.

To appreciate your role in this beautiful universe, you must learn more than just the facts of astronomy. You must understand what we are and how we know. Every page of this book reflects that ideal.

Mike Seeds
mike.seeds@fandm.com

Key Content and Pedagogical Changes to the Ninth Edition

- Chapter 8, "The Sun," has been thoroughly rewritten and reorganized. The new presentation is designed to help you see the sun as a physical object in which energy is generated in its core, flows outward through the interior, and generates the activity we see on the sun's surface.
- The Concept Art Portfolios were redesigned to aid you in navigating through these key visual summaries. The flow through the material has been clarified by adding numerical markers and arrows to better guide you to a stronger conceptual understanding.
- The Guideposts and Summaries were redesigned to make them a more effective study tool. The chapter-opening Guideposts contain a short set of Essential Questions focused on the key ideas in each chapter. The questions are answered in the narrative and then are presented as bulleted insights in the Summary.
- All figures have been updated, redesigned, and carefully annotated to help you pick out the critical details related to the discussions. In particular, the H–R diagrams have been fully revitalized to help you better understand the nature of stars.
- A new type of question has been added to the end-of-chapter material. "Learning to Look" questions give you a chance to critically examine images to test your understanding of fundamental concepts.
- Of course, I have updated the newest exciting discoveries in astronomy, including images from the surface of Titan, gamma ray bursts, volcanism on icy moons, the Kuiper belt and the status of Pluto, dark energy, the search for water on Mars, supernova explosions, extrasolar planets, and much more.

Special Features

- **Special two-page Concept Art Portfolios** provide an opportunity for you to create your own understanding and share in the satisfaction that scientists feel as they uncover the secrets of nature.
- **Guided discovery figures** illustrate important ideas visually and guide you to understand relationships and contrasts interactively.
- **Windows on Science** provide a parallel commentary on how all of science works. For example, the Windows point out where astronomers use statistical evidence, where they are reasoning by analogy, and where they are building a scientific model rather than a scientific hypothesis.
- **Focus on Fundamentals** will help you understand five concepts from physics that are critical to understanding modern astronomy.
- **Guideposts** on the opening page of each chapter help you see the organization of the book. Cast as a series of Essential Questions that are answered within the text narrative, each Guidepost further connects the chapter with preceding and following chapters to provide an overall organizational guide.
- **Building Scientific Arguments** at the end of each text section are carefully designed questions to help you review and synthesize concepts from the section. A short answer follows to show how scientists construct scientific arguments from observations, evidence, theories, and natural laws that lead to a conclusion. A further question then gives you a chance to construct your own argument on a related issue.
- **End-of-Chapter Review Questions** are designed to help you review and test your understanding of the material.

- **End-of-Chapter Discussion Questions** go beyond the text and invite you to think critically and creatively about scientific questions. You can think about these questions yourself or discuss them in class.
- **Media Clusters** list all of the Web resources you can call on to illustrate, review, and expand the material in each chapter.
- **Critical Inquiries for the Web** challenge you to use the World Wide Web to explore further and to think creatively and analytically about astronomy.
- **Exploring *TheSky*** are experiments you can perform with the software *TheSky* or, by making slight modifications, with any planetarium software. These experiments will allow you to see the sky and celestial bodies in new ways on your own computer screen.

As additional aids, notice that most versions of this book also include access to the following electronic enhancements:

- **Virtual Astronomy Laboratories.** This set of online lab exercises covers 20 of the most important concepts of introductory astronomy in an interactive online environment. You may be asked to run simulations, observe animations, perform calculations, or collect data to help you master these fundamental concepts. The labs cover topics from helioseismology to dark matter and allow you to submit your results electronically to your instructor or print them out to hand in. The Media Cluster at the end of each chapter in this textbook notes which labs correlate to that chapter.
- **AceAstronomy.** Take charge of your learning with the first assessment-centered student learning tool for astronomy. AceAstronomy uses a series of diagnostic tests to provide you with a personalized learning plan, so you can begin to maximize your study time with a host of interactive tutorials and quizzes that help you focus on what you need to learn to master astronomy.

Acknowledgments

I started writing astronomy textbooks in 1973, and over the years I have had the guidance of a great many people who care about astronomy and teaching.

I would like to thank all of the students and teachers who have responded so enthusiastically to *Foundations of Astronomy.* Their comments and suggestions have been very helpful in shaping this book.

I would especially like to thank the reviewers whose careful analysis and thoughtful suggestions have been invaluable in completing this new edition.

Many observatories, research institutes, laboratories, and individual astronomers have supplied figures and diagrams for this edition. Their names are given on the page with the illustrations they provided, and I would like to thank them specifically for their generosity.

Writing about every branch of astronomy is a daunting task, and I could not do it without helpful contributions from experts in various fields. I especially want to think Dana Backman, George Jacoby, Victoria Kaspi, and William Keel for their helpful guidance on technical issues.

Certain unique diagrams in Chapters 11, 12, 13, and 14 are based on figures I designed for my article "Stellar Evolution," which appeared in *Astronomy,* February 1979.

I am happy to acknowledge the use of images and data from a number of important programs. In preparing materials for this book I used NASA's Sky View facility located at NASA Goddard Space Flight Center. I have used atlas images and mosaics obtained as part of the Two Micron All Sky Survey (2MASS), a joint project of the University of Massachusetts and the Infrared Processing and Analysis Center/California Institute of Technology, funded by the National Aeronautics and Space Administration and the National Science Foundation. A number of solar images are used by the courtesy of the SOHO consortium, a project of international cooperation between ESA and NASA.

It is always a pleasure to work with the Thomson Learning team at Brooks/Cole and Wadsworth. Special thanks go to all of the people who have contributed to this project, including Teri Hyde, Kathleen Olson, Sam Subity, Lisa Torri, Vernon Boes, Mark Santee, and Brandi Kirksey. I have enjoyed working on production with Margaret Pinette and Bill Heckman of Heckman & Pinette, and I appreciate their understanding and goodwill. I would especially like to thank my editor Chris Hall and my development editor Alyssa White for their help and guidance on this project.

Most of all, I would like to thank my wife, Janet, and my daughter, Kate, for putting up with "the books." They know all too well that textbooks are made of time.

Mike Seeds

Manuscript Reviewers

Steve Desch, Arizona State University
Michael Frey, California State University, Long Beach
Marc Gagné, West Chester University
Paul Hintzen, California State University, Long Beach
Don McCarthy, University of Arizona
Scott Miller, Penn State University
Sharon L. Montgomery, Clarion University of Pennsylvania
Richard Scarborough, University of Nebraska, Lincoln

1 The Scale of the Cosmos

The longest journey begins with a single step.

CONFUCIUS

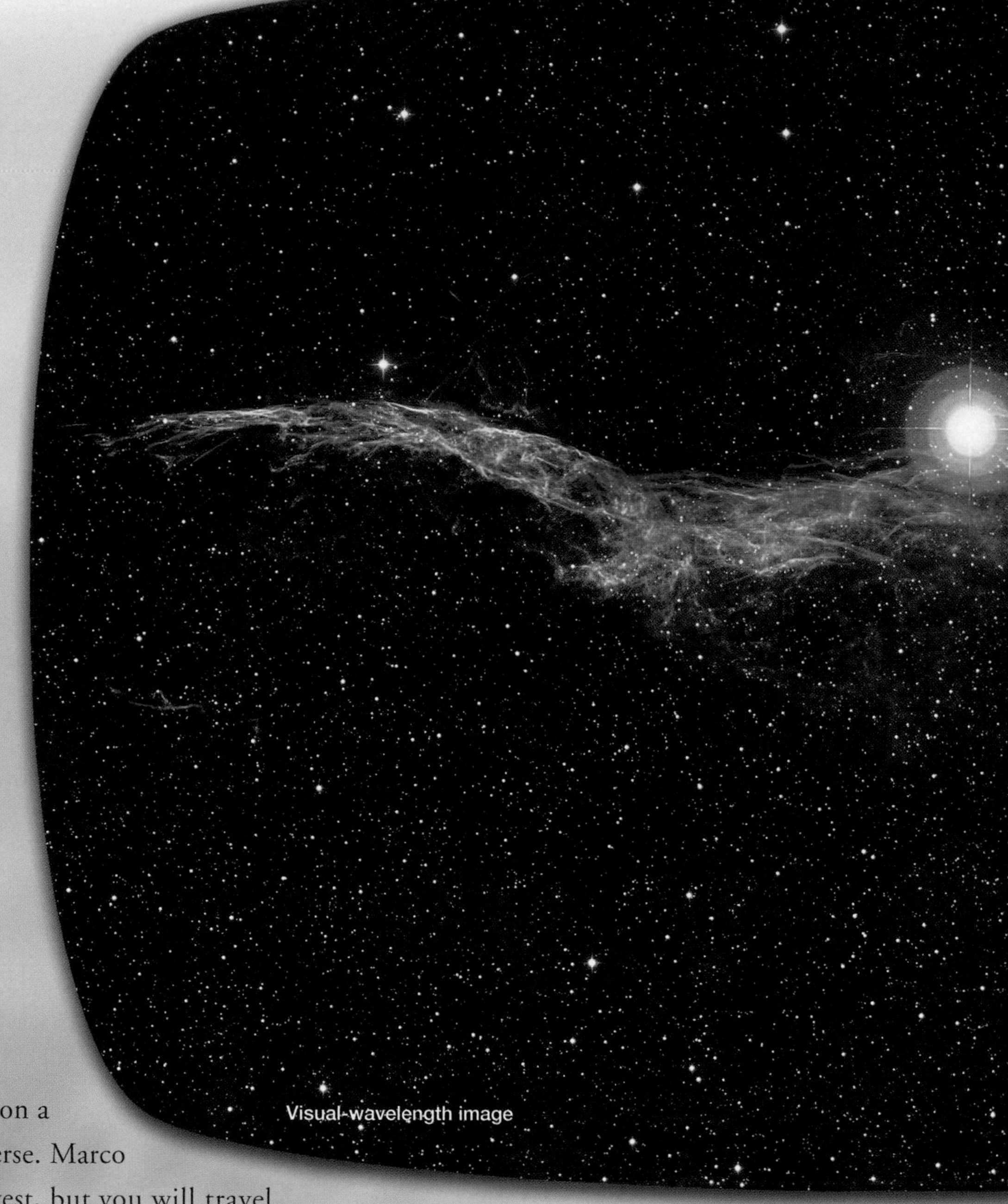
Visual-wavelength image

YOU ARE ABOUT to embark on a voyage out to the end of the universe. Marco Polo journeyed east, Columbus west, but you will travel outward away from your home on Earth, out past the moon, sun, and other planets, past the stars you see in the evening sky, and past billions more that can be seen only with the aid of the largest telescopes. You will journey through great whirlpools of stars to the most distant galaxies visible from Earth—and then you will continue on, carried only by experience and imagination looking for the structure of the universe itself. Imagination is your key to discovery; it will be your scientific time machine transporting you into the past and into the future. Go back to watch the birth of the universe, the formation of the first stars, and the origin of the sun and Earth. Then rush

Continued on page 4

So distant that light has taken 2,500 years to reach Earth, the Veil Nebula was produced by the explosion of a star 15,000 years ago. It now drifts in front of a more distant star not related to the nebula.
(T. Rector, University of Alaska, and WIYN/NOAO/AURA/NSF)

Guidepost

Getting Started

You are already an expert on astronomy. You have enjoyed sunsets and moonrises. You have admired the stars and may know a few constellations. You have probably read about Mars and eclipses, and you may have read about distant galaxies. That is more than most Earthlings know about astronomy. But you have a right to know more.

You are a planetwalker, and you should understand what it means to live on a planet that whirls around a star drifting through a universe of stars and galaxies. You owe it to yourself to know where you are. That is the first step to knowing what you are.

This Chapter

Here you will take a quick trip, a cosmic zoom, from objects you know up to the largest things in the universe. That quick survey will answer four essential questions:

Where is Earth in relation to the sun, the planets, and the galaxies?

How do astronomers talk about these huge astronomical distances?

Which objects are big, and which are small?

Are there other worlds like Earth?

This chapter will give you a sense of the scale of the universe that will carry you to new discoveries.

Looking Ahead

It is easy to learn a few facts, but it is the relationships among facts that are important. This chapter will give you the sense of scale that you need to understand where you are in the universe. The remaining chapters of this book will fill in the details, cite evidence and theories, and illustrate the wonderful intricacy and beauty of the universe. It is an exciting journey, and it begins here.

AceAstronomy™ The AceAstronomy icon throughout the text indicates an opportunity for you to test yourself on key concepts and to explore animations and interactions on the AceAstronomy website at: http://ace.brookscole.com/sf9

into the future to see what will happen when the sun dies and Earth withers.

Although you will discover a beginning to the universe, you will not find an edge in space. No matter how far you voyage, you will not run into a wall or limit beyond which you cannot go. Rather you will discover evidence that our universe may be infinite, that it may extend in all directions without limit. Such vastness may dwarf our earthly dimensions but not our human curiosity and imagination.

Astronomy is more than the study of stars and planets. It is the study of the universe in which we humans exist. You and I live on a small planet circling a small sun drifting through the universe, but astronomy can take us beyond these boundaries and help us not only see where we are in the universe but understand what we are. You have a right to know these things. Perhaps you have a duty to know them.

Do not be humble. Although astronomical sizes and distance may dwarf you, remember that you are an intelligent creature, and you are capable of understanding your universe. It is, after all, yours.

Astronomy will introduce you to sizes, distances, and times far beyond your usual experience on Earth. Your task in this chapter is to grasp the meaning of these unfamiliar sizes, distances, and times. Believe it or not, the solution lies in a single word, *scale.* In this chapter, you will compare objects of different sizes to grasp the scale of the universe.

Let's begin with something familiar. ■ Figure 1-1 shows a region about 52 feet across occupied by a human being, a sidewalk, and a few trees—all objects whose size you can understand. Each successive picture in this chapter will show you a region of the universe that is 100 times wider than the preceding picture. That is, each step will widen your field of view by a factor of 100.

In ■ Figure 1-2 your field of view widens by a factor of 100, and you can see an area 1 mile in diameter. The arrow points to the scene shown in the preceding photograph. People, trees, and sidewalks have vanished, but now you can see a college campus and the surrounding streets and houses. The dimensions of houses and streets are familiar. This is the world you know, and you can relate such objects to the scale of your body.

You started your adventure using feet and miles, but you should use the metric system of units. Not only is it used by all scientists around the world, but it makes calculations much easier. If you are not already familiar with the metric system, or if you need a review, study Appendix A before reading on.

■ Figure 1-1

(Michael A. Seeds)

■ Figure 1-2

(USGS)

■ **Figure 1-3**
(NASA infrared photograph)

range of "colors," from X rays to radio waves, to reveal sights invisible to unaided human eyes.

At the next step in your journey, you will see our entire planet (■ Figure 1-4), which is 12,756 km in diameter. Earth rotates on its axis once a day, exposing half of its surface to daylight at any particular moment. The photo shows most of the daylight side of the planet. The blurriness at the extreme right is the sunset line. The rotation of Earth carries you eastward, and as you cross the sunset line into darkness, you say the sun has set. It is the rotation of the planet that causes the cycle of day and night.

Earth's interior is made mostly of iron and nickel, and its crust is mostly silicate rocks. Only a thin layer of water makes up

■ **Figure 1-4**
(NASA)

The photo in Figure 1-2 is 1 mile in diameter. A mile equals 1.609 kilometers, so you can see in the photo that a kilometer is a bit over two-thirds of a mile—a short walk across a neighborhood.

The view in ■ Figure 1-3 spans 160 kilometers. In this infrared photo, the green foliage shows up as various shades of red. The college campus is now invisible, and the patches of gray are small cities, with the suburbs of Philadelphia visible at the lower right. At this scale, you see the natural features of Earth's surface. The Allegheny Mountains of southern Pennsylvania cross the image in the upper left, and the Susquehanna River flows southeast into Chesapeake Bay. What look like white bumps are a few puffs of clouds.

These features are a reminder that you live on the surface of a changing planet. Forces in Earth's crust pushed the mountain ranges up into parallel folds, like a rug wrinkled on a polished floor. The clouds tell you that Earth's atmosphere is rich in water, which falls as rain and erodes the mountains, washing material down the rivers and into the sea. Mountains and valleys are only temporary features on Earth; they are constantly changing. As you explore the universe, you will come to see that it, like Earth's surface, is always evolving.

Take a closer look at Figure 1-3 and notice the red color. This is an infrared photograph in which healthy green leaves and crops show up as red. Human eyes are sensitive to only a narrow range of colors. As you explore the universe, you will learn to use a wide

the oceans, and the atmosphere is only a few hundred kilometers deep. On the scale of this photograph, the depth of the atmosphere on which life depends is less than the thickness of a piece of thread.

Enlarge your field of view by a factor of 100, and you will see a region 1,600,000 km wide (■ Figure 1-5). Earth is the small blue dot in the center, and the moon, whose diameter is only one-fourth that of Earth, is an even smaller dot along its orbit 380,000 km from Earth.

These numbers are so large that it is inconvenient to write them out. Astronomy is the science of big numbers, and you will use numbers much larger than these to discuss the universe. Rather than writing out these numbers as in the previous paragraph, it is convenient to write them in **scientific notation.** This is nothing more than a simple way to write numbers without writing lots of zeros. In scientific notation, you would write 380,000 as 3.8×10^5. If you are not familiar with scientific notation, read the section on powers of 10 notation in the Appendix. The universe is too big to discuss without using scientific notation.

When you once again enlarge your field of view by a factor of 100 (■ Figure 1-6), Earth, the moon, and the moon's orbit all lie in the small red box at lower left. But now you can see the sun and two other planets that are part of our solar system. Our **solar system** consists of the sun, its family of planets, and some smaller bodies such as moons and comets.

Like Earth, Venus and Mercury are **planets,** small, nonluminous bodies that shine by reflected light. Venus is about the size of Earth, and Mercury is a bit larger than Earth's moon. On this diagram, they are both too small to be seen as anything but tiny dots. The sun is a **star,** a self-luminous ball of hot gas that generates its own energy. The sun is 109 times larger in diameter than Earth (inset), but it too is nothing more than a dot in this diagram.

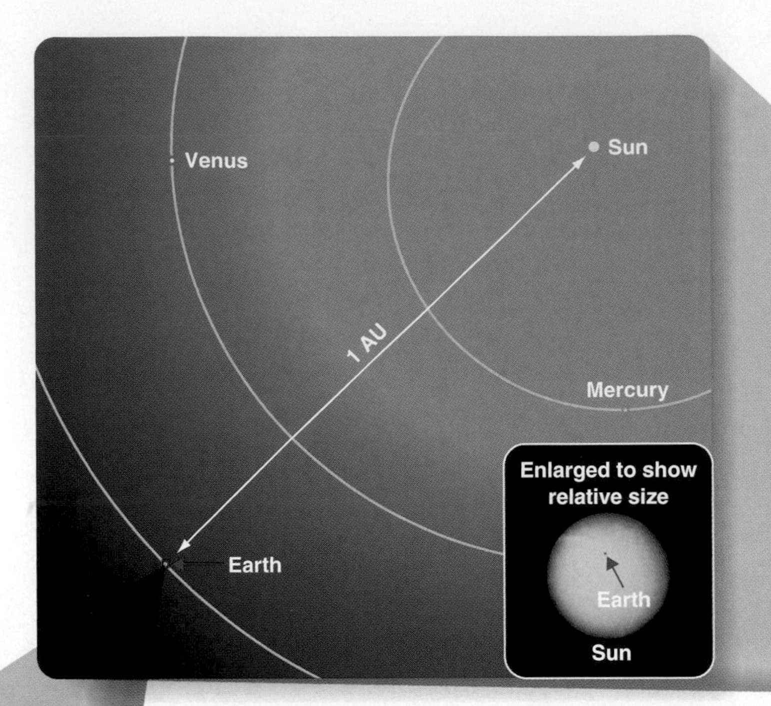

■ **Figure 1-6**
(NOAO)

■ **Figure 1-5**
(NASA)

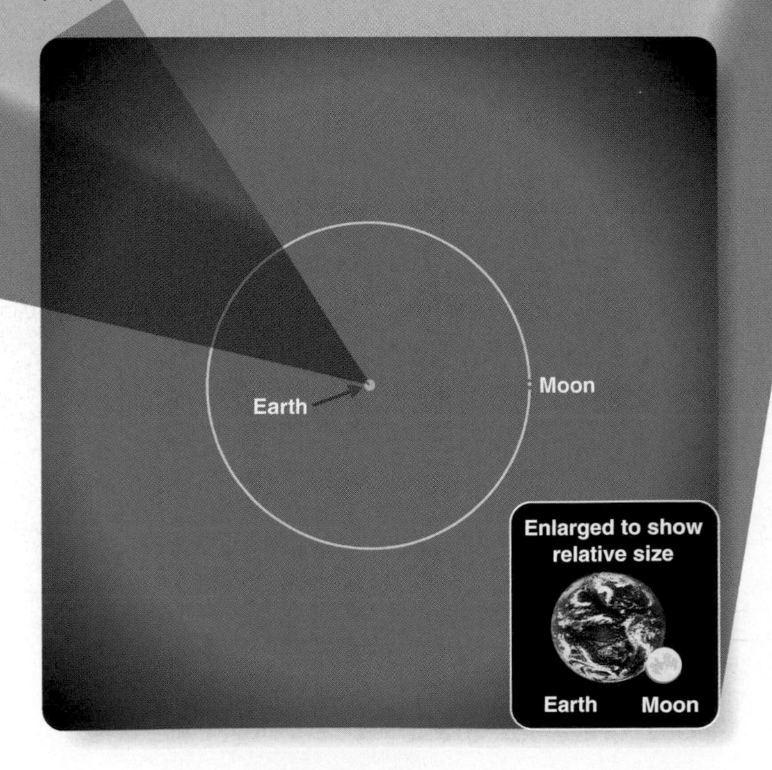

This diagram has a diameter of 1.6×10^8 km. One way astronomers deal with large numbers is to define new units. The average distance from Earth to the sun is a unit of distance called the **astronomical unit (AU),** a distance of 1.5×10^{11} m. Using this unit, you can say that the average distance from Venus to the sun is about 0.7 AU. The average distance from Mercury to the sun is about 0.39 AU.

The orbits of the planets are not perfect circles, and this is particularly apparent for Mercury. Its orbit carries it as close to the sun as 0.307 AU and as far away as 0.467 AU. You can see this variation in the distance from Mercury to the sun in Figure 1-6. Earth's orbit is more circular, and its distance from the sun varies by only a few percent.

Your first field of view was only 52 feet (about 16 m) in width. After only six steps of enlarging by a factor of 100, you can now see the entire solar system (■ Figure 1-7). Your field of view is 1 trillion (10^{12}) times wider than in your first view. The details of the preceding figure are now lost in the red square at the center of this diagram. You see only the brighter, more widely separated objects as you enlarge your view. The sun, Mercury, Venus, and Earth lie so close together that you cannot separate them at this scale. Mars, the next outward planet, lies only 1.5 AU from the sun. In contrast, Jupiter, Saturn, Uranus, Neptune, and Pluto are so far from the sun that they are easy to place in this diagram. These are cold worlds far from the sun's warmth. Light from the sun reaches Earth in only 8 minutes, but it takes over 4 hours to reach Neptune. Notice that Pluto's orbit is so elliptical that Pluto can come closer to the sun than Neptune does, as Pluto did between 1979 and 1999.

When you again enlarge your field of view by a factor of 100, the solar system vanishes (■ Figure 1-8). The sun is only a point of light, and all the planets and their orbits are now crowded into the small red square at the center. The planets are too small and reflect too little light to be visible so near the brilliance of the sun.

Nor are any stars visible except for the sun. The sun is a fairly typical star, and it seems to be located in a fairly average neighborhood in the universe. Although there are many billions of stars like the sun, none is close enough to be visible in this diagram, which shows an area only 11,000 AU in diameter. The stars are typically separated by distances about 10 times larger than the diameter of this diagram. You will see stars in the next field of view, but, except for the sun at the center, this diagram is empty.

It is difficult to grasp the isolation of the stars. If the sun were represented by a golf ball in New York City, the nearest star would be another golf ball in Chicago. Except for the widely scattered stars and a few atoms of gas drifting between the stars, the universe is nearly empty.

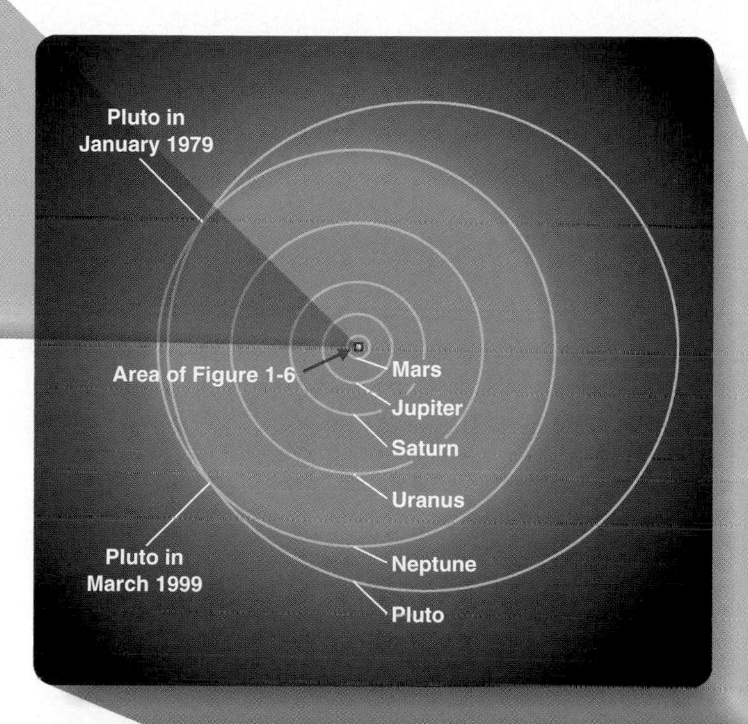

■ Figure 1-7

In ■ Figure 1-9, your field of view has expanded to a diameter a bit over 1 million AU. The sun is at the center, and you can see a few of the nearest stars. These stars are so distant that it is not reasonable to give their distances in astronomical units. To express distances so large, astronomers define a new unit of distance, the light-year. One **light-year (ly)** is the distance that light travels in one year, roughly 10^{13} km or 63,000 AU. The diameter of your field of view is 17 ly. The nearest star to the sun, Alpha Centauri, is 4.2 ly from Earth. In other words, light from Alpha Centauri takes 4.2 years to reach Earth.

Do you remember that Alpha Centauri is one of our Favorite Stars? (See p. xi for a list of Favorite Stars.) It is in the southern sky, so it is invisible from all but the southernmost parts of the United States where it occasionally peeks above the southern horizon. If you ever have the chance, you should locate the sun's nearest companion in space. In a later chapter, you will discover that Alpha Centauri is actually three stars orbiting around each other. Our Favorite Stars have interesting secrets.

Although stars are roughly the same size as the sun, they are so far away that you cannot see them as anything but points of light. Even with the largest telescopes on Earth, you would see only points of light when you looked at stars, and any planets that might circle those stars are much too small and too faint to be visible. Using indirect methods, astronomers have found nearly 200 planets orbiting other stars.

■ Figure 1-8

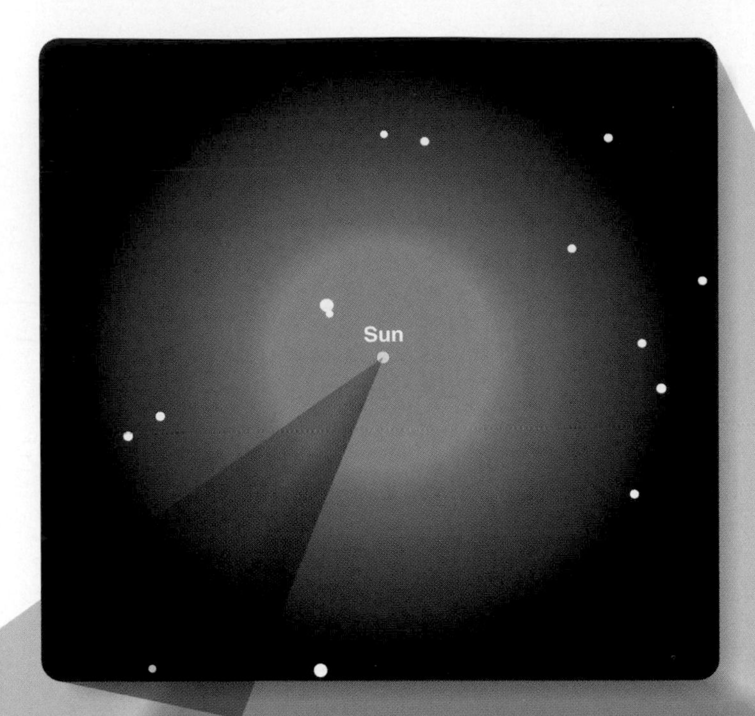

■ Figure 1-9

Milky Way, and our galaxy is called the **Milky Way Galaxy.** Of course, no one can journey far enough into space to look back and photograph our home galaxy, so the photo in Figure 1-11 shows a galaxy similar to our own. Our sun would be invisible in such a photo, but if you could see it, you would find it in the disk of the galaxy about two-thirds of the way out from the center.

Our galaxy, like many others, has graceful **spiral arms** winding outward through the disk. You will discover that stars are born in great clouds of gas and dust as they pass through the spiral arms.

In Figure 1-9, the sizes of the dots represent not the sizes of the stars but their brightness. This is the custom in astronomical diagrams, and it is also how star images are recorded on photographs. Bright stars make larger spots on a photograph than faint stars. The size of a star image in a photograph tells you not how big the star is but only how bright it looks.

In ■ Figure 1-10, you expand your field of view by another factor of 100, and the sun and its neighboring stars vanish into the background of thousands of other stars. The field of view is now 1700 ly in diameter. Of course, no one has ever journeyed thousands of light-years from Earth to look back and photograph the solar neighborhood, so this is a representative photograph of the sky. The sun is a relatively faint star that would not be easily located in a photo at this scale.

■ Figure 1-10

This box ■ represents the relative size of the previous frame. (NOAO)

What you do not see in this photograph is critically important. You do not see the thin gas that fills the spaces between the stars. Although those clouds of gas are thinner than the best vacuum on Earth, it is those clouds that give birth to new stars. Our sun formed from such a cloud about 5 billion years ago. You will see evidence of star formation in your next field of view.

If you expand your field of view by a factor of 100, you see our galaxy (■ Figure 1-11). A **galaxy** is a great cloud of stars, gas, and dust bound together by the combined gravity of all the matter. Galaxies range from 1500 to over 300,000 ly in diameter and can contain over 100 billion stars. In the night sky, you see our galaxy as a great, cloudy wheel of stars ringing the sky as the

Ours is a fairly large galaxy, roughly 75,000 ly in diameter. Only a century ago astronomers thought it was the entire universe—an island universe of stars in an otherwise empty vastness. Now they know that our galaxy is not unique. Indeed ours is only one of many billions of galaxies scattered throughout the universe.

As you expand your field of view by another factor of 100, our galaxy appears as a tiny luminous speck surrounded by other specks (■ Figure 1-12). This diagram includes a region 17 million ly in diameter, and each of the dots represents a galaxy. Notice that our galaxy is part of a cluster of a few dozen galaxies.

Galaxies are commonly grouped together in such clusters. Some of these galaxies have beautiful spiral patterns like our own galaxy, but others do not. Some are strangely distorted. One of the mysteries of modern astronomy is what produces these differences among the galaxies. Astronomers have found some important clues and have developed a fascinating theory.

If you again expand your field of view, you see that the clusters of galaxies are connected in a vast network (■ Figure 1-13). Clusters are grouped into superclusters—clusters of clusters—and the superclusters are linked to form long filaments and walls outlining voids that seem nearly empty of galaxies. These appear to be the largest structures in the universe. Were you to expand your field of view another time, you would probably see a uniform fog of filaments and voids. When you puzzle over the origin of these structures, you are at the frontier of human knowledge.

■ Figure 1-13

This box ▪ represents the relative size of the previous frame. (Detail from galaxy map from M. Seldner, B. L. Siebers, E. J. Groth, and P. J. E. Peebles, *Astronomical Journal 82* [1977])

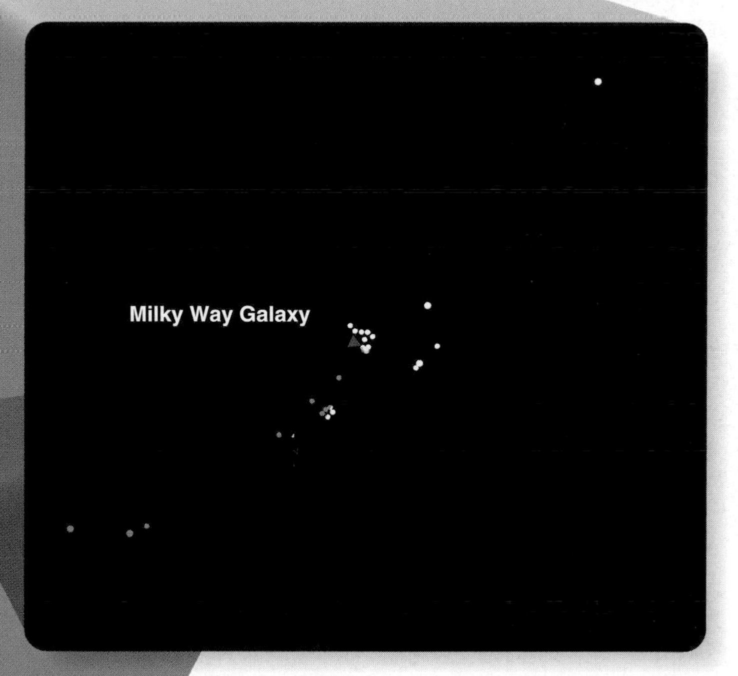

■ Figure 1-12

■ Figure 1-11

(© Anglo-Australian Telescope Board)

The first hurdle in studying astronomy is keeping a proper sense of scale. Remember that each of the billions of galaxies contains billions of stars. Later you will see clear evidence that many of those stars have families of planets like our solar system, and on some of those planets liquid-water oceans and protective atmospheres may have sheltered the spark of life. It is possible that at least a few other planets are inhabited by intelligent creatures who share our curiosity and our wonder at the scale of the cosmos.

Study and Review Tools

Summary

Your goal in this chapter is to preview the scale of astronomical objects. To do so, you journeyed outward from a familiar campus scene by expanding your field of view by factors of 100. Only 12 such steps took you to the largest structures in the universe.

Where is Earth in relation to the sun, the planets, and the galaxies?

- You live on planet Earth, which orbits our star, the sun, once a year. As Earth rotates once a day, you see the sun rise and set.
- The other planets in our solar system, Mercury, Venus, Mars, Jupiter, Saturn, Uranus, Neptune, and Pluto, orbit the sun in ellipses that are nearly circular.
- The sun is just one out of the billions of stars that fill our home galaxy, the Milky Way spiral galaxy.
- Our galaxy is just one of billions of galaxies that fill the universe in great clusters, clouds, filaments, and walls.

How do astronomers talk about these huge astronomical distances?

- Astronomers use the metric system because it simplifies calculations and scientific notation for very large or very small numbers.
- The astronomical unit (AU) is the average distance from Earth to the sun. Mars, for example, orbits 1.5 AU from the sun. The light-year (ly) is the distance light can travel in one year. The nearest star is 4.2 ly from the sun.

Which objects are big, and which are small?

- The moon is only a fourth the diameter of Earth, but the sun is 109 times larger in diameter than Earth—a typical size for a star.
- The solar system includes the sun at the center and all of the planets that orbit around it.
- Galaxies contain many billions of stars. Our galaxy is about 75,000 ly in diameter and contains over 100 billion stars.
- The largest things in the universe are the vast filaments and walls containing many clusters of galaxies.

Are there other worlds like Earth?

- Many stars seem to have planets, but such small, distant worlds are difficult to detect. Fewer than 200 or so have been found so far, but planets seem to be common, so you can probably trust that there are other worlds like Earth.

New Terms

scientific notation (p. 6)	light-year (ly) (p. 7)
solar system (p. 6)	galaxy (p. 8)
planet (p. 6)	Milky Way (p. 8)
star (p. 6)	Milky Way Galaxy (p. 8)
astronomical unit (AU) (p. 6)	spiral arm (p. 8)

Review Questions

Ace Astronomy™ Assess your understanding of this chapter's topics with additional quizzing and animations at **http://ace.brookscole.com/sf9**

1. What is the largest dimension you have personal knowledge of? Have you run a mile? Hiked 10 miles? Run a marathon?
2. In Figure 1-4, the division between daylight and darkness is at the right on the globe of Earth. How do you know this is the sunset line and not the sunrise line?
3. What is the difference between our solar system, our galaxy, and the universe?
4. Look at Figure 1-6. How can you tell that Mercury follows an elliptical orbit? Can you detect the elliptical shape of any other orbits in this figure or the next?
5. Which is the outermost planet in our solar system? Why does that change?
6. Why are light-years more convenient than miles, kilometers, or astronomical units for measuring certain distances?
7. Why is it difficult to detect planets orbiting other stars?
8. What does the size of the star image in a photograph tell you?
9. What is the difference between the Milky Way and the Milky Way Galaxy?
10. What are the largest known structures in the universe?
11. Of the objects listed here, which would be contained inside the object shown in the photograph at the right? Which would contain the object in the photo? Stars, planets, galaxy clusters, filaments, spiral arms

(Bill Schoening/NOAO/AURA/NSF)

Problems

1. The diameter of Earth is 7928 miles. What is its diameter in inches? In yards? If the diameter of Earth is expressed as 12,756 km, what is its diameter in meters? In centimeters?
2. If a mile equals 1.609 km and the moon is 2160 miles in diameter, what is its diameter in kilometers?
3. One astronomical unit is about 1.5×10^8 km. Explain why this is the same as 150×10^6 km.
4. Venus orbits 0.7 AU from the sun. What is that distance in kilometers?
5. Light from the sun takes 8 minutes to reach Earth. How long does it take to reach Mars?
6. The sun is almost 400 times farther from Earth than is the moon. How long does light from the moon take to reach Earth?
7. If the speed of light is 3×10^5 km/s, how many kilometers are in a light-year? How many meters?
8. How long does it take light to cross the diameter of our Milky Way Galaxy?
9. The nearest galaxy to our own is about 2 million light-years away. How many meters is that?
10. How many galaxies like our own would it take laid edge to edge to reach the nearest galaxy? (*Hint:* See Problem 9.)

Media Cluster

Critical Inquries for the Web

1. Locate photographs of Earth taken from space. What do cities look like? Can you see highways? Is the presence of our civilization detectable from space?
2. Locate photographs of nearby galaxies and compare them with photos of very distant galaxies. What kind of detail is invisible for distant galaxies?
3. One of the biggest clusters of galaxies is the Virgo cluster. Find out how many and what kind of galaxies are in the cluster. Is it nearby or far away?

Exploring *TheSky*

1. Locate and center one example of each of three different types of objects:
   a. A planet, such as Saturn. Find its rising and setting time. Such objects have distances measured in astronomical units (AU).
   How to proceed: Decide on the object you want to locate. Then find and center the object by pressing the **Find** button on the **Object Toolbar.** The second method is to press the **F** key. The third is to click **Edit,** then **Find.** Once you have the **Object Information** window, press the **center** button.
   b. A star. All stars in *TheSky* belong to our Milky Way Galaxy. Give the star's name, its magnitude, and its distance in light-years.
   How to proceed: Click on any star, which brings up an **Object Information** window containing a variety of information about the star.
   c. A galaxy; give its name and/or its designation.
   How to proceed: To show galaxies, click on the **Galaxies** button in the **Object Toolbar,** then click on any galaxy. Distances to galaxies are on the order of millions and billions of light years.
2. Look at the solar system from beyond Pluto by clicking on **View** and then on **3D Solar System Mode.** Tip the solar system edge-on and then face-on. Zoom in to see the inner planets. Under **Tools,** set the **Time Skip Increment** to 1 day and then go **forward in time** to watch the planets move.
3. Identify some of the brightest constellations located along the Milky Way. (*Hint:* See **View, Reference Lines.**)

- Go to the Brooks/Cole Astronomy Resource Center **(http://astronomy.brookscole.com)** for critical thinking exercises, articles, and additional readings from InfoTrac College Edition, Brooks/Cole's online student library.

2 The Sky

The Southern Cross I saw every night abeam. The sun every morning came up astern; every evening it went down ahead. I wished for no other compass to guide me, for these were true.

CAPTAIN JOSHUA SLOCUM, *SAILING ALONE AROUND THE WORLD*

Visual-wavelength image

THE NIGHT SKY is the rest of the universe as seen from our planet. When you look up at the stars, you look out through a layer of air only a few hundred kilometers deep. Beyond that, space is nearly empty, with stars scattered light-years apart. Here you will begin your search for the natural laws that govern the universe by trying to understand what the universe looks like. As you read this chapter, keep in mind that you live on a planet. The stars are scattered into the void all around you, most very distant and some closer. Our planet rotates on its axis once a day and revolves around the sun in its orbit once a year, and that makes the sky appear to revolve around you in a daily and yearly cycle. Not only does the sun rise and set, but so also do the stars. The entire sky seems to whirl around you because our planet whirls through the universe.

Continued on page 14

The sky above mountaintop observatories far from city lights is the same sky you see from your window. The stars above you are other suns scattered through the universe. (Kris Koenig/Coast Learning Systems)

Guidepost

Looking Back

Astronomy is about us. As we learn about astronomy, we learn about ourselves. We search for an answer to the question, "What are we?" The cosmic zoom in the previous chapter showed you Earth in relation to other objects in the universe. Even in that quick preview, as you learned about stars and galaxies, you were also learning about yourself.

This Chapter

Now it is time to return to Earth and look closely at the sky. You have noticed stars and may know a few constellations and star names, but now you need to refine your terms and answer a few essential questions:

How do astronomers refer to stars?

How can you compare the brightness of the stars?

How does the sky move as Earth moves?

What causes the seasons?

How do astronomical cycles affect Earth's climate?

Answering these questions will tell you a great deal about yourself and your home on planet Earth.

Looking Ahead

In the next chapter, you will learn more about the beautiful cycles of the sky, and in this chapter and the next it is important to remember that the motions you see in the heavens are produced in part by the motions of Earth. Chapter 4 will show you how hard it was for humanity to understand that we live on a moving planet. In fact, you will discover that modern science was born when people tried to understand the motions in the sky.

Ace Astronomy™ The AceAstronomy icon throughout the text indicates an opportunity for you to test yourself on key concepts and to explore animations and interactions on the AceAstronomy website at: http://ace.brookscole.com/sf9

2-1 The Stars

ON A DARK NIGHT far from city lights, you can see a few thousand stars in the sky. As you begin your study of the sky, the first step is to organize what you see by naming groups of stars and individual stars and by specifying the brightness of individual stars. That will make the sky familiar territory, and you will be ready to explore further.

Constellations

All around the world, ancient cultures celebrated heroes, gods, and mythical beasts by naming groups of stars—**constellations** (■ Figure 2-1). You should not be surprised that the star patterns do not look like the creatures they represent any more than Columbus, Ohio, looks like Christopher Columbus. The constellations "celebrate" the most important mythical figures in each culture. The constellations named within Western culture originated in Mesopotamia over 5000 years ago, with other constellations added by Babylonian, Egyptian, and Greek astronomers during the classical age. Of these ancient constellations, 48 are still in use today.

Different cultures grouped stars and named constellations differently. The constellation you probably know as Orion was known as Al Jabbar, the giant, to the ancient Syrians, as the White Tiger to the Chinese, and as Prajapati in the form of a stag in ancient India. The Pawnee Indians knew the constellation Scorpius as two groupings. The long tail of the scorpion was the Snake, and the two bright stars at the tip of the scorpion's tail were the Two Swimming Ducks.

Many ancient cultures around the world, including the Greeks, northern Asians, and Native Americans, associated the stars of the Big Dipper with a bear. The concept of the celestial bear may have crossed the land bridge into North America with the first Americans over 10,000 years ago. Some of the constellations you see in the sky may be among the oldest surviving traces of human culture.

To the ancients, a constellation was a loose grouping of stars. Many of the fainter stars were not included in any constellation, and regions of the southern sky, not being visible to the ancient astronomers of northern latitudes, were not organized into constellations. Constellation boundaries, when they were defined at all, were only approximate (■ Figure 2-2a), so a star like Alpheratz could be thought of as part of Pegasus or part of Andromeda. In recent centuries, astronomers have added 40 modern constellations to fill gaps, and in 1928 the International Astronomical Union established 88 official constellations with clearly defined boundaries (Figure 2-2b). Consequently, a constellation now represents not a group of stars but an area of the sky, and any star within the boundaries of the area belongs to the constellation. Because the entire sky is covered by constellations, every star is a member of one and only one constellation.

In addition to the 88 official constellations, the sky contains a number of less formally defined groupings called **asterisms.** The Big Dipper, for example, is a well-known asterism that is part of the constellation Ursa Major (the Great Bear). Another asterism is the Great Square of Pegasus (Figure 2-2b), which includes three stars from Pegasus and one from Andromeda. The star charts at the end of this book will introduce you to the brighter constellations and asterisms.

Although you refer to constellations and asterisms by name, most are made up of stars that are not physically associated with one another. Some stars may be many times farther away than others and moving through space in different directions. The only thing they have in common is that they lie in approximately the same direction from Earth (■ Figure 2-3).

■ Figure 2-1

The constellations are an ancient heritage handed down for thousands of years as celebrations of great heroes and mythical creatures. Here Sagittarius and Scorpius hang above the southern horizon.

The Names of the Stars

In addition to naming groups of stars, ancient astronomers named the brighter stars,

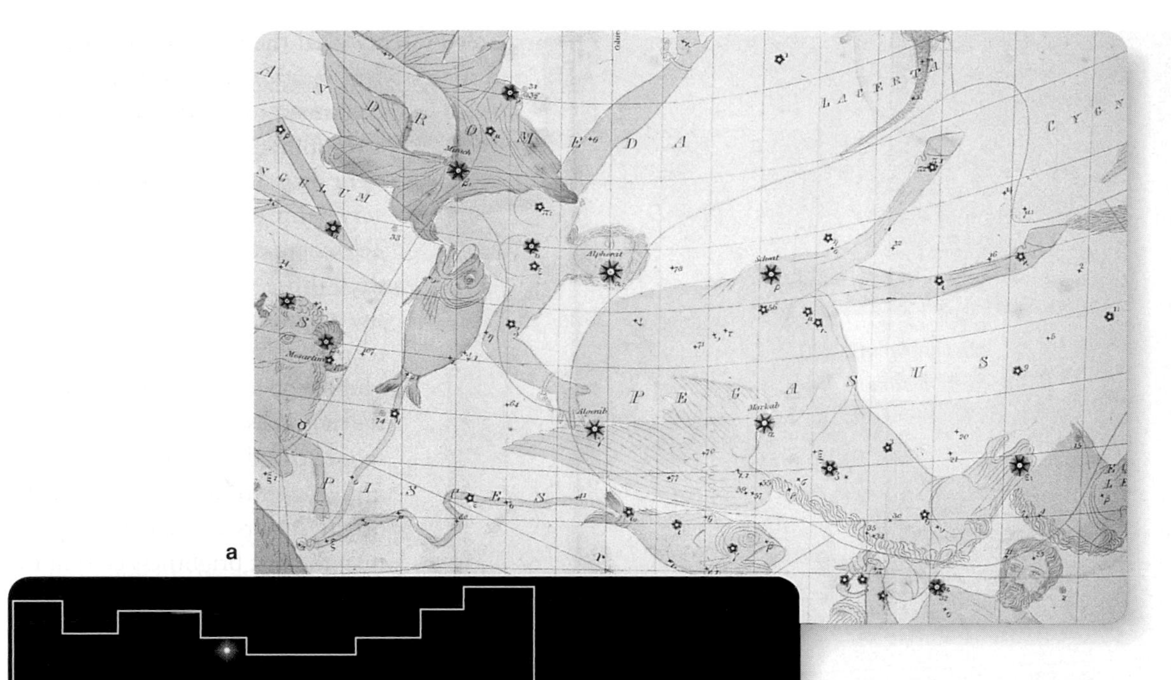

■ Figure 2-2

(a) In antiquity, constellation boundaries were poorly defined, as shown on this map by the curving dotted lines that separate Pegasus from Andromeda. (From Duncan Bradford, *Wonders of the Heavens,* Boston: John B. Russell, 1837) (b) Modern constellation boundaries are precisely defined by international agreement.

and modern astronomers still use many of those names. The constellation names come from Greek versions translated into Latin—the language of science from the fall of Rome to the 19th century—but most star names come from ancient Arabic, though much altered by the passing centuries. The name of Betelgeuse, the bright red star in Orion, for example, comes from the Arabic *yad al jawza,* meaning "armpit of Jawza [Orion]." Names such as Sirius (the Scorched One), Capella (the Little She Goat), and Aldebaran (the Follower of the Pleiades) are beautiful additions to the mythology of the sky.

Giving the stars individual names is not very helpful, because you can see thousands of stars, and these names do not help you locate the star in the sky. In which constellation is Antares, for example? In 1603, Bavarian lawyer Johann Bayer published an atlas of the sky called *Uranometria* in which he assigned lowercase Greek letters to the brighter stars of each constellation in approximate order of brightness. Astronomers have used those Greek letters ever since. (See the Appendix table with the Greek alphabet.) In this way, the brightest star is usually designated α (alpha), the second-brightest β (beta), and so on (■ Figure 2-4). To identify a star in this way, give the Greek letter followed by the possessive form of the constellation name, such as α Scorpii (sometimes written alpha Scorpii) for Antares. That designation reveals that Antares is in the constellation Scorpius and that it is probably the brightest star in the constellation.

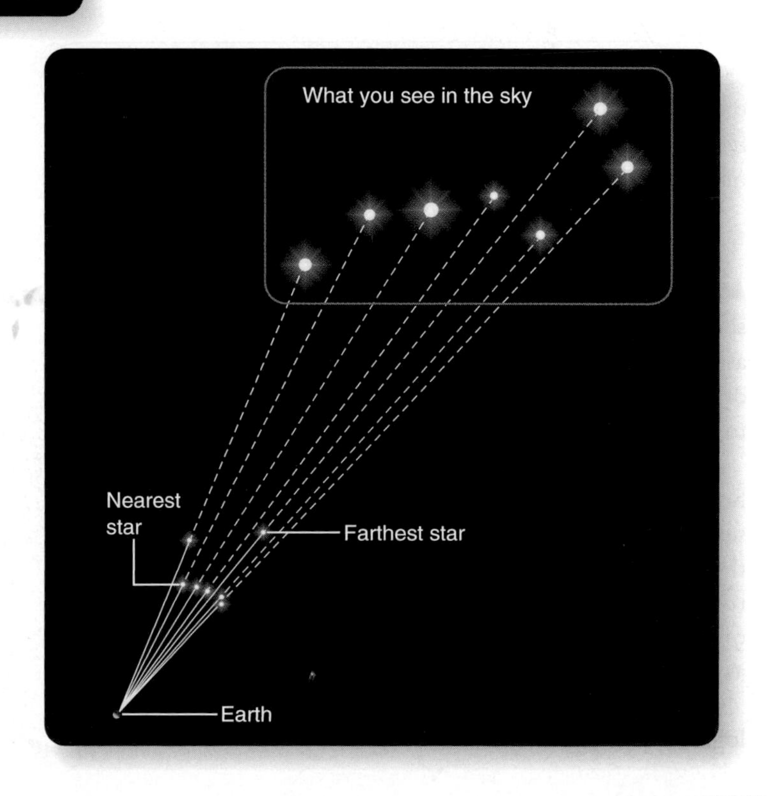

■ Figure 2-3

The stars you see in the Big Dipper—the brighter stars of the constellation Ursa Major, the Great Bear—are not at the same distance from Earth. You see the stars in a group in the sky because they lie in the same general direction as seen from Earth. The size of the star dots in the star chart represents the apparent brightness of the stars.

Window on Science 2-1

Scientific Arguments: The Structure of Science

An argument can be a shouting match, but another definition is "a discourse intended to persuade." Scientists construct arguments as part of the business of science not because they want to persuade others they are right, but because they want to test their ideas. For example, a team of biologists might construct a scientific argument to show that ants communicate by smell. If they do their best and cite all the evidence and all the theories and the argument seems convincing, they gain confidence that they do understand how ants "talk" to each other. If the scientists discover a hole in their argument, they know they have more work to do.

A scientific argument is a logical presentation of evidence and theory with interpretations and explanations that help you understand some aspect of nature. Scientists would be free to include any evidence or theory that helps persuade, but they must observe one fundamental rule of science: They must be totally honest—they must include all of the evidence and all of the theories. The purpose of a scientific argument is to test understanding, not to win votes or sell soap. Dishonesty in a scientific argument is self-deluding, and scientists consider such dishonesty disgraceful.

In the chapters ahead, you will meet lots of scientific arguments about sophisticated theories and detailed evidence involving everything from the formation of Earth to the fate of the universe, but you can begin using scientific arguments now as a way to review your reading. If you can organize what you know about naming stars, for example, and explain why most star names come from Arabic but most constellation names are Latin, and why Greek letters give you clues to the brightness of stars, you will be building a scientific argument that will help you line up the facts you know and convert them to real understanding.

Scientists think in scientific arguments not to convince others, but to test their own understanding. (Roger Ressmeyer/Corbis)

Each section of this book ends with a review tool called "Building Scientific Arguments." Use these as a model to organize your thoughts as you review each concept in a section, and they will help you understand nature the way scientists do.

measurements of the past, right back to the first star catalogs over 2000 years ago.

It's time for you to review the preceding section on constellations, star names, and magnitudes. Before you begin, read **Window on Science 2-1** and notice how scientists organize information. How you review is critical when you study a science. Memorizing facts won't help you much, but organizing your understanding into scientific arguments will line the facts up in a meaningful way. The review tool that follows, "Building Scientific Arguments," will help you review an important concept from the section, but you should use the same process to review each concept from the beginning of the section to the end.

Building Scientific Arguments

Nonastronomers sometimes complain that the magnitude scale is awkward. Why would they think it is awkward, and how did it get that way?

Two things might make the magnitude scale seem awkward. First, it is backward; the bigger the magnitude number, the fainter the star. Of course, that arose because ancient astronomers were not measuring the brightness of stars but rather classifying them, and first-class stars would be brighter than second-class stars. The second awkward feature of the magnitude scale is its mathematical relation to intensity. If two stars differ by one magnitude, one is about 2.5 times brighter than the other. But if they differ by two magnitudes, one is 2.5×2.5 times brighter. This mathematical relationship arises because of the way human eyes perceive brightness as ratios of intensity.

Now build your own scientific argument to analyze the following question: **If the magnitude scale is so awkward, why do you suppose astronomers have used it for over two millennia?**

■ ■ ■

Connections: In this section, you have found a way to identify stars by name and by brightness. Now it is time to look at the sky as a whole and notice its motion.

2-2 The Sky and Its Motion

THE SKY ABOVE seems to be a great blue dome in the daytime and a sparkling ceiling at night. Learning to look at the sky requires that you begin thousands of years ago.

Window on Science 2-2

Frameworks for Thinking about Nature: Scientific Models

In everyday language, you use the word *model* in various ways—fashion model, model airplane, model student—but scientists use the word in a very specific way. A **scientific model** is a carefully devised mental conception of how something works, a framework that helps scientists think about some aspect of nature. For example, astronomers use the celestial sphere as a way to think about the motions of the sky, sun, moon, and stars.

Although a scientific model is a mental conception, it can take many forms. Some models are quite abstract—the psychologist's model of how the human mind processes visual information into images, for instance. But other models are so specific that they can be expressed as a set of mathematical equations. For example, an astronomer might use a set of equations to describe in detail how gas falls into a black hole. You could refer to such a calculation as a model. Of course, you could use metal and plastic to build a celestial globe, but the thing you built wouldn't really be the model any more than the equations are a model. A scientific model is a mental conception, an idea that helps you think about nature.

On the other hand, a model is not meant to be a statement of truth. The celestial sphere is not real; you know the stars are scattered through space at various distances, but you can imagine a celestial sphere and use it to help think about the sky. A scientific model does not have to be true to be useful. Chemists, for example, think about the atoms in molecules by visualizing them as balls joined together by rods. This model of a molecule is not fully correct, but it is a helpful way to think about molecules; it gives chemists a framework within which to organize their ideas.

Because scientific models are not meant to be totally true, you must always remember the assumptions on which they are based. If you begin to think a model is true, it can be misleading instead of helpful. The celestial sphere, for instance, can help you think about the sky, but you must remember that it is only a mental crutch. The universe is much larger and much more interesting than this ancient scientific model of the heavens.

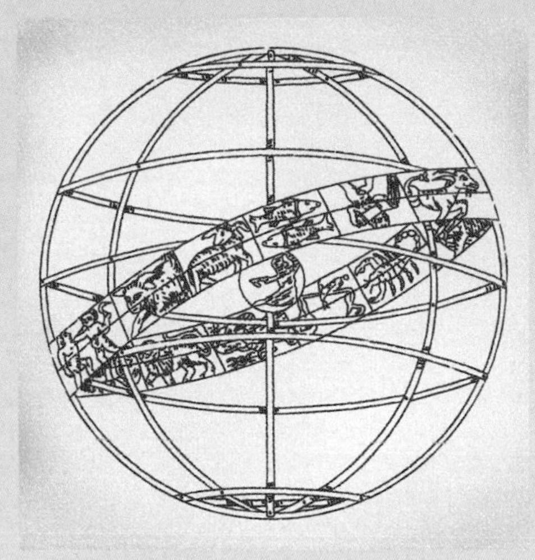

The ancient celestial sphere is a useful model of the sky.

The Celestial Sphere

Ancient astronomers believed the sky was a great sphere surrounding Earth with the stars stuck on the inside like thumbtacks in the ceiling. Modern astronomers know that the stars are scattered through space at different distances, but it is still convenient to think of the sky as a great celestial sphere.

As you study **The Sky around Us** on pages 20–21, notice three important points:

1. The sky appears to rotate westward around Earth each day, but that is a consequence of the eastward rotation of Earth. That produces day and night.
2. Astronomers measure distance across the sky as angles and express them as degrees, minutes, and seconds.
3. What you can see of the sky depends on where you are on Earth. If you lived in Australia, you would see many constellations and asterisms invisible from North America, but you would never see the Big Dipper. Remember our Favorite Star Alpha Centauri? It is in the southern sky and isn't visible from most of the United States. You could just glimpse it above the southern horizon if you were in Miami, but you could see it easily from Australia.

The celestial sphere is an example of a scientific model, a common feature of scientific thought (**Window on Science 2-2**). Notice that a scientific model does not have to be true to be useful. Many scientific models are discussed in the chapters that follow, and you will discover that the most useful models are often quite obviously flawed descriptions of the true facts.

In addition to the daily motion of the sky, Earth's rotation adds a second motion to the sky that can be detected only over centuries.

Precession

Over 2000 years ago, Hipparchus compared a few of his star positions with those made nearly two centuries before and realized that the celestial poles and equator were slowly moving across the sky. Later astronomers understood that this motion is caused by Earth's toplike motion.

If you have ever played with a gyroscope or top, you have seen how the spinning mass resists any change in the direction of its axis of rotation. The more massive the top and the more rapidly it spins, the more difficult it is to change the direction of its axis of rotation. But you probably recall that the axis of even the most

The Sky around Us

1 The eastward rotation of Earth causes the sun, moon, and stars to move westward in the sky as if the **celestial sphere** were rotating westward around Earth. From any location on Earth you see only half of the celestial sphere, the half above the **horizon.** The **zenith** marks the top of the sky above your head, and the **nadir** marks the bottom of the sky directly under your feet. The drawing at right shows the view for an observer in North America. An observer in South America would have a dramatically different horizon, zenith, and nadir.

The apparent pivot points are the **north celestial pole** and the **south celestial pole** located directly above Earth's north and south poles. Halfway between the celestial poles lies the **celestial equator.** Earth's rotation defines the directions you use every day. The **north point** and **south point** are the points on the horizon closest to the celestial poles. The **east point** and the **west point** lie halfway between the north and south points. The celestial equator always touches the horizon at the east and west points.

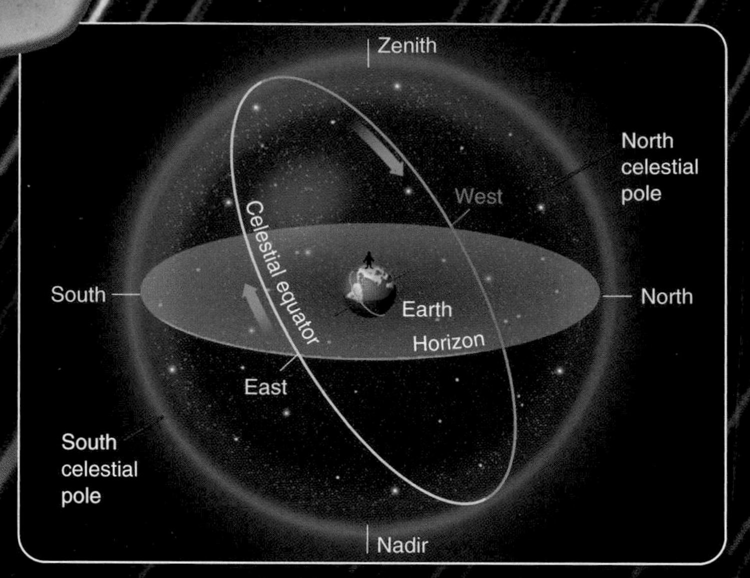

Ace Astronomy™ Log into AceAstronomy and select this chapter to see Active Figure "Celestial Sphere." Notice how each location on Earth has its unique horizon.

AURA/NOAO/NSF

1a This time exposure of about 30 minutes shows stars as streaks, called star trails, rising behind an observatory dome. The camera was facing northeast to take this photo. The motion you see in the sky depends on which direction you look, as shown at right. Looking north, you see the star Polaris, the North Star, located near the north celestial pole. As the sky appears to rotate westward, Polaris hardly moves, but other stars circle the celestial pole. Looking south from a location in North America, you can see stars circling the south celestial pole, which is invisible below the southern horizon.

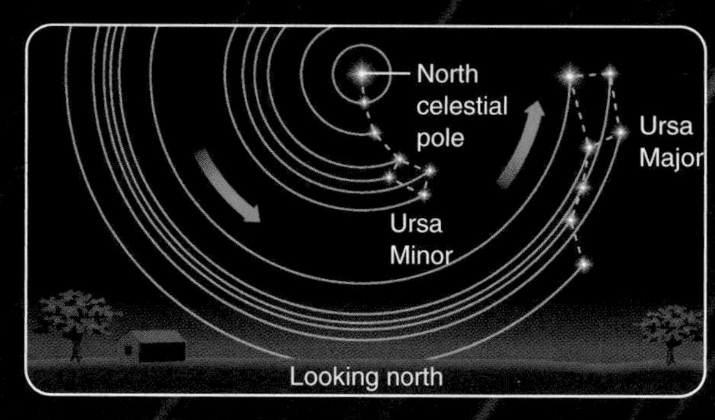

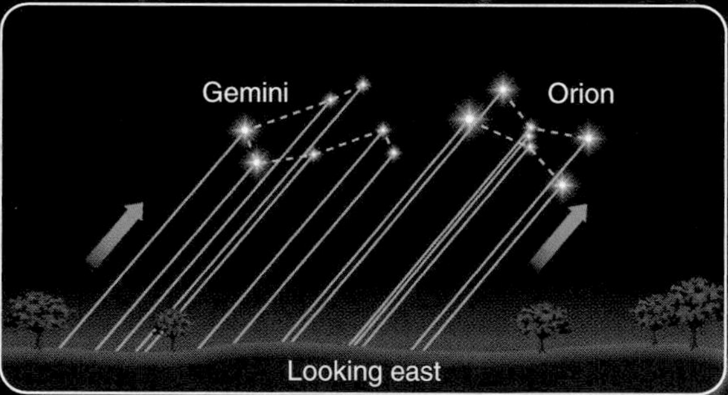

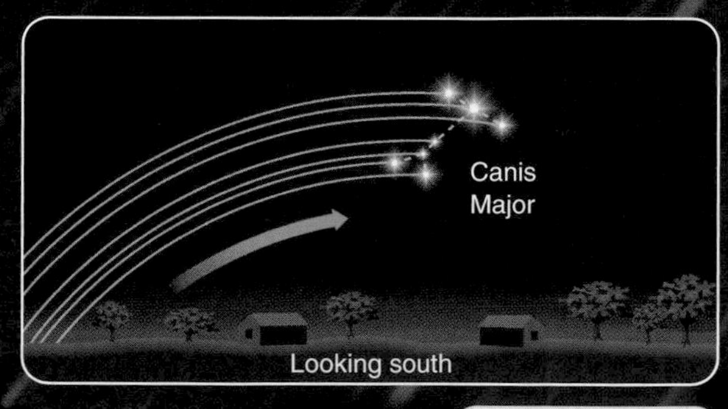

Ace Astronomy™ Log into AceAstronomy and select this chapter to see Active Figure "Rotation of the Sky." Look in different directions and compare the motions of the stars.

Astronomers measure distance across the sky as angles.

Angular distance

2 Astronomers might say, "The star was only 2 degrees from the moon." Of course, the stars are much farther away than the moon, but when you think of the celestial sphere, you can measure distances *on the sky* as **angular distances** in degrees, minutes of arc, and seconds of arc. A **minute of arc** is 1/60th of a degree, and a **second of arc** is 1/60th of a minute of arc. Then the **angular diameter** of an object is the angular distance from one edge to the other. The sun and moon are each about half a degree in diameter, and the bowl of the Big Dipper is about 10° wide.

3 What you see in the sky depends on your latitude as shown at right. Imagine that you begin a journey in the ice and snow at Earth's North Pole with the north celestial pole directly overhead. As you walk southward, the celestial pole moves toward the horizon, and you can see further into the southern sky. The angular distance from the horizon to the north celestial pole always equals your latitude (L)—the basis for celestial navigation. As you cross Earth's equator, the celestial equator would pass through your zenith, and the north celestial pole would sink below your northern horizon.

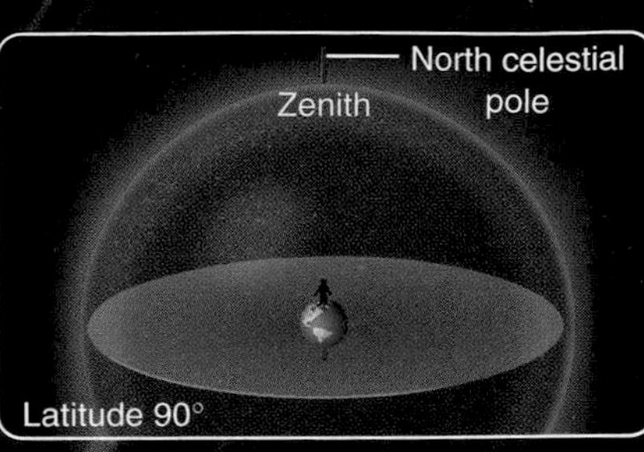

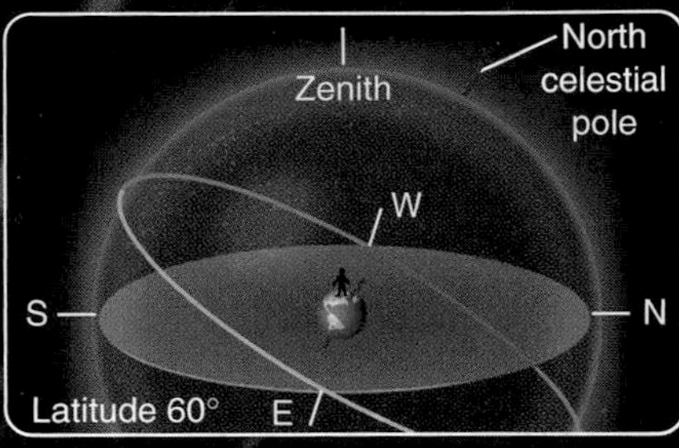

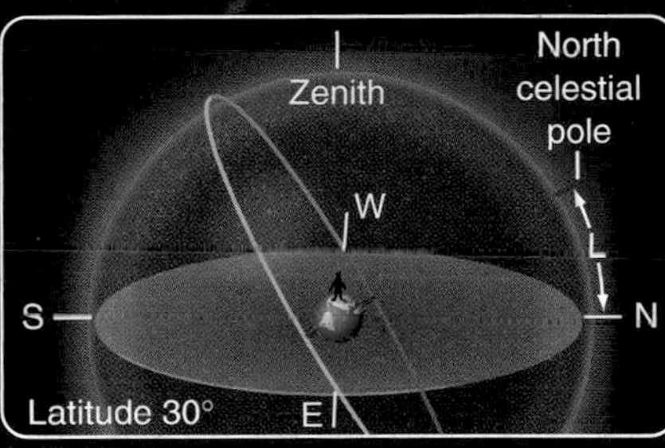

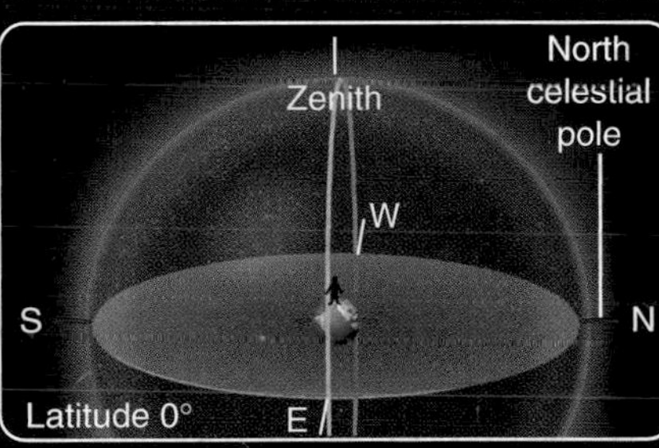

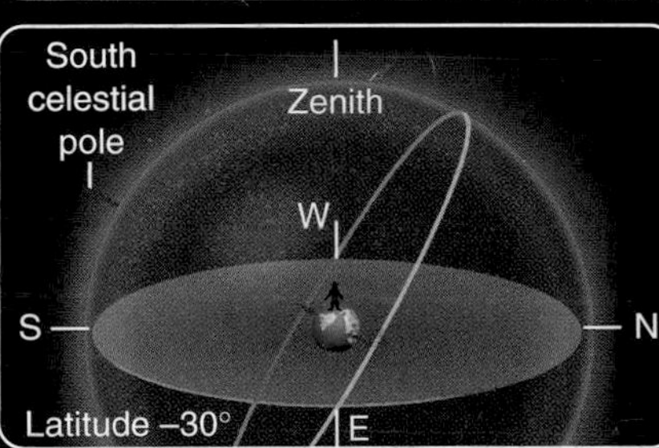

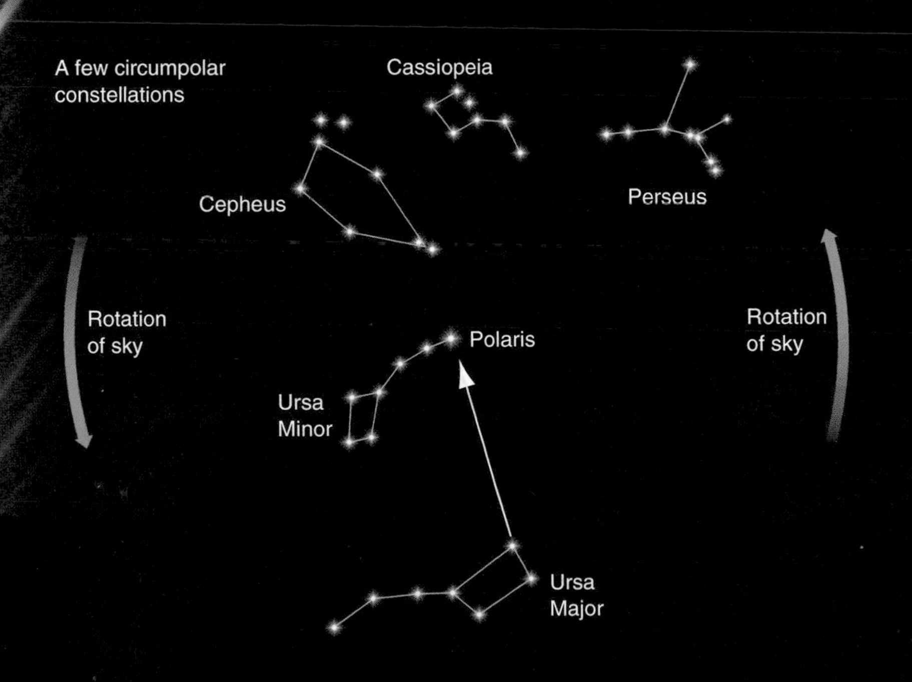

3a **Circumpolar constellations** are those that never rise or set. From mid-northern latitudes, as shown at left, you see a number of familiar constellations circling Polaris and never dipping below the horizon. As the sky rotates, the pointer stars at the front of the Big Dipper always point toward Polaris. Circumpolar constellations near the south celestial pole never rise as seen from mid-northern latitudes. From a high latitude such as Norway, you would have more circumpolar constellations, and from Quito, Ecuador, located on Earth's equator, you would have no circumpolar constellations at all.

Log into AceAstronomy and select this chapter to see Active Figure "Constellations from Different Latitudes."

Ace Astronomy™

The Cycle of the Seasons

1 You can use the celestial sphere to help you think about the seasons. The celestial equator is the projection of Earth's equator on the sky, and the ecliptic is the projection of Earth's orbit on the sky. Because Earth is tipped in its orbit, the ecliptic and equator are inclined to each other by 23.5° as shown at right. As the sun moves eastward around the sky, it spends half the year in the southern half of the sky and half of the year in the northern half. That causes the seasons.

The sun crosses the celestial equator going northward at the point called the **vernal equinox.** The sun is at its farthest north at the point called the **summer solstice.** It crosses the celestial equator going southward at the **autumnal equinox** and reaches its most southern point at the **winter solstice.**

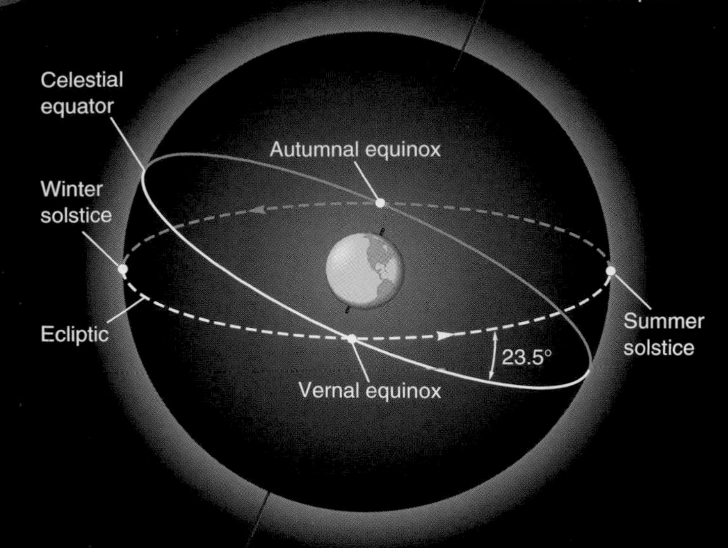

1a The seasons are defined by the dates when the sun crosses these four points, as shown in the table at the right. *Equinox* comes from the word for "equal"; the day of an equinox has equal amounts of daylight and darkness. *Solstice* comes from the words meaning "sun" and "stationary." *Vernal* comes from the word for "green." The "green" equinox marks the beginning of spring.

Event	Date*	Season
Vernal equinox	March 20	Spring begins
Summer solstice	June 22	Summer begins
Autumnal equinox	September 22	Autumn begins
Winter solstice	December 22	Winter begins

* Give or take a day due to leap year and other factors.

AceAstronomy™ Log into AceAstronomy and select this chapter to see Active Figure "Seasons" and watch Earth orbiting the sun.

1b On the day of the summer solstice in late June, Earth's northern hemisphere is inclined toward the sun, and sunlight shines almost straight down at northern latitudes. At southern latitudes, sunlight strikes the ground at an angle and spreads out. North America has warm weather, and South America has cool weather.

Earth's axis of rotation points toward Polaris, and, like a top, the spinning Earth holds its axis fixed as it orbits the sun. On one side of the sun, Earth's northern hemisphere leans toward the sun; on the other side of its orbit, it leans away. However, the direction of the axis of rotation does not change.

23.5°
To Polaris
40° N latitude
Equator
40° S latitude
Sunlight nearly direct on northern latitudes
To sun
Sunlight spread out on southern latitudes

Earth at summer solstice

1c Light striking the ground at a steep angle spreads out less than light striking the ground at a shallow angle. Light from the summer-solstice sun strikes northern latitudes from nearly overhead and is concentrated.

Light from the winter-solstice sun strikes northern latitudes at a much steeper angle and spreads out. The same amount of energy is spread over a larger area, so the ground receives less energy from the winter sun.

2 The two causes of the seasons are shown at right for someone in the northern hemisphere. First, the noon summer sun is higher in the sky and the winter sun is lower, as shown by the longer winter shadows. Thus winter sunlight is more spread out. Second, the summer sun rises in the northeast and sets in the northwest, spending more than 12 hours in the sky. The winter sun rises in the southeast and sets in the southwest, spending less than 12 hours in the sky. Both of these effects mean that northern latitudes receive more energy from the summer sun, and summer days are warmer than winter days.

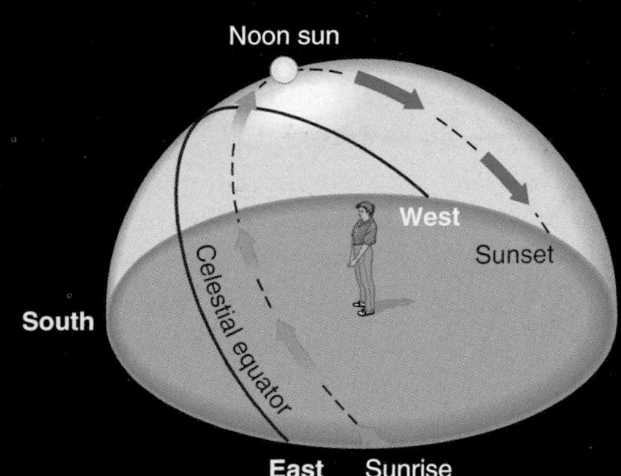

At summer solstice

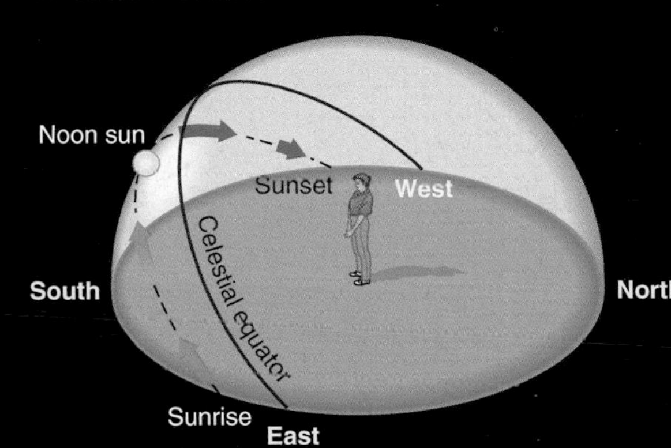

At winter solstice

Ace Astronomy™

Log into AceAstronomy and select this chapter to see Active Figure "Path of the Sun" and see this figure from the inside.

23.5°
To Polaris
Sunlight spread out on northern latitudes
40° N latitude
Equator
To sun
Sunlight nearly direct on southern latitudes
40° S latitude

Earth at winter solstice

1d On the day of the winter solstice in late December, Earth's northern hemisphere is inclined away from the sun, and sunlight strikes the ground at an angle and spreads out. At southern latitudes, sunlight shines almost straight down and does not spread out. North America has cool weather and South America has warm weather.

Earth's orbit is only very slightly elliptical. About January 4, Earth is at **perihelion**, its closest point to the sun, when it is only 1.7 percent closer than average. About July 4, Earth is at **aphelion**, its most distant point from the sun, when it is only 1.7 percent farther than average. This small variation does not significantly affect the seasons.

When you look for planets in the sky, you will always find them near the ecliptic, because their orbits lie in nearly the same plane as the orbit of Earth. As they orbit the sun, they appear to move generally eastward along the ecliptic. (In Chapter 4, you will discover an exception to this general eastward motion.) In fact, the word *planet* comes from a Greek word meaning "wanderer." Mars moves completely around the ecliptic in slightly less than two years, but Saturn, being farther from the sun, takes nearly 30 years.

As seen from Earth, Venus and Mercury can never move far from the sun because their orbits are inside Earth's orbit. They sometimes appear near the western horizon just after sunset or near the eastern horizon just before sunrise. Venus is easier to locate because its larger orbit carries it higher above the horizon than Mercury (■ Figure 2-11). Mercury's orbit is so small that it can never get farther than about 28° from the sun. Consequently, it is usually hard to see against the sun's glare and is often hidden in the clouds and haze near the horizon. At certain times when it is farthest from the sun, however, Mercury shines brightly and can be located near the horizon in the evening or morning sky. (See the Appendix for the best times to observe Venus and Mercury.)

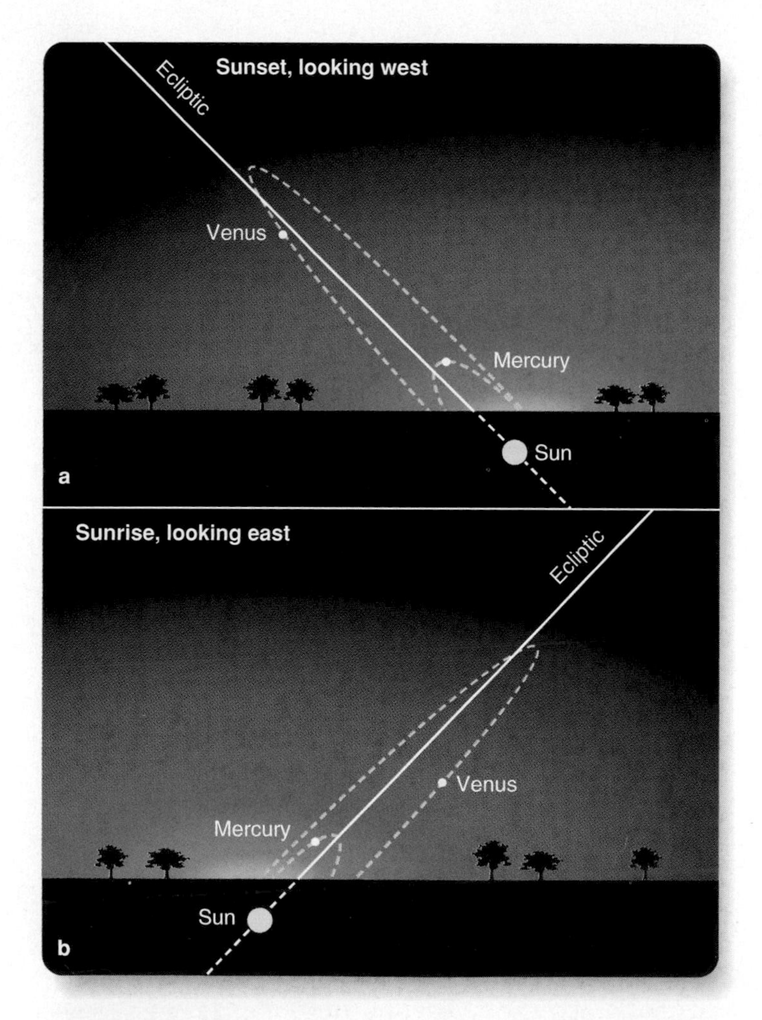

■ Figure 2-11

Mercury and Venus follow orbits that keep them near the sun, and they are visible only soon after sunset or before sunrise. Venus takes 584 days to move from morning sky to evening sky and back again, but Mercury zips around in only 116 days.

By tradition, any planet visible in the evening sky is called an **evening star,** although planets are not stars. Any planet visible in the sky shortly before sunrise is called a **morning star.** Perhaps the most beautiful is Venus, which can become as bright as minus fourth magnitude. As Venus moves around its orbit, it can dominate the western sky each evening for about half a year, but eventually its orbit carries it back toward the sun, and it is lost in the haze near the horizon. In a few weeks it reappears in the dawn sky as a brilliant morning star.

Astrology

Seen from Earth, the planets move gradually eastward along the ecliptic, but they don't follow the ecliptic exactly. Also, each travels at its own pace and seems to speed up and slow down at various times. To the ancients, this complex motion reflected the moods of the sky gods, and astrology was born.

Ancient astrologers defined a **zodiac,** a band 18° wide centered on the ecliptic, as the highway the planets follow. They divided this band into 12 segments named for the constellations along the ecliptic—the signs of the zodiac. A **horoscope** shows the location of the sun, moon, and planets among the zodiacal signs with respect to the horizon at the moment of a person's birth as seen from that longitude and latitude. Even if astrology worked, the generalized horoscopes published in newspapers and tabloids can't have been calculated accurately for individual readers.

Astrology buffs argue that a person's personality, life history, and fate are revealed in his or her horoscope, but the evidence contradicts this belief. Astrology has been tested many times over the centuries, and it just doesn't work. Believers, however, don't give up on it. Thus, astrology is a superstition that depends on blind belief and not a science that depends on evidence (**Window on Science 2-4**).

One reason astronomers find astrology irritating is that it has no link to the physical world. For example, precession has moved the constellations so that they no longer match the zodiacal signs. Whatever sign you were "born under," the sun was probably in the previous zodiacal constellation. In fact, if you were born on or between November 30 and December 17, the sun was passing through a corner of the nonzodiacal constellation Ophiuchus, and you have no official zodiacal sign.* Furthermore, astronomers like to point out, there is no mechanism by which

* The author of this book was born on December 14 and thus has no astrological sign. An astronomer friend claims that the author must therefore have no personality.

Window on Science 2-4

Astrology and Pseudoscience

Astronomers have a low opinion of astrology, not so much because it is groundless but because it pretends to be a science. It is a pseudoscience, from the prefix *pseudo,* meaning false. There are many examples of pseudoscience, and it is illuminating to consider the difference between pseudoscience and science.

A pseudoscience is a set of beliefs that appear to be based on scientific ideas but that fail to obey the most basic rules of science. For example, some years ago a fad arose in which people placed objects under pyramids made of paper, plastic, wire, and so on. The claim was that the pyramidal shape would focus cosmic forces on anything inside and so preserve fruit, sharpen razor blades, and do other miraculous things. Many books promoted this idea, but simple experiments showed that any shape would protect a piece of fruit from airborne spores and allow it to dry without rotting. Likewise, any shape would allow oxidation to improve the cutting edge of a razor blade. In short, experimental evidence contradicted the claim. Nevertheless, supporters of the theory declined to abandon or revise their claims. Thus, the fad of pyramid power was a pseudoscience.

One characteristic of a pseudoscience is that it appeals to our needs and desires. Thus, some pseudoscientific claims are self-fulfilling. For example, some people bought pyramidal tents to put over their beds and thus improve their rest. While there is no logical mechanism by which such a tent could affect a sleeper, because people wanted and expected the claim to be true they slept more soundly. Many pseudoscientific claims involve medical cures, ranging from magnetic bracelets and crystals to focus spiritual power to astonishingly expensive and illegal treatments for cancer. Logic is a stranger to pseudoscience, but human fears and needs are not.

Astrology is a pseudoscience. Over the centuries, astrology has been tested repeatedly, and no correlation has been found. But it survives, and its supporters disregard any evidence that it doesn't work. Like all pseudosciences, astrology is not open to revision in the face of contradictory evidence. Furthermore, astrology fulfills our human need to believe that there is order and meaning to our lives. It may comfort us to believe that our sweetheart has rejected us because of the motions of the planets rather than to admit that we behaved badly on our last date. Comfort aside, astrology is a poor basis for life decisions.

Human nature and human needs probably ensure that pseudoscientific beliefs will continue to plague us, like emotional viruses propagating from person to person. But if we recognize them for what they are we can more easily guide our lives by rational principles and not by giving credit for our successes and blame for our failings to the stars.

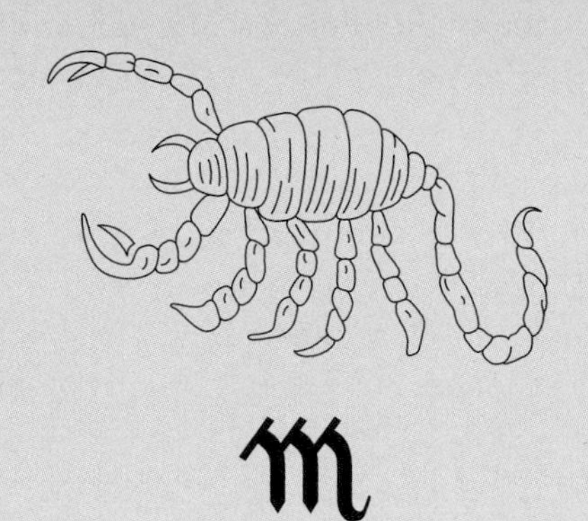

Astrology may be the oldest pseudoscience.

the planets could influence us. The gravitational influence of a doctor who is delivering a baby is many times more powerful than the gravitational influence of the planets.

The arguments in the preceding paragraph actually miss the point. Astrology is not related to the physical world at all. It does not matter what constellation the sun occupies, because astrology divides the zodiac into equal mathematical sections, sometimes called houses, and it does not matter to the believer in astrology that the constellations don't match. Furthermore, the physical mechanism is beside the point for the true believer. Astrology is not so much an astronomical superstition as it is a mathematical superstition.

Astrology makes sense only when you think of the world as the ancients did. They believed in multiple sky gods whose moods altered events on Earth. They believed in supernatural influences between natural events and human events. The ancients did not understand natural forces such as gravity as the mechanisms that cause things to happen, and thus they did not believe in cause and effect as modern people do. If a house burned, they might conclude that it caught fire because it was cursed and not because someone was careless with an oil lamp.

Modern science left astrology behind centuries ago, but astrology survives as a fascinating part of human history—an early attempt to understand the meaning of the sky.

Building Scientific Arguments

Planets like Mars can sometimes be seen rising in the east as the sun sets. Why is the same not true for Mercury or Venus? In this case, the argument hinges on simple geometry. Mars has an orbit outside the orbit of Earth, and that means it can reach a location opposite the sun in the sky. As the sun sets in the west, Mars rises in the east. But Mercury and Venus follow orbits that are smaller than the orbit of Earth, so they can never reach a point opposite the sun in the sky.

As they follow their orbits around the sun, Earthlings see them gradually swing out on the east side of the sun, reach a maximum distance from the sun, swing back toward the sun, move out to the west side of the sun, reach a maximum distance west, and then move back toward the sun again.

Because they never get very far from the sun in the sky, they can never be seen rising in the east as the sun sets.

The planetary motions in the sky are produced by the orbital motions of the planets around the sun. Build a geometrical argument to analyze the following question: **What would you see if the planets followed orbits that were all in exactly the same plane as the orbit of Earth?**

■ ■ ■

Connections: Modern astronomers have a low tolerance for astrological superstition, yet the motions of the heavenly bodies do affect human lives. The motion of the sun produces the seasons, and, as you will discover in Chapter 5, the moon governs the tides. In addition, small changes in Earth's orbit may partially control global climate.

2-5 Astronomical Influences on Earth's Climate

WEATHER IS WHAT HAPPENS TODAY; climate is the average of what happens over decades. Earth has gone through past episodes, called ice ages, when the worldwide climate was cooler and dryer and thick layers of ice covered northern latitudes.

The earliest known ice age occurred about 570 million years ago and the next about 280 million years ago. The most recent ice age began only about 3 million years ago and is still going on. You are living in one of the periodic episodes when the glaciers melt and Earth grows slightly warmer. The current warm period began about 20,000 years ago.

Ice ages seem to occur with a period of roughly 250 million years, and cycles of glaciation within ice ages occur with a period of about 40,000 years. These cyclic changes have an astronomical origin.

The Hypothesis

Sometimes a theory or hypothesis is proposed long before scientists can find the critical evidence to test it. That happened in 1920 when Yugoslavian meteorologist Milutin Milankovitch proposed what became known as the **Milankovitch hypothesis**—that changes in the shape of Earth's orbit, in precession, and in inclination affect Earth's climate and trigger ice ages. Let's examine each of these three motions in turn.

First, astronomers know that the elliptical shape of Earth's orbit varies slightly over a period of about 100,000 years. At present, Earth's orbit carries it 1.7 percent closer than average to the sun during northern-hemisphere winters and 1.7 percent farther away in northern-hemisphere summers. This makes the northern climate very slightly less extreme, and that is critical—most of the landmass where ice can accumulate is in the northern hemisphere. If Earth's orbit became more elliptical, northern summers might be too cool to melt all of the snow and ice from the previous winter. That would make glaciers grow larger.

A second factor is also at work. Precession causes Earth's axis to sweep around a cone with a period of about 26,000 years, and that changes the location of the seasons around Earth's orbit. Northern summers now occur when Earth is 1.7 percent farther from the sun, but in 13,000 years northern summers will occur on the other side of Earth's orbit where Earth is slightly closer to the sun. Northern summers will be warmer, which could melt all of the previous winter's snow and ice and prevent the growth of glaciers.

The third factor is the inclination of Earth's equator to its orbit. Currently at 23.5°, this angle varies from 22° to 24° with a period of roughly 41,000 years. When the inclination is greater, seasons are more severe.

In 1920, Milankovitch proposed that these three factors cycled against each other to produce complex periodic variations in Earth's climate and the advance and retreat of glaciers (■ Figure 2-12a). But no evidence was available to test the theory in 1920, and scientists treated it with skepticism. Many thought it was laughable.

The Evidence

By the middle 1970s, Earth scientists could collect the data that Milankovitch needed. Oceanographers could drill deep into the seafloor and collect samples, and geologists could determine the age of the samples from the natural radioactive atoms they contained. From all this, scientists constructed a history of ocean temperatures that convincingly matched the predictions of the Milankovitch hypothesis (Figure 2-12b).

The evidence seemed very strong, and, by the 1980s, the Milankovitch hypothesis was widely discussed as the leading hypothesis. But science follows a mostly unstated set of rules that holds that a hypothesis must be tested over and over against all available evidence (**Window on Science 2-5**). In 1988, scientists discovered contradictory evidence.

A water-filled crack in Nevada called Devil's Hole contains deposits of the mineral calcite. Diving with scuba gear, scientists drilled out samples of the calcite and analyzed the oxygen atoms found there. For 500,000 years, layers of calcite have built up in Devil's Hole, recording in their oxygen atoms the temperature of the atmosphere when rain fell there. Finding the ages of the mineral samples was difficult, but the results seemed to show that the previous ice age ended thousands of years too early to have been caused by Earth's motions.

These contradictory findings are irritating because we naturally prefer certainty, but such circumstances are common in science. The disagreement between ocean floor samples and Devil's Hole samples triggered a scramble to understand the problem. Were the ages of one or the other set of samples wrong? Were the ancient temperatures wrong? Or were scientists misunderstanding the significance of the evidence?

In 1997, a new study of the ages of the samples confirmed that those from the ocean floor are correctly dated. This seems to

Window on Science 2-5

The Foundation of Science: Evidence

Science is based on evidence. Every theory and conclusion must be supported by evidence obtained from experiments or from observation. If a theory is supported by many pieces of evidence but is clearly contradicted by a single experiment or observation, scientists quickly abandon it. For a theory to be true, there can be no contradictory evidence.

Of course, scientists argue about the significance of particular evidence and often disagree on the interpretation of evidence. Some observations may seem significant at first glance, but upon closer examination you may find that the procedure was flawed and so the piece of evidence is not important. Or you might conclude that the observational fact is correct but is being misinterpreted. The observation may not mean what it seems. Some of the most famous disagreements in science, such as those surrounding Galileo and Darwin, have arisen over the interpretation of well-established factual evidence.

Furthermore, scientists are not allowed to be selective in considering evidence. A lawyer in court can call a certain witness and intentionally fail to ask a critical question that would reveal evidence harmful to the lawyer's case. But a scientist may not ignore any known evidence. The difference in their methods is revealing. The lawyer is attempting to prove only one side of the case and rightly may ignore contradictory evidence. The scientist, however, is searching for the truth and so must test any theory against all available evidence. In a sense, the scientist, in dealing with evidence, must act as both the prosecution and the defense.

As you read about any science, look for the evidence in the form of measurements or observations. Every theory or conclusion should have supporting evidence. If you can find and understand the evidence, the science will make sense. All scientists, from astronomers to zoologists, demand evidence. You should, too.

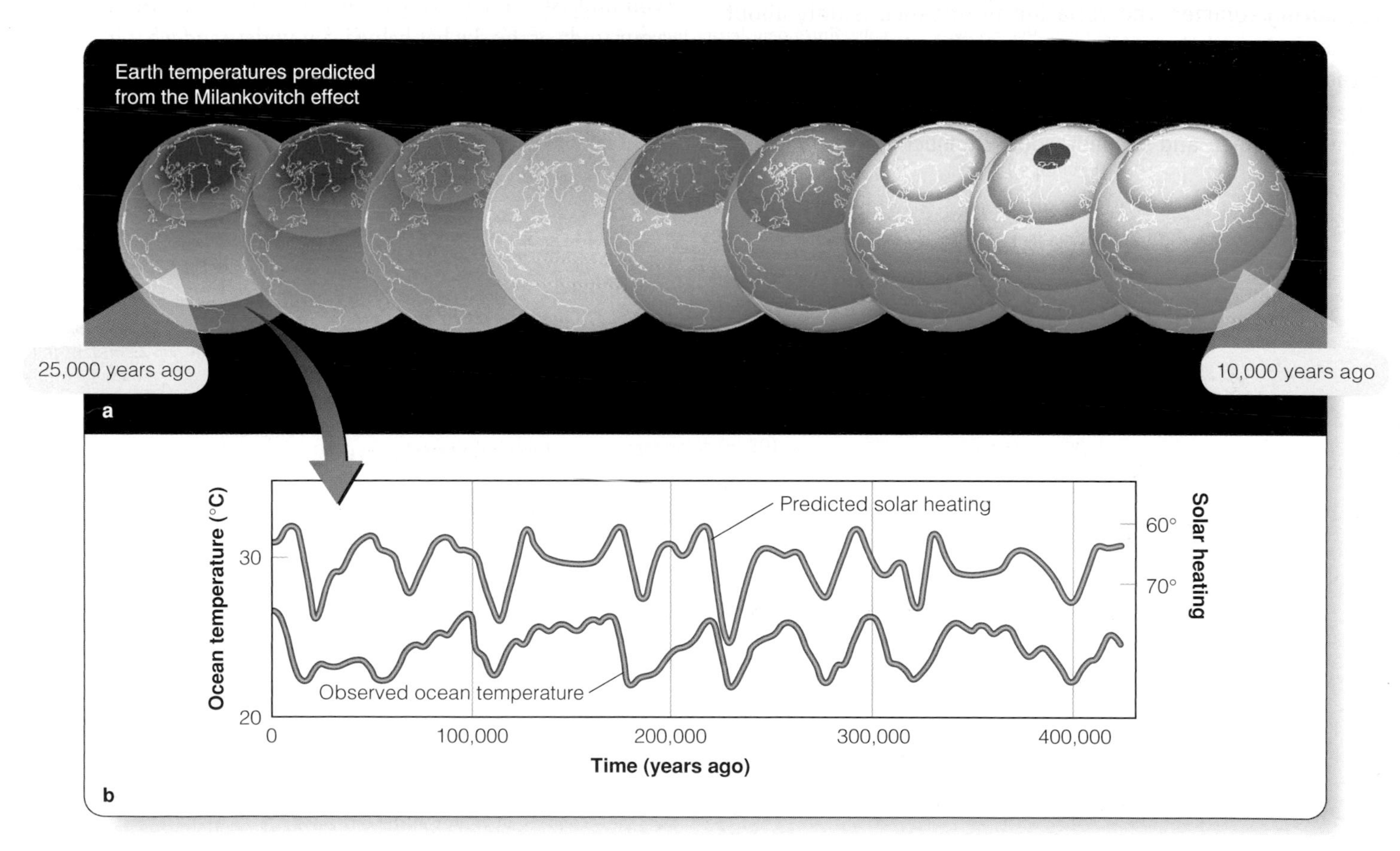

■ Figure 2-12

(a) Mathematical models of the Milankovich effect can be used to predict temperatures on Earth over time. In these Earth globes, cool temperatures are represented by violet and blue and warm temperatures by yellow and red. These globes show the warming that occurred beginning 25,000 years ago, which ended the last ice age. (Courtesy Arizona State University, Computer Science and Geography Departments) (b) Over the last 400,000 years, changes in ocean temperatures measured from fossils found in sediment layers from the seabed match calculated changes in solar heating. (Adapted from Cesare Emiliani)

3 Cycles of the Moon

Even a man who is
pure in heart
And says his prayers by night
May become a wolf when
the wolfbane blooms
And the moon shines
full and bright.

PROVERB FROM OLD WOLFMAN MOVIES

Visual-wavelength image

DID ANYONE EVER WARN YOU, "Don't stare at the moon—you'll go crazy"?* For centuries, the superstitious have associated the moon with insanity. The word *lunatic* comes from a time when even doctors thought that the insane were "moonstruck." A "mooncalf" is someone who has been crazy since birth, and the word is probably related to the belief that moonlight can harm unborn children. Everyone "knows" that people act less rationally when the moon is bright, but everyone is

Continued on page 38

* When I was very small, my grandmother told me if I gazed at the moon, I might go crazy. But it was too beautiful, and I ignored her warning. I secretly watched the moon from my window, became fascinated by the sky, and grew up to be an astronomer.

A total solar eclipse is a lunar phenomenon. It occurs when the moon crosses in front of the sun and hides its brilliant surface. Then you can see the sun's extended atmosphere. (NSO/AURA/NSF)

Guidepost

Looking Back

In the preceding chapter, you watched Earth rotating on its axis, which makes the sun rise and set. You watched Earth revolving along its orbit around the sun, which produces the cycle of the seasons. These two cycles so completely dominate your life on Earth that you hardly notice them. But there are other cycles in the sky, and now you are ready to study one of the most dramatic and beautiful actors on the celestial stage.

This Chapter

The moon is the brightest object in the night sky, and it moves rapidly against the background of stars, changing its shape and occasionally producing strange events called eclipses. You may feel you know the moon well, but you will probably find some surprises as you answer four essential questions about Earth's satellite:

Why does the moon go through phases?

What causes a lunar eclipse?

What causes a solar eclipse?

How can eclipses be predicted?

This chapter illustrates two powerful ways to analyze certain kinds of problems. Many problems in astronomy depend on seeing how light and shadow move through space in three dimensions. Phases and eclipses are good case studies. Also, if a process repeats, you can analyze it by searching for cycles. The seemingly complex motions of the sun and moon become simple and elegant when you see them as cycles.

Looking Ahead

Once you have a 21st-century understanding of your world and its motion, you will be ready to read the next chapter, where you will see how Renaissance astronomers analyzed what they saw in the sky and came to a revolutionary conclusion—that we live on a planet.

Ace Astronomy™ The AceAstronomy icon throughout the text indicates an opportunity for you to test yourself on key concepts and to explore animations and interactions on the AceAstronomy website at: http://ace.brookscole.com/sf9

The Phases of the Moon

1 As the moon orbits Earth, it rotates to keep the same side facing Earth as shown at right. Consequently you always see the same features on the moon, and you never see the far side of the moon. A mountain on the moon that points at Earth will always point at Earth as the moon revolves and rotates.

(Not to scale)

Ace Astronomy™
Log into AceAstronomy and select this chapter to see Active Figure "Lunar Phases" and take control of this diagram.

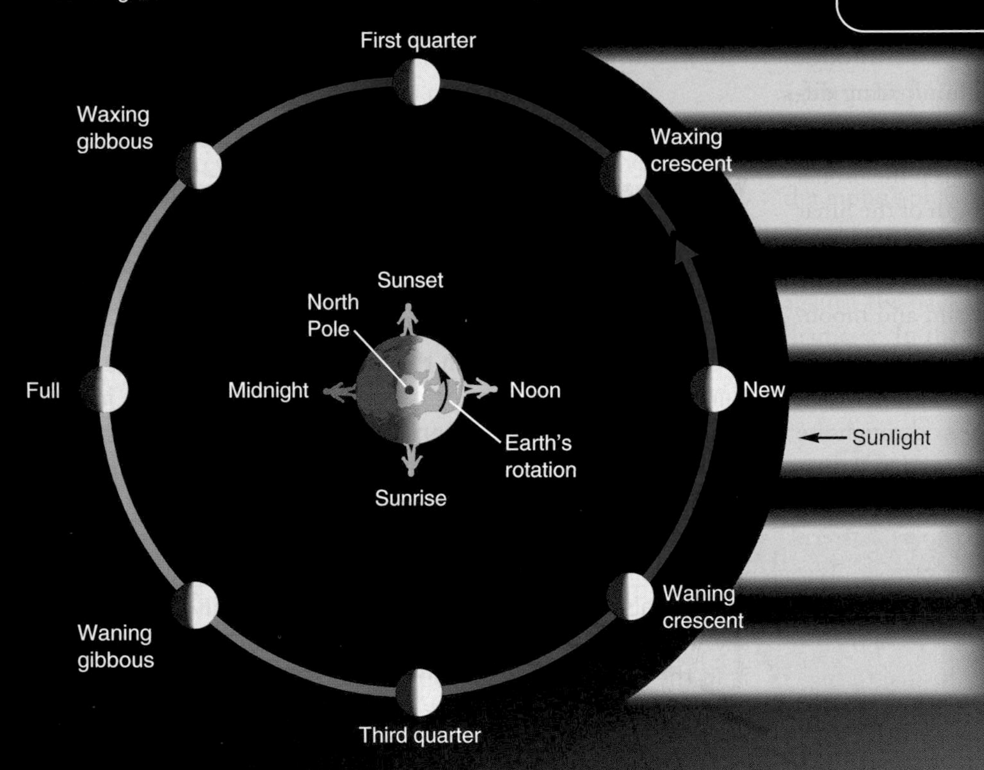

2 As seen at left, sunlight always illuminates half of the moon. Because you see different amounts of this sunlit side, you see the moon cycle through phases. At the phase called "new moon," sunlight illuminates the far side of the moon, and the side you see is in darkness. At new moon you see no moon at all. At full moon, the side you see is fully lit, and the far side is in darkness. How much you see depends on where the moon is in its orbit.

Notice that there is no such thing as the "dark side of the moon." All parts of the moon experience day and night in a month-long cycle.

In the diagram at the left, you see that the new moon is close to the sun in the sky, and the full moon is opposite the sun. The time of day depends on the observer's location on Earth.

2a The first two weeks of the cycle of the moon are shown below by its position at sunset on 14 successive evenings. As the moon grows fatter from new to full, it is said to wax.

The first quarter moon is one week through its 4-week cycle.

Gibbous comes from the Latin word for humpbacked.

Waxing gibbous

Waxing crescent

The full moon is two weeks through its 4-week cycle.

New moon is invisible near the sun

Full moon rises at sunset

THE SKY AT SUNSET

East

South

West

3 The moon orbits eastward around Earth in 27.32 days, its **sidereal period.** This is how long the moon takes to circle the sky once and return to the same position among the stars.

A complete cycle of lunar phases takes 29.53 days, the moon's **synodic period.** (Synodic comes from the Greek words for "together" and "path.")

To see why the synodic period is longer than the sidereal period, study the star charts at the right.

Although you think of the lunar cycle as being about 4 weeks long, it is actually 1.53 days longer than 4 weeks. The calendar divides the year into 30-day periods called months (literally "moonths") in recognition of the 29.53 day synodic cycle of the moon.

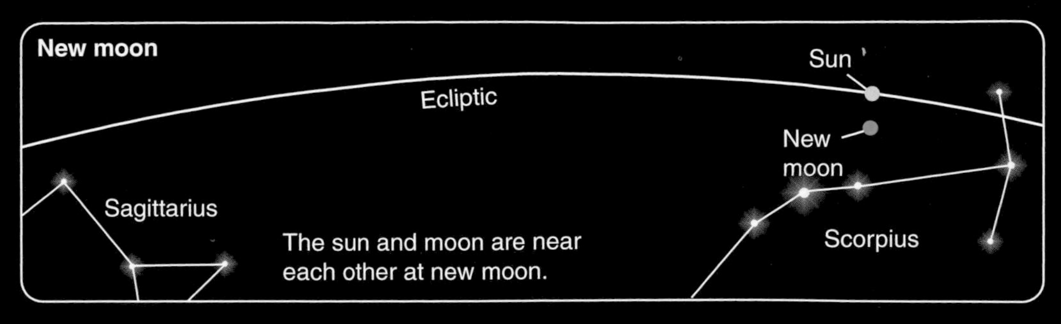
New moon
Sun
Ecliptic
New moon
Sagittarius
Scorpius
The sun and moon are near each other at new moon.

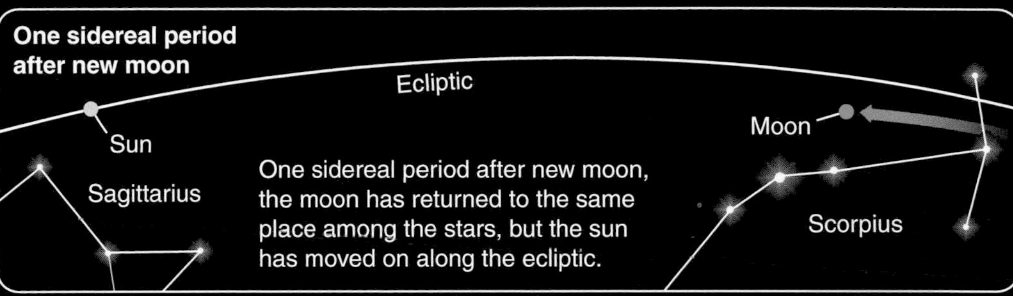
One sidereal period after new moon
Ecliptic
Sun
Moon
Sagittarius
Scorpius
One sidereal period after new moon, the moon has returned to the same place among the stars, but the sun has moved on along the ecliptic.

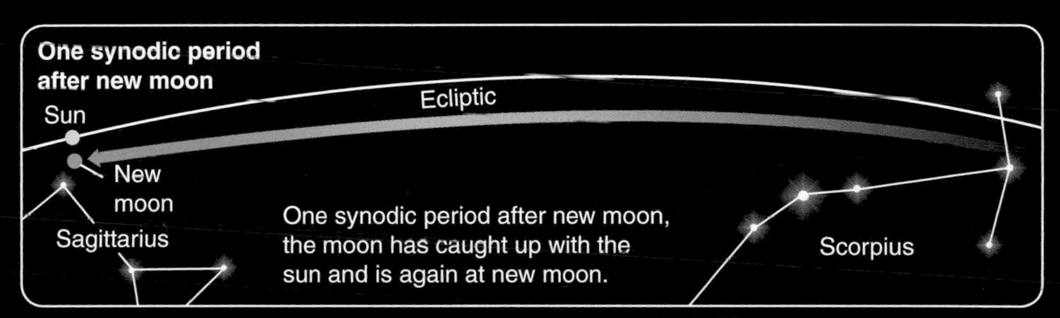
One synodic period after new moon
Sun
Ecliptic
New moon
Sagittarius
Scorpius
One synodic period after new moon, the moon has caught up with the sun and is again at new moon.

You can use the diagram on the opposite page to determine when the moon rises and sets at different phases.

TIMES OF MOONRISE AND MOONSET

Phase	Moonrise	Moonset
New	Dawn	Sunset
First quarter	Noon	Midnight
Full	Sunset	Dawn
Third quarter	Midnight	Noon

2b The last two weeks of the cycle of the moon are shown below by its position at sunrise on 14 successive mornings. As the moon shrinks from full to new, it is said to wane.

The first quarter moon is 3 weeks through its 4-week cycle.

Waning crescent

Waning gibbous

New moon is invisible near the sun

< two weeks >

THE SKY AT SUNRISE

Full moon sets at sunrise

East

South

West

A simple scientific argument analyzing the motion of the moon can explain a lot about what you see and what you don't see. **Why is it extremely difficult to see the crescent moon in the daytime?**

■ ■ ■

Connections: The phases of the moon don't affect you directly—you don't act crazier than usual at full moon. The phases are a decoration in the sky, but occasionally, the moon provides a more dramatic treat.

3-2 Lunar Eclipses

A **LUNAR ECLIPSE** occurs at full moon when the moon moves through Earth's shadow. Because the moon shines only by reflected sunlight, it gradually darkens as it enters the shadow.

Earth's Shadow

Earth's shadow consists of two parts. The **umbra** is the region of total shadow. If you were floating in space in the umbra of Earth's shadow, you would see no portion of the sun; it would be completely hidden behind Earth. However, if you moved into the **penumbra,** you would be in partial shadow and would see part of the sun peeking around Earth's edge. In the penumbra, the sunlight is dimmed but not extinguished.

You can construct a model of Earth's shadow by pressing a map tack into the eraser of a pencil and holding the tack between a lightbulb a few feet away and a white cardboard screen (■ Figure 3-3). The lightbulb represents the sun, and the map tack represents Earth. If you hold the screen close to the tack, you will see that the umbra is nearly as large as the tack and that the penumbra is only slightly larger. However, if you move the screen away from the tack, the umbra shrinks and the penumbra expands. Beyond a certain point, the shadow has no dark core at all, indicating that the screen is beyond the end of the umbra.

The umbra of Earth's shadow is about 1.4 million km (860,000 mi) long and points directly away from the sun. A giant screen placed in the shadow at the average distance of the moon would reveal a dark umbra about 9000 km (5700 mi) in diameter, and the faint outer edges of the penumbra would mark a circle about 16,000 km (10,000 mi) in diameter. For comparison, the moon's diameter is only 3476 km (2160 mi). Consequently, when the moon's orbit carries it through the umbra, it has plenty of room to become completely immersed in shadow.

Total Lunar Eclipses

A lunar eclipse occurs when the moon passes through Earth's shadow and grows dark. If the moon passes through the umbra and no part of the moon remains outside the umbra in the partial sunlight of the penumbra, the eclipse is a **total lunar eclipse.**

■ Figure 3-4 illustrates the stages of a total lunar eclipse as a three-dimensional diagram (**Window on Science 3-1**) showing Earth, its shadows, and the path of the moon. As the moon begins to enter the penumbra, it is only slightly dimmed, and a casual observer may not notice anything odd. After an hour, the moon is deeper in the penumbra and dimmer; and, once it begins to enter the umbra, you see a dark bite on the edge of the lunar disk. The moon travels its own diameter in an hour, so it takes about an hour to enter the umbra completely and become totally eclipsed.

Even when the moon is totally eclipsed, it does not disappear completely. Sunlight, bent by Earth's atmosphere, leaks into the umbra and bathes the moon in a faint glow. Because blue

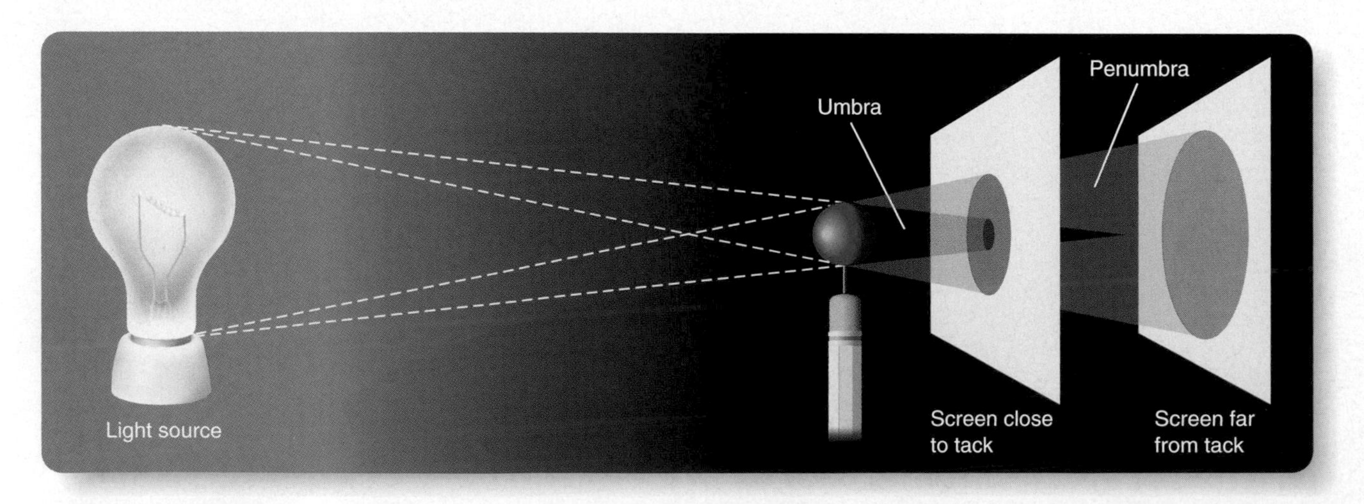

■ Figure 3-3

The shadows cast by a map tack resemble the shadows of Earth and the moon. The umbra is the region of total shadow; the penumbra is the region of partial shadow.

Window on Science | 3-1

3-D Relationships

Much of science is explained in diagrams; one particular type of diagram can be confusing. When artists draw three-dimensional diagrams on flat sheets of paper, they use lots of artistic clues that have been discovered over the centuries. Perspective, shading, color, and shadows help you see the three-dimensional figure if the drawing is familiar—a house, for example. But when a drawing shows something that is unfamiliar, as is often the case in science, the artist's clues don't work as well. When you look at drawings of a molecule, a nerve cell, layers of rock under the ocean, or Earth and its shadows, you must pay special attention to the three-dimensional nature of the figure.

When you see a three-dimensional diagram in any science book, it helps to decide what the point of view is. For example, if the diagram were a photograph, where was the camera? In the drawing in Figure 3-7, the camera would have to have been in space looking back at Earth and the moon. Of course, astronauts have never been that far from Earth, but you can make such a voyage in your imagination, and that helps you understand the geometry of eclipses. Other three-dimensional diagrams have other points of view, so always be sure to first determine the point of view when you look at any three-dimensional diagram.

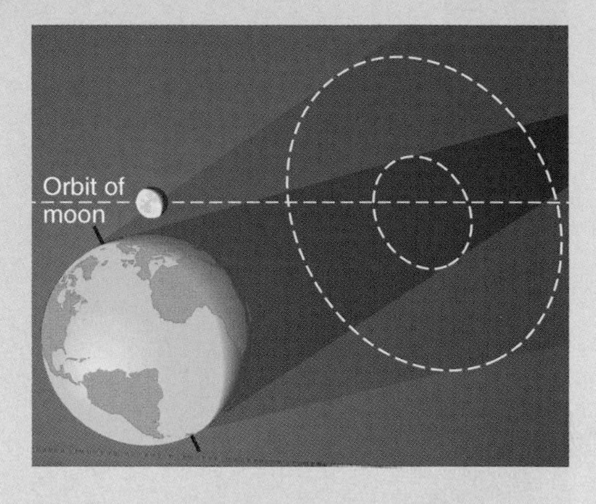
Orbit of
moon

If this were a photograph, where would the photographer have been located?

light is scattered more easily than red light, it is red light that penetrates through Earth's atmosphere to illuminate the moon in a coppery glow. If you were on the moon during totality and looked back at Earth, you would not see any part of the sun because it would be entirely hidden behind Earth. However, you would see Earth's atmosphere illuminated from behind by the sun in a spectacular sunset completely ringing Earth. It is the red glow from this sunset that gives the totally eclipsed moon its reddish color.

How dim the totally eclipsed moon becomes depends on a number of things. If Earth's atmosphere is especially cloudy in those regions that must bend light into the umbra, the moon will be darker than usual. An unusual amount of dust in Earth's atmosphere (from volcanic eruptions, for instance) also causes a dark eclipse. Also, total lunar eclipses tend to be darkest when the moon's orbit carries it through the center of the umbra (■ Figure 3-5).

As the moon moves through Earth's umbral shadow, you can see that the shadow is circular. From this the Greek philosopher Aristotle (384–322 BC) concluded that Earth had to be a sphere, because only a sphere could cast a shadow that was always circular.

Depending on the geometry of the eclipse, the moon can take as long as 1 hour 40 minutes to cross the umbra and another hour to emerge into the penumbra. Still another hour passes as it emerges into full sunlight. A total eclipse of the moon, including the penumbral stage, can take almost six hours from start to finish.

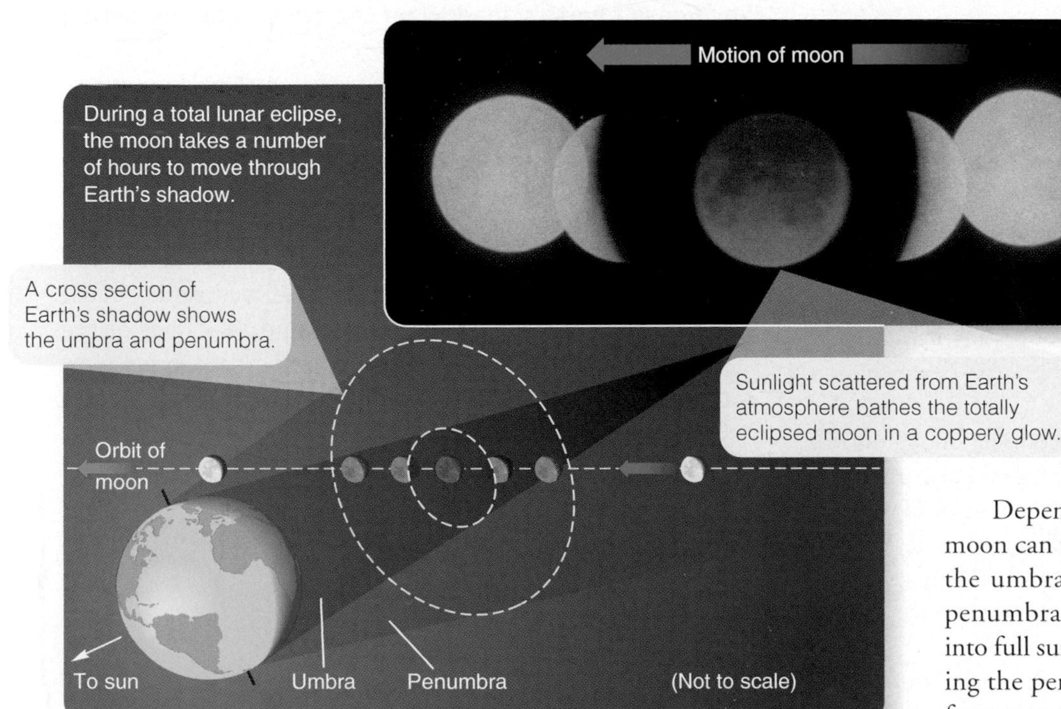
Motion of moon
During a total lunar eclipse, the moon takes a number of hours to move through Earth's shadow.
A cross section of Earth's shadow shows the umbra and penumbra.
Sunlight scattered from Earth's atmosphere bathes the totally eclipsed moon in a coppery glow.
Orbit of moon
To sun
Umbra
Penumbra
(Not to scale)

■ Figure 3-4

During a total lunar eclipse, the moon passes through Earth's shadow, as shown in this multiple-exposure photograph. A longer exposure was used to record the moon while it was totally eclipsed. The moon's path appears curved in the photo because of photographic effects. (© 1982 Dr. Jack B. Marling)

■ **Figure 3-5**

During a total lunar eclipse, the moon turns coppery red. In this photo, the moon is darkest toward the lower right, the direction toward the center of the umbra. The edge of the moon at upper left is brighter because it is near the edge of the umbra. (Celestron International)

■ **Table 3-1 | Total and Partial Eclipses of the Moon, 2006 to 2013**

Year	Time* of Mideclipse (GMT)	Length of Totality (Hr:Min)	Length of Eclipse† (Hr:Min)
2006 Sept. 7	18:52	Partial	1:30
2007 Mar. 3	23:22	1:14	3:40
2007 Aug. 28	10:38	1:30	3:32
2008 Feb. 21	3:27	0:50	3:24
2008 Aug. 16	21:11	Partial	3:08
2009 Dec. 31	19:24	Partial	1:00
2010 June 26	11:40	Partial	2:42
2010 Dec. 21	8:18	1:12	3:28
2011 June 15	20:13	1:40	3:38
2011 Dec. 10	14:33	0:50	3:32
2012 June 4	11:03	Partial	2:08

*Times are Greenwich Mean Time. Subtract 5 hours for Eastern Standard Time, 6 hours for Central Standard Time, 7 hours for Mountain Standard Time, and 8 hours for Pacific Standard Time. From your time zone, lunar eclipses that occur between sunset and sunrise will be visible, and those at midnight will be best placed.

†Does not include penumbral phase.

Partial and Penumbral Lunar Eclipses

Not all eclipses of the moon are total. Because the moon's orbit is inclined a bit over 5° to the plane of Earth's orbit, the moon might not pass through the center of the umbra.

If the moon's orbit carries the full moon too far north or south of the umbra, the moon may only partially enter the umbra. The resulting **partial lunar eclipse** is usually not as dramatic as a total lunar eclipse. Because part of the moon remains outside the umbra, it receives some direct sunlight and looks bright in contrast with the dark part of the moon inside the umbra. Unless the moon almost completely enters the umbra, the glare from the illuminated part of the moon drowns out the fainter red glow inside the umbra. Partial lunar eclipses are interesting because part of the full moon is darkened, but they are not as beautiful as a total lunar eclipse.

If the orbit of the moon carries the moon far enough north or south of the umbra, the moon may only pass through the penumbra and never reach the umbra. Such **penumbral eclipses** are not dramatic at all. In the partial shadow of the penumbra, the moon is only partially dimmed. Most people glancing at a penumbral eclipse would not notice any difference from a full moon.

Total, partial, or penumbral, lunar eclipses are interesting events in the night sky and are not difficult to observe. When the moon passes through Earth's shadow, the eclipse is visible from anywhere on Earth's dark side. One or two lunar eclipses occur in most years. Consult ■ Table 3-1 to find the next lunar eclipse visible in your part of the world.

Building Scientific Arguments

Why doesn't Earth's shadow on the moon look red during a partial lunar eclipse?

Once again, you need to build an argument based on geometry, but now you must consider the direction of sunlight. During a partial lunar eclipse, part of the moon protrudes from Earth's umbral shadow into sunlight. This part of the moon is very bright compared to the fainter red light inside Earth's shadow, and the glare of the reflected sunlight makes it difficult to see the red glow. If a partial eclipse is almost total, so that only a small sliver of moon extends out of the shadow into sunlight, you can sometimes detect the red glow in the shadow.

Of course, this red glow does not happen for every planet–moon combination in the universe. Adapt your argument for a new situation. **Would a moon orbiting a planet that had no atmosphere glow red during a total eclipse? Why or why not?**

■ ■ ■

Connections: Lunar eclipses are slow and stately. For drama and excitement, there is nothing like a solar eclipse, as you will discover in the next section.

3-3 Solar Eclipses

A SOLAR ECLIPSE OCCURS when the moon moves between Earth and the sun. If the moon covers the disk of the sun completely, the eclipse is a **total solar eclipse.** If the moon covers only part of the sun, the eclipse is a **partial solar eclipse.** During a particular solar eclipse, people in one place on Earth may see a total eclipse, while people only a few hundred kilometers away see a partial eclipse.

These spectacular sights are possible because we on Earth are very lucky. Our moon has the same angular diameter as our sun, so it can cover the sun almost exactly. That lucky coincidence allows you to see total solar eclipses.

The Angular Diameter of the Sun and Moon

You learned about the angular diameter of an object in Chapter 2; now you need to think carefully about how the size and distance of an object like the moon determine its angular diameter. This is the key to understanding solar eclipses.

Linear diameter is simply the distance between an object's opposite sides. You use linear diameter when you order a 16-inch pizza—the pizza is 16 inches in diameter. The linear diameter of the moon is 3476 km. The angular diameter of an object is the angle formed by lines extending toward you from opposite sides of the object and meeting at your eye (■ Figure 3-6). Clearly, the farther away an object is, the smaller its angular diameter.

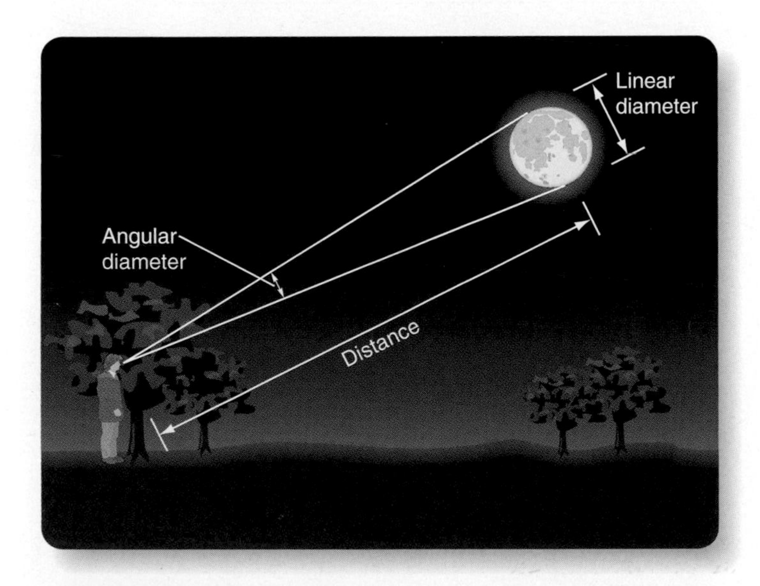

■ Active Figure 3-6

The angular diameter of an object is related to its linear diameter and also to its distance.

AceAstronomy™ Log into AceAstronomy and select this chapter to see the Active Figure called "Small-Angle Formula." Notice that both distance and linear diameter affect an object's angular diameter.

The **small-angle formula** gives you a way to figure out the angular diameter of any object, whether it is a pizza, the moon, or a galaxy. In the small-angle formula, you should always express angular diameter in seconds of arc* and always use the same units for distance and linear diameter:

$$\frac{\text{angular diameter}}{206{,}265''} = \frac{\text{linear diameter}}{\text{distance}}$$

You can use this formula to find any one of these three quantities if you know the other two; here you are interested in finding the angular diameter of the moon.

The moon has a linear diameter of 3476 km and a distance from Earth of about 384,000 km. What is its angular diameter? The moon's linear diameter and distance are both given in the same units, kilometers, so you can put them directly into the small-angle formula:

$$\frac{\text{angular diameter}}{206{,}265''} = \frac{3476 \text{ km}}{384{,}000 \text{ km}}$$

To solve for angular diameter, you can multiply both sides by 206,265 and find that the angular diameter is 1870 seconds of arc. If you divide by 60, you get 31 minutes of arc or, dividing by 60 again, about 0.5°. The moon's orbit is slightly elliptical, so it can sometimes look a bit larger or smaller, but its angular diameter is always close to 0.5°.

You can repeat this calculation for the angular diameter of the sun. The sun is 1.39×10^6 km in linear diameter and 1.50×10^8 km from Earth. If you put these numbers into the small-angle formula, you will discover that the sun has an angular diameter of 1900 seconds of arc, which is 32 minutes of arc or about 0.5°. Earth's orbit is slightly elliptical, and consequently the sun can sometimes look slightly larger or smaller, but it, like the moon, is always close to 0.5° in angular diameter.

By fantastic good luck, you live on a planet with a moon that is almost exactly the same angular diameter as your sun. When the moon passes in front of the sun, it is almost exactly the right size to block the brilliant surface of the sun. Then you see the most exciting sight in astronomy—a total solar eclipse. There are few other worlds where this can happen, because the angular diameters of the sun and a moon rarely match so closely. To see this beautiful sight, all you have to do is arrange to be in the moon's shadow when the moon crosses in front of the sun. That means you have to chase the moon's shadow wherever it sweeps over Earth's surface.

The Moon's Shadow

Like Earth's shadow, the moon's shadow consists of a central umbra of total shadow and a penumbra of partial shadow. What you see when the moon crosses in front of the sun depends on where you

* The number 206,265″ is the number of seconds of arc in a radian. When you divide by 206,265″, you convert the angle from seconds of arc to radians.

are in the moon's shadow. The moon's umbral shadow produces a spot of darkness roughly 269 km (167 mi) in diameter on Earth's surface (■ Figure 3-7). (The exact size of the umbral shadow depends on the location of the moon in its elliptical orbit and the angle at which the shadow strikes Earth.) If you are in this spot of total shadow, you see a total solar eclipse. If you are just outside the umbral shadow but in the penumbra, you see part of the sun peeking around the moon, and the eclipse is partial. Of course, if you are outside the penumbra, you see no eclipse at all.

Because of the orbital motion of the moon, its shadow sweeps across Earth at speeds of at least 1700 km/h (1060 mph). To be sure of seeing a total solar eclipse, you must select an appropriate eclipse (■ Table 3-2), plan far in advance, travel to the right place on Earth, and place yourself in the **path of totality,** the path swept out by the umbral spot.

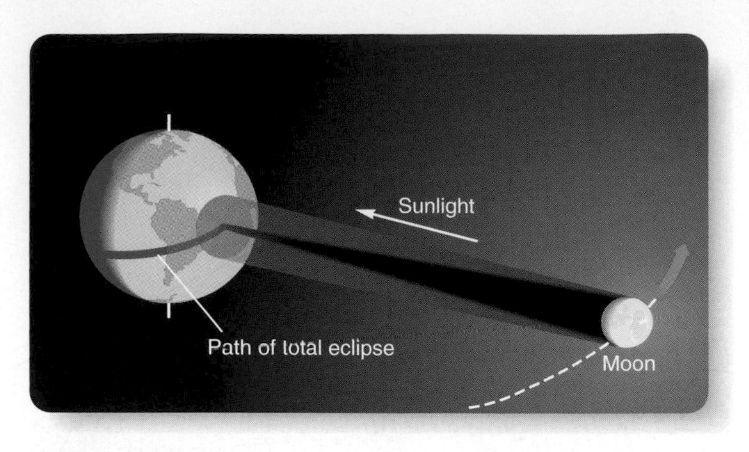

■ **Figure 3-7**

Observers in the path of totality see a total solar eclipse when the umbral shadow sweeps over them. Those in the penumbra see a partial eclipse.

Total Solar Eclipses

A total solar eclipse begins when you first see the edge of the moon encroaching on the sun's disk. This is the moment when the edge of the penumbra sweeps over your location.

During the partial phase, part of the sun remains visible, and it is hazardous to look at the eclipse without protection. Dense filters and exposed film do not necessarily provide protection, because some filters do not block the invisible heat radiation (infrared) that can burn the retina of your eyes. This has led officials to warn the public not to look at solar eclipses and has even frightened some people into locking themselves and their children into windowless rooms during eclipses. In fact, the sun is a bit less dangerous than usual during an eclipse because part of the brilliant surface is covered by the moon. But an eclipse is dangerous in that it can tempt people to look at the sun directly and burn their eyes.

The safest and simplest way to observe the partial phases of a solar eclipse is to use pinhole projection. Poke a small pinhole

■ **Table 3-2** | **Total and Annular Eclipses of the Sun, 2006 to 2016****

Date	Total/Annular (T/A)	Time of Mideclipse* (GMT)	Maximum Length of Total or Annular Phase (Min:Sec)	Area of Visibility
2006 Mar. 29	T	10^h	4:07	Atlantic, Africa, Turkey
2006 Sept. 22	A	12^h	7:09	N.E. of S. America, Atlantic
2008 Feb. 7	A	4^h	2:14	S. Pacific, Antarctica
2008 Aug. 1	T	10^h	2:28	Canada, Arctic, Siberia
2009 Jan. 26	A	8^h	7:56	S. Atlantic, Indian Ocean
2009 July 22	T	3^h	6:40	Asia, Pacific
2010 Jan. 15	A	7^h	11:10	Africa, Indian Ocean
2010 July 11	T	20^h	5:20	Pacific, S. America
2012 May 20	A	23^h	5:46	Japan, N. Pacific, W. U.S.
2012 Nov. 13	T	22^h	4:02	Australia, S. Pacific
2013 May 10	A	0^h	6:04	Australia, Pacific
2013 Nov. 3	AT	13^h	1:40	Atlantic, Africa
2015 March 20	T	10^h	2:47	N. Atlantic, Arctic
2016 March 9	T	2^h	4:10	Borneo, Pacific
2016 Sept. 1	A	9^h	3:06	Atlantic, Africa, Indian Oc.

The next major total solar eclipse visible from the United States will occur on August 21, 2017.

*Times are Greenwich Mean Time. Subtract 5 hours for Eastern Standard Time, 6 hours for Central Standard Time, 7 hours for Mountain Standard Time, and 8 hours for Pacific Standard Time.

^{h}hours.

**There are no total or annular eclipses of the sun during 2014.

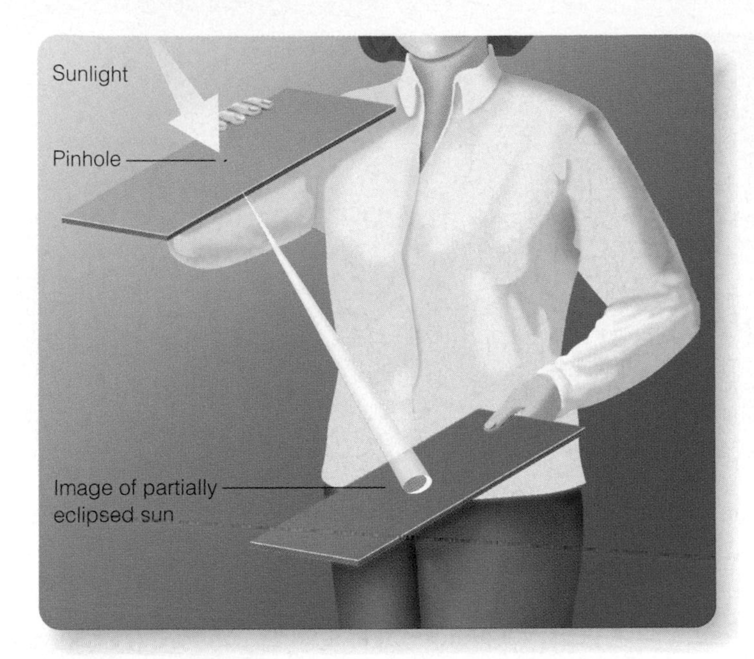

Sunlight
Pinhole
Image of partially eclipsed sun

■ **Figure 3-8**

A safe way to view the partial phases of a solar eclipse. Use a pinhole in a card to project an image of the sun on a second card. The greater the distance between the cards, the larger (and fainter) the image will be.

A Total Solar Eclipse

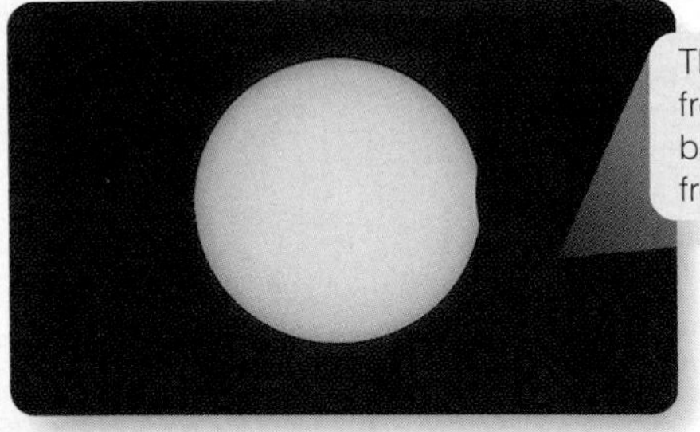

The moon moving from the right just begins to cross in front of the sun.

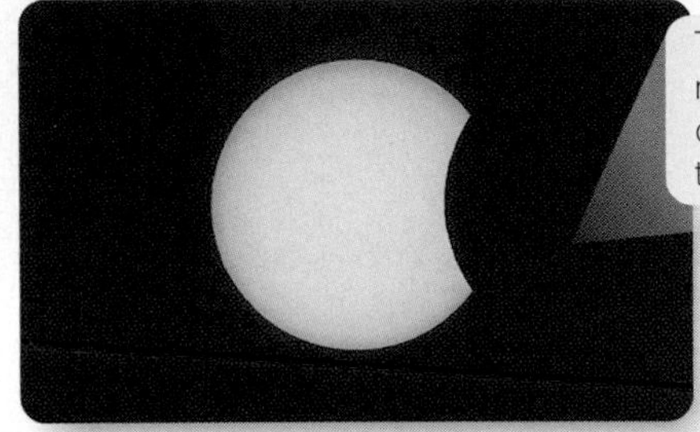

The disk of the moon gradually covers the disk of the sun.

Sunlight begins to dim as more of the sun's disk is covered.

During totality, pink prominences are often visible.

A longer-exposure photograph during totality shows the fainter corona.

■ **Figure 3-9**

This sequence of photos shows the first half of a total solar eclipse. (Daniel Good)

in a sheet of cardboard. Hold the sheet with the hole in sunlight and allow light to pass through the hole to a second sheet of cardboard (■ Figure 3-8). On a day when there is no eclipse, the result is a small, round spot of light that is an image of the sun. During the partial phases of a solar eclipse, the image shows the dark silhouette of the moon obscuring part of the sun. These pinhole images of the partially eclipsed sun can also be seen in the shadows under trees as sunlight peeks through the tiny openings between the leaves and branches. This can produce an eerie effect just before totality as the remaining sliver of sun produces thin crescents of light on the ground under trees.

Throughout the partial phases of a solar eclipse, the moon gradually covers the bright disk of the sun (■ Figure 3-9). Totality begins as the last sliver of the sun's bright surface disappears behind the moon. This is the moment when the edge of the umbra sweeps over your location. So long as any of the sun is visible, the countryside is bright; but, as the last of the sun disappears, dark falls in a few seconds. Automatic streetlights come on, car drivers switch on their headlights, and birds go to roost. The darkness of totality depends on a number of factors, including the weather at the observing site, but it is usually dark enough to make it difficult to read the settings on cameras.

The totally eclipsed sun is a spectacular sight. With the moon covering the bright disk of the sun, called the **photosphere,*** you can see the sun's faint outer atmosphere, the **corona,** glowing with a pale, white light so faint you can safely look at it directly. This corona is made of low-density, hot gas, which is given a wispy appearance by the solar magnetic field, as shown in the last frame of Figure 3-9. Also visible just above the photosphere is a thin

* The photosphere, corona, chromosphere, and prominences will be discussed in detail in Chapter 8. Here the terms are used as the names of features you see during a total solar eclipse.

■ Figure 3-10

(a) During a total solar eclipse, the moon covers the photosphere, and the white corona and pink prominences are visible. Note the streamers in the corona caused by the sun's magnetic field. (Daniel Good) (b) The diamond ring effect can sometimes occur momentarily at the beginning or end of totality if a small segment of the photosphere peeks out through a valley at the edge of the lunar disk. (National Optical Astronomy Observatory)

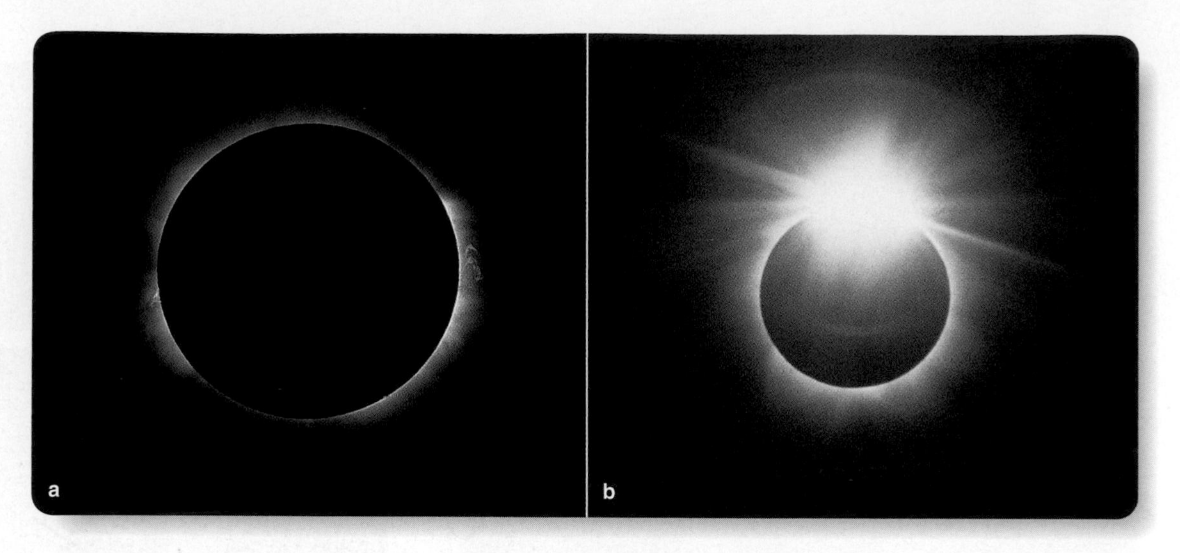

layer of bright gas called the **chromosphere.** The chromosphere is often marked by eruptions on the solar surface called **prominences** (■ Figure 3-10a), which glow with a clear, pink color due to the high temperature of the gases involved. The small-angle formula reveals that a large prominence is about 3.5 times the diameter of Earth.

Totality cannot last longer than 7.5 minutes under any circumstances, and the average is only 2 to 3 minutes. Totality ends when the sun's bright surface reappears at the trailing edge of the moon. This corresponds to the moment when the trailing edge of the moon's umbra sweeps over the observer.

Just as totality begins or ends, a small part of the photosphere can peek out from behind the moon through a valley at the edge of the lunar disk. Although it is intensely bright, such a small part of the photosphere does not completely drown out the fainter corona, which forms a silvery ring of light with the brilliant spot of photosphere gleaming like a diamond (Figure 3-10b). This **diamond ring effect** is one of the most spectacular of astronomical sights, but it is not visible during every solar eclipse. Its occurrence depends on the exact orientation and motion of the moon.

Once totality is over, daylight returns quickly, and the corona and chromosphere vanish. Not too many years ago astronomers traveled great distances to forbidding places to get their instruments into the path of totality and study the faint outer corona visible only during the few minutes of a total solar eclipse. Now many of those observations can be made every day by solar telescopes in space, but eclipse fans still journey to exotic places for the thrill of seeing a total solar eclipse.

Not every solar eclipse is total. Sometimes when the moon crosses in front of the sun, it is too small to fully cover the sun. That can happen because the orbit of the moon is elliptical and its distance from Earth varies. When the moon is at **perigee,** its closest point to Earth, the moon is almost 12 percent larger in angular diameter than when it is at **apogee,** its most distant point from Earth. Because Earth's orbit around the sun is very slightly elliptical, the sun's angular diameter can vary by a total of about 3.4 percent. If a solar eclipse occurs when the moon's angular diameter is too small, you would see an **annular eclipse,** a solar eclipse in which a ring (or annulus) of the photosphere is visible around the disk of the moon (■ Figure 3-11). With a portion of the photosphere visible, the eclipse never becomes total; it never quite gets dark; and you can't see the prominences, chromosphere, and corona. An annular eclipse swept across the United States on May 10, 1994.

Building Scientific Arguments

If you were on Earth watching a total solar eclipse, what would astronauts on the moon see when they looked at Earth?

Building this argument requires that you change your point of view and imagine seeing the geometry from a new direction. Astronauts on the moon could see Earth only if they were on the side that faces Earth. Because solar eclipses always happen at new moon, the near side of the moon would be in darkness, and the far side of the moon would be in full sunlight. The astronauts would be standing in darkness, and they would be looking at the fully illuminated side of Earth. They would see a "full Earth." The moon's shadow would be crossing Earth; and, if the astronauts looked closely, they might be able to see the spot of darkness where the moon's umbral shadow touched Earth. It would take hours for the shadow to cross Earth.

Standing on the moon and watching the moon's umbral shadow sweep across Earth would be a cold, tedious assignment. Perhaps you can imagine a more interesting assignment for the astronauts. **What would astronauts on the moon see while people on Earth were seeing a total lunar eclipse?**

■ ■ ■

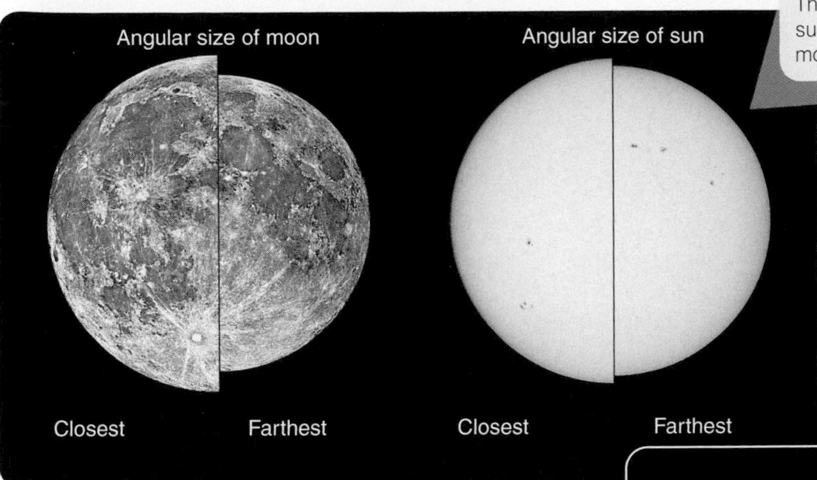

Angular size of moon
Angular size of sun
The angular diameters of the moon and sun vary slightly because the orbits of the moon and Earth are slightly eliptical.
Closest
Farthest
Closest
Farthest

If the moon is too far from Earth during a solar eclipse, the umbra does not reach Earth's surface.
Sunlight
Path of annular eclipse
Moon

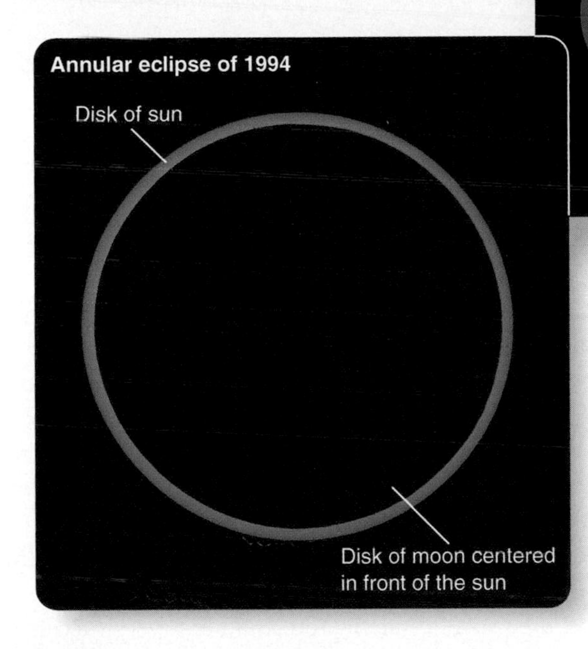

Annular eclipse of 1994
Disk of sun
Disk of moon centered in front of the sun

■ Figure 3-11

An annular eclipse occurs when the moon is far enough from Earth that its umbral shadow does not reach Earth's surface. From Earth, you see an annular eclipse because the moon's angular diameter is smaller than the angular diameter of the sun. In the photograph of the annular eclipse of 1994, the dark disk of the moon is almost exactly centered on the bright disk of the sun. (Moon: UCO/Lick Observatory; Sun: Daniel Good)

Connections: Eclipses of the sun and moon are often dramatic and mysterious. Ancient astronomers studied them and found ways to predict the coming of an eclipse. In the section that follows, you will see how eclipses can be predicted from a basic understanding of the motion of the sun and the moon.

3-4 Predicting Eclipses

TO MAKE EXACT ECLIPSE PREDICTIONS, you would need to calculate the precise motions of the sun and moon, and that requires a computer and proper software. Such software is available for desktop computers, but it isn't necessary if you are satisfied with making less exact predictions. In fact, many primitive peoples, such as the builders of Stonehenge and the ancient Maya, are believed to have made eclipse predictions.

You should think about eclipse prediction for three reasons. First, it is an important part of the history of science. Second, it illustrates how apparently complex phenomena can be analyzed in terms of cycles. Third, eclipse prediction will exercise your mental muscles and force you to see Earth, the moon, and the sun as objects moving through space.

Conditions for an Eclipse

You can predict eclipses by understanding the conditions that make them possible. As you begin to think about these conditions, be sure you are aware of your point of view. (See Window on Science 3-1.) Later you will change your point of view, but to begin you must imagine that you can look up into the sky from your home on Earth and see the sun moving along the ecliptic and the moon moving along its orbit.

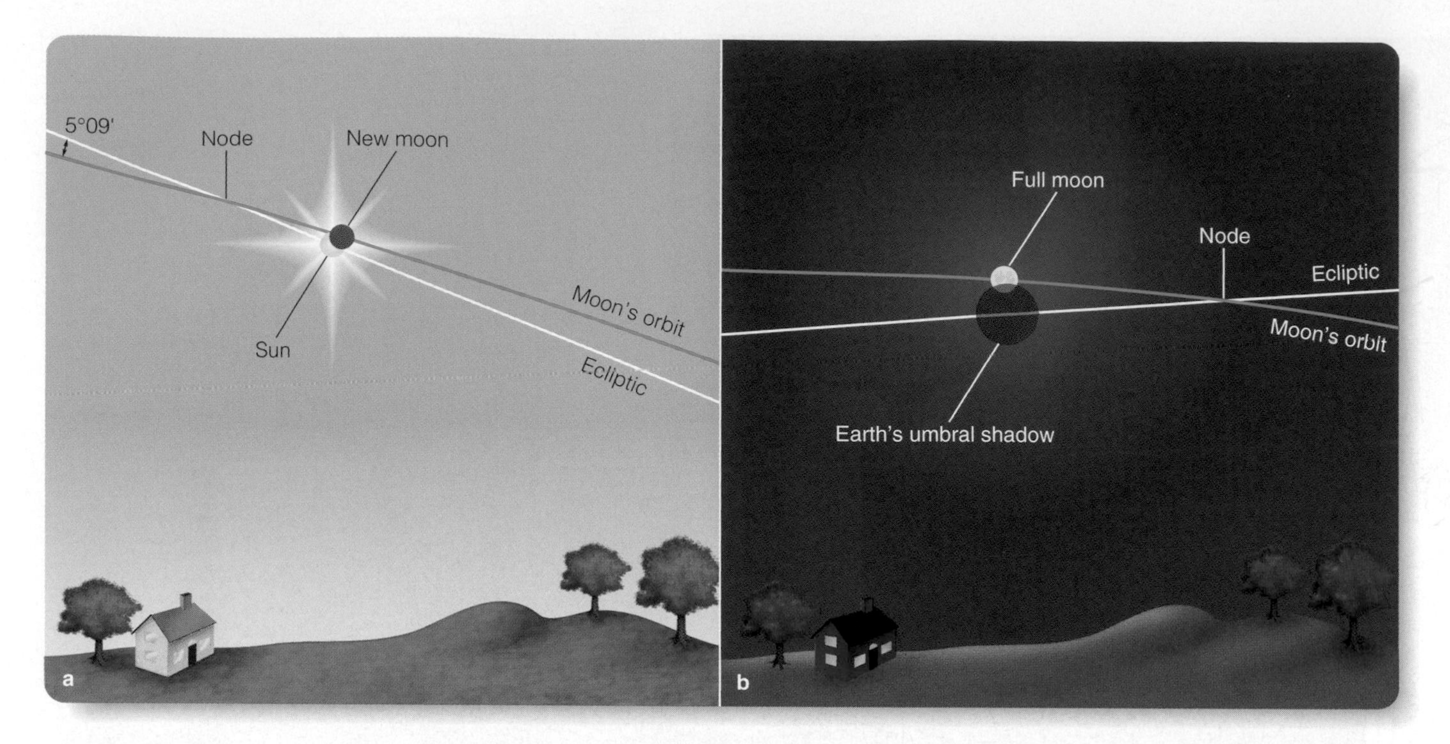

■ Figure 3-12

Eclipses can occur only near the nodes of the moon's orbit. (a) A solar eclipse occurs when the moon meets the sun near a node. (b) A lunar eclipse occurs when the sun and moon are near opposite nodes. Partial eclipses are shown here for clarity.

The orbit of the moon is tipped 5°8'43" to the plane of Earth's orbit, so you see the moon follow a path tipped by that angle to the ecliptic. Each month, the moon crosses the ecliptic at two points called **nodes.** It crosses at one node going southward, and two weeks later it crosses at the other node going northward.

Eclipses can only occur when the sun is near one of the nodes of the moon's orbit. A solar eclipse happens at new moon if the moon passes in front of the sun. Most new moons pass too far north or too far south of the sun to cause an eclipse. Only when the sun is near a node in the moon's orbit can the moon cross in front of the sun, as shown in ■ Figure 3-12a. A lunar eclipse doesn't happen at every full moon because most full moons pass too far north or too far south of the ecliptic and miss Earth's shadow. The moon can enter Earth's shadow only when the shadow is near a node in the moon's orbit, and that means the sun must be near the other node. This is shown in Figure 3-12b.

So there are two conditions for an eclipse: The sun must be crossing a node, and the moon must be crossing either the same node (solar eclipse) or the other node (lunar eclipse). That means, of course, that solar eclipses can occur only when the moon is new, and lunar eclipses can occur only when the moon is full.

Now you know the secret of predicting eclipses. An eclipse can only occur during a period called an **eclipse season** during which the sun is close to a node in the moon's orbit. For solar eclipses, an eclipse season is about 32 days long. Any new moon during this period will produce a solar eclipse. For lunar eclipses, the eclipse season is a bit shorter, about 22 days. Any full moon in this period will encounter Earth's shadow and be eclipsed.

This makes eclipse prediction easy. All you have to do is keep track of where the moon crosses the ecliptic (where the nodes of its orbit are). When the sun is near one of these nodes, you can predict that the nearest new moon will cause a solar eclipse and the nearest full moon will cause a lunar eclipse. This system works fairly well, and ancient astronomers such as the Maya may have used such a system. You could be a very successful ancient Mayan astronomer with what you know about eclipse seasons, but you can do even better if you change your point of view.

The View from Space

Change your point of view and imagine that you are looking at the orbits of Earth and the moon from a point far away in space. You would see the moon's orbit as a small disk tipped at an angle to the larger disk of Earth's orbit. As Earth orbits the sun, the moon's orbit remains fixed in direction. The nodes of the moon's orbit are the points where it passes through the plane of Earth's orbit; an eclipse season occurs each time the line connecting these nodes, the **line of nodes,** points toward the sun. Study ■ Figure 3-13 and notice that the line of nodes does not point at the sun in the example at lower left, and no eclipses are possible. At lower right, the line of nodes points toward the sun, and the shadows produce eclipses.

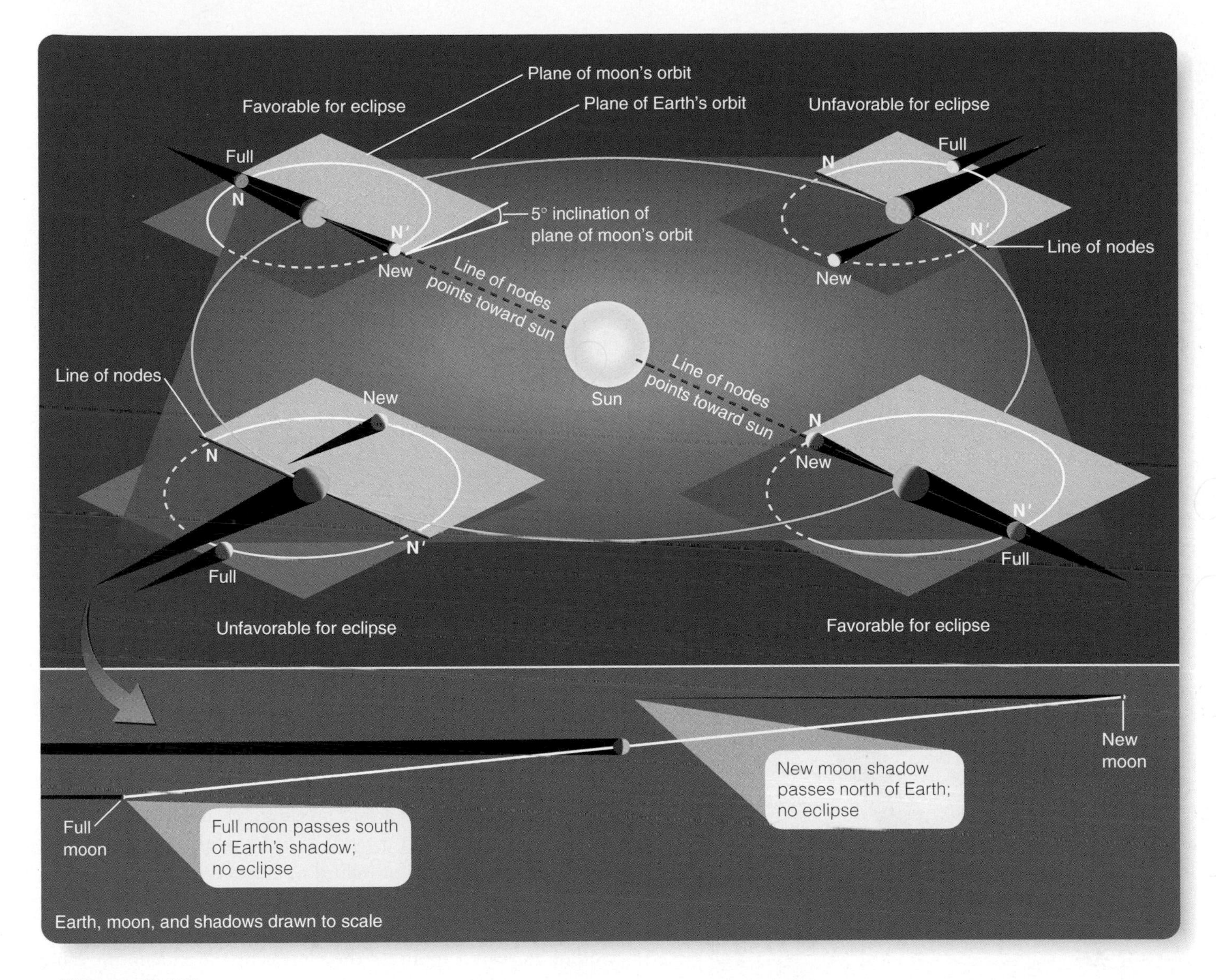

■ Figure 3-13

The moon's orbit is tipped about 5° to Earth's orbit. The nodes N and N′ are the points where the moon passes through the plane of Earth's orbit. If the line of nodes does not point at the sun, the long narrow shadows miss, and there are no eclipses at new moon and full moon. At those parts of Earth's orbit where the line of nodes points toward the sun, eclipses are possible at new moon and full moon.

The shadows of Earth and moon, seen from space, are very long and thin, as shown in the lower part of Figure 3-13. It is easy for them to miss their mark at new moon or full moon and fail to produce an eclipse. Only during an eclipse season, when the line of nodes points toward the sun, do the long, skinny shadows produce eclipses.

If you watched for years from your point of view in space, you would see the orbit of the moon precess like a hubcap spinning on the ground. This precession is caused mostly by the gravitational influence of the sun, and it makes the line of nodes rotate once every 18.6 years. People back on Earth see the nodes slipping westward along the ecliptic 19.4° per year, and the sun takes only 346.62 days (an **eclipse year**) to return to a node. This means that, according to the calendar, the eclipse seasons begin about 19 days earlier every year (■ Figure 3-14).

The cyclic pattern of eclipses shown in Figure 3-14 should give you another clue how to predict eclipses. Eclipses follow a pattern, and if you were an ancient astronomer and understood the pattern, you could predict eclipses without ever knowing what the moon was or how an orbit works. All you need know is the cycle.

The Saros Cycle

Ancient astronomers could predict eclipses in a crude way using the eclipse seasons, but they could have been much more accurate if they recognized that eclipses occur following certain patterns.

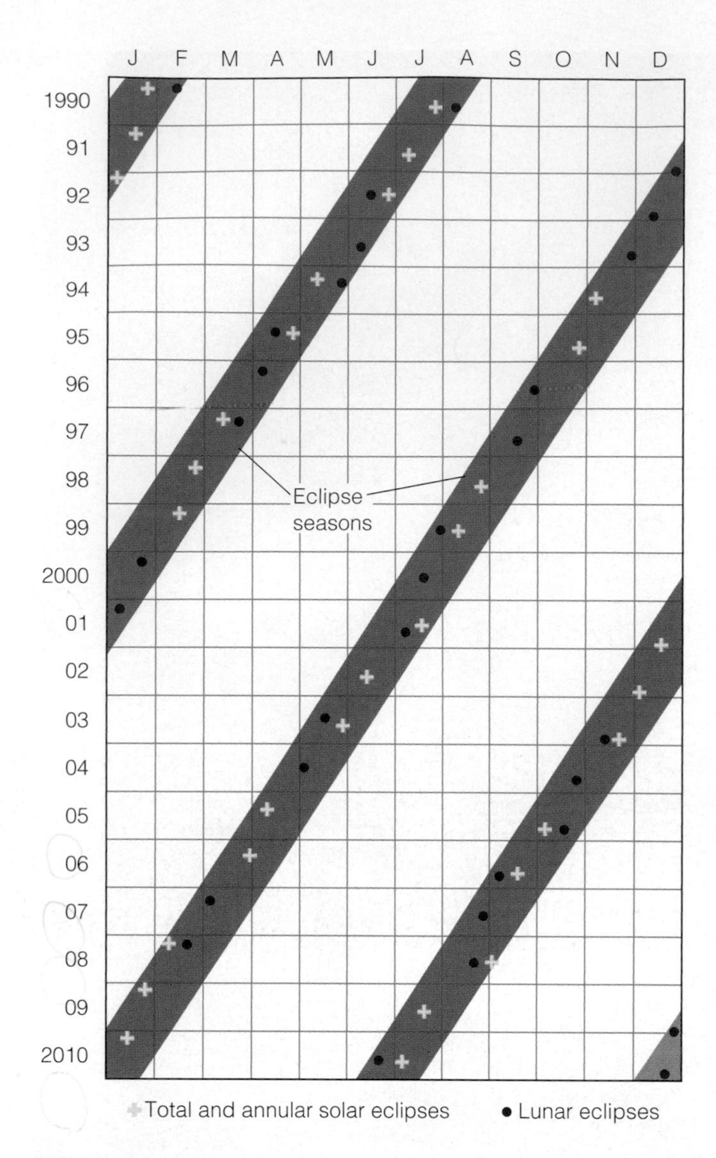

■ **Figure 3-14**

A calendar of eclipse seasons. Each year the eclipse seasons begin about 19 days earlier. Any new moon or full moon that occurs during an eclipse season results in an eclipse. Only total and annular eclipses are shown here.

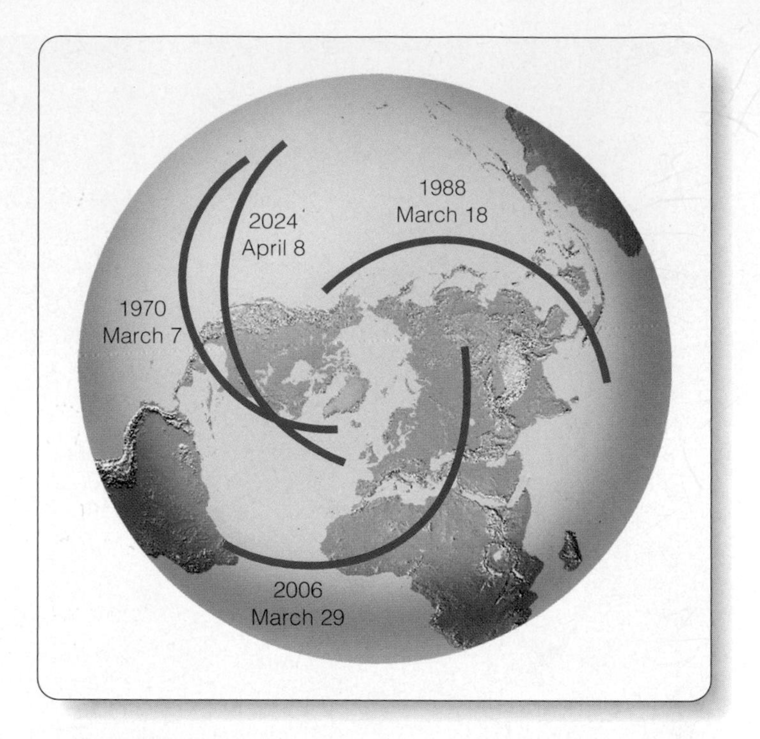

■ **Figure 3-15**

The saros cycle at work. The total solar eclipse of March 7, 1970, recurred after 18 years $11\frac{1}{3}$ days over the Pacific Ocean. After another interval of 18 years $11\frac{1}{3}$ days, the same eclipse was visible from Asia and Africa. After a similar interval, the eclipse will again be visible from the United States.

The most important of these is the **saros cycle** (sometimes referred to simply as the saros). After one saros cycle of 18 years $11\frac{1}{3}$ days, the pattern of eclipses repeats. In fact, *saros* comes from a Greek word that means "repetition."

The eclipses repeat because, after one saros cycle, the moon and the nodes of its orbit return to the same place with respect to the sun. One saros contains 6585.321 days, which is equal to 223 lunar months. Therefore, after one saros cycle the moon is back to the same phase it had when the cycle began. But one saros is also equal to 19 eclipse years. After one saros cycle, the sun has returned to the same place it occupied with respect to the nodes of the moon's orbit when the cycle began. If an eclipse occurs on a given day, then 18 years $11\frac{1}{3}$ days later the sun, the moon, and the nodes of the moon's orbit return to nearly the same relationship, and the eclipse occurs all over again.

Although the eclipse repeats almost exactly, it is not visible from the same place on Earth. The saros cycle is one-third of a day longer than 18 years 11 days. When the eclipse happens again, Earth will have rotated one-third of a turn farther east, and the eclipse will occur one-third of the way westward around Earth (■ Figure 3-15). That means that after three saros cycles—a period of 54 years 1 month—the same eclipse occurs in the same part of Earth.

One of the most famous predictors of eclipses was the Greek philospher Thales of Miletus (about 640–546 BC), who supposedly learned of the saros cycle from the Chaldeans, who had discovered it. No one knows which eclipse Thales predicted, but some scholars suspect the eclipse of May 28, 585 BC. In any case, the eclipse occurred at the height of a battle between the Lydians and the Medes, and the mysterious darkness in midafternoon so startled the two factions that they concluded a truce.

In fact, many historians doubt that Thales actually predicted the eclipse. It would have been very difficult to gather enough information about past eclipses; total solar eclipses are rare as seen from any one place. If you stay in one city, you will see a total solar eclipse about once in 360 years. Also, 585 BC is very early for the Greeks to have known of the saros cycle. The important point is not that Thales did it, but that he could have done it. If he had had records of past eclipses of the sun visible from the area, he could have discovered that they tended to recur with a period of 54 years 1 month (three saros cycles). Indeed, he could have predicted the eclipse without ever understanding what the sun and moon were or how they moved.

Building Scientific Arguments

Why can't two successive full moons be totally eclipsed? Most people suppose that eclipses occur at random or in some pattern so complex you need a big computer to make predictions. You know the geometry is fairly simple, so you can build a simple argument to analyze this question. Remember that a total lunar eclipse occurs when the moon passes through Earth's shadow. Earth's shadow always points toward the ecliptic exactly opposite the sun, and most of the time the moon will pass north or south of Earth's shadow, and there will be no eclipse. A lunar eclipse can happen only when the sun is near one node and the moon crosses Earth's shadow at the other node.

Now you can apply what you know about the moon's phases. An eclipse season for a total lunar eclipse is only 22 days long, but the moon takes 29.5 days to go from one full moon to the next. If one full moon is totally eclipsed, the next full moon 29.5 days later will occur long after the end of the eclipse season and there will be no eclipse.

Now use your knowledge of the cycles of the sun and moon to revise your argument. **Why can the sun be eclipsed by two successive new moons?**

■ ■ ■

Connections: Predicting eclipses isn't very hard, and many ancient peoples were probably familiar enough with the cycles of the sun and moon to know when eclipses were likely. Some people call those first predictions early science, but science involves understanding nature, not just predicting events. In the next chapter, you will see how modern science was born when astronomers tried to understand the cycles they saw in the sky.

Study and Review Tools

Summary

3-1 | The Changeable Moon

Why does the moon go through phases?

- The moon orbits eastward around Earth once a month.
- The moon rotates on its axis as it orbits around Earth and keeps the same side facing Earth throughout the month.
- Because you see the moon by reflected sunlight, its shape appears to change as it orbits Earth and sunlight illuminates different amounts of the side facing Earth.
- The lunar phases wax from new moon to first quarter to full moon and wane from full moon to third quarter to new moon.
- A complete cycle of lunar phases takes 29.53 days.

3-2 | Lunar Eclipses

What causes a lunar eclipse?

- If the full moon passes through Earth's shadow, sunlight is cut off, and the moon darkens in a lunar eclipse. If the moon only grazes the shadow, the eclipse is partial or penumbral and not total.
- The totally eclipsed moon looks copper-red because of sunlight scattered through Earth's atmosphere.

3-3 | Solar Eclipses

What causes a solar eclipse?

- A total solar eclipse occurs if the new moon passes directly between the sun and Earth and the moon's shadow sweeps over Earth's surface. Observers inside the path of totality see a total eclipse, and those just outside the path of totality see a partial eclipse if the penumbra sweeps over their location.
- During a total eclipse of the sun, the bright photosphere of the sun is covered, and the fainter corona, chromosphere, and prominences become visible.
- If during a solar eclipse the moon is in the farther part of its orbit, near apogee rather than perigee, its angular diameter is not large enough to cover the entire photosphere, and you see only an annular eclipse.

3-4 | Predicting Eclipses

How can eclipses be predicted?

- Solar eclipses must occur at new moon, and lunar eclipses must occur at full moon.
- The moon's orbit crosses the ecliptic at two locations called nodes, and eclipses can occur only when the sun is crossing a node. During these periods, called eclipse seasons, a new moon will cause a solar eclipse, and a full moon can cause a lunar eclipse. Knowing when the eclipse seasons occur would allow you to guess which new moons and full moons could cause eclipses.
- Eclipses follow a pattern called the saros cycle. After one saros of 18 years $11\frac{1}{3}$ days, the pattern of eclipses repeats. Some ancient astronomers knew of the saros cycle and used it to predict eclipses.

New Terms

sidereal period (p. 41)
synodic period (p. 41)
lunar eclipse (p. 42)
umbra (p. 42)
penumbra (p. 42)
total eclipse (lunar or solar) (pp. 42, 45)

- partial eclipse (lunar or solar) (pp. 44, 45)
- penumbral eclipse (p. 44)
- small-angle formula (p. 45)
- path of totality (p. 46)
- photosphere (p. 47)
- corona (p. 47)
- chromosphere (p. 48)
- prominences (p. 48)
- diamond ring effect (p. 48)
- perigee (p. 48)
- apogee (p. 48)
- annular eclipse (p. 48)
- nodes (p. 50)
- eclipse season (p. 50)
- line of nodes (p. 50)
- eclipse year (p. 51)
- saros cycle (p. 52)

Review Questions

Ace Astronomy™ Assess your understanding of this chapter's topics with additional quizzing and animations at **http://ace.brookscole.com/sf9**

1. Which lunar phases would be visible in the sky at dawn? at midnight?
2. If you looked back at Earth from the moon, what phase would Earth have when the moon was full? new? a first-quarter moon? a waxing crescent?
3. If a planet has a moon, must that moon go through the same phases that Earth's moon displays?
4. Could a solar-powered spacecraft generate any electricity while passing through Earth's umbral shadow? through Earth's penumbral shadow?
5. Draw the umbral and penumbral shadows onto the diagram in the middle of page 40. Explain why lunar eclipses can occur only at full moon and solar eclipses can occur only at new moon.
6. How did lunar eclipses lead Aristotle to conclude that Earth was a sphere?
7. Why isn't the corona visible during partial or annular solar eclipses?
8. Why can't the moon be eclipsed when it is halfway between the nodes of its orbit?
9. Why aren't solar eclipses separated by one saros cycle visible from the same location on Earth?
10. How could Thales of Miletus have predicted the date of a solar eclipse without observing the location of the moon in the sky?
11. The stamp at right shows a crescent moon. Explain why the moon could never look this way.

12. The photo at right shows the annular eclipse of May 30, 1984. How is it different from the annular eclipse shown in Figure 3-11?

(Laurence Marschall)

Discussion Questions

1. How would eclipses be different if the moon's orbit were not tipped with respect to the plane of Earth's orbit?
2. Are there other planets in our solar system from whose surface you could see a lunar eclipse? a total solar eclipse?
3. Can you detect the saros cycle in Figure 3-14?

Problems

1. Identify the phases of the moon if on March 21 the moon is located at the point on the ecliptic called (a) the vernal equinox, (b) the autumnal equinox, (c) the summer solstice, (d) the winter solstice.
2. Identify the phases of the moon if at sunset the moon is (a) near the eastern horizon, (b) high in the southern sky, (c) in the southeastern sky, (d) in the southwestern sky.
3. About how many days must elapse between first-quarter moon and third-quarter moon?
4. If on March 1 the full moon is near the star Spica, when will the moon next be full? When will it next be near Spica?
5. How many times larger than the moon is the diameter of Earth's umbral shadow at the moon's distance? (*Hint:* See the photo in Figure 3-4.)
6. Use the small-angle formula to calculate the angular diameter of Earth as seen from the moon.
7. During solar eclipses, large solar prominences are often seen extending 5 minutes of arc from the edge of the sun's disk. How far is this in kilometers? in Earth diameters?
8. If a solar eclipse occurs on October 3: (a) Why can't there be a lunar eclipse on October 13? (b) Why can't there be a solar eclipse on December 28?
9. A total eclipse of the sun was visible from Canada on July 10, 1972. When did this eclipse occur next? From what part of Earth was it total?
10. When will the eclipse described in Problem 9 next be total as seen from Canada?
11. When will the eclipse seasons occur during the current year? What eclipse(s) will occur?

Media Cluster

ACTIVE FIGURES

Ace Astronomy™ To access the resources in the Media Cluster, log into AceAstronomy at **http://ace.brookscole.com/sf9** and select Chapter 3.

Small-Angle Formula
The small-angle formula animation illustrates how the size and distance of an object affect its angular size and how well this simple animation works.

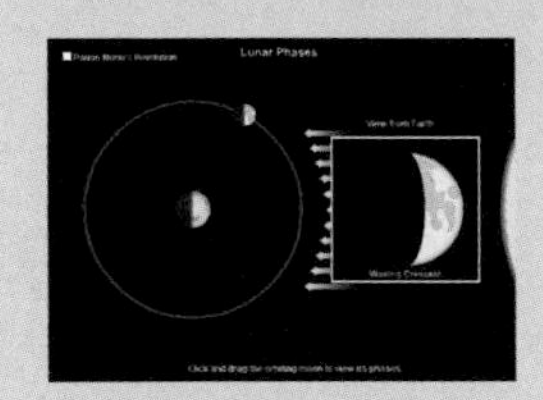

Lunar Phases
Take control of the position of the moon in its orbit and notice how its phases change as seen from Earth.

ASTRONOMY EXERCISES

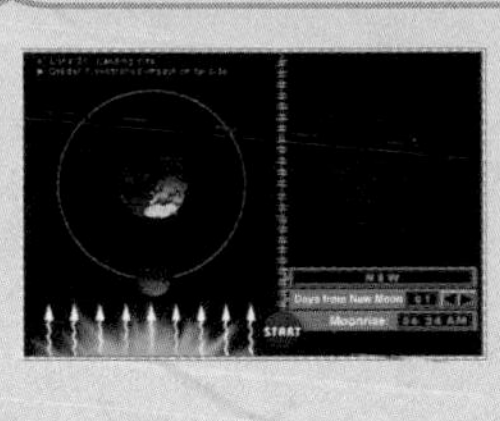

Phases of the Moon
Sunlight always illuminates half of the moon. Because you see different amounts of this sunlit side, you see the moon cycle through phases. Explore this relationship further with this interactive exercise.

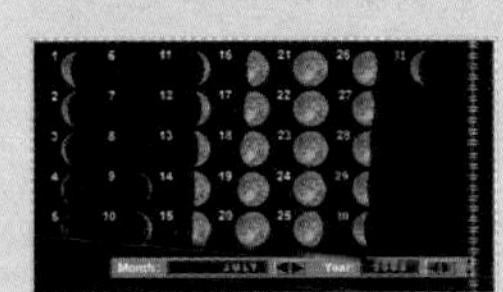

Moon Calendar
This exercise lets you select any month or year in the past, present, or future and see the moon phases for that month.

Critical Inquiries for the Web

1. Search the web for myths about the moon. What common themes do different cultures ascribe to the moon and its cycle of phases?
2. Search for web pages showing photos and observations of a recent total solar eclipse. Can you find web pages that show cultural responses to the eclipse such as celebrations or religious ceremonies?
3. Most people see more total lunar eclipses in a lifetime than total solar eclipses. Why is this so? Compare the regions of visibility for a number of past and upcoming eclipses and determine which future events will be visible from your area.

Exploring *TheSky*

1. How many days must elapse between new moons? Use the **Moon Phase Calendar** under the **Tools** menu to locate new moons.
2. Where would you have to go to see a total solar eclipse in the next year or two? Use the **Eclipse Finder** under the **Tools** menu to select a total solar eclipse. Check **Show Path of Totality** to see a map of Earth.
3. View the total solar eclipse you located in activity 2. In the **Eclipse Finder,** click **View** to see the sun and moon from your present location. Then use **Site Information** under the **Data** menu to change your location to a spot in the path of totality. Set the **Time Step** to 5 minutes and click the **arrows** to move the moon across the sun.
4. Locate and observe an annular eclipse.
5. Locate and observe a total lunar eclipse.

- Go to the Brooks/Cole Astronomy Resource Center **(http://astronomy.brookscole.com)** for critical thinking exercises, articles, and additional readings from InfoTrac College Edition, Brooks/Cole's online student library.

4 The Origin of Modern Astronomy

How you would burst out laughing, my dear Kepler, if you would hear what the greatest philosopher of the Gymnasium told the Grand Duke about me . . .

FROM A LETTER BY GALILEO GALILEI

GALILEO LAUGHED AT JOKES just as you do. We tend to forget that famous scientists in history laughed and joked, felt satisfaction, and experienced frustration day by day. The astronomers you are about to meet loved their work, and they must have laughed often and taken great satisfaction when their observations went well. But there were moments of frustration too. The history of astronomy is the story of great scientific advances made by astronomers working in many lands, but it is also the story of people who loved the sky and felt joy in their work.

As you read, notice how two subplots twist through the story. The first plot involves astronomers trying to understand the place of the Earth. Is it the center of the universe? Does it move? This is not an idle question of philosophy; it led Galileo before the Inquisition.

Continued on page 58

Astronomers like Galileo Galilei and Johannes Kepler struggled to understand the place of the Earth and the motion of the planets.

Guidepost

Looking Back

The previous chapters gave you a modern view of the heavens. You are now familiar with what you see in the sky, and you also can imagine how the Earth, moon, and sun move through space. Now you are ready to understand one of the most sweeping revolutions in human thought: the realization that we live on a planet.

This Chapter

All cultures have tried to explain their place under the sky, and the ancient Greek philosophers developed an elaborate theory. By the 16th century, however, many astronomers were uncomfortable with the ancient theory that Earth sat unmoving at the center of a spherical universe. In this chapter, you will discover how an astronomer named Copernicus changed the ancient theory, how Galileo Galilei changed the rules of the debate, and how astronomers changed the way they thought about motion in the heavens. Here you will find answers to four essential questions:

How did the ancients describe the place of the Earth?

How did Copernicus change the place of the Earth?

Why was Galileo condemned by the Inquisition?

How did Copernican astronomers solve the puzzle of planetary motion?

This chapter is not just about the history of astronomy. As they struggled to understand Earth and the heavens, the astronomers of the Renaissance invented a new way of understanding nature—a way of thinking that is now called science.

Looking Ahead

This is a book of science, and every chapter that follows will use the methods that were invented when Copernicus tried to repair the ancient theory that Earth was the center of the universe. One of the greatest adventures in that story began when an Englishman wondered why objects fall. That key to the universe is the subject of the next chapter.

Ace Astronomy™ The AceAstronomy icon throughout the text indicates an opportunity for you to test yourself on key concepts and to explore animations and interactions on the AceAstronomy website at: http://ace.brookscole.com/sf9

The second plot, the puzzle of planetary motion, reveals astronomers observing the sky and trying to understand how the sun, moon, and planets move. That puzzle led eventually to the discovery of gravity, understanding orbital motion, and putting astronauts in space.

This story goes far beyond the birth of astronomy. The quest for the place of the Earth and the puzzle of planetary motion became, in the 16th and 17th centuries, the center of a storm of controversy over the best way to understand our world and ourselves. That conflict is now known as the scientific revolution. As you read about the birth of astronomy, notice that you are also reading about the birth of modern science.

4-1 The Roots of Astronomy

ASTRONOMY HAS ITS ORIGIN in that most noble of all human traits, curiosity. Just as modern children ask their parents what the stars are and why the moon changes, so did ancient peoples ask themselves those same questions. The answers, often couched in mythical or religious terms, reveal a great reverence for the order of the heavens.

Archaeoastronomy

Most of the history of astronomy is lost forever. You can't go to a library or search the Internet to find out what the first astronomers thought about their world because they left no written record. The study of the astronomy of ancient peoples has been called **archaeoastronomy,** and, in spite of the fog of eons gone, it tells you one thing: trying to understand the heavens is part of human nature.

Perhaps the best-known example of archaeoastronomy is also a huge tourist attraction. Stonehenge, standing on Salisbury Plain in southern England, was built in stages from about 3000 BC to about 1800 BC, a period extending from the late Stone Age into the Bronze Age. Though the public is most familiar with the massive stones of Stonehenge, those were added late in its history. In its first stages, Stonehenge consisted of a circular ditch slightly larger in diameter than the length of a football field, with a concentric bank just inside the ditch and a long avenue leading away toward the northeast. A massive stone, the Heelstone, stood then, as it does now, outside the ditch in the opening of the avenue.

As early as AD 1740, the English scholar W. Stukely suggested that the avenue pointed toward the rising sun at the summer solstice, but few accepted the idea. More recently, astronomers have recognized significant astronomical alignments at Stonehenge. Seen from the center of the monument, the summer-solstice sun rises behind the Heelstone. Other sight lines point toward the most northerly and most southerly risings of the moon (■ Figure 4-1).

The significance of these alignments has been debated. Some have claimed that the Stone Age people who built Stonehenge were using it as a device to predict lunar eclipses. After studying eclipse prediction in the previous chapter, you understand that predicting eclipses is easier than most people assume. You could use Stonehenge to predict eclipses, but was that the intention of the people who built it? Some experts doubt that eclipse prediction was the main use of the monument. The truth may never be known. The builders of Stonehenge had no written language and left no records of their intentions. Nevertheless, the presence of solar and lunar alignments at Stonehenge and at many other Stone Age monuments dotting England and continental Europe shows that so-called primitive peoples were paying detailed attention to the sky. The roots of astronomy lie not in sophisticated science and logic but in human curiosity and wonder.

Astronomical alignments in sacred structures are common all around the world. For example, many tombs are oriented toward the rising sun, and Newgrange, a 5000-year-old passage-grave in Ireland (■ Figure 4-2), faces southeast so that, at dawn on the day of the winter solstice, light from the rising sun shines into its long passageway and illuminates the central chamber. No one today knows what the alignment meant to the builders of Newgrange, and many experts doubt that it was actually intended to be a tomb. Whatever its original purpose, Newgrange is clearly a sacred site linked by its alignment to the order and power of the sky. Astronomical alignments in sacred structures are not just an ancient custom. Many cathedrals are carefully oriented to face east.

Building astronomical alignments into structures gives them meaning by connecting them with the heavens. Navajo Indians of the American southwest, for example, have for centuries built their hogans with the door facing east so they can greet the rising sun each morning with prayers and offerings.

Some alignments may have served calendrical purposes. The 2000-year-old Temple of Isis in Dendera, Egypt, was build to align with the rising point of the bright star Sirius. The first appearance of this star in the dawn twilight marked the flooding of the Nile, so it was an important calendar indicator. The symbolism goes further. According to Egyptian mythology, the goddess Isis was associated with the star Sirius, and her husband, Osirus, was linked to the constellation you know as Orion and also to the Nile, the source of Egypt's agricultural fertility.

An American site in New Mexico is known as the Sun Dagger, but there is no surviving mythology to help you understand it. At noon on the day of the summer solstice, a narrow dagger of sunlight shines across the center of a spiral carved on a cliff face high above the desert floor (■ Figure 4-3). The purpose of the

■ **Figure 4-1** ▶

The central horseshoe of upright stones is only the most obvious part of Stonehenge. The best-known astronomical alignment at Stonehenge is the summer solstice sun rising over the Heelstone. Although a number of astronomical alignments have been found at Stonehenge, experts debate their significance. (Photo: Jamie Backman)

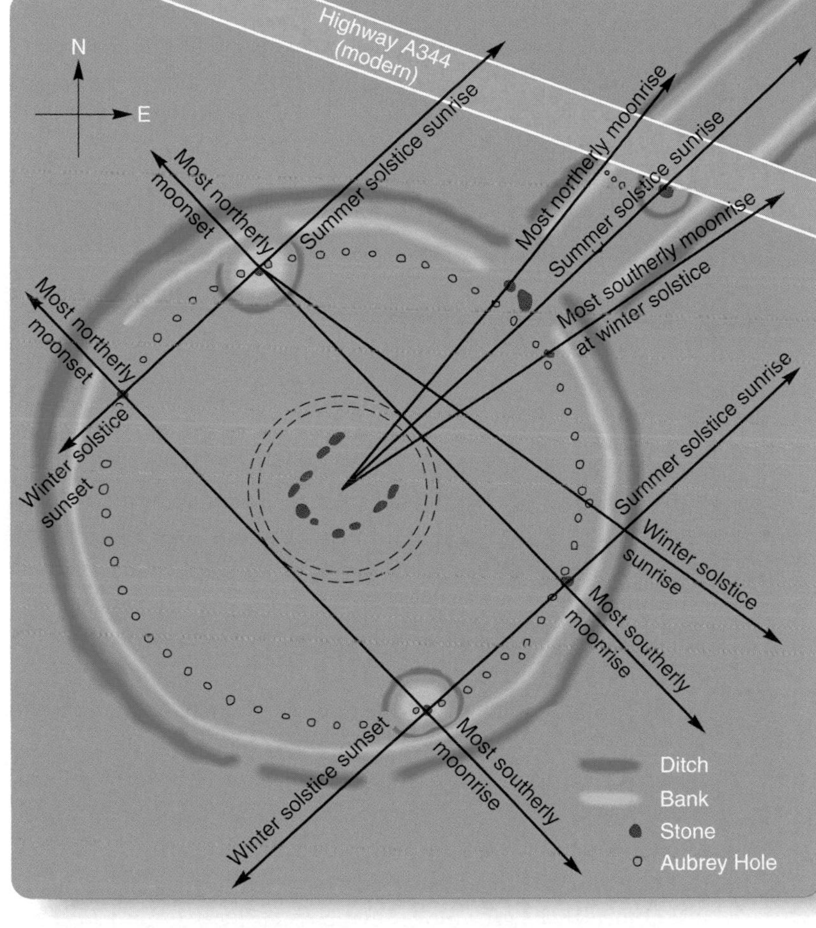

Sun Dagger is open to debate, but similar examples have been found throughout the American southwest. It may have been more a symbolic and ceremonial marker than a precise calendrical indicator. In any case, it is just one of the many astronomical alignments that ancient people built into their structures to link themselves with the sky.

Some scholars are looking not at temples but at small artifacts from thousands of years ago. Scratches on certain bone and stone implements seem to follow a pattern and may be an attempt to keep a record of the phases of the moon (■ Figure 4-4). Some scientists contend that humanity's first attempts at writing were stimulated by a desire to record and predict lunar phases.

Archaeoastronomy is uncovering the earliest roots of astronomy and simultaneously revealing some of the first human efforts at systematic inquiry. The most important lesson of archaeoastronomy is that humans don't have to be technologically sophisticated to admire and study the universe.

One thing about archaeoastronomy is especially sad. Although archaeoastronomy can show how ancient people observed the sky, their thoughts about their universe are in many cases lost. Many had no written language. In other cases, the written record has been lost. Dozens, perhaps hundreds, of beautiful Mayan manuscripts, for instance, were burned by Spanish missionaries who believed that the books were the work of Satan. Only four of these books have survived, and all four contain astronomical references. One contains sophisticated tables that allowed the Maya to predict the motion of Venus and eclipses of the moon. No one will ever know what was burnt.

The fate of the Mayan books illustrates one reason why histories of astronomy usually begin with the Greeks. Some of their writing has survived, and you can discover what they thought about the shape and motion of the heavens.

■ Figure 4-2

Newgrange was built on a small hill in Ireland about 3200 BC. A long passageway extends from the entryway back to the center of the mound, and sunlight shines down the passageway into the central chamber at dawn on the day of the winter solstice. Other passage graves have similar alignments, but their purpose is unknown. (bottom, Newgrange: Benelux Press/Index Stock Imagery)

mind can understand why the universe works the way it does. This view contrasts sharply with those of earlier cultures, which believed that the ultimate causes of things are mysteries beyond human understanding. To Thales and his followers, the mysteries of the universe are mysteries because they are unknown, not because they are unknowable.

High on Fajada Butte, the Sun Dagger is off limits to visitors.
The spiral pattern is the size of a dinner plate.

■ **Figure 4-3**

In the ancient Native American settlement known as Chaco Canyon, New Mexico, sunlight shines between two slabs of stone high on the side of 440-foot-high Fajada Butte to form a dagger of light on the cliff face. About noon on the day of the summer solstice, the dagger of light slices through the center of a spiral pecked into the sandstone. (NPS Chaco Culture National Historic Park)

The second philosopher who made the new scientific attitude possible was Pythagoras (c. 570–500 BC). He and his students noticed that many things in nature seem to be governed by geometrical or mathematical relations. Musical notes, for example, are related in a regular way to the lengths of plucked strings. This led Pythagoras to propose that all nature was underlain by musical principles, by which he meant mathematics. One result of this philosophy was the later belief that the harmony of the celestial movements produced actual music, the music of the spheres. But, at a deeper level, the teachings of Pythagoras made Greek astronomers look at the universe in a new way. Thales said that the universe could be understood, and Pythagoras said that the underlying rules were mathematical.

In trying to understand the universe, Greek astronomers did something that Babylonian astronomers had never done—they tried to construct descriptions based on geometrical forms. Anaximander (c. 611–546 BC) described a universe made up of wheels filled with fire: The sun and moon are holes in the wheels through which the flames can be seen. Philolaus (fifth century BC) argued that Earth moves in a circular path around a central fire (not the sun), which is always hidden behind a counterearth located between the fire and Earth. This was the first theory to suppose that Earth is in motion.

Plato (428–347 BC) was not an astronomer, but his teachings influenced astronomy for 2000 years. Plato argued that the reality humans see is only a distorted shadow of a perfect, ideal form. If human observations are distorted, then observation can be misleading, and the best path to truth is through pure thought on the ideal forms that underlie nature.

Plato also argued that the most perfect geometrical form was the sphere, and therefore, he said, the perfect havens must be made up of spheres rotating at constant rates carrying objects around in circles. Consequently, later astronomers tried to describe the motions of the heavens by imagining multiple, rotating spheres. This became known as the principle of **uniform circular motion.**

The Astronomy of Greece

Greek astronomy was based on the astronomy of Babylon and Egypt, but these astronomies were heavily influenced by religion and astrology. The astronomers of that time studied the motions of the heavens as a way of worship and divination. The Greek astronomers studied astronomy in an entirely new way—they tried to understand the universe.

This new attitude toward the heavens, a truly scientific attitude, was made possible by two early Greek philosophers. Thales of Miletus (c. 624–547 BC) lived and worked in what is now Turkey. He taught that the universe is rational and that the human

■ **Figure 4-4**

A fragment of a 27,000-year-old mammoth tusk found at Gontzi in Ukraine contains scribe marks on its edge, simplified in this drawing. These markings have been interpreted as a record of four cycles of lunar phases. Although controversial, such finds suggest that some of the first human attempts at recording events in written form were stimulated by astronomical phenomena.

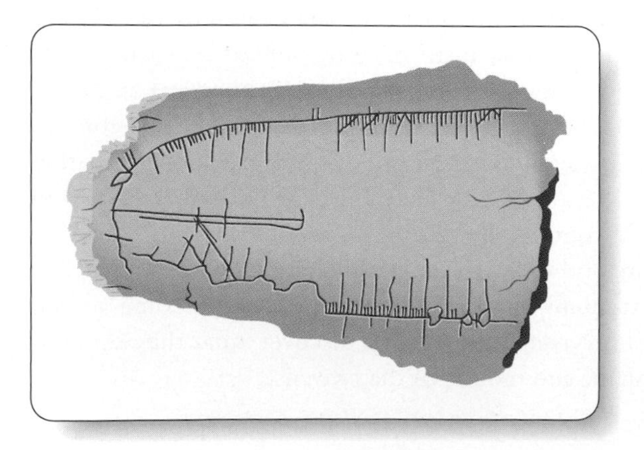

■ Figure 4-5

The spheres of Eudoxus explain the motions in the heavens by means of nested spheres rotating about various axes at different rates. Earth is located at the center.

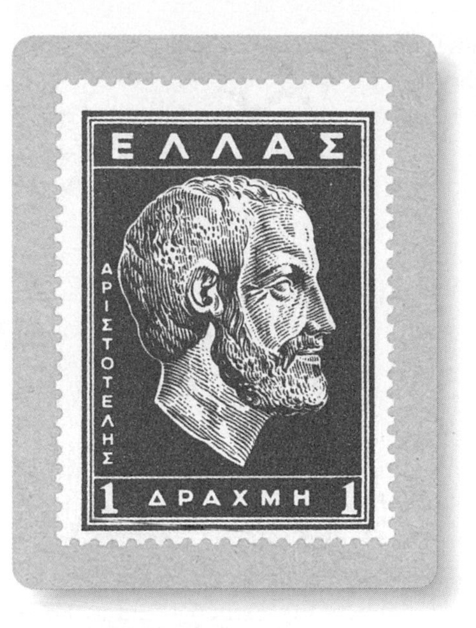

■ Figure 4-6

Aristotle, honored on this Greek stamp, wrote on such a wide variety of subjects and with such deep insight that he became the great authority on all matters of learning. His opinions on the nature of Earth and the sky were widely accepted for almost two millennia.

Pythagoras had taught that Earth is a sphere and that the other heavenly bodies were divine, perfect spheres moving in perfect circles. Eudoxus of Cnidus (409–356 BC), a student of Plato, combined a system of 27 nested spheres rotating at different rates about different axes at constant rates to produce a mathematical description of the motions of the universe (■ Figure 4-5).

At the time of the Greek philosophers, it was common to refer to systems such as that of Eudoxus as descriptions of the world, where the word *world* included not only Earth but all of the heavenly spheres. The reality of these spheres was open to debate. Some thought of the spheres as nothing more than mathematical ideas that described motion in the world model, while others began to think of the spheres as real objects made of perfect celestial material. Aristotle, for example, seems to have thought of the spheres as real.

Aristotle and the Nature of Earth

Aristotle (384–322 BC), one of Plato's students, taught and wrote on philosophy, history, politics, ethics, poetry, drama, and so on (■ Figure 4-6). Because of his sensitivity and insight, he became the great authority of antiquity, and astronomers for almost 2000 years cited him as their authority in adopting the Greek model of the universe.

Much of what Aristotle wrote about scientific subjects was wrong, but he was not a scientist. He was trying to understand the universe by creating a system of formal philosophy. Modern science depends on evidence and hypothesis, but scientific methods had not been invented at the time of Aristotle. Rather, he attempted to use the most basic observations combined with first principles (ideas that he believed were obviously true) to understand the world. The perfection of the heavens was, for Aristotle, a first principle.

Aristotle believed that the universe was divided into two parts—Earth, corrupt and changeable; and the heavens, perfect and unchanging. Like most of his predecessors, he believed that Earth was the center of the universe, so his model is called a **geocentric universe.** The heavens surrounded Earth, and he devised 55 crystalline spheres turning at different rates and at different angles to carry the sun, moon, and planets across the sky. The lowest sphere, that of the moon, marked the boundary between the changeable imperfect region of Earth and the unchanging perfection of the celestial realm above the moon.

Because he believed Earth to be immobile, Aristotle had to make these spheres whirl westward around Earth to produce day and night and move more slowly with respect to one another to produce the motions of the sun, moon, and planets against the background of the stars. Because his model was geocentric, he taught that Earth could be the only center of motion. All of his whirling spheres had to be centered on Earth. Like most other Greek philosophers, Aristotle viewed the universe as a perfect heavenly machine that was not many times larger than Earth itself.

About a century after Aristotle, the Alexandrian philosopher Aristarchus proposed a theory that Earth rotated on its axis and revolved around the sun. This theory is, of course, correct, but most of the writings of Aristarchus were lost, and his theory was not well known. Later astronomers rejected any suggestion that

Earth could move because it conflicted with the teachings of the great philosopher Aristotle.

Aristotle had taught that Earth had to be a sphere because it always casts a round shadow during lunar eclipses, but he could only estimate its size. About 200 BC, Eratosthenes, working in the great library in Alexandria, found a way to calculate Earth's radius. He learned from travelers that the city of Syene (Aswan) in southern Egypt contained a well into which sunlight shone vertically on the day of the summer solstice. This told him that the sun was at the zenith at Syene; but, on that same day in Alexandria, he noted that the sun was $\frac{1}{50}$ of the circumference of the sky (about 7°) south of the zenith.

Because sunlight comes from such a great distance, its rays arrive at Earth traveling almost parallel. That allowed Eratosthenes to use simple geometry to find that the distance from Alexandria to Syene was $\frac{1}{50}$ of Earth's circumference (■ Figure 4-7).

To find Earth's circumference, Eratosthenes had to know the distance from Alexandria to Syene. Travelers told him it took 50 days to cover the distance, and he knew that a camel can travel about 100 stadia per day. That meant the total distance was about 5000 stadia. If 5000 stadia is $\frac{1}{50}$ of Earth's circumference, then Earth must be 250,000 stadia around, and, dividing by 2π, Eratosthenes found Earth's radius to be 40,000 stadia.

How accurate was Eratosthenes? The stadium (singular of *stadia*) had different lengths in ancient times. If you assume 6 stadia to the kilometer, then Eratosthenes's result was too big by only 4 percent. If he used the Olympic stadium, his result was 14 percent too big. In any case, this was a much better measurement of Earth's radius than Aristotle's estimate, which was much too small, about 40 percent of the true radius.

You might think this is just a disagreement between two ancient philosophers, but it had an interesting consequence. At the time of Christopher Columbus, all educated people knew the world was round and not flat, but they weren't sure how big it was. Columbus, like many others, adopted Aristotle's diameter for Earth, so he thought Earth was small enough that he could sail west and reach Japan and the Spice Islands of the East Indies. If he had accepted Eratosthenes' diameter, Columbus would never have risked the voyage. As it turned out, he and his crew were lucky that North America was in the way. If there had been open ocean all the way, they would have starved to death long before they reached Japan.

Aristotle was a philosopher and, following Plato's recommendation, made few direct observations. Eratosthenes made some measurements to find the diameter of Earth, but the next person you need to meet was a real astronomer who observed the sky in detail. Little is known about him, but he is clearly the first great observer. Hipparchus (see Figure 2-5) lived during the second century BC, about two centuries after Aristotle. He is usually credited with the invention of trigonometry, he compiled the first star catalog, and he discovered precession (Chapter 2). Instead of describing the motion of the sun and moon using nested spheres, as most Greek philosophers did, Hipparchus proposed that the sun and moon traveled around circles with Earth near, but not at, their centers. These off-center circles are now known as **eccentrics.** Hipparchus recognized that he could produce the same motion by having the sun, for instance, travel around a small circle that followed a larger circle around Earth. The compounded circular motion that he devised became the key element in the masterpiece of the last great astronomer of classical times, Ptolemy.

■ Active Figure 4-7

On the day of the summer solstice, sunlight fell to the bottom of a well at Syene, but the sun was about $\frac{1}{50}$ of a circle (7°) south of the zenith at Alexandria. From this, Eratosthenes was able to calculate Earth's radius.

Ace Astronomy™ Log into AceAstronomy and select this chapter to see the Active Figure "Eratosthenes' Experiment."

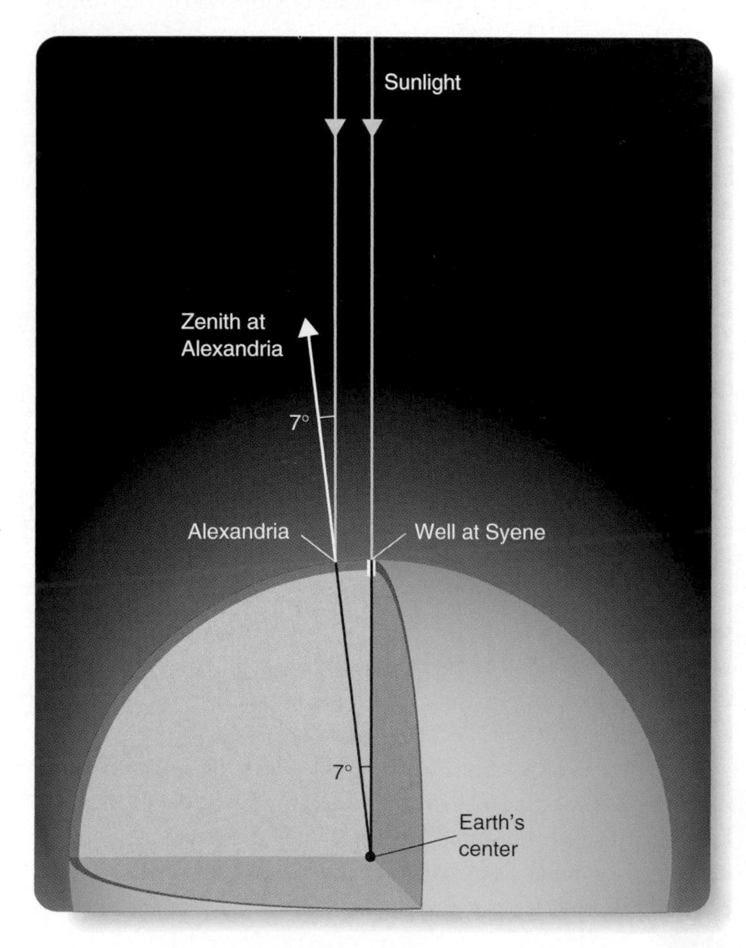

Ace Astronomy™ Log into AceAstronomy and select this chapter to see Astronomy Exercise "Eratosthenes' Calculations." Try his experiment on worlds of different diameters.

The Ptolemaic Universe

Claudius Ptolemaeus was one of the great astronomer-mathematicians of antiquity. His nationality and birth date are unknown, but he lived and worked in the Greek settlement at

Alexandria in what is now Egypt about AD 140. He ensured the survival of Aristotle's universe by transforming it into a sophisticated mathematical model.

Study **The Ancient Universe** on pages 64–65 and notice three important ideas that show how first principles influenced early descriptions of the universe and its motions:

1. Ancient philosophers and astronomers accepted as first principles that Earth was located at the center of the universe and that the heavens moved in uniform circular motion.

2. Second, notice how the observed motion of the planets, the evidence, did not fit the theory very well. The retrograde loops the planets made were very difficult to explain using geocentrism and uniform circular motion.

3. Finally, notice how Ptolemy attempted to explain the motion of the planets by devising a small circle rotating along the edge of a larger circle, which enclosed a slightly off-center Earth. He even allowed the speed of the planets to vary slightly as they circled Earth. Ptolemy lived roughly five centuries after Aristotle, and although Ptolemy believed in the Aristotelian universe, he was interested in a different problem—the motion of the planets. He was a brilliant mathematician, and he was mainly interested in creating a mathematical description of the motions he saw in the heavens. For him, first principles took second place to mathematical precision.

Aristotle's universe, as embodied in the mathematics of Ptolemy, dominated ancient astronomy, but it was wrong. The planets don't follow circles at uniform speeds. At first the Ptolemaic system predicted the positions of the planets well; but, as centuries passed, errors accumulated. If your watch gains only one second a year, it will keep time well for many years, but the error gradually accumulates. After a century your watch will be 100 seconds fast. So, too, did the errors in the Ptolemaic system gradually accumulate as the centuries passed. Islamic and later European astronomers tried to update the system, computing new constants and adjusting epicycles. In the middle of the 13th century, a team of astronomers supported by King Alfonso X of Castile studied the *Almagest* for 10 years. Although they did not revise the theory very much, they simplified the calculation of the positions of the planets using the Ptolemaic system and published the result as *The Alfonsine Tables,* the last great attempt to make the Ptolemaic system of practical use.

AceAstronomy™ Log into AceAstronomy and select this chapter to see Astronomy Exercise "Parallax I." Notice that the parallax angle depends on the length of the baseline.

Building Scientific Arguments

How did the astronomy of Hipparchus and Ptolemy violate the principles of the early Greek philosophers Plato and Aristotle?

Today, scientific arguments depend on evidence and theory; but, in classical times, they started from first principles. Hipparchus and Ptolemy lived very late in the history of classical astronomy, and they concentrated more on the mathematical problems and less on philosophical principles. They replaced the perfect spheres of Plato with nested circles in the form of epicycles and deferents. Earth was moved slightly away from the center of the deferent, so their models of the universe were not exactly geocentric, and the epicycles moved uniformly only as seen from the equant. The celestial motions were no longer precisely uniform, and the principles of geocentrism and uniform circular motion were weakened.

The work of Hipparchus and Ptolemy led eventually to a new understanding of the heavens, but first astronomers had to abandon uniform circular motion. Construct a scientific argument in the classical style based on first principles to answer the question: **Why did Plato argue for uniform circular motion?**

■ ■ ■

Connections: Ptolemy was the last of the great classical astronomers, and his work dominated astronomical thought for almost 1500 years. The collapse of the Ptolemaic model and the rise of a new model of the universe make an exciting story, but it is not just the story of an astronomical idea. It includes the invention of science as a new way of knowing and understanding what we are and where we are.

■ Figure 4-8

Copernicus proposed that the sun and not Earth was the center of the universe. Notice the heliocentric model on this stamp issued in 1973 to commemorate the 500th anniversary of his birth.

4-2 The Copernican Revolution

NICOLAUS COPERNICUS (■ Figure 4-8) triggered an Earthshaking revision in human thought by proposing that the universe is not centered on Earth but rather on the sun. Such a model is called a **heliocentric universe.** That idea eventually brought

1 For 2000 years, the minds of astronomers were shackled by a pair of ideas. The Greek philosopher Plato argued that the heavens were perfect. Because the only perfect geometrical shape is a circle and the only perfect motion is uniform motion, Plato concluded that all motion in the heavens must be made up of combinations of circles turning at uniform rates. This was called **uniform circular motion.**

Plato's student Aristotle argued that Earth was imperfect and lay at the center of the universe. That is, he argued for a **geocentric universe.** He devised a model universe with 55 spheres turning at different rates and at different angles to carry the seven known planets (the moon, Mercury, Venus, the sun, Mars, Jupiter, and Saturn) across the sky.

Aristotle was known as the greatest philosopher in the ancient world, and his authority, lasting for 2000 years, chained the minds of astronomers and forced them to expect the universe to be geocentric and to move in uniform circular motion. See model at right by Peter Apian.

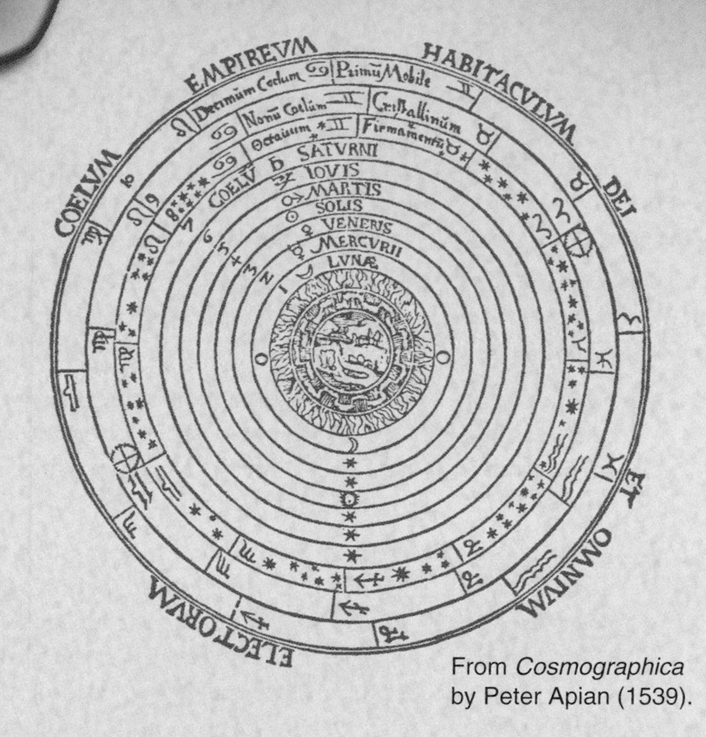

From *Cosmographica* by Peter Apian (1539).

Seen by left eye

1a Ancient astronomers believed that Earth did not move because they saw no **parallax,** the apparent motion of an object because of the motion of the observer. To demonstrate parallax, close one eye and cover a distant object with your thumb held at arm's length. Switch eyes, and your thumb appears to shift position as shown at left. If Earth moves, ancient astronomers reasoned, you should see the sky from different locations at different times of the year, and you should see parallax distorting the shapes of the constellations. They saw no parallax, so they concluded Earth could not move. Actually, the parallax of the stars is too small to see with the unaided eye.

2 Planetary motion was a big problem for ancient astronomers. In fact, the word *planet* comes from the Greek word for "wanderer," referring to the eastward motion of the planets against the background of the fixed stars. The planets did not, however, move at a constant rate, and they could occasionally stop and move westward for a few months before resuming their eastward motion. This backward motion is called **retrograde motion.**

Jan. 29, 2008
East
Nov. 15, 2007
Position of Mars at 5-day intervals
Ecliptic
Dec. 12, 2005
Oct. 3, 2005
Taurus
West
Betelgeuse
Orion
Rigel

Every 2.14 years, Mars passes through a retrograde loop. Two successive loops are shown here. Each loop occurs further east along the ecliptic and has its own shape.

2a Simple uniform circular motion centered on Earth could not explain retrograde motion, so ancient astronomers combined uniformly rotating circles much like gears in a machine to try to reproduce the motion of the planets.

3 Uniformly rotating circles were key elements of ancient astronomy. Claudius Ptolemy created a mathematical model of the Aristotelian universe in which the planet followed a small circle called the **epicycle** that slid around a larger circle called the **deferent.** By adjusting the size and rate of rotation of the circles, he could approximate the retrograde motion of a planet. See illustration at right.

To adjust the speed of the planet, Ptolemy supposed that Earth was slightly off center and that the center of the epicycle moved such that it appeared to move at a constant rate as seen from the point called the **equant.**

To further adjust his model, Ptolemy added small epicycles riding on top of larger epicycles (not shown here), producing a highly complex model.

3a Ptolemy's great book *Mathematical Syntaxis* (c. AD 140) contained the details of his model. Islamic astronomers preserved and studied the book through the Middle Ages, and they called it *Al Magisti* (The Greatest). When the book was found and translated from Arabic to Latin in the 12th century, it became known as *Almagest*.

3b The Ptolemaic model of the universe shown below was geocentric and based on uniform circular motion. Note that Mercury and Venus were treated differently from the rest of the planets. The centers of the epicycles of Mercury and Venus had to remain on the Earth–Sun line as the sun circled Earth through the year.

Equants and smaller epicycles are not shown here. Some versions contained nearly 100 epicycles as generations of astronomers tried to fine-tune the model to better reproduce the motion of the planets.

Notice that this modern illustration shows rings around Saturn and sunlight illuminating the globes of the planets, features that could not be known before the invention of the telescope.

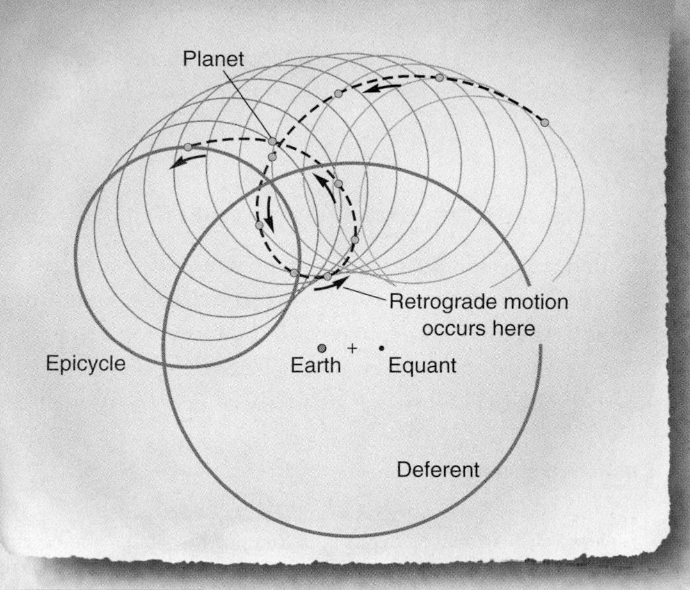

Ace Astronomy™ Go to AceAstronomy and click Active Figures to see "Epicycles." Notice how the counter-clockwise rotation of the epicycle produces retrograde motion.

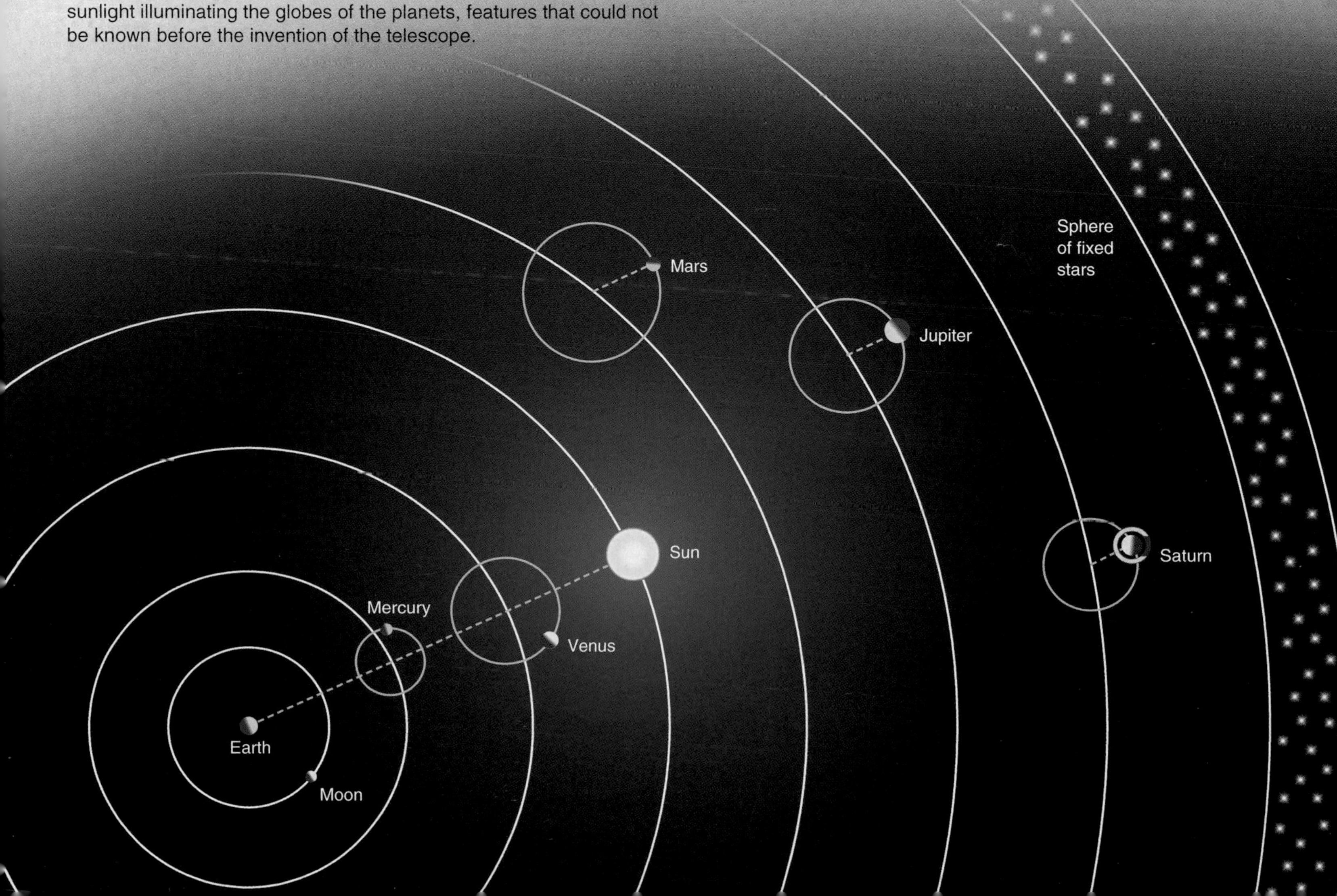

Galileo before the Inquisition and changed forever how we think of our world and ourselves. This Copernican Revolution was much more than an upheaval in astronomy; it marks the birth of modern science.

Copernicus the Revolutionary

When Copernicus proposed that the universe was heliocentric, he risked controversy. According to Aristotle, the most perfect region was the starry sphere, and the most imperfect was Earth's center. In this way the classical geocentric universe matched the commonly held Christian geometry of heaven and hell, and anyone who criticized the geometry of the Aristotelian universe challenged Christian belief and thus risked a charge of heresy.

Throughout his life, Copernicus was associated with the Church. His uncle, by whom he was raised and educated, was an important bishop in Poland; and, after studying canon law and medicine in some of the finest universities in Europe, Copernicus became a canon of the Church at the unusually young age of 24. He was secretary and personal physician to his powerful uncle for 15 years. When his uncle died, Copernicus moved to quarters adjoining the cathedral in Frauenburg, where he was a canon. No doubt his long association with the Church added to his reluctance to publish controversial ideas.

■ Figure 4-9

(a) The Copernican universe as reproduced in *De Revolutionibus*. Earth and all the known planets revolve in separate circular orbits about the sun (Sol) at the center. The outermost sphere carries the immobile stars of the celestial sphere. Notice the orbit of the moon around Earth (Terra). (Yerkes Observatory) (b) The model was elegant not only in its arrangement of the planets but also in their motions. Orbital speed (blue arrows) decreased from Mercury, the fastest, to Saturn, the slowest. Compare the elegance of this model with the complexity of the Ptolemaic model on page 65.

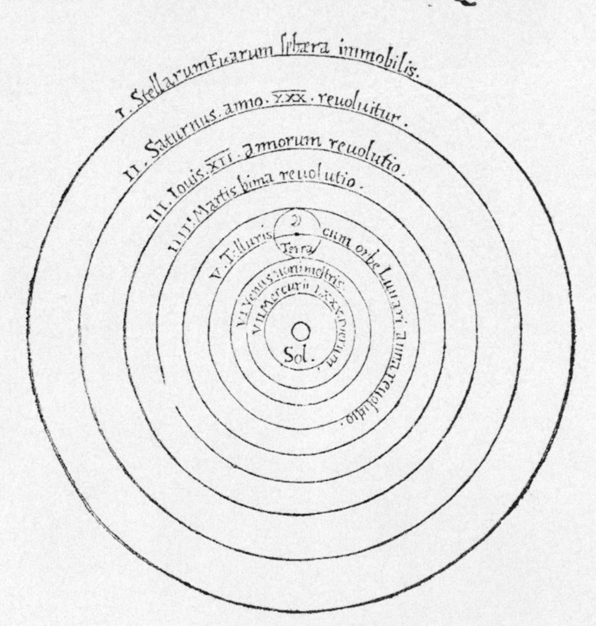

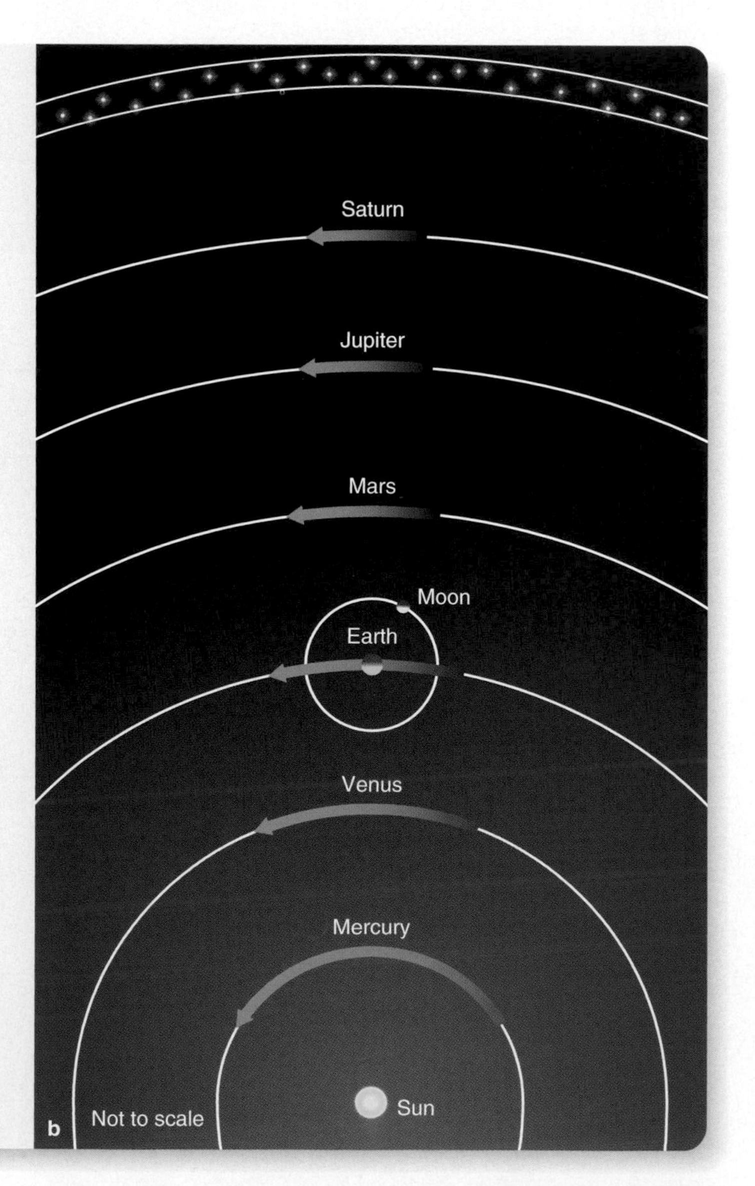

He first wrote about a heliocentric universe sometime before 1514 in a short pamphlet that he distributed in handwritten form and in some cases anonymously. Over the years, Copernicus worked on his book *De Revolutionibus Orbium Coelestium (On the Revolutions of the Celestial Spheres)* (■ Figure 4-9a), but he hesitated to publish it even though other astronomers knew of his theories.

This was a time of rebellion—Martin Luther was speaking harshly about fundamental Church teachings, and others, scholars and scoundrels, questioned the authority of the Church. Even astronomy could stir argument; moving Earth from its central place was a controversial and perhaps heretical idea. Copernicus was concerned about how his ideas would be received, but in 1540 he allowed the visiting astronomer Joachim Rheticus (1514–1576) to publish an account of the Copernican universe in Rheticus's book *Prima Narratio (First Narrative)*. In 1542, Copernicus sent the manuscript for *De Revolutionibus* off to be printed. He died in the spring of 1543 before the printing was completed.

The Copernican Model

You should notice the difference between the Copernican hypothesis and the Copernican model. The Copernican hypothesis was that the universe was heliocentric, and that was correct. The Copernican model was the arrangement and motion of the planets that Copernicus devised to explain the motions in the sky. The model won gradual acceptance, but, as you will discover, it was flawed.

The most important idea in *De Revolutionibus* was the placement of the sun at the center of the universe. That single innovation had an astonishing consequence—the retrograde motion of the planets was immediately explained in a straightforward way without the large epicycles that Ptolemy used. In the Copernican system, Earth moves faster along its orbit than the planets that lie further from the sun. Consequently, Earth periodically overtakes and passes these planets, and they appear to slow and fall behind (■ Figure 4-10). Because the planetary orbits do not lie in precisely the same plane, a planet does not resume its eastward motion in precisely the same path it followed earlier. Consequently, it describes a loop whose shape depends on the angle between the orbital planes.

Copernicus could explain retrograde motion without epicycles, and that was impressive. But he could not do away with epicycles completely. Copernicus, a classical astronomer, had tremendous respect for the old concept of uniform circular motion. In fact, he objected strongly to Ptolemy's use of the equant. It seemed arbitrary to Copernicus, a direct violation of the elegance of Aristotle's philosophy of the heavens. Copernicus called equants "monstrous" in that they violated uniform circular motion. In his model, Copernicus returned to a strong belief in uniform circular motion. Although he did not need epicycles to explain retrograde motion, Copernicus discovered that the sun, moon, and planets suffered small variations in their motions—variations that he could not explain with the concept of uniform circular motion centered on the sun. Today, we recognize those variations as typical of objects following elliptical orbits, but Copernicus held firmly to uniform circular motion, so he had to use small epicycles to produce these minor variations in the motions of the sun, moon, and planets.

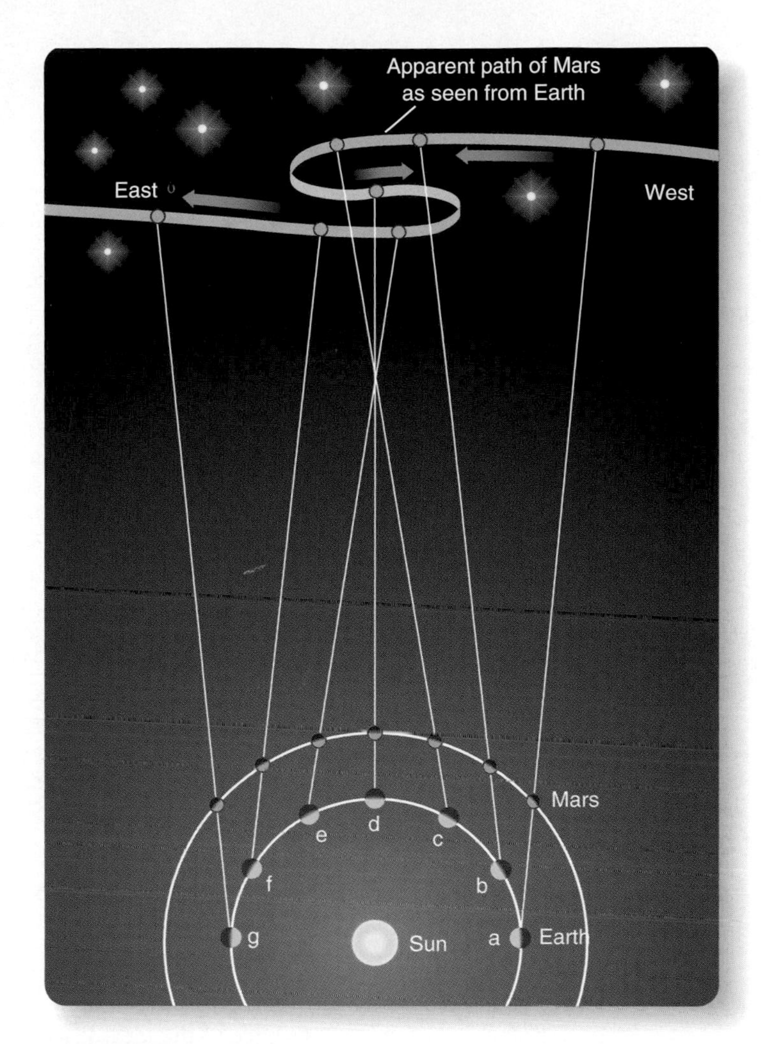

■ Figure 4-10

The Copernican explanation of retrograde motion: As Earth overtakes Mars (a–c), Mars appears to slow its eastward motion. As Earth passes Mars (d), Mars appears to move westward. As Earth draws ahead of Mars (e–g), Mars resumes its eastward motion against the background stars. Compare with the illustration of retrograde motion on page 64. The positions of Earth and Mars are shown at equal intervals of one month.

Because Copernicus imposed uniform circular motion on his model, it could not accurately predict the motions of the planets. *The Prutenic Tables* (1551) were based on the Copernican model, and they were not significantly more accurate than *The Alfonsine Tables* (1251), which were based on Ptolemy's model. Both could be in error by as much as 2°, four times the angular diameter of the full moon.

The Copernican *model* was inaccurate, but the Copernican *hypothesis* that the universe is heliocentric was correct. There are probably a number of reasons why the hypothesis gradually won

Window on Science 4-1

Creating New Ways to See Nature: Scientific Revolutions

The Copernican Revolution is often cited as the perfect example of a scientific revolution. Over a few decades, astronomers abandoned a way of thinking about the universe that was almost 2000 years old and adopted a new set of ideas and assumptions. The American philosopher of science Thomas Kuhn has referred to a commonly accepted set of scientific ideas and assumptions as a scientific **paradigm.** The pre-Copernican astronomers had a geocentric paradigm that included uniform circular motion and the perfection of the heavens. That paradigm survived for many centuries until a new generation of astronomers was able to overthrow the old paradigm and establish a new paradigm that included heliocentrism and Earth's motion.

A scientific paradigm is powerful because it shapes your perceptions. It determines what you judge to be important questions and what you judge to be significant evidence. Consequently, it is often difficult to recognize how your paradigms limit what you can understand.

For example, the geocentric paradigm contained problems that seem obvious today, but because astronomers before Copernicus lived and worked inside that paradigm, they had difficulty seeing the problems. Overthrowing an outdated paradigm is not easy, because you must learn to see nature in an entirely new way. Galileo and Kepler saw nature from a new paradigm that would have been almost incomprehensible to astronomers of earlier centuries.

You can find examples of scientific revolutions in many fields, including biology, geology, genetics, and psychology. These scientific revolutions have been difficult and controversial because they have involved the overthrow of accepted paradigms. But that is why scientific revolutions are exciting. They provide not just a new idea or a new theory, but an entirely new insight into how nature works—a new way of seeing the world.

The ancients believed the stars were attached to a starry sphere. (NOAO and Nigel Sharp)

acceptance in spite of the inaccuracy of the model's predictions. The most important factor may be the elegance of the idea. Placing the sun at the center of the universe produced a symmetry among the motions of the planets that was pleasing to the eye and to the intellect (Figure 4-9b). All of the planets moved in the same direction at speeds that were simply related to their distance from the sun. In the Ptolemaic model, Venus and Mercury were treated differently from the other planets. In the Copernican model, all planets were treated the same. In this way, the model may have won support not for its accuracy, but for its elegance.

The most astonishing consequence of the Copernican hypothesis was not what it said about the sun, but what it said about Earth. By placing the sun at the center, Copernicus made Earth move along a path around the sun just as the other planets did, and that made Earth a planet. By revealing that Earth is a planet, Copernicus revolutionized humanity's view of our place in the universe and triggered a controversy that would eventually bring Galileo before the Inquisition.

Although astronomers throughout Europe read and admired *De Revolutionibus,* they did not usually accept the Copernican hypothesis. The mathematics was elegant, and the astronomical observations and calculations were of tremendous value. Yet few astronomers believed, at first, that the sun actually was the center of the planetary system and that Earth moved. How the Copernican hypothesis became gradually recognized as correct has been named the Copernican Revolution because it involved not just the adoption of a new idea but a total revolution in the way astronomers thought about the place of the Earth (**Window on Science 4-1**).

Galileo the Defender

Most people know two facts about Galileo, and both facts are wrong. You should begin by getting those facts right: Galileo did not invent the telescope, and he was not condemned by the Inquisition for believing that Earth moved around the sun. Then why is Galileo so important that in 1979, almost 400 years after his trial, the Vatican reopened his case? As you read about Galileo, you will discover that his trial concerned not just the place of the Earth but a new way of finding truth, a new way of knowing about the world, a method known today as science.

Galileo Galilei (■ Figure 4-11) was born in Pisa, a city in what is now Italy, in 1564, and he studied medicine at the university there. His true love, however, was mathematics; and, although he had to leave school early because of financial difficulties, he returned only four years later as a professor of mathematics. Three years later, he became professor of mathematics at the university at Padua. He remained there for 18 years.

During this time, Galileo seems to have adopted the Copernican model, although he admitted in a 1597 letter to the Ger-

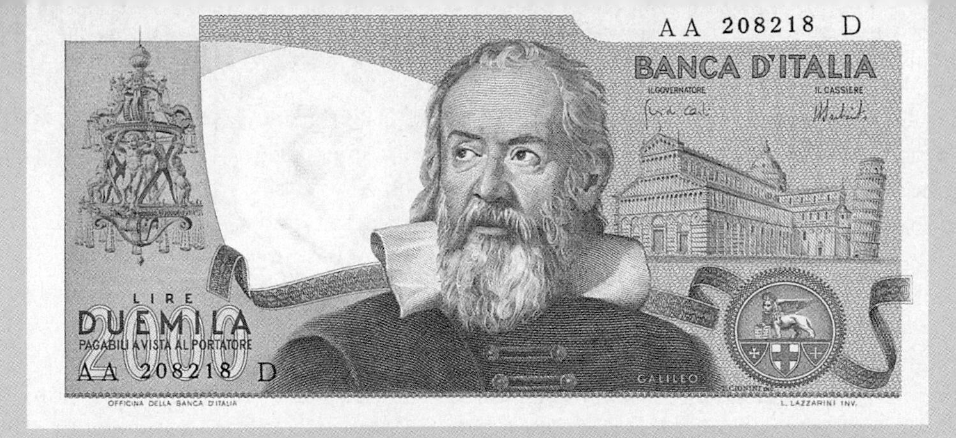

■ Figure 4-11

Galileo, remembered as the great defender of Copernicanism, also made important discoveries in the physics of motion. He is honored here on an Italian 2000-lira note. The reverse side shows one of his telescopes at lower right and a modern observatory above it.

man astronomer Kepler that he did not support Copernicanism publicly because of the criticism such a declaration would bring. It was the telescope that drove Galileo to publicly defend the heliocentric model.

Galileo did not invent the telescope. It was apparently invented around 1608 by lens makers in Holland. Galileo, hearing descriptions in the fall of 1609, was able to build telescopes in his workshop. Galileo was not the first person to look at the sky through a telescope, but he was the first person to observe the sky systematically and apply his observations to the theoretical problem of the day—the place of the Earth.

What Galileo saw through his telescopes was so amazing he rushed a small book into print. *Sidereus Nuncius (The Sidereal Messenger)* reported three major discoveries. First, the moon was not perfect. It had mountains and valleys on its surface, and Galileo used the shadows to calculate the height of the mountains. The Ptolemaic model held that the moon was perfect, but Galileo showed that it was not only imperfect, it was a world like Earth.

The second discovery reported in the book was that the Milky Way was made up of a myriad of stars too faint to see with the unaided eye. While intriguing, this discovery could not match the third discovery. Galileo's telescope revealed four new "planets" circling Jupiter, planets known today as the Galilean moons of Jupiter (■ Figure 4-12).

The moons of Jupiter supported the Copernican model over the Ptolemaic model. Critics of Copernicus had said Earth could not move because the moon would be left behind. But Jupiter moved yet kept its satellites, so Galileo's discovery suggested that Earth, too, could move and keep its moon. Also, the Ptolemaic model included the Aristotelian belief that all heavenly motion was centered on Earth at the center of the universe. Galileo showed that Jupiter's moons revolve around Jupiter, so there could be other centers of motion.

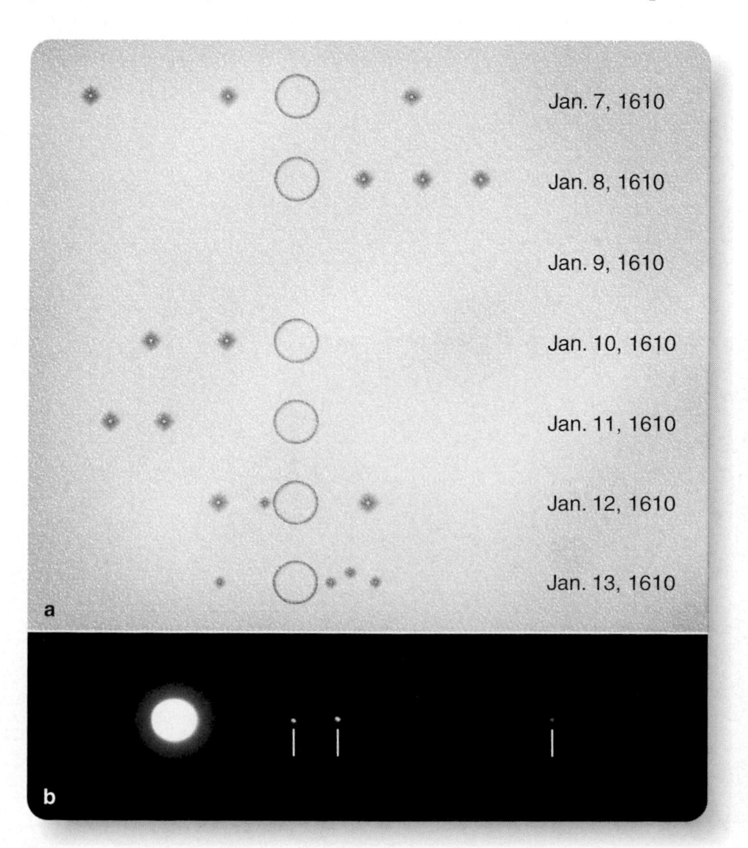

■ Figure 4-12

(a) On the night of January 7, 1610, Galileo saw three small "stars" near the bright disk of Jupiter and sketched them in his notebook. On subsequent nights (excepting January 9, which was cloudy), he saw that the stars were actually four moons orbiting Jupiter. (b) This photo taken through a modern telescope shows the overexposed disk of Jupiter and three of the four Galilean moons. (Grundy Observatory)

Sometime after *Sidereus Nuncius* was published, Galileo made another discovery that made Jupiter's moons even stronger evidence supporting the Copernican universe. Eventually, he was able to measure the orbital periods of the four moons, and he found that the innermost moon moved fastest and that the moons further from Jupiter moved more slowly. In this way, Jupiter's moons made up a harmonious system ruled by Jupiter, just as the planets in the Copernican universe were a harmonious system ruled by the sun (Figure 4-9b). The similarity isn't proof, but Galileo saw it as an argument that the solar system was sun centered and not Earth centered.

Also soon after *Sidereus Nuncius* was published, Galileo made two additional discoveries. When he observed the sun, he discovered sunspots, raising the suspicion that the sun was less than perfect. Further, by noting the movement of the spots, he concluded that the sun was a sphere and that it rotated on its axis. When he observed Venus, Galileo saw that it was going through phases like those of the moon. In the Ptolemaic model, Venus moves around an epicycle centered on a line between Earth and the sun. In that epicycle, it would always be seen as a crescent (■ Figure 4-13a). But Galileo saw Venus go through a complete set of phases, proving that it did indeed revolve around the sun (Figure 4-13b).

Sidereus Nuncius was very popular and made Galileo famous. He became chief mathematician and philosopher to the Grand Duke of Tuscany in Florence. In 1611, Galileo visited Rome and was treated with great respect. He had long, friendly discussions with the powerful Cardinal Barberini, a man with a deep interest in the new discoveries being made. But Galileo also made enemies. Personally, Galileo was outspoken, forceful, and sometimes tactless. He enjoyed debate, but most of all he enjoyed being right. In lectures, debates, and letters he offended important people who questioned his telescopic discoveries.

By 1616, Galileo was the center of a storm of controversy. Some critics said he was wrong, and others said he was lying. Some refused to look through a telescope lest it mislead them, and others looked and claimed to see nothing (hardly surprising given the awkwardness of those first telescopes) (■ Figure 4-14). Pope Paul V decided to end the disruption, so when Galileo visited Rome in 1616 Cardinal Bellarmine interviewed him privately and ordered him to cease debate. There is some controversy today about the nature of Galileo's instructions, but he did not pursue astronomy for some years after the interview. Books relevant to Copernicanism were banned, including *De Revolutionibus,* although owners were allowed to keep their books if they changed certain phrases to make it clear that the central place of the sun was only a theory and not a fact.

In 1621 Pope Paul V died, and his successor, Pope Gregory XV, died in 1623. The next pope was Galileo's friend Cardinal Barberini, who took the name Urban VIII. Galileo rushed to Rome hoping to have the prohibition of 1616 lifted; and, although the new pope did not revoke the orders, he did encourage Galileo. When he got back home, Galileo began to write his great defense of the Copernican model, finally completing it on December 24, 1629. After some delay, the book was approved by both the local censor in Florence and the head censor of the Vatican in Rome. It was printed in February 1632.

Called *Dialogo Dei Due Massimi Sistemi (Dialogue Concerning the Two Chief World Systems),* it confronts the ancient astronomy of Aristotle and Ptolemy with the Copernican model and with telescopic observations as evidence. Galileo wrote the book as a debate among three friends. Salviati, a swift-tongued defender of Copernicus, dominates the book; Sagredo is intelligent but largely uninformed. Simplicio is the dismal defender of Ptolemy. In fact, he does not seem very bright.

■ Figure 4-13

(a) If Venus moved in an epicycle centered on the Earth–sun line (see page 65), it would always appear as a crescent. (b) Galileo's telescope showed that Venus goes through a full set of phases, proving that it must orbit the sun.

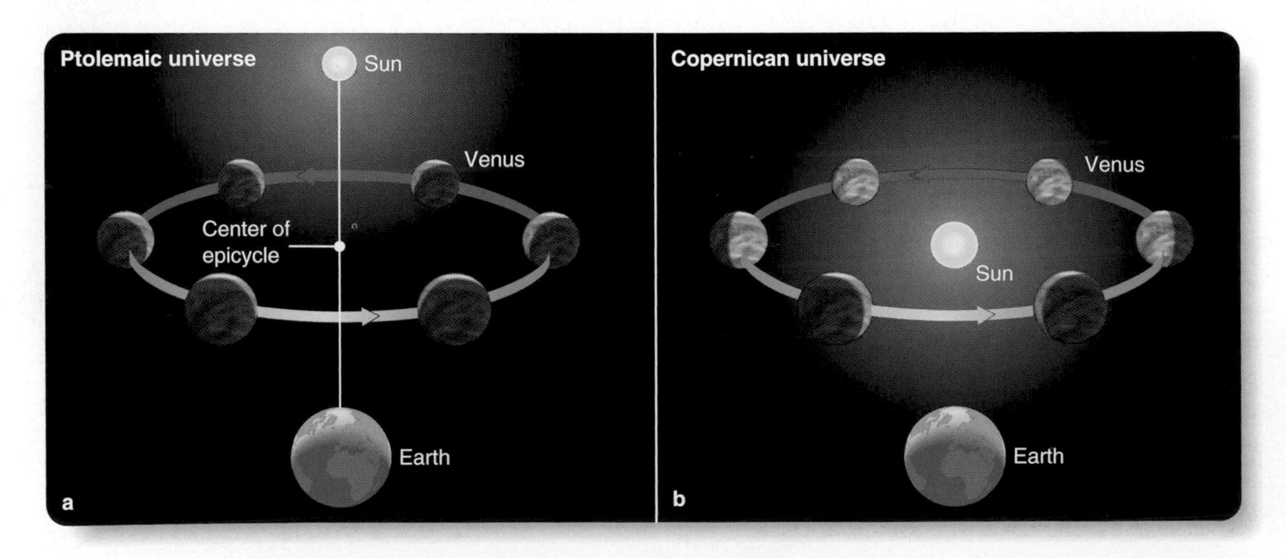

■ Figure 4-14

Galileo's telescope made him famous, and he demonstrated his telescope and discussed his observations with powerful people. Some thought the telescope was the work of the devil and would deceive anyone who looked. In any case, Galileo's discoveries produced intense, and in some cases, angry debate. (Yerkes Observatory)

The publication of *Dialogo* created a sensation, and it was sold out by August 1632, when the Inquisition ordered sales stopped. The book was a clear defense of Copernicus, and, either intentionally or unintentionally, Galileo exposed the pope's authority to ridicule. Urban VIII was fond of arguing that, as God was omnipotent, He could construct the universe in any form while making it appear to humans to have a different form, and thus its true nature could not be deduced by mere observation. Galileo placed the pope's argument in the mouth of Simplicio, and Galileo's enemies showed the passage to the pope as an example of Galileo's disrespect. The pope thereupon ordered Galileo to face the Inquisition.

The Trial of Galileo

The trial of Galileo was one of the turning points in the history of science and human learning, but historians still argue about what happened and why. The trial involved the highest religious principles and the lowest of behind-the-scenes political maneuvering. One thing you can be sure of: The trial changed the way humanity thought about the world and marked the beginning of modern science as a way to understand nature.

Galileo was interrogated by the Inquisition four times and was threatened with torture. He must have thought often of Giordano Bruno, tried, condemned, and burned at the stake in Rome in 1600. One of Bruno's offenses had been Copernicanism. But the trial did not center on Galileo's belief in Copernicanism. After all, *Dialogo* had been approved by two censors. Rather, the trial centered on the instructions given Galileo in 1616. From his file in the Vatican, his accusers produced a record of the meeting between Galileo and Cardinal Bellarmine that included the statement that Galileo was "not to hold, teach, or defend in any way" the principles of Copernicus. Some historians believe that this document, which was signed neither by Galileo nor by Bellarmine nor by a legal secretary, was a forgery. Others suspect it may be a draft that was never used. In any case, it is possible that Galileo's true instructions were much less restrictive. But Bellarmine was dead and could not testify at Galileo's trial.

The Inquisition condemned him not for heresy but for disobeying the orders given him in 1616. On June 22, 1633, at the age of 70, kneeling before the Inquisition, Galileo read a recantation admitting his errors. Tradition has it that as he rose he whispered, *"E pur si muove"* ("Still it moves"), referring to Earth. Although he was sentenced to life imprisonment, he was actually confined at his villa for the next 10 years, perhaps through the intervention of the pope. He died there on January 8, 1642, 99 years after the death of Copernicus.

Galileo was not condemned for heresy, nor was the Inquisition interested when he tried to defend Copernicanism. He was tried and condemned on a charge you might call a technicality. Then why is his trial so important that historians have studied it for almost four centuries? Why have some of the world's greatest authors, including Bertolt Brecht, written about Galileo's trial? Why in 1979 did Pope John Paul II create a commission to reexamine the case against Galileo?

To understand the trial, you must recognize that it was the result of a conflict between two ways of understanding our universe. Plato had argued that observation was deceptive, and that the only way to find truth was through pure thought. Since the Middle Ages, scholars had taught that the only path to true understanding was through religious faith. St. Augustine (AD 354–430) wrote *"Credo ut intelligame,"* which can be translated as, "Believe in order to understand." But Galileo and other scientists of the Renaissance used their own observations to try to understand the universe; and, when their observations contradicted Scripture, they assumed their observations of reality were correct and that Scripture was not being correctly understood (■ Figure 4-15). Galileo paraphrased Cardinal Baronius in saying, "The Bible tells us how to go to heaven, not how the heavens go."

Galileo's discoveries produced intense, and in some cases angry, debate. Various passages of Scripture seemed to contradict observation. For example, Joshua is said to have commanded the sun to stand still, not Earth to stop rotating (Joshua 10:12–13). In response to such passages, Galileo argued that you should "read the book of nature"—that is, you should observe the universe with your own eyes.

The trial of Galileo was not about the place of the Earth. It was not about Copernicanism. It wasn't really about the instructions Galileo received in 1616. It was about the birth of modern science as a rational way to understand our universe. The

■ **Figure 4-15**

Although he did not invent it, Galileo will always be associated with the telescope because it was the source of the observational evidence he used to try to understand the universe. By depending on evidence instead of first principles, Galileo led the way to the invention of modern science as a way to know about the natural world.

commission appointed by John Paul II in 1979, reporting its conclusions in October 1992, said of Galileo's inquisitors, "This subjective error of judgment, so clear to us today, led them to a disciplinary measure from which Galileo 'had much to suffer.'" Galileo was not found innocent in 1992 so much as the Inquisition was forgiven for having charged him in the first place.

The gradual change from reliance on personal faith to reliance on scientific observation came to a climax with the trial of Galileo. Since that time, scientists have increasingly reserved Scripture and religious faith for ethical guidance and personal comfort and have depended on systematic observation to describe the physical world. The final verdict of 1992 was an attempt to bring some balance to this conflict. In his remarks on the decision, Pope John Paul II said, "A tragic mutual incomprehension has been interpreted as the reflection of a fundamental opposition between science and faith. . . . this sad misunderstanding now belongs in the past." Galileo's trial is over, but it continues to echo through history as part of humanity's struggle to understand the place of the Earth in our universe.

Building Scientific Arguments

How were Galileo's observations of the moons of Jupiter evidence against the Ptolemaic model?

Modern scientific arguments depend critically on evidence, and that started with Galileo. He presented his arguments in the form of evidence and conclusions, and the moons of Jupiter were key evidence. Ptolemaic astronomers argued that Earth could not move or it would lose its moon, but even in the Ptolemaic universe Jupiter moved, and Galileo's telescope showed that it kept its moons. Evidently, Earth could move and not leave its moon behind. Furthermore, moons circling Jupiter did not fit the classical belief that all motion was centered on Earth. Obviously there could be other centers of motion. Finally, the orbital periods of the moons were related to their distance from Jupiter, just as the orbital periods of the planets were, in the Copernican system, related to their distance from the sun. This similarity suggested that the sun rules its harmonious family of planets just as Jupiter rules its harmonious family of moons.

Of all of Galileo's telescopic observations, the moons of Jupiter caused the most debate. But there was more. Use the evidence to build an argument to answer the following: **How did craters on the moon and the phases of Venus weigh against the Ptolemaic model?**

■ ■ ■

Connections: Galileo faced the Inquisition because of a conflict between two ways of knowing and understanding the world. The Church taught faith and understanding through revelation, but the scientists of the age were inventing a new way to understand nature that relied on evidence, a way of knowing now called science. In astronomy, evidence means observations, so it is time to turn your attention from the philosophical problem of the place of the Earth to the observational problem of the motion of the planets.

4-3 The Puzzle of Planetary Motion

WHILE GALILEO WAS TEACHING mathematics at Pisa and Padua, two other astronomers were working in northern Europe, beyond the sway of the Inquisition. Although they, too, struggled to understand the place of the Earth, they approached the problem differently. They used the most accurate observations of the positions of the planets to try to discover the rules that govern planetary motion. When that puzzle was finally solved, astronomers understood why the Ptolemaic model of the universe does not work well and how to make the Copernican universe a precise predictor of planetary motion.

Tycho the Observer

The great observational astronomer of this story is a Danish nobleman, Tycho Brahe, born December 14, 1546, only three years after the publication of *De Revolutionibus* (■ Figure 4-16). As a son of a powerful noble family, he was well known for his vanity and lordly manners.

Tycho's college days were eventful. He was officially studying law with the expectation that he would enter Danish politics, but he made it clear to his family that his real interest was astronomy and mathematics. It was also during his college days that he fought a duel and received a wound that disfigured his nose. For the rest of his life, he wore false noses made of gold and silver and stuck on with wax. The disfigurement probably did little to improve his disposition.

Tycho's first astronomical observations were made while he was a student. In 1563, Jupiter and Saturn passed very near each other in the sky, nearly merging into a single point on the night of August 24. Tycho found that the *Alfonsine Tables* were a full month in error and that the *Prutenic Tables* were in error by a number of days. These discrepancies dismayed Tycho and sparked his interest in the motions of the planets.

In 1572, a brilliant "new star" (now called Tycho's supernova) appeared in the sky. Such changes in the sky puzzled classically trained astronomers. Aristotle had argued that the starry sphere was perfect and unchanging, and therefore such new stars had to lie closer to Earth than the moon. Layers below the moon were thought to be less perfect and thus more changeable than the starry sphere.

Tycho carefully measured the position of the star time after time through the night and found that it displayed no parallax. To understand the significance of this observation, you must note that Tycho, like most astronomers of his time, believed Earth was fixed at the center of the universe. Consequently, he believed the heavens rotated westward around Earth once a day. The new star, according to classical astronomy, represented a change in the heavens and therefore had to lie below the sphere of the moon. In that case, Tycho reasoned, the new star would appear slightly too far east when it was in the eastern sky and slightly too far west later in the night when it was carried into in the western sky (■ Figure 4-17). That is an example of parallax. Tycho could detect no parallax in the position of the new star, so he concluded that it must lie above the sphere of the moon and was probably on the starry sphere itself. That was an astonishing discovery because it contradicted Aristotle's belief that the starry sphere was perfect and unchanging.

No one before Tycho could have made this discovery because no one had ever measured the positions of stars so accurately. Tycho had great confidence in the precision of his measurements; so, when he failed to detect parallax for the new star, he knew it was important evidence against the Ptolemaic theory. He announced his discovery in a small book, *De Stella Nova (The New Star)*, published in 1573.

The book attracted the attention of astronomers throughout Europe, and soon Tycho was famous. Although Tycho had finished at the university, he was unemployed, idling at the family home and embarrassing his powerful family. His fame allowed his family to introduce him to the court of the Danish King Frederik II, and there the king offered Tycho funds to build an observatory on the island of Hveen just off the Danish coast. Tycho also received a steady source of income as landlord of a coastal

■ Figure 4-16

Tycho Brahe (1546–1601) was, during his lifetime, the most famous astronomer in the world. Proud of his noble rank, he wears the elephant medal awarded him by the king of Denmark. His artificial nose is suggested in this engraving.

■ Figure 4-17

If an object lay lower than the celestial sphere, reasoned Tycho Brahe, then it should be seen east of its average position as it was rising and west of its average position when it was setting. Because he did not detect this daily parallax, he concluded the new star of 1572 had to lie on the celestial sphere.

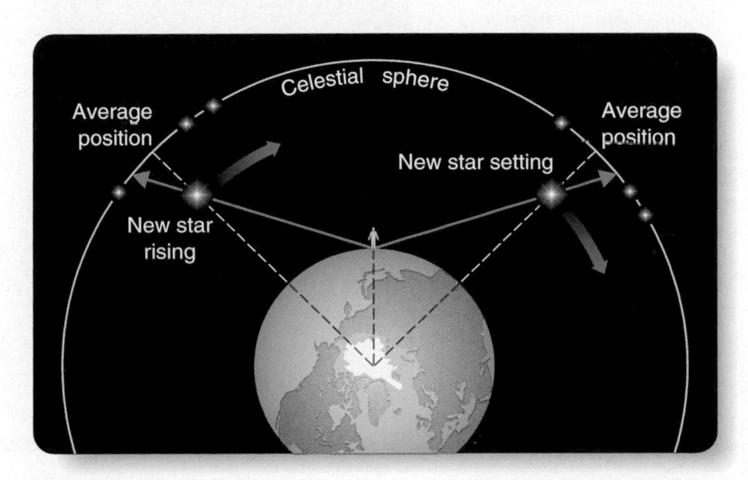

district from which he collected rents. (He was not a popular landlord.) On Hveen, Tycho constructed a luxurious home with six towers specially equipped for astronomy and populated it with servants, assistants, and a dwarf to act as jester. Soon Hveen was an international center of astronomical study.

Tycho Brahe's Legacy

Tycho Brahe made no lasting contribution to astronomical theory. Because he could measure no parallax for the stars, he concluded that Earth had to be stationary, and that led him to reject the Copernican hypothesis. However, he also rejected the Ptolemaic model because of its inaccurate predictions. Instead, he devised a complex model in which Earth was the immobile center of the universe around which the sun and moon moved. The other planets circled the sun. You might find this arrangement familiar; it is really the Copernican model with Earth held stationary and the sun allowed to move around Earth. In this way, Tycho preserved the central, immobile Earth that most astronomers believed was described in scripture (■ Figure 4-18a). Although Tycho's model, the Tychonic Universe, was very popular at first, the Copernican model replaced it within a century.

The true value of Tycho's work was observational. Because he was able to devise new and better instruments, he was able to make highly accurate observations of the positions of the stars, sun, moon, and planets. Tycho had no telescopes—they were not invented until the next century—so his observations were made by the naked eye peering along sights. All of his instruments were designed to measure angles in the sky. For example, his quadrant could measure angles up to 90° above the horizon (Figure 4-18b). By designing and building large instruments with great care, he was able to measure angles to high precision. He measured the positions of 777 stars to better than 4 minutes of arc and regularly measured the positions of the sun, moon, and planets during the 20 years he stayed on Hveen.

Unhappily for Tycho, King Frederik II died in 1588, and his young son took the throne. Suddenly Tycho's temper, vanity, and noble presumptions threw him out of favor. In 1596, taking most of his instruments and books of observations, he went to Prague, the capital of Bohemia, and became imperial mathematician to the Holy Roman Emperor Rudolph II. His goal was to revise *The Alfonsine Tables* and publish the revision as a monument to his new patron. It would be called *The Rudolphine Tables.*

Tycho did not intend to base *The Rudolphine Tables* on the Ptolemaic system but rather on his own Tychonic system, proving once and for all the validity of his hypothesis. To assist him, he hired a few mathematicians and astronomers, including one Johannes Kepler. Then in November 1601, Tycho collapsed at a nobleman's home. Before he died, 11 days later, he asked Rudolph II to make Kepler imperial mathematician. Thus the newcomer, Kepler, became Tycho's replacement (though at one-sixth Tycho's salary).

Kepler the Analyst

No one could have been more different from Tycho Brahe than Johannes Kepler (■ Figure 4-19). He was born December 27, 1571, to a poor family in a region now included in southwestern Germany. His father was unreliable and shiftless, principally employed as a mercenary soldier fighting for whoever paid enough. He finally failed to return from a military expedition, either because he was killed or because he found other circum-

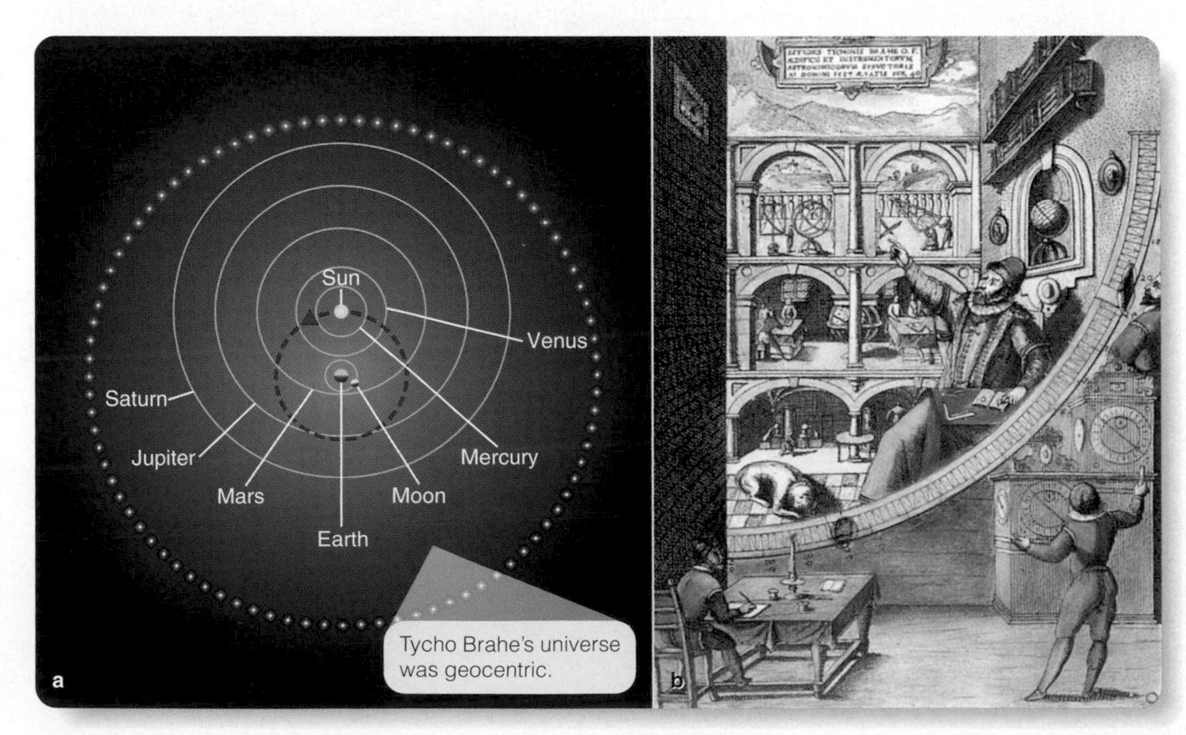

■ Figure 4-18

(a) Tycho Brahe's model of the universe held that Earth was fixed at the center of the starry sphere. The moon and sun circled Earth, while the planets circled the sun. (b) Much of Tycho's success was due to his skill in designing large, accurate instruments. In this engraving of his mural quadrant, the figure of Tycho, his dog, and the background scene are painted on the wall within the arc of the quadrant. The observer (Tycho himself) at the extreme right peers through a sight out the loophole in the wall at the upper left to measure an object's angular distance above the horizon. (Granger Collection, New York)

stances more to his liking. Kepler's mother was apparently an unpleasant and unpopular woman. She was accused of witchcraft in her later years, and Kepler had to defend her in a trial that dragged on for three years. She was finally acquitted but died the following year.

Kepler was the oldest of six children, and his childhood was no doubt unhappy. The family was not only poor but suffered from an absentee father whom Kepler described as "vicious, inflexible, quarrelsome, and doomed to a bad end." In addition, Kepler was never healthy, even as a child, so it is surprising that he did well in school, eventually winning promotion to a Latin school and finally a scholarship to the university at Tübingen, where he studied to become a Lutheran pastor.

During his last year of study, Kepler accepted a job in Graz teaching mathematics and astronomy. Evidently, he was not a good teacher. He had few students his first year and none at all his second. His superiors put him to work teaching a few introductory courses and preparing an annual almanac that contained astronomical, astrological, and weather predictions. Through good luck, in 1595 some of his weather predictions were fulfilled, and he gained a reputation as an astrologer and seer; even in later life he earned money by publishing almanacs.

While still a college student, Kepler had become a believer in the Copernican theory, and at Graz he used his extensive spare time to study astronomy. By 1596, the same year Tycho left Hveen, Kepler was ready to solve the mystery of the universe. That year he published a book called *The Forerunner of Dissertations on the Universe, Containing the Mystery of the Universe.* Like nearly all scientific works of the time, the book was in Latin, and it is now known as *Mysterium Cosmographicum.*

The book begins with a long appreciation of Copernicanism and then goes on to speculate on the spacing of the planetary orbits. Kepler, as a Copernican, knew of six planets circling the sun—Mercury, Venus, Earth, Mars, Jupiter, and Saturn. According to his model, the spheres containing the six planets were separated from one another, and their relative sizes were fixed by the five regular solids*—the cube, tetrahedron, dodecahedron, icosahedron, and octahedron (Figure 4-19). Kepler gave astrological, numerological, and musical arguments for his theory and so followed the tradition of Pythagoras in believing that the order of the universe is underlain by musical (meaning mathematical) principles.

In the second half of the book, Kepler tried to fit the five solids to the planetary circles of the Copernican theory, a complex problem involving three-dimensional trigonometry and geometry. He sent copies to Tycho and to Galileo, and neither seemed very impressed with his theory, but his mathematical abilities shown brightly in his calculations.

■ Figure 4-19

Johannes Kepler (1571–1630) was Tycho Brahe's sucessor. This diagram, based on one drawn by Kepler, shows how he believed the sizes of the celestial spheres carrying the outer three planets are determined by spacers (blue) consisting of the regular solids. Inside the sphere of Mars, the remaining regular solids (not shown here) separate the spheres of the Earth, Venus, and Mercury. The sun lay at the very center of this Copernican universe based on geometrical spacers.

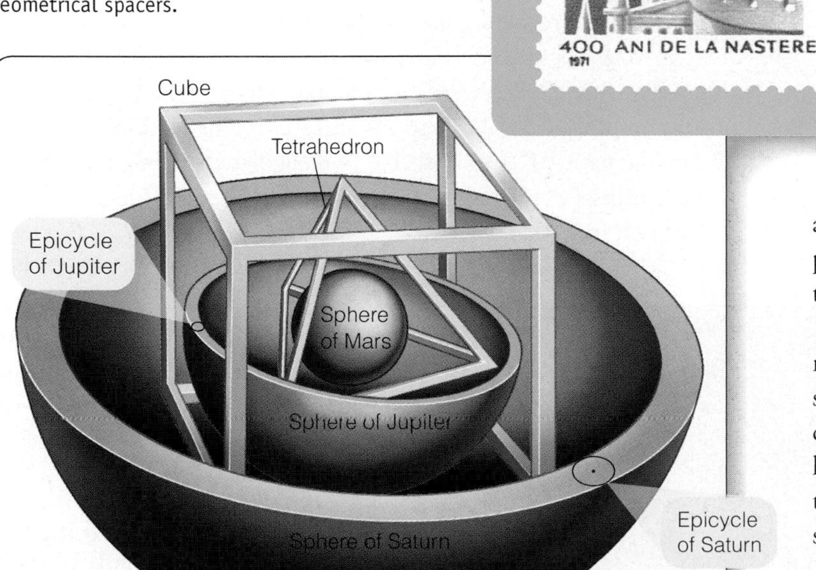

Life was unsettled for Kepler because of the persecution of Protestants in the region, so when Tycho invited him to Prague in 1600, he went readily, eager to work with the famous astronomer. Tycho's sudden death in 1601 left Kepler in a position to use the observations from Hveen to analyze the motions of the planets and complete *The Rudolphine Tables.* Tycho's family, recognizing that Kepler was a Copernican and guessing that he would not follow the Tychonic system in completing *The Rudolphine Tables,* sued to recover the instruments and books of observations. The legal wrangle went on for years. The family did recover the instruments Tycho had brought to Prague; Kepler seems to have had little interest in them, perhaps because

*A regular solid is a three-dimensional body each of whose faces is the same. For example, a cube is a regular solid each of whose faces is a square.

of his poor eyesight. However, Kepler had the books of observations, and he kept them.

Whether Kepler had any legal right to Tycho's records is debatable, but he put them to good use. He began by studying the motion of Mars, trying to deduce from the observations how the planet moves. By 1606, he had solved the puzzle of planetary motion. The orbit of Mars is an ellipse and not a circle, he said, and with that he abandoned the 2000-year-old belief in the circular motion of the planets. But he discovered that the mystery was even more complex: The planets do not move at a uniform speed along their elliptical orbits. Kepler's analysis showed that they move faster when closer to the sun and slower when farther away. With those two brilliant discoveries, Kepler abandoned both circular motion and uniform motion.

Kepler published his results in 1609 in a book called *Astronomia Nova (The New Astronomy).* Like Copernicus's book, *Astronomia Nova* did not become an instant best seller. It is written in Latin for other scientists and is highly mathematical. In some ways, the book is surprisingly advanced. For instance, Kepler discusses the force that holds the planets in their orbits, a question that Isaac Newton considered later in his recognition of gravity as a natural force.

Despite the abdication of Kepler's patron Rudolph II in 1611, Kepler continued his astronomical studies. He wrote about a supernova that had appeared in 1604 (now known as Kepler's supernova) and about comets, and he wrote a textbook about Copernican astronomy. In 1619, he published *Harmonice Mundi (The Harmony of the World),* in which he returned to the cosmic mysteries of *Mysterium Cosmographicum.* The main thing of note in *Harmonice Mundi* is his discovery that the radii of the planetary orbits are related to the planets' orbital periods. That and his two previous discoveries are now recognized as Kepler's three laws of planetary motion.

Kepler's Three Laws of Planetary Motion

Although Kepler dabbled in the philosophical arguments of the day, he was a mathematician, and his triumph was the solution of the problem of the motion of the planets. The key to his solution was the ellipse.

An **ellipse** is a figure drawn around two points called the foci in such a way that the distance from one focus to any point on the ellipse and back to the other focus equals a constant. You can easily draw ellipses with two thumbtacks and a loop of string. Press the thumbtacks into a board, loop the string about the tacks, and place a pencil in the loop. If you keep the string taut as you move the pencil, it traces out an ellipse (■ Figure 4-20a).

The geometry of an ellipse is described by two simple numbers. The **semimajor axis,** ***a,*** is half of the longest diameter. The **eccentricity** of an ellipse, ***e,*** is the distance from either focus to the center of the ellipse divided by the semimajor axis. If you want to draw a circle with the string and tacks as shown in Figure 4-20a, you would move the two thumbtacks together, which shows that a circle is really just an ellipse with eccentricity equal to zero. As you move the thumbtacks farther apart, the ellipse becomes flatter, and the eccentricity moves closer to 1.

Kepler used ellipses to describe the motion of the planets in three fundamental rules that have been tested and confirmed so many times that astronomers now refer to them as natural laws Window on Science 4-2). They are commonly called Kepler's laws of planetary motion (■ Table 4-1).

Kepler's first law states that the orbits of the planets around the sun are ellipses with the sun at one focus. Thanks to the precision of Tycho's observations and the sophistication of Kepler's mathematics, Kepler was able to recognize the elliptical shape of the orbits, even though they are nearly circular. Of the planets known to Kepler, Mercury has the most elliptical orbit, but even it deviates only slightly from a circle (■ Figure 4-21).

Kepler's second law states that a line from the planet to the sun sweeps over equal areas in equal intervals of time. This means

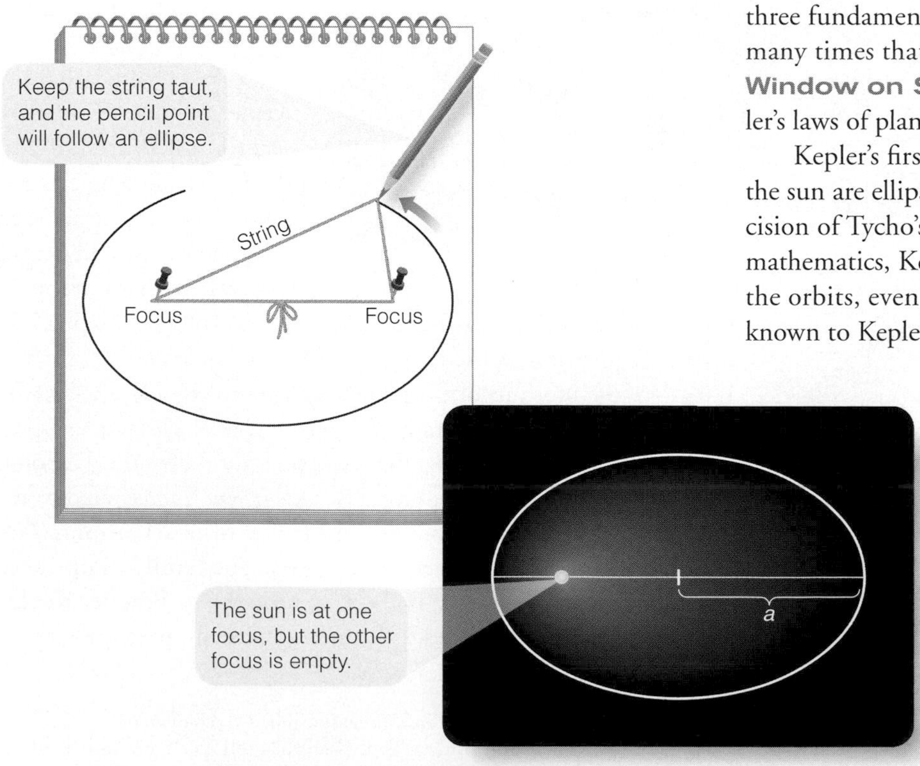

■ Figure 4-20

The geometry of elliptical orbits: Drawing an ellipse with two tacks and a loop of string is easy. The semimajor axis, *a,* is half of the longest diameter. The sun lies at one of the foci of the elliptical orbit of a planet.

Window on Science 4-2

Hypothesis, Theory, and Law: Levels of Confidence

Even scientists misuse the words *hypothesis, theory,* and *law.* You must try to distinguish these terms from one another because they are key elements in science.

A **hypothesis** is a single assertion or conjecture that must be tested. It could be true or false. "All Texans love chili" is a hypothesis. To know whether it is true or false, you need to test it against reality by making observations or performing experiments. Copernicus asserted that the universe was heliocentric; his assertion was a hypothesis subject to testing.

In Chapter 2, you saw that a model (Window on Science 2-1) is a description of some natural phenomenon; it can't be right or wrong. A model is not a conjecture of truth but merely a convenient way to think about a natural phenomenon. Consequently, a model such as the celestial sphere cannot be a hypothesis. Copernicus used his hypothesis to build a model, but they are not the same thing.

A **theory** is a system of rules and principles that can be applied to a wide variety of circumstances. A theory may have begun as one or more hypotheses, but it has been tested, expanded, and generalized. Many textbooks refer to the "Copernican theory," but some historians argue that it was not complete and had not been tested enough to be a theory. It is probably better to call it the Copernican hypothesis.

A **natural law** is a theory that has been refined, tested, and confirmed so often scientists have great confidence in it. Natural laws are the most fundamental principles of scientific knowledge. Kepler's laws are good examples.

Confidence is the key to understanding these terms. Scientists have more confidence in a theory than in a hypothesis and great confidence in a natural law. Nevertheless, scientists are not always consistent about these words. For example, Einstein's theory of relativity is much more accurate than Newton's laws of gravity and motion, but tradition dictates that no one refers to Einstein's laws of relativity and Newton's theories of gravity and motion. Darwin's theory of evolution has been tested many times, and scientists have great confidence in it, but no one refers to Darwin's law of evolution. These distinctions are subtle and sometimes depend more on custom than on levels of confidence.

A fossil of a 500-million-year-old trilobite: Darwin's theory of evolution has been tested many times and is widely accepted, but by custom it is called a theory and not a law. (From the collection of John Coolidge III)

Table 4-1 | Kepler's Laws of Planetary Motion

I. The orbits of the planets are ellipses with the sun at one focus.
II. A line from a planet to the sun sweeps over equal areas in equal intervals of time.
III. A planet's orbital period squared is proportional to its average distance from the sun cubed:

$$P^2_{yr} = a^3_{AU}$$

that when the planet is closer to the sun and the line connecting it to the sun is shorter, the planet moves more rapidly to sweep over the same area that is swept over when the planet is farther from the sun. So the planet in Figure 4-21 (inset) would move from point A to point B in one month, sweeping over the area shown. But when the planet is farther from the sun, one month's motion would be shorter, from A' to B'.

The time that a planet takes to travel around the sun once is its orbital period, P, and its average distance from the sun equals the semimajor axis of its orbit, a. Kepler's third law says these two quantities are related: The orbital period squared is proportional to the semimajor axis cubed. If you measure P in years and a in astronomical units, you can summarize the third law as:

$$P^2_{yr} = a^3_{AU}$$

The subscripts are reminders that you must measure the period in years (yr) and the semimajor axis in astronomical units (AU).

You can use Kepler's third law to make simple calculations. For example, Jupiter's average distance from the sun is 5.20 AU. What is its orbital period? If a equals 5.20, then a^3 equals 140.6. The orbital period must be the square root of 140.6, which equals about 11.8 years.

You should note that Kepler's three laws are empirical. That is, they describe a phenomenon without explaining why it occurs. Kepler derived them from Tycho's extensive observations, not from any fundamental assumption or theory. In fact, Kepler never knew

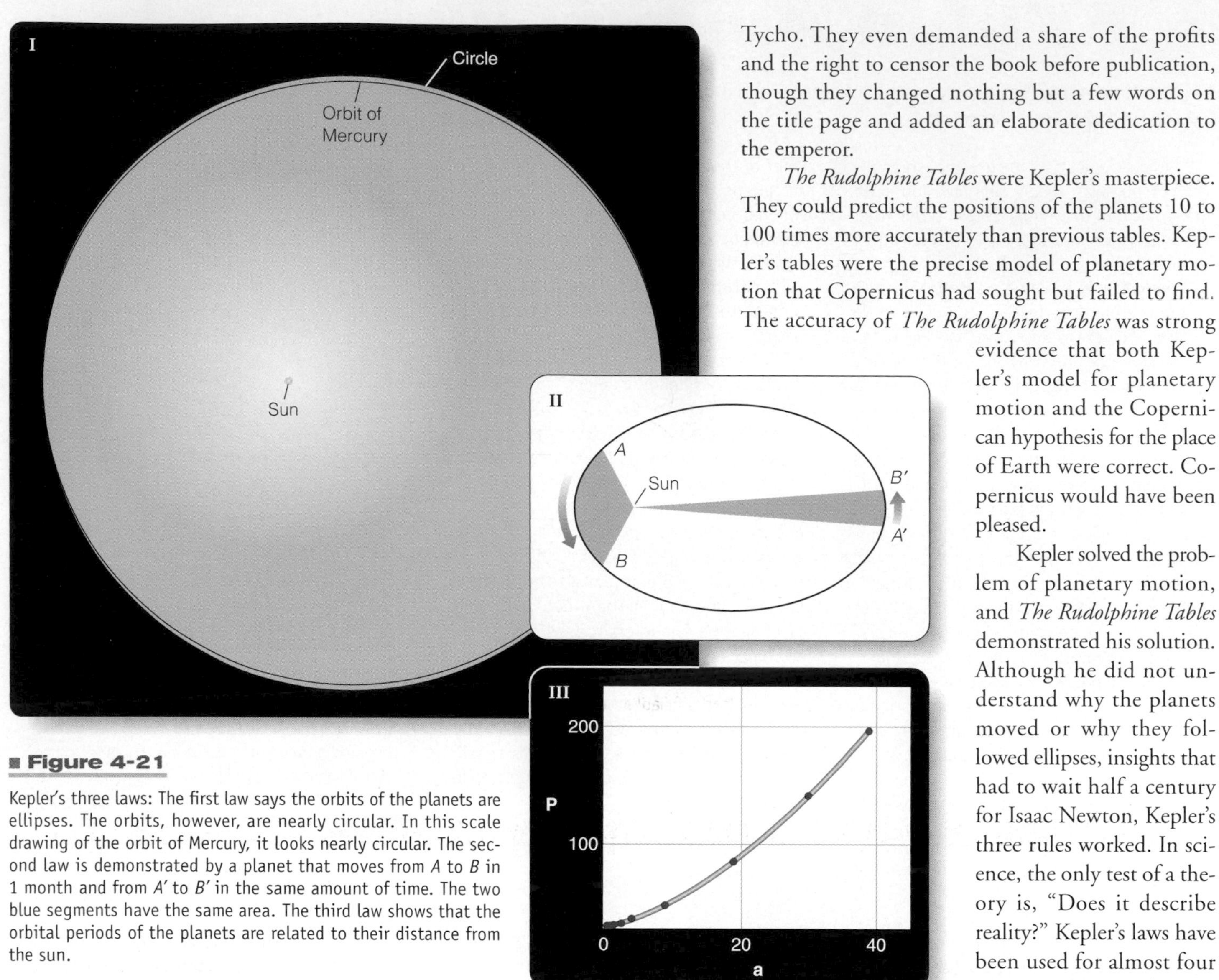

■ **Figure 4-21**

Kepler's three laws: The first law says the orbits of the planets are ellipses. The orbits, however, are nearly circular. In this scale drawing of the orbit of Mercury, it looks nearly circular. The second law is demonstrated by a planet that moves from *A* to *B* in 1 month and from *A′* to *B′* in the same amount of time. The two blue segments have the same area. The third law shows that the orbital periods of the planets are related to their distance from the sun.

what held the planets in their orbits or why they continued to move around the sun. His books are a fascinating blend of careful observation, mathematical analysis, and mystical theory.

The Rudolphine Tables

In spite of Kepler's recurrent involvement with astrology and numerology, he continued to work on *The Rudolphine Tables.* At last, in 1627, they were ready, and he financed their printing himself, dedicating them to the memory of Tycho Brahe. In fact, Tycho's name appears in larger type on the title page than Kepler's own. This is especially surprising when you recall that the tables were based on the heliocentric model of Copernicus and the elliptical orbits of Kepler and not on the Tychonic system. The reason for Kepler's evident deference was Tycho's family, still powerful and still intent on protecting the memory of Tycho. They even demanded a share of the profits and the right to censor the book before publication, though they changed nothing but a few words on the title page and added an elaborate dedication to the emperor.

The Rudolphine Tables were Kepler's masterpiece. They could predict the positions of the planets 10 to 100 times more accurately than previous tables. Kepler's tables were the precise model of planetary motion that Copernicus had sought but failed to find. The accuracy of *The Rudolphine Tables* was strong evidence that both Kepler's model for planetary motion and the Copernican hypothesis for the place of Earth were correct. Copernicus would have been pleased.

Kepler solved the problem of planetary motion, and *The Rudolphine Tables* demonstrated his solution. Although he did not understand why the planets moved or why they followed ellipses, insights that had to wait half a century for Isaac Newton, Kepler's three rules worked. In science, the only test of a theory is, "Does it describe reality?" Kepler's laws have been used for almost four centuries as a true description of orbital motion.

Building Scientific Arguments

What were the main differences among *The Alfonsine Tables, The Prutenic Tables,* and *The Rudolphine Tables?*

Each of these tables was an expression of a different theory, and comparing theory with evidence is the heart of scientific arguments. All three of these tables predicted the motions of the sun, moon, and planets, but only *The Rudolphine Tables* proved accurate. *The Alfonsine Tables,* produced in Toledo around AD 1250, were based on the Ptolemaic model, so they were geocentric and used uniform circular motion; consequently, they were not very accurate. *The Prutenic Tables,* published in 1551, were based on the Copernican model, and so were heliocentric. But because *The Prutenic Tables* included the classical principle of uniform circular motion, they were no more accurate than *The Alfonsine Tables.* Kepler's *Rudol-*

phine Tables were Copernican in that they were heliocentric, but they included Kepler's three laws of planetary motion and consequently could predict the positions of the planets 10 to 100 times more accurately than previous tables.

One of the main reasons for the success of the Copernican hypothesis was not that it was accurate but that it was elegant. Compare the motions of the planets and the explanation of retrograde motion in the Copernican model with those in the Ptolemaic and Tychonic models. **How was the Copernican model more elegant?**

■ ■ ■

Connections: Kepler died on November 15, 1630. During his life he had been an astrologer, a mystic, a numerologist, and a seer, but he became one of the world's great astronomers. His work not only overthrew the concept of uniform circular motion that had shackled the minds of astronomers for over 2000 years but also opened the way to the empirical study of planetary motion, a study that lies at the heart of modern astronomy.

4-4 Modern Astronomy

YOU CAN DATE THE ORIGIN of modern astronomy from the 99 years between the deaths of Copernicus and Galileo (1543 to 1642); it was an age of transition. That period marked the change from the Ptolemaic model of the universe to the Copernican model with the attendant controversy over the place of the Earth. But that same period also marked a transition in the nature of astronomy in particular and science in general, a transition illustrated in the resolution of the puzzle of planetary motion. The puzzle was not solved by philosophical arguments about the perfection of the heavens or by debate over the meaning of scripture. It was solved by precise observation and careful computation, techniques that are the foundation of modern science.

The discoveries of Kepler and Galileo found acceptance in the 1600s because the world was in transition. Astronomy was not the only thing changing during this period. The Renaissance is commonly taken to be the period between 1300 and 1600, and these 99 years of astronomical history lie at the culmination of the reawakening of learning in all fields (■ Figure 4-22). Ships

■ Figure 4-22

The 99 years between the death of Copernicus in 1543 and the death of Galileo in 1642 marked the transition from the ancient astronomy of Ptolemy and Aristotle to the revolutionary theory of Copernicus, and, simultaneously, the invention of science as a way of understanding the world.

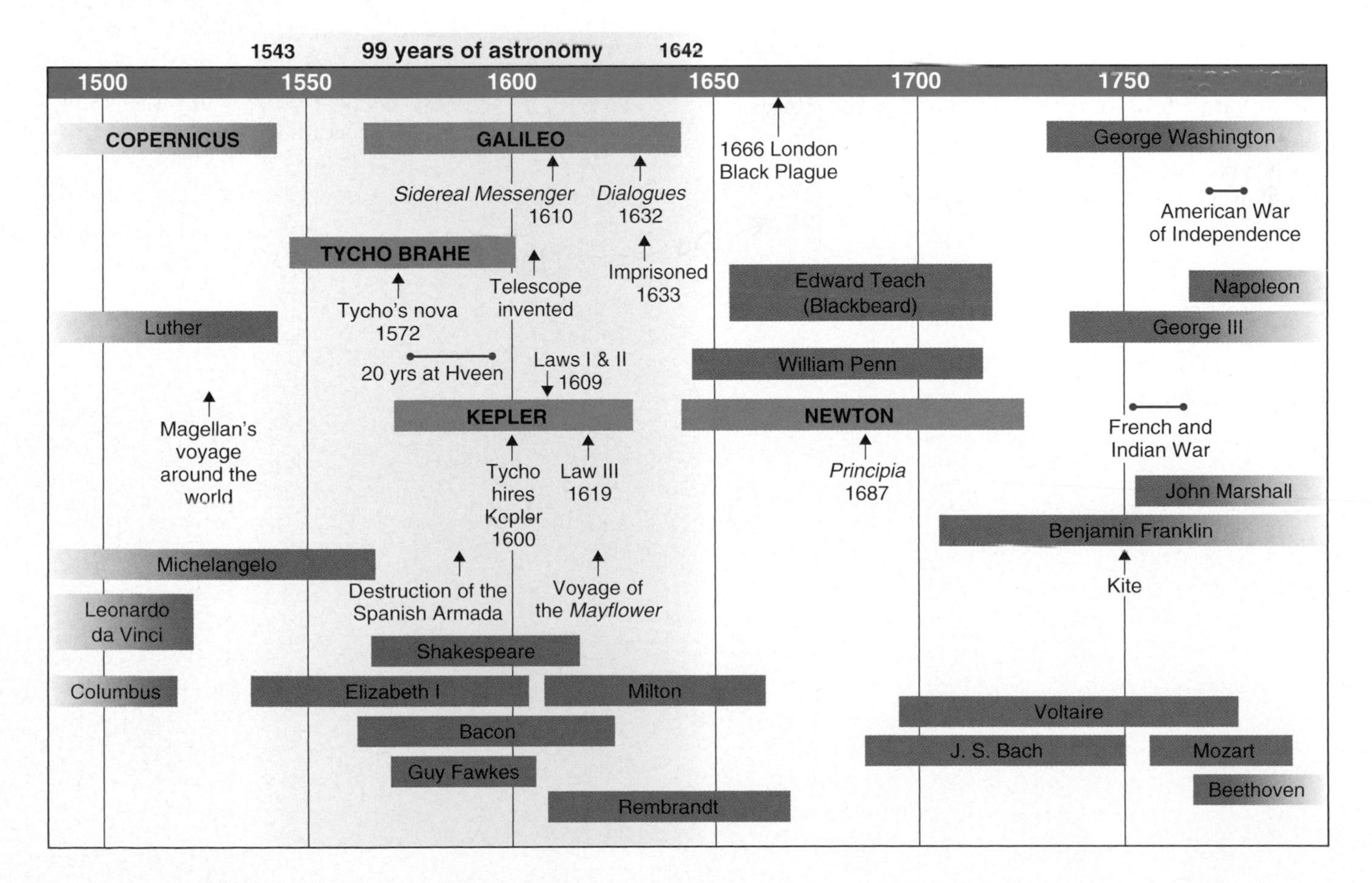

were sailing to new lands and encountering new cultures. The world was open to new ideas and new observations. Martin Luther remade religion, and other philosophers and scholars reformed their areas of human knowledge. Had Copernicus not published his hypothesis, someone else would have suggested that the universe is heliocentric. History was ready to shed the Ptolemaic system.

In addition, this period marks the beginning of the modern scientific method. Beginning with Copernicus, scientists such as Tycho, Kepler, and Galileo depended more and more on evidence, observation, and measurement. This, too, is coupled to the Renaissance and its advances in metalworking and lens making. Before the story told in this chapter began, no astronomer had looked through a telescope, because one could not be made. By 1642, not only telescopes but also other sensitive measuring instruments had transformed science into something new and precise. As you can imagine, scientists were excited by these discoveries, and they founded scientific societies that increased the exchange of observations and hypotheses and stimulated more and better work. The most important advance, however, was the application of mathematics to scientific questions. Kepler's work demonstrated the power of mathematical analysis; and, as the quality of these numerical tools improved, the progress of science accelerated. This story of the birth of modern astronomy is actually the story of the birth of modern science as well.

Building Scientific Arguments

Why was it so hard for astronomers to abandon the Ptolemaic model?

The central position of Earth and uniform circular motion were part of a paradigm—a set of ideas that people accepted as obvious. Because they all worked within that paradigm, they could not see its faults as well as you can today. It did not occur to them to test obvious ideas. Also, astronomers before the time of the Copernican Revolution were not accustomed to thinking in scientific arguments based on theory and evidence. Observations that did not fit the paradigm were not taken as seriously as they would be today. Only after the time of Galileo did scholars begin to think and reason like modern scientists. Scientific thinking is so common today, you take it for granted. When you read about new music players in a magazine like *Consumer Reports,* hear about blood tests at a trial, or try two different kinds of toothpaste to see which you like best, you are thinking scientifically and depending on evidence. Perhaps you are surprised that something as obvious as scientific thinking had to be invented.

There was yet another reason why astronomers found it hard to let go of the Ptolemaic model. The central place of the Earth had serious theological meaning. **Why did moving Earth away from the center make some astronomers hesitate to adopt the Copernican theory?**

■ ■ ■

Connections: You have read that Kepler solved the puzzle of planetary motion, but his solution seems incomplete. Why do the planets move as they do? The story will continue in the next chapter with the life of Isaac Newton. You will see how his accomplishments had their origins in the work of Galileo and led to Einstein's deeper understanding of gravity.

Study and Review Tools

Summary

4-1 | The Roots of Astronomy

How did the ancients describe the place of the Earth?

- Archaeoastronomy is the study of the astronomy of ancient peoples.
- Many cultures around the world observed the sky and marked important alignments. Structures such as Stonehenge, Newgrange, the Sun Dagger, and Egyptian temples have astronomical alignments.
- In most cases, ancient cultures, having no written language, left no detailed records of their astronomical beliefs.
- Greek astronomy, derived in part from Babylon and Egypt, is better known because written documents have survived.
- Classical philosophers accepted as a first principle that Earth was the unmoving center of the universe. Another first principle was that the heavens were perfect, so philosophers such as Plato argued that, because the sphere was the most perfect geometrical form, the heavens must be made up of spheres in uniform rotation. This led to the belief in uniform circular motion.
- The geocentric universe became part of the teachings of the great philosopher Aristotle, who argued that the sun, moon, and stars were carried around Earth on rotating crystalline spheres.
- Aristotle's estimate for the size of Earth was only about one-third of its true size. Eratosthenes used the well at Syene to measure the diameter of Earth and got an accurate estimate.
- About 140 AD, Ptolemy gave mathematical form to Aristotle's model in the *Almagest*. Ptolemy preserved the principles of geocentrism and uniform circular motion, but he added epicycles, deferents, and equants to better predict the motions of the planets.

4-2 | The Copernican Revolution

How did Copernicus change the place of the Earth?

- Copernicus devised a heliocentric model. He preserved the principle of uniform circular motion, but he argued that Earth rotates on its axis and revolves around the sun once a year. His theory was controversial because it contradicted Church teaching.
- Copernicus published his theory in his book *De Revolutionibus* in 1543, the same year he died.
- Because Copernicus kept uniform circular motion as part of his theory, his model did not predict the motions of the plants well, but it did offer a simple explanation of retrograde motion without using big epicycles.
- One reason the Copernican model won converts was that it was more elegant. Venus and Mercury were treated the same as all the other planets, and the velocity of each planet was related to its distance from the sun.

Why was Galileo condemned by the Inquisition?

- Galileo used the newly invented telescope to observe the heavens, and he recognized the significance of what he saw there. His discoveries of the phases of Venus, the satellites of Jupiter, the mountains of the moon, and other phenomena helped undermine the Ptolemaic universe.
- Galileo based his analysis on observational evidence. In 1633, he was condemned before the Inquisition for refusing to halt his defense of Copernicanism.

4-3 | The Puzzle of Planetary Motion

How did Copernican astronomers solve the puzzle of planetary motion?

- Tycho Brahe developed his own model in which the sun and moon circled Earth and the planets circled the sun.
- Tycho's great contribution was to compile detailed observations over a period of 20 years, observations that were later used by Kepler.
- Kepler inherited Tycho's books of observations in 1601 and used them to uncover three laws of planetary motion. He found that the planets follow ellipses with the sun at one focus, that they move faster when near the sun, and that a planet's orbital period squared is proportional to its orbital radius cubed.
- Kepler's final book, *The Rudolphine Tables* (1627), combined heliocentrism with elliptical orbits and predicted the positions of the planets well.

4-4 | Modern Astronomy

- The 99 years from the death of Copernicus to the death of Galileo marked the birth of modern science. From that time on, science depended on evidence to test theories and relied on the analytic methods first demonstrated by Kepler.

New Terms

archaeoastronomy (p. 58)
uniform circular motion (p. 60)
geocentric universe (p. 61)
eccentric (p. 62)
heliocentric universe (p. 63)
parallax (p. 64)
retrograde motion (p. 64)
epicycle (p. 65)
deferent (p. 65)
equant (p. 65)
paradigm (p. 68)
ellipse (p. 76)
semimajor axis, *a* (p. 76)
eccentricity, *e* (p. 76)
hypothesis (p. 77)
theory (p. 77)
natural law (p. 77)

Wednesday

Review Questions

Ace Astronomy™ Assess your understanding of this chapter's topics with additional quizzing and animations at **http://ace.brookscole.com/sf9**

1. What evidence is there that early human cultures observed astronomical phenomena?
2. Why did Plato propose that all heavenly motion was uniform and circular?
3. How do the epicycles of Mercury and Venus differ from those of Mars, Jupiter, and Saturn?
4. Why did Copernicus have to keep small epicycles in his model?
5. Explain how each of Galileo's telescopic discoveries contradicted the Ptolemaic theory.
6. Galileo was condemned, but Kepler, also a Copernican, was not. Why not?
7. When Tycho observed the new star of 1572, he could detect no parallax. Why did that result undermine belief in the Ptolemaic system?
8. Does Tycho's model of the universe explain the phases of Venus that Galileo observed? Why or why not?

9. How do the first two of Kepler's three laws overthrow one of the basic beliefs of classical astronomy?
10. What is the difference between a hypothesis, a theory, and a law?
11. How did *The Alfonsine Tables, The Prutenic Tables,* and *The Rudolphine Tables* differ?
12. What three astronomical objects are represented here? What are the two rings?

13. Whose observatory is shown here? Why are there no telescopes?

Discussion Questions

1. Historian of science Thomas Kuhn has said that *De Revolutionibus* was a revolution-making book but not a revolutionary book. How was it an old-fashioned, classical book?
2. Why might Tycho Brahe have hesitated to hire Kepler? Why do you suppose he appointed Kepler his scientific heir?
3. How does the modern controversy over creationism and evolution reflect two ways of knowing about the physical world?

Problems

1. Draw and label a diagram of the eastern horizon from northeast to southeast and label the rising point of the sun at the solstices and equinoxes. (See page 27 and Figure 4-1).
2. If you lived on Mars, which planets would exhibit retrograde motion? Which would never be visible as crescent phases?
3. Galileo's telescope showed him that Venus has a large angular diameter (61 seconds of arc) when it is a crescent and a small angular diameter (10 seconds of arc) when it is nearly full. Use the small-angle formula to find the ratio of its maximum distance to its minimum distance. Is this ratio compatible with the Ptolemaic universe shown on page 65?
4. Galileo's telescopes were not of high quality by modern standards. He was able to see the moons of Jupiter, but he never reported seeing features on Mars. Use the small-angle formula to find the angular diameter of Mars when it is closest to Earth. How does that compare with the maximum angular diameter of Jupiter?
5. If a planet has an average distance from the sun of 4 AU, what is its orbital period?
6. If a space probe is sent into an orbit around the sun that brings it as close as 0.5 AU and as far away as 5.5 AU, what will be its orbital period?
7. Pluto orbits the sun with a period of 247.7 years. What is its average distance from the sun?

Media Cluster

ACTIVE FIGURES

Ace Astronomy™ To access the resources in the Media Cluster, log into AceAstronomy at **http://ace.brookscole.com/sf9** and select Chapter 4.

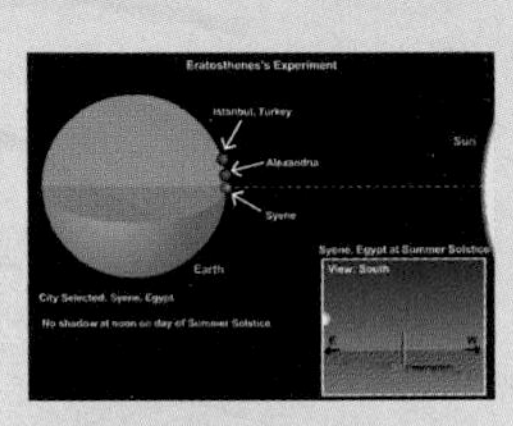

Eratosthenes' Experiment
Instead of light in a well, watch the shadow of a vertical stick from three different places on Earth. From that, you could calculate the size of Earth.

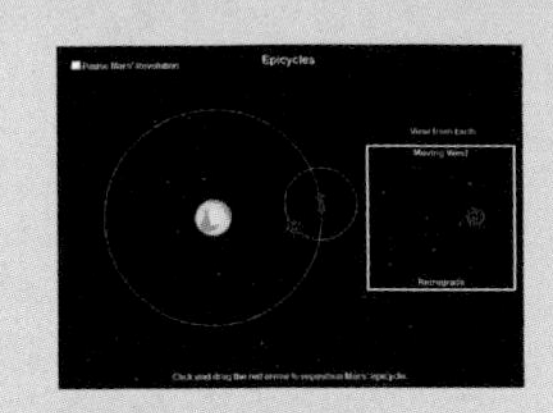

Epicycles
Watch Mars move against the background of stars as its epicycle carries it closest to Earth. You can see the classical explanation of retrograde motion.

ASTRONOMY EXERCISES

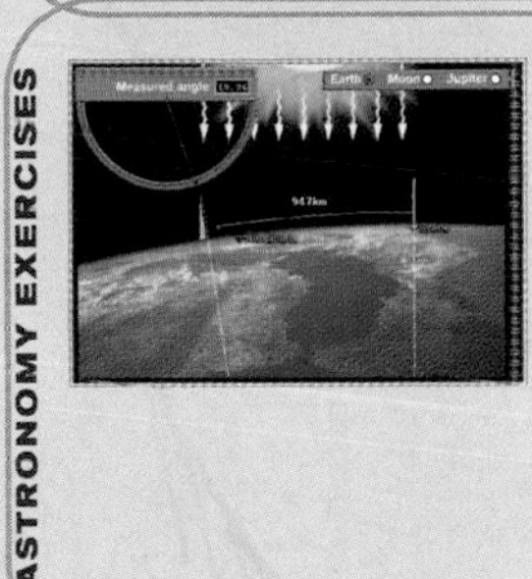

Eratosthenes' Calculations
Reproduce Eratosthenes' experiment and find the size of Earth. Then take the experiment to Earth's moon and to Jupiter.

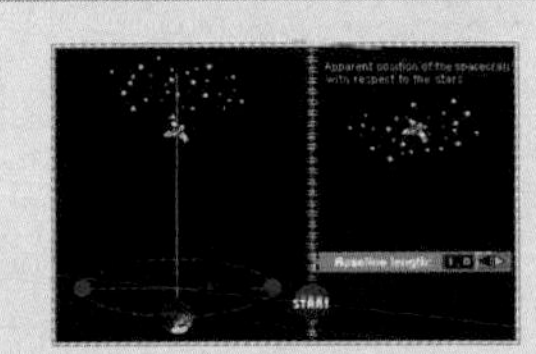

Parallax I
As you learned in this chapter, parallax is the apparent motion of an object because of the motion of the observer. In this animation, see how a spaceship appears to move against a background of stars due to Earth's motion.

Critical Inquiries for the Web

1. The trial of Galileo is an important event in the history of science. You now know, and the Roman Catholic Church now recognizes, that Galileo's view was correct, but what were the arguments on both sides of the issue as it was unfolding? Research the Internet for documents that chronicle the trial, Galileo's observations and publications, and the position of the Church. Use this information to outline the case for and against Galileo in the context of the times in which the trial occurred.
2. Take a "virtual tour" of Stonehenge by browsing the Web for information related to this megalithic site and the possibility that it was used for astronomical purposes. (Be careful to use legitimate scientific and historical websites in your survey.) Summarize the various astronomical alignments evident in the layout of the site.

Exploring *TheSky*

1. Go to Stonehenge in southern England and watch the sunrise on the morning of the summer solstice. Where does it rise on the morning of the winter solstice?
2. Observe Mars going through its retrograde motion. (*Hint:* Use **Reference Lines** under the **View** menu to turn on the ecliptic. Be sure you are in **Free Rotation** under the **Orientation** menu. Locate Mars and use the time skip arrows to watch it move.)
3. Compare the size of the retrograde loops made by Mars, Jupiter, and Saturn.
4. Can you recognize the effects of Kepler's second law in the orbital motion of any of the planets? (*Hint:* Use **3D Solar System Mode** under the **View** menu.)
5. Can you recognize the effects of Kepler's third law in the orbital motion of the planets?

- Go to the Brooks/Cole Astronomy Resource Center **(http://astronomy.brookscole.com)** for critical thinking exercises, articles, and additional readings from InfoTrac College Edition, Brooks/Cole's online student library.

5 Newton, Einstein, and Gravity

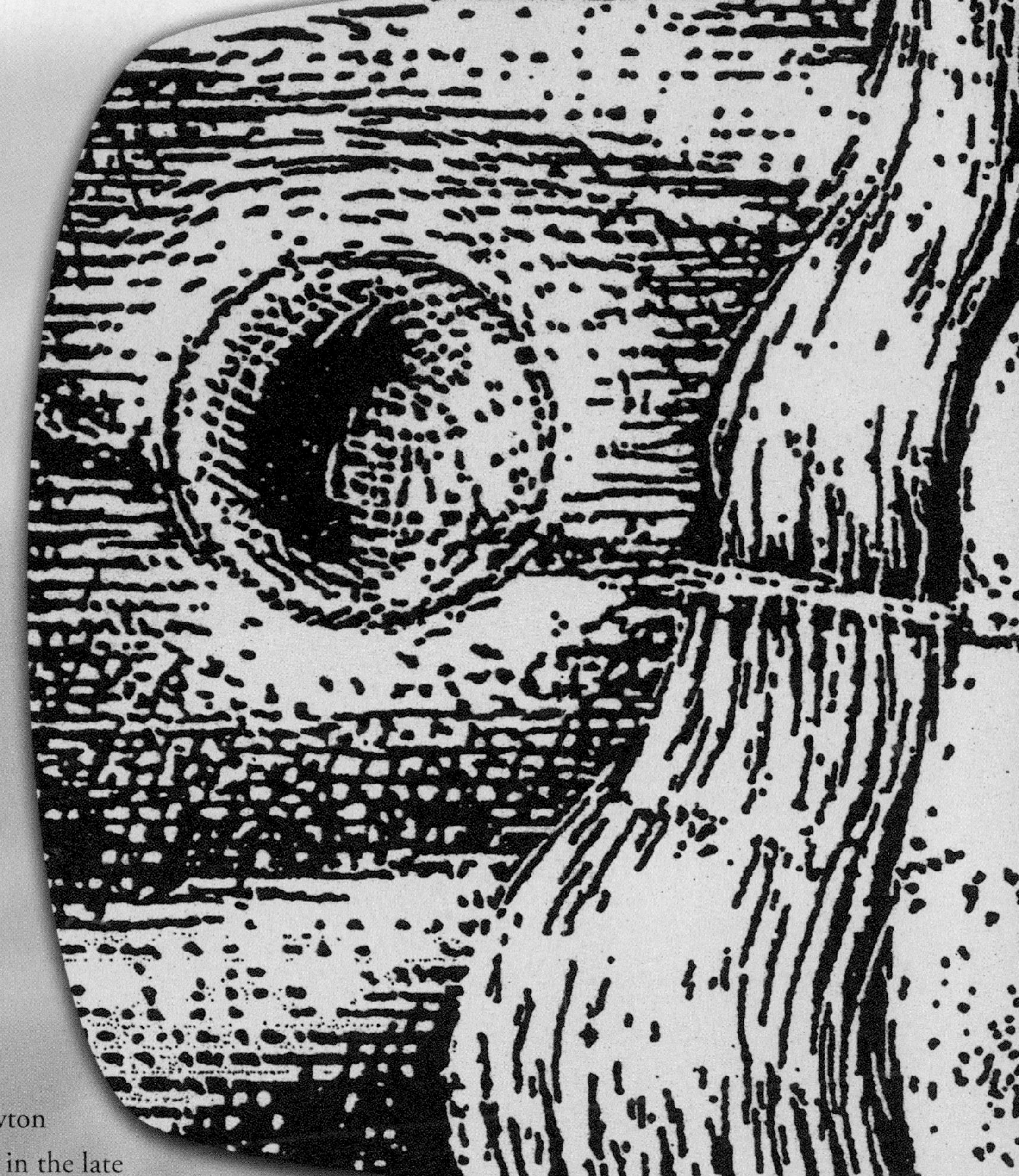

Nature and Nature's laws
lay hid in night:
God said, "Let Newton be!"
and all was light.

ALEXANDER POPE

ISN'T IT WEIRD THAT Isaac Newton is said to have "discovered" gravity in the late 17th century—as if people didn't have gravity before that, as if they floated around holding onto tree branches? Newton's accomplishment is more impressive than just a discovery. Everyone experienced gravity without noticing it. Newton realized that a force had to exist that made things fall, and that changed the way people thought about nature (■ Figure 5-1).

Isaac Newton was born in Woolsthorpe, England, on December 25, 1642, and on January 4, 1643. This was not a biological anomaly but a calendrical quirk. Most of Europe, following the lead of the Catholic countries, had adopted the Gregorian calendar, but Protestant England continued to use the Julian calendar. So December 25 in England was January 4 in Europe. If you take the

Continued on page 86

The motion of the planets was not well understood until Newton discovered gravity. Much later, Einstein found an even better way to understand gravity.

Guidepost

Looking Back

In the last chapter, you saw how Renaissance astronomers struggled to understand the place of the Earth and the motion of the planets. Although Johannes Kepler seemed to solve the puzzle of planetary motion, his three laws only *described* how the planets moved. He never really understood why the planets moved in elliptical orbits.

This Chapter

Now you are ready to finish the adventure. In this chapter you will see how Galileo tried to understand motion and how Newton discovered a wonderful way to understand gravity. But then you will see how Einstein found an even better explanation for motion and gravity. Here you will find answers to five essential questions:

What happens when an object falls?

How did Newton discover gravity?

How does gravity explain orbital motion?

How does gravity explain the tides?

How did Einstein refine our understanding of motion and gravity?

Looking Ahead

Gravity rules. Every object in the universe attracts every other object. The moon orbiting Earth, matter falling into black holes, and the overall structure of the universe are dominated by gravity. The rest of this book will tell the story of matter and gravity, and you will see how modern astronomers have carried on the discoveries made by Galileo, Newton, and Einstein. The universe is a swirling waltz of matter dancing to the music of gravity, and you are going along for the ride.

AceAstronomy™ The AceAstronomy icon throughout the text indicates an opportunity for you to test yourself on key concepts and to explore animations and interactions on the AceAstronomy website at: http://ace.brookscole.com/sf9

■ Figure 5-1

Space stations and astronauts, as well as planets, moons, stars, and galaxies, follow paths called orbits that are described by three simple laws of motion and a theory of gravity first understood by Isaac Newton (1642–1727). Newtonian physics is adequate to send astronauts to the moon and analyze the rotation of the largest galaxies. (NASA/JSC)

English date, then Newton was born in the same year that Galileo Galilei died.

Newton went on to become one of the greatest scientists who ever lived, but even he admitted the debt he owed to those who had studied nature before him. He said, "If I have seen farther than other men, it is because I stood on the shoulders of giants." One of those giants was Galileo. Although Galileo is remembered as the defender of Copernicanism, he was also a talented scientist who studied the motions of falling bodies.

For over two centuries following the publication of Newton's works, astronomers used Newton's laws to describe the universe. Then, early in the 20th century, Albert Einstein proposed a new way to describe gravity. The new theory did not replace Newton's laws but rather showed that they were only approximately correct and could be seriously in error under special circumstances. Einstein's theories further extend the scientific understanding of the nature of gravity. Just as Newton had stood on the shoulders of Galileo, Einstein stood on the shoulders of Newton.

5-1 Galileo and Newton

JOHANNES KEPLER DISCOVERED three laws of planetary motion, but he never understood why the planets move along their orbits. At one place in his writings, he wonders if they are pulled along by magnetic forces emanating from the rotating sun. At another place, he considers and dismisses the idea that the planets are pushed along their orbits by angels.

Newton refined Kepler's model of planetary motion but did not perfect it. In science, a model is an intellectual conception of how nature works (Window on Science 2-1). No model is perfect. Kepler's model was better than Aristotle's, but Newton improved Kepler's model by expanding it into a general theory of motion and gravity. In fact, most scientists now refer to Newton's *law* of gravity. Whether it is called a model, a theory, or a law, it is not perfect. Newton never understood what gravity was. It was as mysterious as an angel pushing the moon inward toward Earth instead of forward along the moon's orbit.

To understand science, you must understand the nature of scientific descriptions. The scientist studies nature by either creating new theories or refining old theories. Yet a theory can never be perfect, because it can never represent the universe in all its intricacies. Instead, a theory must be a limited approximation of a single phenomenon, such as orbital motion. It is fitting that Newton's discoveries all began with Kepler's fellow Copernican, Galileo.

■ Figure 5-2

Although Galileo is often associated with the telescope, as on this Italian stamp, he also made systematic studies of the motion of falling bodies and made discoveries that led to the law of inertia.

Galileo and Motion

Even before Galileo built his first telescope, he had begun studying the motion of freely moving bodies (■ Figure 5-2). After the Inquisition condemned and imprisoned him in 1633, he continued his study of motion. He seems to have realized that he would have to understand motion before he could truly understand the Copernican system. That he was eventually able to formulate principles that later led Newton to the laws of motion and the theory of gravity is a tribute to Galileo's ability to set aside the authority of the ancients and think for himself.

The authority of the age was Aristotle, whose ideas on motion were hopelessly confused. Aristotle said that the world is made up of four classical elements: earth, water, air, and fire, each located in its proper place. The proper place for the element earth was the center of the universe, and the proper place for water was just above earth. Air and then fire formed higher layers and above them was the realm of the planets and stars. (You can see the four layers of the classical elements in the diagram at the top of page 64.) If displaced, the four elements were believed to have a natural tendency to move toward their proper place in the cosmos. Things made up mostly of air or fire—smoke, for instance—tend to move upward. Things composed mostly of earth and water—wood, rock, flesh, bone, and so on—tend to move downward toward the proper place of earth and water. According to Aristotle, objects fall downward because they are moving toward their proper place.*

Aristotle called these motions **natural motions** to distinguish them from **violent motions** produced, for instance, when you push on an object and make it move other than toward its proper place. According to Aristotle, such motions stop as soon as the force is removed. To explain how an arrow could continue to move upward even after it had left the bowstring, he said currents in the air around the arrow carried it forward even though the bowstring was no longer pushing it.

These ideas about natural and violent motion, and the necessity of a force to preserve motion, were still accepted theory in Galileo's time. In fact, in 1590, when Galileo was 26, he wrote a short work called *De Motu (On Motion)* that deals with the proper places of objects and their natural motions.

In Galileo's time and for the two preceding millennia, scholars had commonly tried to resolve problems of science by referring to authority. To analyze the flight of a cannonball, for instance, they would turn to the writings of Aristotle and other classical philosophers and try to deduce what those philosophers would have said on the subject. This generated a great deal of discussion but little real progress. Galileo broke with this tradition and conducted his own experiments.

He began by studying the motions of falling bodies, but he quickly discovered that the velocities were so great and the times so short that he could not measure them accurately. Consequently, he began using polished bronze balls rolling down gently sloping inclines. In that instance, the velocity is lower and the time longer. Using an ingenious water clock, he was able to measure the time the balls took to roll given distances down the incline, and he correctly recognized that these times are proportional to the times taken by falling bodies.

He found that falling bodies do not fall at constant rates, as Aristotle had said, but are accelerated. That is, they move faster with each passing second. Near Earth's surface, a falling object will have a velocity of 9.8 m/s (32 ft/s) at the end of 1 second, 19.6 m/s (64 ft/s) after 2 seconds, 29.4 m/s (96 ft/s) after 3 seconds, and so on. Each passing second adds 9.8 m/s (32 ft/s) to the object's velocity (■ Figure 5-3). In modern terms, this is called the **acceleration of gravity** at Earth's surface.

Galileo also discovered that the acceleration does not depend on the weight of the object. This, too, is contrary to the teachings of Aristotle, who believed that heavy objects, containing more earth and water, fall with higher velocity. Galileo found that the acceleration of a falling body is the same whether it is heavy or light. According to some accounts, he demonstrated this by dropping balls of iron and wood from the top of the Leaning Tower of Pisa to show that they would fall together and hit the ground at the same time (■ Figure 5-4a). In fact, he probably didn't perform this experiment. It would not have been conclusive anyway because of air resistance. More than 300 years later, Apollo 15 astronaut David Scott, standing on the airless moon, demonstrated

* This is one reason why Aristotle had to have a geocentric universe. If Earth's center had not also been the center of the cosmos, his explanation of gravity would not have worked.

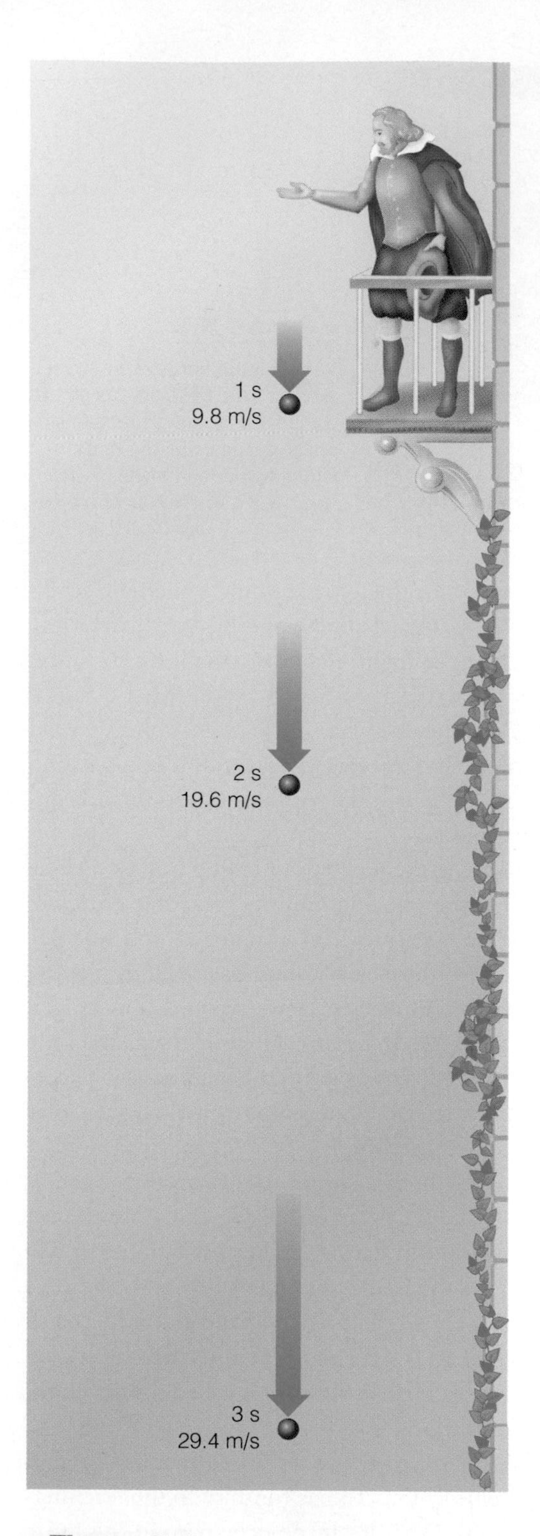

■ **Figure 5-3**

Galileo found that a falling object is accelerated downward. Each second, its velocity increases by 9.8 m/s (32 ft/s).

Air resistance would have slowed the wooden ball more and ruined Galileo's demonstration.
a
On the airless moon, there is no air resistance to slow the feather.
Hammer
Feather
b

■ **Figure 5-4**

(a) According to tradition, Galileo demonstrated that the acceleration of a falling body is independent of its weight by dropping balls of iron and wood from the Leaning Tower of Pisa. In fact, air resistance would have confused the result. (b) In a historic television broadcast from the moon on August 2, 1971, David Scott dropped a hammer and a feather at the same instant. They fell with the same acceleration and hit the surface together. (NASA)

Galileo's discovery by dropping a feather and a steel geologist's hammer. They hit the lunar surface at the same time (Figure 5-4b).

Having described natural motion, Galileo turned his attention to violent motion—that is, motion directed other than toward an object's proper place in the cosmos. He pointed out that an object rolling down an incline is accelerated and that an object rolling up the same incline is decelerated. If the incline were perfectly horizontal and frictionless, he reasoned, there could be no acceleration or deceleration to change the object's velocity, and, in the absence of friction, the object would continue to move forever. In his own words, "any velocity once imparted to a moving body will be rigidly maintained as long as the external causes of acceleration or retardation are removed."

This is contrary to Aristotle's belief that motion can continue only if a force is present to maintain it. In fact, Galileo's statement is a perfectly valid summary of the law of inertia, which became Newton's first law of motion.

Galileo published his work on motion in 1638, two years after he had become entirely blind and only four years before his death. The book was called *Mathematical Discourses and Demonstrations Concerning Two New Sciences, Relating to Mechanics and to Local Motion.* It is known today as *Two New Sciences.*

The book is a brilliant achievement for a number of reasons. To understand motion, Galileo had to abandon the authority of the ancients, devise his own experiments, and draw his own conclusions. In a sense, this was the first example of experimental science. But Galileo also had to generalize his experiments to discover how nature worked. Though his apparatus was finite and plagued by friction, he was able to imagine an infinite, frictionless plane on which a body moves at constant velocity. In his workshop, the law of inertia was obscure; but, in his imagination, it was clear and precise.

Table 5-1 | Newton's Three Laws of Motion

I. A body continues at rest or in uniform motion in a straight line unless acted upon by some force.
II. The acceleration of a body is inversely proportional to its mass, directly proportional to the force, and in the same direction as the force.
III. To every action, there is an equal and opposite reaction.

AceAstronomy™ Log into AceAstronomy and select this chapter to see Astronomy Exercise "Falling Bodies." Try one of the most famous experiments Galileo never did.

Newton and the Laws Of Motion

Newton's three laws of motion (■ Table 5-1) are critical to understanding gravity and orbital motion. They apply to any moving object, from an automobile driving along a highway to galaxies colliding with each other.

The first law is really a restatement of Galileo's law of inertia. An object continues at rest or in uniform motion in a straight line unless acted upon by some force. Astronauts drifting in space will travel at constant rates in straight lines forever if no forces act on them (■ Figure 5-5a).

Newton's first law also explains why a projectile continues to move after all forces have been removed—for instance, why an arrow continues to move after leaving the bowstring. The object continues to move because it has momentum. You can think of an object's **momentum** as a measure of its amount of motion.

Momentum equals velocity times mass. Obviously velocity is important. A low-velocity object such as a paper clip tossed across a room has little momentum, and you could easily catch it in your hand. But the same paper clip fired at the speed of a rifle bullet would have tremendous momentum, and you would not dare try to catch it. Momentum also depends on the mass of an object (**Focus on Fundamentals 1**). To see how, imagine that, instead of tossing a paper clip, someone tosses you a bowling ball. A bowling ball contains much more mass than a paper clip and therefore has much greater momentum at the same velocity.

Newton's first law explains the consequences of the conservation of momentum. When Newton said that momentum is conserved, he meant that it remains constant until something acts to change it. A moving object has a given amount of momentum. To change that momentum, you must exert some force on the object to change either the speed or the direction. Newton's first law and the concept of momentum came from the work of Galileo.

Newton's second law of motion discusses forces, and Galileo did not talk about forces. Galileo spoke instead of accelerations. Newton saw that an acceleration is the result of a force acting on a mass (Figure 5-5b). Newton's second law is commonly written as:

$$F = ma$$

As always, you must carefully define terms when you look at an equation. An **acceleration** is a change in velocity, and a **velocity** is a directed rate of motion. Rate of motion means, of course, a speed, but the word *directed* has a special meaning. Speed itself does not have any direction associated with it, but velocity does. If you drive a car in a circle at 55 mph, your speed is constant, but your velocity is changing because your direction of motion is changing. So an object experiences an acceleration if its speed changes or if its direction of motion changes. Every

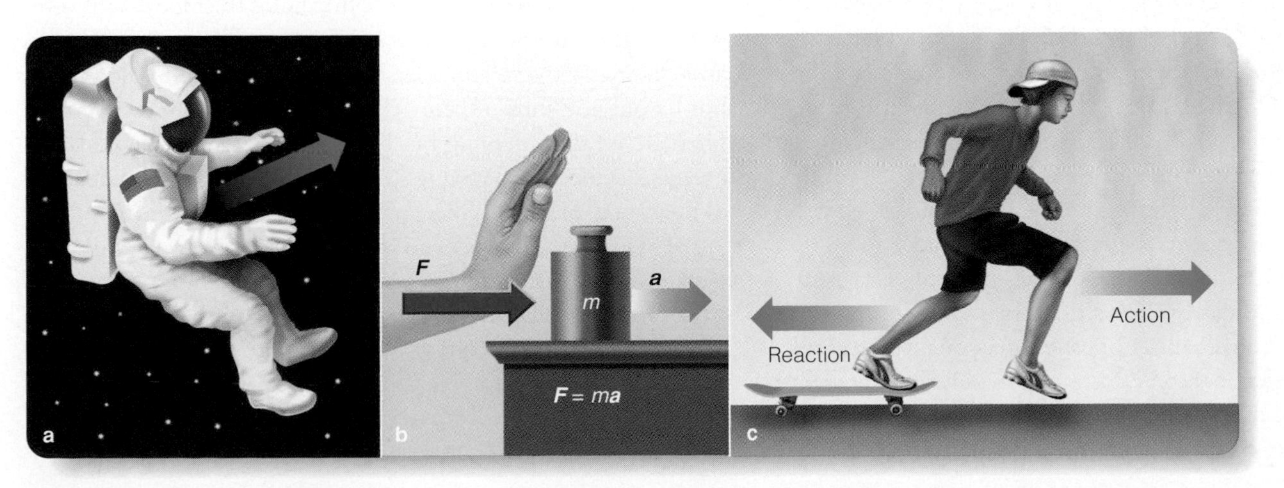

■ Figure 5-5

Newton's three laws of motion.

Focus on Fundamentals 1

Mass

One of the most fundamental parameters in science is **mass,** a measure of the amount of matter in an object. A bowling ball, for example, contains a large amount of mass, but a child's rubber ball contains less matter than the bowling ball, and it is said to be less massive.

Mass is not the same as weight. Your weight is the force that Earth's gravity exerts on the mass of your body. Because gravity pulls you downward, you press against the bathroom scale, and you can measure your weight. Floating in space, you would have no weight at all; a bathroom scale would be useless. But your body would still contain the same amount of matter, so you would still have mass.

Sports analogies illustrate the importance of mass in dramatic ways. A bowling ball, for example, must be massive in order to have a large effect on the pins it strikes. Imagine trying to knock down all the pins with a bowling ball no more massive than a balloon. Even in space, where the bowling ball would be weightless, a low-mass bowling ball would have little effect on the pins. On the other hand, runners want track shoes that have low mass and thus are easy to move. Imagine trying to run a 100-meter dash wearing track shoes that were as massive as bowling balls. They would be very hard to move, and it would be difficult to accelerate away from the starting block. The shot put takes muscle because the shot is massive, not because it is heavy. Imagine throwing the shot in space where it would have no weight. It would still be massive, and it would take great effort to start it moving.

Mass is a unique measure of the amount of material in an object. Using the metric system (Appendix A), mass is measured in kilograms.

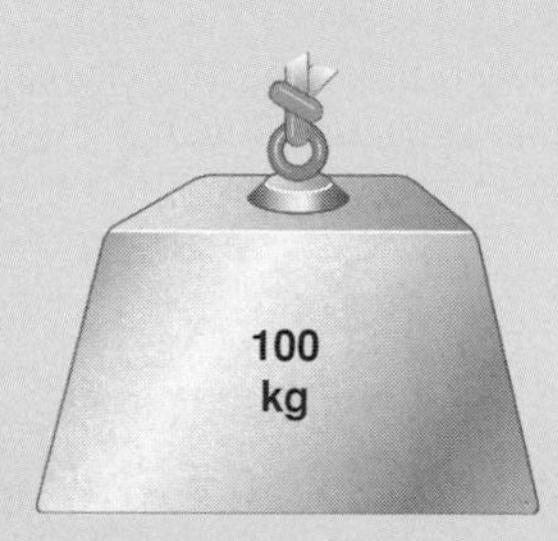

Mass is not the same as weight.

PRESSURE | MASS | ENERGY | TEMPERATURE AND HEAT | DENSITY

automobile has three accelerators—the gas pedal, the brake pedal, and the steering wheel. All three change the car's velocity.

In a way, the second law is just common sense; you experience it every day. The acceleration of a body is proportional to the force applied to it. This is reasonable. If you push gently against a grocery cart, you expect a small acceleration. The second law of motion also says that the acceleration depends on the mass of the body. This, too, is reasonable. If your grocery cart were filled with bricks and you pushed it gently, you would expect very little result. If it were full of inflated balloons, however, it would move easily in response to a gentle push. Finally, the second law says that the resulting acceleration is in the direction of the force. This is also what you would expect. If you push on a cart that is not moving, you expect it to begin moving in the direction you push.

The second law of motion is important because it establishes a precise relationship between cause and effect (**Window on Science 5-1**). Objects do not just move. They accelerate due to the action of a force. Moving objects do not just stop. They decelerate due to a force. Also, moving objects don't just change direction for no reason. Any change in direction is a change in velocity and requires the presence of a force. Aristotle said that objects move because they have a tendency to move. Newton said that objects move due to a specific cause, a force.

Newton's third law of motion specifies that for every action there is an equal and opposite reaction. In other words, forces must occur in pairs directed in opposite directions. For example, if you stand on a skateboard and jump forward, the skateboard will shoot away backward. As you jump, your feet exert a force against the skateboard, which accelerates it toward the rear. But forces must occur in pairs, so the skateboard must exert an equal but opposite force on your feet that accelerates your body forward (Figure 5-5c).

Mutual Gravitation

The three laws of motion led Newton to consider the force that causes objects to fall. The first and second laws tell you that falling bodies accelerate downward because some force must be pulling downward on them. Newton wondered what that force could be.

Newton was also aware that some force has to act on the moon. The moon follows a curved path around Earth, and motion along a curved path is accelerated motion. The second law says that an acceleration requires a force, so a force must be making the moon follow that curved path.

Newton wondered if the force that holds the moon in its orbit could be the same force that causes apples to fall—gravity. He was aware that gravity extends at least as high as the tops of mountains, but he did not know if it could extend all the way to the moon. He believed that it could, but he thought it would be weaker at greater distances, and he guessed that its strength would decrease as the square of the distance increased.

This relationship, the **inverse square law,** was familiar to Newton from his work on optics, where it applied to the intensity of light. A screen set up 1 meter from a candle flame receives a certain amount of energy on each square meter. However, if that screen is moved to a distance of 2 meters, the light that originally illuminated one square meter must cover four square me-

Window on Science | 5-1

The Unstated Assumption of Science: Cause and Effect

One of the most often used and least often stated principles of science is cause and effect, and you could argue that Newton's second law of motion was the first clear statement of the principle.

Ancient philosophers such as Aristotle argued that objects moved because of tendencies. They said that earth and water, and objects made mostly of earth and water, had a natural tendency to move toward the center of the universe. This natural motion had no cause but was inherent in the nature of the objects. But Newton's second law says $F = ma$. If an object (of mass m) changes its motion (a in the equation), then it must be acted on by a force (F in the equation). Any effect (a) must be the result of a cause (F).

The principle of cause and effect goes far beyond motion. The principle of cause and effect gives scientists confidence that every effect has a cause. Hearing loss in certain laboratory rats, color changes in certain chemical dyes, and explosions on certain stars are all effects that must have causes. All of science is focused on understanding the causes of the effects you see around you. If the universe were not rational, then you could never expect to discover causes. Newton's second law of motion was arguably the first clear statement that the behavior of the universe depends rationally on causes.

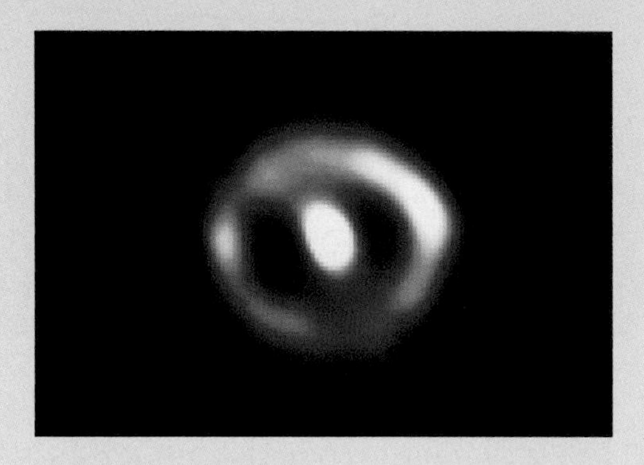

Cause and effect: Why did this star explode in 1992? There must have been a cause. (ESA/STScI and NASA)

ters (■ Figure 5-6). Consequently, the intensity of the light is inversely proportional to the square of the distance to the screen.

Newton made two assumptions that enabled him to predict the strength of Earth's gravity at the distance of the moon. He assumed that the strength of gravity follows the inverse square law and that the critical distance is not the distance from Earth's surface but the distance from Earth's center. Because the moon is about 60 Earth radii away, Earth's gravity at the distance of the moon should be about 60^2 times less than at Earth's surface. Instead of being 9.8 m/s^2 at Earth's surface, it should be about 0.0027 m/s^2 at the distance of the moon.

Now, Newton wondered, could this acceleration keep the moon in orbit? He knew the moon's distance and its orbital period, so he could calculate the actual acceleration needed to keep it in its curved path. The answer is 0.0027 m/s^2. To the accuracy of Newton's data for Earth's radius, it was exactly what his assumptions predicted. The moon is held in its orbit by gravity, and gravity obeys the inverse square law.

Newton's third law says that forces always occur in pairs, and this leads to the conclusion that gravity is mutual. If Earth pulls on the moon, then the moon must pull on Earth. Gravitation is a general property of the universe. The sun, the planets, and all their moons must also attract each other by mutual gravitation. In fact, every particle of mass in the universe must attract every other particle, and Newtonian gravity is often called universal mutual gravitation.

Clearly the force of gravity depends on mass. Your body is made of matter, and you have your own personal gravitational field. But your gravity is weak and does not attract personal satellites orbiting around you. Larger masses have stronger gravity. From an analysis of the third law of motion, Newton realized that the mass that resists acceleration in the first law must be the same as the mass associated with gravity. Then when two bodies attract one another and accelerate toward each other, the two masses must be equally involved. Newton performed precise experiments with pendulums and confirmed this equivalence between the mass that resists acceleration and the mass that causes gravity.

■ Figure 5-6

As light radiates away from a source, it is spread thinner and becomes less intense. Here the light falling on one square meter on the inner sphere must fall on four square meters on a sphere twice as big. This shows how the intensity of light is inversely proportional to the square of the distance.

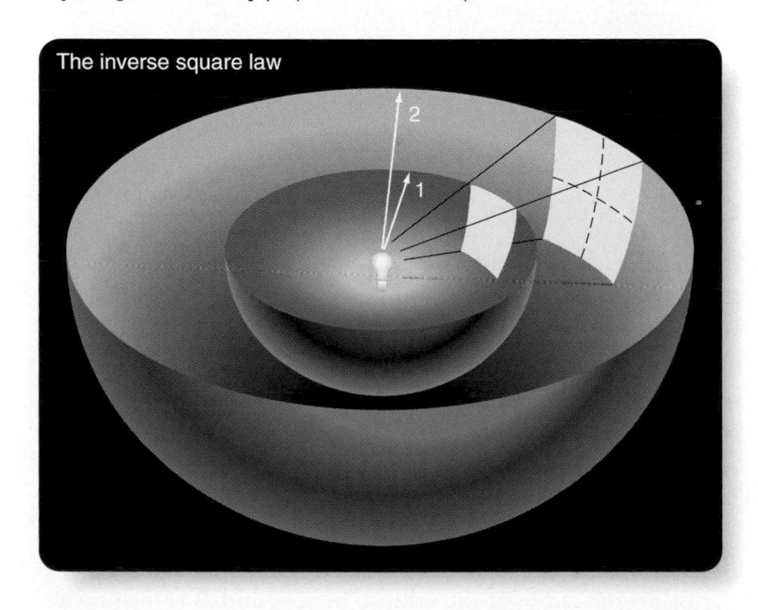

From this, combined with the inverse square law, he was able to write the famous formula for the gravitational force between two masses, M and m:

$$F = -\frac{GMm}{r^2}$$

The constant G is the gravitational constant, and r is the distance between the masses. The negative sign means that the force is attractive, pulling the masses together and making r decrease. In plain English, Newton's law of gravitation says: The force of gravity between two masses M and m is proportional to the product of the masses and inversely proportional to the square of the distance between them.

Newton's description of gravity was a difficult idea for physicists of his time to accept because it is an example of action at a distance. Earth and moon exert forces on each other although there is no physical connection between them. Modern scientists resolve this problem by referring to gravity as a **field.** Earth's presence produces a gravitational field directed toward Earth's center. The strength of the field decreases according to the inverse square law. Any particle of mass in that field experiences a force that depends on the mass of the particle and the strength of the field at the particle's location. The resulting force is directed toward the center of the field.

The field is an elegant way to describe gravity, but it does not say what gravity is. Later in this chapter, when you learn about Einstein's theory of curved space-time, you will get a better idea of what gravity really is.

Building Scientific Arguments

What do the words *universal* and *mutual* mean when you say "universal mutual gravitation"?

Newton argued that the force that makes an apple accelerate downward is the same as the force that accelerates the moon and holds it in its orbit. You can learn more by thinking about Newton's third law of motion, which says that forces always occur in pairs. If Earth attracts the moon, then the moon must attract Earth. That is, gravitation is *mutual* between any two objects.

Furthermore, if Earth's gravity attracts the apple and the moon, then it must attract the sun, and the third law says that the sun must attract Earth. But if the sun attracts Earth, then it must also attract the other planets and even distant stars, which, in turn, must attract the sun and each other. Step by step, Newton's third law of motion leads to the conclusion that gravitation must apply to all masses in the universe. That is, gravitation must be *universal.*

Aristotle explained gravity in a totally different way. Use Aristotle's argument that accounted for a falling apple: **Could that explanation account for a hammer falling on the surface of the moon?**

■ ■ ■

Connections: Newton first came to understand gravity by thinking about the orbital motion of the moon. Gravitation is tremendously important in astronomy. It can explain two different kinds of motions.

5-2 Orbital Motion and Tides

ORBITAL MOTION AND TIDES are two different kinds of gravitational phenomena. As you think about the orbital motion of the moon and planets, you need to think about how gravity pulls on an object. When you think about tides, you must think about how gravity pulls on different parts of an object. Analyzing these two kinds of phenomena will give you a deeper insight into how gravity works.

Orbits

To understand how an object can orbit another object, you must see orbital motion as Newton did. Objects in orbit are falling. You can explore Newton's insight by analyzing the motion of objects orbiting Earth.

Study **Orbiting Earth** on pages 94–95 and notice three important ideas:

1. An object orbiting Earth is actually falling (being accelerated) toward Earth's center. The object continuously misses Earth because of its orbital velocity.
2. Also, notice that objects orbiting each other actually revolve around their center of mass.
3. Finally, notice the difference between closed orbits and open orbits. If you want to leave Earth never to return, you must give your spaceship a high enough velocity so it will follow an open orbit.

AceAstronomy™ Log into AceAstronomy and select this chapter to see Astronomy Exercise "Orbital Motion." Experiment with an object in orbit.

Orbital Velocity

If you were about to ride a rocket into orbit, you would have a critical question. "How fast must I go to stay in orbit?" An object's circular velocity is the lateral velocity the object must have to remain in a circular orbit. If you assume that the mass of your spaceship is small compared with the mass of the object you expect to orbit, Earth in this case, then the circular velocity is:

$$V_c = \sqrt{\frac{GM}{r}}$$

In this formula, M is the mass of the central body in kilograms, r is the radius of the orbit in meters, and G is the gravitational constant, 6.67×10^{-11} m^3/s^{2}kg. This formula is all you need to calculate how fast an object must travel to stay in a circular orbit.

For example, how fast does the moon travel in its orbit? Earth's mass is 5.98×10^{24} kg, and the radius of the moon's orbit is 3.84×10^8 m. Then the moon's velocity is:

$$V_c = \sqrt{\frac{6.67 \times 10^{-11} \times 5.98 \times 10^{24}}{3.84 \times 10^8}} = \sqrt{\frac{39.9 \times 10^{13}}{3.84 \times 10^8}}$$

$$= \sqrt{1.04 \times 10^6} = 1020 \text{ m/s} = 1.02 \text{ km/s}$$

This calculation shows that the moon travels 1.02 km along its orbit each second. That is the circular velocity at the distance of the moon.

A satellite just above Earth's atmosphere is only about 200 km above Earth's surface, or 6578 km from Earth's center, so Earth's gravity is much stronger, and the satellite must travel much faster to stay in a circular orbit. You can use the formula above to find that the circular velocity just above Earth's atmosphere is about 7790 m/s, or 7.79 km/s. This is about 17,400 miles per hour and shows why putting satellites into Earth orbit takes such large rockets. Not only must the rocket lift the satellite above Earth's atmosphere, but the rocket must tip over and accelerate the satellite to circular velocity.

Calculating Escape Velocity

If you launch a rocket upward, it will consume its fuel in a few moments and reach its maximum speed. From that point on, it will coast upward. How fast must a rocket travel to coast away from Earth and escape? Of course, no matter how far it travels, it can never escape from Earth's gravity. The effects of Earth's gravity extend to infinity. It is possible, however, for a rocket to travel so fast initially that gravity can never slow it to a stop. That means it could leave Earth.

The escape velocity is the velocity required to escape from the surface of an astronomical body. Here you are interested in escaping from Earth or a planet; later chapters will consider the escape velocity from stars, galaxies, and even a black hole.

The escape velocity, V_e, is given by a simple formula:

$$V_e = \sqrt{\frac{2GM}{r}}$$

Here G is the gravitational constant 6.67×10^{-11} m^3/s^{2}kg, M is the mass of the astronomical body in kilograms, and r is its radius in meters. (Notice that this formula is very similar to the formula for circular velocity.)

You can find the escape velocity from Earth by looking up its mass, 5.98×10^{24} kg, and its radius, 6.38×10^6 m. Then the escape velocity is:

$$V_e = \sqrt{\frac{2 \times 6.67 \times 10^{-11} \times 5.98 \times 10^{24}}{6.38 \times 10^6}} = \sqrt{\frac{7.98 \times 10^{14}}{6.38 \times 10^6}}$$

$$= \sqrt{1.25 \times 10^8} = 11{,}200 \text{ m/s} = 11.2 \text{ km/s}$$

This is equal to about 25,000 miles per hour.

Notice that the formula says that the escape velocity from a body depends on both its mass and radius. A massive body might have a low escape velocity if it has a very large radius. You will meet such objects in the discussion of giant stars. On the other hand, a rather low-mass body could have a very large escape velocity if it had a very small radius, a condition you will meet in the discussion of black holes.

Circular velocity and escape velocity are two aspects of Newton's laws of gravity and motion. Once Newton understood gravity and motion, he could do what Kepler had failed to do—he could explain why the planets obey Kepler's laws of planetary motion.

AceAstronomy™ Log into AceAstronomy and select this chapter to see Astronomy Exercise "Escape Velocity." Try to escape from a planet.

Kepler's Laws Reexamined

Now that you understand Newton's laws, gravity, and orbital motion, you can understand Kepler's laws of planetary motion in a new way.

Kepler's first law says that the orbits of the planets are ellipses with the sun at one focus. Kepler wondered why the planets keep moving along these orbits, and now you know the answer. They move because there is nothing to slow them down. Newton's first law says that a body in motion stays in motion unless acted on by some force. The gravity of the sun accelerates the planets inward toward the sun and holds them in their orbits, but it doesn't pull backward on the planets, so they don't slow to a stop. With no friction, they must continue to move.

The orbits of the planets are ellipses because gravity follows the inverse square law. In one of his most famous problems, Newton proved that if a planet moves in a closed orbit under the influence of an attractive force that follows the inverse square law, then the planet must follow an elliptical path.

Kepler's second law says that a planet moves faster when it is near the sun and slower when it is farther away. Once again, Newton's discoveries explain why. Earlier you saw that a body moving on a frictionless surface will continue to move in a straight line until it is acted on by some force; that is, the object has momentum. In a similar way, an object set rotating on a frictionless surface will continue rotating until something acts to speed it up or slow it down. Such an object has **angular momentum,** a measure of the rotation of the body about some point. A planet circling the sun has a given amount of angular momentum; and, with no outside influences to alter its motion, it must conserve its angular momentum. That is, its angular momentum must remain constant.

Mathematically, a planet's angular momentum is the product of its mass, velocity, and distance from the sun. This explains why a planet must speed up as it comes closer to the sun along an elliptical orbit. Its angular momentum is conserved, so as its distance from the sun decreases, its velocity must increase. In the

Orbiting Earth

1 You can understand orbital motion by thinking of a cannonball falling around Earth in a circular path. Imagine a cannon on a high mountain aimed horizontally as shown at right. A little gunpowder gives the cannonball a low velocity, and it doesn't travel very far before falling to Earth. More gunpowder gives the cannonball a higher velocity, and it travels farther. With enough gunpowder, the cannonball travels so fast it never strikes the ground. Earth's gravity pulls it toward Earth's center, but Earth's surface curves away from it at the same rate it falls. It is in orbit. The velocity needed to stay in a circular orbit is called the **circular velocity.** Just above Earth's atmosphere, circular velocity is 7790 m/s or about 17,400 miles per hour, and the orbital period is about 90 minutes.

A satellite above Earth's atmosphere feels no friction and will fall around Earth indefinitely.

North Pole

Earth satellites eventually fall back to Earth if they orbit too low and experience friction with the upper atmosphere.

AceAstronomy™

Go to AceAstronomy and click Active Figures to see "Newton's Cannon" and fire your own version of Newton's cannon.

1a A **geosynchronous satellite** orbits eastward with the rotation of Earth and remains above a fixed spot — ideal for communications and weather satellites.

A Geosynchronous Satellite

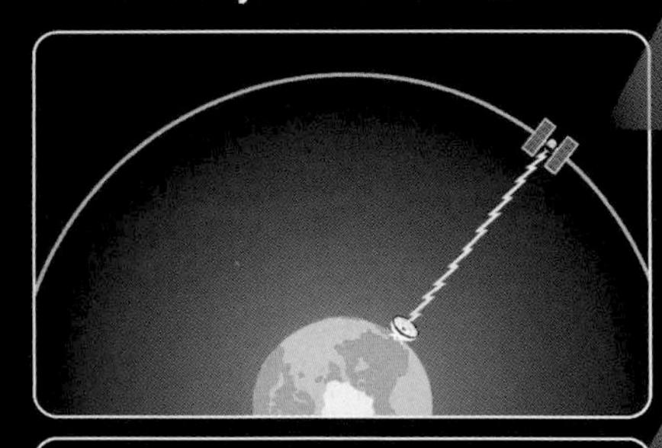

At a distance of 42,250 km (26,260 miles) from Earth's center, a satellite orbits with a period of 24 hours.

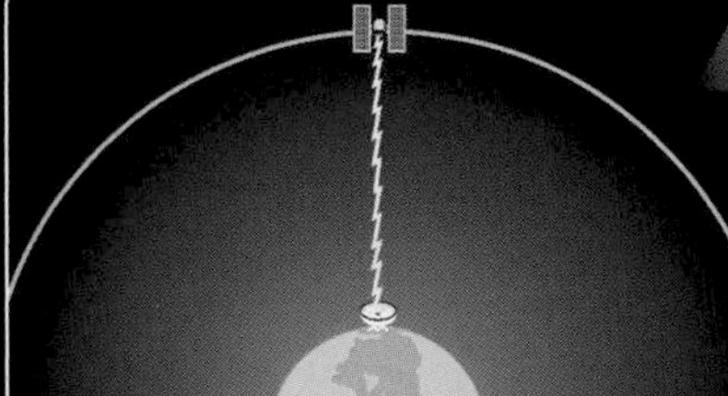

The satellite orbits eastward, and Earth rotates eastward under the moving satellite.

The satellite remains fixed above a spot on Earth's equator.

AceAstronomy™

Go to AceAstronomy and click Active Figures to see "Geosynchronous Orbit" and place your own satellite into geosynchronous orbit.

1b According to Newton's first law of motion, the moon should follow a straight line and leave Earth forever. Because it follows a curve, Newton knew that some force must continuously accelerate it toward Earth — gravity. Each second the moon moves 1020 m (3350 ft) eastward and falls about 1.6 mm (about 1/16 inch) toward Earth. The combination of these motions produces the moon's curved orbit. The moon is falling.

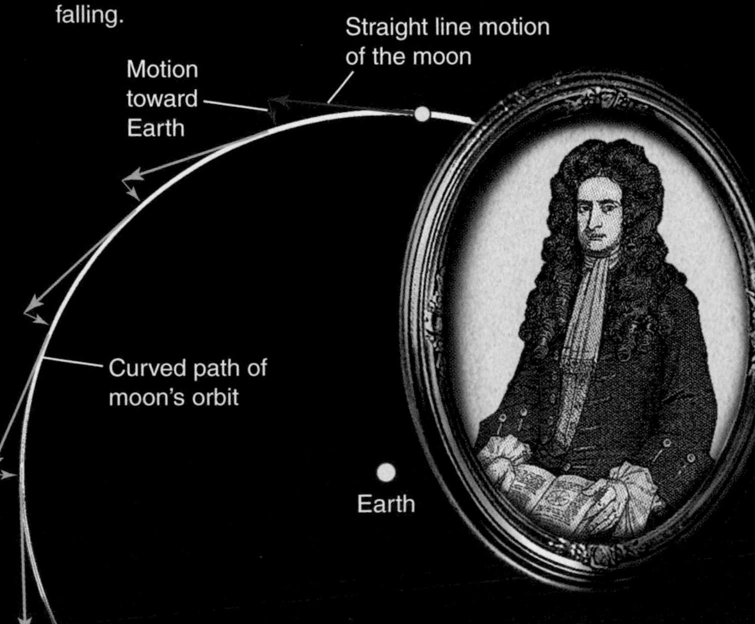

1c Astronauts in orbit around Earth feel weightless, but they are not "beyond Earth's gravity," to use a term from old science fiction movies. Like the moon, the astronauts are accelerated toward Earth by Earth's gravity, but they travel fast enough along their orbits that they continually "miss the Earth." They are literally falling around Earth. Inside or outside a spacecraft, astronauts feel weightless because they and their spacecraft are falling at the same rate. Rather than saying they are weightless, you should more accurately say they are in free fall.

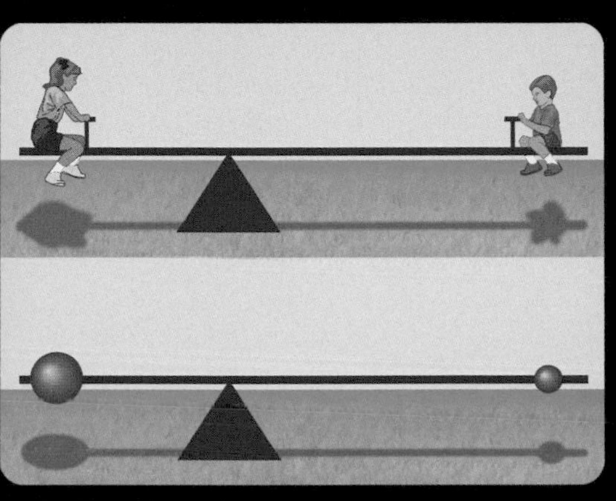

2 To be precise you should not say that an object orbits Earth. Rather the two objects orbit each other. Gravitation is mutual, and if Earth pulls on the moon, the moon pulls on Earth. The two bodies revolve around their common **center of mass**, the balance point of the system.

2a Two bodies of different mass balance at the center of mass, which is located closer to the more massive object. As the two objects orbit each other, they revolve around their common center of mass as shown at right. The center of mass of the Earth–moon system lies only 4708 km (2926 miles) from the center of Earth — inside the Earth. As the moon orbits the center of mass on one side, the Earth swings around the center of mass on the opposite side.

Center of mass

3 **Closed orbits** return the orbiting object to its starting point. The moon and artificial satellites orbit Earth in closed orbits. Below, the cannonball could follow an elliptical or a circular closed orbit. If the cannonball travels as fast as **escape velocity**, the velocity needed to leave a body, it will enter an open orbit. An **open orbit** does not return the cannonball to Earth. It will escape.

Ace Astronomy™ Go to AceAstronomy and click Active Figures to see "Center of Mass." Change the mass ratio to move the center of mass.

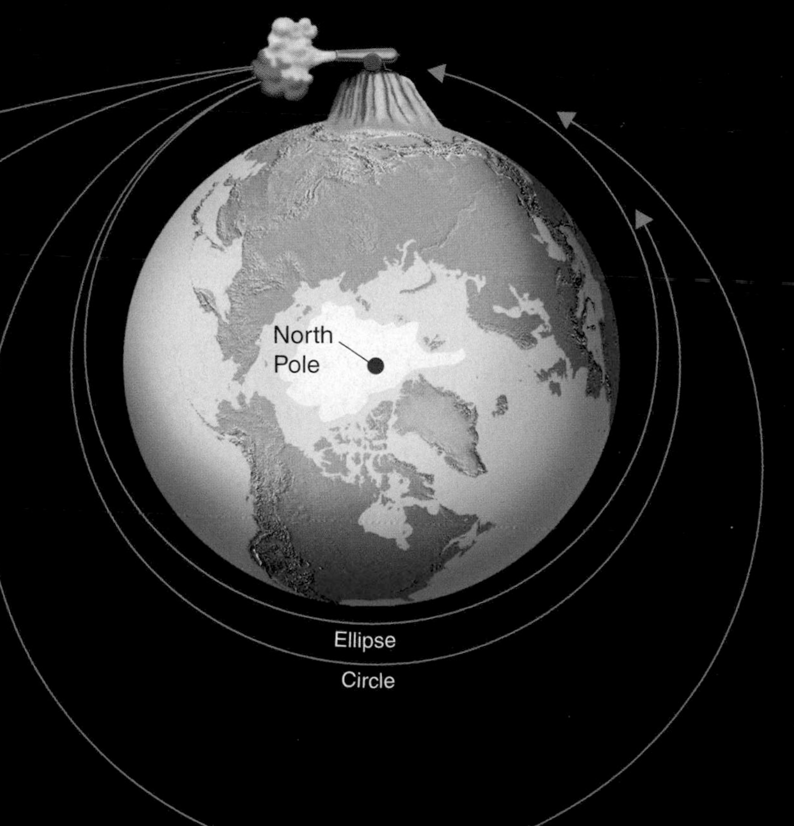

Hyberbola
Parabola
A cannonball with a velocity greater than escape velocity will follow a hyperbola and escape from Earth.
A cannonball with escape velocity will follow a parabola and escape.
North Pole
Ellipse
Circle
Ellipse

3a As described by Kepler's Second Law, an object in an elliptical orbit has its lowest velocity when it is farthest from Earth (apogee), and its highest velocity when it is closest to Earth (perigee). Perigee must be above Earth's atmosphere, or friction will rob the satellite of energy and it will eventually fall back to Earth.

Energy

One of the most fundamental ideas in science is **energy.** Physicists define energy as the ability to do work, but you might paraphrase that definition as the ability to produce a change. A moving body has energy called **kinetic energy.** A planet moving along its orbit, a cement truck rolling down the highway, and a golf ball sailing down the fairway all have the ability to produce a change. Imagine colliding with any of these objects!

Energy need not be represented by motion. Sunlight falling on a green plant, on photographic film, or on unprotected skin can produce chemical changes, and thus light is a form of energy. Batteries and gasoline are examples of chemical energy, and uranium fuel rods contain nuclear energy. A tank of hot water contains thermal energy.

Potential energy is the energy an object has because of its position in a gravitational field. A bowling ball on a shelf above your desk has potential energy. It is only potential, however, and doesn't produce any changes until the bowling ball descends onto your desk. Obviously the higher the shelf, the more potential energy the ball has.

Much of science is the study of how energy flows from one place to another place producing changes. A biologist might study the way a nerve cell transmits energy along its length to a muscle, while a geologist might study how energy flows as heat from Earth's interior and deforms Earth's surface. In such processes, you see energy being transformed from one state to another. Sunlight (energy) is absorbed by ocean plants and stored as sugars and starches (energy). When the plant dies, it and other ocean life are buried and become oil (energy), which gets pumped to the surface and burned in automobile engines to produce motion (energy).

Aristotle believed that all change originated in the motion of the starry sphere and flowed down to Earth. Modern science has found a more sophisticated description of the continual change you see around you. In a way, science is the study the way energy flows through the world and produces change. Energy is the pulse of the natural world.

Using the metric system (Appendix A), you express energy in **joules** (abbreviated **J**). One joule is about as much energy as that given up when an apple falls from a table to the floor.

Energy is the ability to cause change.

PRESSURE | MASS | **ENERGY** | TEMPERATURE AND HEAT | DENSITY

same way, the planet's velocity must decrease as its distance from the sun increases.

This conservation of angular momentum is actually a common human experience. Skaters spinning slowly can draw their arms and legs closer to their axis of rotation and, through conservation of angular momentum, spin faster (■ Figure 5-7). To slow their rotation, they again extend their arms. Similarly, divers can spin rapidly in the tuck position and slow their rotation by stretching into the extended position.

■ Figure 5-7

Skaters demonstrate conservation of angular momentum when they spin faster by drawing their arms and legs closer to their axis of rotation.

Do you like this explanation of the second law? You may not be very comfortable with it if you are not fully familiar with angular momentum. Scientists often use a well-understood idea like the conservation of angular momentum as a stepping-stone to understand a new concept, but you should hop back one stepping-stone and look at the second law again. Imagine you are in an elliptical orbit around the sun. As you round the most distant part of the ellipse, aphelion, you begin to move back closer to the sun, and the sun's gravity pulls you slightly forward in your orbit. You pick up speed as you fall closer to the sun, so, of course, you go faster as you approach the sun. As you round the closest point to the sun, perihelion, you begin to move away from the sun, and the sun's gravity pulls slightly backward on you, slowing you down as you climb away from the sun. So the second law and the conservation of angular momentum make sense when you analyze them in terms of forces and motions.

Kepler's third law is also explained by a conservation law, but in this case it is the law of conservation of energy (**Focus on Fundamentals 2**). A planet orbiting the sun has a specific amount of energy that depends only on its average distance from the sun. That energy can be divided between energy of motion and energy stored in the gravitational attraction between the planet and the sun. The energy of motion depends on how fast the planet moves, and the stored energy depends on the size of its orbit. The relation between these two kinds of energy is fixed by Newton's laws. That means there has to be a fixed relationship between the rate at which a planet moves around its orbit and the size of the orbit—between its orbital period P and the orbit's semi-major axis a. This is just Kepler's third law.

Newton's Version of Kepler's Third Law

The equation for circular velocity is actually a version of Kepler's third law, as you can prove with three lines of simple algebra. The result is one of the most useful formulas in astronomy.

The equation for circular velocity, as you have seen, is:

$$V_c = \sqrt{\frac{GM}{r}}$$

The orbital velocity of a planet is simply the circumference of its orbit divided by the orbital period:

$$V = \frac{2\pi r}{P}$$

If you substitute this for V in the equation for circular velocity and solve for P^2, you will get:

$$P^2 = \frac{4\pi^2}{GM} r^3$$

Here M is just the total mass of the system in kilograms. For a planet orbiting the sun, you can use the mass of the sun for M, because the mass of the planet is negligible compared to the mass of the sun. (In a later chapter, you will apply this formula to two stars orbiting each other, and then the mass M will be the sum of the two masses.) For a circular orbit, r equals the semi-major axis a, so this formula is a general version of Kepler's third law, $P^2 = a^3$. In Kepler's version, you used the units AU and years, but in Newton's version of the formula, you should use units of meters, seconds, and kilograms. G, of course, is the gravitational constant.

This is a powerful formula. Astronomers use it to find the masses of bodies by observing orbital motion. If, for example, you observed a moon orbiting a planet and you could measure the size of the moon's orbit, r, and the orbital period, P, you could use this formula to solve for M, the total mass of the planet plus the moon. There is no other way to find masses in astronomy, and, in later chapters, you will see this formula used to find the masses of stars, galaxies, and planets.

This discussion is a good illustration of the power of Newton's work. By carefully defining motion and gravity and by giving them mathematical expression, Newton was able to derive new truths, among them Newton's version of Kepler's third law. His work transformed the mysterious wanderings of the planets into understandable motions that follow simple rules. In fact, his discovery of gravity explained something else that had mystified philosophers for millennia—the ebb and flow of the oceans.

Tides and Tidal Forces

Newton understood that gravity is mutual—Earth attracts the moon, and the moon attracts Earth—and that means the moon's gravity can explain the ocean tides. But Newton also realized that gravitation is universal, and that means there is much more to tides than just Earth's oceans.

Tides are caused by small differences in gravitational forces. For example, Earth's gravity attracts your body downward with a force equal to your weight. The moon is less massive than Earth and more distant, so the moon attracts your body with a force equal to roughly 0.0003 percent of your weight. You don't notice that tiny force, but Earth's oceans respond dramatically.

The side of Earth facing the moon is about 4000 miles closer to the moon than is the center of Earth. Consequently, the moon's gravity, tiny though it is at the distance of Earth, is just a bit stronger when it acts on the near side of Earth than on the center. It pulls on the oceans on the near side of Earth a bit more strongly than on Earth's center, and the oceans respond by flowing into a bulge of water on the side of Earth facing the moon. There is also a bulge on the side away from the moon, which develops because the moon pulls more strongly on Earth's center than on the far side. Thus the moon pulls Earth away from the far-side oceans, which flow into a bulge on the far side as shown at the top of ■ Figure 5-8.

You might wonder: If Earth and moon accelerate toward each other, why don't they smash together? In fact, they would smash together in just a couple weeks, except that they are moving sideways, and they keep missing. That is, they are orbiting around their common center of mass. The ocean tides are caused by the accelerations Earth and the oceans feel as they move around that center of mass.

The moon's tidal forces are not confined to water. The rocky bulk of Earth also responds to these tidal forces. Although you don't notice, Earth flexes, and the mountains and plains rise and fall by a few centimeters in response to the moon's gravitational acceleration.

You can see dramatic evidence of this effect if you watch the ocean shore for a few hours. Though Earth rotates on its axis, the tidal bulges remain fixed with respect to the moon. As the turning Earth carries you and your beach into a tidal bulge, the ocean water deepens, and the tide crawls up the sand. The tide does not so much "come in" as you are carried into the tidal bulge. Later, when Earth's rotation carries you out of the bulge, the ocean becomes shallower, and the tide falls. Because there are two bulges on opposite sides of Earth, the tides rise and fall twice a day on an ideal coast.

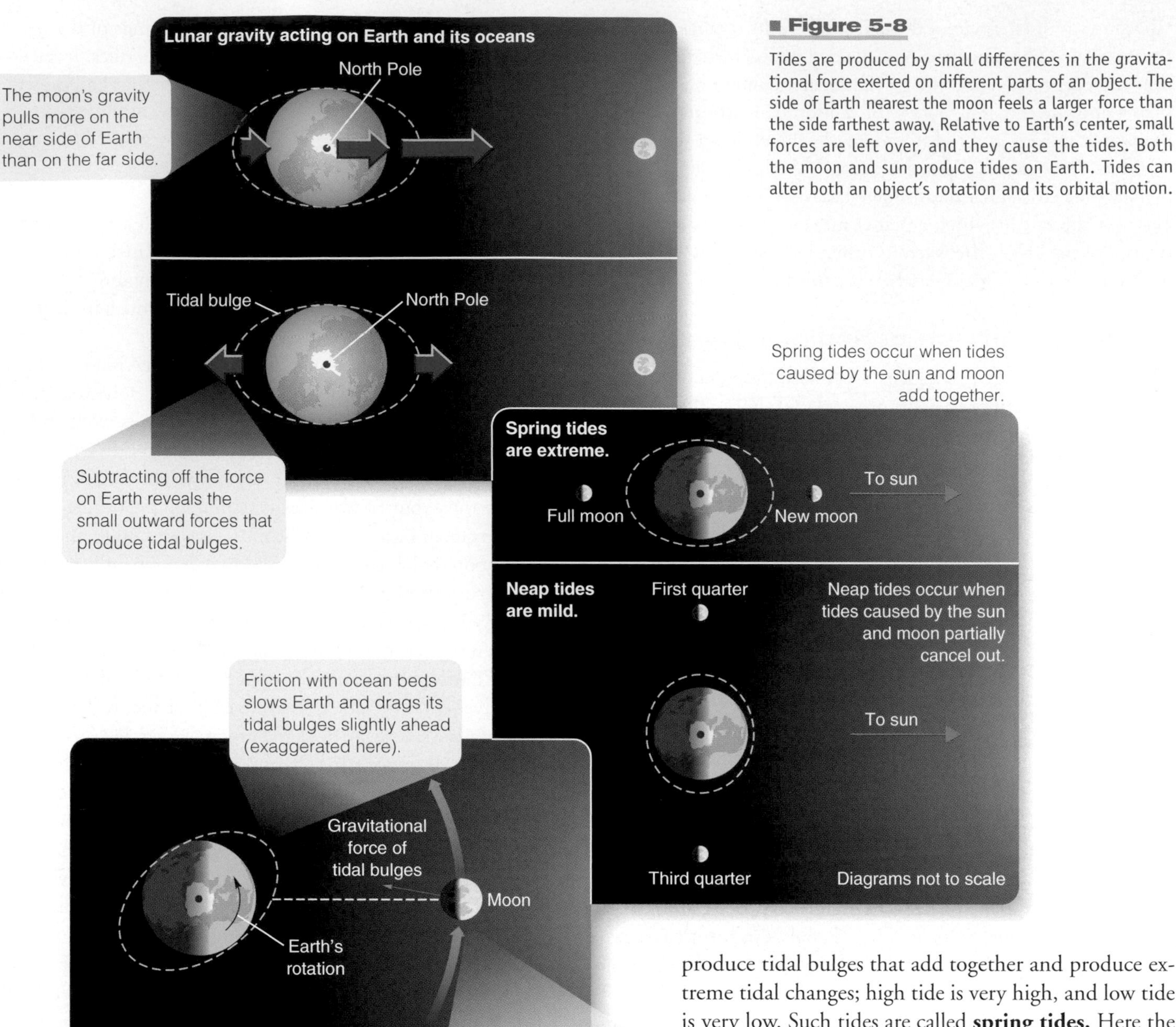
Lunar gravity acting on Earth and its oceans
North Pole
The moon's gravity pulls more on the near side of Earth than on the far side.
Tidal bulge
North Pole
Subtracting off the force on Earth reveals the small outward forces that produce tidal bulges.
Spring tides occur when tides caused by the sun and moon add together.
Spring tides are extreme.
Full moon
New moon
To sun
Neap tides are mild.
First quarter
Neap tides occur when tides caused by the sun and moon partially cancel out.
To sun
Third quarter
Diagrams not to scale
Friction with ocean beds slows Earth and drags its tidal bulges slightly ahead (exaggerated here).
Gravitational force of tidal bulges
Moon
Earth's rotation
Gravity of tidal bulges pulls the moon forward and alters its orbit.

Figure 5-8

Tides are produced by small differences in the gravitational force exerted on different parts of an object. The side of Earth nearest the moon feels a larger force than the side farthest away. Relative to Earth's center, small forces are left over, and they cause the tides. Both the moon and sun produce tides on Earth. Tides can alter both an object's rotation and its orbital motion.

In reality, the tidal cycle at any given location can be quite complex because of the latitude of the site, shape of the shore, winds, and so on. Tides in the Bay of Fundy (New Brunswick, Canada), for example, occur twice a day and can exceed 40 feet. In contrast, the northern coast of the Gulf of Mexico has only one tidal cycle a day of roughly 1 foot.

Gravity is universal, so the sun, too, produces tides on Earth. The sun is roughly 27 million times more massive than the moon, but it lies almost 400 times farther from Earth. Consequently, tides on Earth caused by the sun are less than half those caused by the moon. At new moon and at full moon, the moon and sun produce tidal bulges that add together and produce extreme tidal changes; high tide is very high, and low tide is very low. Such tides are called **spring tides.** Here the word "spring" does not refer to the season of the year but to the rapid welling up of water. Spring tides occur twice a month, at new and full moon. At first- and third-quarter moons, the sun and moon pull at right angles to each other, and the sun's tides cancel out some of the moon's tides. These less-extreme tides are called **neap tides,** and they do not rise very high or fall very low. The word *neap* comes from an obscure Old English word, *nep,* that seems to have meant "lacking power to advance." Spring tides and neap tides are illustrated in Figure 5-8.

Galileo tried to understand tides, but not until Newton described gravity could astronomers analyze tidal forces and recognize their surprising effects. For example, the friction of the ocean waters with the ocean beds slows Earth's rotation and makes the length of a day grow by 0.0023 seconds per century. Fossils of an-

Window on Science 5-2

Testing a Theory by Prediction

When you read about any science, you should notice that scientific theories face in two directions. They look back into the past and explain phenomena previously observed. For example, Newton's laws of motion and gravity explained how the planets moved. But theories also face forward in that they make predictions about what you should find as you explore further. In this way, Newton's laws allowed astronomers to calculate the orbits of comets, predict their return, and eventually understand their origin.

Scientific predictions are important in two ways. First, if a theory leads to a prediction and scientists later discover the prediction was true, the theory is confirmed, and scientists gain confidence that it is a true description of nature. But predictions are important in science in a second way. Using an existing theory to make a prediction may lead to an unexplored avenue of knowledge. For example, the first theories of genetics made predictions that confirmed the genetic theory of inheritance, but those predictions also created a new understanding of how living creatures evolve.

As you read about any scientific theory, think about both what it can explain and what it can predict.

The motion of comets can be predicted using Newton's laws. (Nigel Sharp/NOAO/AURA/NSF)

cient corals confirm that only 900 million years ago Earth's day was 18 hours long. In addition, Earth's gravitation exerts tidal forces on the moon, and although there are no bodies of water on the moon, friction within the flexing rock has slowed the moon's rotation to the point that it now keeps the same face toward Earth.

Tidal forces can also affect orbital motion. Friction with the ocean beds drags the tidal bulges eastward out of a direct Earth–moon line. These tidal bulges contain a large amount of mass, and their gravitational field pulls the moon forward in its orbit, as shown at the bottom of Figure 5-8. As a result, the moon's orbit is growing larger by about 4 cm a year, an effect that astronomers can measure by bouncing laser beams off reflectors left on the lunar surface by the Apollo astronauts.

Newton's gravitation is much more than just the force that makes apples fall. In later chapters, you will see how tides can pull gas away from stars, rip galaxies apart, and melt the interiors of small moons orbiting near massive planets. Tidal forces produce some of the most surprising and dramatic processes in the universe.

Astronomy after Newton

Newton published his work in July 1687 in a book called *Philosophiae Naturalis Principia Mathematica* (*Mathematical Principles of Natural Philosophy*), now known simply as *Principia* (■ Figure 5-9). It is one of the most important books ever written. *Principia* (pronounced *Prin KIP ee uh*) changed astronomy, changed science, and changed the way people think about nature.

■ Figure 5-9

Newton, working from the discoveries of Galileo and Kepler, derived three laws of motion and the principle of mutual gravitation. He and some of his discoveries are honored on this English pound note. Notice the diagram of orbital motion in the background and the open copy of *Principia* in Newton's hands.

Principia changed astronomy and ushered in a new age. No longer did astronomers appeal to the whim of the gods to explain things in the heavens. No longer did they speculate on why the planets wander across the sky. *Principia* says that the motions of the heavenly bodies are governed by simple, universal rules that describe the motions of everything from planets to falling apples. Suddenly the universe was understandable in simple terms.

Newton's laws of motion and gravity made it possible for astronomers to calculate the orbits of planets and moons. Not only could they explain how the heavenly bodies move, they could predict future motions (Window on Science 5-2). This subject, known as gravitational astronomy, dominated astronomy for almost 200 years and is still important. It included the calculation of the orbits of comets and asteroids and the theoretical prediction of the existence of two planets, Neptune and Pluto.

Principia also changed science in general. The works of Copernicus and Kepler had been mathematical, but no book before had so clearly demonstrated the power of mathematics as a language of precision. Newton's arguments in *Principia* were so powerful an illustration of the quantitative study of nature that scientists around the world adopted mathematics as their most powerful tool.

Also, *Principia* changed the way people thought about nature. Newton showed that the rules that govern the universe are simple. Particles move according to three rules of motion and attract each other with a force called gravity. These motions are predictable, and that makes the universe a vast machine based on a few simple rules. It is complex only in that it contains a vast number of particles. In Newton's view, if he knew the location and motion of every particle in the universe, he could, in principle, derive the past and future of the universe in every detail. This mechanical determinism has been undermined by modern quantum mechanics, but it dominated science for more than two centuries during which scientists thought of nature as a beautiful clockwork that would be perfectly predictable if they knew how all the gears meshed.

Most of all, Newton's work broke the last bonds between science and formal philosophy. Newton did not speculate on the good or evil of gravity. He did not debate its meaning. Not more than a hundred years before, scientists would have argued over the "reality" of gravity. Newton didn't care for these debates. He wrote, "It is enough that gravity exists and suffices to explain the phenomena of the heavens."

Building Scientific Arguments

How do Newton's laws of motion explain the orbital motion of the moon?

Natural laws are powerful because they explain so much. The key is often to build an argument step by step. If Earth and the moon did not attract each other, the moon would move in a straight line in accord with Newton's first law of motion and vanish into space in a few months. Instead, gravity pulls the moon toward Earth's center, and the moon accelerates toward Earth. This acceleration is just enough to pull the moon away from its straight-line motion and cause it to follow a curve around Earth. In fact, it is correct to say that the moon is falling, but because of its lateral motion it continuously misses Earth.

Every orbiting object is falling toward the center of its orbit but is moving laterally fast enough to compensate for the inward motion, and it follows a curved orbit. That is an elegant argument, but it raises a question: **How can astronauts float inside spacecraft in a "weightless" state? Why might "free fall" be a more accurate term?**

■ ■ ■

Connections: Newton's laws of motion and gravitation are critical in astronomy not only because they describe orbital motion and tides but also because they describe nearly all interactions between astronomical bodies. Newton made other discoveries, and you will meet those in later chapters. Now you are ready to jump forward a bit more than two centuries to see how Einstein described gravity in a new and powerful way.

5-3 Einstein and Relativity

IN THE EARLY YEARS of the last century, Albert Einstein (1879–1955) (■ Figure 5-10) began thinking about how motion and gravity interact. He soon gained international fame by showing that Newton's laws of motion and gravity were only partially correct. The revised theory became known as the theory of relativity. As you will see, there are really two theories of relativity.

Special Relativity

Einstein began by thinking about how moving observers see events around them. His analysis led him to the first postulate of relativity, also known as the principle of relativity:

> **First postulate** (the principle of relativity): Observers can never detect their *uniform* motion except relative to other objects.

You may have experienced the first postulate while sitting on a train in a station. You suddenly notice that the train on the next track has begun to creep out of the station. However, after several moments you realize that it is your own train that is moving and that the other train is still motionless on its track.

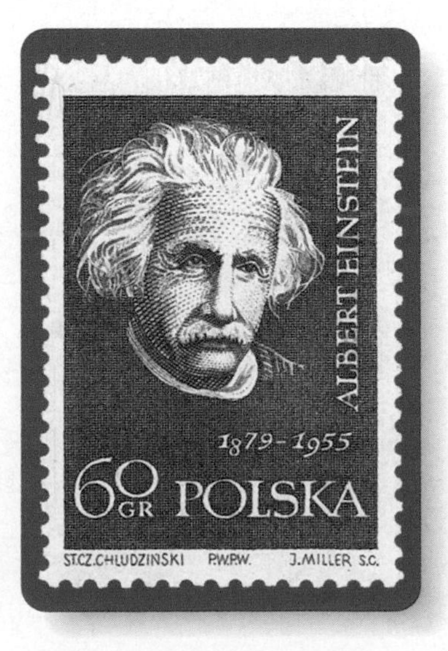

■ **Figure 5-10**
Einstein has become a symbol of the brilliant scientist, but his fame began when he was a young man and thought deeply about the nature of motion. That led him to revolutionary insights into the meaning of space and time and a new understanding of gravity.

Consider another example. Suppose you are floating in a spaceship in interstellar space and another spaceship comes coasting by (■ Figure 5-11a). You might conclude that it is moving and you are not, but someone in the other ship might be equally sure that you are moving and it is not. Of course, you could just look out a window and compare the motion of your spaceship with a nearby star, but that just expands the problem. Which is moving, your spaceship or the star? The principle of relativity says that there is no experiment you can perform to decide which ship is moving and which is not. This means that there is no such thing as absolute rest—all motion is relative.

Because neither you nor the people in the other spaceship could perform any experiment to detect your absolute motion through space, the laws of physics must have the same form in both spaceships. Otherwise, experiments would give different results in the two ships, and you could decide who was moving. So, a more general way of stating the first postulate refers to these laws of physics:

> **First postulate** (alternate version):
> The laws of physics are the same for all observers, no matter what their motion, so long as they are not *accelerated.*

The words *uniform* and *accelerated* are important. If either spaceship were to fire its rockets, then its velocity would change. The crew of that ship would know it because they would feel the acceleration pressing them into their couches. Accelerated motion, therefore, is different—the pilots of the spaceships can always tell which ship is accelerating and which is not. The postulates of relativity discussed here apply only to observers in *uniform* motion, which means *unaccelerated* motion. That is why the theory is called **special relativity.**

The first postulate led to the conclusion that the speed of light must be constant for all observers. No matter how you are moving, your measurement of the speed of light has to give the same result (Figure 5-11b). This became the second postulate of special relativity:

> **Second postulate:** The velocity of light is constant and will be the same for all observers independent of their motion relative to the light source.

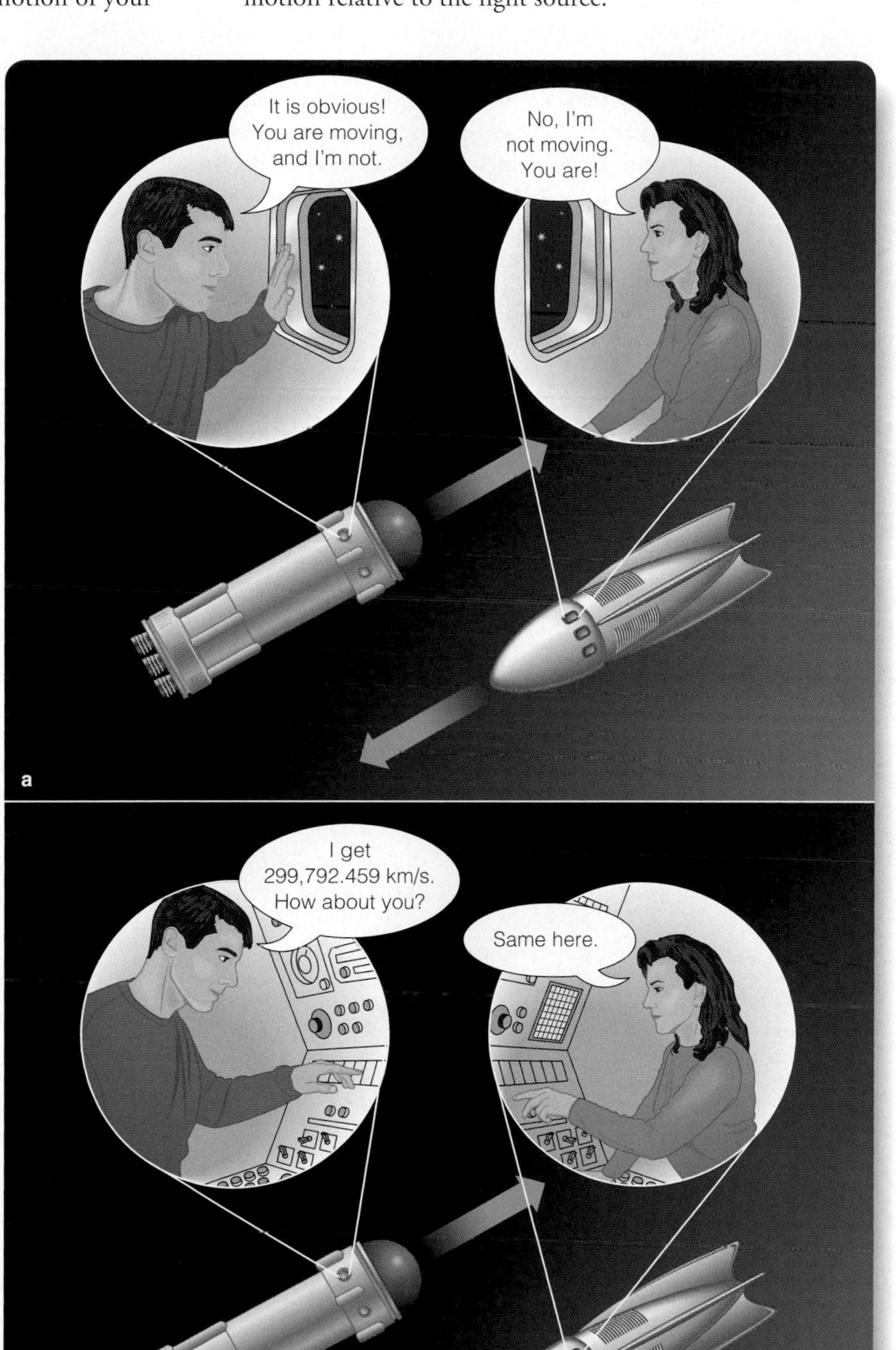

■ Figure 5-11

(a) The principle of relativity says that observers can never detect their uniform motion, except relative to other objects. Thus, neither of these travelers can decide who is moving and who is not. (b) If the velocity of light depended on the motion of the observer through space, then these travelers could perform measurements inside their spaceships to discover who was moving. If the principle of relativity is correct, then the velocity of light must be a constant for all observers.

Remember, this is required by the first postulate; if the velocity of light were not constant, then the pilots of the spaceships could decide who was moving.

Once Einstein had accepted the basic postulates of relativity, he was led to some startling discoveries. Newton's laws of motion and gravity worked well as long as distances were small and velocities were low. But when you begin to think of very large distances or very high velocities, Newton's laws are no longer adequate to describe what happens. Instead, you must use relativistic physics. For example, special relativity shows that the observed mass of a moving particle depends on its velocity. The higher the velocity, the greater the mass of the particle. This is not significant at low velocities, but it becomes very important as the velocity approaches the velocity of light. Such increases in mass are observed whenever physicists accelerate atomic particles to high velocities (■ Figure 5-12).

This discovery led to yet another insight. The relativistic equations that describe the energy of a moving particle predict that the energy of a motionless particle is not zero. Rather, its energy at rest is m_0c^2. This is of course the famous equation:

$$E = m_0c^2$$

The c is the speed of light, and the m_0 is the mass of the particle when it is at rest. This simple formula suggests that mass and energy are related, and you will see in later chapters how nature can convert one into the other inside stars.

For example, suppose that you convert 1 kg of matter into energy. You must express the velocity of light as 3×10^8 m/s, and your result is 9×10^{16} joules (J) (approximately equal to a 20-megaton nuclear bomb). (Recall that a joule is a unit of energy roughly equivalent to the energy given up when an apple falls from a table to the floor.) This simple calculation shows that the energy equivalent of even a small mass is very large.

■ Figure 5-12

The observed mass of moving electrons depends on their velocity. As the ratio of their velocity to the velocity of light, v/c, gets larger, the mass of the electrons in terms of their mass at rest, m/m_0, increases. Such relativistic effects are quite evident in particle accelerators, which accelerate atomic particles to very high velocities.

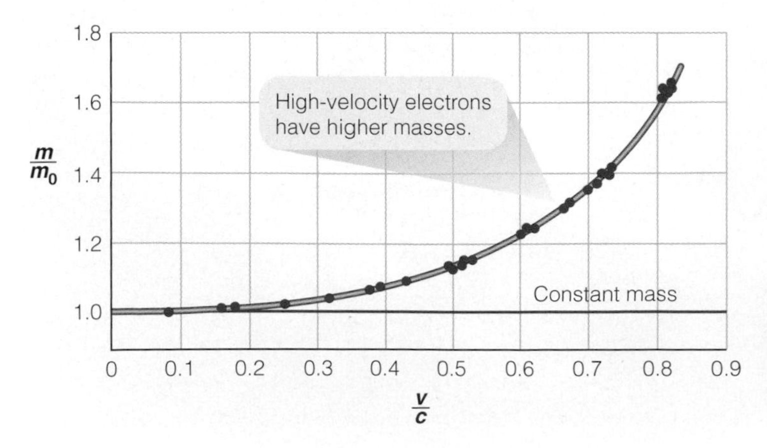

Other relativistic effects include the slowing of moving clocks and the shrinkage of lengths measured in the direction of motion. A detailed discussion of the major consequences of the special theory of relativity is beyond the scope of this book. Instead, let's consider Einstein's second advance, the general theory.

The General Theory of Relativity

In 1916, Einstein published a more general version of the theory of relativity that dealt with accelerated as well as uniform motion. This **general theory of relativity** contained a new description of gravity.

Einstein began by thinking about observers in accelerated motion. Imagine an observer sitting in a windowless spaceship. Such an observer cannot distinguish between the force of gravity and the inertial forces produced by the acceleration of the spaceship (■ Figure 5-13). This led Einstein to conclude that gravity and acceleration are related, a conclusion now known as the equivalence principle:

> **Equivalence principle:** Observers cannot distinguish locally between inertial forces due to acceleration and uniform gravitational forces due to the presence of a massive body.

This should not surprise you. Earlier in this chapter you read that Newton concluded that the mass that resists acceleration is the same as the mass that exerts gravitational forces. He even performed an elegant experiment with pendulums to test the equivalence of the mass related to motion and the mass related to gravity.

The importance of the general theory of relativity lies in its description of gravity. Einstein concluded that gravity, inertia, and acceleration are all associated with the way space is related to time. This relation is often referred to as curvature, and a one-line description of general relativity explains a gravitational field as a curved region of space-time:

> **Gravity according to general relativity:** Mass tells space-time how to curve, and the curvature of space-time (gravity) tells mass how to accelerate.

So you feel gravity because Earth's mass causes a curvature of space-time. The mass of your body responds to that curvature by accelerating toward Earth's center. According to general relativity, all masses cause curvature, and the larger the mass, the more severe the curvature. That's gravity.

Confirmation of the Curvature of Space-Time

Einstein's general theory of relativity has been confirmed by a number of experiments, but two are worth mentioning here because they were among the first tests of the theory. One involves Mercury's orbit, and the other involves eclipses of the sun.

Johannes Kepler understood that the orbit of Mercury is elliptical, but only since 1859 have astronomers known that the

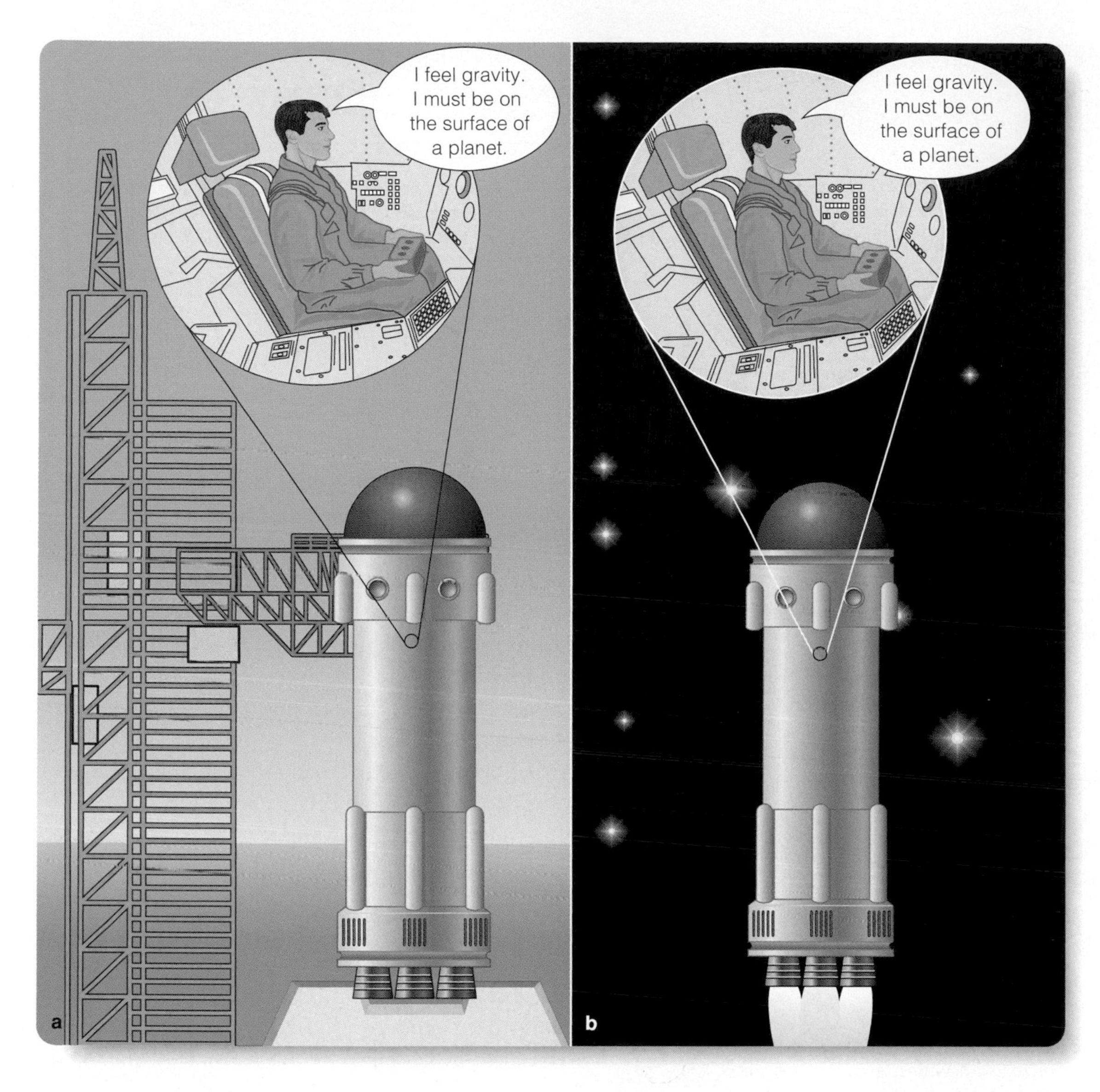

■ **Figure 5-13**

(a) An observer in a closed spaceship on the surface of a planet feels gravity. (b) In space, with the rockets smoothly firing and accelerating the spaceship, the observer feels inertial forces that are equivalent to gravitational forces.

long axis of the orbit sweeps around the sun in a motion called precession (■ Figure 5-14). The total observed precession is 5600.73 seconds of arc per century (as seen from Earth), which equals about 1.5° per century. This precession is produced by the gravitation of Venus, Earth, and the other planets. However, when astronomers used Newton's description of gravity, they calculated that the precession should amount to only 5557.62 seconds of arc per century. So Mercury's orbit is advancing 43.11 seconds of arc per century faster than Newton's law predicts.

This is a tiny effect. Each time Mercury returns to perihelion, its closest point to the sun, it is about 29 km (18 miles) past the position predicted by Newton's laws. This is such a small distance compared with the planet's diameter of 4850 km that it could never have been detected had it not been cumulative. Each orbit, Mercury gains 29 km, and in a century it gains over 12,000 km—more than twice its own diameter. This tiny effect, called the advance of perihelion of Mercury's orbit, accumulated into a serious discrepancy in the Newtonian description of the universe.

The advance of perihelion of Mercury's orbit was one of the first problems to which Einstein applied the principles of general relativity. First he calculated how much the sun's mass curves space-time in the region of Mercury's orbit, and then he calculated how Mercury moves through the space-time. The theory predicted that the curved space-time should cause Mercury's orbit to advance by 43.03 seconds of arc per century, well within the observational accuracy of the excess (Figure 5-14b).

Einstein was elated with this result, and he would be even happier with modern studies that have shown that Mercury, Venus, Earth, and even Icarus, an asteroid that comes close to the sun, have orbits observed to be slipping forward due to the curvature of space-time near the sun (■ Table 5-2). This same effect has been detected in pairs of stars that orbit each other.

A second test of the curvature of space-time was directly related to the motion of light through the curved space-time near the sun. The equations of general relativity predicted that light would be deflected by curved space-time, just as a rolling golf

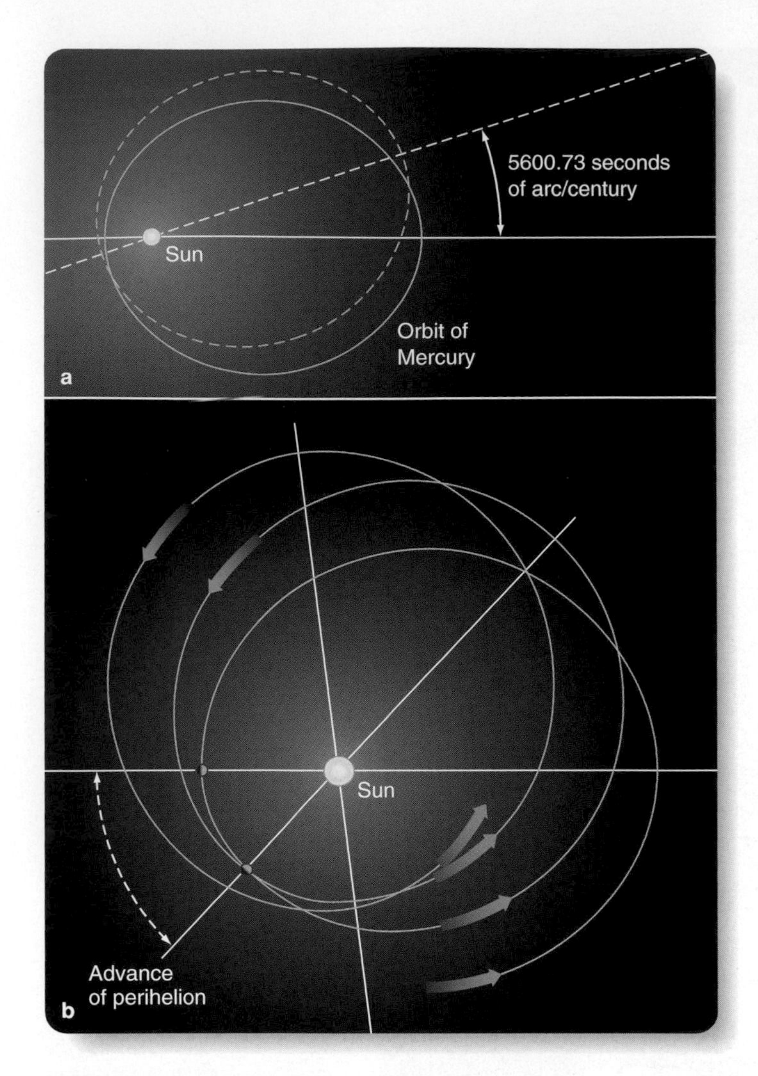

■ Figure 5-14

(a) Mercury's orbit precesses 5600.73 seconds of arc per century—43.11 seconds of arc per century faster than predicted by Newton's laws. (b) Even when you ignore the influences of the other planets, Mercury's orbit is not a perfect ellipse. Curved space-time near the sun distorts the orbit from an ellipse into a rosette. The advance of Mercury's perihelion is exaggerated about a million times in this figure.

ball is deflected by undulations in a putting green. Einstein predicted that starlight grazing the sun's surface would be deflected by 1.75 seconds of arc (■ Figure 5-15). Starlight passing near the sun is normally lost in the sun's glare, but during a total solar eclipse stars beyond the sun could be seen. As soon as Einstein published his theory, astronomers rushed to observe such stars and test the curvature of space-time.

The first solar eclipse following Einstein's announcement in 1916 was June 8, 1918. It was cloudy at some observing sites, and results from other sites were inconclusive. The next occurred on May 29, 1919, only months after the end of World War I, and was visible from Africa and South America. British teams went to both Brazil and Príncipe, an island off the coast of Africa. Months before the eclipse, they photographed that part of the sky where the sun would be located during the eclipse and measured the positions of the stars on the photographic plates. Then, during the eclipse, they photographed the same star field with the eclipsed sun located in the middle. After measuring the plates, they found slight changes in the positions of the stars. During the eclipse, the positions of the stars on the plates were shifted outward, away from the sun (■ Figure 5-16). If a star had been located at the edge of the solar disk, it would have been shifted outward by about 1.8 seconds of arc. This represents good agreement with the theory's prediction.

This test has been repeated at many total solar eclipses since 1919, with similar results. The most accurate results were obtained in 1973 when a Texas–Princeton team measured a deflection of 1.66 ± 0.18 seconds of arc—good agreement with Einstein's theory.

The general theory of relativity is critically important in modern astronomy. You will meet it again in the discussion of black holes, distant galaxies, and the big bang universe. The theory revolutionized modern physics by providing a theory of gravity based on the geometry of curved space-time. Thus,

■ Table 5-2 | Precession in Excess of Newtonian Physics

PLANET	Observed Excess Precession (Sec of arc per century)	Relativistic Prediction (Sec of arc per century)
Mercury	43.11 ± 0.45	43.03
Venus	8.4 ± 0.48	8.6
Earth	5.0 ± 1.2	3.8
Icarus	9.8 ± 0.8	10.3

■ Figure 5-15

Like a depression in a putting green, the curved space-time near the sun deflects light from distant stars and makes them appear to lie slightly farther from the sun than their true positions.

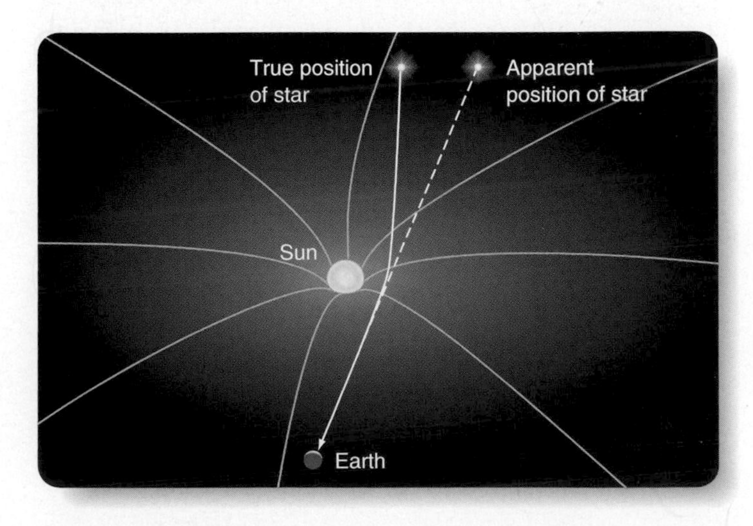

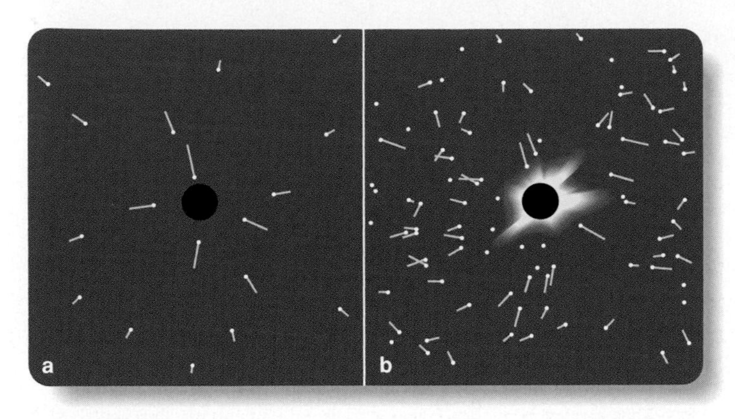

Figure 5-16

(a) Schematic drawing of the deflection of starlight by the sun's gravity. Dots show the true positions of the stars as photographed months before the eclipse. Lines point toward the positions of the stars during the eclipse. (b) Actual data from the eclipse of 1922. Random uncertainties of observation cause some scatter in the data, but in general the stars appear to move away from the sun by 1.77 seconds of arc at the edge of the sun's disk. The deflection of stars is magnified by a factor of 2300 in both (a) and (b).

Galileo's inertia and Newton's mutual gravitation are shown to be not just descriptive rules but fundamental properties of space and time.

Building Scientific Arguments

What does the equivalence principle tell you?

The equivalence principle says that there is no observation you can make inside a closed spaceship to distinguish between uniform acceleration and gravitation. Of course, you could open a window and look outside, but then you would no longer be in a closed spaceship. As long as you make no outside observations, you can't tell whether your spaceship is firing its rockets and accelerating through space or resting on the surface of a planet where gravity gives you weight.

Einstein took the equivalence principle to mean that gravity and acceleration through space-time are somehow related. The general theory of relativity gives that relationship mathematical form and shows that gravity is really a distortion in space-time that physicists refer to as curvature. Consequently, you can say "mass tells space-time how to curve, and space-time tells mass how to move." The equivalence principle led Einstein to an explanation for gravity.

Einstein began his work by thinking carefully about common things such as what you feel when you are moving uniformly or accelerating. This led him to deep insights now called postulates. Special relativity sprang from two postulates. **Why does the second postulate have to be true if the first postulate is true?**

■ ■ ■

Connections: Your study of the origin of astronomy began with the builders of Stonehenge and reaches the modern day with Einstein's general theory of relativity. Now that you have seen where astronomy came from, you are ready to see how it helps you understand your place the universe. Your first question could be "How do astronomers get information?" The answer involves the astronomer's most basic tool, the telescope, and that is the subject of the next chapter.

Study and Review Tools

Summary

5-1 | Galileo and Newton

What happens when an object falls?

- Galileo found that a falling object is accelerated; that is, it falls faster and faster with each passing second. The rate at which it accelerates, termed the acceleration of gravity, is 9.8 m/s^2 (32 ft/s^2) at Earth's surface and does not depend on the weight of the object.
- According to tradition, Galileo demonstrated this by dropping balls of iron and wood from the Leaning Tower of Pisa to show that they would fall together.
- Galileo stated the law of inertia. In the absence of friction, a moving body on a horizontal plane will continue moving forever.

How did Newton discover gravity?

- Newton adopted Galileo's law of inertia as the first of three laws of motion. The second law says that a change in motion, an acceleration, must be caused by a force, and the third law says that forces occur in pairs acting in opposite directions.
- Newton realized that the curved path of the moon meant that it was being accelerated away from a straight-line path, and that required the presence of a force—gravity.
- From his mathematical analysis, Newton was able to show that the force of gravity between two masses is proportional to the product of their masses and inversely proportional to the square of the distance between them.

5-2 | Orbital Motion and Tides

How does gravity explain orbital motion?

- An object in space near Earth would move along a straight line and quickly leave Earth were it not for Earth's gravity accelerating the object toward Earth's center and forcing it to follow a curved path, an orbit. Objects in orbit around Earth are falling (being accelerated) toward Earth's center.

- If there is no friction, the object will fall around its orbit forever.
- Newton's laws explain Kepler's three laws of planetary motion. The planets follow elliptical orbits because gravity follows the inverse square law. The planets move faster when closer to the sun and slower when farther away because they conserve angular momentum. A planet's orbital period squared is proportional to its orbital radius cubed because the moving planet conserves energy.
- An object in a closed orbit follows an elliptical path. A circle is just a special ellipse of zero eccentricity.
- If a body's velocity equals or exceeds the escape velocity, V_e, it will follow a parabola or hyperbola. These orbits are termed "open" because the object never returns to its starting place.

How does gravity explain the tides?

- Tides are caused by differences in the force of gravity acting on different parts of a body.
- Tides on Earth occur because the moon's gravity pulls more strongly on the near side of Earth than on the center of Earth. A tidal bulge occurs on the far side because the moon's gravity is slightly weaker there than at the center of Earth.
- Tides produced by the moon combine with tides produced by the sun to cause extreme tides (called spring tides) at new and full moons. The moon and sun work against each other to produce less-extreme tides (neap tides) at quarter moons.
- Friction from tides can slow the rotation of a rotating world, and the gravitational pull of tidal bulges can make orbits change slowly.

5-3 | Einstein and Relativity

How did Einstein refine the understanding of motion and gravity?

- Einstein published two theories that extended Newton's laws of motion and gravity, the special theory of relativity and the general theory of relativity.
- The special theory of relativity says that uniform (unaccelerated) motion is relative. Observers cannot detect their uniform motion through space except relative to outside objects. This is known as the first postulate.
- This leads to the second postulate: the speed of light is a constant for all observers.
- A consequence of special relativity is that mass and energy are related.
- The general theory of relativity says that a gravitational field is a curvature of space-time caused by the presence of a mass. For example, Earth's mass curves space-time, and the mass of your body responds to that curvature by accelerating toward Earth's center.
- The curvature of space-time was confirmed by the slow advance in perihelion of the orbit of Mercury and by the deflection of starlight observed during a 1919 total solar eclipse.

New Terms

natural motion (p. 87)
violent motion (p. 87)
acceleration of gravity (p. 87)
momentum (p. 89)
acceleration (p. 89)
velocity (p. 89)
mass (p. 90)
inverse square law (p. 90)
field (p. 92)
angular momentum (p. 93)
circular velocity (p. 94)
geosynchronous satellite (p. 94)
center of mass (p. 95)
closed orbit (p. 95)
escape velocity (p. 95)
open orbit (p. 95)
energy (p. 96)
kinetic energy (p. 96)
potential energy (p. 96)
joule (J) (p. 96)
spring tide (p. 98)
neap tide (p. 98)
special relativity (p. 101)
general theory of relativity (p. 102)

Review Questions

Ace Astronomy™ Assess your understanding of this chapter's topics with additional quizzing and animations at **http://ace.brookscole.com/sf9**

1. Why wouldn't Aristotle's explanation of gravity work if Earth was not the center of the universe?
2. According to the principles of Aristotle, what part of the motion of a baseball pitched across the home plate is natural motion? What part is violent motion?
3. If you drop a feather and a steel hammer at the same moment, they should hit the ground at the same instant. Why doesn't this work on Earth, and why does it work on the moon?
4. What is the difference between mass and weight? between speed and velocity?
5. Why did Newton conclude that some force had to pull the moon toward Earth?
6. Why did Newton conclude that gravity has to be mutual and universal?
7. How does the concept of a field explain action at a distance? Name another kind of field also associated with action at a distance.
8. Why can't a spacecraft go "beyond Earth's gravity"?
9. What is the center of mass of the Earth–moon system? Where is it?
10. How do planets orbiting the sun and skaters doing a spin conserve angular momentum?
11. Why is the period of an open orbit undefined?
12. How does the first postulate of special relativity imply the second?
13. When you ride a fast elevator upward, you feel slightly heavier as the trip begins and slightly lighter as the trip ends. How is this phenomenon related to the equivalence principle?
14. From your knowledge of general relativity, would you expect radio waves from distant galaxies to be deflected as they pass near the sun? Why or why not?
15. Why can the object shown at the right be bolted in place and used 24 hours a day without adjustment?
16. Why is it a little bit misleading to say that this astronaut is weightless?

Larry Mulvehill/The Image Works

(NASA/JSC)

Discussion Questions

1. How did Galileo idealize his inclines to conclude that an object in motion stays in motion until it is acted on by some force?
2. Give an example from everyday life to illustrate each of Newton's laws.
3. People who lived before Newton may not have believed in cause and effect as strongly as you do. How do you suppose they saw their daily lives?

Problems

1. Compared with the strength of Earth's gravity at its surface, how much weaker is gravity at a distance of 10 Earth radii from Earth's center? at 20 Earth radii?
2. Compare the force of lunar gravity on the surface of the moon with the force of Earth's gravity at Earth's surface.
3. If a small lead ball falls from a high tower on Earth, what will be its velocity after 2 seconds? after 4 seconds?
4. What is the circular velocity of an Earth satellite 1000 km above Earth's surface? (*Hint:* Earth's radius is 6380 km.)

5. What is the circular velocity of an Earth satellite 36,000 km above Earth's surface? What is its orbital period? (*Hint:* Earth's radius is 6380 km.)
6. What is the orbital period of an imaginary satellite orbiting just above Earth's surface? Ignore friction with the atmosphere.
7. Repeat the previous problem for Mercury, Venus, the moon, and Mars.
8. Describe the orbit followed by the slowest cannonball on page 94 on the assumption that the cannonball could pass freely through Earth. (Newton got this problem wrong the first time he tried to solve it.)
9. If you visited an asteroid 30 km in radius with a mass of 4×10^{17} kg, what would be the circular velocity at its surface? A major league fastball travels about 90 mph. Could a good pitcher throw a baseball into orbit around the asteroid?
10. What is the orbital period of a satellite orbiting just above the surface of the asteroid in Problem 9?
11. What would be the escape velocity at the surface of the asteroid in Problem 9? Could a major league pitcher throw a baseball off of the asteroid?

Media Cluster

ACTIVE FIGURES

Ace Astronomy™ To access the resources in the Media Cluster, log into AceAstronomy at **http://ace.brookscole.com/sf9** and select Chapter 5.

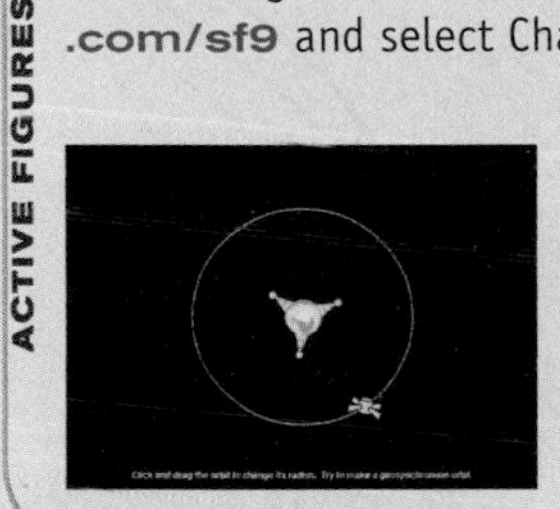

Geosynchronous Orbit
Can you put your satellite into geosynchronous orbit so that it can communicate with a ground antenna 24 hours a day? Can you move it from one antenna to the next?

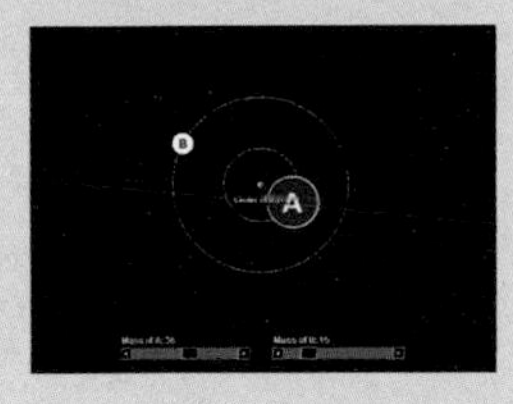

Center of Mass
Adjust the masses of the stars in a binary star system and see how the orbits change. Do the sizes of the orbits correctly reflect the ratio of the masses?

ASTRONOMY EXERCISES

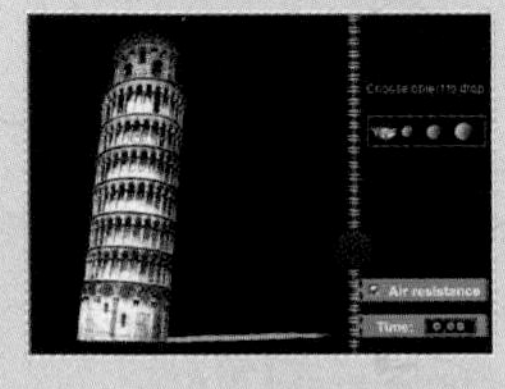

Falling Bodies
According to tradition, Galileo demonstrated that the acceleration of a falling body is independent of its weight by dropping various objects from the Leaning Tower of Pisa. You can repeat his experiment with this interactive exercise.

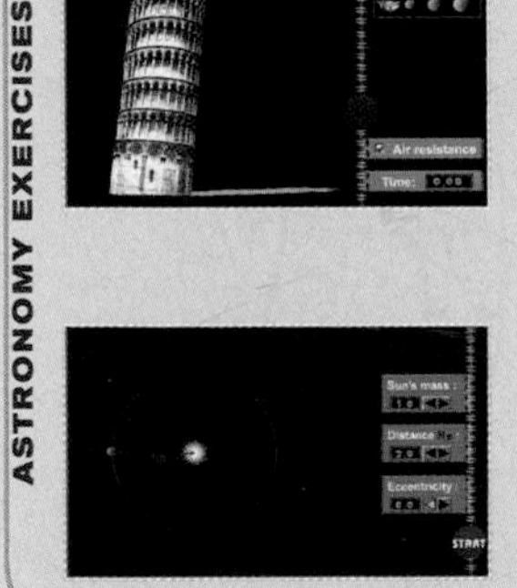

Orbital Motion
Explore how the orbit of a hypothetical planet is affected by changing the mass of the sun, distance to the sun, and eccentricity of orbit in this animation.

Escape Velocity
Escape velocity is the initial velocity an object needs to escape from the surface of a celestial body. See if you can determine the correct escape velocity for the rocket in this animation.

Critical Inquiries for the Web

1. Einstein's general theory of relativity predicts the curvature of space-time, but here on Earth you have little opportunity to observe such effects. Find an astronomical situation in which space-time curvature is evident from observations and describe the effect of the curvature on what astronomers see when they view these objects.
2. Communications satellites are obvious uses of the geosynchronous orbit, but can you think of other uses for such orbits? Find an Internet site that uses or displays information gleaned from geosynchronous orbit that provides a useful service.

- Go to the Brooks/Cole Astronomy Resource Center **(http://astronomy.brookscole.com)** for critical thinking exercises, articles, and additional readings from InfoTrac College Edition, Brooks/Cole's online student library.

6 Light and Telescopes

The strongest thing that's
given us to see with's
A telescope. Someone in
every town
Seems to me owes it to the
town to keep one.

ROBERT FROST, "THE STAR-SPLITTER," *NEW HAMPSHIRE: A POEM WITH NOTES AND GRACE NOTES* (NEW YORK: HENRY HOLT AND CO., 1923), pp. 27–30.

DO YOU ENJOY BEING outdoors on a clear, starry night? Astronomers appreciate the beauty of the night like everyone else, but they have an additional insight. They know that starlight is going to waste. Every night, light from the stars falls on trees, oceans, roofs, and parking lots, and it is all wasted. To an astronomer, nothing is so precious as starlight. It is the only link to the sky, and the astronomer's quest is to gather as much starlight as possible and extract from it the secrets of the stars. ▌ The telescope is the emblematic tool of the astronomer because its purpose is to gather and concentrate light for analysis. Nearly all of the interesting objects in the sky are faint sources of light, so modern astronomers are driven to build the largest possible telescopes to gather the maximum amount of light (■ Figure 6-1).

▌ Continued on page 110 ▌

Every sunset atop a Hawaiian mountain finds astronomers preparing for a night's work using giant telescopes to gather faint starlight and search out the secrets of the stars. (NOAO/AURA/NSF)

Guidepost

Looking Back

In the early chapters of this book, you looked at the sky the way ancient astronomers did, with the unaided eye. In the last chapter, you got a glimpse through Galileo's telescope, and it revealed astonishing things about Jupiter, the moon, and Venus. Now it is time to examine the tools of the modern astronomer.

This Chapter

You should begin by noting that these tools gather and focus light and its related forms of radiation. For that reason, the first of the essential questions is about mysterious light:

What is light?

How do telescopes work, and how are they limited?

What kind of instruments do astronomers use to record and analyze light?

Why do astronomers use radio telescopes?

Why must some telescopes go into space?

As you answer these questions, you will discover how important it is to understand light in all its forms.

Looking Ahead

Astronomy is almost entirely an observational science. Astronomers cannot visit distant galaxies and far-off worlds, so they must observe using astronomical telescopes. Twenty chapters remain, and every one will discuss information gathered by telescopes.

Ace Astronomy™ The AceAstronomy icon throughout the text indicates an opportunity for you to test yourself on key concepts and to explore animations and interactions on the AceAstronomy website at: http://ace.brookscole.com/sf9

■ Figure 6-1

Astronomical telescopes are often very large to gather large amounts of starlight. The Northern Gemini Telescope stands over 19 m (60 ft) high when pointed straight up. Its main mirror is 8.1 m (26.5 ft) in diameter—larger than some classrooms. The dome of this telescope is shown at the left of the photo on the opening page of this chapter. (NOAO/AURA/NSF)

If you wish to gather visible light, a normal telescope will do; but, as you will soon see, visible light is only one kind of radiation from the stars. Astronomers can extract information from other forms of radiation by using specialized telescopes. Radio telescopes, for example, provide an entirely different view of the sky. Some of these specialized telescopes can be used from Earth's surface, but some must go into orbit above Earth's atmosphere. Telescopes that observe X rays, for instance, must be placed in space.

As you study the sophisticated telescopes and instruments that modern astronomers use, keep in mind Robert Frost's suggestion: In every town, someone should keep a telescope. Astronomy is more than technology and scientific analysis. It tells us what we are, and every town should have a telescope to keep us looking upward.

6-1 Radiation: Information from Space

JUST AS A BOOK on bread baking might begin with a discussion of flour, this chapter on telescopes begins with a discussion of light—not just visible light, but the entire range of radiation from the sky.

Light as a Wave and a Particle

If you have admired the colors in a soap bubble, you have seen light behave as a wave. But when light enters the light meter on your camera, it behaves as a particle. How it behaves depends on how you observe it—it is both wave and particle.

You experience waves whenever you hear sound. Sound waves are a mechanical disturbance that travels through the air from source to ear. Sound requires a medium; so, on the moon, where there is no air, there can be no sound. In contrast, light is made up of electric and magnetic fields that can travel through empty space. Unlike sound, light does not require a medium, and so it can travel through a perfect vacuum. There is no sound on the moon, but there is plenty of sunlight.

Light is merely one form of radiation, called **electromagnetic radiation** because it is associated with changing electric and magnetic fields that travel through space and transfer energy from one place to another. When light enters your eye, the fluctuating electric and magnetic fields carry energy that stimulates nerve endings, and you see light.

The oscillating electric and magnetic fields that constitute electromagnetic radiation move through space at about 300,000 km/s (186,000 mi/s). This speed is commonly referred to as the speed of light, *c,* but it is in fact the speed of all such radiation in a vacuum.

It may seem odd to use the word *radiation* when you speak of light. The word can be used to refer to high-energy particles emitted from radioactive atoms, and you have learned to be a little bit concerned when you see the word *radiation.* But it really refers to anything that spreads outward from a source. Light radiates from a source, so you can correctly refer to light as a form of radiation.

Electromagnetic radiation is a wave phenomenon; that is, it is associated with a periodically repeating disturbance, or wave. You are familiar with waves in water. If you disturb a quiet pool of water, waves spread across the surface. Imagine that you use a meter stick to measure the distance between the successive peaks of a wave. This distance is the **wavelength,** usually represented by the Greek letter lambda (λ). If you were measuring ripples in a pond, you might find that the wavelength is a few centimeters, whereas the wavelength of ocean waves might be a hundred meters or more. There is no restriction on the wavelength of electromagnetic radiation. Wavelengths can range from smaller than the diameter of an atom to larger than that of Earth.

Because all electromagnetic radiation travels at the speed of light, wavelength is related to **frequency,** the number of cycles that pass in one second. Short-wavelength radiation has a high frequency; long-wavelength radiation has a low frequency. To understand this, imagine watching an electromagnetic wave race past while you count its peaks (■ Figure 6-2). If the wavelength is short, you will count many peaks in one second; if the wavelength is long, you will count few peaks per second. The dials on radios are marked in frequency, but they could just as easily be marked in wavelength. The relation between wavelength and frequency is a simple one:

$$\lambda = \frac{c}{f}$$

That is, the wavelength equals the speed of light c divided by the frequency f. Notice that the larger (higher) the frequency, the smaller (shorter) the wavelength. In most cases, astronomers use wavelength rather than frequency.

Radio waves can have wavelengths from a few millimeters for microwaves to kilometers. In contrast, the wavelength of light is so short that you will need more convenient units. This book uses **nanometers (nm)** because this unit is consistent with the International System of units. One nanometer is 10^{-9} meter, and visible light has wavelengths that range from about 400 nm to about 700 nm. Another unit that astronomers commonly use, and a unit that you will see in many references on astronomy, is the **Angstrom (Å).** One Angstrom is 10^{-10} meter, and visible light has wavelengths between 4000 Å and 7000 Å.

You may find radio astronomers describing wavelengths in centimeters or millimeters, and infrared astronomers often refer to wavelengths in micrometers (or microns). One micrometer (μm) is 10^{-6} meter. Whatever unit is used to describe the wavelength, you must keep in mind that all electromagnetic radiation is the same phenomenon.

■ Figure 6-2

All electromagnetic waves travel at the speed of light. The wavelength is the distance between successive peaks. The frequency of the wave is the number of peaks that pass you in one second.

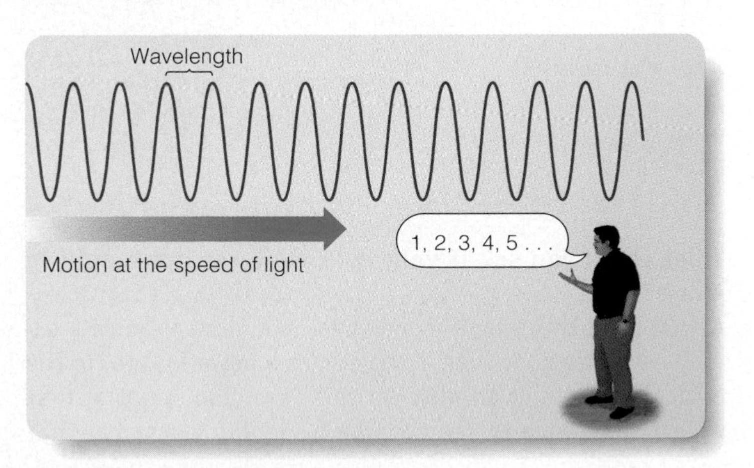

What exactly is electromagnetic radiation? Is it a particle or a wave? Throughout his life, Newton believed that light was made up of particles, but modern physicists now recognize that light can behave as both particle and wave. The modern model of light is more complete than Newton's, and it refers to "a particle of light" as a **photon.** You can recognize its dual nature by thinking of it as a bundle of waves.

The amount of energy a photon carries depends on its wavelength. The shorter the wavelength, the more energy the photon carries; the longer the wavelength, the less energy it contains. This is easy to remember because short wavelengths have high frequencies, and you would naturally expect rapid fluctuations to be more energetic. A simple formula expresses the relationship between energy and wavelength:

$$E = \frac{hc}{\lambda}$$

Here h is Planck's constant (6.6262×10^{-34} joule s), c is the speed of light (3×10^8 m/s), and λ is the wavelength in meters. A photon of visible light carries a very small amount of energy, but a photon with a very short wavelength can carry much more.

The Electromagnetic Spectrum

A spectrum is an array of electromagnetic radiation in order of wavelength. You are most familiar with the spectrum of visible light, which you see in rainbows. The colors of the spectrum differ in wavelength, with red having the longest wavelength and violet the shortest. The visible spectrum is shown at the top of ■ Figure 6-3.

The average wavelength of visible light is about 0.00005 cm. You could put 50 light waves end to end across the thickness of a sheet of household plastic wrap. Measured in nanometers, the wavelength of visible light ranges from about 400 to 700 nm. Just as you sense the wavelength of sound as pitch, you sense the wavelength of light as color. Light near the short-wavelength end of the visible spectrum (400 nm) looks violet to your eyes, and light near the long-wavelength end (700 nm) looks red.

Figure 6-3 shows how the visible spectrum makes up only a small part of the entire electromagnetic spectrum. Beyond the red end of the visible spectrum lies **infrared radiation,** where wavelengths range from 700 nm to about 0.1 cm. Your eyes are not sensitive to this radiation, but your skin senses it as heat. A "heat lamp" is just a bulb that gives off principally infrared radiation.

Beyond the infrared part of the electromagnetic spectrum lie radio waves. Microwaves have wavelengths of a millimeter to a few centimeters and are used for radar and long-distance telephone communication. Longer wavelengths are used for UHF and VHF television transmissions. FM, military, governmental, and ham radio signals have wavelengths up to a few meters, and AM radio waves can have wavelengths of kilometers.

The distinction between the wavelength ranges is not sharp. Long-wavelength infrared radiation and the shortest microwave

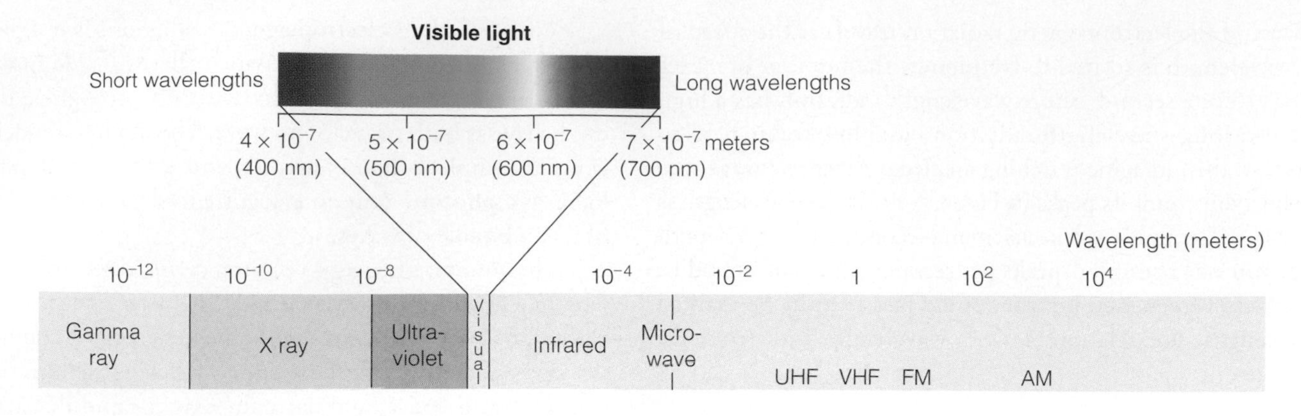

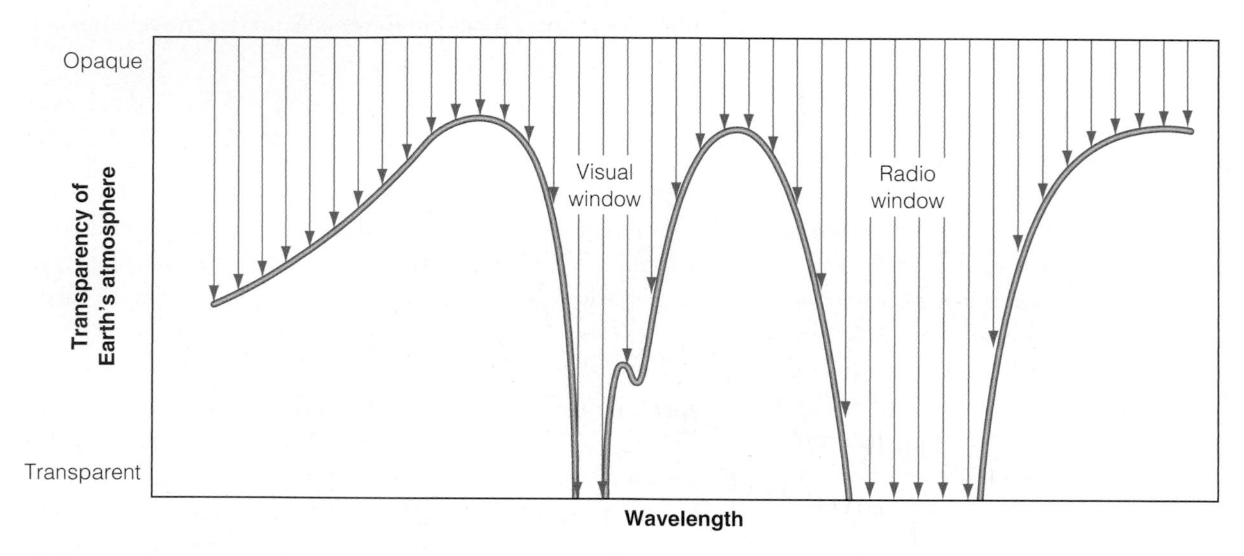

■ Figure 6-3

The spectrum of visible light, extending from red to violet, is only part of the electromagnetic spectrum. Most radiation is absorbed in Earth's atmosphere, and only radiation in the visual window and the radio window can reach Earth's surface.

radio waves are the same. Similarly, there is no clear division between the short-wavelength infrared and the long-wavelength part of the visible spectrum. It is all electromagnetic radiation.

Look once again at the electromagnetic spectrum in Figure 6-3 and notice that electromagnetic waves shorter than violet are called **ultraviolet.** Electromagnetic waves even shorter are called X rays, and the shortest are gamma rays. Again, the boundaries between these wavelength ranges are not clearly defined.

X rays and gamma rays can be dangerous, and even ultraviolet photons have enough energy to do harm. Small doses of ultraviolet produce a suntan and larger doses sunburn and skin cancers. Contrast this to the lower-energy infrared photons. Individually they have too little energy to affect skin pigment, a fact that explains why you can't get a tan from a heat lamp. Only by concentrating many low-energy photons in a small area, as in a microwave oven, can you transfer significant amounts of energy.

Astronomers are interested in electromagnetic radiation because it carries clues to the nature of stars, planets, and other celestial objects. Earth's atmosphere is opaque to most electromagnetic radiation, as shown by the graph at the bottom of Figure 6-3. Gamma rays, X rays, and some radio waves are absorbed high in Earth's atmosphere, and a layer of ozone (O_3) at an altitude of about 30 km absorbs ultraviolet radiation. Water vapor in the lower atmosphere absorbs the longer wavelength infrared radiation. Only visible light, some shorter wavelength infrared, and some radio waves reach Earth's surface through two wavelength regions called **atmospheric windows.** Obviously, if you wish to study the sky from Earth's surface, you must look out through one of these windows.

AceAstronomy™ Log into AceAstronomy and select this chapter to see Astronomy Exercise "The Electromagnetic Spectrum." Explore different wavelength regions.

Building Scientific Arguments

What could you see if your eyes were sensitive only to X rays?

As you build this scientific argument, you must imagine a totally new situation. That is sometimes a powerful tool in the critical analysis of an idea. In this case, you might at first expect to be able to see through walls, but remember that your eyes detect only light that already exists. There are almost no X rays bouncing around at Earth's surface, so if you had X-ray eyes, you would be in the dark and would be unable to see anything. Even when you looked up at the sky, you would see nothing because Earth's atmosphere is not transparent to X rays. If Superman can see through walls, it is not because his eyes can detect X rays.

But now imagine a slightly different situation and modify your argument. **Would you be in the dark if your eyes were sensitive only to radio wavelengths?**

■ ■ ■

Connections: Now that you know something about electromagnetic radiation, you are ready to meet the tools astronomers use to gather and analyze that radiation.

6-2 Optical Telescopes

ASTRONOMERS BUILD OPTICAL TELESCOPES to gather light and focus it into sharp images. This requires sophisticated optical and mechanical designs, and it leads astronomers to build gigantic telescopes on the tops of high mountains. As you begin, you will need to learn some technical terms that describe telescopes, but remember your real goal. You want to understand how different kinds of telescopes work and why some are better than others.

Two Kinds of Telescopes

Astronomical telescopes focus light into an image in one of two ways, as shown in ■ Figure 6-4. A lens bends (refracts) the light as it passes through the glass and brings it to a focus to form a small

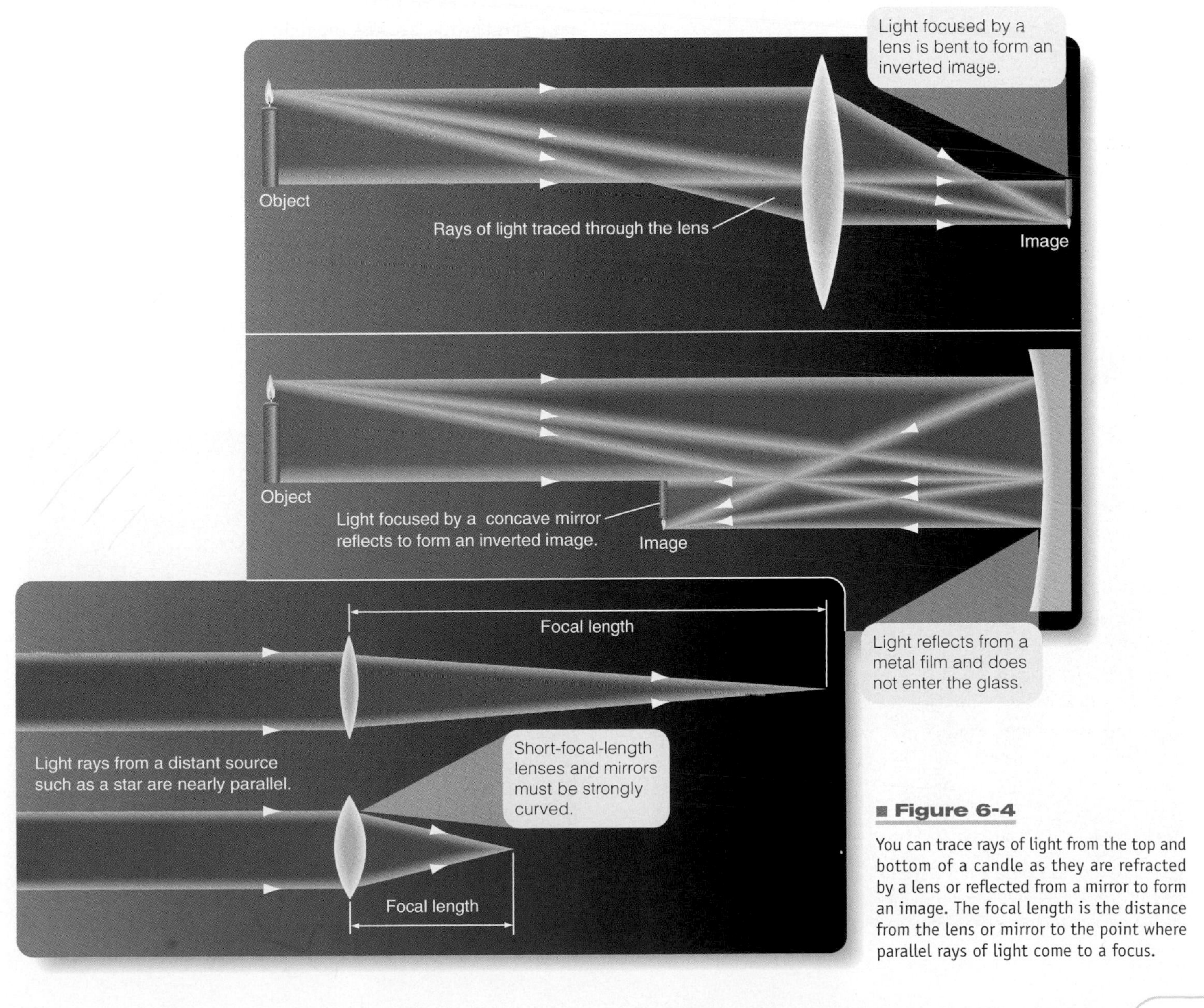

■ Figure 6-4

You can trace rays of light from the top and bottom of a candle as they are refracted by a lens or reflected from a mirror to form an image. The focal length is the distance from the lens or mirror to the point where parallel rays of light come to a focus.

inverted image. A mirror—a concave piece of glass with a reflective surface—forms an image by reflecting the light. In either case, the **focal length** is the distance from the lens or mirror to the image formed of a distant light source, such as a star. Short-focal-length lenses and mirrors must be strongly curved, and long-focal-length lenses and mirrors are less strongly curved. Grinding the proper shape on a lens or mirror is a delicate, time-consuming, and expensive process.

Because there are two ways to focus light, there are two kinds of astronomical telescopes. **Refracting telescopes** use a large lens to gather and focus the light, whereas **reflecting telescopes** use a concave mirror. The advantages of the reflecting telescope have made it the preferred design for modern observatories.

The main lens in a refracting telescope is called the **primary lens,** and the main mirror in a reflecting telescope is called the **primary mirror.** These are also called the **objective lens** and **mirror.** Both kinds of telescopes form a very small, inverted image that is difficult to observe directly, so astronomers use a small lens called the **eyepiece** to magnify the image and make it convenient to view (■ Figure 6-5).

Refracting telescopes suffer from a serious optical distortion that limits their usefulness. When light is refracted through glass, shorter wavelengths bend more than longer wavelengths, and blue light, having shorter wavelengths, comes to a focus closer to the lens than does red light (■ Figure 6-6a). If you focus the eyepiece on the blue image, the red light is out of focus, and you see a red blur around the image. If you focus on the red image, the blue light blurs. The color separation is called **chromatic aberration.** Telescope designers can grind a telescope lens of two components made of different kinds of glass and so bring two different wavelengths to the same focus (Figure 6-6b). This does improve the image, but these **achromatic lenses** are not totally free of chromatic aberration, because other wavelengths still blur. Telescopes made with such lenses were popular until the end of the 19th century.

The primary lens of a refracting telescope is very expensive to make because it must be achromatic, and the glass must be pure and flawless because the light passes through the lens. The four surfaces must be ground precisely, and the lens can be supported only along its edge. The largest refracting telescope in the world was completed in 1897 at Yerkes Observatory in Wisconsin. Its lens is 1 m (40 in.) in diameter and weighs half a ton. Larger refracting telescopes are prohibitively expensive.

Reflecting telescopes are much less expensive because the light reflects from the front surface of the mirror. Consequently

■ Active Figure 6-5

(a) A refracting telescope uses a primary lens to focus starlight into an image that is magnified by a lens called an eyepiece. The primary lens has a long focal length, and the eyepiece has a short focal length. (b) A reflecting telescope uses a primary mirror to focus the light by reflection. A small secondary mirror reflects the starlight back down through a hole in the middle of the primary mirror to the eyepiece.

Ace Astronomy™ Log into AceAstronomy and select this chapter to see the Active Figure "Refractors and Reflectors." Watch light pass through the optics.

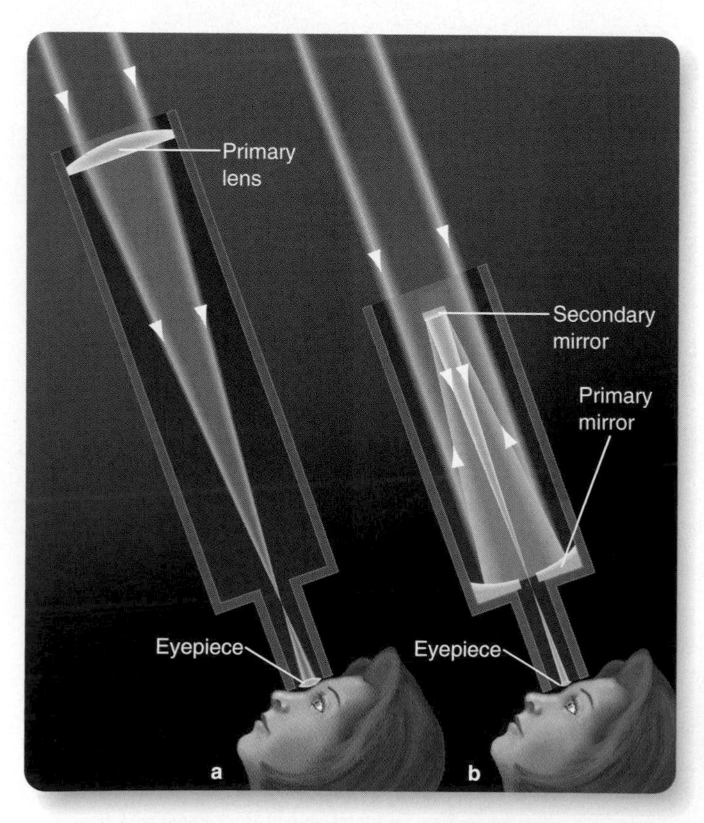

■ Figure 6-6

(a) A normal lens suffers from chromatic aberration because short wavelengths bend more than long wavelengths. (b) An achromatic lens, made in two pieces of two different kinds of glass, can bring any two colors to the same focus, but other colors remain slightly out of focus.

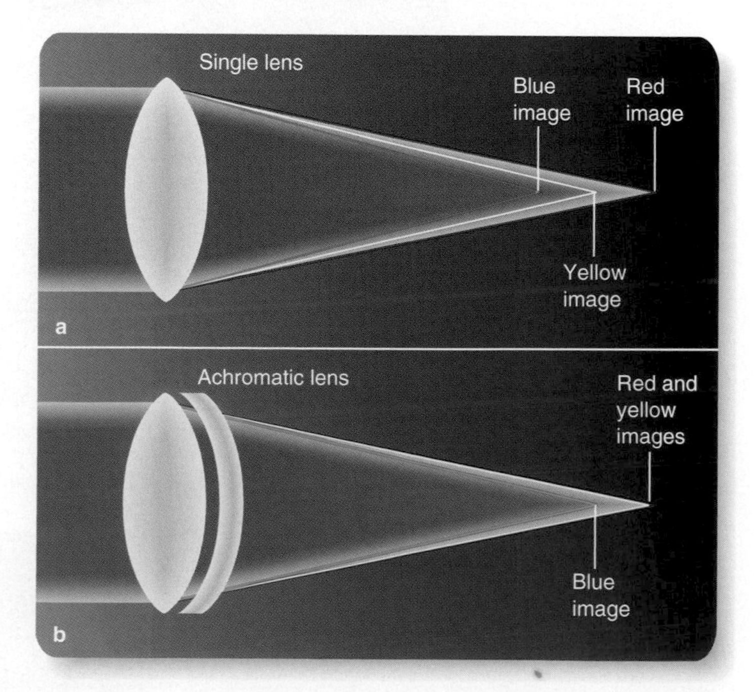

only the front surface need be ground to precise shape. Also, the glass of the mirror need not be perfectly transparent, and the mirror can be supported over its back surface to reduce sagging. Most important, reflecting telescopes do not suffer from chromatic aberration because the light is reflected from the metallic film on the front surface of the mirror and never enters the glass. For these reasons, every large astronomical telescope built since the beginning of the 20th century has been a reflecting telescope.

Ace Astronomy™ Log into AceAstronomy and select this chapter to see Astronomy Exercise "Lenses: Focal Length." Create different lenses and observe their focal lengths.

Ace Astronomy™ Log into AceAstronomy and select this chapter to see Astronomy Exercise "Telescopes: Objective Lens and Eyepiece." See how the two lenses interact to produce an image.

The Powers of a Telescope

A telescope can aid your eyes in only three ways—the three powers of a telescope. They make images brighter, more detailed, and larger.

Most interesting celestial objects are faint sources of light, so you need a telescope that can gather large amounts of light to produce a bright image. **Light-gathering power** refers to the ability of a telescope to collect light. Catching light in a telescope is like catching rain in a bucket—the bigger the bucket, the more rain it catches (■ Figure 6-7). Light-gathering power is proportional to the area of the telescope objective. A lens or mirror with a large area gathers a large amount of light. The area of a circular lens or mirror of diameter D is just πr^2 or, written in terms of the diameter, the area is $\pi(D/2)^2$. To compare the relative light-gathering powers *(LGP)* of two telescopes *A* and *B,* you can calculate the ratio of the areas of their objectives, which reduces to the ratio of their diameters *(D)* squared.

$$\frac{LGP_A}{LGP_B} = \left(\frac{D_A}{D_B}\right)^2$$

For example, suppose you compared a telescope 24 cm in diameter with a telescope 4 cm in diameter. The ratio of the diameters is 24/4, or 6, but the larger telescope does not gather 6 times as much light. Light-gathering power increases as the ratio of diameters squared, so it gathers 36 times more light than the smaller telescope. This example shows the importance of diameter in astronomical telescopes. Even a small increase in diameter produces a large increase in light-gathering power and allows astronomers to study much fainter objects.

The second power, **resolving power,** refers to the ability of the telescope to reveal fine detail. Because light acts as a wave, it produces a small **diffraction fringe** around every point of light in the image, and you cannot see any detail smaller than the fringe (■ Figure 6-8). Astronomers can't eliminate diffraction fringes,

■ **Figure 6-7**

Gathering light is like catching rain in a bucket. A large-diameter telescope gathers more light and has a brighter image than a smaller telescope of the same focal length.

but the larger a telescope is in diameter, the smaller the diffraction fringes are. That means the larger the telescope, the better its resolving power.

As you will see later, resolving power is worse for longer wavelengths, but if you consider only optical telescopes, you can estimate the resolving power by calculating the angular distance between two stars that are just barely visible through the telescope as two separate images. Astronomers say the two images are "resolved," meaning they are separated from each other. The resolving power, α, in seconds of arc, equals 11.6 divided by the diameter of the telescope in centimeters:

$$\alpha = \frac{11.6}{D}$$

For example, the resolving power of a 25-cm telescope is 11.6 divided by 25, or 0.46 second of arc. No matter how perfect the telescope optics, this is the smallest detail you can see through that telescope.

In addition to resolving power, two other factors—lens quality and atmospheric conditions—limit the detail you can see through a telescope. A telescope must contain high-quality optics to achieve its full potential resolving power. Even a large telescope reveals little detail if its optics are marred with imperfections. Also, when you look through a telescope, you are looking up through miles of turbulent air in Earth's atmosphere, which makes the image dance and blur, a condition called **seeing.** On a

Window on Science 6-1

Resolving Power and the Resolution of a Measurement

Have you ever seen a movie in which the hero magnifies a newspaper photo and reads some tiny detail? It isn't really possible, because newspaper photos are made up of tiny dots of ink, and no detail smaller than a single dot will be visible no matter how much you magnify the photo. In fact, all images are made up of elements of some sort, and that means there is a limit to the amount of detail you can see in an image. In an astronomical image, the resolution is often set by seeing. It is foolish to attempt to see a detail in the image that is smaller than the resolution.

This limitation is true of all measurements in science. A zoologist might be trying to measure the length of a live snake, or a sociologist might be trying to measure the attitudes of people toward drunk driving, but both face limits to the resolution of their measurements. The zoologist might specify that the snake was 43.28932 cm long, and the sociologist might say that 98.2491 percent of people oppose drunk driving, but a critic might point out that it isn't possible to make these measurements that accurately. The resolution of the techniques does not justify the accuracy implied by all of those digits.

Science is based on measurement, and whenever you make a measurement you should ask yourself how accurate that measurement can be. The accuracy of the measurement is limited by the resolution of the measurement technique, just as the amount of detail in a photograph is limited by its resolution.

A high-resolution image of Mars reveals details such as mountains, craters, and the southern polar cap. (NASA)

night when the atmosphere is unsteady and the images are blurred, the seeing is bad (■ Figure 6-9). Even under good seeing conditions, the detail visible through a large telescope is limited, not by its diffraction fringes, but by the air through which the telescope must look. A telescope performs better on a high mountaintop where the air is thin and steady, but even there Earth's atmosphere limits the detail the best telescopes can reveal to about 0.5 second of arc.

This limitation on the amount of information in an image is related to the limitation on the accuracy of a measurement. All measurements have some built-in uncertainty (**Window on Science 6-1**), and scientists must learn to work within those limitations.

The third and least important power of a telescope is **magnifying power,** the ability to make the image bigger. Because the amount of detail you can see is limited by the seeing conditions and the resolving power, very high magnification does not necessarily show more detail. Also, you can change the magnification by changing the eyepiece, but you cannot alter the telescope's light-gathering power or resolving power without changing the diameter of the objective lens or mirror, and that would be so expensive you might as well build a whole new telescope.

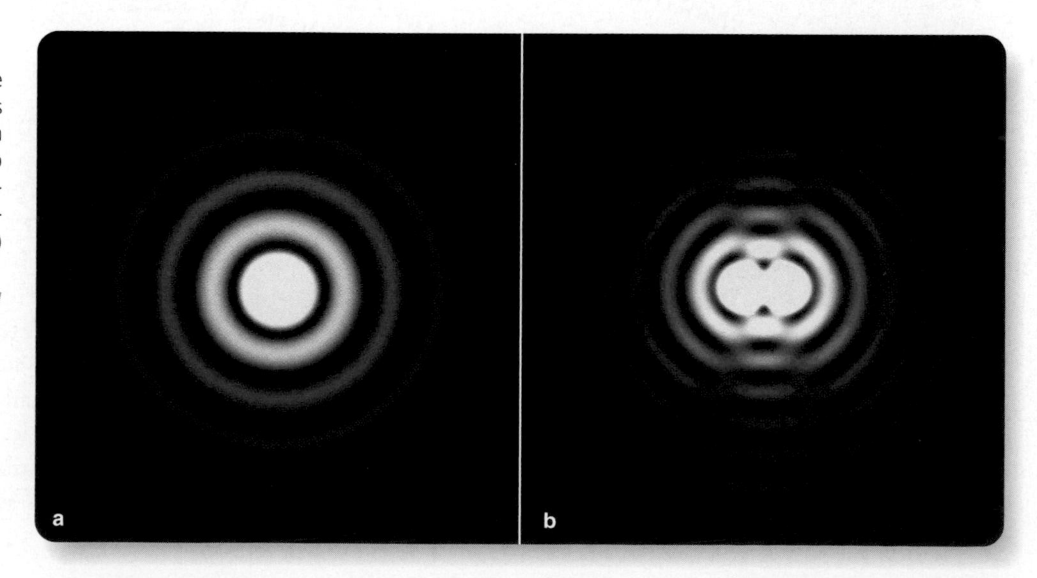

■ Active Figure 6-8

(a) Stars are so far away that their images are points, but the wave nature of light surrounds each star image with diffraction fringes (much magnified in this computer model). (b) Two stars close to each other have overlapping diffraction fringes and become impossible to detect separately. (Computer model by M. A. Seeds)

AceAstronomy™ Log into AceAstronomy and select this chapter to see Active Figure "Resolution and Telescopes." You can control telescope diameter and watch resolution change.

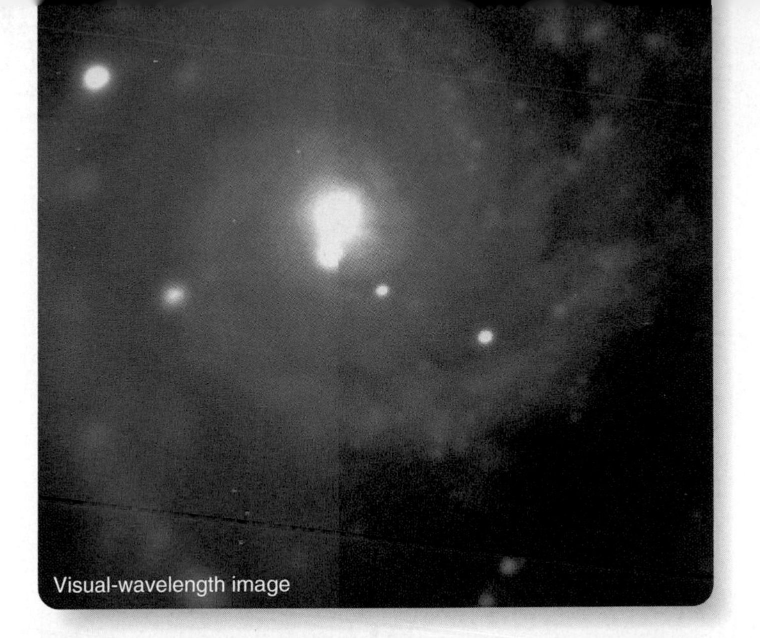

■ Figure 6-9

The left half of this photograph of a galaxy is from an image recorded on a night of poor seeing. Small details are blurred. The right half of the photo is from an image recorded on a night when Earth's atmosphere above the telescope was steady and the seeing was better. Much more detail is visible under good seeing conditions. (Courtesy William Keel)

You can calculate the magnification of a telescope by dividing the focal length of the objective by the focal length of the eyepiece:

$$M = \frac{F_o}{F_e}$$

For example, if a telescope has an objective with a focal length of 80 cm and you use an eyepiece whose focal length is 0.5 cm, the magnification is 80/0.5, or 160 times.

Notice that the two most important powers of the telescope, light-gathering power and resolving power, depend on the diameter of the telescope. This explains why astronomers refer to telescopes by diameter and not by magnification. Astronomers will refer to a telescope as an 8-meter telescope or a 10-meter telescope, but they would never identify a telescope as a 200-power telescope.

The search for light-gathering power and high resolution explains why nearly all major observatories are located far from big cities and usually on high mountains. Astronomers avoid cities because **light pollution,** the brightening of the night sky by light scattered from artificial outdoor lighting, can make it impossible to see faint objects (■ Figure 6-10). In fact, many residents of cities are unfamiliar with the beauty of the night sky because they can

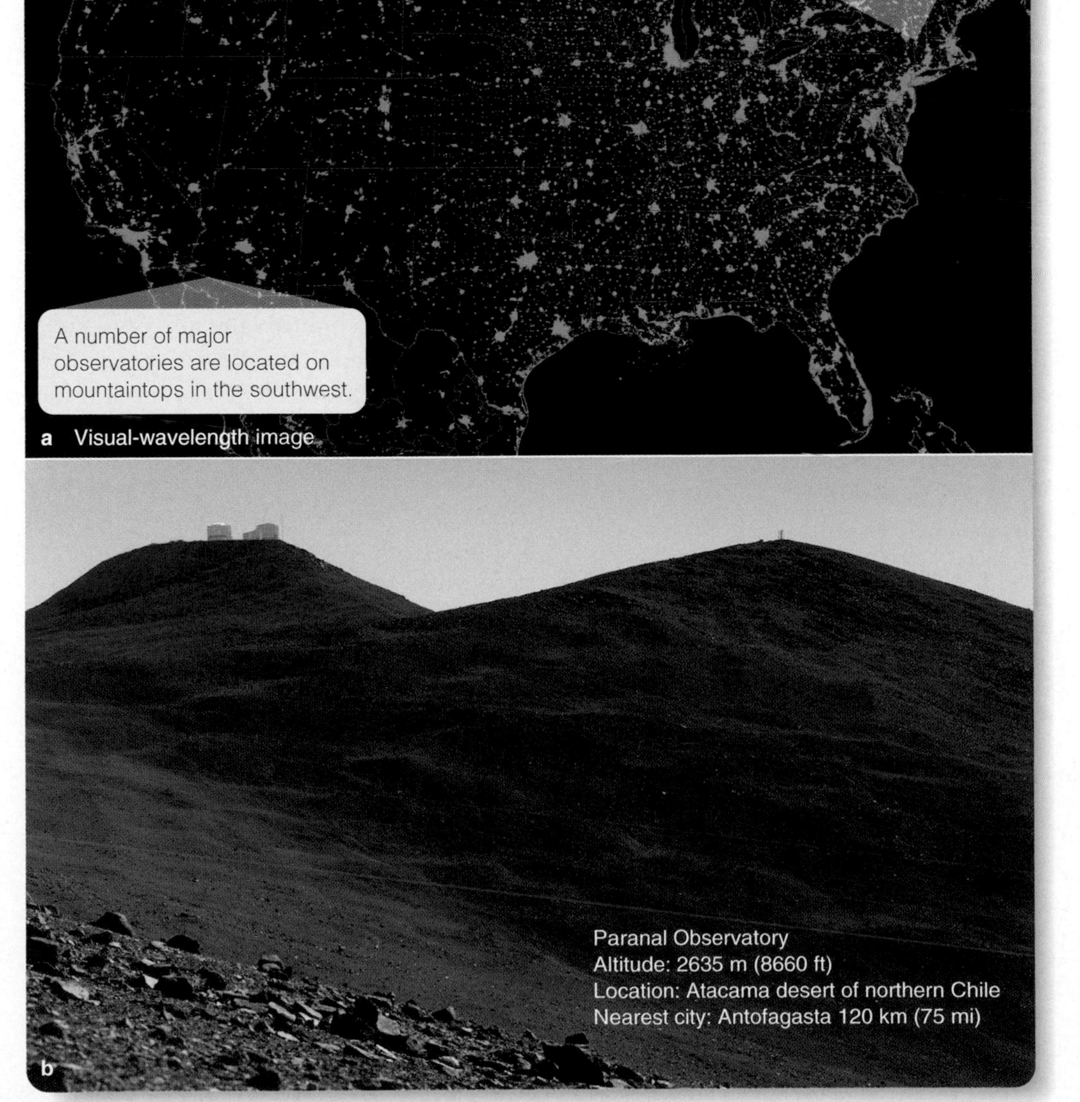

■ Figure 6-10

(a) This satellite view of the continental United States at night shows the light pollution and energy waste produced by outdoor lighting. Observatories cannot be located near large cities. (NOAA) (b) The domes of four giant telescopes are visible at upper left at Paranal Observatory, built by the European Southern Observatory. The Atacama Desert is believed to be the driest place on Earth. (ESO)

see only the brightest stars. Nevertheless, nature's own light pollution, the moon, is so bright it drowns out fainter objects, and astronomers are often unable to observe on the nights near full moon when faint objects cannot be observed even with the largest telescopes on high mountains.

Astronomers prefer to place their telescopes on carefully selected high mountains. The air there is thin and more transparent. The air is very dry at high altitudes and is more transparent to infrared radiation. Most important, astronomers select mountains where the air flows smoothly and is not turbulent. This produces the best seeing. Building an observatory on top of a high mountain far from civilization is difficult and expensive, as you can imagine from the photo in Figure 6-10, but the dark sky and steady seeing make it worth the effort.

Ace Astronomy™ Log into AceAstronomy and select this chapter to see Astronomy Exercise "Telescopes and Resolution I." What affects resolution most in the design of a telescope?

Ace Astronomy™ Log into AceAstronomy and select this chapter to see Astronomy Exercise "Telescopes and Resolution II." How does wavelength affect the resolution of a telescope?

Ace Astronomy™ Log into AceAstronomy and select this chapter to see Astronomy Exercise "Particulate, Heat, and Light Pollution." What happens when the air near a city is polluted by dust, heat, or light?

Buying a Telescope

Thinking about how to shop for a new telescope will not only help you if you decide to buy one but will also illustrate some important points about astronomical telescopes.

Assuming you have a fixed budget, you should buy the highest-quality optics and the largest-diameter telescope you can afford. Of the two things that limit what you see, optical quality is under your control. You can't make the atmosphere less turbulent, but you should buy good optics. If you buy a telescope from a toy store and it has plastic lenses, you shouldn't expect to see very much. Also, you want to maximize the light-gathering power of your telescope, so you want to purchase the largest-diameter telescope you can afford. Given a fixed budget, that means you should buy a reflecting telescope rather than a refracting telescope. Not only will you get more diameter per dollar, but your telescope will not suffer from chromatic aberration.

You can safely ignore magnification. Department stores and camera stores may advertise telescopes by quoting their magnification, but it is not an important number. What you can see is fixed by light-gathering power, optical quality, and Earth's atmosphere. Besides, you can change the magnification by changing eyepieces.

Other things being equal, you should choose a telescope with a solid mounting that will hold the telescope steady and allow it to point at objects easily. Computer-controlled pointing systems are available for a price on many small telescopes. A good telescope on a poor mounting is almost useless.

You might be buying a telescope to put in your backyard, but you must think about the same issues astronomers consider when they design giant telescopes to go on mountaintops. Designing new, giant telescopes has led astronomers to solve some traditional problems in new ways, as you will see in the next section.

New-Generation Telescopes

For most of the 20th century, astronomers faced a serious limitation on the size of astronomical telescopes. Traditional telescope mirrors were made thick to avoid sagging that would distort the reflecting surface, but those thick mirrors were heavy. The 5-m (200-in.) mirror on Mount Palomar weighs 14.5 tons. These traditional telescopes were big, heavy, and expensive.

Modern astronomers have solved these problems in a number of ways. Study **Modern Astronomical Telescopes** on pages 120–121 and notice four important advances in telescope design made possible by high-speed computers:

1. Traditional telescopes use large, solid, heavy mirrors to focus starlight.
2. Astronomers can now build simpler, lighter-weight telescope mountings and depend on computers to move the telescope and follow the westward motion of the stars as Earth rotates.
3. Notice that computer control of the shape of telescope mirrors allows the use of thin, lightweight mirrors—either "floppy" mirrors or segmented mirrors. Lowering the weight of the mirror lowers the weight of the rest of the telescope and makes it stronger and less expensive. Also, thin mirrors cool faster at nightfall and produce better images.
4. Also notice the way astronomers use high-speed computers to reduce seeing distortion caused by Earth's atmosphere. Only a few decades ago, many astronomers argued that it wasn't worth building more large telescopes on Earth's surface because of the limitations set by seeing. Now a number of new giant telescopes have been built, and more are in development that can partially overcome the seeing problem.

Did you notice that astronomical telescopes must be aligned with the north celestial pole? Polaris, the North Star, is one of our favorite stars in the list at the beginning of this book. It marks the location of the north celestial pole. Equatorial mountings have an axis that points toward Polaris, and alt-azimuth telescopes are run by computers, which align their motion with Polaris. Even telescopes in the southern hemisphere, where the north celestial pole lies below the horizon, must tip their hats toward Polaris. That's one reason Polaris deserves to be one of your favorite stars; whenever you notice Polaris in the night sky, think of all the

The LBT has the light-gathering power of a 11.8-m telescope and the resolving power of a 22.8-m telescope.

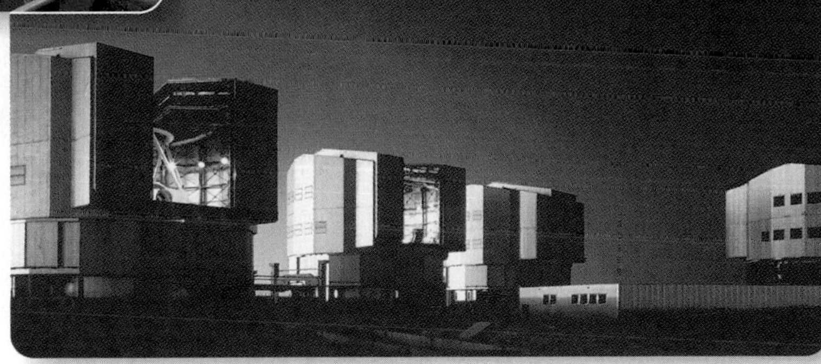

The mirrors in the VLT telescopes are 8.2 m in diameter.
The two Gemini telescopes have 8.1-m mirrors with active optics.

■ **Figure 6-11**

Astronomers have begun building multiple telescopes to solve different problems. Two Gemini telescopes have been built, one in Hawaii and one in Chile, to observe the entire sky. The Large Binocular Telescope (LBT) carries two 8.4-m mirrors that combine their light. The four telescopes of the VLT are housed in separate domes, but they can combine to work as a very large telescope. (Gemini: NOAO/AURA/NSF; LBT: Large Binocular Telescope Project and European Industrial Engineer; VLT: ESO)

astronomical telescopes in backyards and observatories all over the world that bow toward Polaris.

High-speed computers have allowed astronomers to build new, giant telescopes as shown in ■ Figure 6-11. An international collaboration of astronomers have built the Gemini telescopes with 8.1-m thin mirrors. One is located in the northern hemisphere and one in the southern hemisphere to cover the entire sky. The European Southern Observatory has built the Very Large Telescope (VLT) high in the remote Andes Mountains of northern Chile. The VLT consists of four telescopes with computer-controlled mirrors 8.2 m in diameter and only 17.5 cm (6.9 in.) thick. The four telescopes can work singly or can combine their light to work as one large telescope. Italian and American astronomers are building the Large Binocular Telescope, which carries a pair of 8.4-m mirrors on a single mounting.

High-speed computers have improved astronomical telescopes in another way that might surprise you. Computer control and data handling have made possible huge surveys of the sky in which millions of objects are observed. The Sloan Digital Sky Survey, for example, is mapping the sky, measuring the position and brightness of 100 million stars and galaxies at a number of wavelengths. The Two-Micron All Sky Survey (2MASS) has mapped the entire sky at three wavelengths in the infrared. Other surveys are being made at many other wavelengths. Every night large telescopes scan the sky, and billions of bytes of data are compiled automatically in immense sky atlases. Astronomers will study those data banks for decades to come.

The days when astronomers worked beside their telescopes through long, dark, cold nights are nearly gone. The complexity and sophistication of telescopes require a battery of computers, and almost all research telescopes are run from control rooms that astronomers call "warm rooms." Astronomers don't need to be kept warm, but computers demand comfortable working conditions (■ Figure 6-12).

■ **Figure 6-12**

In the control room of the 4-meter telescope atop Kitt Peak National Observatory, the telescope operator at left manages the operation and safety of the telescope. The astronomer at right operates the instruments, records data, and makes decisions on the observing program. Astronomers work through the night controlling the computers that control the telescope and its instruments. (NOAO/AURA/NSF)

Modern Astronomical Telescopes

1 The traditional telescopes described on this page are limited by complexity, weight, and Earth's atmosphere. Modern solutions are shown on the opposite page.

In larger telescopes the light can be focused to a **prime focus** position high in the telescope tube as shown at the right. Although it is a good place to image faint objects, the prime focus is inconvenient for large instruments. A **secondary mirror** can reflect the light through a hole in the primary mirror to a **Cassegrain focus.** This focal arrangement may be the most common form of astronomical telescope.

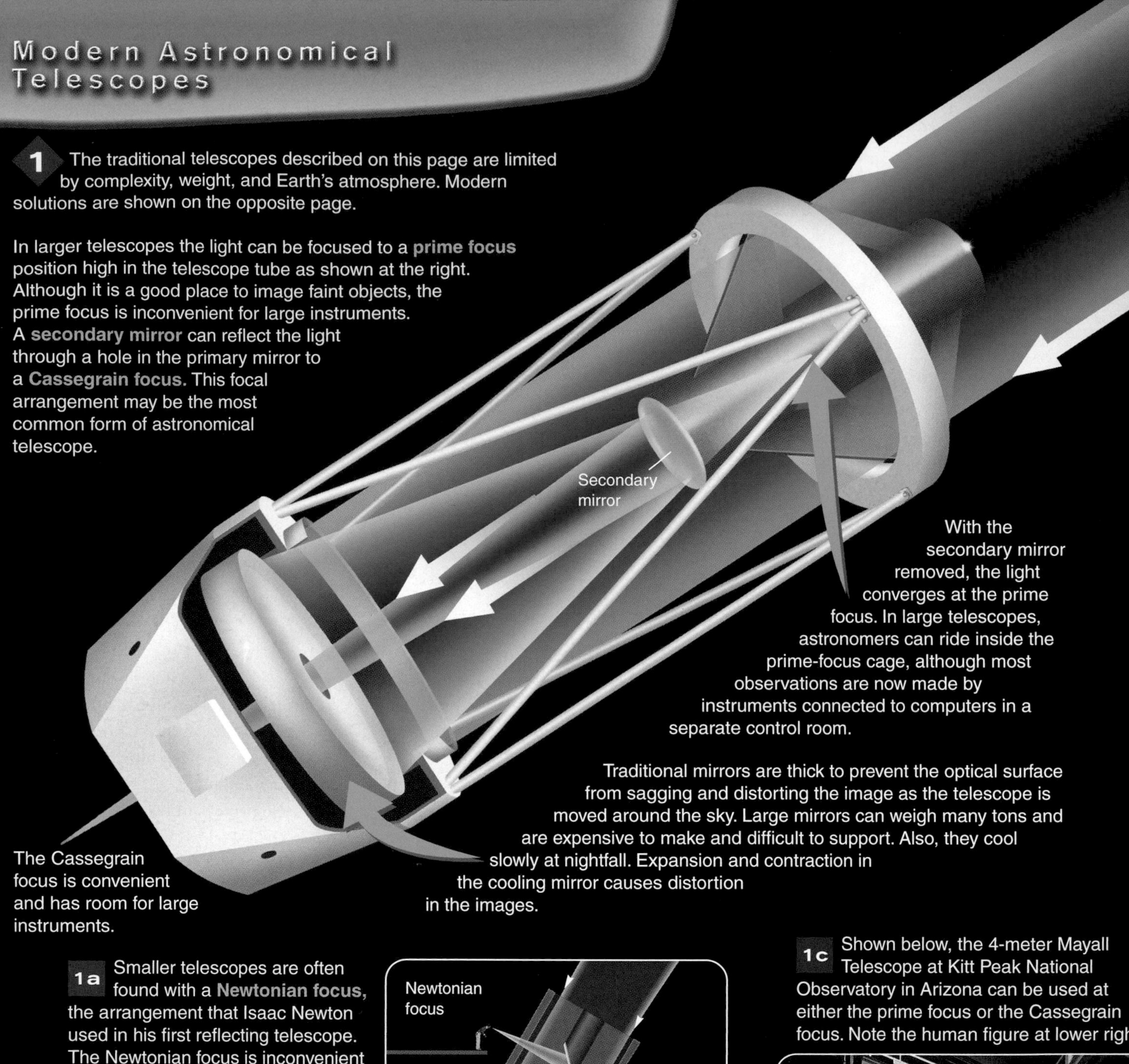

With the secondary mirror removed, the light converges at the prime focus. In large telescopes, astronomers can ride inside the prime-focus cage, although most observations are now made by instruments connected to computers in a separate control room.

Traditional mirrors are thick to prevent the optical surface from sagging and distorting the image as the telescope is moved around the sky. Large mirrors can weigh many tons and are expensive to make and difficult to support. Also, they cool slowly at nightfall. Expansion and contraction in the cooling mirror causes distortion in the images.

The Cassegrain focus is convenient and has room for large instruments.

1a Smaller telescopes are often found with a **Newtonian focus,** the arrangement that Isaac Newton used in his first reflecting telescope. The Newtonian focus is inconvenient for large telescopes as shown at right.

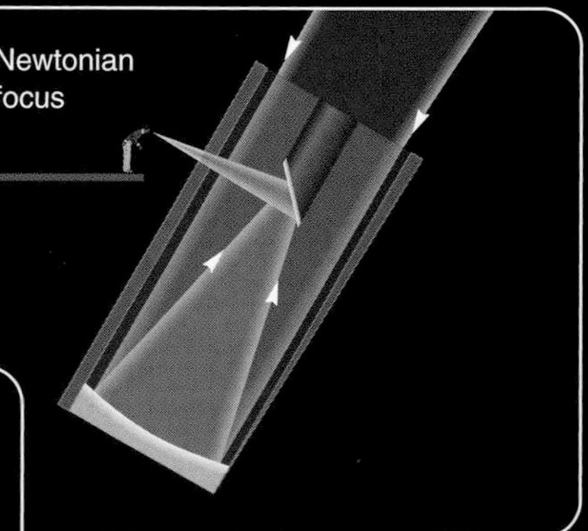

Thin correcting lens

Schmidt-Cassegrain telescope

1b Many small telescopes such as the one on your left use a **Schmidt-Cassegrain focus.** A thin correcting plate improves the image but is too slightly curved to introduce serious chromatic aberration.

1c Shown below, the 4-meter Mayall Telescope at Kitt Peak National Observatory in Arizona can be used at either the prime focus or the Cassegrain focus. Note the human figure at lower right.

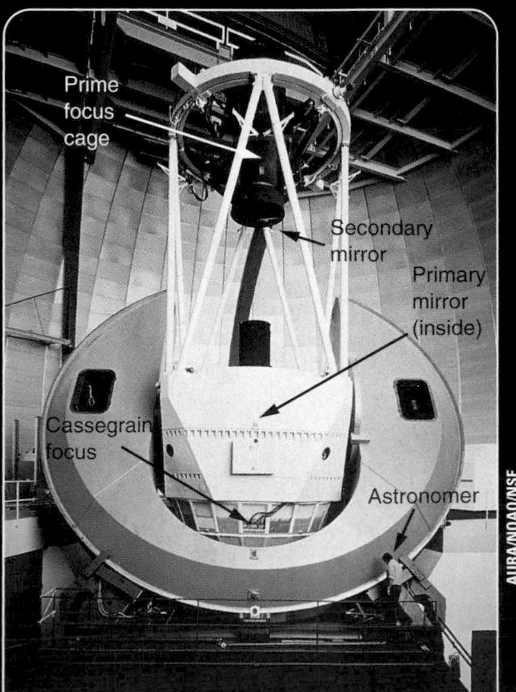

AURA/NOAO/NSF

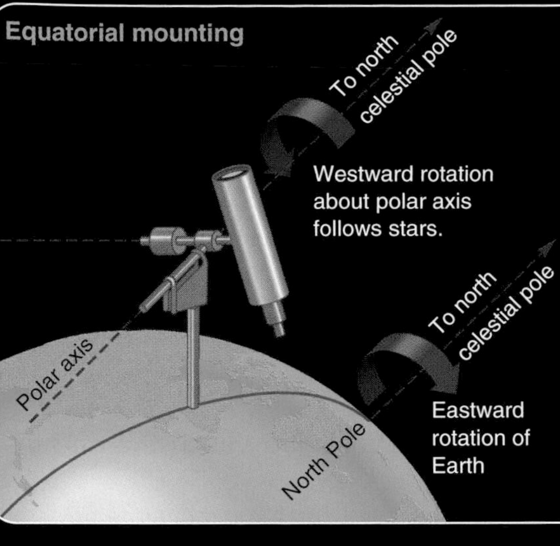

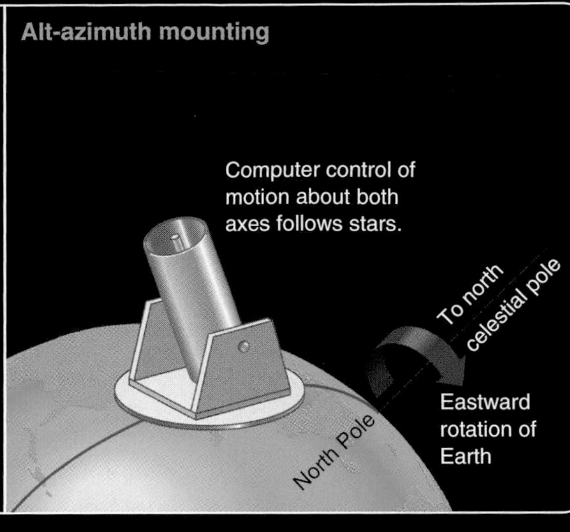

2 Telescope mountings must contain a **sidereal drive** to move smoothly westward and counter the eastward rotation of Earth. The traditional **equatorial mounting** (far left) has a **polar axis** parallel to Earth's axis, but the modern **alt-azimuth mounting** (near left) moves like a cannon — up and down and left to right. Such mountings are simpler to build but need computer control to follow the stars.

3 Unlike traditional thick mirrors, thin mirrors, sometimes called floppy mirrors as shown at right, weigh less and require less massive support structures. Also, they cool rapidly at nightfall and there is less distortion from uneven expansion and contraction.

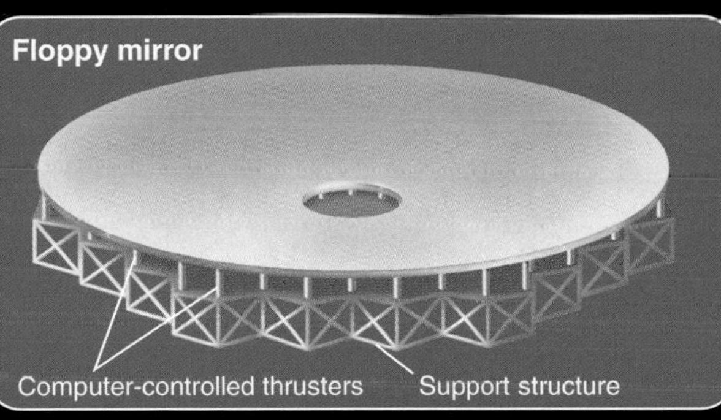

3a Grinding a large mirror may remove tons of glass and take months, but new techniques speed the process. Some large mirrors are cast in a rotating oven that causes the molten glass to flow to form a concave upper surface. Grinding and polishing such a preformed mirror is much less time consuming.

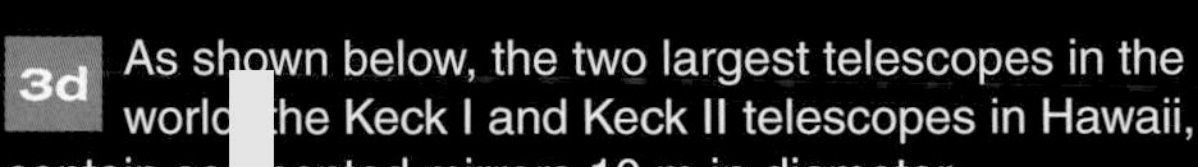

3b Mirrors made of segments are economical because the segments can be made separately. The resulting mirror weighs less and cools rapidly. See image at right.

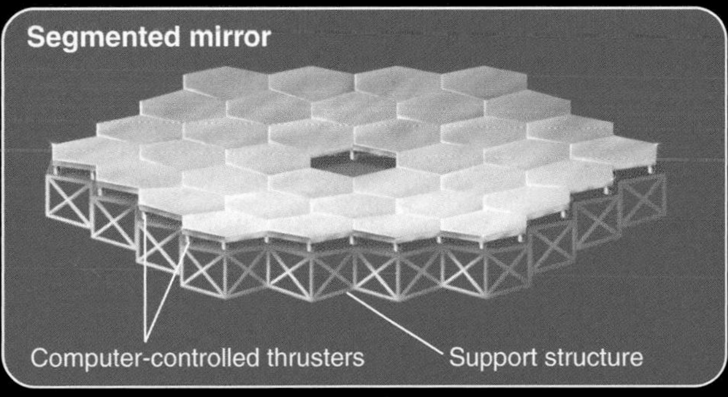

3c Both floppy mirrors and segmented mirrors sag under their own weight. Their optical shape must be controlled by computer-driven thrusters under the mirror in what is called **active optics.**

3d As shown below, the two largest telescopes in the world, the Keck I and Keck II telescopes in Hawaii, contain segmented mirrors 10 m in diameter.

W. M. Keck Observatory

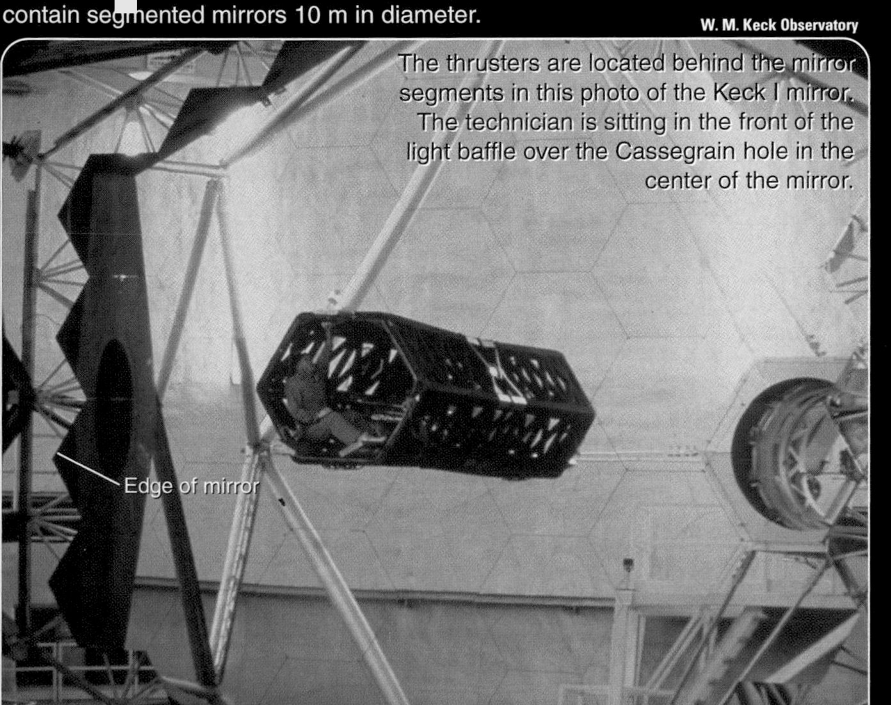

The thrusters are located behind the mirror segments in this photo of the Keck I mirror. The technician is sitting in the front of the light baffle over the Cassegrain hole in the center of the mirror.

Adaptive optics in telescopes

Adaptive optics off — Object appears to be a single star.

Adaptive optics on — Object revealed as a pair of stars.

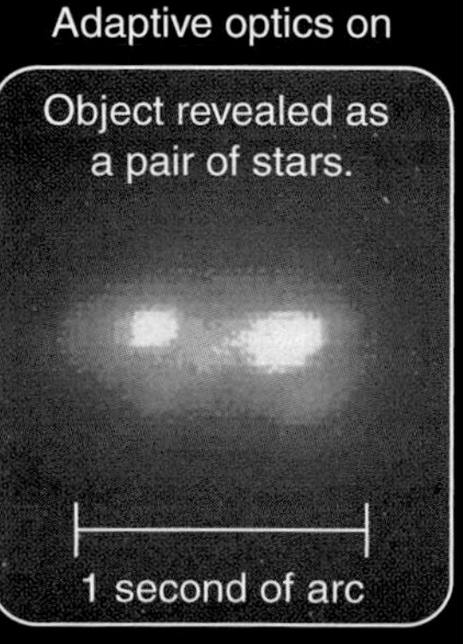

Paul Kalas

4 **Adaptive optics** uses high-speed computers to monitor the image distortion caused by Earth's atmosphere and adjust the optics many times a second to compensate. This can reduce the blurring due to seeing and dramatically improve image quality in Earth based telescopes.

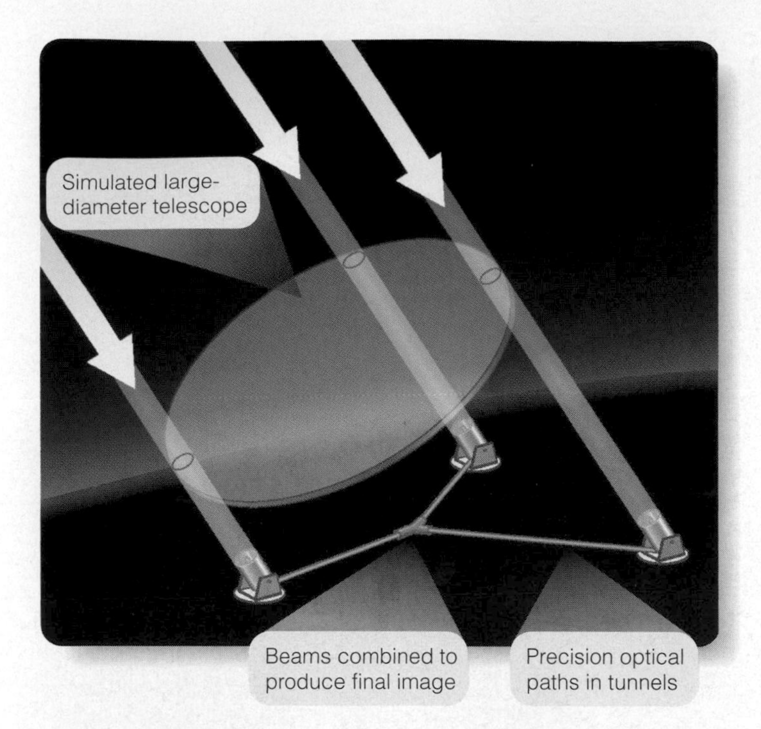

■ **Figure 6-13**

In an astronomical interferometer, smaller telescopes can combine their light through specially designed optical tunnels to simulate a larger telescope with a resolution set by the separation of the smaller telescopes.

Interferometry

One of the reasons astronomers build big telescopes is to increase resolving power, and astronomers have been able to achieve very high resolution by connecting multiple telescopes together to work as if they were a single telescope. This method of synthesizing a larger telescope is known as **interferometry** (■ Figure 6-13).

To work as an interferometer, the separate telescopes must combine their light through a network of mirrors, and the path that each light beam travels must be controlled so that it does not vary more than some small fraction of the wavelength. Turbulence in Earth's atmosphere constantly distorts the light, and high-speed computers must continuously adjust the light paths. Recall that the wavelength of light is very short, roughly 0.0005 mm, so building optical interferometers is one of the most difficult technical problems that astronomers face. Infrared- and radio-wavelength interferometers are slightly easier to build because the wavelengths are longer. In fact, as you will discover later in this chapter, the first astronomical interferometers worked at radio wavelengths.

The VLT shown in Figure 6-11 consists of four 8.2-m telescopes that can operate separately, but they can be linked together through underground tunnels with three 1.8-m telescopes on the same mountaintop. The resulting optical interferometer provides the resolution of a telescope 200 meters in diameter. Other telescopes can work as interferometers. The two Keck 10-m telescopes can be used as an interferometer. The CHARA array on Mt. Wilson combines six 1-meter telescopes to create the equivalent of a telescope one-fifth of a mile in diameter. The Large Binocular Telescope shown in Figure 6-11 can be used as an interferometer.

Although turbulence in Earth's atmosphere can be partially averaged out in an interferometer, plans are being made to put interferometers in space. The Space Interferometry Mission, for example, will work at visual wavelengths and study everything from the cores of erupting galaxies to planets orbiting nearby stars.

Building Scientific Arguments

Why do astronomers build observatories at the tops of mountains?

To build this argument you need to think about the powers of a telescope. Astronomers have joked that the hardest part of building a new observatory is constructing the road to the top of the mountain. It certainly isn't easy to build a large, delicate telescope at the top of a high mountain, but it is worth the effort. A telescope on top of a high mountain is above the thickest part of Earth's atmosphere. There is less air to dim the light, and there is less water vapor to absorb infrared radiation. Even more important, the thin air on a mountaintop causes less disturbance to the image, and consequently the seeing is better. A large telescope on Earth's surface has a resolving power much better than the distortion caused by Earth's atmosphere. So it is limited by seeing, not by its own diffraction. It really is worth the trouble to build telescopes atop high mountains.

Astronomers not only build telescopes on mountaintops, they also build gigantic telescopes many meters in diameter. Revise your argument to focus on telescope design. **What are the problems and advantages in building such giant telescopes?**

■ ■ ■

Connections: Astronomers sometimes refer to a telescope that produces distorted images as a "light bucket." In a sense, all astronomical telescopes are light buckets because the light they focus into images reveals very little until it is recorded and analyzed by special instruments attached to the telescopes.

6-3 Special Instruments

LOOKING THROUGH A TELESCOPE doesn't tell you much. A star looks like a point of light. A planet looks like a little disk. A galaxy looks like a hazy patch. To use an astronomical telescope to learn about the universe, you must be able to analyze the light the telescope gathers. Special instruments attached to the telescope make that possible.

Galaxy NGC 891 in true color. It is edge-on and contains thick dust clouds.
In this image, color shows brightness. White and red are brightest, and yellow and green are dimmer.
Visual-wavelength image
Visual image in false color
In these negative images of NGC 891, the sky is white and the stars are black.
Visual-wavelength negative images

■ Figure 6-14

Astronomical images can be manipulated in many ways to bring out details. The photo of the galaxy at upper left is dark, and the details of the dust clouds in the disk of the galaxy do not show well. The two negative images of the galaxy have been produced to show the dust clouds more clearly. (C. Hawk, B. Savage, N. A. Sharp NOAO/WIYN/NSF) The image at upper right shows two interacting galaxies known as Arp 273. The visual-wavelength image has been given false color according to brightness. (NOAO/WIYN/NSF)

Imaging Systems

The original imaging device in astronomy was the photographic plate. It could record faint objects in long time exposures and could be stored for later analysis. But photographic plates have been almost entirely replaced in astronomy by electronic imaging systems.

Most modern astronomers use **charge-coupled devices (CCDs)** to record images. A CCD is a specialized computer chip containing roughly a million microscopic light detectors arranged in an array about the size of a postage stamp. These devices can be used like small photographic plates, but they have dramatic advantages. They can detect both bright and faint objects in a single exposure, are much more sensitive than photographic plates, and can be read directly into computer memory for later analysis. Although CCDs for astronomy are extremely sensitive and therefore expensive, less sophisticated CCDs are used in video and digital cameras.

The image from a CCD is stored as numbers in computer memory, so it is easy to manipulate the image to bring out details that would not otherwise be visible. For example, astronomical images are often reproduced as negatives with the sky white and the stars dark. This makes the faint parts of the image easier to see (■ Figure 6-14). Astronomers also manipulate images to produce **false-color images** in which the colors represent different levels of intensity and are not related to the true colors of the object. You can see an example in Figure 6-14. In fact, false-color images are common in many fields (**Window on Science 6-2**).

Measurements of intensity and color were made in the past using a photometer, a highly sensitive light meter attached to a telescope. Today, however, most such measurements are made directly on CCD images. Because the CCD image is easily digitized, brightness and color can be measured more easily and more accurately than on photographic plates.

The Spectrograph

To analyze light in detail, astronomers need to spread the light out according to wavelength to form a spectrum, a task performed by a **spectrograph.** You can understand how this works if you imagine reproducing an experiment performed by Isaac Newton in 1666. Newton bored a small hole in the window shutter of his bedroom to admit a thin beam of sunlight. When he placed a prism in the beam, it spread the light into a beautiful spectrum splashed across his bedroom wall. From this Newton concluded that white light was made of a mixture of all the colors.

Newton didn't think in terms of wavelength, but you can use that modern concept to see that the light passing through the prism is bent at an angle that depends on the wavelength. Violet (short wavelength) bends most, and red (long wavelength) least. In this way, the white light entering the prism is spread into a spectrum (■ Figure 6-15). You could build a spectrograph with a prism to spread the light and a lens to guide the light into a camera.

Window on Science 6-2

False-Color Images and Reality

Astronomical images are usually recorded in digital form directly into computer memory, so astronomers often add false color to images to reveal things otherwise hard to see. They might exaggerate colors already present, but more often they add colors to represent levels of brightness. Of course, the colors used are entirely arbitrary. One astronomer might like shades of blue, and another might use all of the colors ranging from violet to red. To understand the image, you need to know the meaning of the colors—that white and red are bright areas, for instance, and that yellow and green are dimmer.

Astronomers also use false color to create images recorded at wavelengths not visible to the human eye. Radio, infrared, ultraviolet, and X-ray images are often displayed as false-color images. You might think of them as maps in which the colors show the relative intensity of the radiation in different parts of the image. Again the choice of colors is up to the astronomer analyzing the image.

False-color images are common in astronomy, but they are also used in other sciences. Doctors often analyze medical X-ray images, CAT scans, and so on by converting the images to false color. Biologists use false color to analyze microscopic photographs, and geologists use false color to study photographs of Earth recorded by a satellite at various wavelengths.

Even a simple photograph of a galaxy or a glowing cloud of gas in space is a kind of false-color image. The object is much too faint to see with the human eye, and even if you flew in a spaceship to within a short distance, it would still be too faint to see with your eyes. The photograph you see is a time exposure in which the light was allowed to accumulate over an extended period ranging from seconds to hours. That makes the galaxy or gas cloud visible. Even a black-and-white photograph of a glowing cloud of gas shows an aspect of the universe you can never see with unaided eyes. In that sense, all astronomical images are false-color images. Their virtue is not that they are colorful but that they are meaningful.

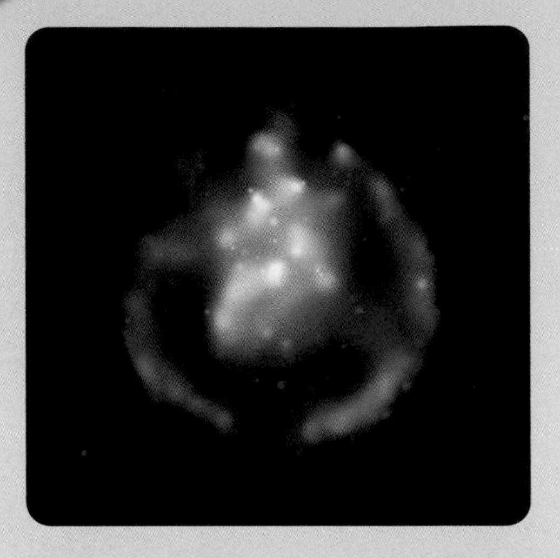

A false-color X-ray image of the expanding cloud of gas left behind by the explosion of a massive star. (NASA/CXC/Rutgers/J. Hughes)

Nearly all modern spectrographs use a grating in place of a prism. A **grating** is a piece of glass with thousands of microscopic parallel grooves scribed onto its surface. Different wavelengths of light reflect from the grating at slightly different angles, so white light is spread into a spectrum. You have probably noticed this effect when you look at the closely spaced lines etched onto a compact disk; as you move the disk about, different colors flash across its surface. You could build a modern spectrograph by using a high-quality grating to spread the light into a spectrum and a CCD camera to record the spectrum.

The spectrum of an astronomical object can contain hundreds of spectral lines produced by the atoms in the object. Because astronomers must identify these lines and measure the wavelengths, they use a **comparison spectrum** as a calibration of their spectrograph. Special bulbs built into the spectrograph produce bright lines given off by such atoms as thorium and argon or neon. The wavelengths of these spectral lines have been measured to high precision in the laboratory, so astronomers can use spectra of these light sources as guides to measure wavelengths and identify spectral lines in the spectrum of a star, galaxy, or planet.

Because astronomers understand how light interacts with matter, a spectrum carries a tremendous amount of information (as you will see in the next chapter), and that makes a spectrograph the astronomer's most powerful instrument. An astronomer recently remarked, "We don't know anything about an object till we get a spectrum," and that is only a slight exaggeration.

Building Scientific Arguments

What is the difference between light going through a lens and light passing through a prism?

When you think about natural processes, it is often helpful to compare similar things and scientific arguments often make such comparisons. A few simple rules explain most natural events, so the similarities are often revealing. A refracting telescope producing chromatic aberration and a prism dispersing light into a spectrum are two examples of the same thing, but one is bad and one is good. When light passes through the curved surfaces of a lens, different wavelengths are bent by slightly different amounts, and the different colors of light come to focus at different focal lengths. This produces the color fringes in an image called chromatic aberration, and that's bad. But the surfaces of a prism are made to be precisely flat, so all of the light enters the prism at the same angle, and any given wavelength is bent by the same amount. Consequently, white light is dispersed into a spectrum. You could call the dispersion of light by a prism "controlled chromatic aberration," and that's good.

Now you can build your own argument comparing similar things. CCDs have been very good for astronomy, and they have almost completely replaced photographic plates. **How are CCD chips similar to photographic plates, and how are they better?**

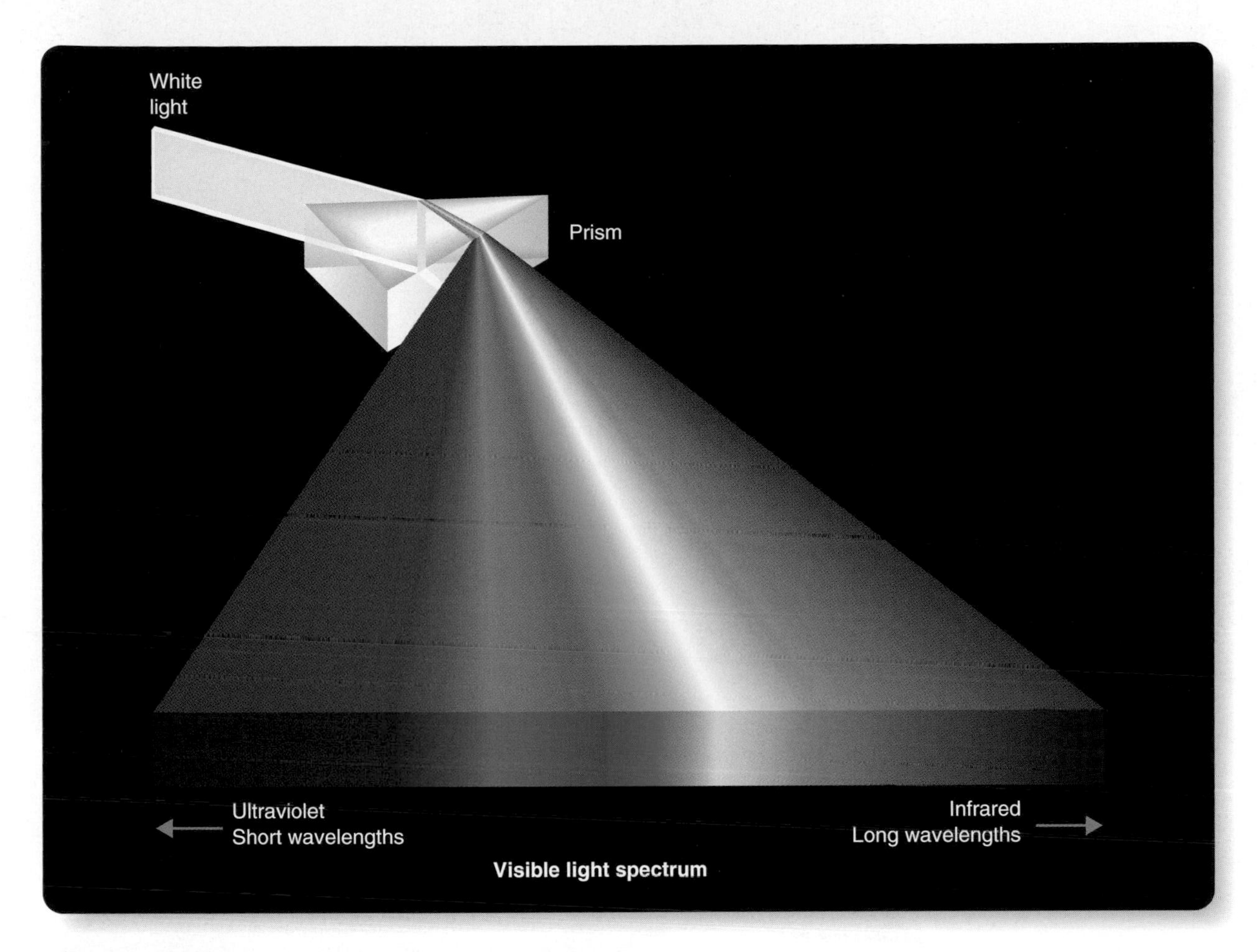

■ Figure 6-15

A prism bends light by an angle that depends on the wavelength of the light. Short wavelengths bend most and long wavelengths least. Thus, white light passing through a prism is spread into a spectrum.

■ ■ ■

Connections: So far, this discussion has been limited to visual wavelengths. Now it is time for you to explore the rest of the electromagnetic spectrum.

6-4 Radio Telescopes

ALL THE TELESCOPES and instruments you have discussed so far look out through the visible light window in Earth's atmosphere, but there is another window running from a wavelength of 1 cm to about 1 m (see Figure 6-3). By building the proper kinds of instruments, astronomers can study the universe through this radio window.

Operation of a Radio Telescope

A radio telescope usually consists of four parts: a dish reflector, an antenna, an amplifier, and a recorder (■ Figure 6-16). The components, working together, make it possible for astronomers to detect radio radiation from celestial objects.

The dish reflector of a radio telescope, like the mirror of a reflecting telescope, collects and focuses radiation. Because radio waves are much longer than light waves, the dish need not be as smooth as a mirror; wire mesh will reflect all but the shortest wavelength radio waves. But don't be surprised if you see a photo of a radio telescope that doesn't have a dish shape. There are a number of ways to collect radio energy from the sky. Nevertheless, dish-shaped reflectors are very common at radio observatories.

Though a radio telescope's dish may be many meters in diameter, the antenna may be as small as your hand. Like the antenna on a TV set, its only function is to absorb the radio energy and direct it along a cable to an amplifier. After amplification, the signal goes to some kind of recording instrument. Most radio observatories record data directly into computer memory.

However it is recorded, an observation with a radio telescope measures the amount of radio energy coming from a specific point on the sky, and that causes two problems. For one thing, the intensity at one spot doesn't tell you much, so the radio telescope must be scanned over an object, a cloud of gas, for example, to produce a map of the radio intensity at different points. The

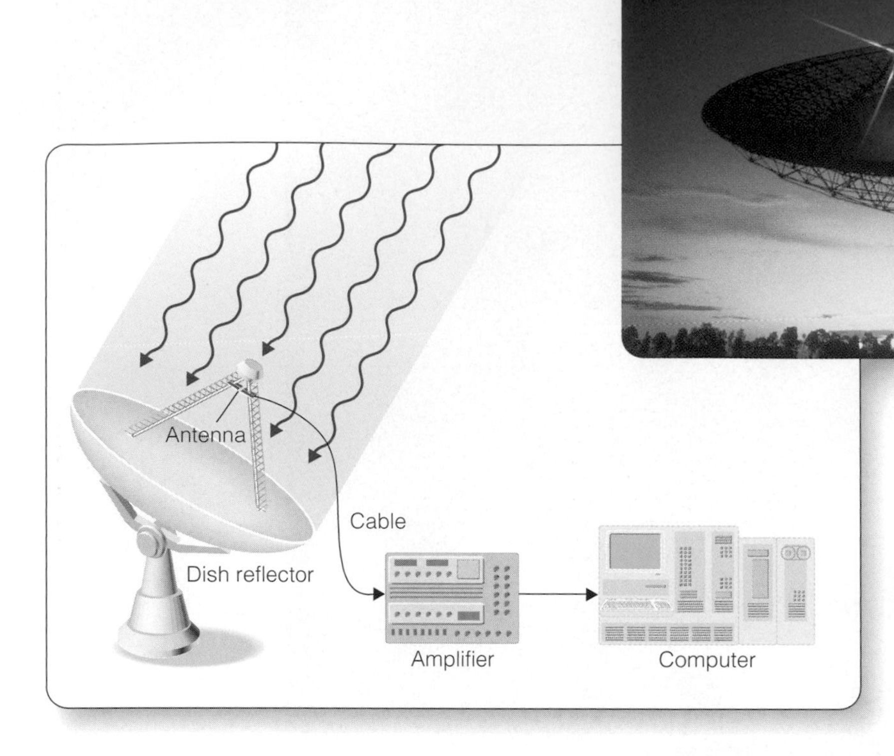

■ Figure 6-16

In most radio telescopes, a dish reflector concentrates the radio signal on the antenna. The signal is then amplified and recorded. For all but the shortest radio waves, wire mesh is an adequate reflector (photo). (Courtesy Seth Shostak/SETI Institute)

Seat prices in a baseball stadium
Red most expensive
Violet least expensive

a

■ Figure 6-17

(a) A contour map of a baseball stadium shows regions of similar admission prices. The most expensive seats are those behind home plate. (b) A false-color-image radio map of Tycho's supernova remnant, the expanding shell of gas produced by the explosion of a star in 1572. The radio contour map has been color-coded to show intensity. (Courtesy NRAO)

Radio energy map
Red strongest
Violet weakest

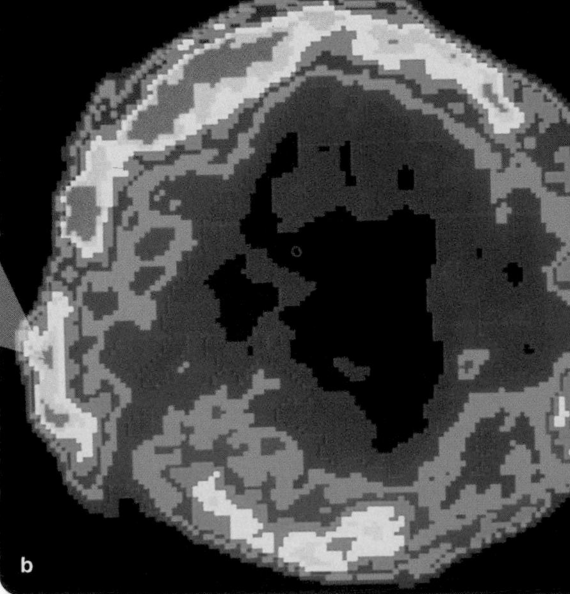

second problem is that humans can't see radio waves, so astronomers draw maps in which contours mark areas of similar radio intensity. You could compare such a map to a seating diagram for a baseball stadium in which the contours mark areas in which the seats have the same price (■ Figure 6-17a). Contour maps are very common in radio astronomy and are often reproduced using false colors (Figure 6-17b)

Limitations of a Radio Telescope

A radio astronomer works under three handicaps: poor resolution, low intensity, and interference. You remember that the resolving power of an optical telescope depends on the diameter of the objective lens or mirror. It also depends on the wavelength of the radiation. At very long wavelengths, like those of radio waves, images become fuzzy because the diffraction fringes are very large. As with an optical telescope, there is no way to improve the resolving power without building a bigger telescope. Consequently, radio telescopes must be quite large.

Even so, the resolving power of a radio telescope is not good. A dish 30 m in diameter receiving radiation with a wavelength of 21 cm has a resolving power of about 0.5°. Such a radio telescope would be unable to reveal any details in the sky smaller than the moon. Fortunately, radio astronomers can combine two or more radio telescopes to form a **radio interferometer** capable of much higher resolution. For example, the Very Large Array (VLA) consists of 27 dish antennas spread in a Y-shape across the New Mexico desert (■ Figure 6-18). In combination, they have the resolving power of a radio telescope 36 km (22 mi) in diameter. The VLA can resolve details smaller than 1 second of arc. Eight new dish antennas being added across New Mexico will give the VLA 10 times better resolving power. Another large radio interferometer, the Very Long Baseline Array (VLBA), consists of matched radio dishes spread from Hawaii to the Virgin Islands and has an effective diameter almost as large as Earth.

■ Figure 6-18

The Very Large Array uses 27 radio dishes, which can be moved to different positions along a Y-shaped set of tracks across the New Mexico desert. They are shown here in the most compact arrangement. Signals from the dishes are combined to create very-high-resolution radio maps of celestial objects. (NRAO/AUI/NSF)

The second handicap radio astronomers face is the low intensity of the radio signals. You saw earlier that the energy of a photon depends on its wavelength. Photons of radio energy have such long wavelengths that their individual energies are quite low. In order to get strong signals focused on the antenna, the radio astronomer must build large collecting dishes.

The largest fully steerable radio telescope in the world is at the National Radio Astronomy Observatory in Green Bank, West Virginia (■ Figure 6-19a). The telescope has a reflecting surface 100 meters in diameter, big enough to hold an entire football field, and can be pointed anywhere in the sky. Its surface consists of 2004 computer-controlled panels that adjust to maintain the shape of the reflecting surface.

The largest radio dish in the world is 300 m (1000 ft) in diameter. So large a dish can't be supported in the usual way, so it is built into a mountain valley in Arecibo, Puerto Rico.

■ Figure 6-19

(a) The largest steerable radio telescope in the world is the GBT located in Green Bank, West Virginia. With a diameter of 100 m, it stands higher than the Statue of Liberty. (Mike Bailey: NRAO/AUII) (b) The 300-m (1000-ft) radio telescope in Arecibo, Puerto Rico, hangs from cables over a mountain valley. The Arecibo Observatory is part of the National Astronomy and Ionosphere Foundation operated by Cornell University and the National Science Foundation. (David Parker/SPL/Photo Researchers, Inc.)

The reflecting dish is a thin metallic surface supported above the valley floor by cables attached near the rim, and the antenna hangs above the dish on cables from three towers built on three mountain peaks that surround the valley (Figure 6-19b).

Although this telescope can look only overhead, the operators can change its aim slightly by moving the antenna and by waiting for Earth's rotation to point the telescope in the proper direction. This may sound clumsy, but the telescope's ability to detect weak radio sources, together with its good resolution, makes it one of the most important radio observatories in the world.

The third handicap the radio astronomer faces is interference. A radio telescope is an extremely sensitive radio receiver listening to radio signals thousands of times weaker than artificial radio and TV transmissions. Such weak signals are easily drowned out by interference. Sources of such interference include everything from poorly designed transmitters in Earth satellites to automobiles with faulty ignition systems. To avoid this kind of interference, radio astronomers locate their telescopes as far from civilization as possible. Hidden deep in mountain valleys, they are able to listen to the sky protected from human-made radio noise.

Advantages of a Radio Telescope

Building large radio telescopes in isolated locations is expensive, but three factors make it all worthwhile. First, and most important, a radio telescope can reveal clouds of cool hydrogen in space. Because 90 percent of the atoms in the universe are hydrogen, that is important information. Large clouds of cool hydrogen are completely invisible to normal telescopes because they produce no visible light of their own and reflect too little to be detected on photographs. However, cool hydrogen emits a radio signal at the specific wavelength of 21 cm. (You will see how the hydrogen produces this radiation in the discussion of the gas clouds in space in Chapter 10.) These hydrogen clouds are important because, for one thing, they are the places where stars are born. The only way astronomers can detect these clouds of gas is with a radio telescope that receives the 21-cm radiation, so that is one reason that radio telescopes are important.

The second reason is related to dust in space. Because radio signals have relatively long wavelengths, they can penetrate the vast clouds of dust that obscure astronomers' view at visual wavelengths. Light waves are short, and they interact with tiny dust grains floating in space; as a result, the light is scattered and never gets through the dust to reach optical telescopes on Earth. However, radio signals from far across the galaxy pass unhindered through the dust, giving radio astronomers an unobscured view.

Finally, radio telescopes are important because they can detect objects that are more luminous at radio wavelengths than at visible wavelengths. This includes everything from the coldest clouds of gas to the hottest stars. Some of the most distant objects in the universe, for instance, are detectable only at radio wavelengths.

Building Scientific Arguments

Why do optical astronomers build big telescopes, while radio astronomers build groups of widely separated smaller telescopes?

Once again you can learn a lot by building a scientific argument based on comparison. Optical astronomers build large telescopes to maximize light-gathering power, but the problem for radio telescopes is resolving power. Because radio waves are so much longer than light waves, a single radio telescope can't see details in the sky much smaller than the moon. By linking radio telescopes miles apart, radio astronomers build a radio interferometer that can simulate a radio telescope miles in diameter and thus increase the resolving power.

The difference between the wavelengths of light and radio waves makes a big difference in building the best telescopes. Keep that difference in mind as you build a new argument: **Why don't radio astronomers want to build their telescopes on mountaintops as optical astronomers do?**

■ ■ ■

Connections: Earth's atmosphere causes trouble for astronomers in two ways. It distorts images, and it absorbs radiation at many wavelengths. The only way to avoid these limitations completely is to send telescopes above the atmosphere, into space.

6-5 Astronomy from Space

YOU HAVE LEARNED about the observations that ground-based telescopes can make through the two atmospheric windows in the visible and radio parts of the electromagnetic spectrum. Most of the rest of the electromagnetic radiation—infrared, ultraviolet, X ray, and gamma ray—never reaches Earth's surface. To observe at these wavelengths, telescopes must fly above the atmosphere in high-flying aircraft, rockets, balloons, and satellites. The only exceptions are observations that can be made in the near-infrared and the near-ultraviolet.

The Ends of the Visual Spectrum

Astronomers can observe in the near-infrared just beyond the red end of the visible spectrum. You can't see this light, but some of it leaks through the atmosphere in narrow, partially open atmospheric windows scattered from 1200 nm to about 40,000 nm. Infrared astronomers usually measure wavelength in micrometers (10^{-6} meters), so they refer to this wavelength range as 1.2 to 40 micrometers (or microns for short). In this range, much of the radiation is absorbed by water vapor, but carbon dioxide and oxygen molecules also absorb infrared. As you saw earlier in this chapter, it is an advantage to place telescopes on mountaintops

Infrared astronomers can often observe with the dome lights on. Their instruments are not usually sensitive to visible light.
SOFIA will fly at roughly 12 km (over 40,000 ft) to get above most of Earth's atmosphere.
NASA DLR
SOFIA
Adding liquid nitrogen to the camera on a telescope is a familiar task for astronomers.

■ Figure 6-20

Comet Hale–Bopp hangs in the sky over the 3-meter NASA Infrared Telescope Facility (IRAF) atop Mauna Kea. The air at high altitudes is so dry that it is transparent to shorter infrared photons. SOFIA will fly so high it will be able to observe infrared wavelengths that cannot be observed from mountaintops. Most astronomical CCD cameras must be cooled to low temperatures, and this is especially true for infrared cameras. (Hale–Bopp over observatory: William Keel/IRAF; SOFIA: NASA; Camera: Kris Koenig/Coast Learning Systems)

where the air is thin and dry. For example, a number of important infrared telescopes observe from the 4150-m (13,600-ft) summit of Mauna Kea in Hawaii. At this altitude, the telescopes are above much of the water vapor in Earth's atmosphere (■ Figure 6-20).

The far-infrared range, which includes wavelengths longer than 40 micrometers, carries clues to the nature of comets, planets, forming stars, and other cool objects, but these wavelengths are absorbed high in Earth's atmosphere—much higher than mountaintops. Infrared telescopes have flown to high altitudes under balloons and in airplanes. NASA is now building the Stratospheric Observatory for Infrared Astronomy (SOFIA), a Boeing 747 that will carry a 2.5-m telescope, control systems, and a team of technicians and astronomers to the fringes of the atmosphere. Once at that altitude, they can open a door above the telescope and make extended infrared observations as the plane flies a precisely calculated path. You can see the door in the photo in Figure 6-20. Some infrared wavelengths are too strongly absorbed by the atmosphere to be observed from even the highest-flying aircraft. The only way to make those infrared observations is to put the telescope in space above the atmosphere.

Wherever infrared telescopes are based, they have one thing in common—cooling. If a telescope observes at far-infrared wavelengths, then it must be cooled to low temperature. Infrared radiation is emitted by heated objects, and if the telescope is warm it will emit many times more infrared radiation than that coming from a distant object. Imagine trying to look for rabbits at night through binoculars that are themselves glowing. In a telescope observing in the near-infrared, only the detector, the element on which the infrared radiation is focused, must be cooled, usually with liquid nitrogen, as shown in Figure 6-20. To observe in the far-infrared, however, the entire telescope must be cooled.

At the short-wavelength end of the spectrum, astronomers can observe in the near-ultraviolet. Your eyes don't detect this radiation, but it can be recorded by photographic plates and CCDs. Wavelengths shorter than about 290 nm, the far-ultraviolet, are completely absorbed by the ozone layer extending from 20 km to about 40 km above Earth's surface. No mountaintop is that high, and no airplane can fly to such an altitude. To observe in the far-ultraviolet or beyond at X-ray or gamma-ray wavelengths, telescopes must be in space above the atmosphere.

Telescopes in Space

To observe far beyond the ends of the visible spectrum, astronomical telescopes must go above Earth's atmosphere into space. This is very expensive and difficult, but it is the only way to study some processes. Stars, for example, are born inside clouds of dust and gas, and visible wavelengths cannot escape from these dust clouds. Only observations in the infrared can reveal the secrets of star formation. Black holes are small and hard to detect, but matter falling into a black hole emits X rays. Telescopes in space can explore such processes as these that are invisible from within Earth's atmosphere.

One of the most successful space telescopes was the International Ultraviolet Explorer (IUE), launched in 1978. It carried a

The Great Observatories in Space

1 The Hubble Space Telescope was carried into orbit by the Space Shuttle in 1990. The telescope contains a 2.4-m (96-in.) mirror and can observe from the near-infrared to the near-ultraviolet.

Orbiting above Earth's blurring atmosphere, Hubble is limited only by diffraction in its optics. It can detect details 10 times smaller than Earth-based telescopes.

The telescope, as big as a large bus, has been visited twice by astronauts, who repaired equipment and installed new instruments. Named after Edwin Hubble, the astronomer who discovered the expansion of the universe, the telescope has been tremendously productive observing everything from the weather on Mars to the most distant galaxies visible in the universe.

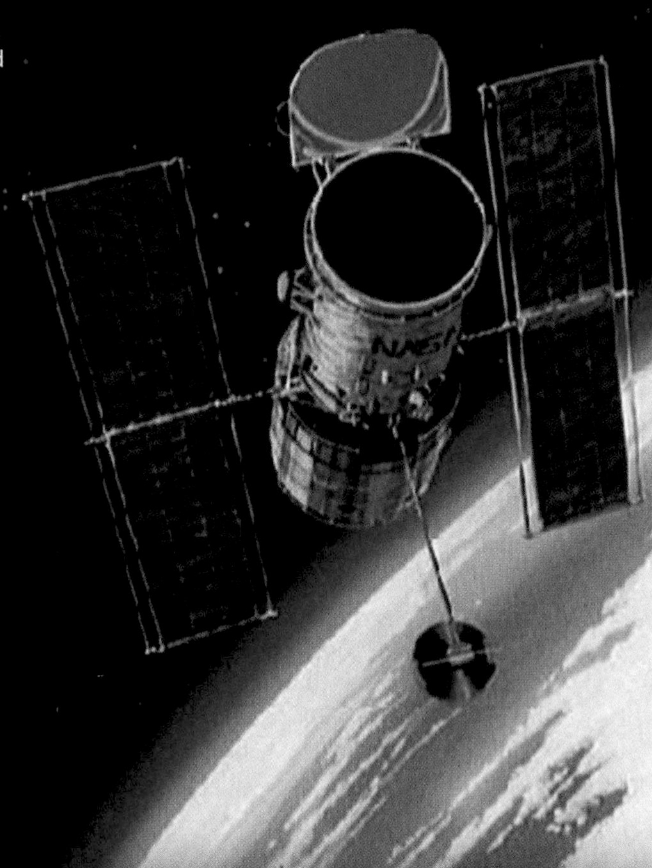

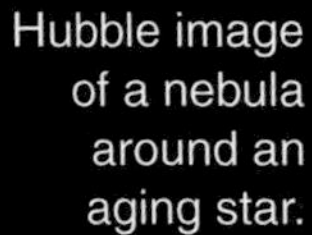

Hubble image of Mars and its polar cap.

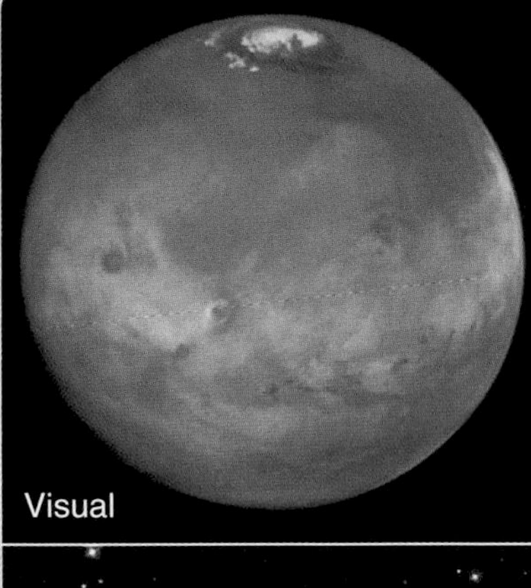

Hubble image of a nebula around an aging star.

Hubble image of a dust-filled galaxy.

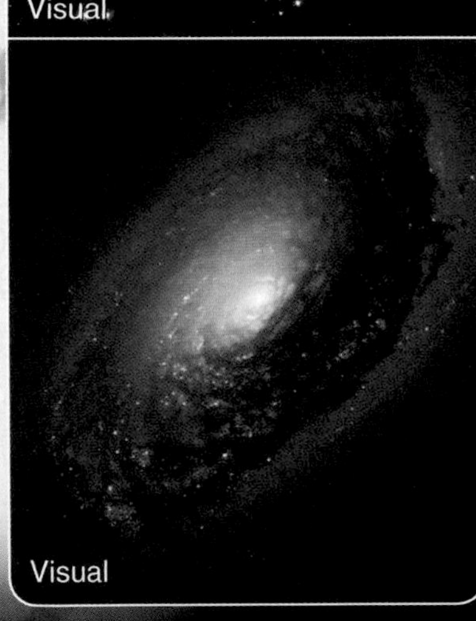

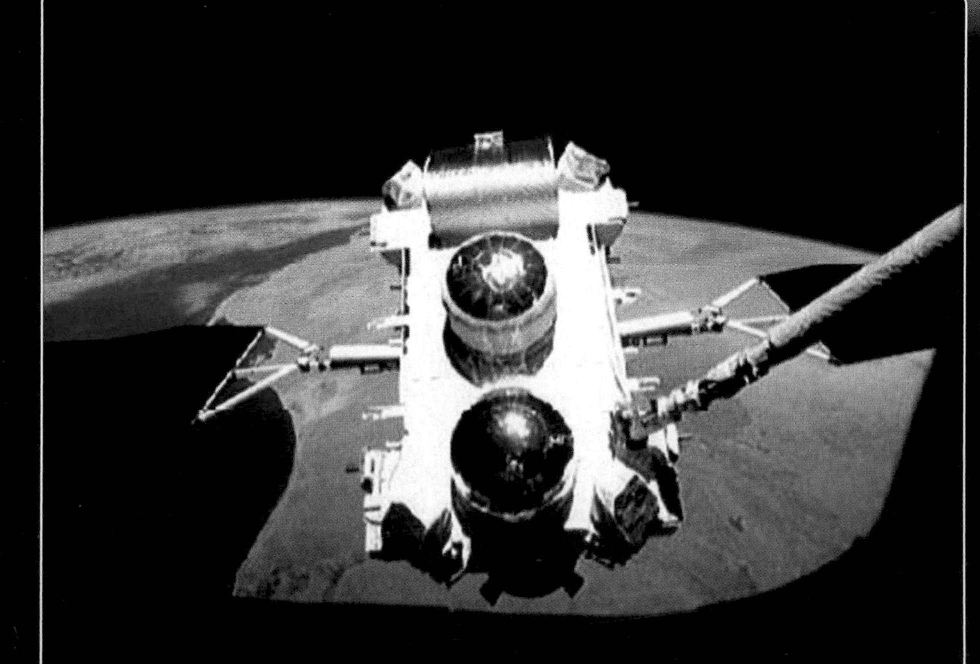

2 The Compton Gamma Ray Observatory at left was in orbit from 1991 to 2000. It made observations of very-high-energy photons, helping astronomers understand such violently active objects as neutron stars and black holes.

2a The Chandra X-Ray Observatory was placed in orbit 1/3 of the way to the moon in 1999. It is nearly 14 m (45 ft) long and carries highly precise mirrors 1.2 m (47 in.) in diameter. X rays would penetrate into regular mirrors, so Chandra's mirrors are designed as cylinders polished on the inside so that X rays just graze the surface and are focused onto detectors. The telescope was named after the late Indian-American Nobel laureate Subrahmanyan Chandrasekhar, who was a pioneer in many branches of theoretical astronomy.

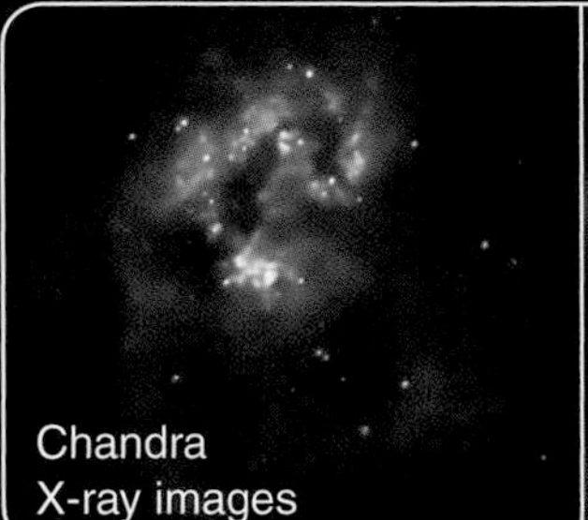

Two galaxies collide and trigger the birth of stars.

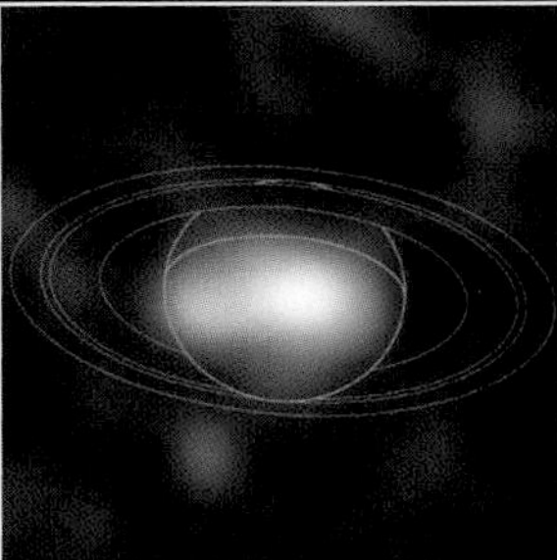

Saturn emits X-rays from near its equator.

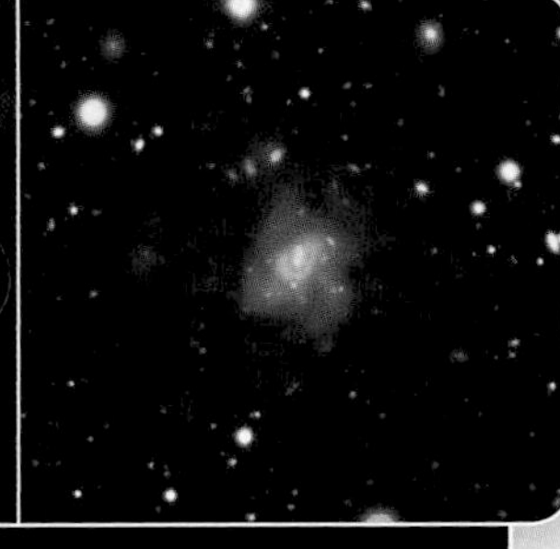

Very hot gas is trapped in a cluster of galaxies.

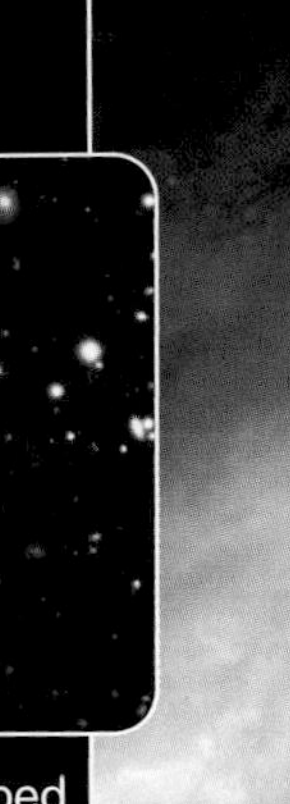

Chandra can detect X-ray emitting objects 50 times fainter and resolve details 10 times smaller than any previous X-ray telescope.

2b The Spitzer Space Telescope (at left) observes in the infrared. It was launched in 2003 and named in honor of the late astronomer Lyman Spitzer, a leader in space astronomy. The telescope is cooled to –273°C (–459°F) so it cannot orbit the warm Earth. Instead it is in an orbit around the sun and will drift slowly away from Earth during its lifetime. Protected from sunlight by a heat screen, it can observe a wide range of astronomical objects.

Spitzer infrared image

A comet glows in the infrared.

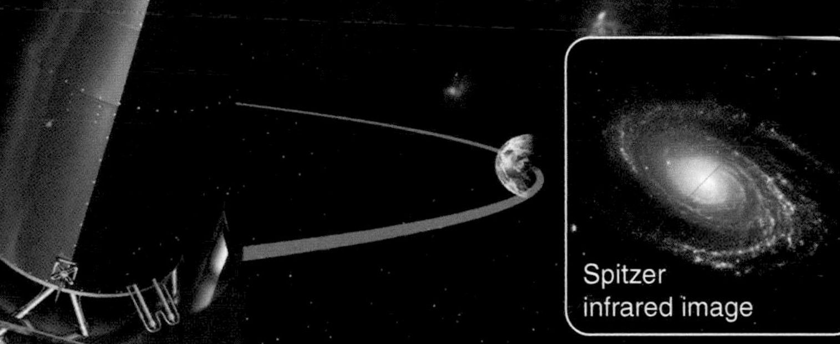

Spitzer infrared image

Dust warmed by hot young stars glows in the disk of this spiral galaxy.

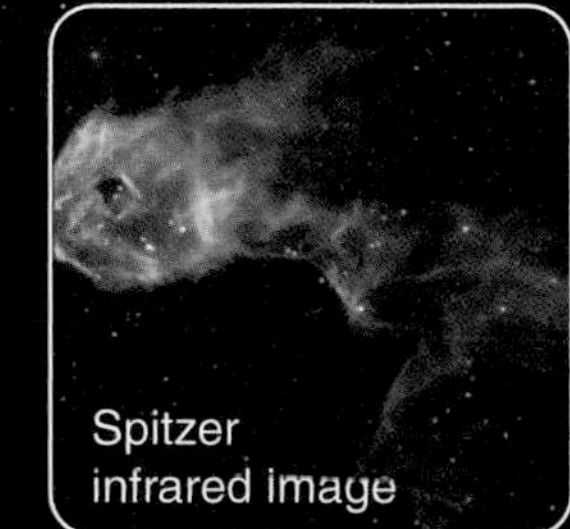

An infrared image penetrates a dusty nebula to reveal newborn stars just beginning to shine.

3 The James Webb Space Telescope (at right) is planned as the next great observatory in space. Named after the early director of NASA who oversaw the planning of the Apollo moon landings, the telescope will carry a 6.5-m (256-in.) segmented mirror made of the metal beryllium. It will observe without a telescope tube from behind a multilayered sunscreen. Launch is planned for 2011.

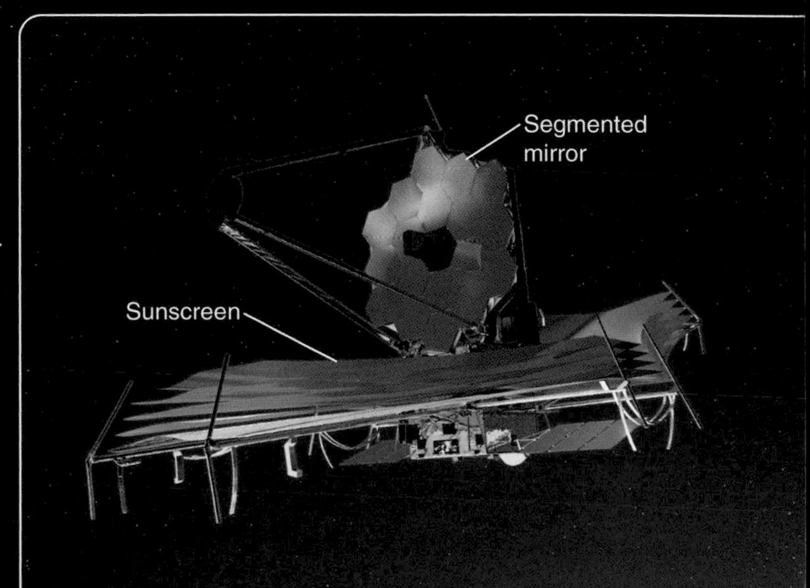

telescope only 45 cm (18 in.) in diameter and was expected to last only a year or two, but it became the little telescope that could. It made many observations and exciting discoveries until it finally failed in 1996.

Many space telescopes are small satellites designed to make specific observations for a short period, but some are large general-purpose telescopes. Over two decades ago, astronomers developed a plan to place a series of great observatories in space. Those space telescopes have revolutionized human understanding of what we are and where we are in the universe. Study **The Great Observatories in Space** on pages 130–131 and notice three points:

1. Not only can a telescope in space observe at a wide range of wavelengths, but it is above the atmospheric blurring called seeing. The Hubble Space Telescope observes mostly at visual wavelengths and has the advantage of sharp images undistorted by seeing.

2. Notice how telescopes must be specialized for their wavelength range. The Compton Gamma Ray Observatory had special detectors, the Chandra X-Ray Observatory must have cylindrical mirrors, and the Spitzer infrared observatory must have cooled optics.

3. The Hubble Space Telescope has been maintained by visits from astronauts, but such visits are expensive, and the future of Hubble is in doubt. Astronauts cannot reach the Chandra and Spitzer telescopes, and the Compton Observatory was removed from orbit in 2000. Space observatories have limited lifetimes, and astronomers are already planning the next great observatory in space. The new James Webb Space Telescope will not be available for many years.

These great observatories in space are controlled from research centers on Earth and are open to proposals from any astronomer with a good idea; but competition is fierce, and only the most worthy projects win approval.

Cosmic Rays

All of the radiation you have read about in this chapter has been electromagnetic radiation. **Cosmic rays,** however, are not really rays; they are subatomic particles traveling at tremendous velocities that strike Earth's atmosphere from space. Almost no cosmic rays reach the ground, but they do smash gas atoms in the upper atmosphere, and fragments of those atoms shower down on you day and night over your entire life. These secondary cosmic rays are passing through you as you read this sentence.

Some cosmic-ray research can be done from high mountains or high-flying aircraft; but, to study cosmic rays in detail, detectors must go into space. A number of cosmic-ray detectors have been carried into orbit, but this area of astronomical research is just beginning to bear fruit.

Astronomers can't be sure what produces cosmic rays. Because they are atomic particles with electric charges, they are deflected by the magnetic fields spread through our galaxy, and that means you can't tell where they are coming from. The space between the stars is a glowing fog of cosmic rays. Some lower-energy cosmic rays come from the sun, but observations show that at least some cosmic rays are produced by the violent explosions of dying stars.

At present, cosmic rays largely remain an exciting mystery. You will meet them again in future chapters.

Building Scientific Arguments

Why can infrared astronomers observe from high mountaintops, while X-ray astronomers must observe from space? Once again, you can analyze this question by building a scientific argument based on comparison. Infrared radiation is absorbed by water vapor in Earth's atmosphere. If you built an infrared telescope on top of a high mountain, you would be above most of the water vapor in the atmosphere, and you could collect some infrared radiation from the stars. The longer-wavelength infrared radiation is absorbed much higher in the atmosphere, so you couldn't observe it from our mountaintop. Similarly, X rays are absorbed in the uppermost layers of the atmosphere, and you would not be able to find any mountain high enough to get an X-ray telescope above those absorbing layers. To observe the stars at X-ray wavelengths, you would need to put your telescope in space, above Earth's atmosphere.

You can see why X-ray and far-infrared telescopes must observe from space. Now build another argument based on comparison: **Why must the Hubble Space Telescope be in space when it observes in the visual-wavelength range?**

■ ■ ■

Connections: The tools of the astronomer are designed to gather radiation from the sky and extract information. Perhaps no tool is as important as the spectrograph because no form of observation is as loaded with information as a spectrum. In the next chapter, you will see how astronomers can harvest the information in a star's spectrum.

Study and Review Tools

Summary

6-1 | Radiation: Information from Space

What is light?

- Light is the visible form of electromagnetic radiation, an electric and magnetic disturbance that transports energy at the speed of light. The electromagnetic spectrum includes gamma rays, X rays, ultraviolet radiation, visible light, infrared radiation, and radio waves.
- You can think of a particle of light, a photon, as a bundle of waves that acts sometimes as a particle and sometimes as a wave.
- The energy a photon carries depends on its wavelength. The wavelength of visible light, usually measured in nanometers (10^{-9} m), ranges from 400 nm to 700 nm. Infrared and radio photons have longer wavelengths and carry less energy. Ultraviolet, X-ray, and gamma-ray photons have shorter wavelengths and carry more energy.

6-2 | Optical Telescopes

How do telescopes work, and how are they limited?

- Astronomers use telescopes to gather light, resolve fine detail, and magnify the image. The first two of these three powers of the telescope depend on the telescope's diameter. Consequently, astronomers strive to build telescopes with large diameters.
- A refracting telescope uses a lens to bend the light and focus it into an image. Because of chromatic aberration, refracting telescopes cannot bring all colors to the same focus, resulting in color fringes around the images. An achromatic lens partially corrects for this, but such lenses are expensive and cannot be made much larger than about 1 m in diameter.
- Reflecting telescopes use a mirror to focus the light and are less expensive than refracting telescopes of the same diameter. Also, reflecting telescopes do not suffer from chromatic aberration. Most recently built large telescopes are reflectors.
- Astronomers build observatories on high mountains for two reasons. Turbulence in Earth's atmosphere blurs the image of an astronomical telescope, a phenomenon that astronomers refer to as seeing. Atop a mountain, the air is steady, and the seeing is better; the air on a mountaintop is also thin and dry and is more transparent, especially in the infrared.
- Sometimes astronomical telescopes can be linked together to form an interferometer, which has a resolution equivalent to that of a telescope as large in diameter as the separation between the telescopes.

6-3 | Special Instruments

What kind of instruments do astronomers use to record and analyze light?

- For many decades astronomers used photographic plates to record images at the telescope, but modern electronic systems such as CCD cameras have replaced photographic plates in most applications.
- Spectrographs using prisms or a grating spread starlight out according to wavelength to form a spectrum revealing hundreds of spectral lines produced by atoms in the object being studied.

6-4 | Radio Telescopes

Why do astronomers use radio telescopes?

- Astronomers use radio telescopes for three reasons: They can detect cool hydrogen in space; they can see through dust clouds that block visible light; and they can detect certain objects invisible at other wavelengths.
- Most radio telescopes contain a dish reflector, an antenna, an amplifier, and a data recorder. Such a telescope can record the intensity of the radio energy coming from a spot on the sky. Scans of small regions are used to produce radio maps.
- Because of the long wavelength, radio telescopes have very poor resolution, and astronomers often link separate radio telescopes together to form a radio interferometer capable of resolving much finer detail.

6-5 | Astronomy from Space

Why must some telescopes go into space?

- Earth's atmosphere is transparent in two wavelength ranges called atmospheric windows, the visual window and the radio window. At other wavelengths, the atmosphere absorbs radiation. To observe at other wavelengths, telescopes must go into space.
- Earth's atmosphere distorts and blurs images. Telescopes in orbit are above this seeing distortion and are limited only by diffraction in their optics.
- Cosmic rays are not electromagnetic radiation; they are subatomic particles such as electrons and protons traveling at nearly the speed of light. They can best be studied from above Earth's atmosphere.

New Terms

electromagnetic radiation (p. 110)
wavelength (p. 110)
frequency (p. 111)
nanometer (nm) (p. 111)
Angstrom (Å) (p. 111)
photon (p. 111)
infrared radiation (p. 111)
ultraviolet radiation (p. 112)
atmospheric window (p. 112)
focal length (p. 114)
refracting telescope (p. 114)
reflecting telescope (p. 114)
primary lens, mirror (p. 114)
objective lens, mirror (p. 114)
eyepiece (p. 114)
chromatic aberration (p. 114)
achromatic lens (p. 114)
light-gathering power (p. 115)
resolving power (p. 115)
diffraction fringe (p. 115)
seeing (p. 115)
magnifying power (p. 116)
light pollution (p. 117)
prime focus (p. 120)
secondary mirror (p. 120)
Cassegrain focus (p. 120)
Newtonian focus (p. 120)
Schmidt–Cassegrain focus (p. 120)
sidereal drive (p. 121)
equatorial mounting (p. 121)
polar axis (p. 121)
alt-azimuth mounting (p. 121)
active optics (p. 121)
adaptive optics (p. 121)
interferometry (p. 122)
charge-coupled device (CCD) (p. 123)
false-color image (p. 123)
spectrograph (p. 123)
grating (p. 124)
comparison spectrum (p. 124)
radio interferometer (p. 127)
cosmic ray (p. 132)

Review Questions

Ace Astronomy™ Assess your understanding of this chapter's topics with additional quizzing and animations at **http://ace.brookscole.com/sf9**

1. Why would you not plot sound waves in the electromagnetic spectrum?
2. If you had limited funds to build a large telescope, which type would you choose, a refractor or a reflector? Why?
3. Why do nocturnal animals usually have large pupils in their eyes? How is that related to astronomical telescopes?
4. Why do optical astronomers sometimes put their telescopes at the tops of mountains, while radio astronomers sometimes put their telescopes in deep valleys?
5. Optical and radio astronomers both try to build large telescopes but for different reasons. How do these goals differ?
6. What are the advantages of making a telescope mirror thin? What problems does this cause?
7. Small telescopes are often advertised as "200 power" or "magnifies 200 times." As someone knowledgeable about astronomical telescopes, how would you improve such advertisements?
8. Not long ago an astronomer said, "Some people think I should give up photographic plates." Why might she change to something else?
9. What purpose do the colors in a false-color image or false-color radio map serve?
10. How is chromatic aberration related to a prism spectrograph?
11. Why would radio astronomers build identical radio telescopes in many different places around the world?
12. Why do radio telescopes have poor resolving power?
13. Why must telescopes observing in the far-infrared be cooled to low temperatures?
14. What might you detect with an X-ray telescope that you could not detect with an infrared telescope?
15. The moon has no atmosphere at all. What advantages would you have if you built an observatory on the lunar surface?
16. The two images at the right show a star before and after an adaptive optics system was switched on. What causes the distortion in the first image, and how does adaptive optics correct the image?

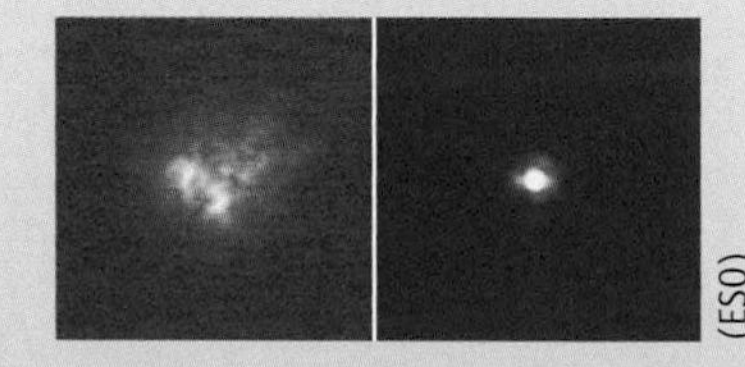
(ESO)

17. The X-ray image at right shows the remains of an exploded star. Explain why images recorded by telescopes in space are often displayed in false color rather than in the "colors" received by the telescope.

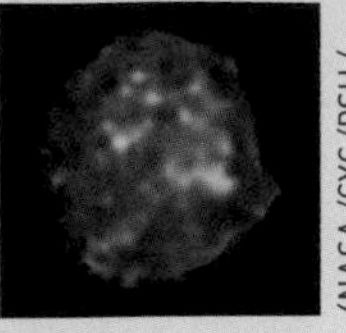
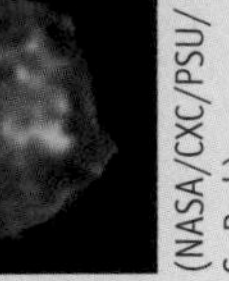
(NASA/CXC/PSU/S. Park)

Discussion Questions

1. Why does the wavelength response of the human eye match so well the visual window of Earth's atmosphere?
2. Most people like beautiful sunsets with brightly glowing clouds, bright moonlit nights, and twinkling stars. Most astronomers don't. Why?

Problems

1. The thickness of the plastic in plastic bags is about 0.001 mm. How many wavelengths of red light is this?
2. What is the wavelength of radio waves transmitted by a radio station with a frequency of 100 million cycles per second?
3. Compare the light-gathering powers of one of the Keck telescopes and a 0.5-m telescope.
4. How does the light-gathering power of one of the Keck telescopes compare with that of the human eye? (*Hint:* Assume that the pupil of your eye can open to about 0.8 cm.)
5. What is the resolving power of a 25-cm telescope? What do two stars 1.5 seconds of arc apart look like through this telescope?
6. Most of Galileo's telescopes were only about 2 cm in diameter. Should he have been able to resolve the two stars mentioned in Problem 5?
7. How does the resolving power of the 5-m telescope compare with that of the Hubble Space Telescope? Why does the HST outperform the 5-m telescope?
8. If you build a telescope with a focal length of 1.3 m, what focal length should the eyepiece have to give a magnification of 100 times?
9. Astronauts observing from a space station need a telescope with a light-gathering power 15,000 times that of the human eye, capable of resolving detail as small as 0.1 second of arc, and having a magnifying power of 250. Design a telescope to meet their needs. Could you test your design by observing stars from Earth?
10. A spy satellite orbiting 400 km above Earth is supposedly capable of counting individual people in a crowd. Roughly what minimum-diameter telescope must the satellite carry? (*Hint:* Use the small-angle formula.)

Media Cluster

Ace Astronomy™ To access the resources in the Media Cluster, log into AceAstronomy at **http://ace.brookscole.com/sf9** and select Chapter 6.

ACTIVE FIGURES

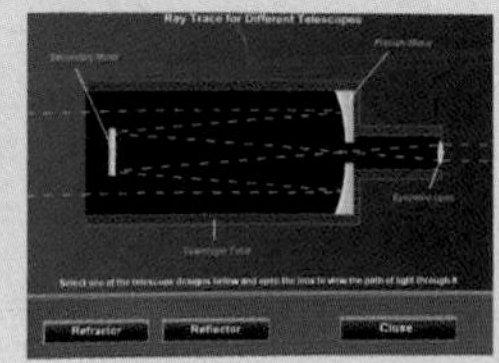

Refractors and Reflectors
Compare the two basic types of telescope designs and the paths light takes through each with this animation.

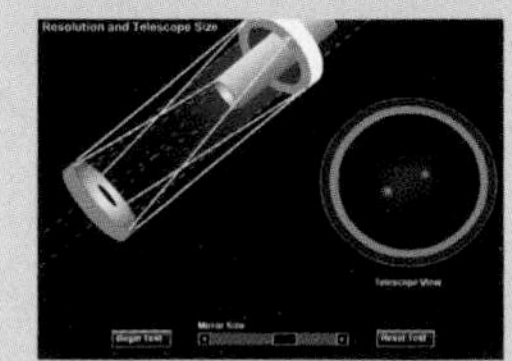

Resolution and Telescopes
This animation allows you to contrast the high resolution of larger-aperture telescopes with the lower resolution of smaller ones.

ASTRONOMY EXERCISES

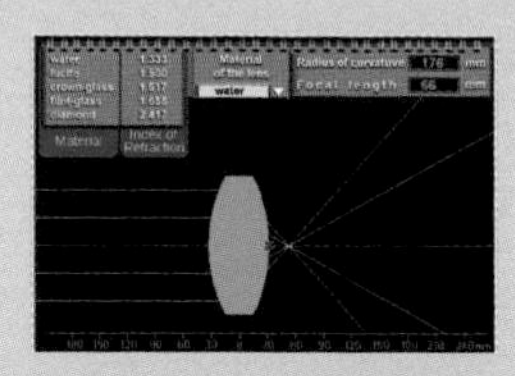

Lenses: Focal Length
Astronomers use lenses to focus starlight into an image. You can change the shape and material of the lens in this animation to study how these parameters affect the focal length.

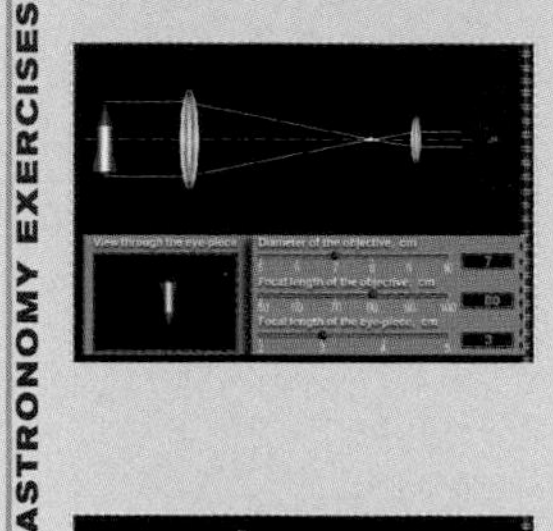

Telescopes: Objective Lens and Eyepiece
Use this animation to study how the objective lens and eyepiece work together to form an image in a telescope. You can change parameters for the diameter of the lens and the focal lengths of the objective lens and eyepiece.

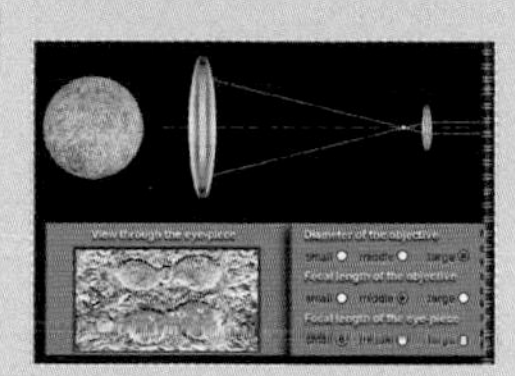

Telescopes and Resolution I
In this simulation, you can vary the focal length of the objective lens, the focal length of the eyepiece lens, and the diameter of the objective lens. Try them all and see which factor (or factors) determines how sharp the image is.

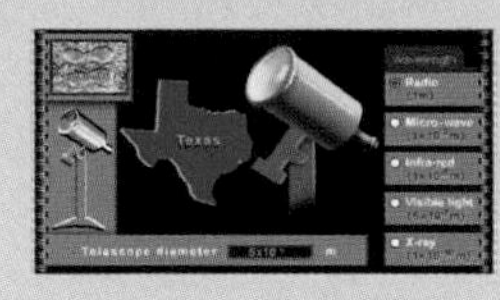

Telescopes and Resolution II
Get a feel for how wavelength affects the resolution of a telescope by comparing the lens diameter needed to observe craters on the moon with a visible-light telescope versus other types such as radio and microwave telescopes.

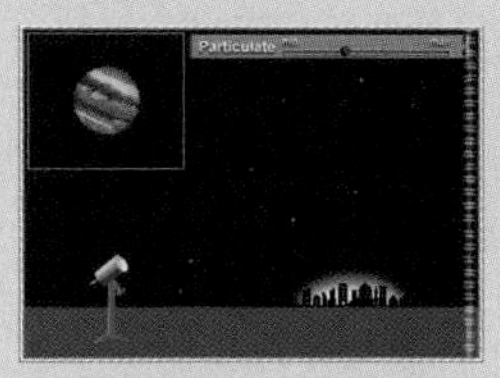

Particulate, Heat, and Light Pollution
These three animations let you study how environmental variables can affect viewing quality. How does this fit in with what you learned in this chapter about the locations of major astronomical telescopes?

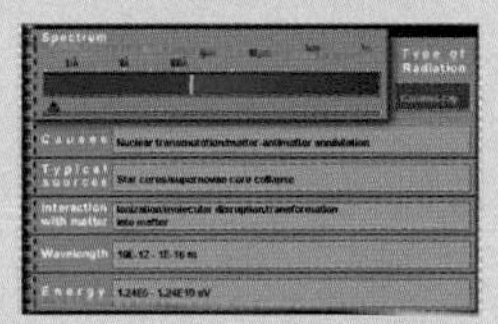

The Electromagnetic Spectrum
In this exercise, study the properties of radiation, such as wavelength and sources of various parts of the electromagnetic spectrum.

VIRTUAL ASTRONOMY LABS

Lab 4: Solar Wind and Cosmic Rays
This lab begins with an overview of the properties of the sun's atmosphere and how energetic particles escape and travel through the solar system. The lab ends with a discussion of cosmic rays.

Critical Inquiries for the Web

1. Research in chemistry, physics, and biology is supported in part by industry. Because astronomy has few industrial applications, it is not well supported by industry. Visit websites for major observatories and find out who pays their bills.
2. Visit websites for the major observatories in space, such as Hubble Space Telescope and Chandra, and check on their latest status. What kinds of observations are they making, and what kinds of discoveries are they making?
3. Astronomers are leaders in efforts to reduce electric power wasted on outdoor lighting because it causes light pollution. Find websites on light pollution and see if your town is making any effort to preserve the beauty of the night sky and simultaneously save electrical power.

Exploring *TheSky*

1. Astronomical telescopes using equatorial mountings must be aligned precisely with the north celestial pole. Locate Polaris and determine how far it is from the north celestial pole. (*Hint:* Use **Reference Lines** under the **View** menu and check **Grid** under Equatorial. Be sure the spacing is set to auto/fine. Then locate the Little Dipper and zoom in on Polaris.)

- Go to the Brooks/Cole Astronomy Resource Center **(http://astronomy.brookscole.com)** for critical thinking exercises, articles, and additional readings from InfoTrac College Edition, Brooks/Cole's online student library.

7 Starlight and Atoms

Awake! for Morning in the
Bowl of Night
Has flung the Stone that puts
the Stars to Flight:
And Lo! the Hunter of the
East has caught
The Sultan's Turret in a
Noose of Light.

THE RUBÁIYÁT OF OMAR KHAYYÁM,
TRANS. EDWARD FITZGERALD

THE UNIVERSE IS FILLED with fabulously beautiful clouds of glowing gas illuminated by brilliant stars, but it is all hopelessly beyond reach. No laboratory jar on Earth holds a sample labeled "star stuff," and no space probe has ever visited the inside of a star. The stars are far away, and the only information you can obtain about them comes hidden in starlight (■ Figure 7-1). Whatever you want to know about stars you must catch in a noose of light. ▌ Earthbound humans knew almost nothing about stars until the early 19th century, when the Munich optician Joseph von Fraunhofer studied the solar spectrum and found it interrupted by some 600 dark lines. As scientists realized that the lines were related to the various atoms in the sun and found that stellar spectra had similar patterns of lines, the door to an understanding of stars finally opened.

Continued on page 138

Clouds of glowing gas illuminated by hot, bright stars lie thousands of light years across space, but clues hidden in starlight tell a story of star birth and star death. (ESO)

Guidepost

Looking Back

In the last chapter you read how hard astronomers work to gather light from the stars using giant telescopes on mountaintops and in space. You also read how spectrographs can spread light out into spectra. Now you are ready to see what all the fuss is about.

This Chapter

This chapter explains how light interacts with matter and how astronomers must understand that interaction to understand stars. Here you will find answers to five essential questions:

How do stars produce light?

What is an atom?

How do atoms interact with light?

What kind of spectra do you see when you look at celestial objects?

What can you learn from a star's spectrum?

This chapter marks a change in the way you will look at nature. Up to this point, you have been thinking about what you can see with your eyes alone or aided by telescopes. In this chapter, you begin using modern astrophysics, the application of physics to study the sky. Now you can search out secrets of the stars that lie beyond what you can see.

Looking Ahead

The analysis of spectra is a powerful tool, and in the next chapter you will use that tool to study the sun. In the chapters that follow, you will study other suns—the stars.

Ace Astronomy™ The AceAstronomy icon throughout the text indicates an opportunity for you to test yourself on key concepts and to explore animations and interactions on the AceAstronomy website at: http://ace.brookscole.com/sf9

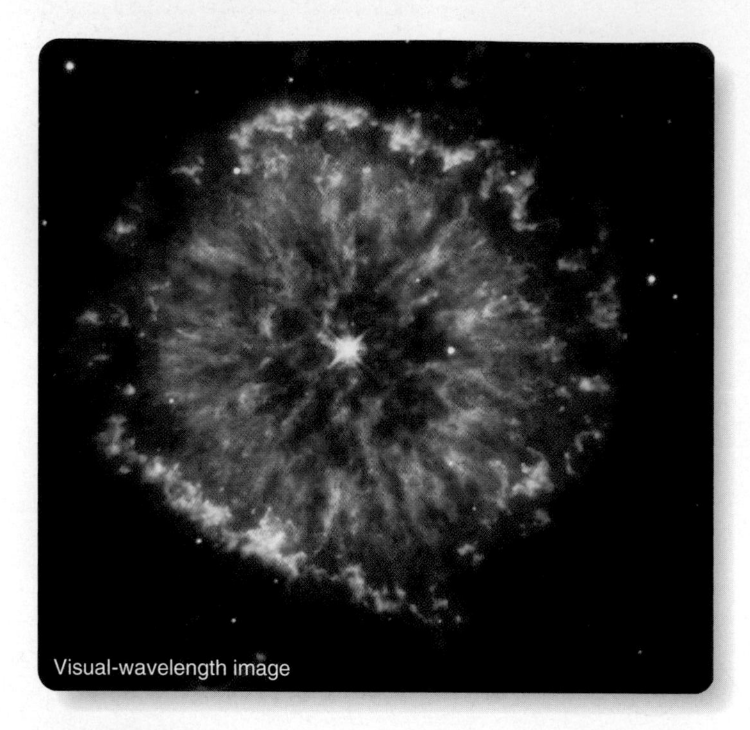

■ **Figure 7-1**

What's going on here? The sky is filled with beautiful and mysterious objects that lie far beyond your reach—in the case of the nebula NGC 6751, about 6500 ly beyond your reach. The only way to understand such objects is by analyzing their light. Such an analysis reveals that this object is a dying star surrounded by the expanding shell of gas it ejected a few thousand years ago. You will learn more about this phenomenon in Chapter 13. (NASA Hubble Heritage Team/STScI/AURA)

In this chapter, you will go through that door by considering how stars produce light and how atoms interact with light to produce spectral lines. The first step is to consider the hydrogen atom because it is the most common atom in the universe as well as the simplest. Other atoms are larger and more complicated, but in many ways their properties resemble those of hydrogen.

Once you understand how an atom's structure can interact with light to produce spectral lines, you will recognize certain patterns in stellar spectra. By classifying the spectra according to these patterns, you can arrange the stars in a sequence according to temperature. One of the most important pieces of information revealed in a star's spectrum is its temperature.

But, properly analyzed, a stellar spectrum can tell you much more. The spectrum contains information about the chemical composition of the star and the star's motion relative to Earth.

7-1 Starlight

IF YOU LOOK at the stars in the constellation Orion, you will notice that they are not all the same color (see Figure 2-4). One of our Favorite Stars, Betelgeuse, in the upper left corner, is quite red; another Favorite Star, Rigel, in the lower right corner, is blue. These differences in color arise from the way the stars produce light, and they provide an important clue to the temperatures of stars.

Temperature and Heat

Temperature is one of the defining characteristics of a star. That is, if you know a star's temperature and a few other properties, such as size, you can understand the star. But if you don't know a star's temperature, you can know almost nothing about it.

A gas is made up of particles—atoms and molecules—that are in constant motion, colliding with one another millions of times a second. The **temperature** of a gas is a measure of the average kinetic energy of the particles. Recall that kinetic energy is energy of motion, so if a gas is hot, the particles must be moving very rapidly; if the gas is cool, the particles are moving more slowly. The motion of the particles represents stored energy, and you feel that energy as heat. Read **Focus on Fundamentals 3** and notice the distinction between the temperature, the stored energy, and the heat you feel.

Astronomers express the temperature of stars on the **Kelvin temperature scale.** The Kelvin scale sets its zero point at **absolute zero,** the temperature at which the particles of a gas have no remaining kinetic energy that could be extracted as heat. Absolute zero is −273.2°C, which is −459.7°F. (See Appendix A.) The temperatures of the stars range from less than 2000 K to 40,000 K or more. To see how astronomers can estimate the temperatures of the stars from starlight, you must understand how stars produce light.

The Origin of Starlight

The starlight you see comes from gases in the outer surface layers of the star—the photosphere. The gases deep inside the star also emit light, but it is absorbed before it can escape, and the low-density gas above the photosphere is too thin to emit significant amounts of light. Starlight comes from the photosphere of a star, that layer of gases dense enough to emit significant amounts of light but thin enough to allow the light to escape. (Recall that you met the photosphere of the sun in Chapter 3.) So when astronomers refer to a star as "hot" they are referring to the temperature of its visible surface—its photosphere.

To see how the photosphere of a star can produce light, you must consider two things, how photons are produced and how the temperature of a material is related to motion among its atoms.

First, a photon can be produced by a changing electric field. An **electron** is a negatively charged subatomic particle, and if you disturb the motion of an electron, the sudden change in the electric field around it can produce an electromagnetic wave. For example, if you run a comb through your hair while standing near a radio tuned to an AM station, you can produce popping noises on the radio. The moving comb disturbs the electrons in both the comb and your hair, building static electricity. The sudden sparks of static electricity produce electromagnetic waves that the radio

Focus on Fundamentals 3

Temperature, Heat, and Thermal Energy

One of the most common misconceptions in science involves temperature. People often say "temperature" when they really mean "heat," and sometimes they say "heat" when they mean something entirely different. This is a fundamental idea, so you need to understand the differences.

When something is hot, the particles in the object, be they atoms or molecules, are moving rapidly. Temperature is a measure of the average motion of the particles. (Mathematically, temperature is proportional to the square of the average velocity.) In a hot object, the particles race around at higher speeds than in a cool object. If you have your temperature taken, it will probably be 98.6°F, an indication that the atoms and molecules in your body are moving about at a normal pace. If you measure the temperature of a month-old baby, the thermometer should register the same temperature, showing that the atoms and molecules in the baby's body are moving at the same average velocity as the atoms and molecules in your body.

The energy of the moving particles in a body is called **thermal energy.** You have much more mass than the baby, so you must contain more thermal energy even though you have the same temperature. The thermal energy in your body and in the baby's body have the same intensity (temperature) but different amounts. People often confuse temperature and thermal energy, so you must be careful to distinguish between them. Temperature is an intensity, and thermal energy is an amount.

Many people say "heat" when they should say thermal energy. Heat is the thermal energy that moves from a hot object to a cool object. If two objects have the same temperature, you and the infant for example, there is no transfer of thermal energy and no heat. This is a fine distinction, but when you hear someone say "heat," check to see if he or she doesn't really mean thermal energy.

What's the difference between temperature and heat?

You may have burned yourself on cheese pizza, but you probably haven't burned yourself on green beans. At the same temperature, cheese holds more thermal energy than green beans. It isn't the temperature that burns your tongue, but the flow of thermal energy, and that's heat.

PRESSURE | MASS | ENERGY | **TEMPERATURE AND HEAT** | DENSITY

picks up as pops and crackles. This illustrates an important principle: Any change in the motion of an electron can generate an electromagnetic wave.

Second, recall that temperature is related to the motion among the particles in a material. In the gases of a hot star, the atoms move faster, on average, than the atoms in a cool star.

Now put these two ideas together, and you can understand why a hot object glows. The hotter an object is, the more violent the motion among its particles. The agitated particles collide with electrons, and when electrons are accelerated, part of the energy is carried away as electromagnetic radiation in the form of a photon. The radiation emitted by a heated object is called **black body radiation,** a name that refers to the way a perfect emitter of radiation would behave. A perfect emitter would also be a perfect absorber and, at room temperature, would look black. You will often find the term *black body radiation* referring to objects that glow brightly.

Black body radiation is quite common. In fact, it is responsible for the light emitted by an incandescent lightbulb. Electricity flowing through the wire filament of the lightbulb heats the wire to high temperature, and it glows. You can also recognize the light emitted by a heated horseshoe in the blacksmith's forge as black body radiation. Many objects in astronomy, including stars, emit radiation approximately as if they were black bodies.

Hot objects emit black body radiation, but so do cold objects. Ice cubes are cold, but their temperature is higher than absolute zero, so they contain some thermal energy and must emit some black body radiation. The coldest gas drifting in space has a temperature only a few degrees above absolute zero, but it too emits black body radiation.

Two important features of black body radiation will help you understand what you see when you look at a star. First, the hotter an object is, the more black body radiation it emits. Hot objects emit more radiation because their agitated particles travel faster and collide more often. So, of course, you expect a glowing coal from a fire to emit more total energy than an ice cube of the same size.

The second feature is the relationship between the temperature of the object and the wavelengths of the photons it emits. The wavelength of a photon emitted when a particle collides with an electron depends on the violence of the collision. Only a violent collision can produce a short-wavelength (high-energy) photon. Because extremely violent collisions don't occur very often, short-wavelength photons are rare. Similarly, most collisions are not extremely gentle, so long-wavelength (low-energy) photons are also rare. Consequently, black body radiation is made up of photons with a distribution of wavelengths, and very short and very long wavelengths are rare. The **wavelength of maximum**

intensity (λ_{max}), the wavelength at which the object emits the most radiation, occurs at some intermediate wavelength.

■ Figure 7-2 shows the intensity of radiation versus wavelength for three objects of different temperatures. The curves are high in the middle and low at either end, which tells you that these objects emit most intensely at intermediate wavelengths. The total area under each curve is proportional to the total energy emitted, and you can see that the hotter object emits more total energy than the cooler objects. Look closely at the curves, and you will notice that the wavelength of maximum intensity depends on temperature. The hotter the object is, the shorter the wavelength of maximum intensity. Notice in this figure how temperature determines the color of a glowing black body. The hotter object emits more blue light than red and thus looks blue, and the cooler object emits more red light than blue and consequently looks red. Now you can understand why two of our Favorite Stars, Betelgeuse and Rigel, have such different colors. Betelgeuse is cool and looks red, but Rigel is hot and looks blue

Notice that cool objects may emit little visible radiation but are still producing black body radiation. For example, the human body has a temperature of 310 K and emits black body radiation mostly in the infrared part of the spectrum. Infrared security cameras can detect burglars by the radiation they emit, and mosquitoes can track you down in total darkness by homing in on your infrared radiation. Although humans emit lots of infrared radiation, you rarely emit higher energy photons; and you almost never emit X-ray or gamma-ray photons. Your wavelength of maximum intensity lies in the infrared part of the spectrum.

■ **Figure 7-2**

Black body radiation from three bodies at different temperatures demonstrates that a hot body radiates more total energy and that the wavelength of maximum intensity is shorter for hotter objects. The hotter object here will look blue to your eyes, while the cooler object will look red.

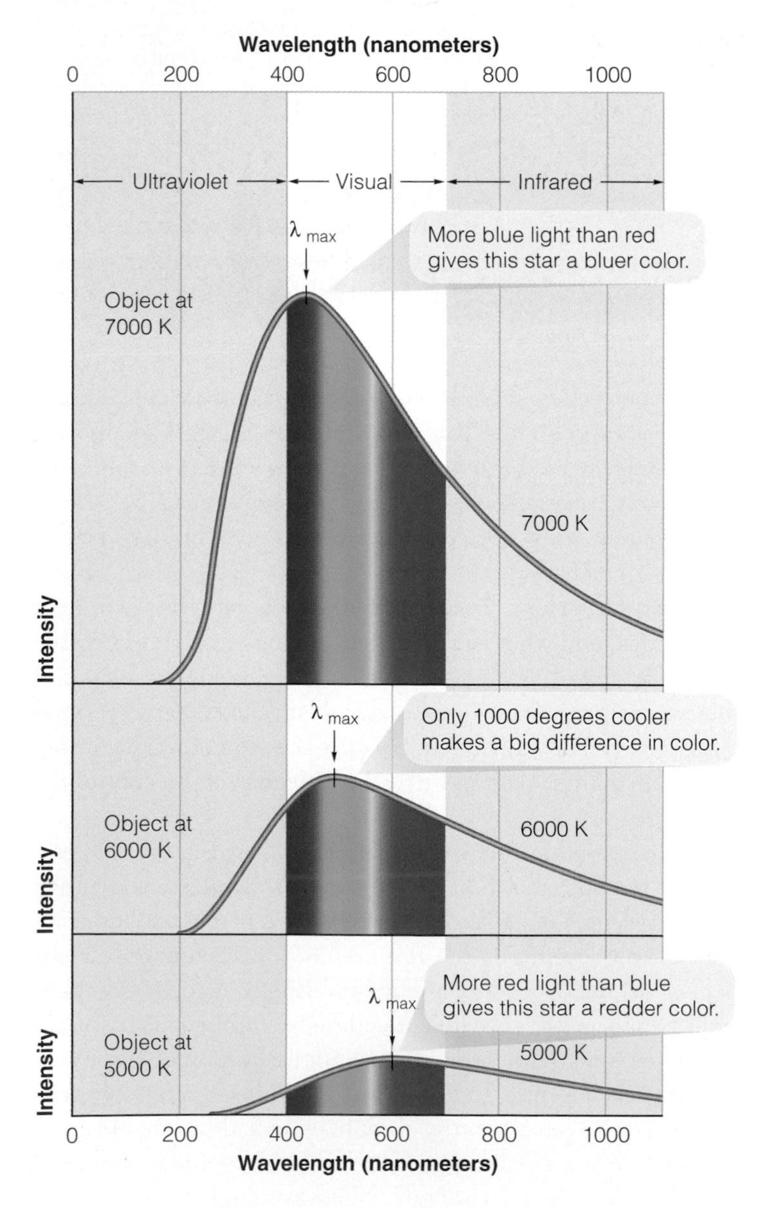

Two Radiation Laws

The two features of black body radiation that you have just considered can be given precise mathematical form, and they have proven so dependable, they are known as laws. One law is related to energy and one to color.

As you saw in the previous section, a hot object emits more black body radiation than a cool object. That is, it emits more energy. Recall from Chapter 5 that energy is expressed in units called joules (J); 1 joule is about the energy of an apple falling from a table to the floor. The total radiation given off by 1 square meter of the object in joules per second equals a constant number, represented by σ, times the temperature raised to the fourth power.* This relationship is called the Stefan–Boltzmann law:

$$E = \sigma T^4 \text{ (J/s/m}^2\text{)}$$

How does this help you understand stars? Suppose a star the same size as the sun had a surface temperature that was twice as hot as the sun's surface. Then each square meter of that star would radiate not twice as much energy but 2^4, or 16, times as much energy. From this law you can see that a small difference in temperature can produce a very large difference in the amount of energy emitted.

The second radiation law is related to the color of stars. In the previous section, you saw that hot stars look blue and cool stars look red. Wien's law tells you that the wavelength at which a star radiates the most energy, its wavelength of maximum intensity (λ_{max}), depends only on the star's temperature:

$$\lambda_{max} = \frac{3{,}000{,}000}{T}$$

*For the sake of completeness, you can note that the constant σ equals 5.67×10^{-8} J/m²s degree⁴.

That is, the wavelength of maximum radiation in nanometers equals 3 million divided by the temperature on the Kelvin scale.

This is a powerful tool in astronomy, because it means you can relate the temperature of a star and its wavelength of maximum intensity. For example, you might find a star that has a surface temperature of 3000 K. Then its wavelength of maximum intensity would be 3,000,000/3000, or 1000 nm—in the near-infrared. Later you will meet objects much hotter than most stars; such objects radiate most of their energy at very short wavelengths. The hottest stars, for instance, radiate most of their energy in the ultraviolet.

Now you can understand how astronomers can estimate the temperature of a star's surface from its color. The astronomers use the same technique that doctors and nurses use to measure a patient's body temperature. Medical personnel use a small device that detects the infrared radiation emerging from the patient's ear. You might suspect the device depends on the Stephan–Boltzmann law and measures the intensity of the infrared radiation. A person with a fever will emit more energy than a healthy person. However, a healthy person with a large ear canal would emit more energy than a person with a small ear canal, so measuring intensity wouldn't be accurate. The device actually depends on Wien's law in that it measures the "color" of the infrared radiation. A patient with a fever will emit at a slightly shorter wavelength of maximum intensity, and the infrared radiation emerging from that person's ear will be a tiny bit "bluer" than normal. Remember that the wavelength of maximum intensity depends on the temperature, so even a small fever produces changes in the color of the infrared radiation. Astronomers use the same technique to find the temperatures of the stars; they observe the color of the starlight.

AceAstronomy™ Log into AceAstronomy and select this chapter to see Astronomy Exercise "Black Body." Change the temperature and watch the black body curve change.

AceAstronomy™ Log into AceAstronomy and select this chapter to see Astronomy Exercise "Stefan-Boltzmann Law." Watch the brightness of a star change as you adjust its temperature.

Building Scientific Arguments

Why does the wavelength of maximum intensity depend on temperature?

Remember as you build any scientific argument that it must proceed logically, step by step. No gaps are allowed. In this case, you can begin with the temperature of an object. The hotter the object is, the more rapidly its particles move. That means that the typical collision between a particle and an electron will be more violent in a hotter body, and more violent collisions will accelerate the electrons more violently and, on average, produce higher-energy, shorter-wavelength photons. So the wavelength of maximum intensity depends on the temperature of the body. That's Wien's law. It all makes sense when you think about it step by step.

Now change the argument to answer a different question: **Why does the Stefan–Bolzmann law work?**

■ ■ ■

Connections: You have been thinking about atoms in a general way, but now it is time to get specific. What is an atom, and how can it interact with light?

7-2 Atoms

THE ATOMS IN THE SURFACE LAYERS of stars leave their marks on the light the stars emit. By understanding what atoms are and how they interact with light, you can decode the spectra of the stars.

A Model Atom

To think about atoms and how they can interact with light, you need a working model of an atom. In Chapter 2, you used a working model of the sky, the celestial sphere. You identified and named the important parts and described how they were located and how they interacted. So you should begin your study of atoms by creating a model of an atom.

Your model atom contains a positively charged **nucleus** at the center; this nucleus consists of two kinds of particles. **Protons** carry a positive electrical charge, and **neutrons** have no charge. That means the nucleus has a net positive charge.

The nucleus in this model atom is surrounded by a whirling cloud of orbiting electrons, low-mass particles with a negative charge. In a normal atom, the number of electrons equals the number of protons, and the positive and negative charges balance to produce a neutral atom. Because protons and neutrons each have a mass 1836 times greater than that of an electron, most of the mass of the atom lies in the nucleus. The hydrogen atom is the simplest of all atoms. The nucleus is a single proton orbited by a single electron, with a total mass of only 1.67×10^{-27} kg, about a trillionth of a trillionth of a gram.

An atom is mostly empty space. To see this, imagine constructing a simple scale model. The nucleus of a hydrogen atom is a proton with a diameter of about 0.0000016 nm, or 1.6×10^{-15} m. If you multiply this by one trillion (10^{12}), you can represent the nucleus of your model atom with a grape seed, which is about 0.16 cm in diameter. The region of a hydrogen atom containing the whirling electron has a diameter of about 0.4 nm, or 4×10^{-10} m. Multiplying by a trillion magnifies the diameter to about 400 m, or about 4.5 football fields laid end to end (■ Figure 7-3). When you imagine a grape seed in the midst of a sphere

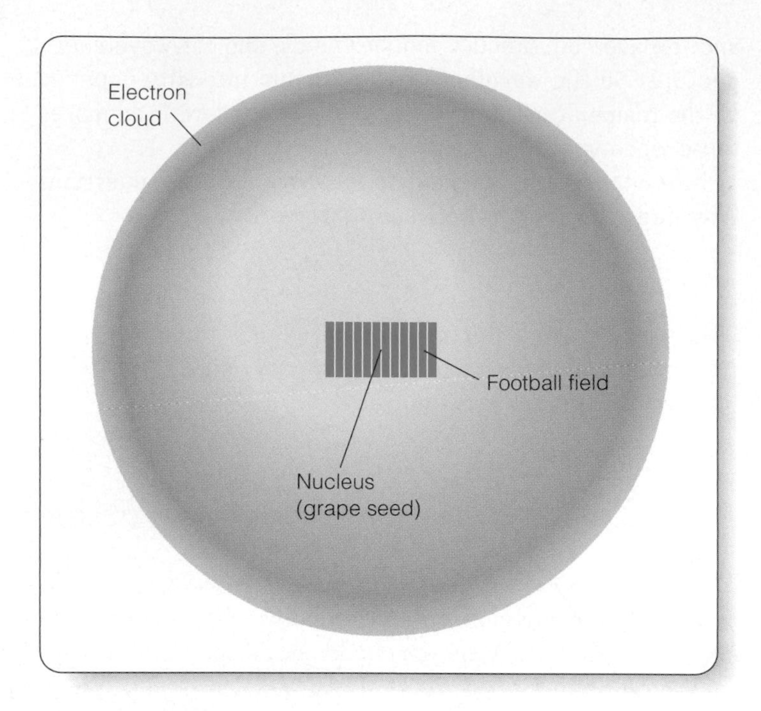

Figure 7-3

Magnifying a hydrogen atom by 10^{12} makes the nucleus the size of a grape seed and the diameter of the electron cloud about 4.5 times longer than a football field. The electron itself is still too small to see.

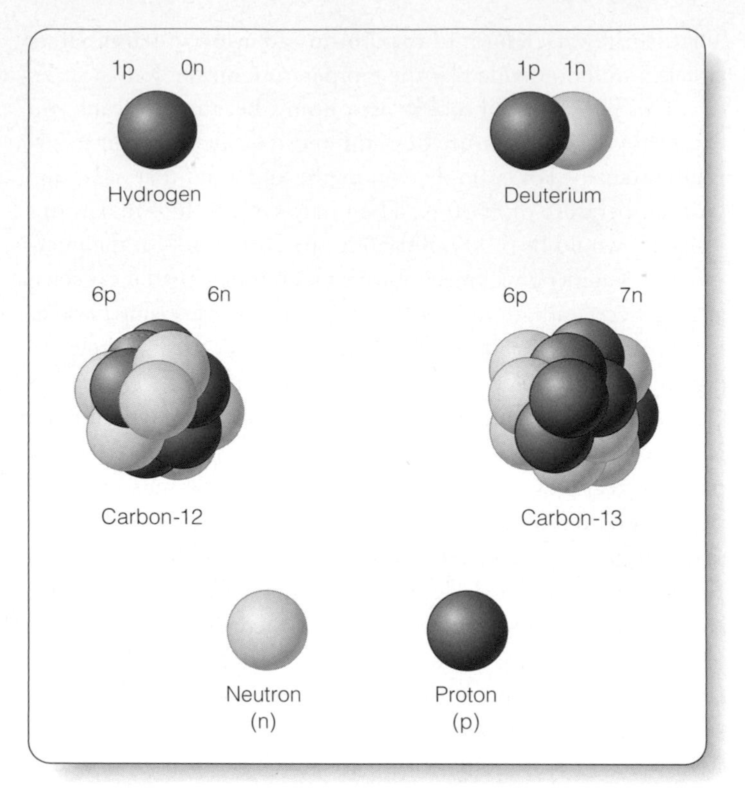

Figure 7-4

Some common isotopes. A rare isotope of hydrogen, deuterium, contains a proton and a neutron in its nucleus. Two isotopes of carbon are carbon-12 and carbon-13.

4.5 football fields in diameter, you can see that an atom is mostly empty space.

Different Kinds of Atoms

There are over a hundred kinds of atoms, called chemical elements. Which element an atom is depends only on the number of protons in the nucleus. For example, carbon has six protons in its nucleus. An atom with one more proton than this is nitrogen, and an atom with one proton fewer is boron.

Although the number of protons in an atom of an element is fixed, you could change the number of neutrons in an atom's nucleus without changing the atom significantly. For instance, if you added a neutron to a carbon nucleus, you would still have carbon, but it would be slightly heavier than normal carbon. Atoms that have the same number of protons but a different number of neutrons are **isotopes.** Carbon has two stable isotopes. One form contains six protons and six neutrons for a total of 12 particles and is thus called carbon-12. Carbon-13 has six protons and seven neutrons in its nucleus (■ Figure 7-4).

Protons and neutrons are bound tightly into the nucleus, but the electrons are held loosely in the electron cloud. Running a comb through your hair creates a static charge by removing a few electrons from their atoms. This process is called **ionization,** and the atom that has lost one or more electrons is an **ion.** A carbon atom is neutral if it has six electrons to balance the positive charge of the six protons in its nucleus. If you ionize the atom by removing one or more electrons, the atom is left with a net positive charge. Under some circumstances, an atom may capture one or more extra electrons, giving it more negative charges than positive. Such a negatively charged atom is also considered an ion.

Atoms that collide may form bonds with each other by exchanging or sharing electrons. Two or more atoms bonded together form a **molecule.** Atoms do collide in stars, but the high temperatures cause violent collisions that are unfavorable for chemical bonding. Only in the coolest stars are the collisions gentle enough to permit the formation of chemical bonds. You will see later that the presence of molecules such as titanium oxide (TiO) in a star is a clue that the star is very cool. In later chapters, you will see that molecules can form in cool gas clouds in space and in the atmospheres of planets.

Electron Shells

So far you have been thinking of the cloud of the whirling electrons in a general way, but now it is time to be more specific as to how the electrons behave within the cloud.

The electrons are bound to the atom by the attraction between their negative charge and the positive charge on the nucleus. This attraction is known as the **Coulomb force,** after the French physicist Charles-Augustin de Coulomb (1736–1806). To ionize

Window on Science 7-1

Quantum Mechanics: The World of the Very Small

Quantum mechanics is the set of rules that describe how atoms and subatomic particles behave. When you think about large objects such as stars, planets, aircraft carriers, and hummingbirds, you don't have to think about quantum mechanics, but on the atomic scale, particles behave in ways that seem unfamiliar.

One of the principles of quantum mechanics specifies that you cannot know simultaneously the exact location and motion of a particle. This is why physicists refer to the electrons in an atom as if they were a cloud of negative charge surrounding the nucleus. Because you can't know the position and motion of the electron, you can't really describe it as a small particle following an orbit. You can use that image as a model to help your imagination, but the reality is much more interesting, and describing the electrons as a charge cloud provides a better and more sophisticated model of an atom.

This raises some serious questions about reality. Is an electron really a particle at all? Quantum mechanics can describe particles as waves and waves as particles. If you can't know simultaneously the position and motion of a specific particle, how can you know how it will react to a collision with a photon or another particle? The answer is that you can't know, and that seems to violate the principle of cause and effect (Window on Science 5-1).

Needless to say, you can't expect to explore these discrepancies here. Scientists and philosophers of science continue to struggle with the meaning of reality on the quantum-mechanical level. Here you should note that the reality you see on the scale of stars and hummingbirds is only part of nature. You have constructed a model to help think about nature on the scale of atoms, but the truth is much more interesting and much more exciting than anything you see on larger scales. Although you will use models of atoms to study stars, there is still much to learn about the atoms themselves.

an atom, you need a certain amount of energy to pull an electron away from the nucleus. This energy is the electron's **binding energy,** the energy that holds it to the atom.

An electron may orbit the nucleus at various distances. If the orbit is small, the electron is close to the nucleus, and a large amount of energy is needed to pull it away. Consequently, its binding energy is large. An electron orbiting farther from the nucleus is held more loosely, and less energy will pull it away. That means it has less binding energy. The size of an electron's orbit is related to the energy that binds it to the atom.

Nature permits atoms only certain amounts (quanta) of binding energy, and the laws that describe how atoms behave are called the laws of **quantum mechanics** (Window on Science 7-1). Much of this discussion of atoms is based on the laws of quantum mechanics.

Because atoms can have only certain amounts of binding energy, your model atom can have orbits of only certain sizes, called **permitted orbits.** These are like steps in a staircase: You can stand on the number-one step or the number-two step, but not on the number-one-and-one-quarter step. The electron can occupy any permitted orbit but not orbits in between.

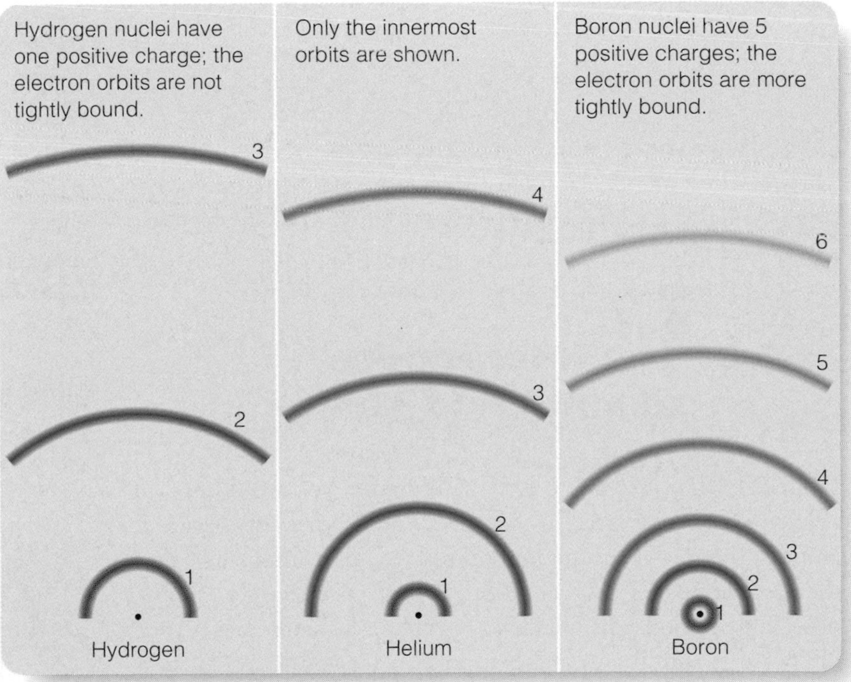

Hydrogen nuclei have one positive charge; the electron orbits are not tightly bound.
Only the innermost orbits are shown.
Boron nuclei have 5 positive charges; the electron orbits are more tightly bound.
3
2
1
Hydrogen
4
3
2
1
Helium
6
5
4
3
2
1
Boron

■ Figure 7-5

The electron in an atom may occupy only certain, permitted orbits. Because different elements have different charges on their nuclei, the elements have different, unique patterns of permitted orbits.

The arrangement of permitted orbits depends primarily on the charge of the nucleus, which in turn depends on the number of protons. Consequently, each kind of element has its own pattern of permitted orbits (■ Figure 7-5). Isotopes of the same elements have nearly the same pattern because they have the same number of protons. However, ionized atoms have orbital patterns that differ from their un-ionized forms. Thus the arrangement of permitted orbits differs for every kind of atom and ion.

Building Scientific Arguments

How many hydrogen atoms would it take to cross the head of a pin?

This is not a frivolous question. In answering it, you will discover how small atoms really are, and you will see how powerful physics and mathematics can be as a way to understand nature. Many scientific arguments are convincing because they have the precision of mathematics. To begin, assume that the head of a pin is about 1 mm in diameter. That is 0.001 m. The size of a hydrogen atom is represented by the diameter of the electron cloud, roughly 0.4 nm. Because 1 nm equals 10^{-9} m, you can multiply and discover that 0.4 nm equals 4×10^{-10} m. To find out how many atoms would stretch 0.001 m, you can divide the diameter of the pinhead by the diameter of an atom. That is, divide 0.001 m by 4×10^{-10} m, and you get 2.5×10^{6}. It would take 2.5 million hydrogen atoms lined up side by side to cross the head of a pin.

Now you can see how tiny an atom is and also how powerful a bit of physics and mathematics can be. It reveals a view of nature beyond the capability of your eyes. Now build an argument using another bit of arithmetic: **How many hydrogen atoms would you need to add up to the mass of a paper clip (1 g)?**

■ ■ ■

Connections: Astronomers are experts on electron orbits because the electrons in those orbits can interact with light. Such interactions fill starlight with clues to the secrets of the stars.

7-3 The Interaction of Light and Matter

IF LIGHT DID NOT INTERACT with matter, you would not be able to see these words. In fact, you would not exist, because, among other problems, photosynthesis would be impossible, and there would be no grass, wheat, bread, beef, cheeseburgers, or any other kind of food. The interaction of light and matter makes your life possible, and it also makes it possible for you to understand the universe.

You should begin your study of light and matter by considering the hydrogen atom. As you read earlier, hydrogen is both simple and common. Roughly 90 percent of all atoms in the universe are hydrogen.

The Excitation of Atoms

Each orbit in an atom represents a specific amount of binding energy, so physicists commonly refer to the orbits as **energy levels.** Using this terminology, you can say that an electron in its smallest and most tightly bound orbit is in its lowest permitted energy level. You could move the electron from one energy level to another by supplying enough energy to make up the difference between the two energy levels. It would be like moving a flowerpot from a low shelf to a high shelf; the greater the distance between the shelves, the more energy you would need to raise the pot. The amount of energy needed to move the electron is the energy difference between the two energy levels.

If you move the electron from a low energy level to a higher energy level, you can call the atom an **excited atom.** That is, you have added energy to the atom in moving its electron. If the electron falls back to the lower energy level, that energy is released.

An atom can become excited by collision. If two atoms collide, one or both may have electrons knocked into a higher energy level. This happens very commonly in hot gas, where the atoms move rapidly and collide often.

Another way an atom can get the energy that moves an electron to a higher energy level is to absorb a photon. Only a photon with exactly the right amount of energy can move the electron from one level to another. If the photon has too much or too little energy, the atom cannot absorb it. Because the energy of a photon depends on its wavelength, only photons of certain wavelengths can be absorbed by a given kind of atom. ■ Figure 7-6 shows the lowest four energy levels of the hydrogen atom along with three photons the atom could absorb. The longest-wavelength photon has only enough energy to excite the electron to the second energy level, but the shorter-wavelength photons can excite the electron to higher levels. A photon with too much or too little energy cannot be absorbed. Because the hydrogen atom has many more energy levels than shown in Figure 7-6, it can absorb photons of many different wavelengths.

■ **Figure 7-6**

A hydrogen atom can absorb only those photons that move the atom's electron to one of the higher-energy orbits. Here three different photons are shown along with the change they would produce if they were absorbed.

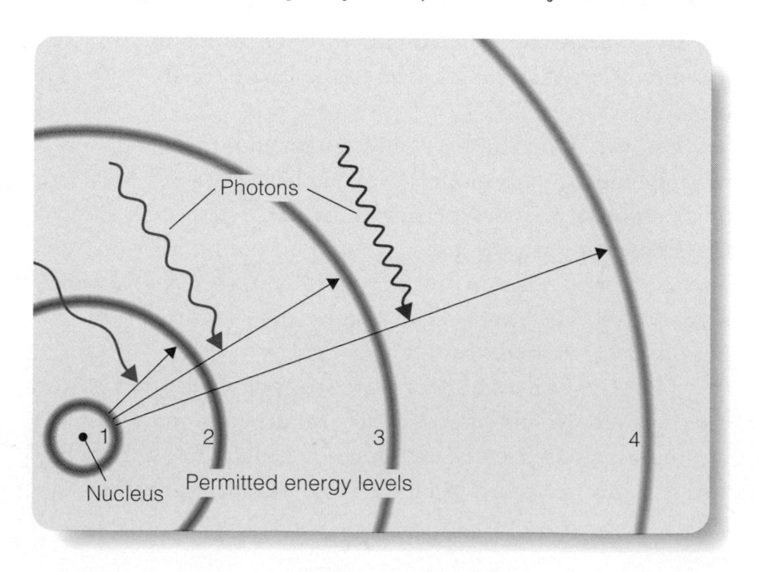

Atoms, like humans, cannot exist in an excited state forever. The excited atom is unstable and must eventually (usually within 10^{-6} to 10^{-9} seconds) give up the energy it has absorbed and return its electron to the lowest energy level. Because the electrons eventually tumble down to this bottom level, physicists call it the **ground state.**

When the electron drops from a higher to a lower energy level, it moves from a loosely bound level to one more tightly bound. The atom then has a surplus of energy—the energy difference between the levels—that it can emit as a photon. Study the sequence of events in ■ Figure 7-7 to see how an atom can absorb and emit photons. Because each type of atom or ion has its unique set of energy levels, each type absorbs and emits photons with a unique set of wavelengths. As a result, you can identify the elements in a gas by studying the characteristic wavelengths of light absorbed or emitted.

The process of excitation and emission is a common sight in urban areas at night. A neon sign glows when atoms of neon gas in the glass tube are excited by electricity flowing through the tube. As the electrons in the electric current flow through the gas, they collide with the neon atoms and excite them. As you have seen, immediately after an atom is excited, its electron drops back to a lower energy level, emitting the surplus energy as a photon of a certain wavelength. The photons emitted by excited neon produce a reddish-orange glow. Signs of other colors, erroneously called "neon," contain other gases or mixtures of gases instead of pure neon.

The Formation of a Spectrum

The spectrum of a star is formed as light passes outward through the gases near its surface. Study **Atomic Spectra** on pages 146–147. Notice three important properties of spectra:

1. There are three kinds of spectra described by three simple rules. When you see one of these types of spectra, you can recognize the kind of matter that emitted the light.
2. Notice that the wavelengths of the photons that are absorbed or emitted are determined by the atomic energy levels in the atoms. The emitted photons coming from a hot cloud of hydrogen gas have the same wavelengths as the photons absorbed by hydrogen atoms in the gases of a star. Although the hydrogen atom produces many spectral lines from the ultraviolet to the infrared, only three are visible to human eyes.
3. Most modern astronomy books display spectra as graphs of intensity versus wavelength. Be sure you see the connection between dark absorption lines and dips in the graphed spectrum.

Spectra are filled to bursting with information about the sources of the light; but, to extract that information, astronomers must be experts on the interaction of light and matter. Electrons moving among the orbits within atoms can reveal the secrets of the stars.

Ace Astronomy™ Log into AceAstronomy and select this chapter to see Astronomy Exercise "Emission and Absorption Spectra." This exercise will give you an inside look at atoms as they interact with photons.

Building Scientific Arguments

What spectrum would you see if you observed molten iron? To analyze this situation, you can begin your argument with the well-understood Kirchhoff's laws. Molten iron is a dense liquid, and the atoms and molecules collide so often they emit all wavelengths, so you would see a continuous spectrum. That is Kirchhoff's first law. But in order to see the molten iron, you would have to look through the hot vapors rising from it. Photons on their way to your spectrograph would pass through these gases, and atoms in the gases would absorb certain wavelengths. So what you would really see would be the continuous spectrum of the molten iron with weak absorption lines caused by the gases above the iron. This is what Kirchhoff's third law describes.

Now expand this argument. Suppose you had a very sensitive spectrograph that could look at the hot gases above the molten iron from the side so as to avoid looking directly at the molten iron. **What kind of spectrum would the hot gases emit?**

■ ■ ■

Connections: Whatever kind of spectrum astronomers look at, the most common spectral lines are the Balmer lines of hydrogen, the only hydrogen lines that can be studied from Earth's surface. In the next section, you will see how the Balmer lines work like a thermometer to measure a star's temperature.

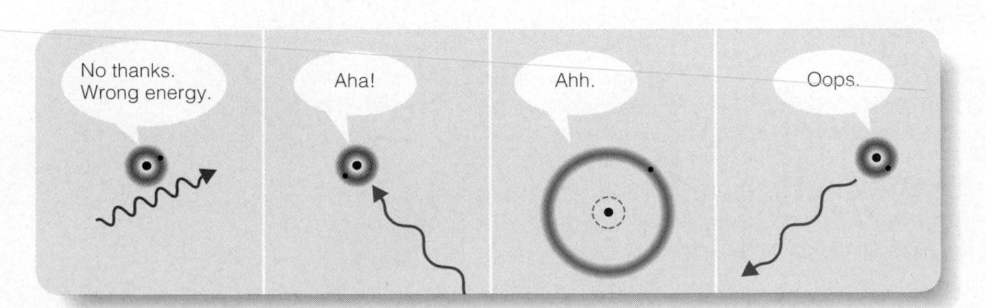

■ **Figure 7-7**

An atom can absorb a photon only if the photon has the correct amount of energy. The excited atom is unstable and within a fraction of a second returns to a lower energy level, reradiating the photon in a random direction.

Atomic Spectra

1 To understand how to analyze a spectrum, begin with a simple incandescent lightbulb. The hot filament emits black body radiation, which forms a **continuous spectrum.**

An **absorption spectrum** results when radiation passes through a cool gas. In this case you can imagine that the lightbulb is surrounded by a cool cloud of gas. Atoms in the gas absorb photons of certain wavelengths, which are missing from the spectrum, and you see their positions as dark **absorption lines.** Such spectra are sometimes called **dark-line spectra.**

An **emission spectrum** is produced by photons emitted by an excited gas. You could see **emission lines** by turning your telescope aside so that photons from the bright bulb did not enter the telescope. The photons you would see would be those emitted by the excited atoms near the bulb. Such spectra are also called **bright-line spectra.**

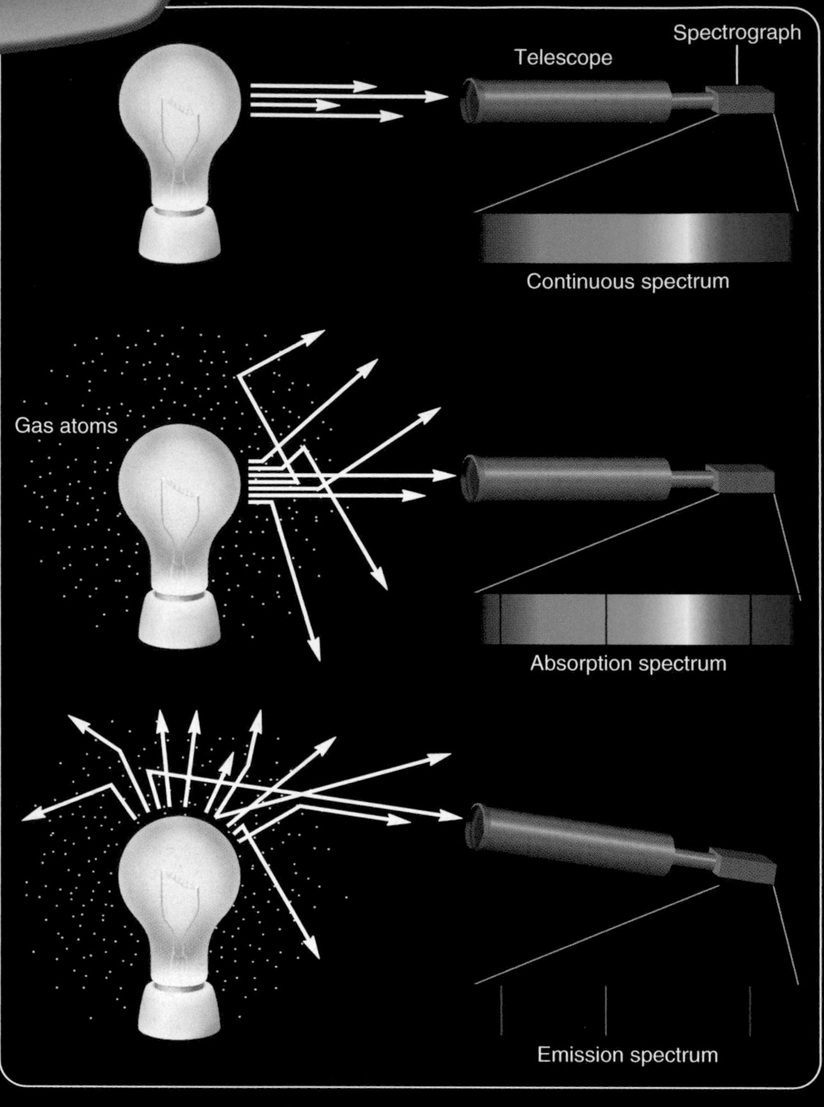

1a The spectrum of a star is an absorption spectrum. The denser layers of the photosphere emit black body radiation. Gases in the atmosphere of the star absorb their specific wavelengths and form dark absorption lines in the spectrum.

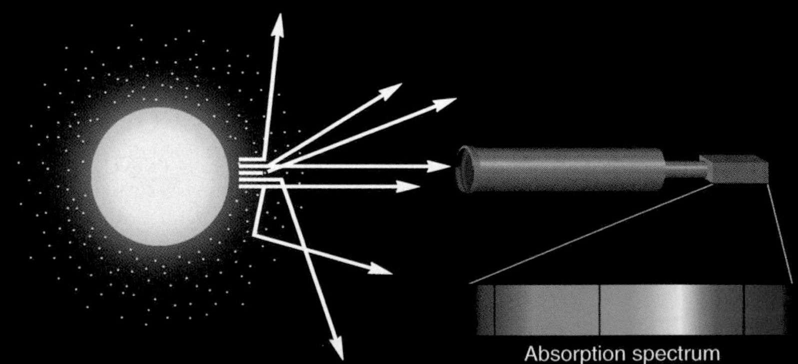

1b In 1859, long before scientists understood atoms and energy levels, the German scientist Gustav Kirchhoff formulated three rules, now known as **Kirchhoff's laws,** that describe the three types of spectra.

KIRCHHOFF'S LAWS

Law I: The Continuous Spectrum

A solid, liquid, or dense gas excited to emit light will radiate at all wavelengths and thus produce a continuous spectrum.

Law II: The Emission Spectrum

A low-density gas excited to emit light will do so at specific wavelengths and thus produce an emission spectrum.

Law III: The Absorption Spectrum

If light comprising a continuous spectrum passes through a cool, low-density gas, the result will be an absorption spectrum.

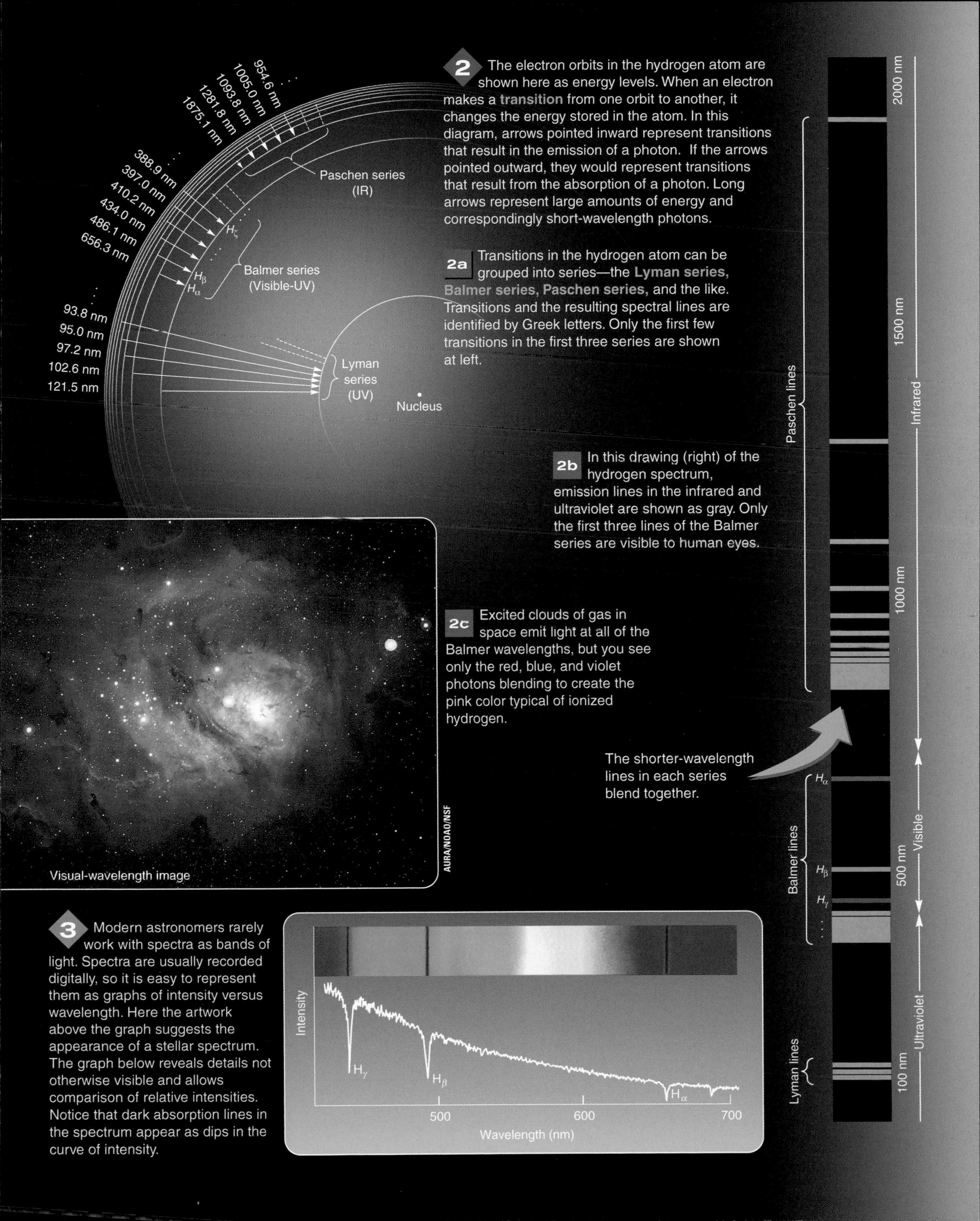

954.6 nm
1005.0 nm
1093.8 nm
1281.8 nm
1875.1 nm
Paschen series
(IR)
388.9 nm
397.0 nm
410.2 nm
434.0 nm
486.1 nm
656.3 nm
Hζ
Hβ
Hα
Balmer series
(Visible-UV)
93.8 nm
95.0 nm
97.2 nm
102.6 nm
121.5 nm
Lyman
series
(UV)
Nucleus
2 The electron orbits in the hydrogen atom are shown here as energy levels. When an electron makes a transition from one orbit to another, it changes the energy stored in the atom. In this diagram, arrows pointed inward represent transitions that result in the emission of a photon. If the arrows pointed outward, they would represent transitions that result from the absorption of a photon. Long arrows represent large amounts of energy and correspondingly short-wavelength photons.
2a Transitions in the hydrogen atom can be grouped into series—the Lyman series, Balmer series, Paschen series, and the like. Transitions and the resulting spectral lines are identified by Greek letters. Only the first few transitions in the first three series are shown at left.
2b In this drawing (right) of the hydrogen spectrum, emission lines in the infrared and ultraviolet are shown as gray. Only the first three lines of the Balmer series are visible to human eyes.
2c Excited clouds of gas in space emit light at all of the Balmer wavelengths, but you see only the red, blue, and violet photons blending to create the pink color typical of ionized hydrogen.
The shorter-wavelength lines in each series blend together.
Visual-wavelength image
AURA/NOAO/NSF
3 Modern astronomers rarely work with spectra as bands of light. Spectra are usually recorded digitally, so it is easy to represent them as graphs of intensity versus wavelength. Here the artwork above the graph suggests the appearance of a stellar spectrum. The graph below reveals details not otherwise visible and allows comparison of relative intensities. Notice that dark absorption lines in the spectrum appear as dips in the curve of intensity.
Intensity
Hγ
Hβ
Hα
500
600
700
Wavelength (nm)
Paschen lines
Balmer lines
Lyman lines
Hα
Hβ
Hγ
2000 nm
1500 nm
Infrared
1000 nm
500 nm
Visible
100 nm
Ultraviolet

7-4 Stellar Spectra

IN LATER CHAPTERS, you will use spectra to study galaxies and planets, but here you can begin by studying the spectra of stars. Such spectra are the easiest to understand, and the nature of stars is central to the study of all celestial objects.

The Balmer Thermometer

You can use the Balmer absorption lines as a thermometer to find the temperatures of stars. From the discussion of black body radiation, you know how to estimate temperature from color—red stars are cool, and blue stars are hot. (Remember that stellar spectra tell you about the surface of the star; that's where the light comes from.) You can estimate temperature from color, but the Balmer lines give you much greater accuracy.

The Balmer thermometer works because the Balmer absorption lines are produced only by atoms whose electrons are in the second energy level. If the star is cool, there are few violent collisions between atoms to excite the electrons, and most atoms have their electrons in the ground state. If most electrons are in the ground state, they can't absorb photons in the Balmer series. As a result, you should expect to find weak Balmer absorption lines in the spectra of cool stars.

In the surface layers of stars hotter than about 20,000 K, on the other hand, there are many violent collisions between atoms, exciting electrons to high energy levels or knocking the electron clear out of some atoms; that is, some atoms are ionized. Consequently, few hydrogen atoms have their electron in the second orbit to form Balmer absorption lines, and you should expect hot stars, like cool stars, to have weak Balmer absorption lines.

At an intermediate temperature, roughly 10,000 K, the collisions are just right to excite large numbers of electrons into the second energy level. With many atoms excited to the second level, the gas absorbs Balmer wavelength photons strongly and produces strong Balmer lines.

To summarize, the strength of the Balmer lines depends on the temperature of the star's surface layers. Both hot and cool stars have weak Balmer lines, but medium-temperature stars have strong Balmer lines.

Theoretical calculations can predict just how strong the Balmer lines should be for stars of various temperatures. Such calculations are the key to finding temperatures from stellar spectra. The curve in ■ Figure 7-8a shows the strength of the Balmer lines for various stellar temperatures. You could use this as a temperature indicator, except that the curve gives two answers. A star with Balmer lines of a certain strength might have either of two temperatures, one high and one low. How do you know which is the right answer? You must examine other spectral lines to choose the correct temperature.

You have seen how the strength of the Balmer lines depends on temperature. Temperature has a similar effect on the spectral lines of other elements, but the temperature at which the lines reach maximum strength differs for each element (Figure 7-8b).

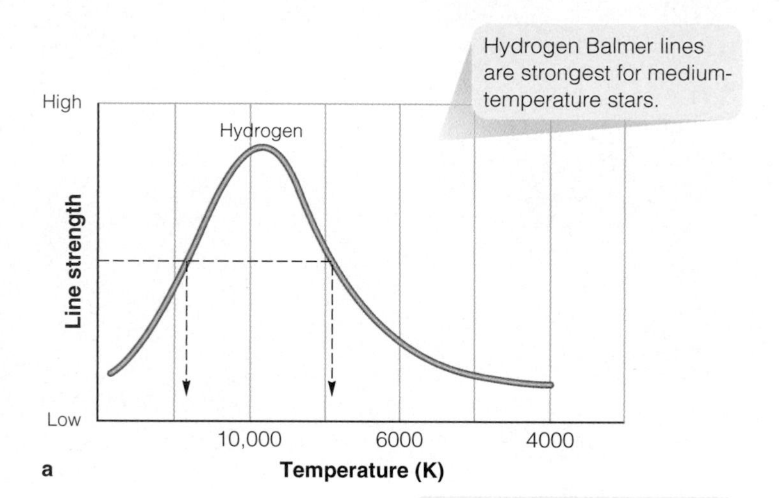

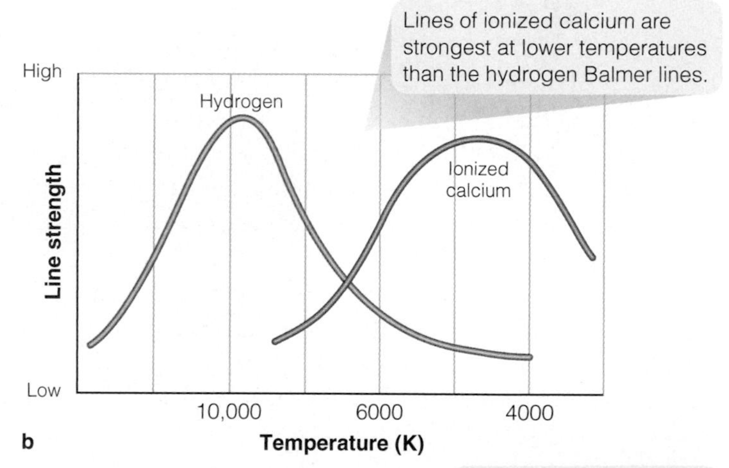

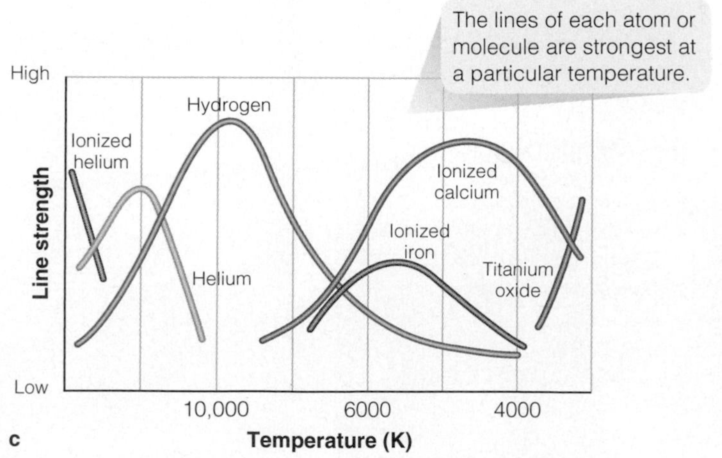

■ **Figure 7-8**

The strength of spectral lines can tell you the temperature of a star. (a) Balmer hydrogen lines alone are not enough because they give two answers. Balmer lines of a certain strength could be produced by a hotter star or a cooler star. (b) Adding another atom to the diagram helps, and (c) adding many atoms and molecules to the diagram creates a precise tool to find the temperatures of stars.

If you add these elements to your graph, you get a powerful tool for finding the stars' temperatures (Figure 7-8c).

How do you use this tool? You determine a star's temperature by comparing the strengths of its spectral lines with your graph. For instance, if you recorded the spectrum of a star and found medium-strength Balmer lines and strong helium lines, you could conclude it had a temperature of about 20,000 K. But if the star had weak hydrogen lines and strong lines of ionized iron, you would assign it a temperature of about 5800 K, similar to that of the sun.

The spectra of stars cooler than about 3000 K contain dark bands produced by molecules such as titanium oxide (TiO). Because of their structure, molecules can absorb photons at many wavelengths, producing numerous, closely spaced spectral lines that blend together to form bands. These molecular bands appear only in the spectra of the coolest stars because, as mentioned before, molecules in cool stars are not subject to the violent collisions that would break up molecules in hotter stars. Consequently, the presence of dark bands in a star's spectrum indicates that the star is very cool.

From stellar spectra, astronomers have found that the hottest stars have surface temperatures above 40,000 K and the coolest about 2000 K. Compare these with the surface temperature of the sun, which is about 5800 K.

Spectral Classification

You have seen that the strengths of spectral lines depend on the surface temperature of the star. From this you can conclude that all stars of a given temperature should have similar spectra. If you learn to recognize the pattern of spectral lines produced by a 6000 K star, for instance, you need not use Figure 7-8c every time you see that kind of spectrum. You can save time by classifying stellar spectra rather than analyzing each one individually.

The first widely used classification system was devised by astronomers at Harvard during the 1890s and 1900s. One of them, Annie J. Cannon, personally inspected and classified the spectra of over 250,000 stars. The spectra were first classified in groups labeled A through Q, but some groups were later dropped, merged with others, or reordered. The final classification includes the seven major **spectral classes, or types,** still used today: O, B, A, F, G, K, M.*

This sequence of spectral types, called the **spectral sequence,** is important because it is a temperature sequence. The O stars are the hottest, and the temperature decreases down to the M stars, the coolest. For maximum precision, astronomers divide each spectral class into 10 subclasses. For example, spectral class A consists of the subclasses A0, A1, A2, . . . A8, A9. Next comes F0, F1, F2, and so on. This finer division gives a star's temperature to an accuracy within about 5 percent. The sun, for example, is not just a G star, but a G2 star, with a temperature of 5800 K.

Astronomers classify a star by the lines and bands in its spectrum, as shown in ■ Table 7-1. For example, if it has weak Balmer lines and lines of ionized helium, it must be an O star. This table is based on the same information shown in Figure 7-8c.

■ Figure 7-9 shows 13 stellar spectra ranging from the hottest at the top to the coolest at the bottom. Notice how special features change gradually from hot to cool stars. Although these spectra are attractive, astronomers rarely work with spectra as color images. Rather, they display spectra as graphs of intensity versus wavelength with dark absorption lines as dips in the graph (■ Figure 7-10). Such graphs show more detail than photographs. Notice also that the overall curves are similar to black body curves. The wavelength of maximum intensity is in the infrared for the coolest stars and in the ultraviolet for the hottest stars.

Compare Figures 7-9 and 7-10 and notice how the strength of spectral lines depends on temperature. Note that the Balmer lines are strongest in A stars, where the temperature is moderate but still high enough to excite the electrons in hydrogen atoms

*Generations of astronomy students have remembered the spectral sequence using the mnemonic "Oh, Be A Fine Girl (Guy), Kiss Me." More recent suggestions from students include, "Oh Boy, An F Grade Kills Me," and "Only Bad Astronomers Forget Generally Known Mnemonics."

■ **Table 7-1 | Spectral Classes**

Spectral Class	Approximate Temperature (K)	Hydrogen Balmer Lines	Other Spectral Features	Naked-Eye Example
O	40,000	Weak	Ionized helium	Meissa (O8)
B	20,000	Medium	Neutral helium	Achernar (B3)
A	10,000	Strong	Ionized calcium weak	Sirius (A1)
F	7,500	Medium	Ionized calcium weak	Canopus (F0)
G	5,500	Weak	Ionized calcium medium	Sun (G2)
K	4,500	Very weak	Ionized calcium strong	Arcturus (K2)
M	3,000	Very weak	TiO strong	Betelgeuse (M2)

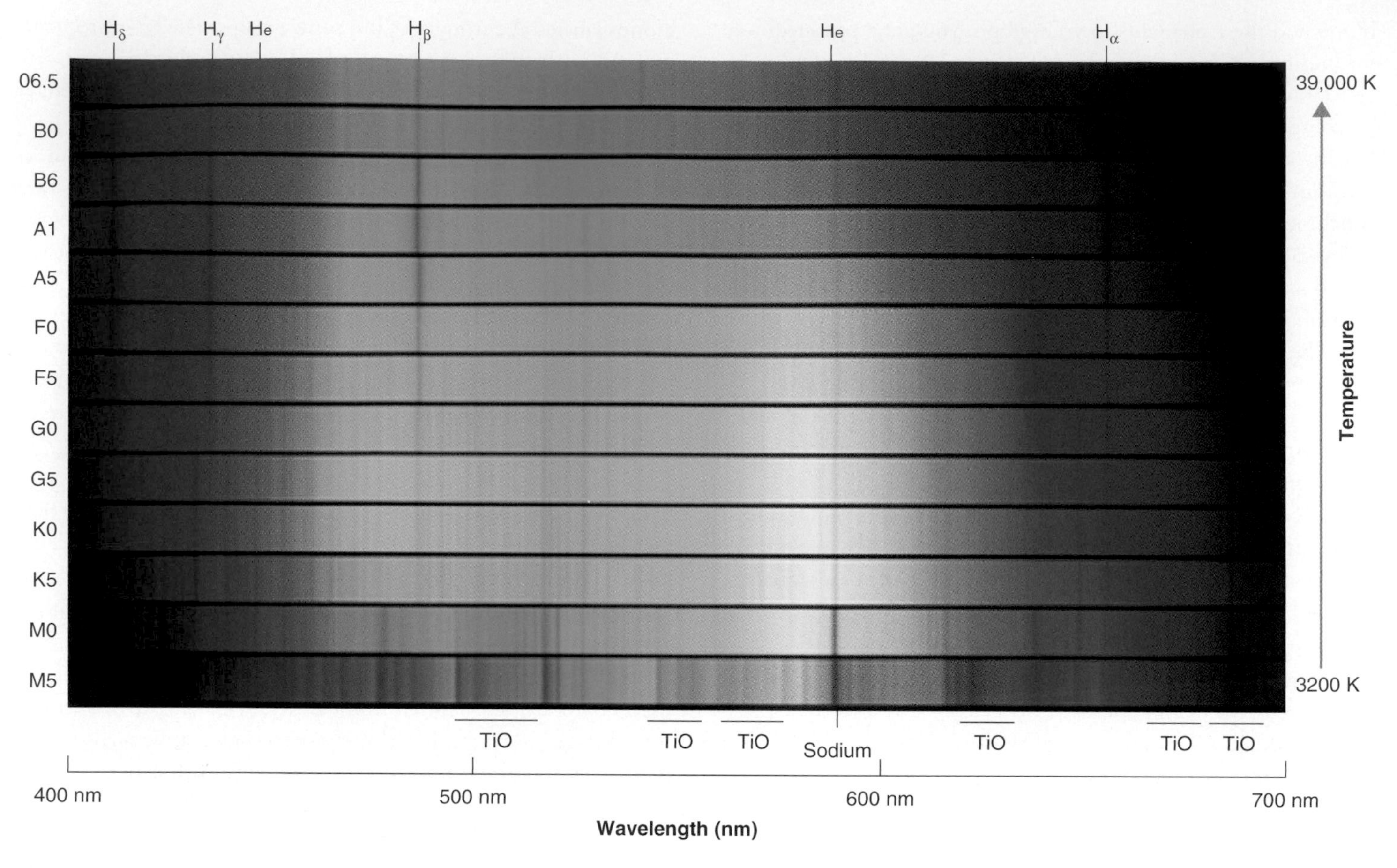

■ Figure 7-9

These spectra show stars from hot O stars at the top to cool M stars at the bottom. The Balmer lines of hydrogen are strongest about A0, but the two closely spaced lines of sodium in the yellow are strongest for very cool stars. Helium lines appear only in the spectra of the hottest stars. Notice that the helium line visible in the top spectrum has nearly but not exactly the same wavelength as the sodium lines visible in cooler stars. Bands produced by the molecule titanium oxide are strong in the spectra of the coolest stars. (AURA/NOAO/NSF)

to the second energy level, where they can absorb Balmer wavelength photons. In the hotter stars (O and B), the Balmer lines are weak because the higher temperature excites the electrons to energy levels above the second or ionizes the atoms. The Balmer lines in cooler stars (F through M) are also weak but for a different reason. The lower temperature cannot excite many electrons to the second energy level, so few hydrogen atoms are capable of absorbing Balmer wavelength photons.

The spectral lines of other atoms also change from class to class. Helium is visible only in the spectra of the hottest classes, and titanium oxide bands only in the coolest. Two lines of ionized calcium increase in strength from A to K and then decrease from K to M. Because the strength of these spectral lines depends on temperature, it requires only a few moments to study a star's spectrum and determine its temperature.

Now you can learn something new about our Favorite Stars. Sirius, brilliant in the winter sky, is an A1 star; and Vega, bright overhead in the summer sky, is an A0 star. They have nearly the same temperature and color and strong Balmer lines in their spectra. The bright red star in Orion is Betelgeuse, a cool M2 star, but blue-white Rigel is a hot B8 star. Polaris, the North Star, is an F8 star a bit hotter than our sun, and Alpha Centauri, the closest star to the sun, seems to be a G2 star just like the sun.

The study of spectral types is a century old, but astronomers continue to discover new types of stars. The **L dwarfs,** found in 1998, are cooler and fainter than M stars. The L dwarfs are

■ Figure 7-10 ▶

Modern digital spectra are often represented by graphs of intensity versus wavelength with dark absorption lines appearing as sharp dips in the curves. The hottest stars are at the top and the coolest at the bottom. Hydrogen Balmer lines are strongest at about A0, while lines of ionized calcium (CaII) are strong in K stars. Titanium oxide (TiO) bands are strongest in the coolest stars. Compare these spectra with Figures 7-8c and 7-9. (Courtesy NOAO, G. Jacoby, D. Hunter, and C. Christian)

clearly a different type of star. The spectra of M stars contain bands produced by metal oxides such as titanium oxide, but L dwarf spectra contain bands produced by molecules such as iron hydride (FeH) (■ Figure 7-11). The **T dwarfs,** discovered in 2000, are even cooler and fainter than L dwarfs. Their spectra show absorption by methane (CH_4) and water vapor. The development of giant telescopes and highly sensitive infrared cameras and spectrographs is allowing astronomers to find and study these coolest of stars.

The Composition of the Stars

It seems as though it should be easy to find the composition of the sun and stars just by looking at their spectra, but it turns out to be a difficult task that wasn't well understood until the 1920s. The story of how astronomers first discovered the composition of the stars is worth telling, not only because it is the story of an important American astronomer who never got proper credit, but also because the story illustrates the temperature dependence of spectral features. The story begins in England.

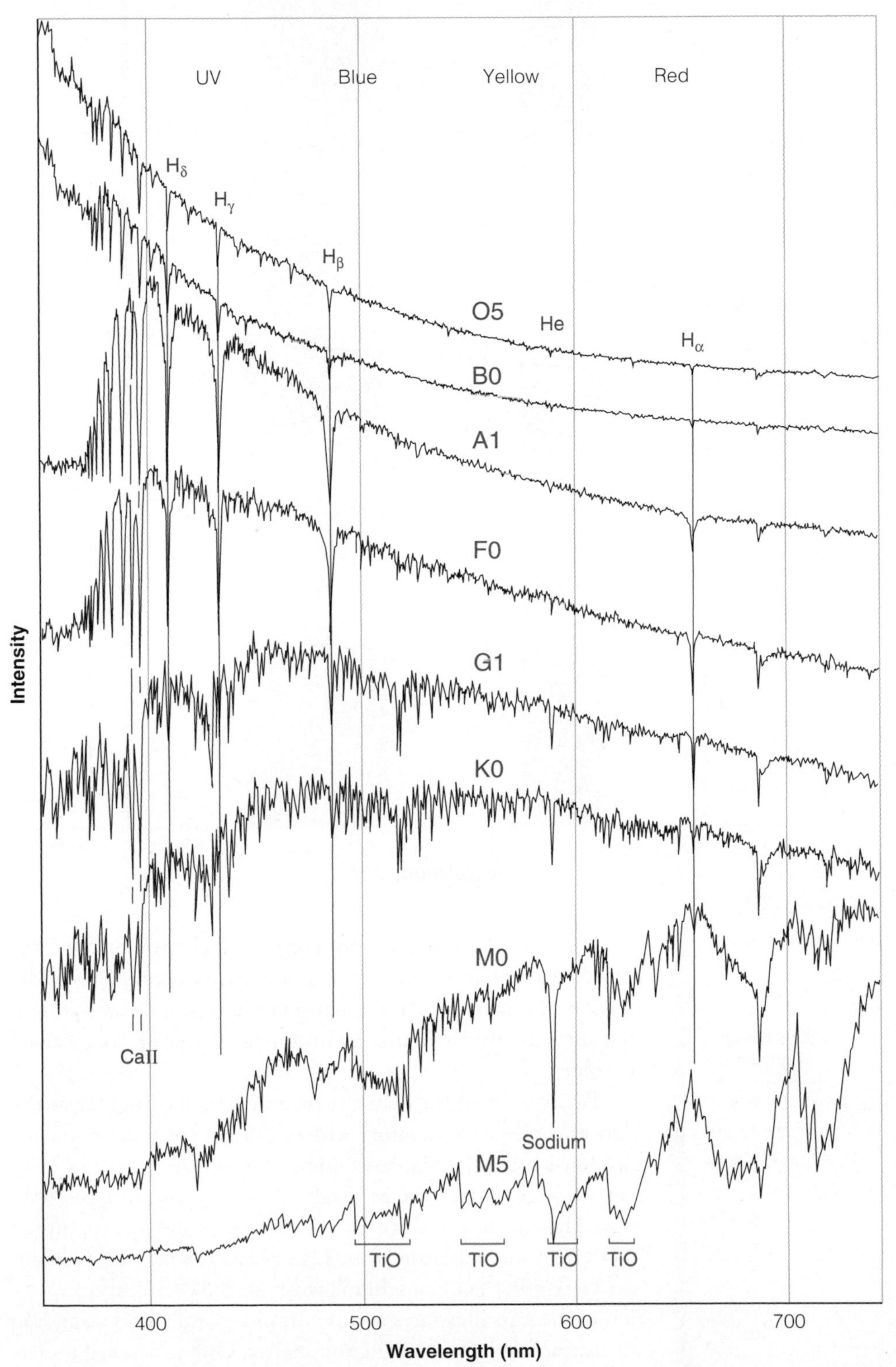

As a child in England, Cecilia Payne (1900–1979) excelled in classics, languages, mathematics, and literature, but her first love was astronomy. After finishing Newnham College in Cambridge, she left England, sensing that there were no opportunities in England for a woman of science. In 1922, Payne arrived at Harvard, where she eventually earned her Ph.D., although her degree was awarded by Radcliffe because Harvard did not then admit women.

In her thesis, Payne attempted to relate the strength of the absorption lines in stellar spectra to the physical conditions in the atmospheres of the stars. This was not easy because, as you have seen in this chapter, a given spectral line can be weak because the atom is rare or because the temperature is too high or too low for that atom to be able to absorb efficiently. If you see sodium lines in a star's spectrum, you can be sure that the star contains sodium atoms, but if you see no sodium lines, you must consider the possibility that the star is too hot or too cool for sodium to produce spectral lines.

Payne's problem was to untangle these two factors and find the true temperatures of the stars and the true abundance of the atoms in their atmospheres. Recent advances in atomic physics gave her the theoretical tools she needed. About the time Payne left Newnham College, Indian physicist Meghnad Saha published his work on the ionization of atoms. Drawing from such theoretical work, Payne was able to show

■ Figure 7-11

These six infrared spectra show the dramatic differences between L dwarfs and T dwarfs. Spectra of M stars show titanium oxide bands (TiO), but L and T dwarfs are so cool that TiO molecules do not form. Other molecules such as iron hydride (FeH), water (H_2O), and methane (CH_4) can form in these very cool stars. (Adapted from Thomas R. Geballe, Gemini Observatory, from a graph that originally appeared in *Sky and Telescope Magazine*, February 2005, p. 37.)

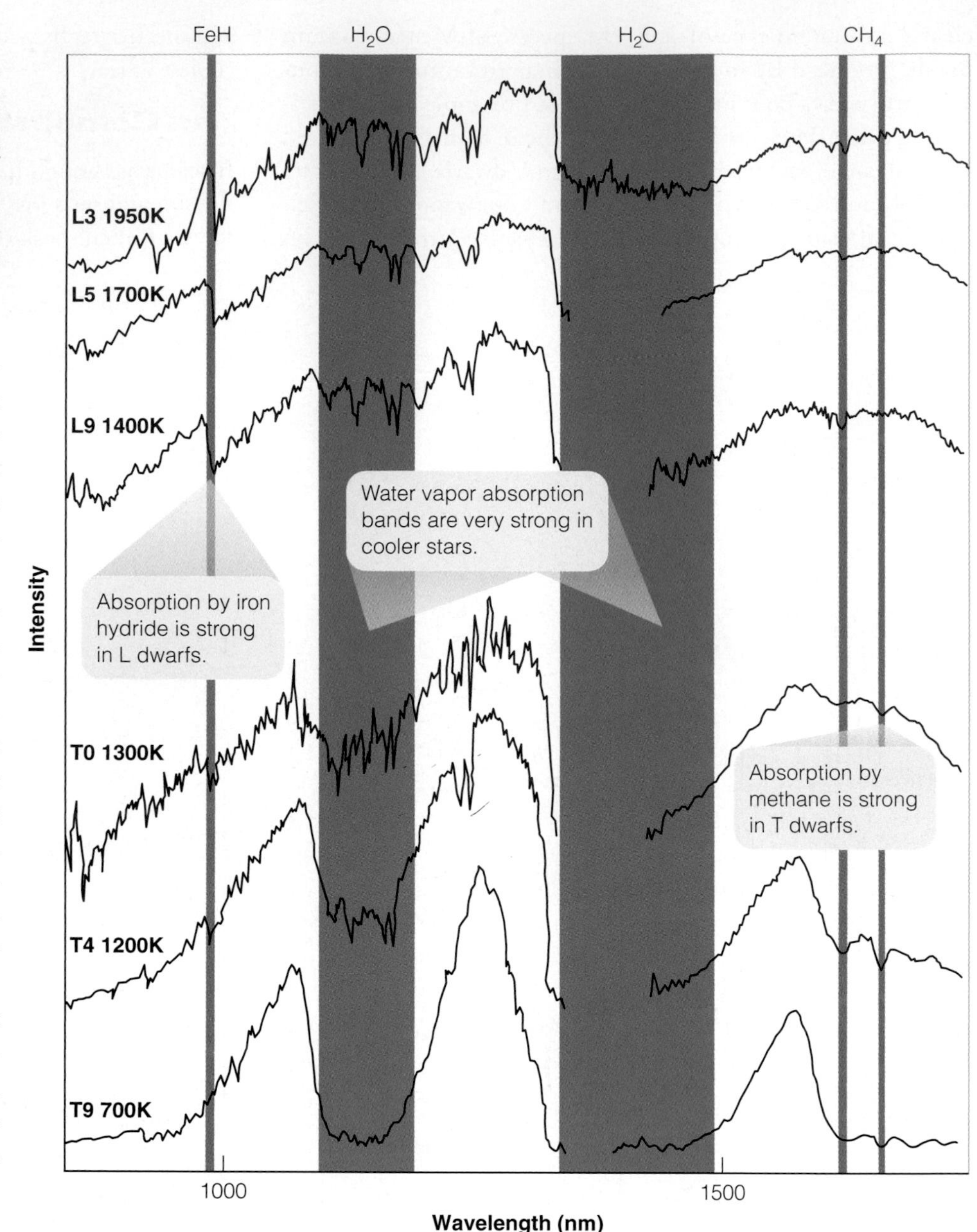

that over 90 percent of the atoms in stars (including the sun) were hydrogen and most of the rest helium (■ Table 7-2). The heavier atoms like calcium, sodium, and iron seem more abundant only because they are better at absorbing photons at the temperatures of stars.

At the time, astronomers found it hard to believe Payne's abundances of hydrogen and helium. They especially found the abundance of helium unacceptable. After all, hydrogen lines are at least visible in most stellar spectra, but helium lines are almost invisible in the spectra of all but the hottest stars. Rather, nearly all astronomers assumed that the stars had roughly the same composition as Earth's surface; that is, they believed that the stars were composed mainly of heavier atoms such as carbon, silicon, iron, aluminum, and so on. Even the most eminent astronomers dismissed Payne's result as illusory. Faced with this pressure and realizing the limited opportunities available to women in science in the 1920s, Payne could not press her discovery.

■ Table 7-2 I The Most Abundant Elements in the Sun

Element	Percentage by Number of Atoms	Percentage by Mass
Hydrogen	91.0	70.9
Helium	8.9	27.4
Carbon	0.03	0.3
Nitrogen	0.008	0.1
Oxygen	0.07	0.8
Neon	0.01	0.2
Magnesium	0.003	0.06
Silicon	0.003	0.07
Sulfur	0.002	0.04
Iron	0.003	0.1

It was 1929 before astronomers generally understood the importance of temperature in measurements of composition derived from stellar spectra. At that point, astronomers recognized that stars are mostly hydrogen and helium, but Payne received no credit.

Payne worked for many years as a staff astronomer at the Harvard College Observatory with no formal position on the faculty. She married Russian astronomer Sergei Gaposchkin in 1934 and was afterward known as Cecilia Payne-Gaposchkin. In 1956, when Harvard accepted women to its faculty, she was appointed a full professor and chair of the Harvard astronomy department.

Cecilia Payne-Gaposchkin's work on the chemical composition of the stars illustrates the importance of fully understanding the interaction between light and matter. Only a detailed understanding of the physics could lead her to the correct composition. As you turn your attention to other information that can be derived from stellar spectra, you will again discover the importance of understanding light.

Ace Astronomy™ Log into AceAstronomy and select this chapter to see Astronomy Exercise "Stellar Atomic Absorption Lines." You can change the chemical composition of a star and watch its spectrum change.

The Doppler Effect

Surprisingly, one of the pieces of information hidden in a spectrum is the velocity of the light source. Astronomers can measure the wavelengths of lines in a star's spectrum and find the velocity of the star. The **Doppler effect** is the apparent change in the wavelength of radiation caused by the motion of the source.

You can detect the Doppler shift in sound. Sound is not electromagnetic radiation, of course; it is a mechanical wave transmitted through air, but because it is a wave, it is subject to the Doppler effect. Sounds with long wavelengths have low pitches, and sounds with short wavelengths have higher pitches. When a car is approaching you on the highway, you hear its sound waves with a slightly shorter wavelength, and they have a higher pitch. As the car passes you, the pitch drops, and you hear slightly longer wavelengths. That's the Doppler effect.

If a star is moving toward Earth, the lines in its spectrum will be shifted slightly toward shorter wavelengths. That is, they are shifted toward the blue end of the spectrum—a **blueshift.** If a star is moving away from Earth, the lines are shifted slightly toward the red end of the spectrum—a **redshift.** These Doppler shifts are small and don't change the color of the star. But it is easy to detect these changes in wavelength in a star's spectrum.

■ Figure 7-12a shows the Doppler effect in two spectra of the star Arcturus. The lines in the top spectrum are slightly blueshifted because the spectrum was recorded when Earth, following its orbit, was moving toward Arcturus. The lines in the bottom spectrum are redshifted because it was recorded six months later, when Earth was moving away from Arcturus. The greater the shift, the greater the velocity, so astronomers can measure velocities this way.

You may be quite familiar with the Doppler effect. Meteorologists use Doppler radar to measure the velocities of clouds and weather systems. A large redshift close to a large blueshift is a warning of a possible tornado. Or perhaps you have had your own velocity measured by police radar as you drive down the highway. Radar can even be used in sports to measure the speed of a baseball or tennis ball. It's just redshifts and blueshifts.

The Doppler effect tells you how rapidly the distance between you and the source of light is increasing or decreasing. It does not matter whether you are moving or the star is moving. Only the relative velocity is important. Also, the Doppler shift is sensitive only to the part of the velocity directed away from you or toward you. This part of the velocity is called the **radial velocity** ***(V_r),*** as shown in ■ Figure 7-13.

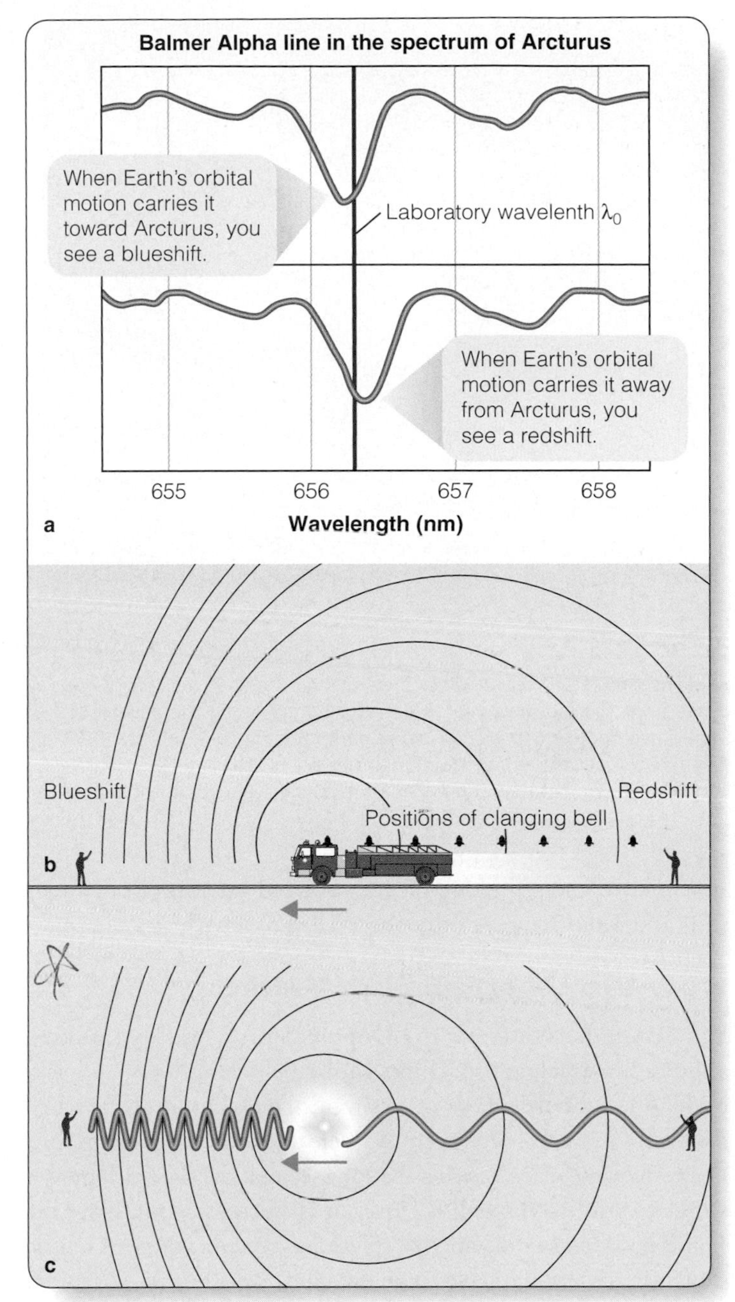

Balmer Alpha line in the spectrum of Arcturus
When Earth's orbital motion carries it toward Arcturus, you see a blueshift.
Laboratory wavelenth λ_0
When Earth's orbital motion carries it away from Arcturus, you see a redshift.
655
656
657
658
Wavelength (nm)
a
Blueshift
Redshift
Positions of clanging bell
b
c

■ Figure 7-12

The Doppler effect. (a) Earth's orbital motion causes small Doppler sifts in a star's spectrum. (b) The clanging bell on a moving fire truck produces sounds that move outward (black circles). An observer ahead of the truck hears the clangs closer together, while an observer behind the truck hears them farther apart. (c) A moving source of light emits waves that move outward (black circles). An observer in front of the light source observes a shorter wavelength (a blueshift), and an observer behind the light source observes a longer wavelength (a redshift).

The Doppler effect occurs for any kind of radiation, not just light. Radio astronomers, for instance, observe the Doppler effect when they measure the wavelength of radio waves coming from objects moving toward or away from Earth. Whatever wavelength range you consider—radio, X rays, gamma rays—you can

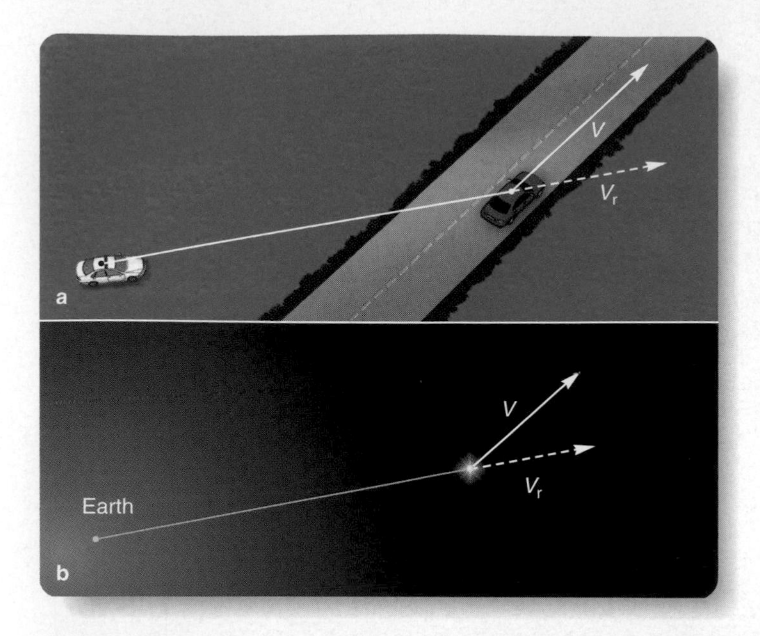

■ **Figure 7-13**

(a) Police radar can measure only the radial part of your velocity *(V_r)* as you drive down the highway, not your true velocity along the pavement *(V)*. That is why police using radar never park far from the highway. (b) From Earth, astronomers can use the Doppler effect to measure the radial velocity *(V_r)* of a star, but they cannot measure its true velocity, *V*, through space.

still refer to a lengthening of the observed wavelength as a redshift and a shortening of the observed wavelength as a blueshift.

How the Doppler Shift Works

You can understand how the Doppler shift works by thinking about a fire truck approaching with a bell clanging once a second. When the bell clangs, the sound travels ahead of the truck to reach your ears. One second later, the bell clangs again, but not at the same place. During that one second, the fire truck moved closer to you, so the bell is closer at its second clang. Now the sound has a shorter distance to travel and reaches your ears a little sooner than it would have if the fire truck were not approaching. The third time the bell clangs, it is even closer. By timing the bell, you could observe that the clangs are slightly less than 1 second apart, all because the fire truck is approaching. If the fire truck were moving away from you, you would hear the clangs sounding more than one second apart, because each successive clang of the bell would occur farther away.

Figure 7-12b shows a fire truck moving toward one observer and away from another observer. The position of the bell at each clang is shown by the small black bells with the sound spreading outward as black circles. You can see that the clangs are squeezed together ahead of the fire truck and stretched apart behind.

Now you can substitute a source of light waves for the clanging of the bell. If the source is approaching, then each time the source emits the peak of a wave it will be slightly closer to you, and you will observe a shorter wavelength. If it is moving away, you will observe a longer wavelength. This is shown in Figure 7-12c, where the peaks of the waves appear compressed in front of the moving light source and stretched out behind the source.

Of course, how great the change in wavelength is depends on the velocity. Just as a slowly moving car has a smaller Doppler effect in sound than does a high-speed airplane, a slowly moving star has a smaller Doppler effect in light than does a high-speed star. Consequently, you can measure the velocity by measuring the amount by which the spectral lines are shifted in wavelength. A large Doppler shift means a high radial velocity.

Calculating the Doppler Velocity

It is easy to calculate the radial velocity of an object from its Doppler shift. The formula is a simple ratio relating the radial velocity V_r divided by the speed of light c to the change in wavelength, λ, divided by the unshifted wavelength, λ_0:

$$\frac{V_r}{c} = \frac{\Delta\lambda}{\lambda_0}$$

For example, suppose you observed a line in a star's spectrum with a wavelength of 600.1 nm. Laboratory measurements show that the line should have a wavelength of 600 nm. That is, its unshifted wavelength is 600 nm. What is the star's radial velocity? First note that the change in wavelength is 0.1 nm:

$$\frac{V_r}{c} = \frac{0.1}{600} = 0.000167$$

Multiplying by the speed of light, 3×10^5 km/s, gives the radial velocity, 50 km/s. Because the wavelength is shifted to the red (lengthened), the star must be receding.

Now that you understand the Doppler shift you can understand a final illustration of the information hidden in stellar spectra. Even the shapes of the spectral lines can reveal secrets about the stars.

Ace Astronomy™ Log into AceAstronomy and select this chapter to see Astronomy Exercise "Doppler Shift." Compare the Doppler shift for a source of sound with that for a source of light.

The Shapes of Spectral Lines

When astronomers refer to the shape of a spectral line, they mean the variation of intensity across the line. An absorption line, for instance, is darkest in the center and brighter to each side. Two examples are shown in ■ Figure 7-14.

The exact shape of a line can reveal a great deal about a star, but the most important characteristic is the width of the line. Spectral lines are not perfectly narrow; if they were, they would be undetectable. They have a natural width because nature allows an atom some leeway in the energy it may absorb or emit. In the absence of all other effects, spectral lines have a natural width of about 0.001 to 0.00001 nm—very narrow indeed.

The natural widths of spectral lines are not important in most branches of astronomy because other effects smear out the

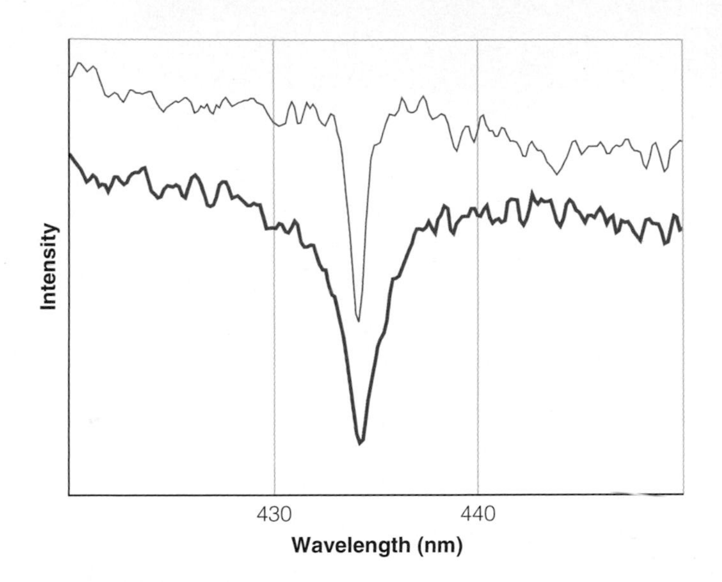

■ Figure 7-14

Here you see two dark absorption lines magnified from the spectra of two A1 stars. The upper line is quite narrow, but the bottom line is much broader. Because the two stars have the same spectral type, you can trust that they must have the same temperature. The stars differ not in temperature but in gas density. The star with the narrow spectral lines has a very low-density atmosphere. Precise observations of the shapes of spectral lines can reveal a great deal about stars. (Courtesy NOAO, G. Jacoby, D. Hunter, and C. Christian)

lines and make them much broader. For example, if a star spins rapidly, the Doppler effect will broaden the spectral lines. As the star rotates, one side will recede from Earth, and the other side will approach Earth. Light from the receding side will be redshifted, and light from the approaching side will be blueshifted, so any spectral lines will be broadened. Astronomers can measure a star's rotation rate from the width of its spectral lines.

Another important process is called **Doppler broadening.** To consider this process, imagine that you photograph the spectrum of a jar full of hydrogen atoms (■ Figure 7-15). Because the gas has some thermal energy (it is not at absolute zero), the gas atoms are in motion. Some will be coming toward your spectrograph, and some will be receding. Most, of course, will not be traveling very fast, but some will be moving very quickly. The photons emitted by the atoms approaching you will have slightly shorter wavelengths because of the Doppler effect, and photons emitted by atoms receding from you will have slightly longer wavelengths. Thus, the Doppler shifts due to the motions of the individual atoms will smear the spectral line out and make it broader. This summary describes the Doppler broadening of an emission line, but the effect is the same for absorption lines.

The extent of Doppler broadening depends on the temperature of the gas. If the gas is cold, the atoms travel at low velocities, and the Doppler shifts are small (Figure 7-15a). If the gas is hot, however, the atoms travel faster, Doppler shifts are larger, and the lines will be wider (Figure 7-15b). Sometimes astronomers will estimate the temperature of a cloud of gas in space by looking at the widths of its spectral lines.

Another form of broadening, **collisional broadening,** is caused by collisions between atoms, and consequently it depends on the density of the gas (**Focus on Fundamentals 4**). Densities in astronomy cover an enormous range, from one atom per cubic centimeter in space to millions of tons of atoms per cubic centimeter inside dead stars. Clearly, you would expect such densities to affect the way atoms collide with one another and how they absorb and emit photons.

Collisional broadening spreads out spectral lines when the atoms absorb or emit photons while they are colliding with other

■ Figure 7-15

Doppler broadening. The atoms of a gas are in constant motion. Photons emitted by atoms moving toward the observer will have slightly shorter wavelengths, and those emitted by atoms moving away will have slightly longer wavelengths. This broadens the spectral line. If the gas is cool (a), the atoms do not move very fast, the Doppler shifts are small, and the line is narrow. If the gas is hot (b), the atoms move faster, the Doppler shifts are larger, and the line is broader.

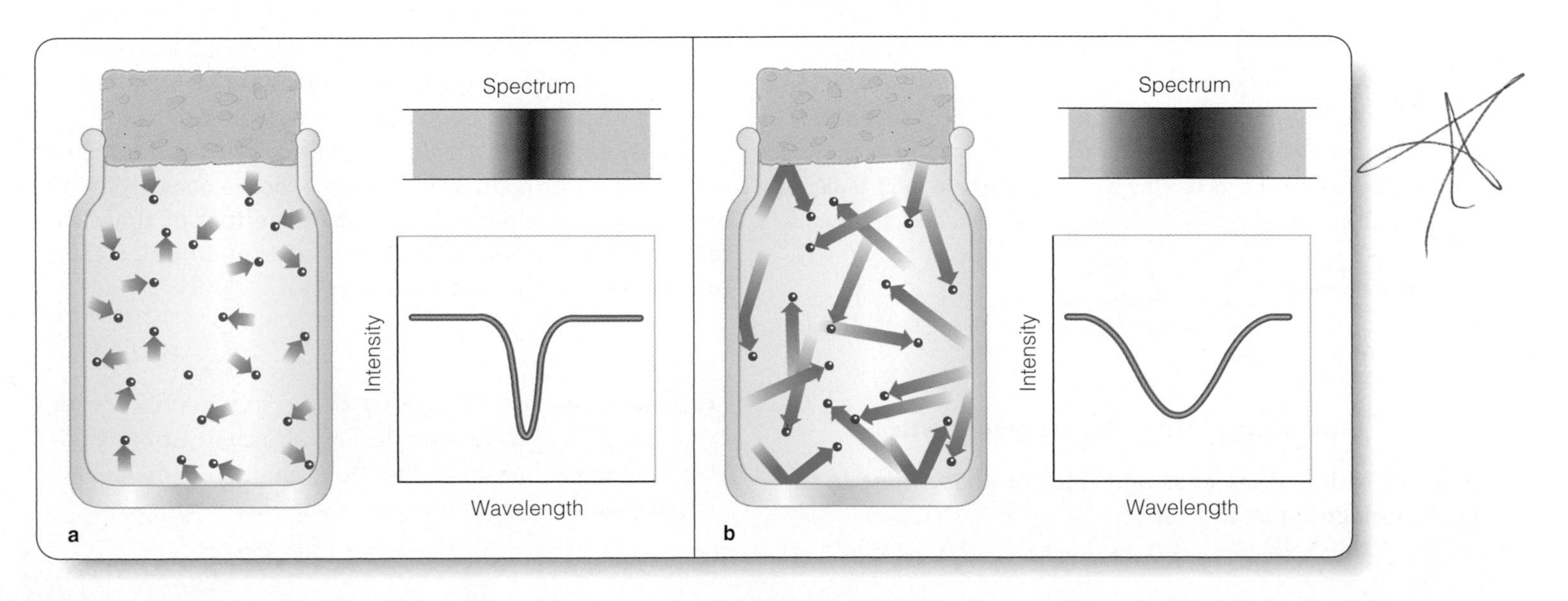

Focus on Fundamentals 4

Density

You are about as dense as an average star. What does that mean? As you study astronomy, you will use the term *density* often, so you should be sure to understand this fundamental concept. **Density** is a measure of the amount of matter in a given volume. Density is expressed as mass per volume, such as grams per cubic centimeter. The density of water, for example, is about 1 g/cm^3, and you are almost as dense as water.

To get a feel for density, imagine holding a brick in one hand and a similar-sized block of Styrofoam in the other hand. You can easily tell that the brick contains more matter than the Styrofoam block, even though both are the same size. The brick weighs more than the Styrofoam, but it isn't really the weight that you should consider. Rather, you should think about the mass of the two objects. In space, where they have no weight, the brick and the Styrofoam would still have mass, and you could tell just by moving them around that the brick contains more mass than the Styrofoam. For example, imagine tapping each object gently against your ear. The massive brick would be easy to distinguish from the low-mass Styrofoam block, even in weightlessness.

When you think of density, you divide mass by volume, and your mind makes that comparison at an instinctive level. You sense the density of an object just by handling it. Gift shops sometimes sell imitation rocks made of Styrofoam as humorous gifts. "Rocks" made of Styrofoam seem odd when you handle them because your brain expects rocks to be dense.

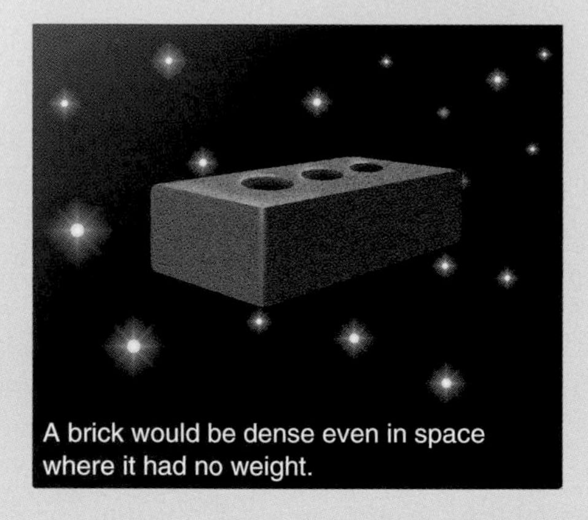
A brick would be dense even in space where it had no weight.

Density is a fundamental idea in science because it is a general property of materials. Metals tend to be dense; lead, for example, has a density of about 7 g/cm^3. Rock, in contrast, has a density of 3 to 4 g/cm^3. Water and ice have densities of about 1 g/cm^3. If you knew that a small moon orbiting Saturn had a density of 1.5 g/cm^3, you could immediately draw some conclusions about what kinds of materials the little moon might be made of—ice and a little rock, but not much metal. The density of an object is a basic clue to its composition.

Astronomical bodies can have dramatically different densities. The gas in a nebula can have a very low density, but the same kind of gas in a star can have a much higher density. The sun, for example, has an average density of about 1 g/cm^3, about the same as your body. As you study astronomical objects, pay special attention to their densities. Density is fundamental.

PRESSURE | MASS | ENERGY | TEMPERATURE AND HEAT | **DENSITY**

atoms, ions, or electrons. The collisions disturb the energy levels in the atoms, making it possible for the atoms to absorb a slightly wider range of wavelengths. Because of this, the spectral lines are wider. Because atoms in a dense gas collide more often than atoms in a low-density gas, collisional broadening depends on the density of the gas. Temperature is also an important factor. Atoms in a hot gas travel faster and collide more often and more violently than atoms in a cool gas. The two spectral lines in Figure 7-14 illustrate the effect of density.

Once again, the physics of the interaction of light and matter provides a tool to understand starlight. In later chapters, you will see how astronomers use the widths of spectral lines to better understand clouds of gas in space, stars, and even distant galaxies.

Ace Astronomy™ Log into AceAstronomy and select this chapter to see Astronomy Exercise "Stellar Rotation." See an interesting application of the Doppler shift.

Building Scientific Arguments

Why are helium lines weak and calcium lines strong in the visible spectrum of the sun?

To analyze this problem, you must recall that the ability of an atom or ion to absorb light depends on temperature. Helium is quite abundant in the sun, but the surface of the sun is too cool to excite helium atoms and enable them to easily absorb visible-wavelength photons. On the other hand, calcium is easily ionized, and calcium atoms that have lost an electron are very good absorbers of photons at temperatures like those at the sun's surface. Consequently, calcium lines are strong in the solar spectrum even though calcium ions are rare, and helium lines are weak even though helium atoms are common.

When you create a scientific argument, you must include all of the important factors. In this case, both composition and temperature are important. **But why is neither of these factors important when astronomers use the Doppler effect to measure a star's radial velocity?**

■ ■ ■

Connections: The spectra of the stars are filled with clues about the gas emitting the light. Much of astronomy is based on unraveling these clues. You will begin in the following chapter by studying the nearest star—the sun.

Study and Review Tools

Summary

7-1 | Starlight

How do stars produce light?

- Stars emit light from the hot gases at their surfaces. Light from deeper layers does not escape.
- Temperature is a measure of the average kinetic energy of the particles in a gas. In a hot gas, the particles move fast: in a cool gas, they move slowly.
- Thermal energy is the total kinetic energy in the particles of a material. Heat is the flow of thermal energy from hot regions to cool regions.
- In a hot material, the rapidly moving atoms or molecules collide with electrons, and when electrons are accelerated, part of the energy is emitted as electromagnetic radiation. This is known as black body radiation.
- The hotter a black body is, the more it radiates and the shorter is its wavelength of maximum intensity, λ_{max}. This allows astronomers to estimate the temperatures of stars from their colors.

7-2 | Atoms

What is an atom?

- An atom consists of a nucleus surrounded by a cloud of electrons. The nucleus is made up of positively charged protons and uncharged neutrons.
- The number of protons in an atom determines which element it is. Atoms of the same element (that is, having the same number of protons) with different numbers of neutrons are called isotopes.
- A neutral atom is surrounded by a number of negatively charged electrons equal to the number of protons in the nucleus. An atom that has lost or gained an electron is called an ion.
- The electrons in an atom may occupy various permitted orbits around the nucleus but not orbits in between.

7-3 | The Interaction of Light and Matter

How do atoms interact with light?

- The size of an electron's orbit depends on the energy stored in the motion of the electron.
- An electron may be excited to a higher orbit during a collision between atoms, or it may move from one orbit to another by absorbing or emitting a photon of the proper energy.
- Kirchhoff's laws describe the formation of spectra. A hot solid, liquid or dense gas will emit black body radiation, producing a continuous spectrum. An excited low-density gas will produce an emission (bright-line) spectrum. Light passing through a cool, low-density gas will form an absorption (dark-line) spectrum.

7-4 | Stellar Spectra

What kind of spectra do you see when you look at celestial objects?

- Because orbits of only certain energies are permitted in an atom, photons of only certain wavelengths can be absorbed or emitted. Each kind of atom has its own characteristic set of spectral lines. The hydrogen atom has the Lyman series of lines in the ultraviolet, the Balmer series in the visible, and the Paschen series (and others) in the infrared.
- If light passes through a low-density gas on its way to your telescope, the gas can absorb photons of certain wavelengths, and you see dark lines in the spectrum at those positions. Such a spectrum is called an absorption spectrum.
- If you look at a low-density gas that is excited to emit photons, you see bright lines in the spectrum. Such a spectrum is called an emission spectrum.

What can you learn from a star's spectrum?

- Nearly all stellar spectra are absorption spectra, and the hydrogen lines seen there are Balmer lines. In cool stars, the Balmer lines are weak because atoms are not excited out of the ground state. In hot stars, the Balmer lines are weak because atoms are excited to higher orbits or are ionized. Only at medium temperatures are the Balmer lines strong.
- The strength of spectral lines in a star's spectrum can tell you its temperature. In its simplest form, this amounts to classifying the star's spectrum in the spectral sequence: O, B, A, F, G, K, M.
- Long after the spectral sequence was created, astronomers found the L and T stars at temperatures even cooler than the M stars.
- A spectrum can reveal the chemical composition of the stars. The presence of spectral lines of a certain element shows that that element must be present in the star. But astronomers must proceed with care. Lines of a certain element may be weak or absent if the star is too hot or too cool.
- The wavelengths of spectral lines provide clues to the motions of the stars. When a star is approaching, you observe slightly shorter wavelengths (a blueshift); when it is receding, you observe slightly longer wavelengths (a redshift). This Doppler effect reveals a star's radial velocity, that part of its velocity directed toward or away from Earth.
- The shape of a spectral line can reveal the rotation of a star because one side of the star approaches Earth while the other side recedes. The Doppler shift broadens the star's spectral lines.
- Doppler broadening can make spectral lines wider because the atoms in a hot gas move very rapidly and at any moment some are receding and some are approaching.
- Collisional broadening occurs in a dense gas where the atoms collide often. This widens spectral lines.

New Terms

temperature (p. 138)
Kelvin temperature scale (p. 138)
absolute zero (p. 138)
electron (p. 138)
thermal energy (p. 139)
black body radiation (p. 139)
wavelength of maximum intensity (λ_{max}) (p. 139)
nucleus (p. 141)
proton (p. 141)
neutron (p. 141)
isotope (p. 142)
ionization (p. 142)
ion (p. 142)
molecule (p. 142)
Coulomb force (p. 142)
binding energy (p. 143)
quantum mechanics (p. 143)
permitted orbit (p. 143)
energy level (p. 144)
excited atom (p. 144)
ground state (p. 145)
continuous spectrum (p. 146)
absorption spectrum (dark-line spectrum) (p. 146)
absorption line (p. 146)
emission spectrum (bright-line spectrum) (p. 146)
emission line (p. 146)
Kirchhoff's laws (p. 146)
transition (p. 147)
Lyman series (p. 147)
Balmer series (p. 147)
Paschen series (p. 147)

spectral class or type (p. 149)
spectral sequence (p. 149)
L dwarf (p. 150)
T dwarf (p. 151)
Doppler effect (p. 153)
blueshift (p. 153)
redshift (p. 153)
radial velocity (V_r) (p. 153)
Doppler broadening (p. 155)
collisional broadening (p. 155)
density (p. 156)

Review Questions

Ace Astronomy™ Assess your understanding of this chapter's topics with additional quizzing and animations at **http://ace.brookscole.com/sf9**

1. Why can a good blacksmith judge the temperature of a piece of heated iron by its color?
2. Why do hot stars look bluer than cool stars?
3. Use black body radiation to explain why you could hold a cigarette between your lips and light it with a match, but you couldn't hold the cigarette in your lips while you light it by sticking the tip into a bonfire. The match and bonfire have about the same temperature. (This is just one of the hazards of smoking.)
4. What is the difference between a neutral atom, an ion, and an excited atom?
5. How do the energy levels in an atom determine which wavelength photons it can absorb or emit?
6. What kind of spectrum would you expect to record if you observed molten lava? In practice, to view molten lava you must look through gases boiling out of the lava. What kind of spectrum might you see in that case?
7. Why do the strengths of the Balmer lines depend on the temperature of the star?
8. Why does a stellar spectrum tell you about the surface layers but not the deeper layers of the star?
9. Explain the similarities among Figure 7-8, Figure 7-9, Figure 7-10, and Table 7-1.
10. Why would you not expect to see TiO bands in the spectra of hot stars?
11. Imagine that you observed a star's spectrum and found that the lines of a certain element were not present. Would you be safe in concluding that the star did not contain this element? Why or why not?
12. If a star moves exactly perpendicular to a line connecting it to Earth, will the Doppler effect change its spectrum? Why or why not?
13. Why would you expect a star that rotates very rapidly to have broad spectral lines? What else could broaden the lines?
14. The nebula shown at right is mostly low-density hydrogen excited to emit photons. What kind of spectrum would you expect this nebula to produce?
15. If the nebula in the image here crosses in front of the star and the nebula and star have different radial velocities, what might the spectrum of the star look like?

(T. Rector, University of Alaska, and WIYN/NURO/AURA/NSF)

Discussion Questions

1. In what ways is the model of an atom used in this chapter a scientific model? How can you use it when it is not a completely correct description of an atom?
2. Can you think of classification systems commonly used to simplify what would otherwise be very complex measurements? Consider foods, movies, cars, grades, clothes, and so on.

Problems

1. Human body temperature is about 310 K (98.6°F). At what wavelength do humans radiate the most energy? What kind of radiation do we emit?
2. If a star has a surface temperature of 20,000 K, at what wavelength will it radiate the most energy?
3. Infrared observations of a star show that it is most intense at a wavelength of 2000 nm. What is the temperature of the star's surface?
4. If you double the temperature of a black body, by what factor will the total energy radiated per second per square meter increase?
5. If one star has a temperature of 6000 K and another star has a temperature of 7000 K, how much more energy per second will the hotter star radiate from each square meter of its surface?
6. Transition A produces light with a wavelength of 500 nm. Transition B involves twice as much energy as A. What wavelength light does it produce?
7. Determine the temperatures of the following stars based on their spectra. Use Figure 7-8.
   a. medium-strength Balmer lines, strong helium lines
   b. medium-strength Balmer lines, weak ionized calcium lines
   c. TiO bands very strong
   d. very weak Balmer lines, strong ionized calcium lines
8. To which spectral classes do the stars in Problem 7 belong?
9. In a laboratory, the Balmer beta line has a wavelength of 486.1 nm. If the line appears in a star's spectrum at 486.3 nm, what is the star's radial velocity? Is it approaching or receding?
10. The highest-velocity stars an astronomer might observe have velocities of about 400 km/s. What change in wavelength would this cause in the Balmer gamma line? (*Hint:* Wavelengths are given on page 147).

ACTIVE FIGURES

Ace Astronomy™ To access the resources in the Media Cluster, log into AceAstronomy at **http://ace.brookscole.com/sf9** and select Chapter 7.

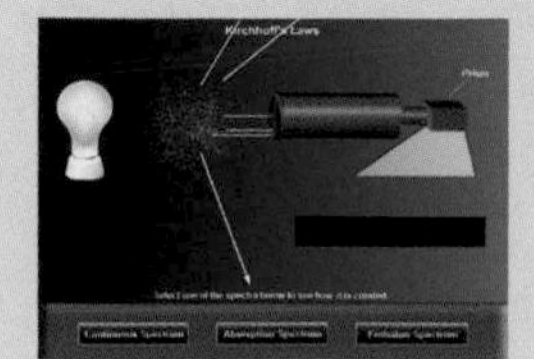

Kirchhoff's Laws
Kirchhoff studied and identified three different spectra in his experiments. Use this animation to study the conditions for producing each of the types of spectra.

ASTRONOMY EXERCISES

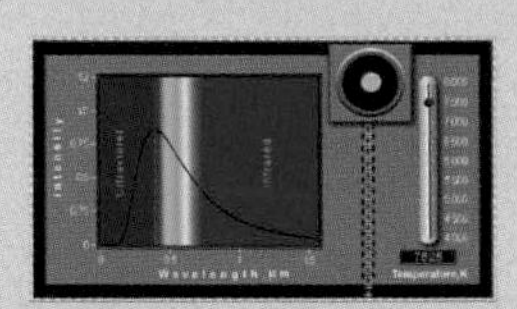

Black Body
In this animation, you can change the temperature of an object and see how its black body curve changes.

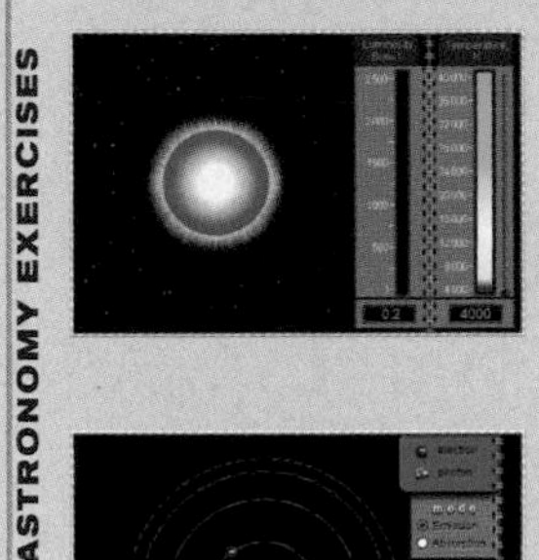

Stefan–Boltzmann Law
In this introduction to the Stefan–Boltzmann law, watch the luminosity of a star change as you adjust its temperature.

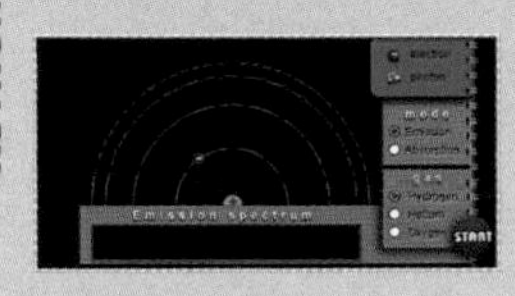

Emission and Absorption Spectra
This exercise will give you an inside look at atoms as they interact with photons. The photons are absorbed by the atom, sending the electron to a higher energy level.

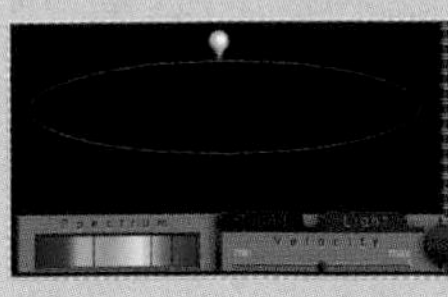

Doppler Shift
Recall that the Doppler effect is the change in wavelength of radiation due to relative radial motion of source and observer. You can compare the Doppler shift for a source of sound with that for a source of light in this animation.

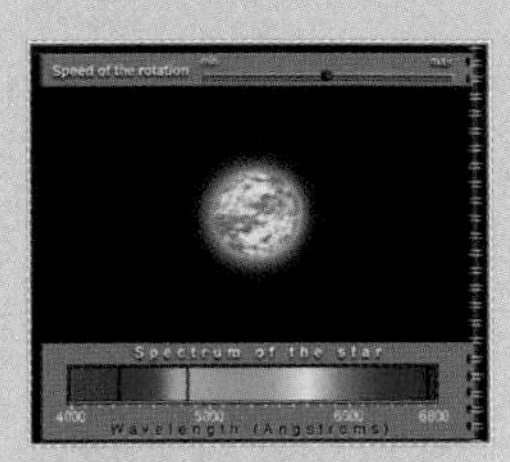

Stellar Rotation
This animation lets you increase or decrease the rotation speed of a star and see the effect on the star's spectrum.

Stellar Atomic Absorption Lines
In this exercise, you can build your own star. Select from the type and ratio of element mixes in your new star.

VIRTUAL ASTRONOMY LABS

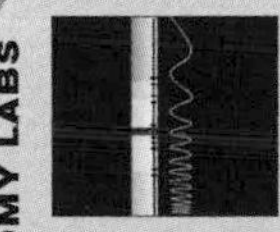

Lab 2: Properties of Light and Its Interaction with Matter
This lab examines the wave properties of electromagnetic radiation and how different regions of the electromagnetic spectrum are related. It ends by looking at the interaction between matter and radiation on the atomic scale.

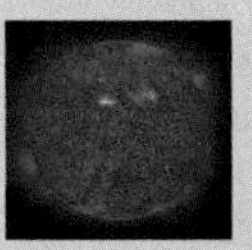

Lab 11: The Spectral Sequence and the H–R Diagram
This lab introduces the tools astronomers use to identify the stars of the main sequence and how that information is used to estimate the distance to the stars and the age of a star cluster.

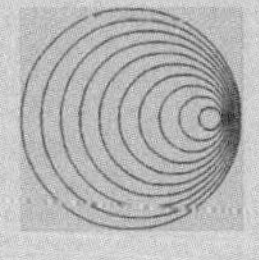

Lab 3: The Doppler Effect
This lab provides a brief review of some basic properties of waves and investigates the shift in observed wavelengths of sound and light waves caused by the motion of an emitting source with respect to an observer.

Critical Inquiries for the Web

1. The name for the element helium has astronomical roots. Search the Internet for information on the discovery of helium. How and when was it discovered, and how did it get its name? Why do you suppose it took so long for helium to be recognized?
2. This chapter contains a model for an atom that may seem familiar to you. How was that model developed? What was the plum pudding model? Search the Web for information on historical models of the atom and compile a time line of important developments leading to our current understanding.
3. What can you find out about Anne J. Cannon and Cecilia Payne-Gaposchkin?

Exploring *TheSky*

1. Locate the following stars, click on them, and determine their spectral types: Antares in Scorpius, Betelgeuse in Orion, Aldebaran in Taurus, Sirius in Canis Major, Rigel in Orion.
2. How are spectral types correlated with the colors of stars? (*Hint:* Locate Orion and choose **Spectral Colors** under the **View** menu.)

- Go to the Brooks/Cole Astronomy Resource Center **(http://astronomy.brookscole.com)** for critical thinking exercises, articles, and additional readings from InfoTrac College Edition, Brooks/Cole's online student library.

8 The Sun

All cannot live on the piazza, but everyone may enjoy the sun.

ITALIAN PROVERB

UV image

A WIT ONCE REMARKED that solar astronomers would know a lot more about the sun if it were farther away, and that contains a grain of truth; the sun is only a humdrum star, and although there are billions like it in the sky, the sun is the only one close enough to show surface detail. Solar astronomers can see so much detail in the swirling currents of gas and arching bridges of magnetic force that present theories seem inadequate to describe it. Yet the sun is not a complicated object. It is just a star. ▌ In their general properties, stars are very simple. They are great balls of hot gas held together by their own gravity. Their gravity would make them collapse into small, dense bodies were they not so hot. The tremendously hot gas inside stars has such a high pressure that the stars would surely explode were it not for their own confining gravity.

▌ Continued on page 162 ▌

This far-ultraviolet image of the sun made from space reveals complex structure on the surface and clouds of gas being ejected into space. (NASA/SOHO)

Guidepost

Looking Back

The interaction of light and matter, which you studied in Chapter 7, is a powerful tool. It can reveal the secrets of the stars. The next step is to apply that tool to the sun.

This Chapter

In this chapter, you will discover how little astronomers could know about the sun were it not for the analysis of its spectrum. Just a bit of spectrographic ingenuity reveals that the brilliance of the sun hides a complex atmosphere of hot gases, powered by nuclear reactions and churned by powerful storms.

This chapter will help you answer four essential questions:

What do you see when you look at the sun?

How does the sun make its energy?

What are the dark sunspots and other forms of solar activity?

Why does the sun go through a cycle of activity?

Perhaps the most important question answered here is more general: How can astronomers learn about the sun and stars?

Looking Ahead

This is the first chapter that applies the methods of science to understand a celestial body. The tools developed in this chapter will help you explore stars, galaxies, and planets through the rest of this book.

AceAstronomy™ The AceAstronomy icon throughout the text indicates an opportunity for you to test yourself on key concepts and to explore animations and interactions on the AceAstronomy website at: http://ace.brookscole.com/sf9

Like soap bubbles, stars are simple structures balanced between opposing forces that individually would destroy them. You can study our sun as a close-up example of a star.

Another reason to study the sun is that life on Earth depends critically on the sun. Very small changes in the sun's luminosity can alter Earth's climate, and a slightly larger change might make Earth uninhabitable. Nearly all of our energy comes from the sun—oil and coal are merely stored sunlight. Furthermore, the sun's atmosphere of very thin gas reaches out past Earth's orbit, and any change in the sun, such as an eruption or a magnetic storm, can have a direct effect on Earth.

Finally, you should study the sun because it is beautiful. Your analysis of sunlight will reveal that the sun is both powerful and delicate. So you should study the sun not only because it is *a star,* not only because it is *our star,* but because it is the sun.

8-1 The Solar Atmosphere

WHEN YOU WATCH the sun set in the west, you see a glowing disk. What are you seeing? Here you will see the sun sketched in its general properties and describe the surface you see. Later you can plunge inside to see how it makes its energy and then return to the surface to study the storms and eruptions not generally visible to the unaided eye. What do you see when you look at the sun? You see a star.

The sun is 109 times Earth's diameter and 333,000 times Earth's mass. This seems dramatic, but look at **Celestial Profile 1** and notice the sun's density. It is only a little bit denser than water. So, although the sun is very large and very massive, it must be a gas from its surface to its center. When you look at the sun you see only the outer layers of this vast sphere of gas. In fact, these outer layers, the solar atmosphere, extend high above the visible surface of the sun.

Heat Flow in the Sun

When you look at the sun you see a hot, glowing surface, and simple logic tells you that energy in the form of heat is flowing outward from the sun's interior. The solar spectrum reveals that the sun is a G2 star with a temperature of about 5800 K. At that temperature, every square millimeter of the sun's surface must be radiating more energy than a 60-watt lightbulb. With all that energy radiating into space, the sun's surface would cool rapidly if energy did not flow up from the interior to keep the surface hot.

Not until the 1930s did astronomers understand how the sun makes its energy. Nuclear reactions occur in the core of the sun and generate energy, which flows upward as heat and keeps the surface hot. These nuclear reactions are discussed in detail later in this chapter, but here you should notice the importance of the energy flowing outward through the sun's surface.

As you study the atmosphere of the sun, you will find many phenomena that are driven by the energy flowing outward. Like a pot of boiling soup on a hot stove, the surface of the sun is in constant activity as the heat flows up from below.

For now, you can consider the sun in its quiescent, unchanging state, and explore the layers in its atmosphere.

The Photosphere

The visible surface of the sun looks like a smooth layer of gas marked by a few dark **sunspots.** Although the photosphere seems to be a distinct surface, it is not solid. In fact, the sun is gaseous from its outer atmosphere right down to its center. The photosphere is the thin layer of gas from which Earth receives most of the sun's light. It is less than 500 km deep and has an average temperature of about 5800 K. If the sun magically shrank to the size of a bowling ball, the photosphere would be no thicker than a layer of tissue paper wrapped around the ball. For comparison, the chromosphere lies above the photosphere and is only a few times thicker in extent, but the corona, beginning above the chromosphere, extends far above the visible surface (■ Figure 8-1). Recall that you first met the terms *photosphere, chromosphere,* and *corona* in connection with solar eclipses in Chapter 3.

Below the photosphere, the gas is denser and hotter and therefore radiates plenty of light, but that light cannot escape from the sun because of the outer layers of gas. So you cannot detect light from these deeper layers. Above the photosphere, the gas is less dense and so is unable to radiate much light. The photosphere is the layer in the sun's atmosphere that is dense enough to emit plenty of light but not so dense that the light can't escape.

One reason the photosphere is so shallow is related to the hydrogen atom. Because the temperature of the photosphere is sufficient to ionize some atoms, there are a large number of free electrons in the gas. Neutral hydrogen atoms can add an extra electron and become an H^- (H-minus) ion, but this extra electron is held so loosely that almost any photon has energy enough to free it. In the process, of course, the photon is absorbed. That makes the H^- ions very good absorbers of photons and makes the gas of the photosphere opaque. Light from below cannot escape easily, and you see a well-defined surface glowing like a layer of hot fog—the photosphere.

Although the photosphere appears to be substantial, it is really a very-low-density gas. Even in the deepest and densest layers visible, the photosphere is 3400 times less dense than the air you breathe. To find gases as dense as the air you breathe, you would have to descend about 70,000 km below the photosphere, about 10 percent of the way to the sun's center. With fantastically efficient insulation, you could fly a spaceship right through the photosphere.

The spectrum of the sun is an absorption spectrum, and that can tell you a great deal about the photosphere. You know from Kirchhoff's third law that an absorption spectrum is produced when a source of a continuous spectrum is viewed through a gas. In the case of the photosphere, the deeper layers are dense enough to produce a continuous spectrum, but atoms in the photosphere

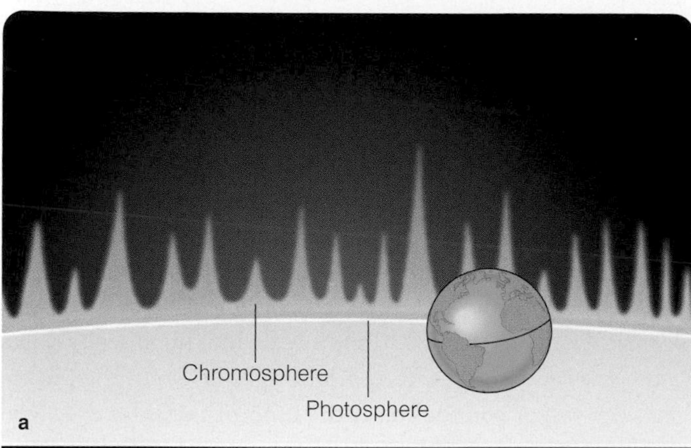

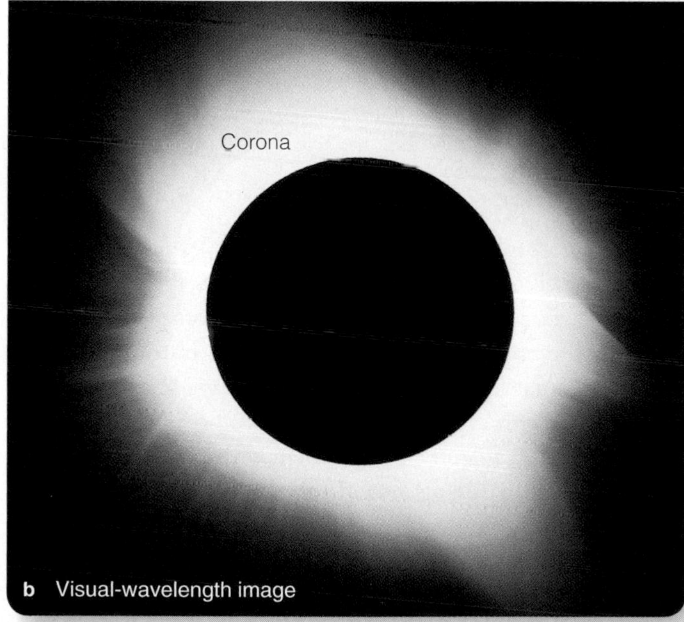

■ Figure 8-1

(a) A cross section at the edge of the sun shows the relative thickness of the photosphere and chromosphere. Earth is shown for scale. On this scale, the disk of the sun would be more than 1.5 m (5 feet) in diameter. (b) The corona extends from the top of the chromosphere to great height above the photosphere. (b) This photograph, made during a total solar eclipse, shows only the inner part of the corona. (Daniel Good)

absorb photons of specific wavelengths, producing the absorption lines you see.

In good photographs, the photosphere has a mottled appearance because it is made up of dark-edged regions called granules, and the visual pattern is called **granulation** (■ Figure 8-2a). Each granule is about the size of Texas and lasts for only 10 to 20 minutes before fading away. Faded granules are continuously replaced by new granules. Spectra of these granules show that the centers are a few hundred degrees hotter than the edges, and Doppler shifts reveal that the centers are rising and the edges are sinking at speeds of about 0.4 km/second.

From this evidence, astronomers recognize granulation as the surface effects of convection just below the photosphere.

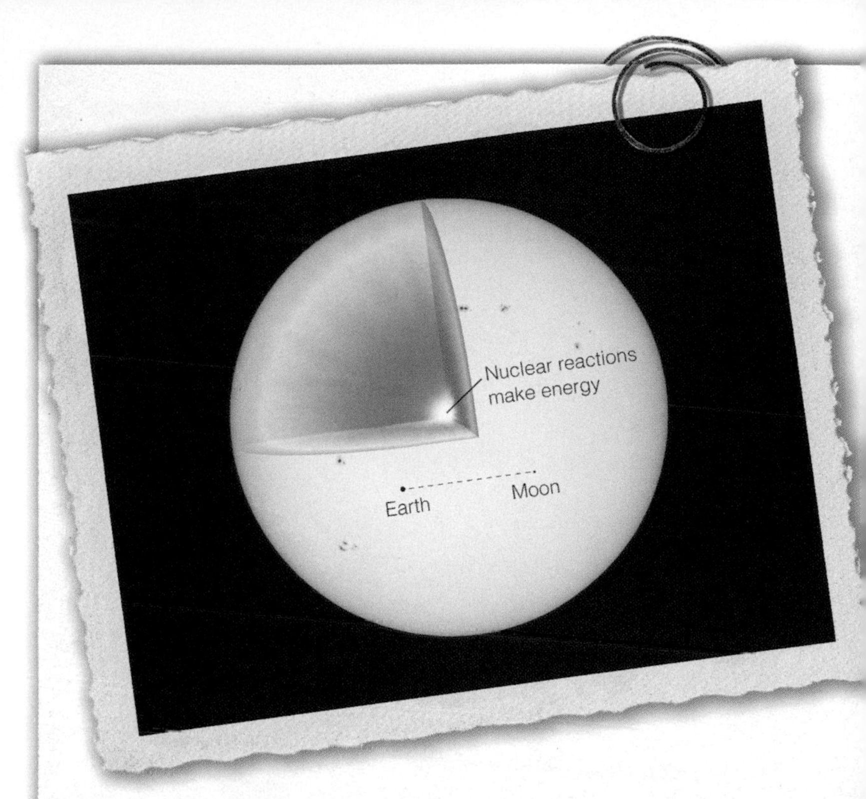

This visual-wavelength image of the sun shows a few sunspots and is cut away to show the location of energy generation at the sun's center. The Earth–moon system is shown for scale. (Dan Good)

Celestial Profile 1: The Sun

From Earth:

Average distance from Earth	1.00 AU (1.495979×10^8 km)
Maximum distance from Earth	1.0167 AU (1.5210×10^8 km)
Minimum distance from Earth	0.9833 AU (1.4710×10^8 km)
Average angular diameter	0.53° (32 minutes of arc)
Period of rotation	25.38 days at equator
Apparent visual magnitude	−26.74

Characteristics:

Radius	6.9599×10^5 km
Mass	1.989×10^{30} kg
Average density	1.409 g/cm^3
Escape velocity at surface	617.7 km/s
Luminosity	3.826×10^{26} J/s
Surface temperature	5800 K
Central temperature	15×10^6 K
Spectral type	G2 V
Absolute visual magnitude	4.83

Personality Profile:

In Greek mythology, the sun was carried across the sky in a golden chariot pulled by powerful horses and guided by the sun-god Helios. When Phaeton, son of Helios, drove the chariot one day, he lost control of the horses, and Earth was nearly set ablaze before Zeus smote Phaeton from the sky. Even in classical times, people understood that life on Earth depends critically on the sun.

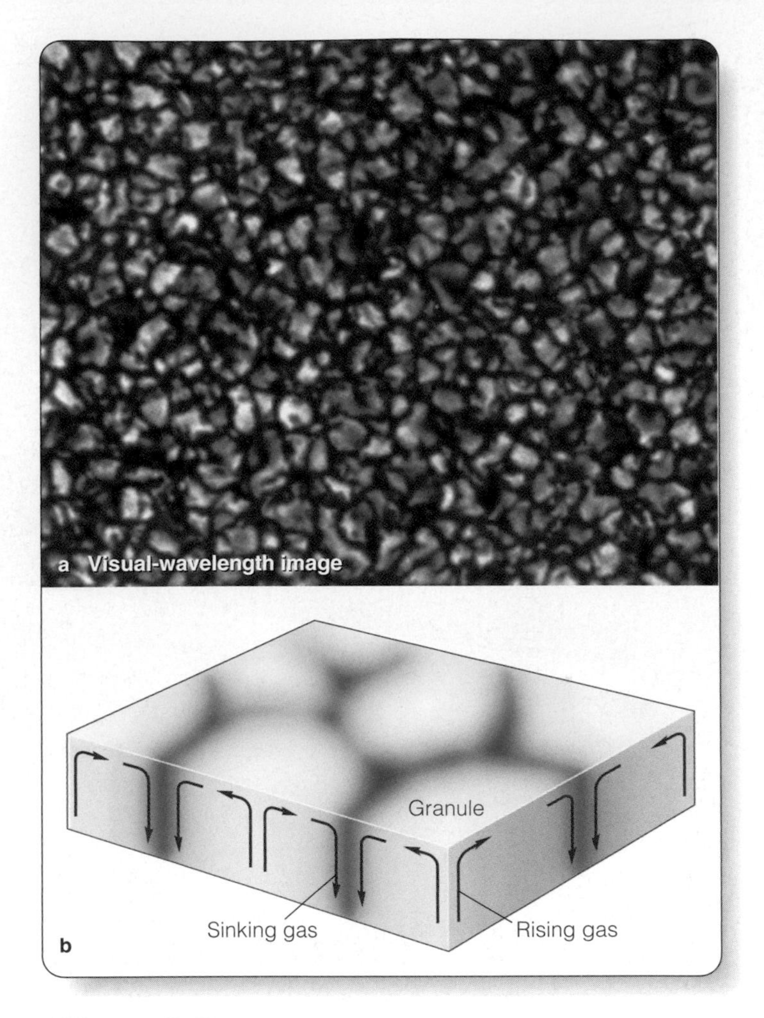

Figure 8-2

(a) This ultra-high-resolution image of the photosphere shows granulation. The largest granules here are about the size of Texas. (P. N. Brandt, G. Scharmer, G. W. Simon, Swedish Vacuum Solar Telescope) (b) This model explains granulation as the tops of rising convection currents just below the photosphere. Heat flows upward as rising currents of hot gas and downward as sinking currents of cool gas. The rising currents heat the solar surface in small regions seen from Earth as granules.

Convection occurs when hot fluid rises and cool fluid sinks, as when, for example, a convection current of hot gas rises above a candle flame. You can create convection in a liquid by adding a bit of cool nondairy creamer to an unstirred cup of hot coffee. The cool creamer sinks, warms, rises, cools, sinks again, and so on, creating small regions on the surface of the coffee that mark the tops of convection currents. Viewed from above, these regions look much like solar granules.

In the sun, rising currents of hot gas heat small regions of the photosphere, which, being slightly hotter, emit more black body radiation and look brighter. The cool sinking gas of the edges emits less light and thus looks darker (Figure 8-2b). The presence of granulation is clear evidence that energy is flowing upward through the photosphere.

Spectroscopic studies of the solar surface have revealed another kind of granulation. **Supergranules** are regions about 30,000 km in diameter (about 2.3 times Earth's diameter) and include about 300 granules. These supergranules are regions of very slowly rising currents that last a day or two. They may be the surface traces of larger currents of rising gas deeper under the photosphere.

The edge, or **limb,** of the solar disk is dimmer than the center (see the figure in Celestial Profile 1). This **limb darkening** is caused by the absorption of light in the photosphere. When you look at the center of the solar disk, you are looking directly down into the sun, and you see deep, hot, bright layers in the photosphere. But when you look near the limb of the solar disk, you are looking at a steep angle and cannot see as deeply. The photons you see come from shallower, cooler, dimmer layers in the photosphere. Limb darkening proves that the temperature in the photosphere increases with depth, as you would expect if energy is flowing up from below.

The Chromosphere

Above the photosphere lies the chromosphere. Solar astronomers define the lower edge of the chromosphere as lying just above the visible surface of the sun with its upper regions blending gradually with the corona. You can think of the chromosphere as being an irregular layer with a depth on average less than Earth's diameter (see Figure 8-1). Because the chromosphere is roughly 1000 times fainter than the photosphere, you can see it with your unaided eyes only during a total solar eclipse when the moon covers the brilliant photosphere. Then, the chromosphere flashes into view as a thin line of pink just above the photosphere. The word *chromosphere* comes from the Greek word *chroma,* meaning "color." The pink color is produced by the combined light of three bright emission lines—the red, blue, and violet Balmer lines of hydrogen.

Astronomers know a great deal about the chromosphere from its spectrum. The chromosphere produces an emission spectrum, and Kirchhoff's second law tells you the chromosphere must be an excited, low-density gas. The density is about 10^8 times less dense than the air you breathe.

Atoms in the lower chromosphere are ionized, and atoms in the higher layers of the chromosphere are even more highly ionized. That is, they have lost more electrons. From this, astronomers can find the temperature in different parts of the chromosphere. Just above the photosphere the temperature falls to a minimum of about 4500 K and then rises rapidly (■ Figure 8-3). The region where the temperature increases fastest is called the **transition region** because it makes the transition from the lower temperatures of the photosphere and chromosphere to the extremely high temperatures of the corona.

Solar astronomers can take advantage of some elegant physics to study the chromosphere. The gases of the chromosphere are transparent to nearly all visible light, but atoms in the gas are

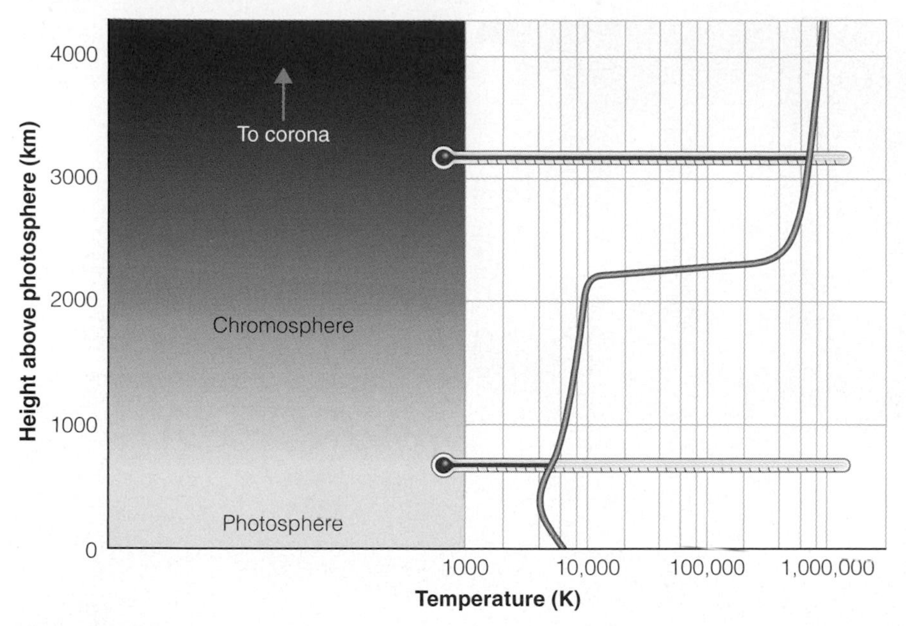

■ **Figure 8-3**

The chromosphere. If you could place thermometers in the sun's atmosphere, you would discover that the temperature increases from 5800 K at the photosphere to 10^6 K at the top of the chromosphere.

very good at absorbing photons of specific wavelengths. This produces certain dark absorption lines in the spectrum of the photosphere. A photon at one of those wavelengths is very unlikely to escape from deep layers. A **filtergram** is a photograph made using light in one of those dark absorption lines. Those photons can only have escaped from higher in the atmosphere. In this way filtergrams reveal detail in the upper layers of the chromosphere. In a similar way, an image recorded in the far-ultraviolet or in the X-ray part of the spectrum reveals other structures in the solar atmosphere.

■ Figure 8-4 shows a filtergram made at the wavelength of the H_α Balmer line. This image shows complex structure in the chromosphere including long, dark **filaments** silhouetted against the brighter surface. **Spicules** are flamelike jets of gas extending upward into the chromosphere and lasting 5 to 15 minutes. Seen at the limb of the sun's disk, these spicules blend together and look like flames covering a burning prairie (Figure 8-1a), but they are not flames at all. Spectra show that spicules are cooler gas from the lower chromosphere extending upward into hotter regions. Images at the center of the solar disk show that spicules spring up around the edge of supergranules like weeds around flagstones (Figure 8-4). Although spicules are not yet well understood, they are clearly driven by the outward flow of energy in the sun.

Spectroscopic analysis of the chromosphere alerts you that it is a low-density gas in constant motion where the temperature increases rapidly with height. Just above the chromosphere lies even hotter gas.

The Solar Corona

The outermost part of the sun's atmosphere is called the corona, after the Greek word for crown. The corona is so dim that it is not visible in Earth's daytime sky because of the glare of scattered light from the brilliant photosphere. During a total solar eclipse, however, when the moon covers the photosphere, you can see the innermost parts of the corona, as shown in Figure 8-1b. Observations made with specialized telescopes called **coronagraphs** on Earth or in space can block the light of the photosphere and image the corona out beyond 20 solar radii, almost 10 percent of the

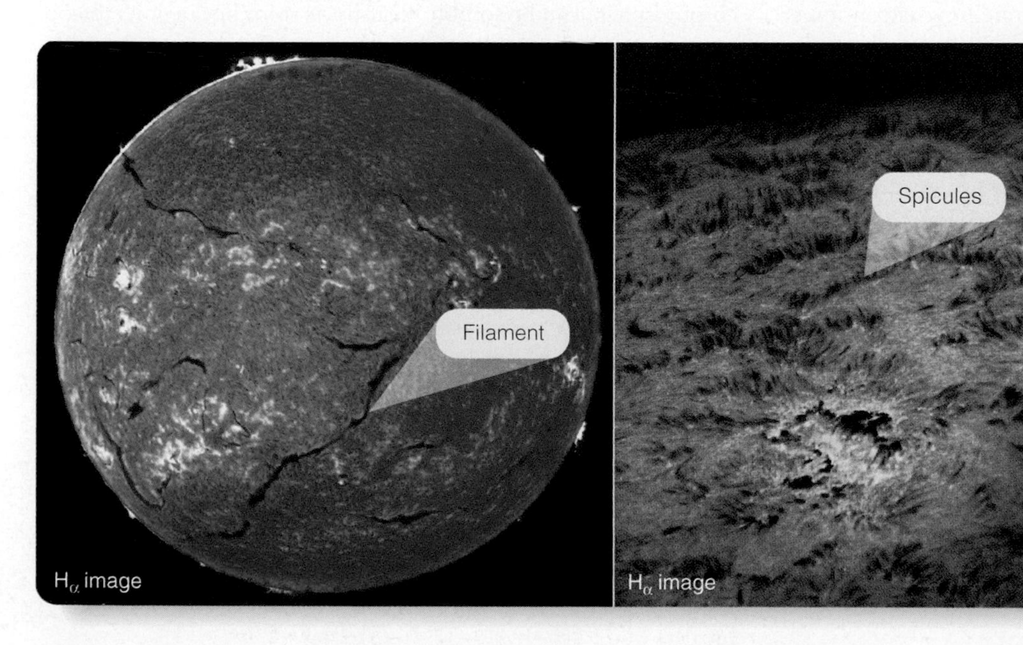

■ **Figure 8-4**

H_α filtergrams reveal complex structure in the chromosphere, including long, dark filaments and spicules springing from the edges of supergranules twice the diameter of Earth. (NOAA/SEL/USAF; NOAO/NSO)

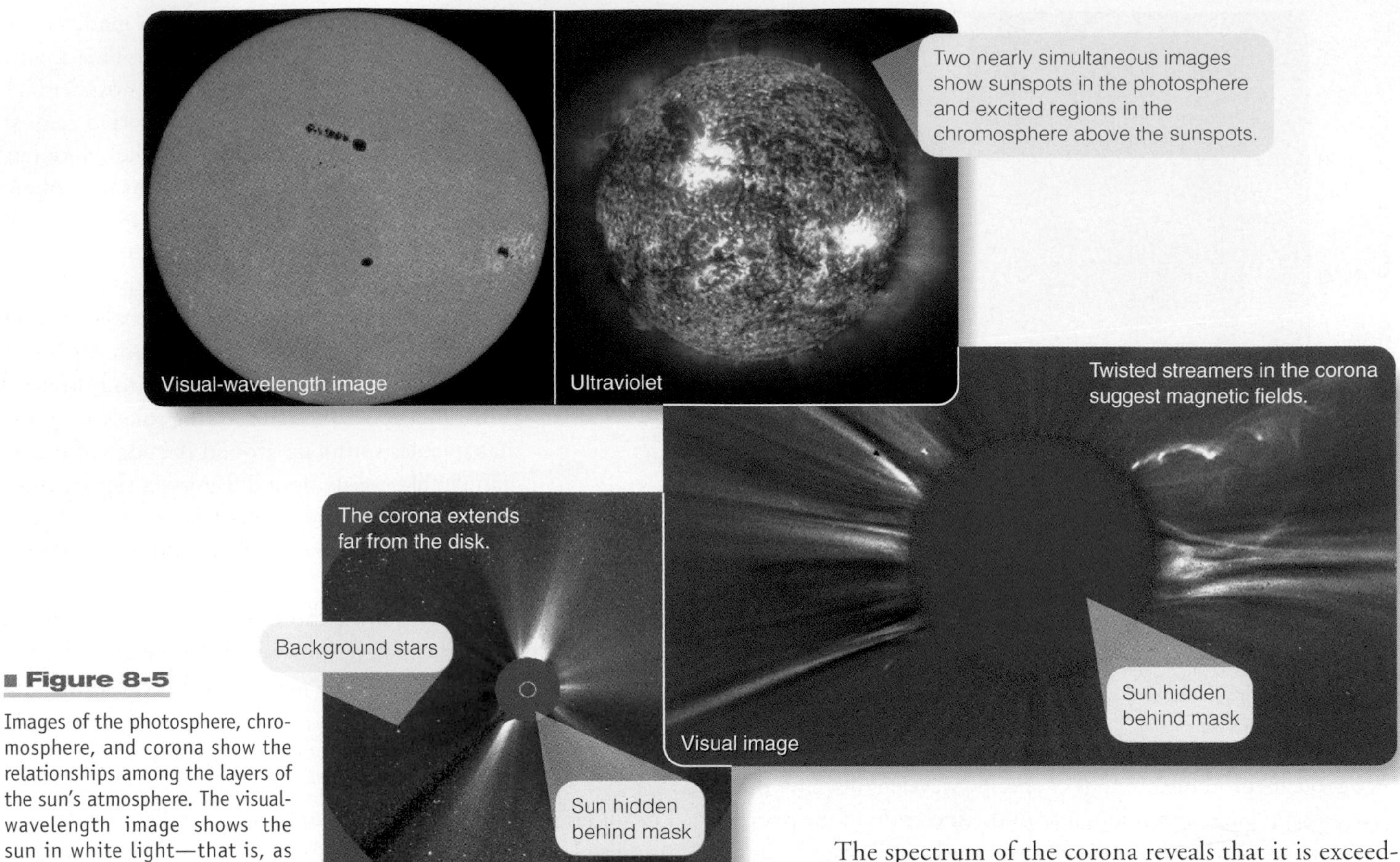
Two nearly simultaneous images show sunspots in the photosphere and excited regions in the chromosphere above the sunspots.
Visual-wavelength image
Ultraviolet
Twisted streamers in the corona suggest magnetic fields.
The corona extends far from the disk.
Background stars
Sun hidden behind mask
Visual image
Sun hidden behind mask
Visual image

■ Figure 8-5

Images of the photosphere, chromosphere, and corona show the relationships among the layers of the sun's atmosphere. The visual-wavelength image shows the sun in white light—that is, as you would see it with your eyes. (SOHO/ESA/NASA)

way to Earth. Such images reveal that magnetic fields link the sunspots with features in the chromosphere and corona (■ Figure 8-5).

The spectrum of the corona can tell you a great deal about the coronal gases and simultaneously illustrate how astronomers can analyze a spectrum. Some of the light from the outer corona produces a spectrum with absorption lines the same as the sun's spectrum. This light is just sunlight reflected from dust particles in the corona. In contrast, some of the light from the corona produces a continuous spectrum that lacks absorption lines, and that happens when sunlight from the photosphere is scattered off free electrons in the ionized coronal gas. Because the coronal gas has a temperature over 1 million K and the electrons travel very fast, the reflected photons suffer large, random Doppler shifts that smear out solar absorption lines to produce a continuous spectrum.

Superimposed on the corona's continuous spectrum are emission lines of highly ionized gases. In the lower corona, the atoms are not as highly ionized as they are at higher altitudes, and this tells you that the temperature of the corona rises with altitude. Just above the chromosphere, the temperature is about 500,000 K; but in the outer corona the temperature can be as high as 2 million K or more.

The spectrum of the corona reveals that it is exceedingly hot gas, but it is not very bright. Its density is very low, only 10^6 atoms/cm^3 in its lower regions. That is about a trillion times less dense than the air you breath. In its outer layers the corona contains only 1 to 10 atoms/cm^3, better than the best vacuum on Earth. Because of this low density, the hot gas does not emit much radiation.

Astronomers have wondered for years how the corona and chromosphere can be so hot. Heat flows from hot regions to cool regions, never from cool to hot. So how can the heat from the photosphere, with a temperature of only 5800 K, flow out into the much hotter chromosphere and corona? Observations made by the SOHO satellite have mapped a **magnetic carpet** of looped magnetic fields extending up through the photosphere (■ Figure 8-6). Turbulence below the surface may be whipping these fields about and heating the gases of the chromosphere and corona. Remember that the gas of the chromosphere and corona has very low density, so it can't resist the moving magnetic fields. The gas gets whipped about as the magnetic fields flick back and forth, and that heats the gas. In this instance, energy appears to flow outward as the agitation of the magnetic fields.

Gas from the solar atmosphere follows the magnetic fields pointing outward and flows away from the sun in a breeze called the **solar wind.** Like an extension of the corona, the low-density gases of the solar wind blow past Earth at 300 to 800 km/s with gusts as high as 1000 km/s. Earth is bathed in the corona's hot breath.

■ Figure 8-6

This extreme-ultraviolet image of a section of the sun's lower corona has been given a green color. White and black areas are regions of opposite magnetic polarity. Computer models show the location of the magnetic fields that connect these regions. The largest loops shown here could encircle Earth. The entire surface of the sun is covered by this magnetic carpet. (Stanford-Lockheed Institute for Space Research, Palo Alto, CA, and NASA GSFC)

Because of the solar wind, the sun is slowly losing mass, but this is only a minor loss for an object as massive as the sun. The sun loses about 10^7 tons per year, but that is only 10^{-14} of a solar mass per year. Later in life, the sun, like many other stars, will lose mass rapidly. You will see in future chapters how this affects stars.

Do other stars have chromospheres, coronae, and stellar winds like the sun? Ultraviolet and X-ray observations suggest that the answer is yes. The spectra of many stars contain emission lines in the far-ultraviolet that could only have formed in the low-density, high-temperature gases of a chromosphere and corona. Also, many stars are sources of X rays, which appear to have been produced by coronae. This observational evidence gives astronomers good reason to believe that the sun, for all its complexity, is a typical star.

Helioseismology

Almost no light emerges from below the photosphere, so you can't see into the solar interior. However, solar astronomers can use the vibrations in the sun to explore its depths in a process called **helioseismology.** Random motions in the sun constantly produce vibrations—rumbles too low to hear with human ears. Some of these vibrations resonate in the sun like sound waves in organ pipes. A vibration with a period of 5 minutes is strongest, but the periods range from 3 to 20 minutes. These are very, very low-pitched rumbles!

Astronomers can detect these vibrations by observing Doppler shifts in the solar surface. As a vibrational wave travels down into the sun, the increasing density and temperature curve its path and it returns to the surface, where it makes the photosphere heave up and down by small amounts—roughly plus or minus 15 km (■ Figure 8-7a). Short-wavelength waves penetrate less deeply and travel shorter distances than longer-wavelength waves. This covers the surface of the sun with a pattern of rising and falling regions that can be mapped using the Doppler effect (Figure 8-7b). By observing these motions, astronomers can determine which vibrations resonate the strongest. Just as

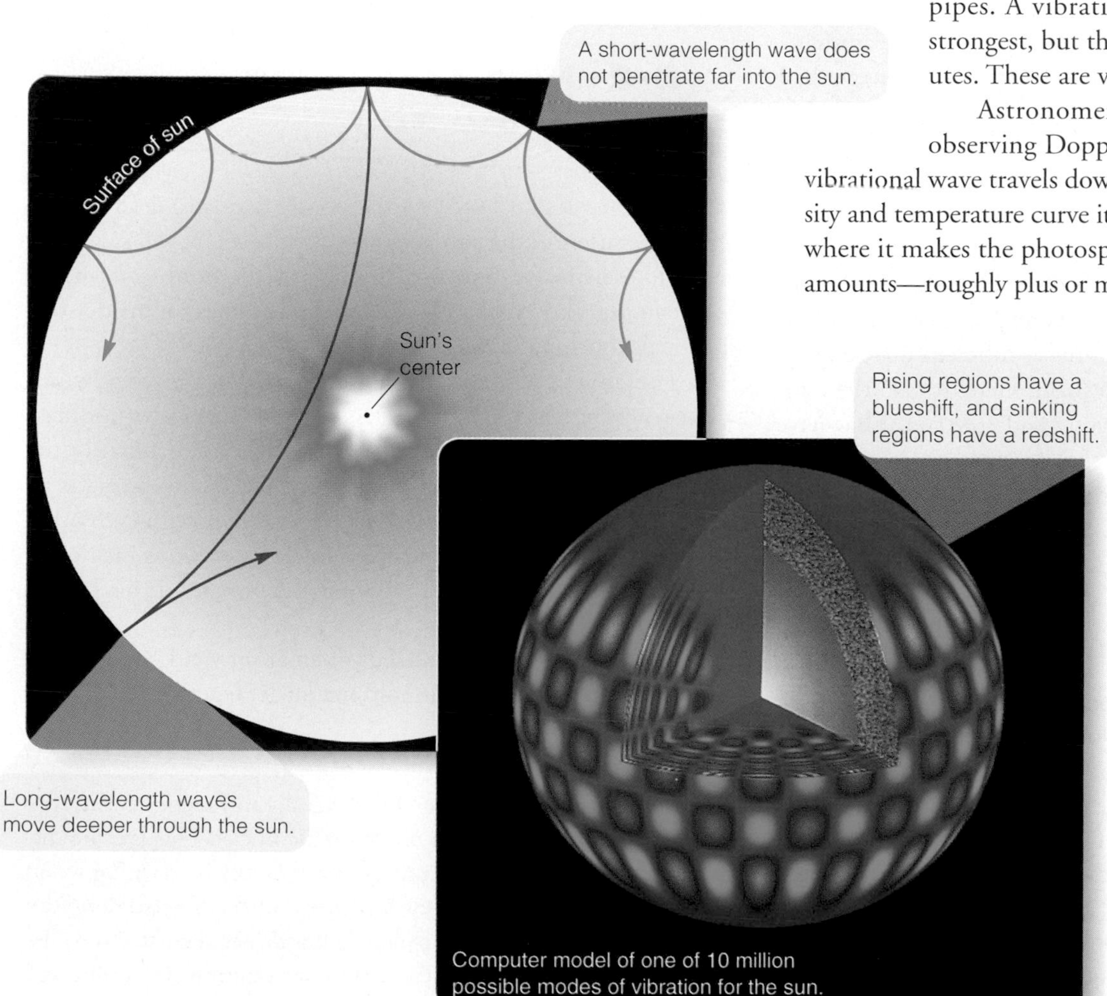
A short-wavelength wave does not penetrate far into the sun.
Surface of sun
Sun's center
Rising regions have a blueshift, and sinking regions have a redshift.
Long-wavelength waves move deeper through the sun.
Computer model of one of 10 million possible modes of vibration for the sun.

■ Figure 8-7

Helioseismology: The sun can vibrate in millions of different patterns or modes, and each mode corresponds to a different wavelength vibration penetrating to a different level. By measuring Doppler shifts as the surface moves gently up and down, astronomers can map the inside of the sun. (AURA/NOAO/NSF)

geologists can study Earth's interior by analyzing vibrations from earthquakes, so solar astronomers can use helioseismology to explore the sun's interior.

Helioseismology sounds almost magical, but you can understand it better if you think of a duck pond. If you stood at the shore of a duck pond and looked down at the water, you would see ripples arriving from all parts of the pond. Because every duck on the pond contributes to the ripples, you could, in principle, study the ripples near the shore and draw a map showing the position and velocity of every duck on the pond. Of course, it would be difficult to untangle the different ripples, so you would need lots of data and a big computer. Nevertheless, all of the information would be there, lapping at the rocks at your feet.

The sun can oscillate in about 10 million different ways, and each mode of oscillation has its own characteristic wavelength and its own unique pattern on the solar surface. The waves producing different modes penetrate to different depths where conditions can weaken or strengthen a wave. By discovering which wavelength waves are actually present, solar astronomers can determine the temperature, density, pressure, composition, and motion at different depths inside the sun.

Of course, with 10 million possible wavelengths, the observations and analysis are difficult. A single wave produces a complicated pattern of motion on the solar surface. Large amounts of data are necessary, so helioseismologists have used a network of telescopes around the world operated by the Global Oscillation Network Group (GONG). The network can observe the sun continuously for weeks at a time as Earth rotates. The sun never sets on GONG. The SOHO satellite in space can observe solar oscillations continuously and can detect motions as slow as 1 mm/s (0.002 mph). Solar astronomers can then use high-speed computers to separate the different patterns on the solar surface and measure the strength of the waves at many different wavelengths.

Helioseismology has allowed astronomers to map the temperature, density, and rate of rotation inside the sun. They have been able to detect great currents of gas flowing below the photosphere and the emergence of sunspots before they appear in the photosphere. Helioseismology can even locate sunspots on the back side of the sun, sunspots that are not yet visible from Earth.

Building Scientific Arguments

How deeply into the sun can you see?

Scientific arguments usually involve observations, and it is always important to know how observations are made. When you look into the layers of the sun, your sight does not really penetrate into the sun. Rather, your eyes record photons that have escaped from the sun and traveled outward through the layers of the sun's atmosphere. If you observe at a wavelength at the center of a dark absorption line, then the photosphere and lower chromosphere are opaque, photons can't escape to your eyes, and the only photons you can see come from the upper chromosphere. What you see are the details of the upper chromosphere. On the other hand, if you observe at a wavelength that is not easily absorbed (a wavelength between spectral lines), the atmosphere is more transparent, and photons from deep inside the photosphere can escape to your eyes. There is a limit, however, set by the H^- ion, a hydrogen atom with an extra electron. At a certain depth, there is so much of this ion that the sun's atmosphere is opaque for almost all wavelengths, few photons can escape, and you can't see deeper.

By choosing the proper wavelength, solar astronomers can observe to different depths. But the corona is so thin and the gas below the photosphere so dense that this method doesn't work in these regions. Now it is time to build a new argument. **How can you observe the corona and the deeper layers of the sun?**

■ ■ ■

Connections: So far you have thought of the sun as a static, unchanging ball of gas with energy flowing outward from the interior and through the atmospheric layers. Now you are ready to plunge into the sun and find out how that energy is made.

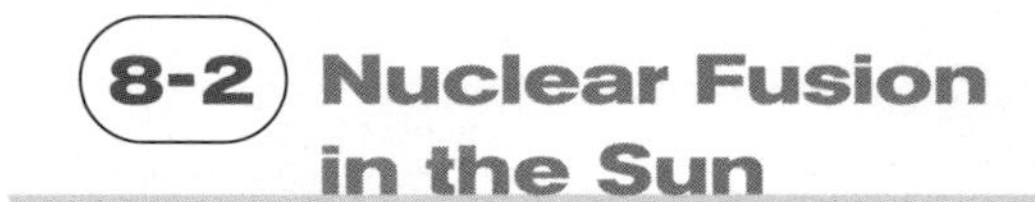

8-2 Nuclear Fusion in the Sun

ASTRONOMERS OFTEN USE the wrong words to describe energy generation in the sun and stars. Astronomers will say, "The star ignites hydrogen burning." In English, the word *ignite* means *catch on fire,* and *burn* means *on fire.* What goes on inside stars isn't really burning in the usual sense.

The sun is a star, and it is not burning. It is powered by nuclear reactions that occur near its center. The energy produced keeps the interior hot, and the gas is totally ionized. That is, the electrons are not attached to the atomic nuclei, and the gas is an atomic soup of rapidly moving particles colliding with each other at high velocity. When you discuss nuclear reactions inside the sun and stars, you should be careful to refer to atomic nuclei and not to atoms.

How exactly can the nucleus of an atom yield energy? The answer lies in the forces that hold the nuclei together.

Nuclear Binding Energy

The sun generates its energy by breaking and reconnecting the bonds between the particles *inside* atomic nuclei. This is quite different from the way you would generate energy by burning wood in a fireplace. The process of burning wood extracts energy by breaking and reconnecting chemical bonds between atoms in the wood. Chemical bonds are formed by the electrons in atoms, and you saw in Chapter 7 that the electrons are bound to the atoms by the electromagnetic force. So chemical energy originates in the electromagnetic force.

There are only four forces in nature: the force of gravity, the electromagnetic force, the **weak force,** and the **strong force.** The weak force is involved in the radioactive decay of certain kinds of nuclear particles, and the strong force binds together atomic nuclei. That means nuclear energy must come from the strong force.

Nuclear power plants on Earth generate energy through **nuclear fission** reactions that split uranium nuclei into less massive fragments. A uranium nucleus contains a total of 235 protons and neutrons, and it splits into a range of fragments containing roughly half as many particles. Because the fragments produced are more tightly bound than the uranium nuclei, binding energy is released during uranium fission.

Stars don't use nuclear fission. They make energy in **nuclear fusion** reactions that combine light nuclei into heavier nuclei. The most common reaction, including that in the sun, fuses hydrogen nuclei (single protons) into helium nuclei (two protons and two neutrons). Because the nuclei produced are more tightly bound than the original nuclei, energy is released. Notice in ■ Figure 8-8 that both fusion and fission reactions move downward in the diagram toward more tightly bound nuclei. They both produce energy by releasing the binding energy of atomic nuclei.

■ Figure 8-8

The red line in this graph shows the binding energy per particle, the energy that holds particles inside an atomic nucleus. The horizontal axis shows the atomic mass number of each element, the number of protons and neutrons in the nucleus. Both fission and fusion nuclear reactions move downward in the diagram (arrows) toward more tightly bound nuclei. Iron has the most tightly bound nucleus, so no nuclear reactions can begin with iron and release energy.

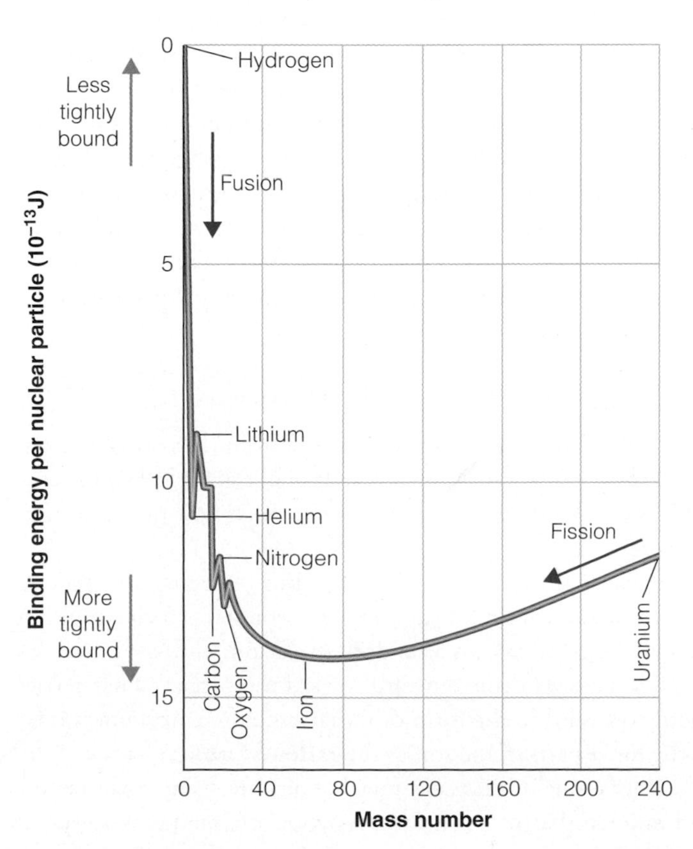

Hydrogen Fusion

Not until the 1930s did astronomers realize how the sun generates energy. The sun fuses together four hydrogen nuclei to make one helium nucleus. Because one helium nucleus has 0.7 percent less mass than four hydrogen nuclei, it seems that some mass vanishes in the process. To see this, subtract the mass of a helium nucleus from the mass of four hydrogen nuclei.

$$\begin{array}{r} \text{4 hydrogen nuclei} = 6.693 \times 10^{-27}\text{ kg} \\ -\ \text{1 helium nucleus} = 6.645 \times 10^{-27}\text{ kg} \\ \hline \text{Difference in mass} = 0.048 \times 10^{-27}\text{ kg} \end{array}$$

It seems from this subtraction that a small amount of mass disappears, but it doesn't really vanish; it merely changes form. The equation $E = m_0c^2$ reminds you that mass and energy are related, and under certain circumstances mass may become energy and vice versa. So the 0.048×10^{-27} kg does not vanish but merely becomes energy. To see how much, you use Einstein's equation:

$$\begin{aligned} E &= m_0c^2 \\ &= (0.048 \times 10^{-27}\text{ kg})(3 \times 10^{8}\text{ m/s})^2 \\ &= 0.43 \times 10^{-11}\text{ J} \end{aligned}$$

This is a very small amount of energy, hardly enough to raise a housefly one-thousandth of an inch. Because one reaction produces such a small amount of energy, it is obvious that many reactions are necessary to supply the energy needs of a star. The sun, for example, needs 10^{38} reactions per second, transforming 5 million tons of mass into energy every second, just to stay hot enough to resist its own gravity.

It seems from this that nuclear fusion is very powerful, especially if you calculate that the fusion of a milligram of hydrogen (roughly the mass of a match head) produces as much energy as burning 30 gallons of gasoline. However, the nuclear reactions in the sun are spread through a large volume in its core, and any single gram of matter produces only a little energy. A person of normal mass eating a normal diet produces about 4000 times more heat per gram than the matter in the core of the sun. The sun produces a lot of energy because it contains a lot of grams of matter in its core.

Fusion reactions can occur only when the nuclei of two atoms get very close to each other. Because atomic nuclei carry positive charges, they repel each other with an electrostatic force called the Coulomb force. Physicists commonly refer to this repulsion between nuclei as the **Coulomb barrier.** To overcome this barrier and get close together, atomic nuclei must collide violently. Violent collisions are rare unless the gas is very hot, in which case the nuclei move at high speeds and collide violently. (Remember, an object's temperature is related to the speed with which its particles move.)

So nuclear reactions in the sun take place only near the center, where the gas is hot and dense. A high temperature insures that some of the collisions between nuclei are violent enough to overcome the Coulomb barrier, and a high density ensures that there are enough collisions, and thus enough reactions, to meet the sun's energy needs.

You can symbolize the fusion reactions in the sun with a simple nuclear reaction:

$$4\,{}^{1}\text{H} \rightarrow {}^{4}\text{He} + \text{energy}$$

In this equation, ^{1}H represents a proton, the nucleus of the hydrogen atom, and ^{4}He represents the nucleus of a helium atom. The superscripts indicate the approximate weight of the nuclei (the number of protons plus the number of neutrons). The actual steps in the process are more complicated than this convenient summary suggests. Instead of waiting for four hydrogen nuclei to collide simultaneously, a highly unlikely event, the process can proceed step by step in a chain of reactions—the proton–proton chain.

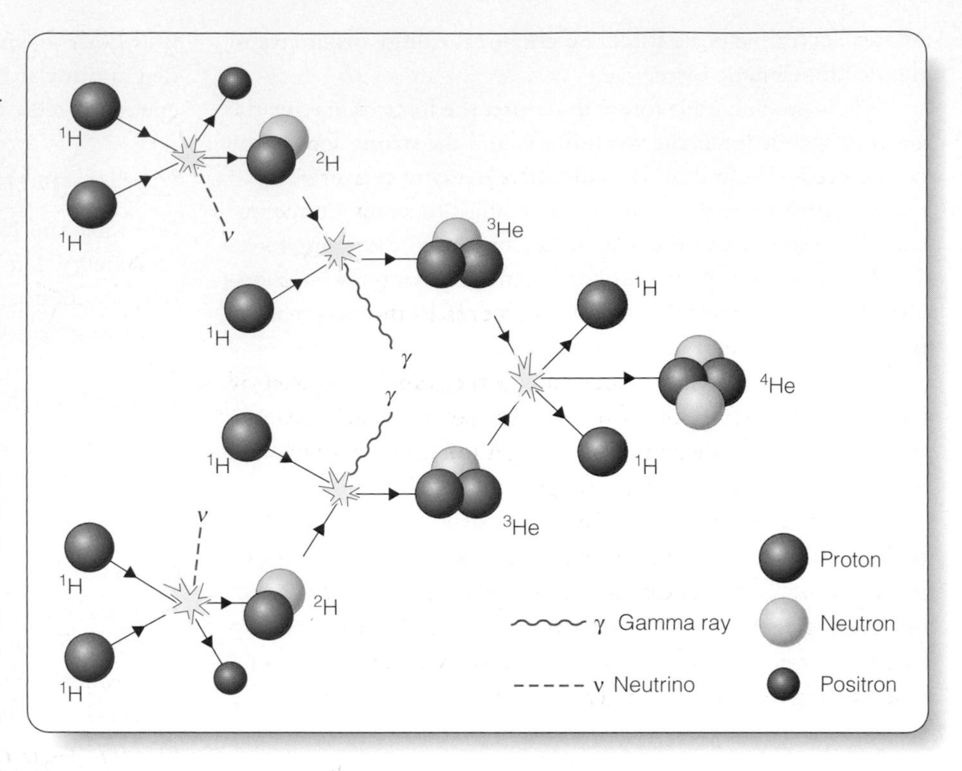

■ **Figure 8-9**

The proton–proton chain combines four protons (at far left) to produce one helium nucleus (at right). Energy is produced mostly as gamma rays and positrons, which combine with electrons and convert their mass into energy. Neutrinos escape, carrying away about 2 percent of the energy produced.

The **proton–proton chain** is a series of three nuclear reactions that builds a helium nucleus by adding together protons. This process is efficient at temperatures above 10,000,000 K. The sun, for example, manufactures over 90 percent of its energy in this way.

The three steps in the proton–proton chain entail these reactions:

$$^{1}\text{H} + {}^{1}\text{H} \rightarrow {}^{2}\text{H} + e^{+} + \nu$$
$$^{2}\text{H} + {}^{1}\text{H} \rightarrow {}^{3}\text{He} + \gamma$$
$$^{3}\text{He} + {}^{3}\text{He} \rightarrow {}^{4}\text{He} + {}^{1}\text{H} + {}^{1}\text{H}$$

In the first reaction, two hydrogen nuclei (two protons) combine to form a heavy hydrogen nucleus called **deuterium,** emitting a particle called a **positron,** e^{+} (a positively charged electron), and a **neutrino,** ν (a subatomic particle having an extremely low mass and a velocity nearly equal to the velocity of light). In the second reaction, the heavy hydrogen nucleus absorbs another proton and, with the emission of a gamma ray, γ, becomes a lightweight helium nucleus. Finally, two lightweight helium nuclei combine to form a common helium nucleus and two hydrogen nuclei. Because the last reaction needs two ^{3}He nuclei, the first and second reactions must occur twice (■ Figure 8-9). The net result of this chain reaction is the transformation of four hydrogen nuclei into one helium nucleus plus energy.

The energy appears in the form of gamma rays, positrons, the energy of motion of the particles, and neutrinos. The gamma rays are photons that are absorbed by the surrounding gas before they can travel more than a fraction of a millimeter. This heats the gas and helps maintain the pressure. The positrons produced in the first reaction combine with free electrons, and both particles vanish, converting their mass into gamma rays, which are absorbed and also help keep the center of the star hot. In addition, when fusion produces new nuclei, they fly apart at high velocity. This energy of motion helps raise the temperature of the gas. The neutrinos resemble photons except that they almost never interact with other particles. The average neutrino could pass unhindered through a lead wall a light-year thick. Consequently, the neutrinos do not help heat the gas but race out of the star at nearly the speed of light, carrying away roughly 2 percent of the energy produced.

Energy Transport in the Sun

Now you are ready to follow the energy from the core of the sun to the surface. The surface is cool, only about 5800 K, and the center is over 10 million Kelvin, so energy must flow outward from the core.

Because the core is so hot, the photons being emitted are gamma rays. Each time a gamma ray encounters an electron, it is deflected or scattered in a random direction; and, as it bounces around, it slowly drifts outward toward the surface. That carries energy outward in the form of radiation, and astronomers refer to the inner parts of the sun as the **radiative zone.**

To examine this process, imagine picking a single gamma ray and following it to the surface. As your gamma ray is scattered over and over by the hot gas, it drifts outward into cooler layers, and the cooler gas tends to emit photons of longer wavelength. Your gamma ray will eventually be absorbed by the gas and reemitted as two X rays. Now you must follow those two X rays as they bounce around, and soon you will see them drifting outward into even cooler gas, where they will become a number of longer wavelength photons. The packet of energy that began as a single gamma ray gets broken down into a large number of lower-energy photons, and it eventually emerges from the sun's surface as about 1800 photons of visible light.

But something else happens along the way. The packet of energy that you began following from the core eventually reaches the outer layers of the sun where the gas is so cool it is not very transparent to radiation. The energy backs up like water behind a dam, and the gas begins to churn in convection. Hot blobs of gas rise, and cool blobs sink. In this region, known as the **convective zone,** the energy is carried outward as circulating gas. These two zones are shown in ■ Figure 8-10. The granulation visible on the photosphere is clear evidence that the sun has a convective zone just below the photosphere carrying energy upward to the surface.

Sunlight is nuclear energy produced in the core of the sun. The energy of a single gamma ray can take a million years to work its way outward, first as radiation and then as convection on its journey to the photosphere.

It is time to ask the critical question that lies at the heart of science. What is the evidence to support this theoretical explanation of how the sun makes its energy? The search for that evidence will introduce you to one of the great problems of modern astronomy.

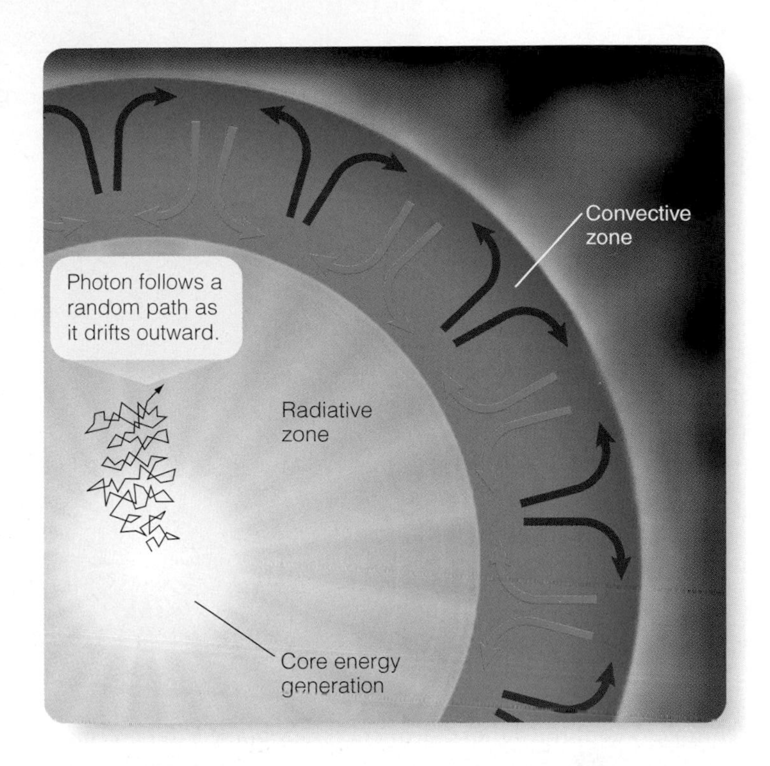

Active Figure 8-10

A cross section of the sun. Near the center, nuclear fusion reactions generate high temperatures. Energy flows outward through the radiative zone as photons are randomly deflected over and over by electrons. In the cooler, more opaque outer layers, the energy is carried by rising convection currents of hot gas (red arrows) and sinking currents of cooler gas (blue arrows).

AceAstronomy™ Log into AceAstronomy and select this chapter to see Active Figure "The Sun." Watch energy flow outward from the core to the surface.

AceAstronomy™ Log into AceAstronomy and select this chapter to see Astronomy Exercise "Nuclear Fusion" and take control of fusion in the sun's core.

The Solar Neutrino Problem

The center of a star seems forever hidden, but the sun is transparent to neutrinos because these subatomic particles almost never interact with normal matter. Nuclear reactions in the sun's core produce floods of neutrinos that rush out of the sun and off into space. If you could detect these neutrinos, you could probe the sun's interior.

Because neutrinos almost never interact with atoms, you never feel the flood of over 10^{12} solar neutrinos that flow through your body every second. Even at night, neutrinos from the sun rush through Earth as if it weren't there, up through your bed, through you, and onward into space. Obviously you are lucky to be transparent to neutrinos, but it means that neutrinos are extremely hard to detect. Certain nuclear reactions, however, can be triggered by a neutrino of the right energy; and, in the late 1960s, chemist Raymond Davis, Jr., began using such a reaction to detect solar neutrinos.

Davis filled a 100,000-gallon tank with the cleaning fluid perchloroethylene (C_2Cl_4). Theory predicts that about once a day, a solar neutrino will convert a chlorine atom in the tank into radioactive argon, which can be detected later by its radioactive decay. To protect the detector from cosmic rays from space, the tank was buried nearly a mile deep in a South Dakota gold mine (■ Figure 8-11a). Of course, the mile of rock overhead had no effect on the neutrinos.

The result of the Davis experiment startled astronomers. The cleaning fluid detected too few neutrinos—not one neutrino per day as predicted by models of the sun but about one every three days. The experiment was refined, tested, and calibrated for three decades; but it did not find the missing neutrinos. Other detectors were built, and they too counted too few neutrinos coming from the sun.

The missing solar neutrinos were one of the great mysteries of modern astronomy. Some scientists argued that astronomers didn't correctly understand how the sun and stars make their energy, but other scientists wondered if there was something about neutrinos that could explain the problem. Astronomers have great confidence in their theories of the sun's interior, and helio-

■ Figure 8-11

(a) The Davis solar neutrino experiment used cleaning fluid and could detect only one of the three flavors of neutrinos. (Brookhaven National Laboratory) (b) The Sudbury Neutrino Observatory is a 12-meter-diameter globe containing water rich in deuterium in place of hydrogen. Buried 6800 feet deep in an Ontario mine, it can detect all three flavors of neutrinos and confirms that neutrinos oscillate. (Photo courtesy of SNO)

seismology confirmed those theories, so astronomers did not abandon their theories immediately (**Window on Science 8-1**).

As the 21st century began, scientists were able to solve the mystery. Physicists know of three kinds of neutrinos, which they call flavors. The Davis experiment could detect (or taste) only one flavor, electron-neutrinos. Theory hinted that the electron-neutrinos produced in the core of the sun might oscillate among the three flavors as they rushed out through the sun and across space to Earth. Observations begun in 2000 confirm this theory (Figure 8-11b). Some of the electron-neutrinos produced in the sun transform into tau- and muon-neutrinos, which most detectors cannot detect.

This solution to the solar neutrino problem is exciting because neutrinos can't oscillate unless they have mass. Neutrinos were long thought to be massless, but if they have even a small mass, they are so common their gravity could affect the evolution of the universe as a whole—something you will read about in Chapter 18. The detection of neutrino oscillation excites astronomers for another reason. It confirms the theories that describe the interior of the sun and stars.

Building Scientific Arguments

Why does nuclear fusion require that the gas be very hot? This argument has to include some basic physics of atoms and thermal energy. Inside a star, the gas is so hot it is ionized, which means the electrons have been stripped off the atoms, and the nuclei are bare and have a positive charge. For hydrogen fusion, the nuclei are single protons. These atomic nuclei repel each other because of their positive charges, so they must collide with each other at high velocity to overcome that repulsion and get close enough together to fuse. If the atoms in a gas are moving rapidly, then it must have a high temperature, and so nuclear fusion requires that the gas have a very high temperature. If the gas is cooler than about 10 million K, hydrogen can't fuse because the protons don't collide violently enough to overcome the repulsion of their positive charges.

It is easy to see why nuclear fusion in the sun requires high temperature, but now expand your argument. **Why does it require high density?**

■ ■ ■

Connections: You have been to the center of the sun and seen how it makes its energy, and you have followed that energy up to the surface. Now it is time for you to see how outrageous the sun can be as that energy bursts through the photosphere.

8-3 Solar Activity

THE SUN IS UNQUIET. It is home to slowly changing spots larger than Earth and vast eruptions that dwarf human imagination. All of these seemingly different forms of solar activity have one thing in common—magnetic fields. The weather on the sun is magnetic.

Observing the Sun

Solar activity is often visible with even a small telescope, but you should be very careful about observing the sun. Sunlight is intense, and when it enters your eye it is absorbed and converted into thermal energy. Equally dangerous is the infrared radiation in sunlight. Your eyes can't detect the infrared, but it is converted to thermal energy in your eyes and can burn and scar the retina.

Window on Science 8-1

Avoiding Hasty Judgments: Scientific Confidence

Scientists like to claim that every scientific understanding is based on evidence, that every theory has been tested, and that the moment a theory fails a test, it is discarded or revised. The truth is much more complicated than that, and the solar neutrino problem is a good illustration. If the detection of too few solar neutrinos contradicted the theory of stellar structure, why wasn't the theory abandoned?

While scientists do indeed have tremendous respect for evidence, they also have confidence in theories that have been tested successfully many times. If a theory has been tested and confirmed over and over, they may even begin to call it a natural law, and that means scientists have great confidence that the theory or law is a good description of how nature works.

Nevertheless, it is not unusual for an experiment or an observation to contradict well-established theories. In many cases, the experiments and observations are simple mistakes, or they have not been interpreted correctly. Scientists resist abandoning a well-tested theory even when an observation continues to contradict it. If confidence in the theory is stronger than the evidence, scientists begin by testing the evidence. Can it be right? Do we understand it correctly? Of course, if the evidence cannot be impeached and it continues to contradict the theory, scientists must eventually abandon or modify the theory no matter how many times it has previously been tested and confirmed. This ultimate reliance on evidence is the distinguishing characteristic of science.

It is only human nature to hang on to the principles you have come to trust, and that confidence in well-tested scientific principles helps scientists avoid rushing to faulty judgments. For example, claims for perpetual motion machines occasionally crop up in the news, but the world's scientists don't instantly abandon the laws of energy and motion pending an analysis of the latest claim. Of course, if such a claim did prove true, the entire structure of scientific knowledge would come crashing down, but because the known laws of energy and motion have been well tested and no perpetual motion machine has ever been successful, scientists know which way to bet. Like the keel on a ship, confidence in well-tested theories and laws keeps the scientific boat from rocking before every little breeze.

Ultimately scientific confidence must be open to change. If, eventually, a single experiment or observation conclusively contradicts a cherished law of nature, scientists must abandon that law and find a new way to understand nature. You can see this scientific confidence at work in many controversies, from the origin of the human race, to the meaning of IQ measurements; but one of the best examples is the solar neutrino problem. While astronomers struggled to understand the origin of solar neutrinos, they continued to have confidence that they do indeed understand how stars make energy.

It is not safe to look directly at the sun, and it is even more dangerous to look at the sun through any optical instrument such as a telescope, binoculars, or even the viewfinder of a camera. The light-gathering power of such an optical system concentrates the sunlight and can cause severe injury. Never look at the sun with any optical instrument unless you are certain it is safe. ■ Figure 8-12 shows a safe way to observe the sun with a small telescope.

In the early 17th century, Galileo observed the sun and saw spots on its surface; day by day he saw the spots moving across the sun's disk. He rightly concluded that the sun was a sphere and was rotating. You could repeat his observations, and you would probably see something that looks like Figure 8-12b. You would see sunspots.

Sunspots

The dark sunspots that you see at visible wavelengths only hint at the complex processes that go on in the sun's atmosphere. To explore those processes, you must turn to the analysis of images and spectra at a wide range of wavelengths.

Study **Sunspots and the Sunspot Cycle** on pages 174–175 and notice five important points:

1. Sunspots are cool spots on the sun's surface caused by strong magnetic fields.
2. Also notice that sunspots follow an 11-year cycle not only in the number of spots visible but in their location on the sun.
3. The Zeeman effect gives astronomers a way to measure the strength of magnetic fields on the sun.
4. Notice that the sunspot cycle can vary over centuries and appears to affect Earth's climate.
5. Finally, notice the clear evidence that sunspots are part of a larger magnetic process that involves all layers of the sun's atmosphere.

The sunspot groups are merely the visible traces of magnetically active regions. But what causes this magnetic activity? The answer appears to be linked to the waxing and waning of the sun's overall magnetic field.

Ace Astronomy™ Log into AceAstronomy and select this chapter to see Astronomy Exercise "Zeeman Effect" and see how astronomers measure magnetic fields.

Ace Astronomy™ Log into AceAstronomy and select this chapter to see Astronomy Exercise "Sunspot Cycle I"; you can explore the link between sunspots and the sun's magnetic field.

Ace Astronomy™ Log into AceAstronomy and select this chapter to see Astronomy Exercise "Sunspot Cycle II." You can observe sunspots over the centuries.

Sunspots and the Sunspot Cycle

1 The dark spots that appear on the sun are only the visible traces of complex regions of activity. Observations over many years and at a range of wavelengths tell you that sunspots are clearly linked to the sun's magnetic field.

Spectra show that sunspots are cooler than the photosphere with a temperature of about 4240 K. The photosphere has a temperature of about 5800 K. Because the total amount of energy radiated by a surface depends on its temperature raised to the fourth power, sunspots look dark in comparison. Actually, a sunspot emits quite a bit of radiation. If the sun were removed and only an average-size sunspot were left behind, it would be brighter than the full moon.

A typical sunspot is about twice the size of Earth, but there is a wide range of sizes. They appear, last a few weeks to as long as two months, and then shrink away. Usually, sunspots occur in pairs or complex groups.

Size of Earth

NASA

Royal Swedish Academy of Sciences

Umbra

Penumbra

Sunspots are not shadows, but astronomers refer to the dark core of a sunspot as its umbra and the outer, lighter region as the penumbra.

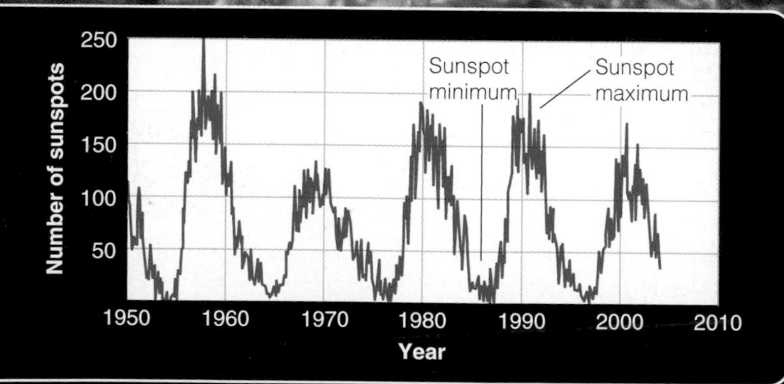

2 The number of spots visible on the sun varies in a cycle with a period of 11 years. At maximum, there are often over 100 spots visible. At minimum, there are very few.

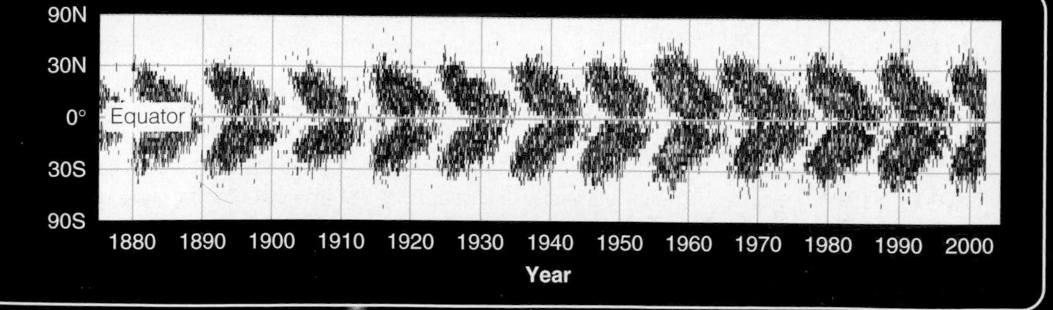

2a Early in the cycle, spots appear at high latitudes north and south of the sun's equator. Later in the cycle, the spots appear closer to the sun's equator. If you plot the latitude of sunspots versus time, the graph looks like butterfly wings, as shown in this **Maunder butterfly diagram**, named after E. Walter Maunder of Greenwich Observatory.

3 Astronomers can measure magnetic fields on the sun using the **Zeeman effect** as shown below. When an atom is in a magnetic field, the electron orbits are altered, and the atom is able to absorb a number of different wavelength photons even though it was originally limited to a single wavelength. In the spectrum, you see single lines split into multiple components, with the separation between the components proportional to the strength of the magnetic field.

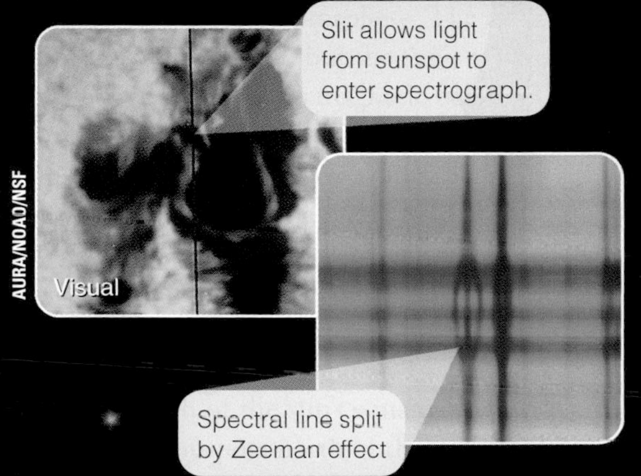

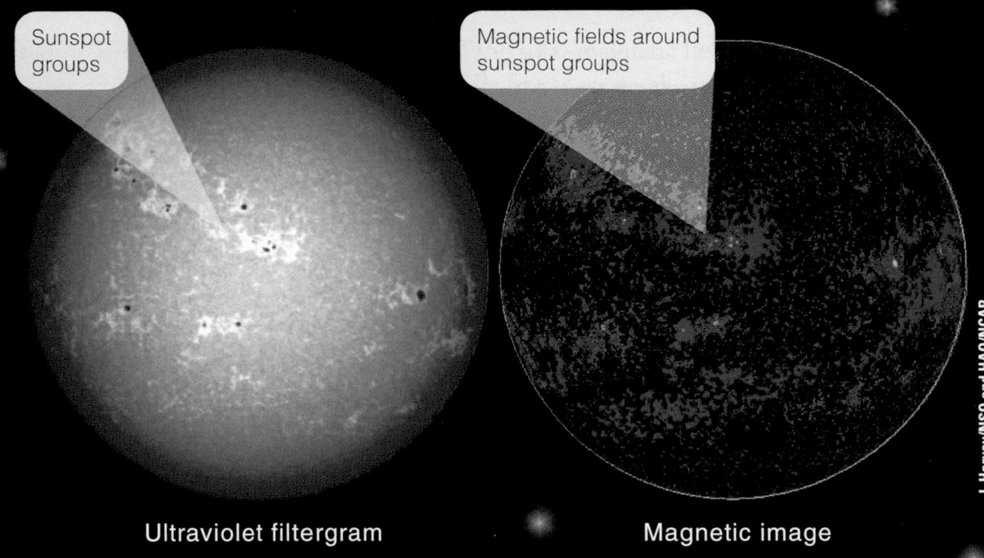

Simultaneous images

3a Images of the sun above show that sunspots contain magnetic fields a few thousand times stronger than Earth's. The strong fields are believed to inhibit gas motion below the photosphere; consequently, convection is reduced below the sunspot, and the surface there is cooler. Heat prevented from emerging through the sunspot is deflected and emerges around the sunspot, which can be detected in infrared images.

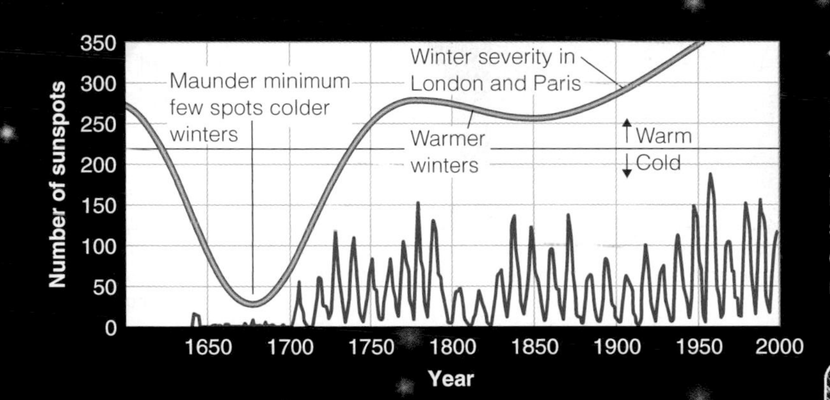

4 Historical records show that there were very few sunspots from about 1645 to 1715, a phenomenon known as the **Maunder minimum.** This coincides with a period called the "little ice age," a period of unusually cool weather in Europe and North America from about 1500 to about 1850, as shown in the graph at left. Other such periods of cooler climate are known. The evidence suggests that there is a link between solar activity and the amount of solar energy Earth receives. This link has been confirmed by measurements made by spacecraft above Earth's atmosphere.

M. Seeds

Magnetic fields can reveal themselves by their shape. For example, iron filings sprinkled over a bar magnet reveal an arched shape.

Far -UV image

SOHO/EIT, ESA and NASA

5 Observations at nonvisible wavelengths reveal that the chromosphere and corona above sunspots are violently disturbed in what astronomers call **active regions.** Spectrographic observations show that active regions contain powerful magnetic fields.

Arched structures above an active region are evidence of gas trapped in magnetic fields.

The complexity of an active region becomes visible at short wavelengths.

Visual-wavelength image

Simultaneous images

Far-UV image

NASA/TRACE

Magnetic Solar Phenomena

1 Magnetic phenomena in the chromosphere and corona, like magnetic weather, result as constantly changing magnetic fields on the sun trap ionized gas to produce beautiful arches and powerful outbursts. Some of this solar activity can affect Earth's magnetic field and atmosphere.

This ultraviolet image of the solar surface was made by the NASA TRACE spacecraft. It shows hot gas trapped in magnetic arches extending above active regions. At visual wavelengths, you would see sunspot groups in these active regions.

The gas in prominences may be 60,000 to 80,000 K, quite cold compared with the low-density gas in the corona, which may be as hot as a million Kelvin.

Trace/NASA

1a A **prominence** is composed of ionized gas trapped in a magnetic arch rising up through the photosphere and chromosphere into the lower corona. Seen during total solar eclipses at the edge of the solar disk, prominences look pink because of the three Balmer emission lines. The image below shows the arch shape suggestive of magnetic fields. Seen from above against the sun's bright surface, prominences form dark filaments.

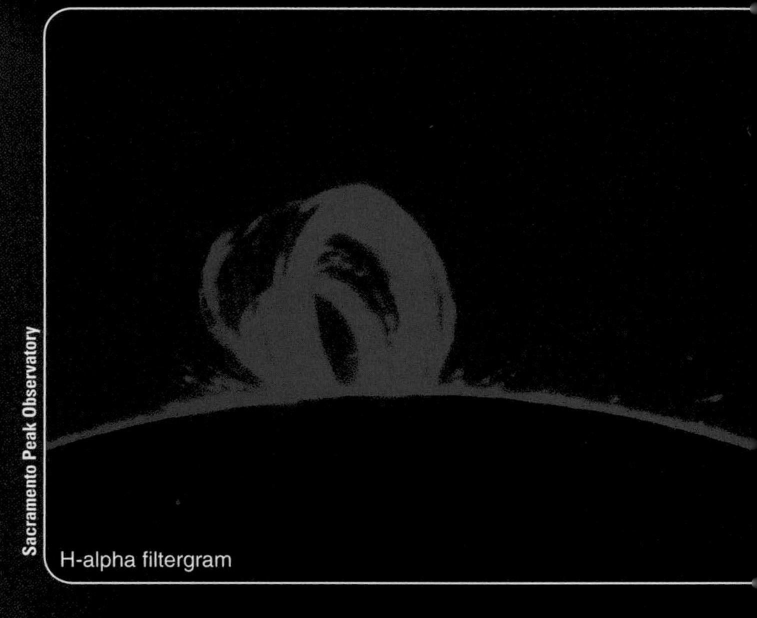

H-alpha filtergram

1b Quiescent prominences may hang in the lower corona for many days, whereas eruptive prominences burst upward in hours. The eruptive prominences below are many Earth diameters long.

Far-UV image

SOHO, EIT, ESA and NASA

2 Solar **flares** rise to maximum in minutes and decay in an hour. They occur in active regions where oppositely directed magnetic fields meet and cancel each other out in what astronomers call **reconnections.** Energy stored in the magnetic fields is released as short-wavelength photons and as high-energy protons and electrons. X-ray and ultraviolet photons reach Earth in 8 minutes and increase ionization in our atmosphere, which can interfere with radio communications. Particles from flares reach Earth hours or days later as gusts in the solar wind, which can distort Earth's magnetic field and disrupt navigation systems. Solar flares can also cause surges in electrical power lines and damage to Earth satellites.

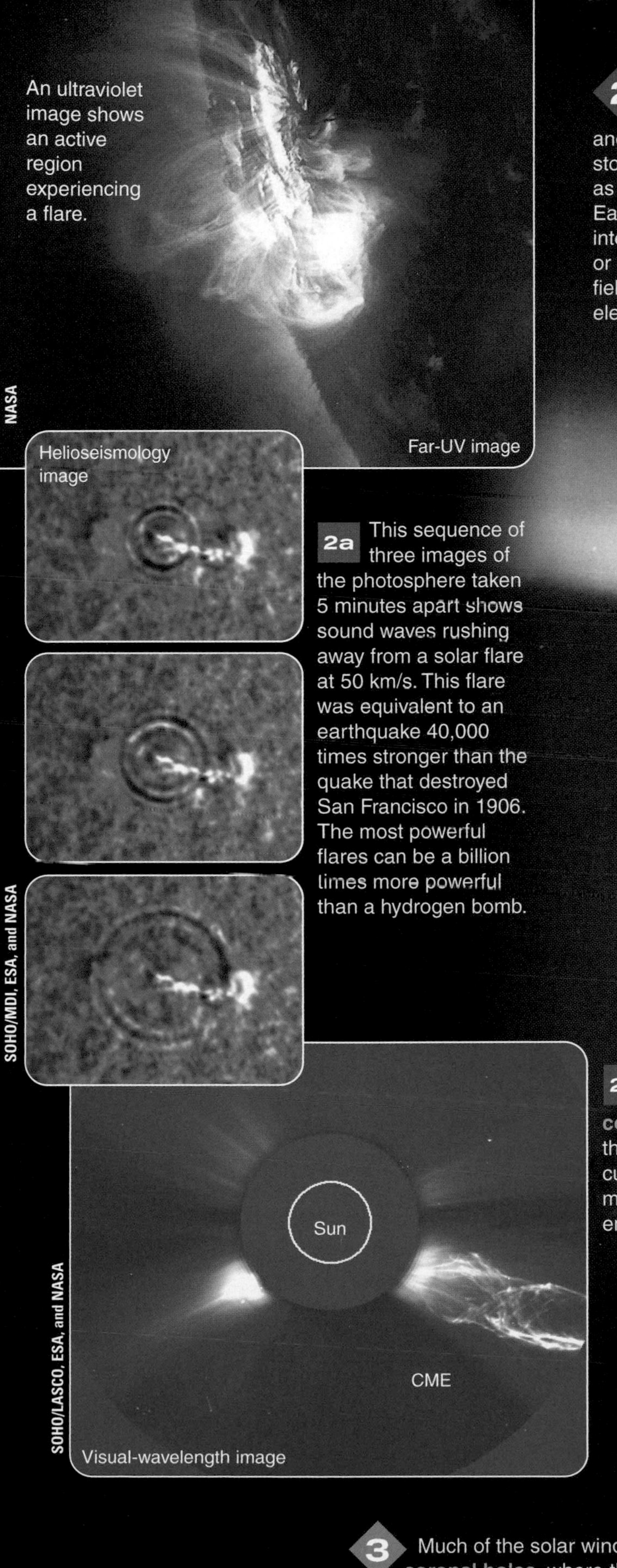

An ultraviolet image shows an active region experiencing a flare.

2a This sequence of three images of the photosphere taken 5 minutes apart shows sound waves rushing away from a solar flare at 50 km/s. This flare was equivalent to an earthquake 40,000 times stronger than the quake that destroyed San Francisco in 1906. The most powerful flares can be a billion times more powerful than a hydrogen bomb.

2b **Auroras** occur about 130 km above Earth's surface when energy in the solar wind guided by Earth's magnetic field excites gases in the upper atmosphere.

Copyright Jan Curtis

2c Magnetic reconnections can release enough energy to blow large amounts of ionized gas outward from the corona in **coronal mass ejections (CMEs).** These produce violent gusts in the solar wind, and if one blows past Earth it can create electrical currents up to a million megawatts, which flow down into Earth's magnetic poles and excite atoms in Earth's upper atmosphere to emit photons, which are seen as an aurora, as shown above.

3 Much of the solar wind comes from **coronal holes,** where the magnetic field does not loop back into the sun. These open magnetic fields allow ionized gas in the corona to flow away as the solar wind. The dark area in this X-ray image at right is a coronal hole.

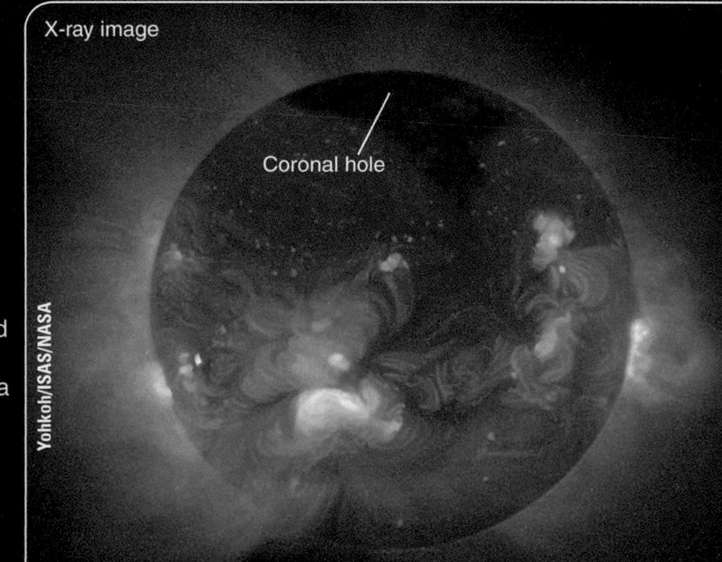

Ace Astronomy™ Log into AceAstronomy and select this chapter to see Astronomy Exercise "Auroras." You can take control of Earth's magnetic field and the solar wind to create your own aurorae.

The Solar Constant

Even a small change in the sun's energy output could produce dramatic changes in Earth's climate. The continued existence of the human species depends on the constancy of the sun, but we humans know very little about the variation of the sun's energy output.

The energy production of the sun can be measured by adding up all of the energy falling on 1 square meter of Earth's surface during 1 second. Of course, some correction for the absorption of Earth's atmosphere is necessary, and you must count all wavelengths from X rays to radio waves. The result, which is called the **solar constant,** amounts to about 1360 joules per square meter per second. A change in the solar constant of only 1 percent could change Earth's average temperature by 1 to 2°C (about 1.8 to 3.6°F). For comparison, during the last ice age Earth's average temperature was about 5°C cooler than it is now.

Some of the best measurements of the solar constant were made by instruments aboard the Solar Maximum Mission satellite. These have shown variations in the energy received from the sun of about 0.1 percent that lasted for days or weeks. Superimposed on that random variation is a long-term decrease of about 0.018 percent per year that has been confirmed by observations made by sounding rockets, balloons, and satellites. This long-term decrease may be related to a cycle of activity on the sun with a period longer than the 22-year magnetic cycle.

Small, random fluctuations will not affect Earth's climate, but a long-term decrease over a decade or more could cause worldwide cooling. History contains some evidence that the solar constant may have varied in the past. As you saw on page 175, the "Little Ice Age" was a period of unusually cool weather in Europe and America that lasted from about 1500 to 1850.* The average temperature worldwide was about 1°C cooler than it is now. This period of cool weather corresponded to the Maunder minimum, a period of reduced solar activity—few sun spots, no auroral displays, and no solar coronae visible during solar eclipses.

In contrast, an earlier period called the Grand Maximum, lasting from about 1100 to about 1250 AD, saw a warming of Earth's climate. The Vikings were able to explore and colonize Greenland, and native communities in parts of North America were forced to abandon their settlements because of long droughts.

Other minima and maxima have been found in climate data taken from studies of the growth rings of trees. In good years, trees add a thicker growth ring than in poor years, so measuring tree rings can reveal the climate in the past. Evidently, solar activity can increase or decrease the solar constant very slightly and affect Earth's climate in dramatic ways. The future of our civilization on Earth may depend on our learning to understand the solar constant.

*Ironically, the Maunder minimum coincides with the reign of Louis XIV of France, the "Sun King."

Building Scientific Arguments

What kind of activity would the sun have if it didn't rotate differentially?

This is a really difficult question because only one star is visible close up. Nevertheless, you can construct a scientific argument by thinking about the Babcock model. If the sun didn't rotate differentially, with its equator traveling faster than do the higher latitudes, then the magnetic field might not get twisted up, and there might not be a solar cycle. Twisted tubes of magnetic field might not form and rise through the photosphere to produce prominences and flares, although convection might tangle the magnetic field and produce some activity. Is the magnetic activity that heats the chromosphere and corona driven by differential rotation or by convection? It is hard to guess, but without differential rotation, the sun might not have a strong magnetic field and high-temperature gas above its photosphere.

This is very speculative, but sometimes in the critical analysis of ideas it helps to imagine a change in a single important factor and try to understand what might happen. For example, redo the argument above. **What do you think the sun would be like if it had no convection inside?**

■ ■ ■

Connections: The sun is beautiful and complex, with great eruptions, prominences, active regions, and dark spots sweeping across its surface like magnetic weather. All of this activity is driven by the outward flow of energy generated by hydrogen fusion in the core. Presumably other stars are like the sun, but other stars are so far away their surfaces are invisible. Through the largest telescopes, they look like nothing more than points of light. How can astronomers learn about the stars? You will begin to answer that question in the next chapter.

Study and Review Tools

Summary

8-1 | The Solar Atmosphere

What do you see when you look at the sun?

- The solar atmosphere consists of three layers of hot, low-density gas: the photosphere, chromosphere, and corona. The photosphere, or visible surface, is the level in the sun from which visible photons most easily escape.
- The granulation of the photosphere is produced by convection currents of gas rising from below.
- The chromosphere is most easily visible during total solar eclipses, when it flashes into view for a few seconds. It is a thin, hot layer of gas just above the photosphere, and its pink color is caused by the Balmer emission lines in its spectrum.
- Filtergrams of the chromosphere reveal spicules and filaments.
- The corona is the sun's outermost atmospheric layer. It is composed of a very-low-density, very hot gas extending many solar radii from the visible sun. Its high temperature—up to three million Kelvin—is believed to be maintained by the magnetic field extending up through the photosphere—the magnetic carpet.
- Parts of the corona give rise to the solar wind, a breeze of low-density ionized gas streaming away from the sun.
- Solar astronomers can study the motion, density, and temperature of gases inside the sun by analyzing the way the solar surface oscillates. Known as helioseismology, this process requires large amounts of data and extensive computer analysis.

8-2 | Nuclear Fusion in the Sun

How does the sun make its energy?

- The sun generates its energy near its center, where the temperature and density are high enough for nuclear fusion reactions to combine hydrogen nuclei to make helium nuclei. This is known as the proton–proton chain.
- Observations of too few neutrinos coming from the sun's core are now explained by the oscillation of neutrinos among three different types. The neutrinos confirm that the sun makes its energy by hydrogen fusion.

8-3 | Solar Activity

What are the dark sunspots and other forms of solar activity?

- Sunspots seem dark because they are slightly cooler than the rest of the photosphere. The average sunspot is about twice the size of Earth and contains magnetic fields a few thousand times stronger than Earth's. Sunspots are thought to form because the magnetic field inhibits rising currents of hot gas and allows the surface to cool.
- Astronomers can use the Zeeman effect to measure magnetic fields on the sun.

Why does the sun go through a cycle of activity?

- The sun is very bright, and its light and infrared radiation can burn your eyes, so you must take great care in observing the sun.
- The average number of sunspots varies over a period of about 11 years and appears to be related to a magnetic cycle. The sunspot cycle does not repeat exactly each cycle, and the decades from 1645 to 1715, known as the Maunder minimum, seem to have been a time when solar activity was very low and Earth's climate was slightly colder.
- The sun rotates differentially, with regions far from the equator rotating slower than equatorial regions.
- Alternate sunspot cycles have reversed magnetic polarity, which has been explained by the Babcock model, in which the differential rotation of the sun winds up the magnetic field. Tangles in the field rise to the surface and cause active regions visible to your eyes as sunspot pairs. When the field becomes strongly tangled, it reorders itself into a simpler but reversed field, and the cycle starts over.
- Other stars are too far away for starspots to be visible, but spectroscopic observations reveal that many other stars have spots and magnetic fields that follow long-term cycles like the sun's.
- Prominences occur in the chromosphere; their arched shapes show that they are formed of ionized gas trapped in the magnetic field. You see prominences in filtergrams as dark filaments silhouetted against the bright chromosphere.
- Flares are sudden eruptions of X-ray, ultraviolet, and visible radiation plus high-energy atomic particles produced when magnetic fields on the sun interact and reconnect. Flares are important because they can have dramatic effects on Earth such as communications blackouts and auroras.
- Spacecraft images show long streamers extending from the corona out into space. Coronal mass ejections occur when magnetic fields on the surface of the sun eject bursts of ionized gas that flow outward in the solar wind. Such bursts can produce auroras and other phenomena if they strike Earth.
- Small changes in the solar constant over decades can affect Earth's climate. The Maunder minimum, a time of low solar activity, coincides with the little ice age. Other climate fluctuations have occurred and appear to be linked to small changes in solar activity.

New Terms

sunspot (p. 162)
granulation (p. 163)
convection (p. 164)
supergranule (p. 164)
limb (p. 164)
limb darkening (p. 164)
transition region (p. 164)
filtergram (p. 165)
filament (p. 165)
spicule (p. 165)
coronagraph (p. 165)
magnetic carpet (p. 166)
solar wind (p. 166)
helioseismology (p. 167)
weak force (p. 169)
strong force (p. 169)
nuclear fission (p. 169)
nuclear fusion (p. 169)
Coulomb barrier (p. 169)
proton–proton chain (p. 170)
deuterium (p. 170)
positron (p. 170)
neutrino (p. 170)
radiative zone (p. 170)
convective zone (p. 171)
Maunder butterfly diagram (p. 174)
Zeeman effect (p. 175)
Maunder minimum (p. 175)
active region (p. 175)
differential rotation (p. 176)
dynamo effect (p. 177)
Babcock model (p. 177)
prominence (p. 180)
flare (p. 181)
reconnection (p. 181)
aurora (p. 181)
coronal hole (p. 181)
coronal mass ejection (CME) (p. 181)
solar constant (p. 182)

Review Questions

Ace Astronomy™ Assess your understanding of this chapter's topics with additional quizzing and animations at **http://ace.brookscole.com/sf9**

1. Why can't you see deeper than the photosphere?
2. What evidence can you give that granulation is caused by convection?
3. How are granules and supergranules related? How do they differ?
4. How can astronomers detect structure in the chromosphere?
5. What evidence can you give that the corona has a very high temperature?
6. What heats the chromosphere and corona to high temperature?
7. How are astronomers able to explore the layers of the sun below the photosphere?
8. Why does nuclear fusion require high temperatures?
9. Why does nuclear fusion in the sun occur only near the center?
10. How can astronomers detect neutrinos from the sun?
11. How can neutrino oscillation explain the solar neutrino problem?
12. What evidence can you give that sunspots are magnetic?
13. How does the Babcock model explain the sunspot cycle?
14. What does the spectrum of a prominence reveal? What does its shape reveal?
15. How can solar flares affect Earth?
16. The upper two images here show two solar phenomena. What are they, and how are they related? How do they differ?
17. The lower image was recorded in the extreme ultraviolet by the SOHO spacecraft. Explain the features you see.

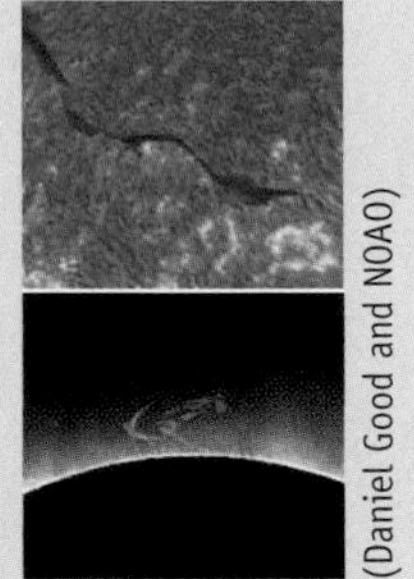

(Daniel Good and NOAO)

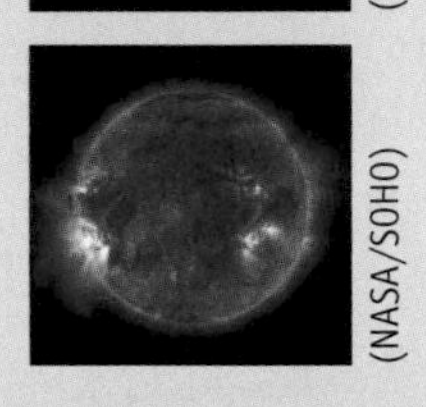

(NASA/SOHO)

Discussion Questions

1. What energy sources on Earth cannot be thought of as stored sunlight?
2. What would the spectrum of an auroral display look like? Why?
3. What observations would you make if you were ordered to set up a system that could warn astronauts in orbit of dangerous solar flares? Such a warning system exists.

Problems

1. The radius of the sun is 0.7 million km. What percentage of the radius is taken up by the chromosphere?
2. The smallest detail visible with ground-based solar telescopes is about 1 second of arc. How large a region does this represent on the sun? (*Hint:* Use the small-angle formula.)
3. What is the angular diameter of a star like the sun located 5 ly from Earth? Is the Hubble Space Telescope able to detect detail on the surface of such a star?
4. How much energy is produced when the sun converts 1 kg of mass into energy?
5. How much energy is produced when the sun converts 1 kg of hydrogen into helium? (*Hint:* How does this problem differ from Problem 4?)
6. A 1-megaton nuclear weapon produces about 4×10^{15} J of energy. How much mass must vanish when a 5-megaton weapon explodes?
7. Use the luminosity of the sun, the total amount of energy it emits each second, to calculate how much mass it converts to energy each second.
8. If a sunspot has a temperature of 4240 K and the solar surface has a temperature of 5800 K, how many times brighter is the surface compared to the sunspot? (*Hint:* Use the Stefan–Boltzmann law, Chapter 7.)
9. A solar flare can release 10^{25} J. How many megatons of TNT would be equivalent? (*Hint:* A 1-megaton bomb produces about 4×10^{15} J.)
10. The United States consumes about 2.5×10^{19} J of energy in all forms in a year. How many years could you run the United States on the energy released by the solar flare in Problem 9?
11. Neglecting energy absorbed or reflected by Earth's atmosphere, the solar energy hitting 1 square meter of Earth's surface is 1360 J/s (the solar constant). How long does it take a baseball diamond (90 ft on a side) to receive 1 megaton of solar energy?

Media Cluster

ACTIVE FIGURES

Ace Astronomy™ To access the resources in the Media Cluster, log into AceAstronomy at **http://ace.brookscole.com/sf9** and select Chapter 8.

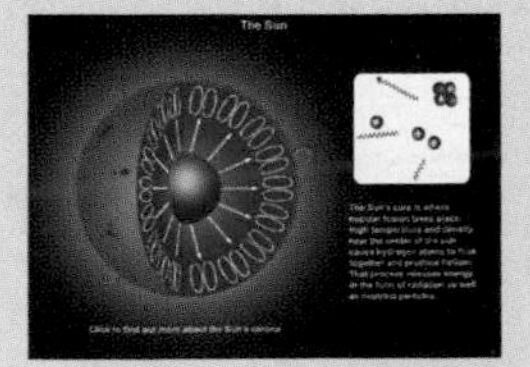

The Sun
Take a look inside the sun in this animation. You can click on various parts of the sun to see things like an animation of nuclear fusion and a movie of a solar flare.

ASTRONOMY EXERCISES

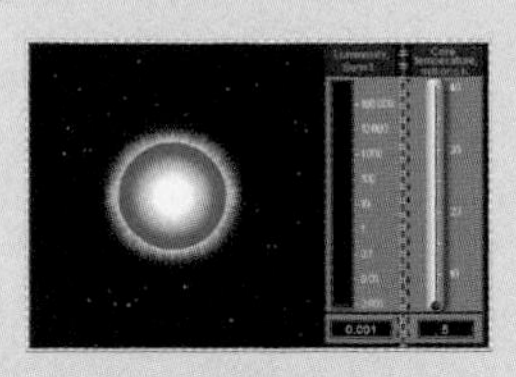

Nuclear Fusion
Nuclear fusion is a reaction that joins the nuclei of atoms to form more massive nuclei. Vary the temperature of the star in this animation to see how this affects the rate of nuclear fusion and luminosity.

Zeeman Effect
The phenomenon known as the Zeeman effect allows astronomers to measure magnetic fields. In this animation, you can adjust the strength of the magnetic field around a star to study the effects on the star's spectrum.

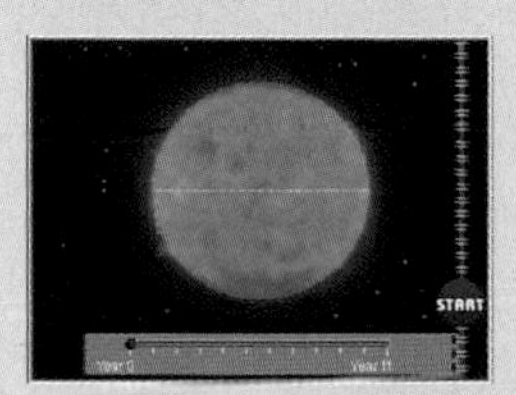

Sunspot Cycle I
Sunspots are relatively dark spots on the sun that contain intense magnetic fields. This animation lets you study the number and location of sunspots occurring over an 11-year cycle.

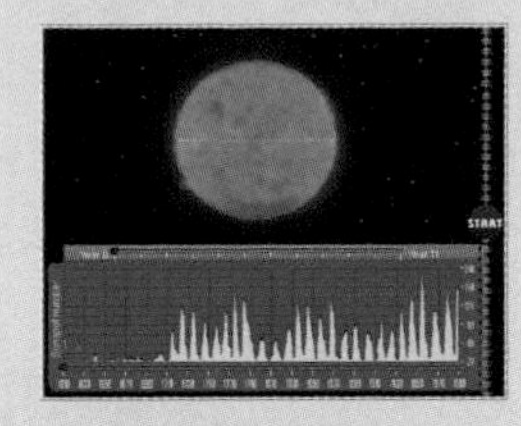

Sunspot Cycle II
In this animation, you can track sunspot activity over the past 300 years.

Convection and Magnetic Fields
You can vary the temperature of the sun's core in this animation. As you do so, notice how the amount of convection changes, as well as what happens to the magnetic field around the sun.

Auroras
This animation lets you vary the strength of Earth's magnetic field as well as the speed and number of particles in the solar wind. See if you can determine what factors cause the most- and least-intensive auroras.

VIRTUAL ASTRONOMY LABS

Lab 4: Solar Wind and Cosmic Rays
This lab begins with an overview of the properties of the sun's atmosphere and how energetic particles escape and travel through the solar system. It ends with a discussion of cosmic rays.

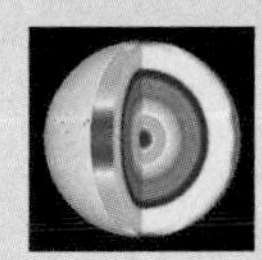

Lab 10: Helioseismology
This lab examines how helioseismology, the study of the sun's vibrations, allows you to obtain detailed information about its interior. Such information allows you to test your understanding of the sun very precisely.

Critical Inquiries for the Web

1. Do disturbances in one layer of the solar atmosphere produce effects in other layers? You have seen that filtergrams are useful in identifying the layers of the solar atmosphere and the structures within them. Visit a website that provides daily solar images. Then, choose today's date (or one near it) and examine the sun in several wavelengths to explore the relation between disturbances in various layers.
2. Explore the web to find out how auroral activity is affected as solar activity rises and falls through the solar cycle. What changes in auroral visibility occur during this cycle? In what other ways can the increased activity associated with a solar maximum affect Earth?
3. Explore the Web to find photos and observations of auroras. From what places on Earth are auroras most often seen?
4. What can you find on the Web about Earth-based efforts to generate energy through nuclear fusion? How do nuclear fusion power experiments attempt to trigger and control nuclear fusion? So-called cold fusion has been abandoned as a false trail. How did it resemble nuclear fusion?

Exploring *TheSky*

1. Locate the six photos of the sun provided in *TheSky* and attempt to draw in the sun's equator in each photo. (*Hint:* In the sun's information box, choose **More Information** and then **Multimedia.** What features are visible in these images that help us recognize the orientation of the sun's equator?)

- Go to the Brooks/Cole Astronomy Resource Center **(http://astronomy.brookscole.com)** for critical thinking exercises, articles, and additional readings from InfoTrac College Edition, Brooks/Cole's online student library.

9 The Family of Stars

Ice is the silent language
of the peak;
and fire the silent language
of the star.

CONRAD AIKEN,
AND IN THE HUMAN HEART

DOES YOUR FAMILY include some characters? The family of stars is amazingly diverse. In a photograph, the stars differ only slightly in color and brightness, but you are going to discover that some are huge and some are tiny, some are astonishingly hot and some are quite cool, some are ponderously massive and some are weeny little stars hardly massive enough to shine. If your family is as diverse as the family of stars, you must have some peculiar relatives. ▌ Unfortunately, finding out what a star is like is quite difficult. When you look at the stars, you look across vast distances and see only bright points of light. Just looking tells you almost nothing about the star's energy production, diameter, or mass. Rather than just looking at stars, you must analyze starlight with great care. This chapter concentrates on finding out three things

Continued on page 188

The stars in this image of the Eagle Nebula look similar at first glance, but look carefully. The stars make up a diverse family of many different types. (Mark McCaughrean and Morten Andersen of the Astrophysical Institute, Potsdam, and the European Southern Observatory)

Guidepost

Looking Back

If you want to study anything scientifically, the first thing you have to do is find a way to measure it. But measurement in astronomy is very difficult. In Chapter 6 you studied the sophisticated telescopes and instruments that astronomers use, and in Chapter 7 you read about the information that can be found in spectra. Chapter 8 showed you how astronomers use their understanding of the interaction of light and matter to study the nearest star, the sun. Now you are ready to expand your universe and study other stars.

This Chapter

Here you will discover how those same tools—telescopes, instruments, and basic physics—can tell you what stars are like. Here you will find answers to five essential questions:

How far away are the stars?

How much energy do stars make?

How big are stars?

How much matter do stars contain?

What is the typical star like?

With this chapter you leave our sun behind and begin your study of the billions of stars that dot the sky. In a sense, the star is the basic building block of the universe. If you hope to understand what the universe is and how it works, you must understand the stars.

Looking Ahead

Once you know how to find the basic properties of stars, you will be ready to trace the history of the stars from birth to death, a story that includes our sun, our Earth, and you and me. That story begins in the next chapter.

Ace Astronomy™ The AceAstronomy icon throughout the text indicates an opportunity for you to test yourself on key concepts and to explore animations and interactions on the AceAstronomy website at: http://ace.brookscole.com/sf9

about stars—how much energy they emit, how big they are, and how much mass they contain. These three properties, combined with stellar temperatures—already discussed in Chapter 7—will give you an insight into the family relations among stars.

Although you begin with three things to learn about stars, you immediately meet a detour. To find out how much energy a star emits, you must know how far away it is. If at night you see lights in the distance, you cannot tell whether you are looking at a powerful searchlight miles away or the fainter but nearby bulb on someone's front porch. Only when you know the distance to the lights can you judge their intrinsic brightness. In the same way, to find the intrinsic brightness of a star—or, more precisely, the amount of energy it emits—you must know its distance from you. A short detour will provide you with a method of measuring stellar distances.

After you reach these three goals, you will be able to put your data together to find out what the average star is like. Which types of stars are most common? Which are rare? Look at the photograph in ■ Figure 9-1. The stars are beautiful, but their light tells a cosmic story of great power and beauty, a story you can understand only if you understand the family of the stars.

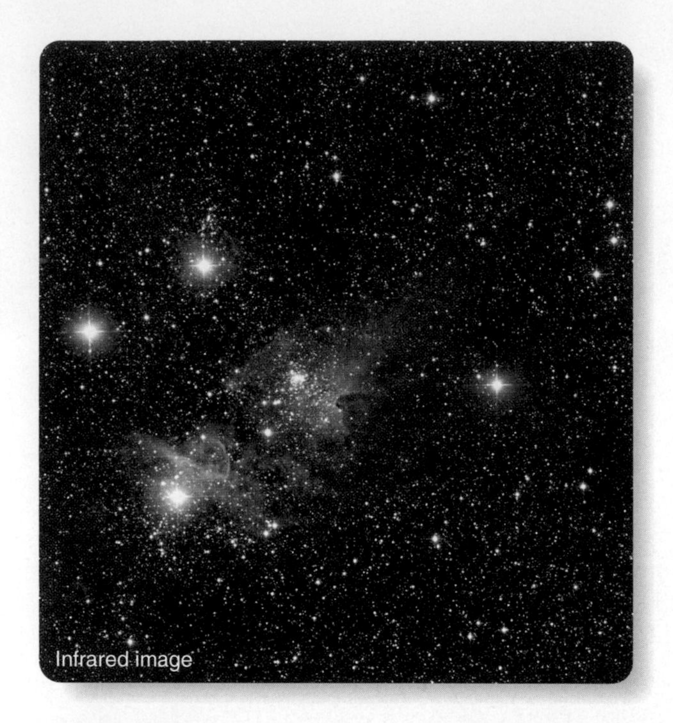

■ Figure 9-1

The modern quest to understand the universe is symbolized in this photo of the Eta Carinae Nebula, roughly 7500 light-years from Earth. Only by the careful analysis of starlight have astronomers learned that the stars in the nebula are only 3 million years old and that the brightest star at lower left contains 100 times more matter than the sun. (2MASS Sky Survey and IPAC)

9-1 Measuring the Distances to Stars

DISTANCE IS THE MOST IMPORTANT and the most difficult measurement in astronomy, and astronomers have found many different ways to estimate the distance to stars. Yet each of those ways depends on a direct geometrical method that is much like the method surveyors would use to measure the distance across a river they cannot cross. You can begin by reviewing the surveyor's method and then applying it to stars.

The Surveyor's Method

To measure the distance across a river, a team of surveyors begins by driving two stakes into the ground. The distance between the stakes is the baseline of the measurement. The surveyors then choose a landmark on the opposite side of the river, a tree perhaps, and that establishes a large triangle marked by the two stakes and the tree. Using their surveyor's instruments, they sight the tree from the two ends of the baseline and measure the two angles on their side of the river.

Knowing two angles of this large triangle and the length of the side between them, the surveyors can then find the distance across the river by simple trigonometry. Another way to find the distance is to construct a scale drawing. For example, if the baseline is 50 m and the angles are 66° and 71°, you can draw a line 50 mm long to represent the baseline. Using a protractor, you can construct angles of 66° and 71° at each end of the baseline, and then, as shown in ■ Figure 9-2, extend the two sides until they meet at *C*. Point *C* on your drawing is the location of the tree. Measuring the height of your triangle, you would find it to be 64 mm, and that would tell you that the distance across the river to the tree is 64 m.

Of course, modern surveyors don't make drawings. They make their trigonometric calculations in computers. The important

■ Figure 9-2

You can find the distance *d* across a river by measuring the baseline and the angles *A* and *B* and then constructing a scale drawing of the triangle.

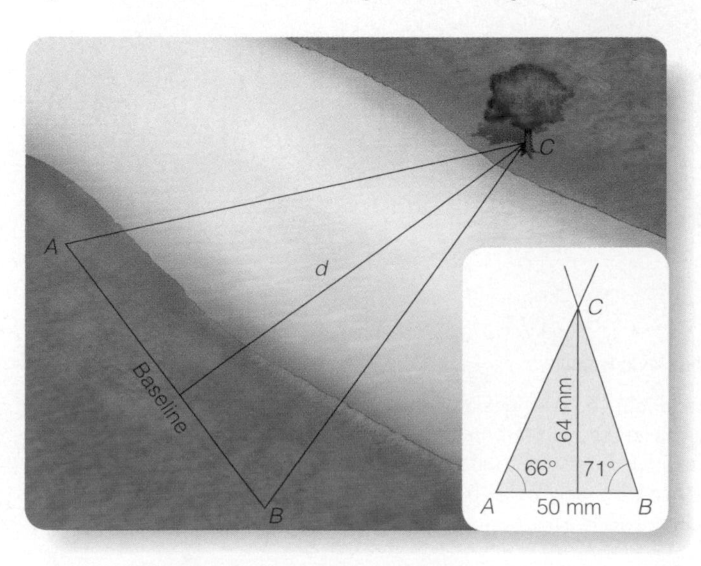

point is that if you measure the baseline and the two angles, you can figure out the distance across the river.

The more distant an object is, the longer the baseline you must use to measure the distance to the object. You could use a baseline 50 m long to find the distance across a river, but to measure the distance to a mountain on the horizon, you might need a baseline 1000 m long. Great distances require very long baselines.

The Astronomer's Method

To find the distance to a star, you must use an extremely long baseline, the diameter of Earth's orbit. If you take a photograph of a nearby star and then wait six months, Earth will have moved halfway around its orbit. You can then take another photograph of the star. This second photograph is taken at a point in space 2 AU (astronomical units) from the point where the first photograph was taken. So your baseline equals the diameter of Earth's orbit, or 2 AU.

You then have two photographs of the same part of the sky taken from slightly different locations in space. When you examine the photographs, you will discover that the nearby star is not in exactly the same place in the two photographs. This apparent shift in the position of the star is called *parallax.*

Parallax is the apparent change in the position of an object due to a change in the location of the observer. You saw in Chapter 4 that parallax is an everyday experience. Your thumb, held at arm's length, appears to shift position against a distant background when you look with first one eye and then with the other. In this case, the baseline is the distance between your eyes, and the parallax is the apparent movement of your thumb when you change eyes. The farther away you hold your thumb, the smaller the parallax.

Did you ever use a stereo viewer to see photographs in three dimensions? That's just parallax at work. Your brain uses the parallax of objects seen through your two eyes to determine distance and create a three-dimensional image.

Because the stars are so distant, their parallaxes are very small angles, usually expressed in seconds of arc. The quantity that astronomers call **stellar parallax *(p)*** is half the total shift of the star, as shown in ■ Figure 9-3. Astronomers measure the parallax, and surveyors measure the angles at the ends of the baseline, but both measurements tell you the same thing—the shape of the triangle and the distance to the object in question.

The distance to a star with parallax p is given by the simple formula

$$d = \frac{1}{p}$$

where the parallax, p, is measured in seconds of arc and the distance, d, is measured in a unit of distance invented by astronomers, the **parsec (pc),*** defined as the distance to an imaginary star with a parallax of 1 second of arc. One parsec turns out to equal 206,265 AU, or 3.26 ly. This makes it very easy to calculate the distance to a star given the parallax. For example, one of our Favorite Stars, Sirius, has a parallax of 0.375 second of arc. Then the distance to Sirius in parsecs is 1 divided by 0.375, which equals 2.7 pc. To convert to light-years, multiply 2.7 pc by 3.26 and discover that Sirius is 8.8 ly away.

Measuring the small angle p is very difficult. The nearest star to the sun is α Centauri, another of our Favorite Stars. It has a parallax of only 0.76 second of arc, and more distant stars have even smaller parallaxes. To see how small these angles are, hold a piece of paper edgewise at arm's length. The thickness of the paper covers an angle of about 30 seconds of arc. You can see that the parallax of a star, smaller than 1 second of arc, must be very difficult to measure accurately.

The blurring caused by Earth's atmosphere smears star images into tiny blobs of light about one second of arc in diameter, and that makes it difficult to measure parallax from Earth's sur-

*The parsec is used throughout astronomy because it simplifies the calculation of distance. However, there are instances when the light-year is also convenient. Consequently, the chapters that follow use either parsecs or light-years as convenience and custom dictate.

■ Figure 9-3

You can measure the parallax of a nearby star by photographing it from two points along Earth's orbit. For example, you might photograph it now and again 6 months from now. Half of the star's total change in position from one photograph to the other is its stellar parallax, p.

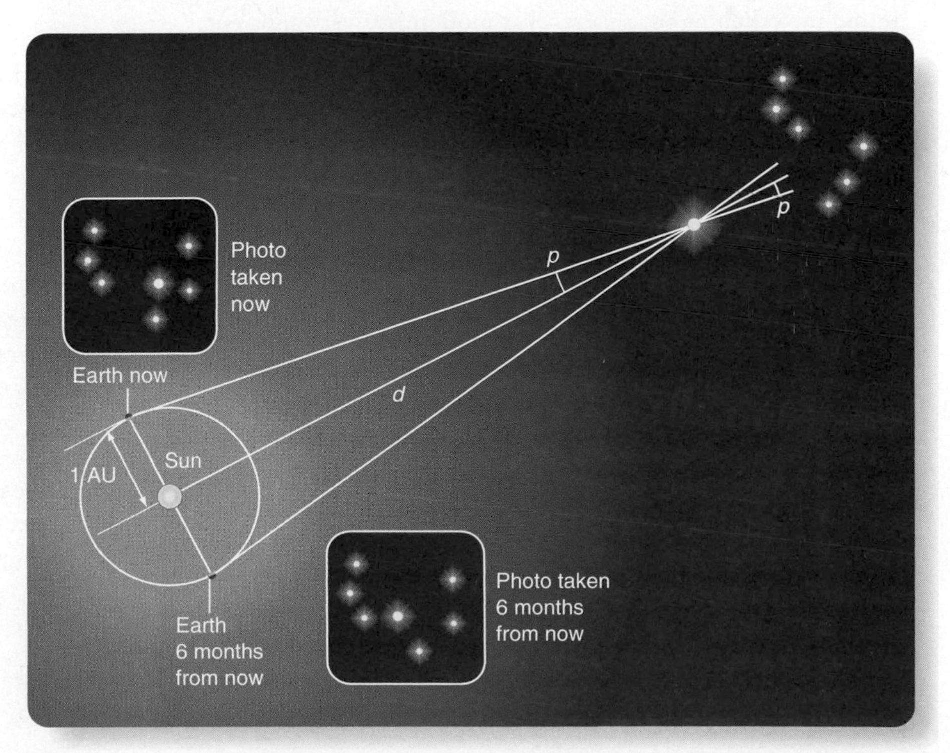

face. Even when astronomers average together many observations, they cannot measure parallax with an uncertainty smaller than about 0.002 second of arc. If you measure a parallax of 0.02 second of arc, the uncertainty is about 10 percent. That means that astronomers observing from Earth's surface can't measure accurate parallaxes smaller than about 0.02 second of arc, which corresponds to a distance of 50 pc; consequently, ground-based parallax measurements are limited to only the closest stars. Since the first stellar parallax was measured in 1838, ground-based astronomers have been able to measure accurate parallaxes for only about 10,000 stars.

In 1989, the European Space Agency launched the satellite Hipparcos to measure stellar parallaxes from orbit above the blurring effects of Earth's atmosphere. The little satellite observed for four years, and the data were reduced by the most sophisticated computers to produce two parallax catalogs in 1997. One catalog contains 120,000 stars with parallaxes 20 times more accurate than ground-based measurements. The other catalog contains over a million stars with parallaxes as accurate as ground-based parallaxes. The Hipparcos data have given astronomers new insights into the nature of stars.

Before dropping the subject of parallax measurement, you should learn about a related observation that reveals the motions of the stars through space.

AceAstronomy™ Log into AceAstronomy and select this chapter to see Astronomy Exercises "Parallax I" and "Parallax II" to experiment with parallax.

The Changing Shape of the Big Dipper

100,000 years ago the Big Dipper had a different shape.

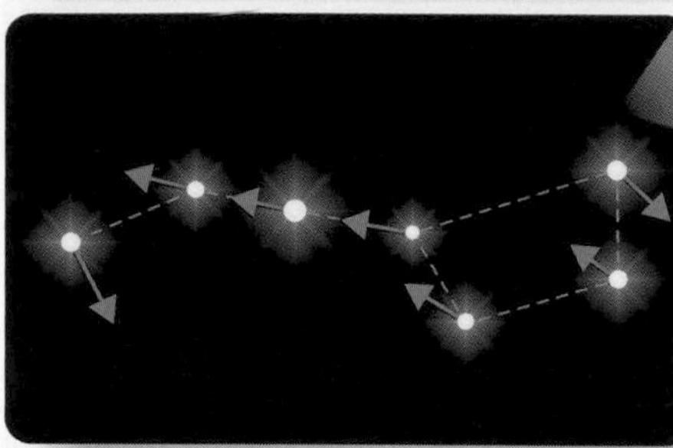

Proper motion is moving the stars of the Big Dipper across the sky.

100,000 years in the future, the Big Dipper will have a distorted shape.

■ **Figure 9-4**

"Proper motion" refers to the slow movement of the stars across the sky.

Proper Motion

All the stars in the sky, including our sun, are moving along orbits around the center of our galaxy. That motion isn't very apparent over periods of years, but over the centuries it can significantly distort the shape of constellations (■ Figure 9-4). Astronomers use proper motion in a number of ways, but one thing it provides is an estimate of distance.

If you photograph a small area of the sky on two dates separated by 10 years or more, you can notice that some of the stars in the photograph have moved very slightly against the background stars. This motion, expressed in units of seconds of arc per year, is the **proper motion** of the stars. As examples, consider two stars from our list of Favorite Stars. Vega is a bright blue-white star in the summer sky, and it has a proper motion of 0.327 second of arc per year. Rigel, a bright blue-white star in the winter sky, has a proper motion of only 0.002 second of arc per year. The two stars are nearly the same brightness in the sky and nearly the same temperature, but the proper motion of Rigel is over a hundred times less than that of Vega. What does that mean?

A star might have a small proper motion if it is moving almost directly toward or away from you; then its position on the sky would change very slowly, so it would have a small proper motion. That is unlikely, but it does happen. Another reason a star might have a small proper motion is that it could be quite far away from you. Then even if the star were moving rapidly through space, it would not have a large proper motion. That explains why Rigel, at a distance of 237 pc, has a smaller proper motion, and Vega, at a distance of only 7.7 pc, has a larger proper motion. By the way, this alerts you to something interesting. Although Rigel is 31 times further away than Vega, they have nearly the same brightness in the sky. Rigel must be emitting a lot more light than Vega.

Astronomers can use proper motion to look for nearby stars. If you see a star with a small (or zero) proper motion, it is probably a distant star, but a star with a large proper motion is probably quite close. You may have seen this effect if you watch birds. Distant geese move slowly across the sky, but a nearby bird flits quickly across your field of view. In this way, proper motions can give astronomers a way to locate nearby stars.

Building Scientific Arguments

Why are parallax measurements made from space better than parallax measurements made from Earth?

At first you might suppose that a satellite in orbit can measure the parallax of the stars better because the satellite is closer to the stars, but that will lead your argument astray. When you recall the immense distances to the stars, you can see that being in space doesn't really put the satellite significantly closer to the stars. Rather, the reason for increased accuracy is that the satellite is above Earth's atmosphere. When astronomers try to measure parallax, the turbulence in Earth's atmosphere blurs the star images and smears them out into blobs roughly 1 second of arc in diameter. It isn't possible to measure the position of these fuzzy images accurately. Astronomers can't measure parallax smaller than about 0.02 second of arc. A parallax of 0.02 second of arc corresponds to a distance of 50 pc, so ground-based astronomers can't measure parallax accurately beyond that distance.

A satellite in orbit, however, is above Earth's atmosphere, so the only blurring in the star images is that produced by diffraction in the optics. In other words, the star images are very sharp, and a satellite in orbit can measure the positions of stars and thus their parallaxes to high accuracy.

Now extend your argument one more step. **If a satellite can measure parallaxes as small as 0.001 second of arc, then how far are the most distant stars it can measure?**

■ ■ ■

Connections: You started this chapter with three things you wanted to know about stars: energy production, size, and mass. Now you have found a way to locate nearby stars and measure their distances, so you are ready to discuss the first of your three goals. You are ready to find out how much energy stars emit.

9-2 Intrinsic Brightness

YOUR EYES TELL YOU that some stars look brighter than others, and in Chapter 2 you used the scale of apparent magnitudes to refer to stellar brightness. The faintest stars you can see with the naked eye are about sixth magnitude, with brighter stars having magnitudes represented by smaller numbers. The brightest stars you see in the sky have negative magnitudes, such as Sirius, whose apparent magnitude is -1.47.

The scale of apparent magnitudes only tells you how bright stars look, however, and you need to know their true, or intrinsic, brightness. *Intrinsic* means "belonging to the thing." When astronomers refer to the intrinsic brightness of a star, they mean a measure of the total amount of light the star emits. Apparent magnitudes don't tell you the intrinsic brightness of the stars, only how bright they look. An intrinsically very bright star might appear faint if it were far away. To find the true brightness of stars, you must correct the apparent magnitudes for the influence of distance.

Brightness and Distance

If you see lights at night, it is difficult to determine which are less powerful but nearby and which are highly luminous but farther away (■ Figure 9-5). You face the same problem when you look at stars, and to resolve that problem, you must think carefully about how brightness depends on distance.

When you look at a bright light, your eyes respond to the visual-wavelength energy falling on the eye's retina, which tells you how bright the object looks. That means the brightness is related to the flux of energy entering your eye. **Flux** is the energy in joules (J) per second falling on one square meter. Recall that a joule is about as much energy as is released when an apple falls from a table onto the floor. Scientists commonly refer to 1 joule per second as 1 watt. The light flux entering your eye is directly related to the intensity of the light. The more flux entering your eye, the brighter the light looks.

If you placed a screen 1 meter square near a lightbulb, a certain amount of flux would fall on the screen. If you moved the screen twice as far from the bulb, the light that previously fell on the screen would be spread to cover an area four times larger, and the screen's surface would receive only one-fourth as much light per square meter. If you tripled the distance to the screen, its surface would receive only one-ninth as much light per square meter. In this way, the flux you receive from a light source is inversely proportional to the square of the distance to the source. This is known as the inverse square relation. (You first encoun-

■ **Active Figure 9-5**

To judge the true brightness of a light source, you need to know how far away it is. With no clues to distance, the distant headlight on a truck might look as bright as the nearby headlight on a bicycle.

AceAstronomy™ Log into AceAstronomy and select this chapter to see Active Figure "Brightness and Distance." You can take control of this figure.

tered the inverse square relation in Chapter 5, where it was applied to the strength of gravity. See Figure 5-6.)

Now you understand how the brightness of a star depends on its distance. If you knew the apparent magnitude of a star and its distance from Earth, you could use the inverse square law to correct for the effect of distance. Astronomers do that using a special kind of magnitude scale as described in the next section.

Absolute Visual Magnitude

If all the stars were the same distance away, you could compare one with another and decide which was emitting more light and which less. Of course, the stars are scattered at different distances, and you can't shove them around to line them up for comparison. If, however, you knew the distance to a star, you could use the inverse square relation to calculate the brightness the star would have at some standard distance. Astronomers take 10 pc as the standard distance and refer to the intrinsic brightness of the star as its **absolute visual magnitude** ***(M_V)***, the apparent visual magnitude the star would have it if were 10 pc away.

The symbol for absolute visual magnitude is an uppercase M with a subscript V. Recall from Chapter 2 that the symbol for apparent visual magnitude is a lowercase m with a subscript V. The subscript tells you that the visual magnitude system is based only on the wavelengths of light human eyes can see. Other magnitude systems are based on other parts of the electromagnetic spectrum such as the infrared, ultraviolet, and so on.

The intrinsically brightest stars known have absolute visual magnitudes of about -8 and the faintest about $+19$. The sun has an absolute magnitude of 4.78. If the sun were only 10 pc away from Earth, it would look no brighter than the faintest star in the handle of the Little Dipper. Look at our list of Favorite Stars. The nearest star to the sun, alpha Centauri, is only 1.4 pc away, and its apparent magnitude is 0.0, indicating that it looks bright in the sky. However, its absolute magnitude is 4.39, about the same as the sun. Remember Vega and Rigel from the previous section? Vega has an absolute magnitude of 0.6, but Rigel has an absolute magnitude of -6.8. The two stars look the same in the sky, but Rigel is producing a lot more energy than Vega.

Calculating Absolute Visual Magnitude

How exactly do astronomers find the absolute visual magnitude of a star? This question leads to one of the most common formulas in astronomy, a formula that relates a star's magnitude and its distance.

The **magnitude–distance formula** relates the apparent magnitude m_V, the absolute magnitude M_V, and the distance d in parsecs:

$$m_V - M_V = -5 + 5 \log_{10}(d)$$

If you know any two of the parameters in this formula, you can easily calculate the third. If you are interested in finding the absolute magnitude of a star, then you should consider an example in which you know the distance and apparent magnitude of a star. Suppose a star has a distance of 50 pc and an apparent magnitude of 4.5. A pocket calculator tells you that the log of 50 is 1.70, and $-5 + 5 \times 1.70$ equals 3.5, so you know that the absolute magnitude is 3.5 magnitudes brighter than the apparent magnitude. That means the absolute magnitude is 1.0, since 4.5 minus 3.5 is 1.0. (Remember that smaller numbers mean brighter magnitudes.) If this star were 10 pc away, it would look bright in the sky, a first-magnitude star.

Astronomers also use the magnitude–distance formula to calculate the distance to a star if the apparent and absolute magnitudes are known. For that purpose, it is handy to rewrite the formula in the following form:

$$d = 10^{(m_V - M_V + 5)/5}$$

If you knew that a star had an apparent magnitude of 7 and an absolute magnitude of 2, then $m_V - M_V$ is 5 magnitudes, and the distance would be 10^2 or 100 parsecs.

The magnitude difference $m_V - M_V$ is known as the **distance modulus,** a measure of how far away the star is. The larger the distance modulus, the more distant the star. You could use the magnitude–distance formula to construct a table of distance and distance modulus (■ Table 9-1).

The magnitude–distance formula may seem awkward at first, but a pocket calculator makes it easy to use. It is important because it performs a critical function in astronomy: It allows astronomers to convert observations of distance and apparent magnitude into absolute magnitude, a measure of the true bright-

■ Table 9-1 | Distance Moduli

$m_V - M_V$	d (pc)
0	10
1	16
2	25
3	40
4	63
5	100
6	160
7	250
8	400
9	630
10	1000
⋮	⋮
15	10,000
⋮	⋮
20	100,000
⋮	⋮

ness of the star. Once you know the absolute magnitude, you can go one step further and figure out the total amount of energy a star is radiating into space.

Ace Astronomy™ Log into AceAstronomy and select this chapter to see Astronomy Exercise "Apparent Brightness and Distance." You can change the distance to a star and see its apparent brightness change.

Luminosity

The first of your three goals for this chapter was to find out how much energy stars emit. With the absolute magnitudes of the stars in hand, you can now compare other stars with the sun. That is easiest if you convert absolute magnitude into luminosity. The **luminosity** ***(L)*** of a star is the total amount of energy the star radiates in 1 second—not just visible light, but all wavelengths. To find a star's luminosity, you begin with its absolute visual magnitude, make a small correction, and compare the star with the sun.

The correction you must make adjusts for the radiation emitted at wavelengths humans cannot see. Absolute visual magnitude includes only visible light. The absolute visual magnitudes of hot stars and cool stars will underestimate their total luminosities because those stars radiate significant amounts of radiation in the ultraviolet or infrared parts of the spectrum. You can correct for the missing radiation because the amount of missing energy depends only on the star's temperature. For hot and cool stars, the correction can be large, but for medium-temperature stars like the sun, the correction is small. Adding the proper correction to the absolute visual magnitude changes it into the **absolute bolometric magnitude**—the absolute magnitude the star would have if you could see all wavelengths.

Once you know a star's absolute bolometric magnitude, you can find its luminosity by comparing it with the sun. The absolute bolometric magnitude of the sun is 4.7. For every magnitude a star is brighter than 4.7, it is 2.512 times more luminous than the sun. (Recall from Chapter 2 that a difference of 1 magnitude corresponds to an intensity ratio of 2.512.) That means that a star with an absolute bolometric magnitude of 2.7 is 2 magnitudes brighter than the sun, which means it is 6.3 times more luminous (6.3 is approximately 2.512×2.512).

Our Favorite Star Aldebaran makes a convenient example. It has an absolute bolometric magnitude of -0.39. That makes it just a bit over 5 magnitudes brighter than the sun. A difference of 5 magnitudes is defined to be a factor of 100 in brightness, so the luminosity of Aldebaran is 100 times the sun's luminosity, or 100 $L_\odot$. Aldebaran is the red eye of Taurus the Bull; next time you see it in the winter sky, nudge your friends and say, "See that star? It emits just over 100 times more energy than the sun."

Remember Favorite Stars Vega and Rigel? Earlier in this chapter you noted that they look the same in the sky, but Rigel is much farther away. If you analyze their brightnesses you will discover that Rigel is a hundred times more luminous than Vega.

The symbol $L_\odot$ represents the luminosity of the sun, a number astronomers can calculate in a direct way. Earth satellites can measure the total solar energy hitting 1 square meter in 1 second just above Earth's atmosphere (the solar constant defined in the previous chapter). The distance from Earth to the sun is known, so it is a simple matter to calculate how much energy the sun must radiate in all directions to provide Earth with the energy it receives per second (see Problem 9 at the end of this chapter). The measured luminosity of the sun is about 4×10^{26} joules/s. You can use that to convert luminosity in terms of the sun into actual joules per second. For example, if Aldebaran is 100 times the luminosity of the sun, it must be emitting a total luminosity of 4×10^{28} joules/s.

Review this point: If you can measure the parallax of a star, you can find its distance, calculate its absolute visual magnitude, correct for the light you can't see to find the absolute bolometric magnitude, and then find the luminosity in terms of the sun. Then you can multiply by the sun's luminosity to find the luminosity of the star in joules per second.

Some stars are a million times more luminous than the sun, and some are almost a million times less luminous. Clearly, the family of stars is filled with interesting characters.

Building Scientific Arguments

How can two stars look the same in the sky but have dramatically different luminosities?

You can answer this question by building a scientific argument that relates three factors: the appearance of a star, its true luminosity, and its distance. The farther away a star is, the fainter it looks, and that is just the inverse square law. If two stars such as Vega and Rigel have the same apparent visual magnitude, then your eyes must be receiving the same amount of light from them. But Rigel is much more luminous than Vega, so it must be farther away. Parallax observations from the Hipparcos satellite confirm that Rigel is 31 times farther away.

Distance is often the key to understanding the brightness of stars, but temperature can also be important. Build a scientific argument to answer the following: **Why must astronomers make a correction in converting the absolute visual magnitude of very hot or very cool stars into luminosities?**

■ ■ ■

Connections: Although you had to make a detour in order to find the distances to the stars, you have reached your first goal. You have found a way to discover the energy emitted by stars, and you have discovered that the range of stellar luminosities is very large. How can that be? The answer lies with your second goal—the diameters of the stars.

9-3 The Diameters of Stars

YOUR SECOND GOAL in this chapter is to find the diameters of stars. You know little about stars until you know their diameters. Are they all the same size as the sun, or are some larger and some smaller? You certainly can't see their diameters through a telescope; the images of stars are much too small for astronomers to resolve their disks and measure their diameters. But there is a way to find out how big stars really are. If you know their temperatures and luminosities, you can find their diameters. That relationship will introduce you to the most important diagram in astronomy, where you will discover more family relations among the stars.

Luminosity, Radius, and Temperature

The luminosity and temperature of a star can tell you its diameter if you understand the two factors that affect a star's luminosity: surface area and temperature. For example, you can eat dinner by candlelight because the candle flame has a small surface area, and, although it is very hot, it cannot radiate much heat; it has a low luminosity. However, if the candle flame were 12 ft tall, it would have a very large surface area from which to radiate, and although it might be no hotter than a normal candle flame, its luminosity would drive you from the table (■ Figure 9-6).

In a similar way, a hot star may not be very luminous if it has a small surface area. It could be highly luminous if it were larger, and even a cool star could be very luminous if it were very large and so had a large surface area from which to radiate. So you can see that both temperature and surface area help determine the luminosity of a star.

■ Figure 9-6

Molten lava pouring from a volcano is not as hot as a candle flame, but a lava flow has more surface area and radiates more energy than a candle flame. Approaching a lava flow without protective gear is dangerous. (Krafft/Photo Researchers, Inc.)

Stars are spheres, and the surface area of a sphere is $4\pi R^2$. If you express the radius in meters, the area is the number of square meters on the surface of the star. Each square meter radiates like a black body, and you will remember from Chapter 7 that the total energy given off each second from each square meter is σT^4. So the total luminosity of the star is the surface area multiplied by the energy radiated per square meter:

$$L = 4\pi R^2 \sigma T^4$$

If you divide by the same quantities for the sun, you can cancel out the constants and get a simple formula for the luminosity of a star in terms of its radius and temperature:

$$\frac{L}{L_\odot} = \left(\frac{R}{R_\odot}\right)^2 \left(\frac{T}{T_\odot}\right)^4$$

Here the symbol $\odot$ stands for the sun, and the formula says that the luminosity of a star in terms of the sun equals the radius of the star in terms of the sun squared times the temperature of the star in terms of the sun raised to the fourth power.

Suppose a star is 10 times the sun's radius but only half as hot. How luminous would it be?

$$\frac{L}{L_\odot} = \left(\frac{10}{1}\right)^2 \left(\frac{1}{2}\right)^4 = \left(\frac{100}{1}\right)\left(\frac{1}{16}\right) = 6.25$$

The formula says that the star would be 6.25 times more luminous than the sun.

How can you use this to find the diameters of the stars? If you see a cool star that is very luminous, you know it must be very large, and if you see a hot star that is not very luminous, you know it must be very small.

The formula allows you to calculate these sizes. For instance, suppose that a star is 40 times the luminosity of the sun and twice as hot. If you put these numbers into the formula, you get:

$$\frac{40}{1} = \left(\frac{R}{R_\odot}\right)^2 \left(\frac{2}{1}\right)^4$$

Solving for the radius, you get:

$$\left(\frac{R}{R_\odot}\right)^2 = \frac{40}{2^4} = \frac{40}{16} = 2.5$$

So the radius is:

$$\frac{R}{R_\odot} = \sqrt{2.5} = 1.58$$

The star is 1.58 times larger in radius than the sun.

Because a star's luminosity depends on its surface area and its temperature, you can use luminosity and temperature to sort the stars and discover which are large and which are small. Astronomers use a special diagram for that sorting.

The H–R Diagram

The **Hertzsprung–Russell (H–R) diagram,** named after its originators, Ejnar Hertzsprung and Henry Norris Russell, is a graph that separates the effects of temperature and surface area on stellar luminosities and sorts the stars according to their sizes. Before you study the details of the H–R diagram (as it is often called), try looking at a similar diagram you might use to sort automobiles.

You could plot a diagram such as ■ Figure 9-7 to show horsepower versus weight for various makes of cars. And in so doing, you would find that, in general, the more a car weighs, the more horsepower it has. Most cars fall somewhere along the sequence of cars running from heavy, high-powered cars to light, low-powered models. You might call this the main sequence of cars. But some cars have much more horsepower than normal for their weight—the sport or racing models—and the economy models have less power than normal for cars of the same weight. Just as this diagram sorts cars into family groups, the H–R diagram sorts the stars into groups based on size.

The H–R diagram is a graph with luminosity on the vertical axis and temperature on the horizontal axis. A star is represented by a point on the graph, which tells you the luminosity of the star and its temperature. The H–R diagram in ■ Figure 9-8 also contains a scale of spectral type across the top. Because a star's spectral type is determined by its temperature, you could use either spectral type or temperature on the horizontal axis. Notice that the vertical axis of the H–R diagram is an exponential scale. That's a convenient way to graph data, but you should make special note of what it means (**Window on Science 9-1**).

In an H–R diagram, the location of a point representing a star tells you a great deal about the star. Points near the top of the diagram represent very luminous stars, and points near the bottom represent very-low-luminosity stars. Also, points near the right edge of the diagram represent very cool stars, and points near the left edge of the diagram represent very hot stars. Notice in Figure 9-8 how the artist has used color to represent temperature. Cool stars are red, and hot stars are blue. You can see that dramatically in the photo of real stars in ■ Figure 9-9.

Astronomers use H–R diagrams so often that they usually skip the words "the point that represents the star." Rather they will say that a star is located in a certain place in the diagram. Of course, they mean the point that represents the luminosity and temperature of the star and not the star itself. The location of a star in the

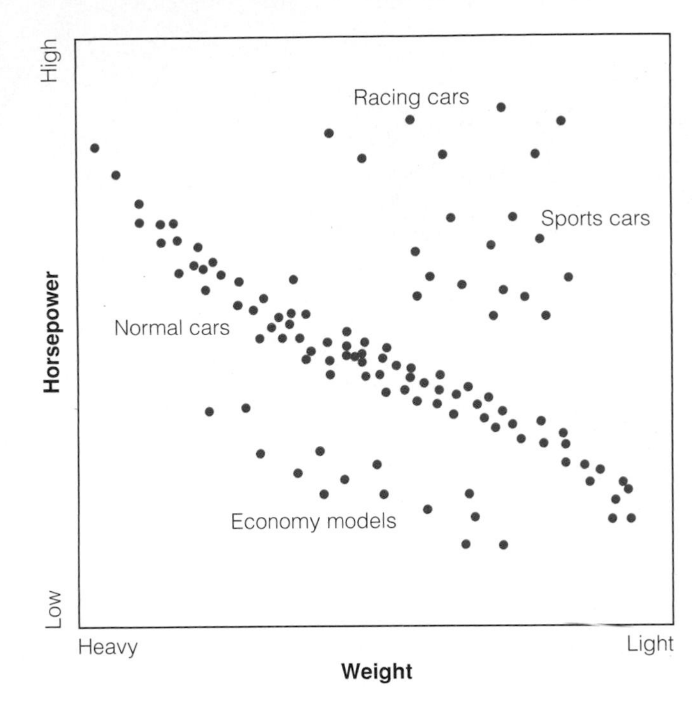

■ Figure 9-7

You could analyze automobiles by plotting their horsepower versus their weight and thus reveal relationships between various models. Most would lie somewhere along the main sequence of "normal" cars.

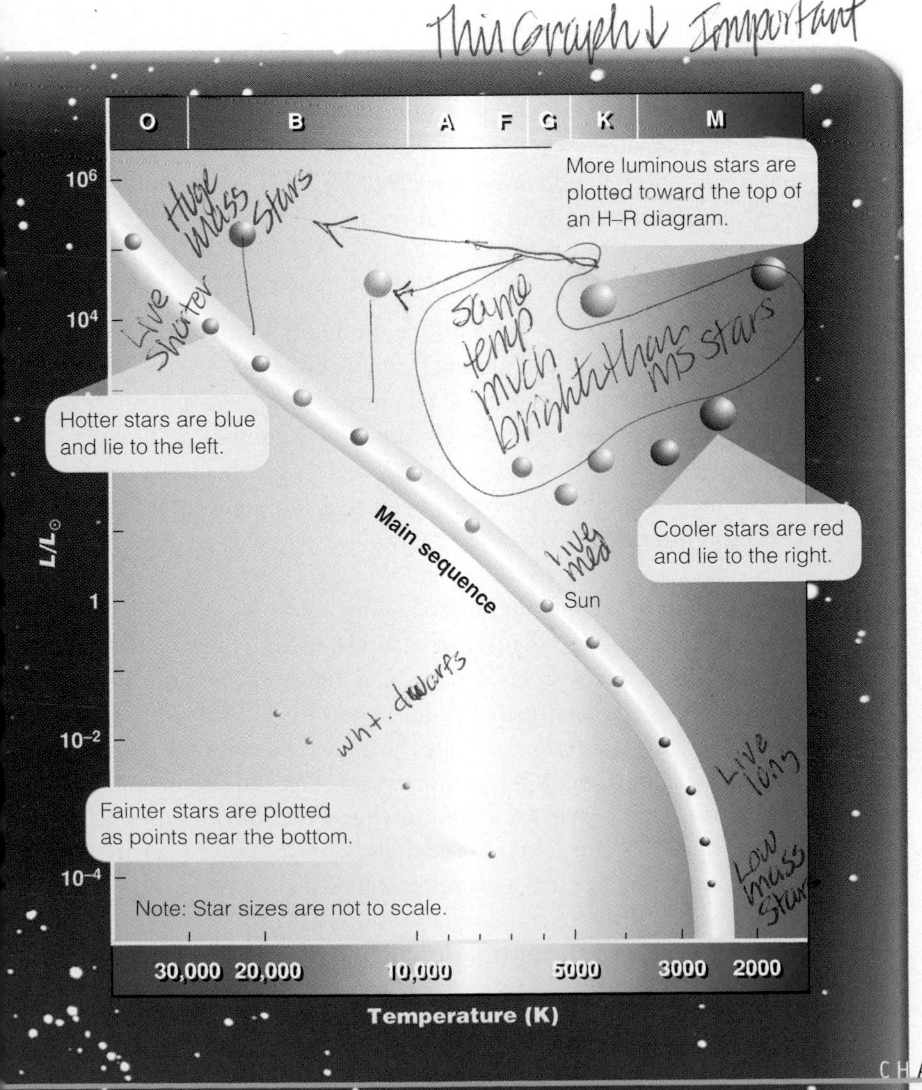

■ Figure 9-8

In an H–R diagram, a star is represented by a dot that shows the luminosity and temperature of the star. The background color in this diagram indicates the temperature of the stars. The sun is a yellow-white G2 star. Most stars fall along the main sequence running from hot luminous stars at upper left to cool low luminosity stars at lower right.

Window on Science 9-1

Exponential Scales

Graphs are important in all branches of science, and some graphs have scales that are exponential (related to the power of some base, such as 10^2, 10^3, 10^4, and so on). Because you are accustomed to thinking about graphs with linear scales, exponential scales can be misleading. H–R diagrams are good examples of graphs with exponential scales, but such scales are common in all the sciences, including chemistry, biology, and geology. They also appear in economics, government, and other fields that deal with quantitative data.

In an exponential scale, each step (tick mark) along the axis corresponds to a constant factor. For example, suppose you plotted a diagram in which the vertical axis showed weight on an exponential scale. At the very bottom of the graph, you could plot the weight of a cat and, above that, your own weight. A little higher in the graph, you could plot the weight of a bull elephant and, just above that, the weight of a whale. In fact, just above that, you could plot the weight of all the whales on Earth plus all the elephants, people, and cats. An exponential scale compresses the high numbers and spreads out the low numbers. That would make your weight on the graph seem much greater than that of a cat and surprisingly (perhaps distressingly) close to that of an elephant. Exponential scales are very helpful for compressing a wide range of numbers into a single graph, but they can be misleading.

For an astronomical example, notice that the luminosity scale on the vertical axis of the H–R diagram in Figure 9-10 has tick marks separated by a constant factor of 10 in stellar luminosity. If a star is just one tick mark higher than another star, it is *10 times* more luminous.

To see how misleading this can be, try drawing an H–R diagram with a linear scale on the vertical axis. Put zero luminosity at the bottom and make each tick mark equal to a step of 100,000 $L_{\odot}$, with 1,000,000 $L_{\odot}$ at the top. You will discover that it is difficult to plot all of the different kinds of stars, but their relative luminosities will be much more clearly illustrated.

You are accustomed to judging the sizes of things from appearances, so exponential scales can be very confusing. (M. Seeds)

■ Figure 9-9

Notice the colors of the stars in the small star cluster M39. The brightest stars are either hot and blue or cool and red. Compare with the most luminous stars in Figure 9-8. (Heidi Schweiker/NAOA/AURA/NSF)

H–R diagram has nothing to do with the location of the star in space. Furthermore, a star may move in the H–R diagram as it ages and its luminosity and temperature change, but such motion in the diagram has nothing to do with the star's motion in space.

Giants, Supergiants, and Dwarfs

Look again at Figure 9-8 and notice the **main sequence,** a region of the H–R diagram running from upper left to lower right. It includes roughly 90 percent of all normal stars. As you might expect, the hot main-sequence stars are more luminous than the cool main-sequence stars. There are, however, stars that don't fall on the main sequence. That alerts you that temperature is not the only thing that determines the luminosity of a star. Size is important too.

The relation among luminosity, radius, and temperature is a precise mathematical relationship that can be used to draw lines of constant radius across an H–R diagram. ■ Figure 9-10 is an H–R diagram on which slanting dashed lines show the location of stars of certain radii. For example, locate the dashed line labeled 1 $R_{\odot}$. That line passes through the point marked "Sun" and represents the location of any star whose radius equals that of the sun. Of course, the line slants down to the right because cooler stars are always fainter than hotter stars of the same size.

The H–R diagram reveals relationships within the family of stars. The stars called **giants** lie at the right above the main sequence. Although these stars are cool, they are luminous because

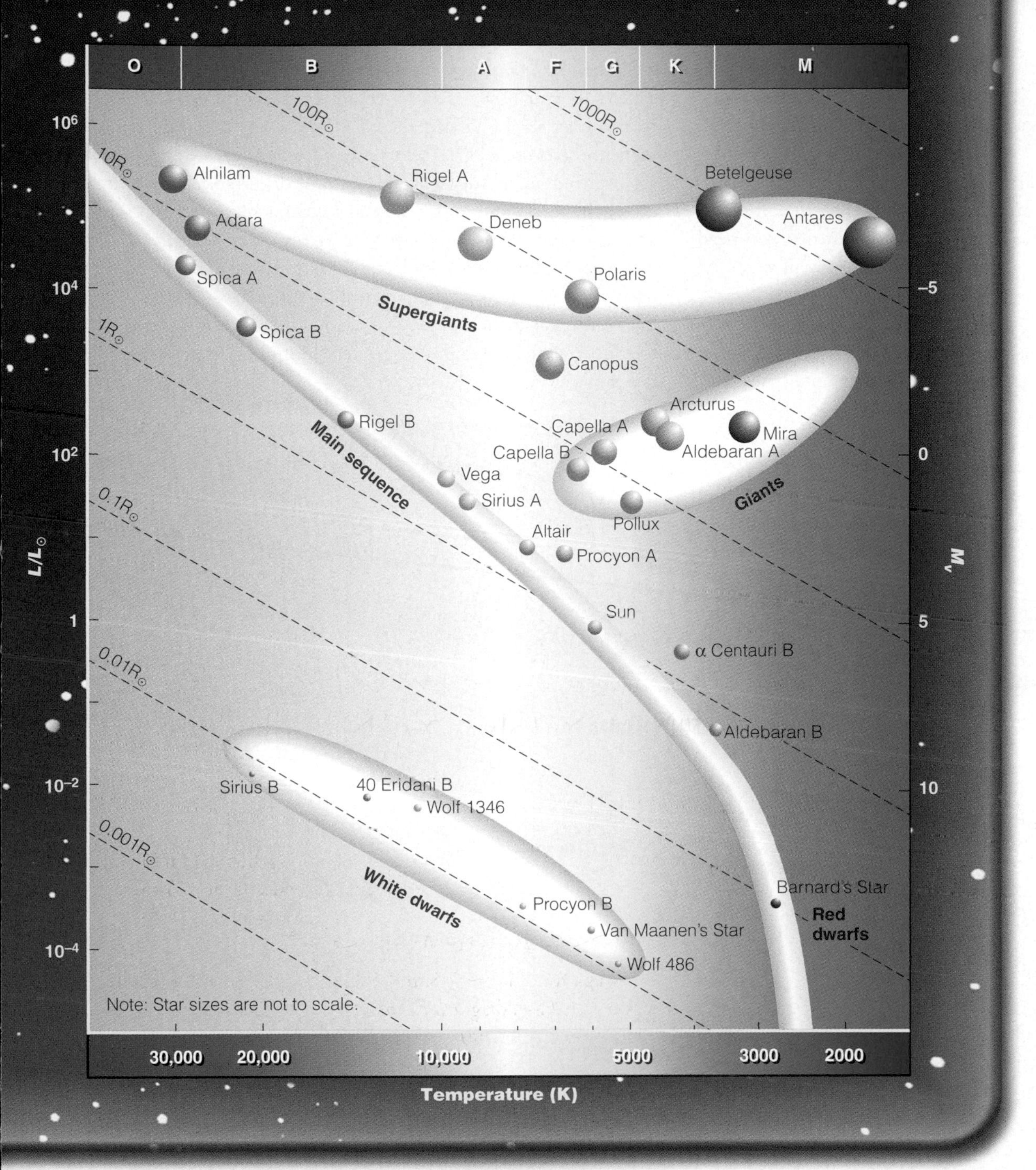

■ **Figure 9-10**

An H–R diagram showing the luminosity and temperature of many well-known stars. The dashed lines are lines of constant radius. The star sizes on this diagram are not to scale. (Individual stars that orbit each other are designated A and B, as in Spica A and Spica B.)

they are 10 to 100 times larger than the sun. The **supergiants** lie near the top of the H–R diagram and are 10 to 1000 times the size of the sun. Now you can understand why Rigel is so much more luminous than Vega. They have nearly the same temperature, but Rigel is a supergiant and has a much larger surface area from which to radiate. Another of our Favorite Stars is Betelgeuse in Orion, also a supergiant. If it magically replaced the sun at the center of our solar system, it would swallow up Mercury, Venus, Earth, and Mars. The largest stars known have radii of about 7 AU; if they replaced the sun, they would extend nearly to the orbit of Saturn.

At the bottom of the H–R diagram lie the economy models, stars that are very low in luminosity because they are very small. At the bottom end of the main sequence, the **red dwarfs** are not only small, but they are also cool, and that means they can't radiate much energy; they have very low luminosities. In contrast, the **white dwarfs** lie in the lower left of the H–R diagram, and, although some are very hot, they are so small they can't be very bright. They are all roughly the size of Earth and can't radiate much energy from their small surface areas.

The H–R diagram shows that there is a great range in the sizes of stars. The largest are 100,000 times larger than the smallest. Notice that the size of the dots in the H–R diagrams here are only symbolic of the true sizes of the stars. If those dots were plotted in true size, your book would need to be as big as a billboard (■ Figure 9-11).

The distribution of stars in the H–R diagram according to size is a clue to how stars are born and how they die. You will follow those clues in the chapters that follow, but first you need to gather more information about the diverse family of stars.

AceAstronomy™ Log into AceAstronomy and select this chapter to see Astronomy Exercise "Stefan-Boltzmann Law II." You can control the size and temperature of a star and watch its luminosity change.

Luminosity Classification

You can tell from a star's spectrum what kind of star it is. Recall from Chapter 7 that collisional broadening can make spectral lines wider when the gas is dense and the atoms collide often.

Main-sequence stars are relatively small and have dense atmospheres in which the gas atoms collide often and distort their electron energy levels. That makes the lines in the spectra of main-sequence stars broad. On the other hand, giant stars are larger, their atmospheres are less dense, and the atoms disturb one another relatively little. Spectra of giant stars have narrower spectral lines, and spectra of supergiants have very narrow lines (■ Figure 9-12).

That means you can look at a star's spectrum and tell roughly how big it is. These are called **luminosity classes,** because the size of the star is the dominating factor in determining luminosity. Supergiants, for example, are very luminous because they are very large.

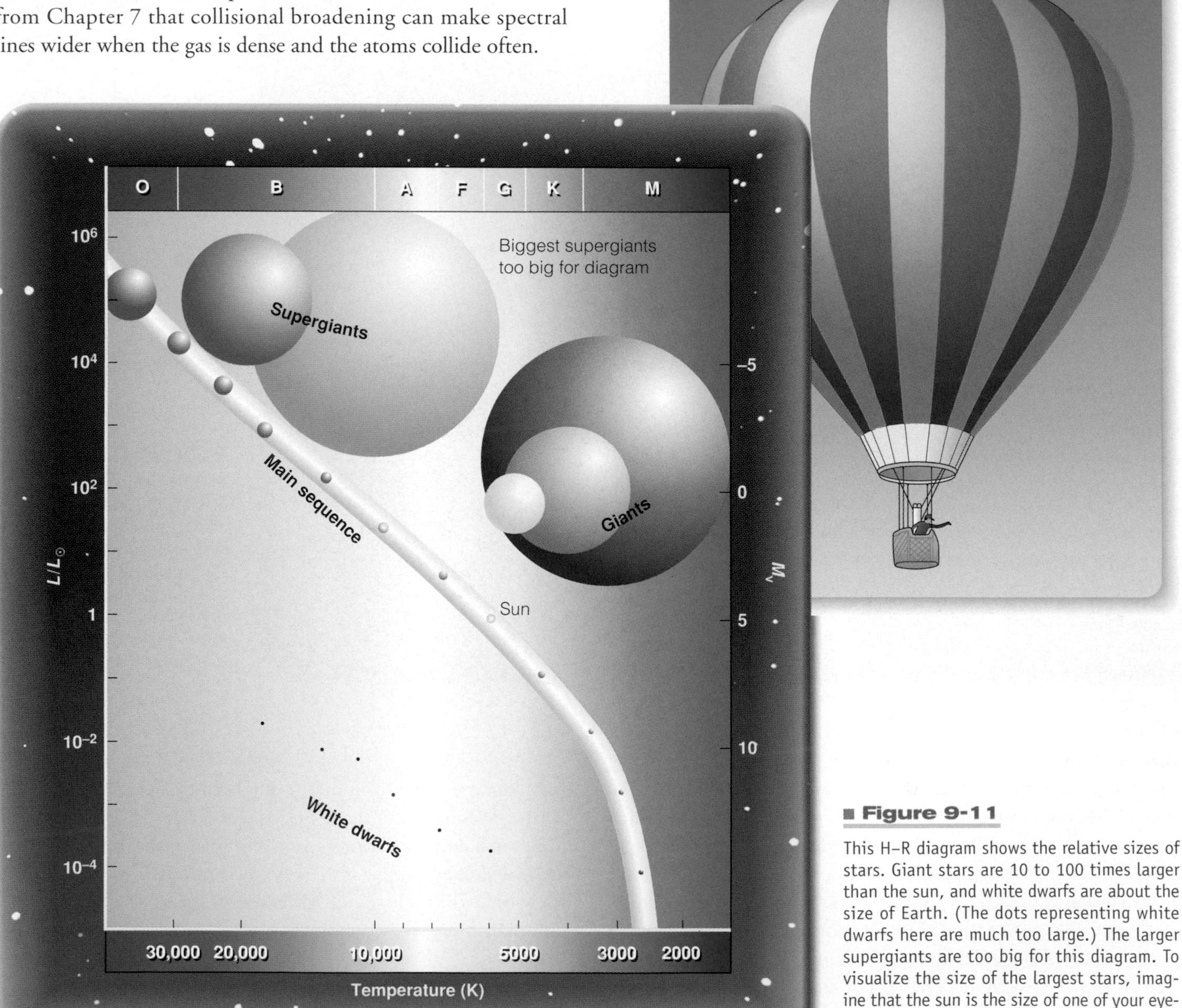

■ **Figure 9-11**

This H–R diagram shows the relative sizes of stars. Giant stars are 10 to 100 times larger than the sun, and white dwarfs are about the size of Earth. (The dots representing white dwarfs here are much too large.) The larger supergiants are too big for this diagram. To visualize the size of the largest stars, imagine that the sun is the size of one of your eyeballs. Then the largest supergiants would be the size of a hot-air balloon.

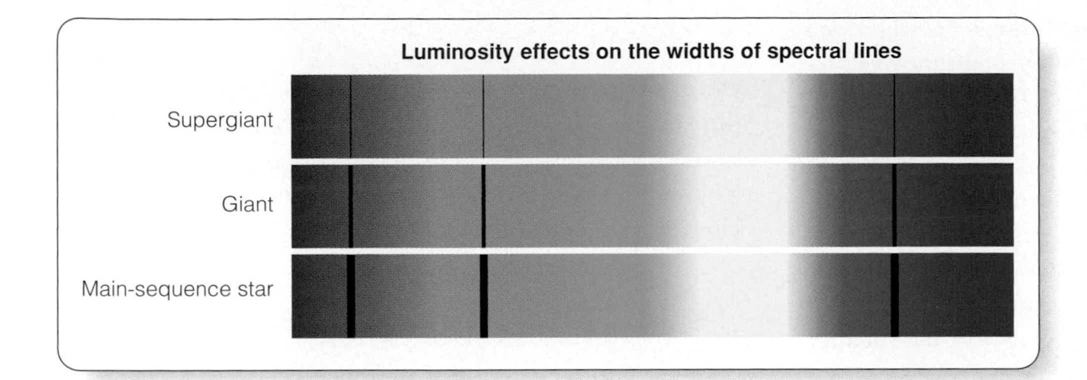

■ Figure 9-12

These schematic spectra show how the widths of spectral lines reveal a star's luminosity classification. Supergiants have very narrow spectral lines, and main-sequence stars have broad lines. In addition, certain spectral lines are more sensitive to this effect than others, so an experienced astronomer can inspect a star's spectrum and determine its luminosity classification.

The luminosity classes are represented by the Roman numerals I through V, with supergiants further subdivided into types Ia and Ib, as follows:

Luminosity Classes

Ia	Bright supergiant
Ib	Supergiant
II	Bright giant
III	Giant
IV	Subgiant
V	Main-sequence star

You can distinguish between the bright supergiants (Ia) such as Rigel and the regular supergiants (Ib) such as Polaris, the North Star. The star Adhara is a bright giant (II), Aldebaran is a giant (III), and Altair is a subgiant (IV). The sun is a main-sequence star (V). The luminosity class is written after the spectral type, as in G2 V for the sun. White dwarfs don't enter into this classification, because their spectra are peculiar. Some of our Favorite Stars are quite unusual. Next time you look at the North Star, remind yourself that it is a supergiant.

If you plot the positions of the luminosity classes on the H–R diagram you get a figure like ■ Figure 9-13. Remember that these are rather broad classifications and that the lines on the diagram are only approximate. A star of luminosity class III may lie slightly above or below the line labeled III. Luminosity classification is subtle and not too accurate, but it is an important tool in modern astronomy. As you will see in the next section, luminosity classification, combined with the H–R diagram, gives astronomers a way to find the distance to stars that are too far away to have measurable parallaxes.

Spectroscopic Parallax

Astronomers can measure the stellar parallax of nearby stars, but most stars are too distant to have measurable parallaxes. The distances to these stars can be estimated from their spectra and apparent magnitude in a process called **spectroscopic parallax.** Spectroscopic parallax is not an actual measure of parallax, but it does reveal the distance to the star.

The method of spectroscopic parallax depends on the H–R diagram. If you photographed the spectrum of a star, you could determine its spectral class, and that would tell you its horizontal location in the H–R diagram. You could also determine its luminosity class by looking at the widths of its spectral lines, and that would tell you the star's vertical location in the diagram. Once you plotted the point that represents the star in an H–R diagram such as Figure 9-13, you could read off its absolute magnitude. As you have seen earlier in this chapter, you can find the distance to a star by comparing its apparent and absolute magnitudes.

For example, our Favorite Star Betelgeuse is classified M2 Ia, and its apparent magnitude is about 0.05. You can plot this

■ Figure 9-13

The approximate location of the luminosity classes on the H–R diagram.

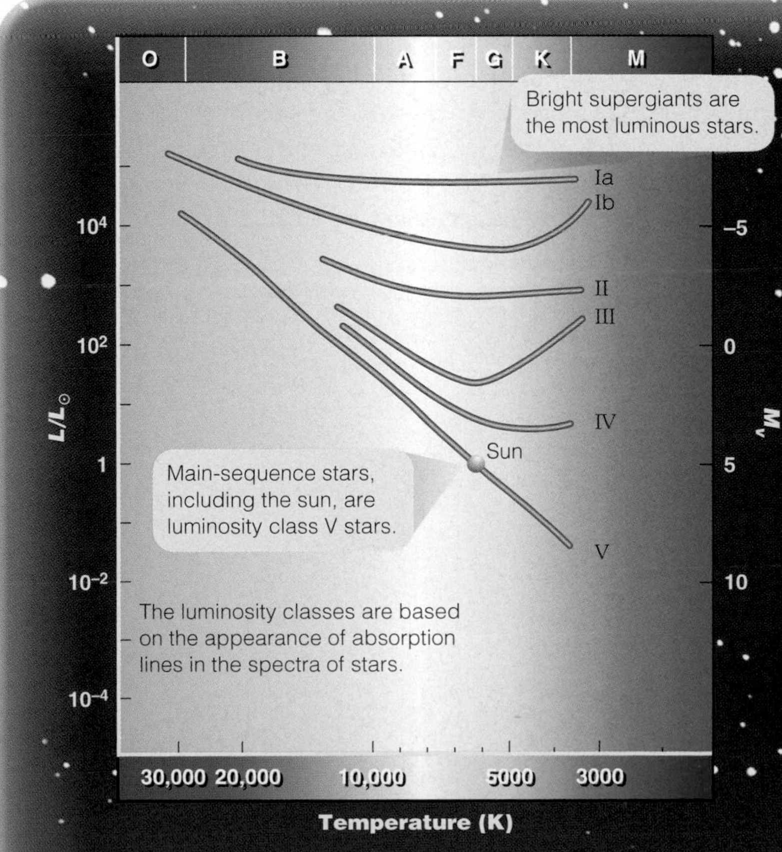

star in an H–R diagram such as that in Figure 9-13, where you would find that it should have an absolute magnitude of about −7.2. That means its distance modulus is 0.05 minus (−7.2), or about 7.7, and the distance (estimated from Table 9-1) is about 350 pc. Parallax from the Hipparcos satellite shows that the true distance to Betelgeuse is 520 pc, so the estimate from spectroscopic parallax is only approximate. An error of 1 magnitude changes the distance by a *factor* of 1.6. That's an error of 60 percent, so spectroscopic parallax isn't very accurate, but it can provide an estimate for stars so distant parallax can't be measured.

Building Scientific Arguments

What evidence can you give that giant stars really are bigger than the sun?

Scientific arguments are based on evidence, so you need to proceed step by step here. Stars exist that have the same spectral type as the sun but are clearly more luminous. Capella, for example, is a G star with an absolute magnitude of 0. Because it is a G star, it must have about the same temperature as the sun, but its absolute magnitude is almost 5 magnitudes brighter than the sun's. A magnitude difference of 5 magnitudes corresponds to an intensity ratio of 100, so Capella must be about 100 times more luminous than the sun. If it has the same surface temperature as the sun but is 100 times more luminous, then it must have a surface area 100 times greater than the sun's. Because the surface area of a sphere is proportional to the square of the radius, Capella must be 10 times larger in radius. That is clear observational evidence that Capella is a giant star.

In Figure 9-10, you can see that Procyon B is a white dwarf only slightly warmer than the sun but about 10,000 times less luminous than the sun. Build a scientific argument based on evidence to resolve this question. **Why do astronomers conclude that white dwarfs must be small stars?**

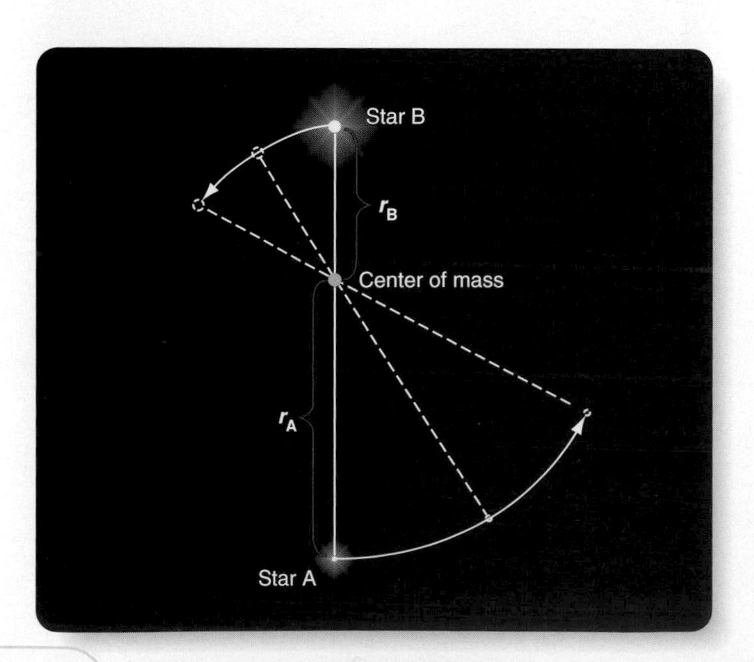

■ ■ ■

Connections: You have achieved the first two goals of this chapter and discovered the luminosities and radii of the stars. That has revealed surprising differences among stars that look similar in the sky. Now it is time to pursue the third goal. How massive are these stars?

9-4 The Masses of Stars

YOUR THIRD GOAL is to find out how much matter stars contain, that is, to know their mass. Do they all contain about the same mass as our sun, or are some more massive and others less? Unfortunately, it's difficult to determine the mass of a star. Looking through a telescope at a star, you see only a point of light that tells you nothing about the star. Gravity is the key. Matter produces a gravitational field, and you can figure out how much matter a star contains if you watch an object move through the star's gravitational field. To find the masses of stars, you must study **binary stars,** pairs of stars that orbit each other.

Binary Stars in General

The key to finding the mass of a binary star is understanding orbital motion. Chapter 5 illustrated orbital motion by imagining a cannonball fired from a high mountain (see page 94). If Earth's gravity didn't act on the cannonball, it would follow a straight-line path and leave Earth forever. Because Earth's gravity pulls it away from its straight-line path, the cannonball follows a curved path around Earth—an orbit. When two stars orbit each other, their mutual gravitation pulls them away from straight-line paths and makes them follow closed orbits around a point between the stars.

Each star in a binary system moves in its own orbit around the system's center of mass, the balance point of the system. If the stars were connected by a massless rod and placed in a uniform gravitational field such as that near Earth's surface, the system would balance at its center of mass like a child's seesaw (see page 95). If one star were more massive than its companion, then the massive star would be closer to the center of mass and would travel in a smaller orbit, while the lower-mass star would whip around in a larger orbit (■ Figure 9-14). The ratio of the masses of the stars M_A/M_B equals r_B/r_A, the inverse of the ratio of the radii of the orbits. If one star has an orbit twice as large as the other star's orbit, then it must be half as massive. Getting the ratio of the masses is easy, but that doesn't tell you the individual masses

■ Active Figure 9-14

As stars in a binary star system revolve around each other, the line connecting them always passes through the center of mass, and the more massive star is always closer to the center of mass.

AceAstronomy™ Log into AceAstronomy and select this chapter to see Active Figure "Center of Mass." Experiment to see how the ratio of the masses affects the relative size of the orbits.

Window on Science 9-2

Learning about Nature through Chains of Inference

Scientists can rarely observe the things they really want to know, so they must construct chains of inference. You can't observe the mass of stars directly, so you must find a way to use what you can observe, orbital period and angular separation, and step by step figure out the parameters you need to reach your goal. In this case, the goal is the masses of the two stars. Scientists follow chains of inference from the observable parameters to the unobservable quantities they want to know.

Chains of inference can be mathematical, as when geologists use earthquakes to calculate the temperature and density of Earth's interior. There is no way to drill a hole to Earth's center and lower a thermometer or recover a sample, so the internal temperature and density are not directly observable. Nevertheless, the speed of vibrations from distant earthquakes depends on the temperature and density of the rock they pass through. Geologists can't measure the speed of the vibrations deep inside Earth; but they can measure the delay in the arrival time at different locations on the surface, and that allows them to work their way back to the speed and, finally, the temperature and density.

Chains of inference can also be nonmathematical. Biologists studying the migration of whales can't follow individual whales for years at a time, but they can observe feeding and mating in different locations; take into consideration food sources, ocean currents, and water temperatures; and construct a chain of inference that leads back to the seasonal migration pattern for whales.

This chapter contains a number of chains of inference because it describes how astronomers know what stars are like. Almost all sciences use chains of inference. If you can link the observable parameters step by step to the final conclusions, you have gained a strong insight into the nature of science.

San Andreas fault: A chain of inference connects earthquakes to conditions inside Earth. (USGS)

of the stars, which is what you really want to know. That takes further analysis.

To find the mass of a binary star system, you must know the size of the orbits and the orbital period—the length of time the stars take to complete one orbit. The smaller the orbits are and the shorter the orbital period is, the stronger the two stars' gravity must be to hold each other in orbit. For example, if two stars whirl rapidly around each other in small orbits, then their gravity must be very strong to prevent their flying apart. Such stars would have to be very massive. From the size of the orbits and the orbital period, astronomers can figure out the masses of the stars from Newton's laws.

Calculating the Masses of Binary Stars

According to Newton's laws of motion and gravity, the total mass of two stars orbiting each other is related to the average distance a between them and their orbital period P. If the masses are M_A and M_B, then

$$M_A + M_B = \frac{a^3}{P^2}$$

In this formula, a is expressed in AU, P in years, and the mass in solar masses.

Notice that this formula is related to Kepler's third law of planetary motion (see Table 4-1). Almost all of the mass of the solar system is in the sun. If you apply this formula to any planet in our solar system, the total mass is 1 solar mass. Then the formula becomes $P^2 = a^3$, which is Kepler's third law.

In other star systems, the total mass is not necessarily 1 solar mass, and this gives you a way to find the masses of binary stars. If you can find the average distance in AU between the two stars, a, and their orbital period in years, P, the sum of the masses of the two stars is just a^3/P^2.

Example 1: If you observe a binary system with a period of 32 years and an average separation of 16 AU, what is the total mass? *Solution:* The total mass equals $16^3/32^2$, which equals 4 solar masses.

Example 2: Let's call the two stars in the previous example A and B. Suppose star A is 12 AU away from the center of mass, and star B is 4 AU away. What are the individual masses? *Solution:* The ratio of the masses must be 12:4, which is the same as 3:1. What two numbers add up to 4 and have the ratio of 3:1? In this case, the answer is easy. Star B must be 3 solar masses, and Star A must be 1 solar mass.

Actually, figuring out the mass of a binary star system is not as easy as it might seem from this discussion. The orbits of the two stars may be elliptical, and, although the orbits lie in the same plane, that plane can be tipped at an unknown angle to your line of sight, further distorting the shapes of the orbits. Astronomers must find ways to correct for these distortions. In addition, astronomers analyzing binary systems must find the distances to the stars so they can estimate the true size of the orbits in astronomical units. Notice that finding the masses of binary stars requires a number of steps to get from what can be observed to what astronomers really want to know, the masses. Constructing such sequences of steps is an important part of science (**Window on Science 9-2**).

Although there are many different kinds of binary stars, three types are especially important for determining stellar masses. These are discussed separately in the next sections.

Visual Binary Systems

In a **visual binary system,** the two stars are separately visible in the telescope. Only a pair of stars with large orbits can be separated visually; if the orbits are small, the star images blend together in the telescope, and you see only a single point of light. Because visual binary systems must have large orbits, they also have long orbital periods. Some take hundreds or even thousands of years to complete a single orbit.

Astronomers study visual binary systems by measuring the position of the two stars directly at the telescope or in images. In either case, the astronomers need measurements over many years to map the orbits. The first frame of ■ Figure 9-15 shows a photograph of one of our Favorite Stars, Sirius, which is actually a visual binary system made up of the bright star Sirius A and its white dwarf companion Sirius B. The photo was taken in 1960. Successive frames in Figure 9-15 show the motion of the two stars as observed since 1960 and the orbits the stars follow. The orbital period is 50 years.

Binary systems are common; more than half of all stars are members of binary star systems. Our Favorite Star Polaris, for example, has a binary companion in an orbit with a period of 29. 59 years. Two smaller stars orbit even farther from Polaris, so it is a four-star binary system. Favorite Star Alpha Centauri, the nearest star to the sun, is a three-star system. Although binary stars are common, few can be analyzed completely. Many visual binaries are so far apart that their periods are much too long for practical mapping of their orbits. Other binary systems are so close together they are not visible as separate stars.

Spectroscopic Binary Systems

If the stars in a binary system are close together, the telescope, limited by diffraction and by seeing, reveals only a single point of light. Only by looking at a spectrum that is formed by light from both stars and contains spectral lines from both can astronomers tell that there are two stars present and not one. Such a system is a **spectroscopic binary system.**

■ Figure 9-16 shows a pair of stars orbiting each other in identical circular orbits. From the diagram you can guess that the two stars have equal masses, and you should notice that the circular orbit appears elliptical because you see it nearly edge-on. Of course, if this were a spectroscopic binary system, you would not see the separate stars. Nevertheless, the Doppler shift would alert you that there were two stars orbiting each other. In the first frame of the figure, star A is approaching Earth while star B recedes. In the spectrum, you see a spectral line from star A blueshifted while the same spectral line from star B is redshifted. As you watch the two stars revolve around their orbits, they alternately approach and recede from Earth, and you can see their spectral lines Doppler shifted first toward the blue and then toward the red. In a real spectroscopic binary, you can't see the individual stars, but the sight of pairs of spectral lines moving back and forth across each other would alert you that you were observing a spectroscopic binary system (■ Figure 9-17).

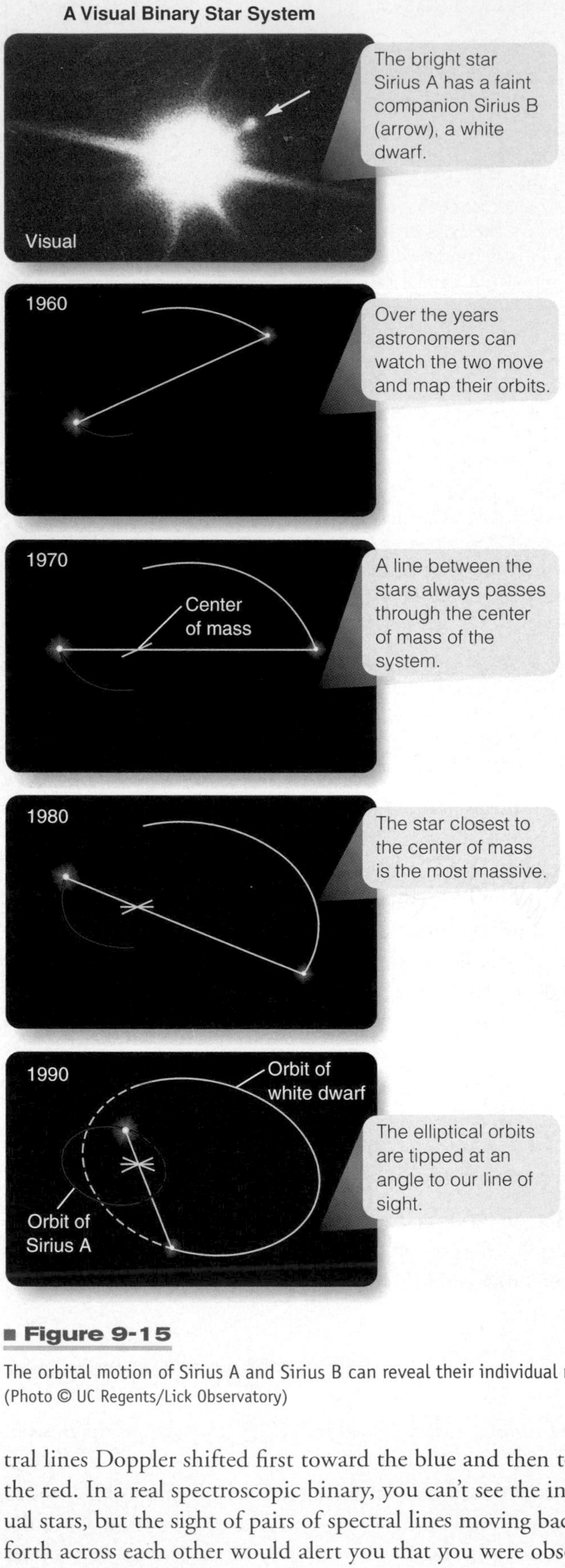

A Visual Binary Star System

The bright star Sirius A has a faint companion Sirius B (arrow), a white dwarf.

Visual

1960

Over the years astronomers can watch the two move and map their orbits.

1970

Center of mass

A line between the stars always passes through the center of mass of the system.

1980

The star closest to the center of mass is the most massive.

1990

Orbit of white dwarf

Orbit of Sirius A

The elliptical orbits are tipped at an angle to our line of sight.

■ Figure 9-15

The orbital motion of Sirius A and Sirius B can reveal their individual masses. (Photo © UC Regents/Lick Observatory)

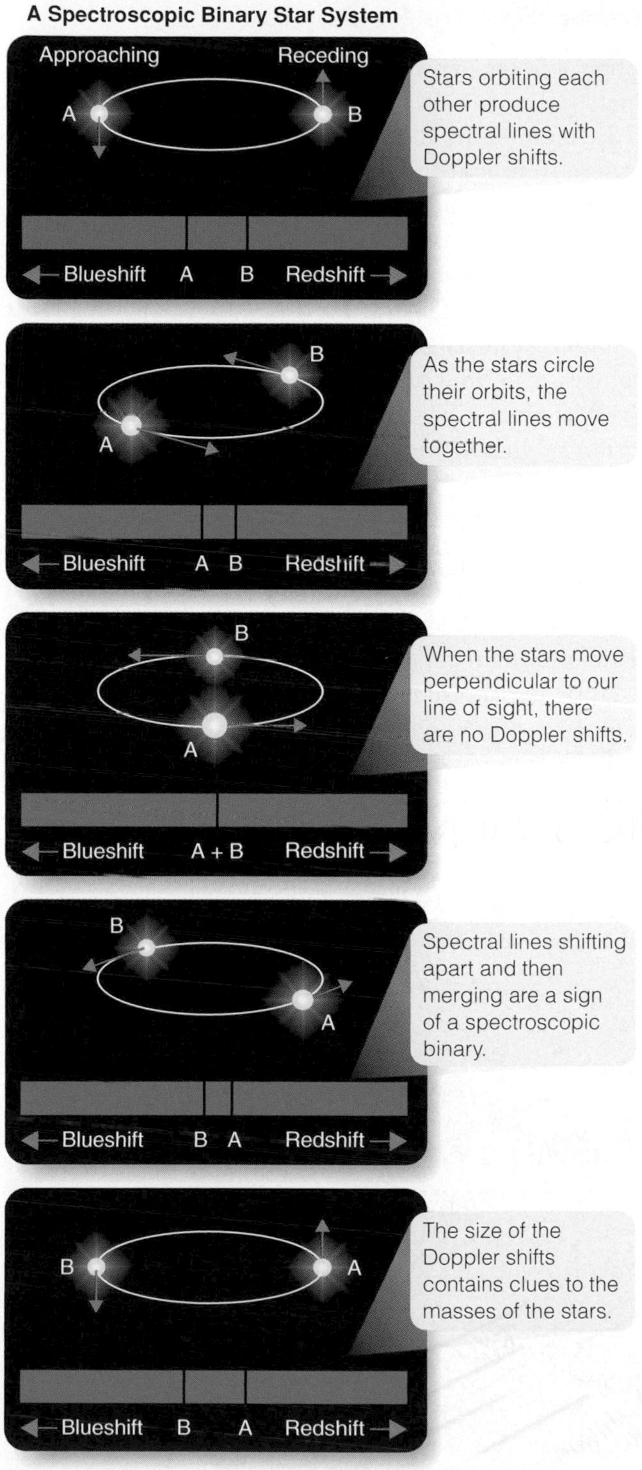

A Spectroscopic Binary Star System

Approaching
Receding
A
B
Stars orbiting each other produce spectral lines with Doppler shifts.
Blueshift A B Redshift

As the stars circle their orbits, the spectral lines move together.
Blueshift A B Redshift

When the stars move perpendicular to our line of sight, there are no Doppler shifts.
Blueshift A + B Redshift

Spectral lines shifting apart and then merging are a sign of a spectroscopic binary.
Blueshift B A Redshift

The size of the Doppler shifts contains clues to the masses of the stars.
Blueshift B A Redshift

■ **Figure 9-16**

From Earth, a spectroscopic binary looks like a single point of light, but the Doppler shifts in its spectrum reveal the orbital motion of the two stars.

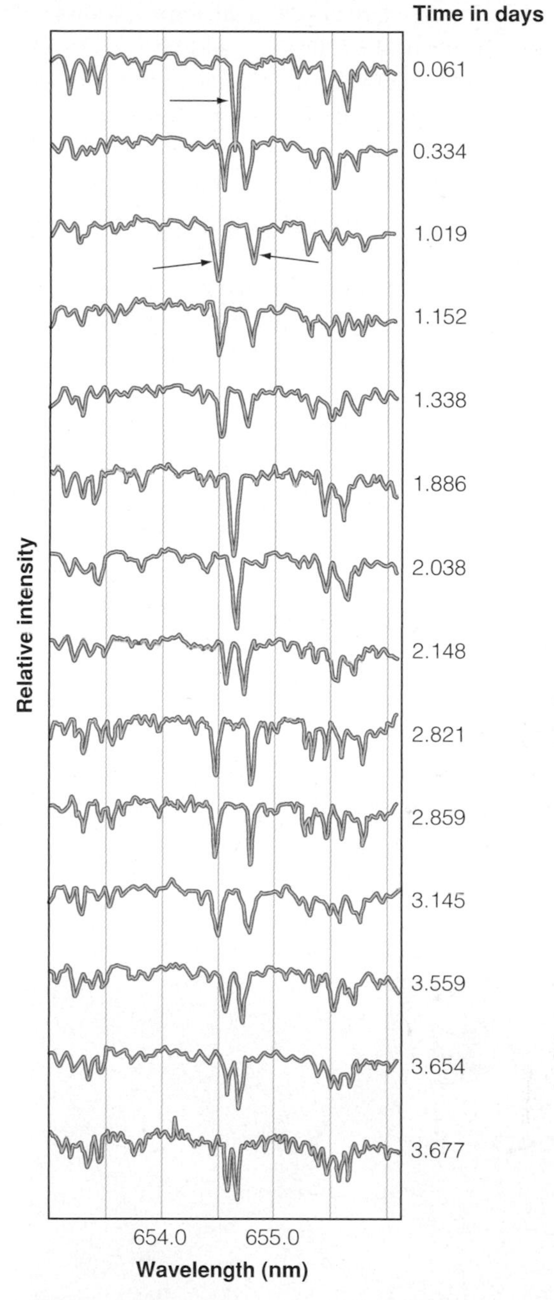

■ **Figure 9-17**

Fourteen spectra of the star HD80715 are shown here as graphs of intensity versus wavelength. A single spectral line (arrow in top spectrum) splits into a pair of spectral lines (arrows in third spectrum), which then merge and split apart again. These changing Doppler shifts reveal that HD80715 is a spectroscopic binary. (Adapted from data courtesy of Samuel C. Barden and Harold L. Nations)

You can find the orbital period of a spectroscopic binary system by waiting to see how long it takes for the spectral lines to return to their starting positions. You can measure the size of the Doppler shifts to find the orbital velocities of the two stars. If you multiply velocity times orbital period, you can find the circumference of the orbit, and from that you can find the radius of the orbit. Of course, if you knew the orbital period and the size of the orbit, you could calculate the mass. One important detail is missing, however. You don't know how much the orbits are inclined to your line of sight.

You can find the inclination of a visual binary system because you can see the shape of the orbits. In a spectroscopic binary system, however, you cannot see the individual stars, so you can't find the inclination or untip the orbits. Recall that the Doppler effect only tells you the radial velocity, the part of the velocity directed toward or away from Earth. Because you cannot find the inclination, you cannot correct these radial velocities to find the true orbital velocities. Consequently, you cannot find the true masses. All you find from a spectroscopic binary system is a lower limit to the masses.

More than half of all stars are in binary systems, and most of those are spectroscopic binary systems. Many of the familiar stars in the sky are actually pairs of stars orbiting each other (■ Figure 9-18).

You might wonder what happens when the orbits of a spectroscopic binary system lie exactly edge-on to Earth. The result is the most useful kind of binary system.

Ace Astronomy™ Log into AceAstronomy and select this chapter to see Astronomy Exercise "Spectroscopic Binaries." You can see how the motions of the two stars affect their shared spectrum.

■ Figure 9-18

(a) At the bend of the handle of the Big Dipper lies a pair of stars, Mizar and Alcor. Through a telescope you will discover that Mizar has a fainter companion and so is a member of a visual binary system. (b) Spectra of Mizar recorded at different times show that it is itself a spectroscopic binary system rather than a single star. In fact, both the faint companion to Mizar and the nearby star Alcor are also spectroscopic binary systems. (The Observatories of the Carnegie Institution of Washington)

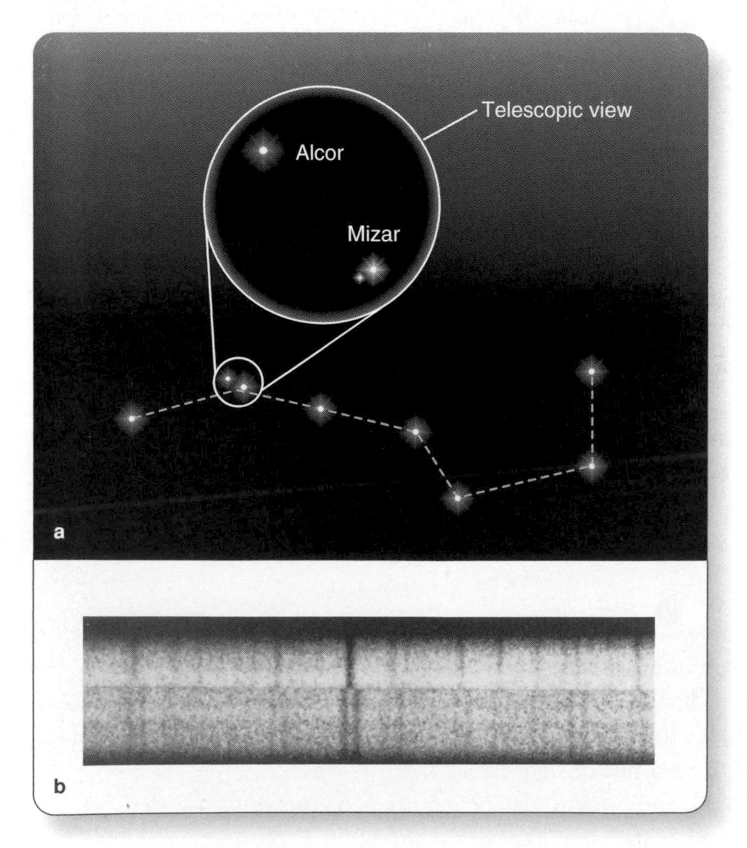

Eclipsing Binary Systems

As mentioned in the previous section, the orbits of the two stars in a binary system always lie in a single plane. If that plane is nearly edge-on to Earth, then the stars can cross in front of each other as seen from Earth. Imagine a model of a binary star system in which a cardboard disk represents the orbital plane and balls represent the stars, as in ■ Figure 9-19. If you see the model from the edge, then the balls that represent the stars can move in front of each other as they follow their orbits. The small star crosses in front of the large star, and then, half an orbit later, the large star crosses in front of the small star. When one star moves in front of the other, it blocks some of the light, and astronomers say the star is eclipsed. Such a system is called an **eclipsing binary system.**

Seen from Earth, the two stars are not visible separately. The system looks like a single point of light. But when one star moves in front of the other star, part of the light is blocked, and the total brightness of the point of light decreases. ■ Figure 9-20 shows a smaller star moving in an orbit around a larger star, first eclipsing the larger star and then being eclipsed as it moves behind. The resulting variation in the brightness of the system is shown as a graph of brightness versus time, a **light curve.**

■ Figure 9-19

Imagine a model of a binary system with balls for stars and a disk of cardboard for the plane of the orbits. Only if you view the system edge-on do you see the stars cross in front of each other.

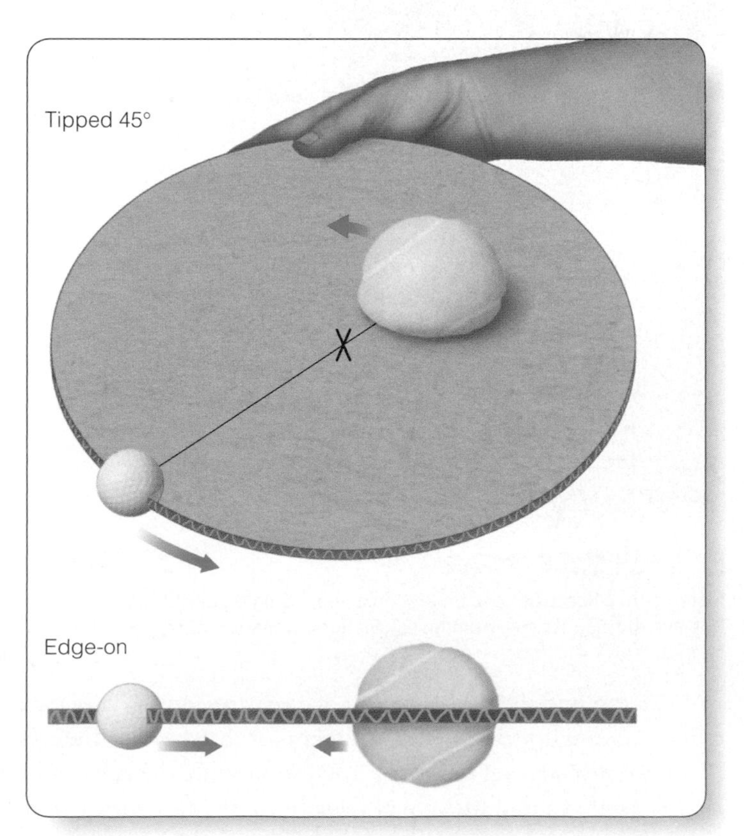

The light curves of eclipsing binary systems contain tremendous amounts of information about the stars, but the curves can be difficult to analyze. Figure 9-20 shows an idealized system.

■ Figure 9-20

From Earth, an eclipsing binary looks like a single point of light, but changes in brightness reveal that two stars are eclipsing each other. Doppler shifts in the spectrum combined with the light curve, shown here as magnitude versus time, can reveal the size and mass of the individual stars.

An Eclipsing Binary Star System

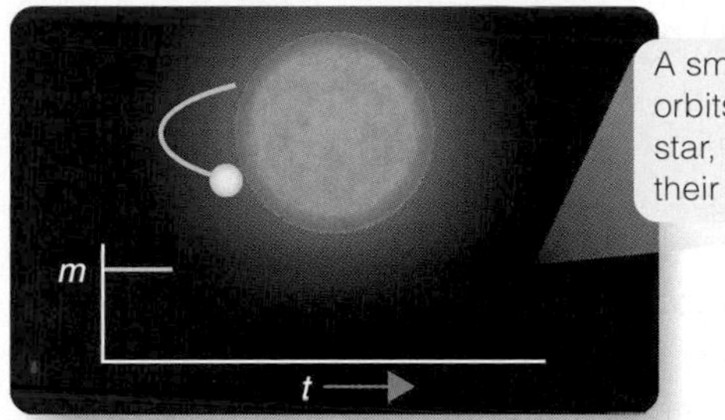
A small, hot star orbits a large, cool star, and you see their total light.
m
t

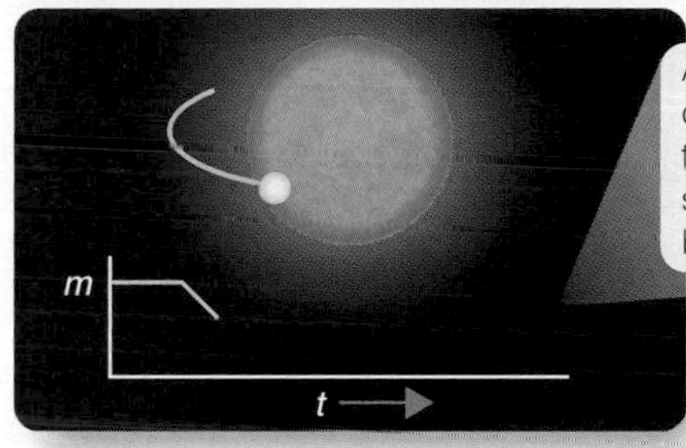
As the hot star crosses in front of the cool star, you see a decrease in brightness.
m
t

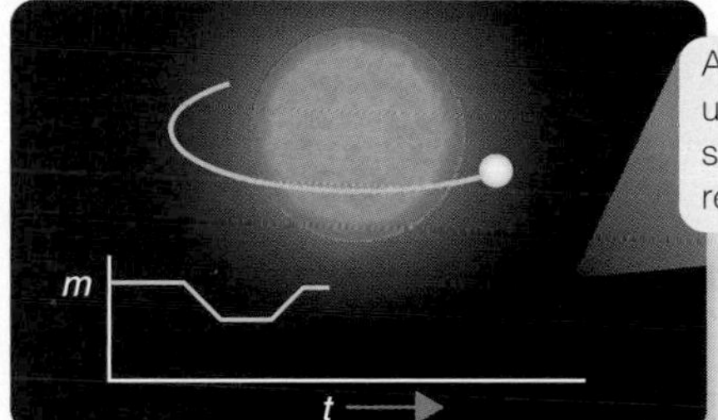
As the hot star uncovers the cool star, the brightness returns to normal.
m
t

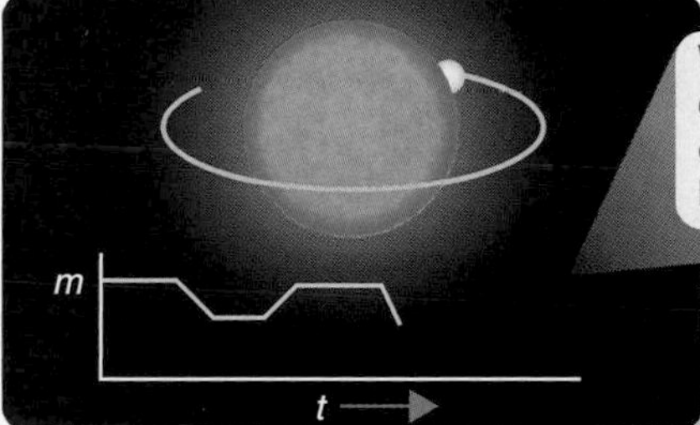
When the hot star is eclipsed behind the cool star, the brightness drops.
m
t

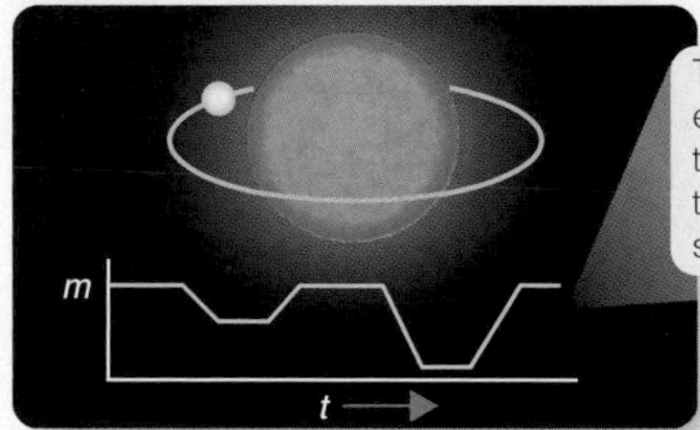
The depth of the eclipses depends on the surface temperatures of the stars.
m
t

■ Figure 9-21 shows the light curve of a real system in which the stars have dark spots on their surfaces and are so close to each other that their shapes are distorted.

Once the light curve of an eclipsing binary system has been accurately observed, you can construct a chain of inference that leads to the masses of the two stars. You could find the orbital period easily, and you could get spectra showing the Doppler shifts of the two stars. You could find the orbital velocity because you don't have to untip the orbits; you know they are nearly edge-on or there would not be eclipses. Then you can find the size of the orbits and the masses of the stars.

Earlier in this chapter, luminosity and temperature were used to calculate the radii of stars, but eclipsing binary systems provide a way to measure the sizes of stars directly. From the light curve you could tell how long it took for the small star to cross the large star. Multiplying this time interval by the orbital velocity of the small star would give you the diameter of the larger star. You could also determine the diameter of the small star by noting how long it took to disappear behind the edge of the large star.

■ Figure 9-21

The observed light curve of the binary star VW Cephei (lower curve) shows that the two stars are so close together their gravity distorts their shapes. Slight distortions in the light curve reveal the presence of dark spots at specific places on the star's surface. The upper curve shows what the light curve would look like if there were no spots. (Graphics created with Binary Maker 2.0)

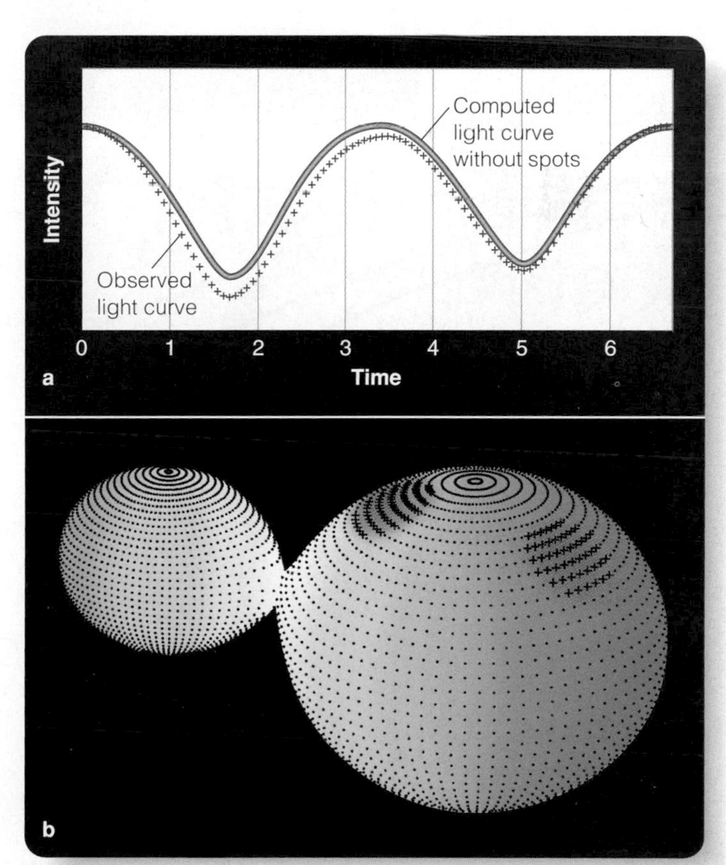
Computed light curve without spots
Observed light curve
Intensity
0
1
2
3
4
5
6
Time
a
b

For example, if it took 300 seconds for the small star to disappear while traveling 500 km/s relative to the large star, then it would have to be 150,000 km in diameter.

Of course, there are complications due to the inclination and eccentricity of orbits, but often these effects can be taken into account, and the system can reveal not only the masses of its stars but also their diameters.

Algol (alpha Persei) is one of the best-known eclipsing binaries, because its eclipses are visible to the naked eye. Normally, its magnitude is about 2.15, but its brightness drops to 3.4 in eclipses that occur every 68.8 hours. Although the nature of the star was not recognized until 1783, its periodic dimming was probably known to the ancients. *Algol* comes from the Arabic for "the demon's head," and it is associated in constellation mythology with the severed head of Medusa, the sight of whose serpentine locks turned mortals to stone (■ Figure 9-22). Indeed, in some accounts, Algol is the winking eye of the demon.

From the study of binary stars, astronomers have found that the masses of stars range from 0.08 solar mass at the low end to as high as 150 solar masses at the high end. The most massive stars ever found in a binary system are a pair of stars with masses of 83 and 82 solar masses. A few other stars are believed to be more massive, but they do not lie in binary systems, so astronomers can only estimate their masses. Stars near the upper limit are very rare, and few are known, so this upper limit is uncertain.

Ace Astronomy™ Log into AceAstronomy and select this chapter to see Astronomy Exercise "Eclipsing Binaries." You can take control of a binary star system.

Building Scientific Arguments

When you look at the light curve for an eclipsing binary system with total eclipses, how can you tell which star is hotter?

This scientific argument brings together a number of things you have learned about light and binary stars. If you assume that the two stars in an eclipsing binary system are not the same size, then you can refer to them as the larger star and the smaller star. When the smaller star moves behind the larger star, you lose the light coming from the total area of the small star. And when the smaller star moves in front of the larger star, it blocks off light from the same amount of area on the larger star. In both cases, the same amount of area, the same number of square meters, is hidden from your sight. Then the amount of

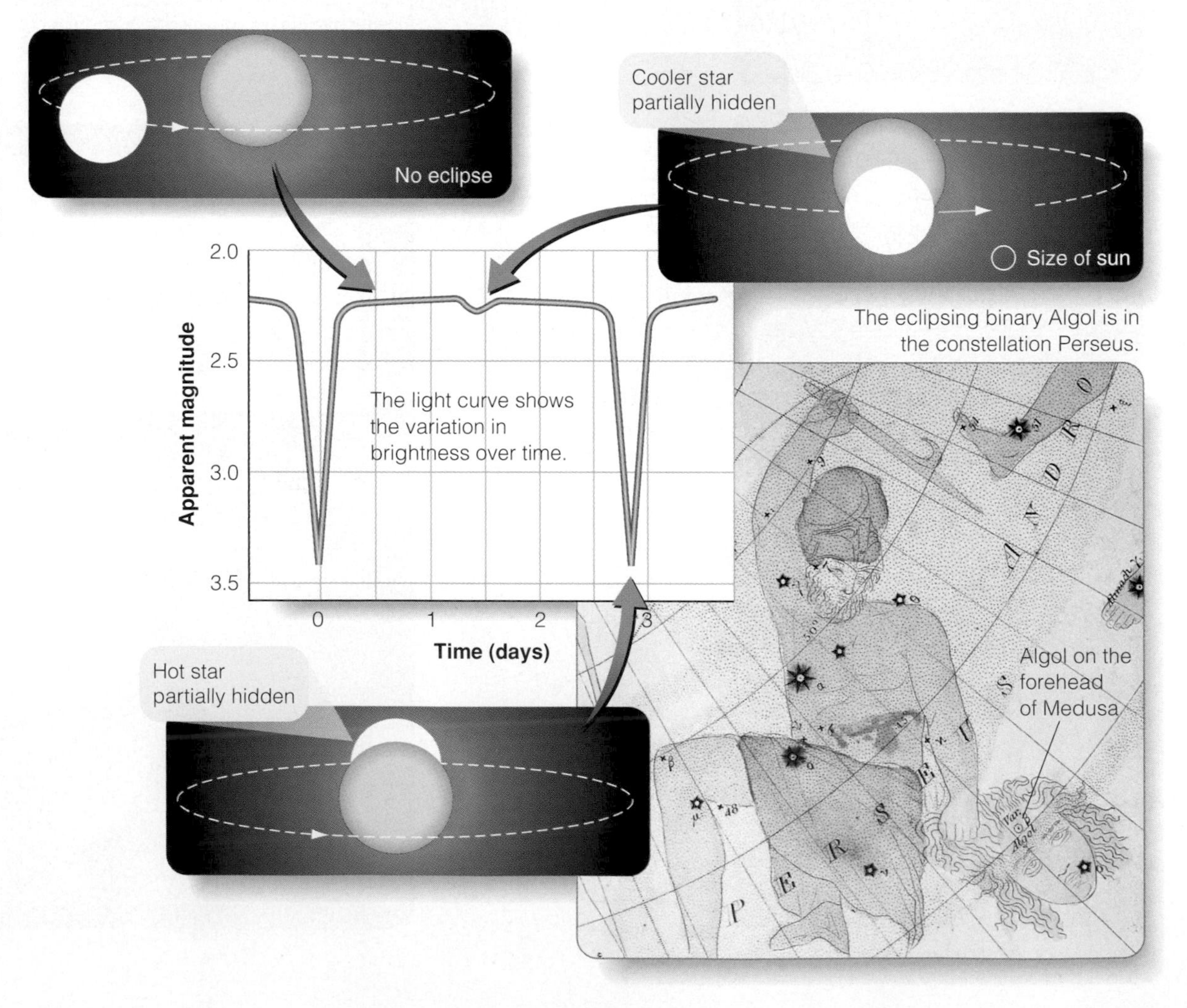

■ **Figure 9-22**

The eclipsing binary Algol consists of a hot B star and a cooler G or K star. The eclipses are partial, meaning that neither star is completely hidden during eclipses. The orbit here is drawn as if the cooler star were stationary.

Window on Science 9-3

Information about Nature: Basic Scientific Data

In a simple sense, science is the process by which scientists look at data and search for relationships that show how nature works. But, as a result, science sometimes requires large amounts of data. For example, astronomers need to know the masses and luminosities of many stars before they can begin looking for relationships.

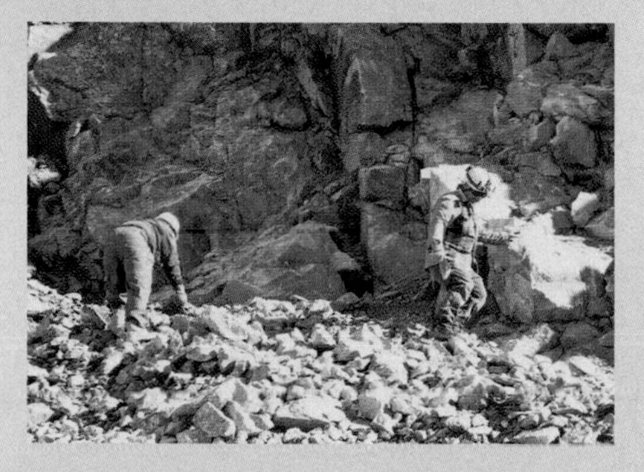

Collecting mineral samples can be hard work, but it is also fun. Scientists sometimes collect large amounts of data because they enjoy the process. (M. Seeds)

Compiling basic data is one of the common forms of scientific study. This work may not seem very exciting to an outsider, but scientists often love their work not so much because they want to know nature's secrets but because they love the process of studying nature. Using a microscope or a telescope is fun. Gathering plants in the rain forest or geological samples from a cliff face can be tremendously exciting and satisfying. It is sometimes the process of science that is most rewarding, and enjoyment of the process can lead scientists to gather significant amounts of information.

Solving a single binary star system to find the masses of the stars does not tell an astronomer a great deal about nature, but solving a binary is like solving a puzzle. It is fun, and it is satisfying. Over the years, many astronomers have added their results to the growing data file on stellar masses. Astronomers can now analyze that data to search for relationships between the masses of stars and other parameters such as diameter and luminosity.

The history of science is filled with stories of hardworking scientists who compiled large amounts of data that later scientists used to make important discoveries. For example, Tycho Brahe spent 20 years recording the positions of the stars and planets (Chapter 4). He must have loved his work to spend 20 years on his windy island, but he did not live to use his data. It was his successor, Johannes Kepler, who used Tycho Brahe's data to discover the laws of planetary motion.

Whatever science you study, you will encounter accumulations of measurements and observations that have been compiled over the years, including everything from the hardness of rocks to the attention span of infants. Determining these basic scientific data is as much a part of science as is testing a hypothesis.

light lost during an eclipse depends only on the temperature of the hidden surface because temperature is what determines how much each square meter can radiate per second. When the surface of the hotter star is hidden, the brightness will fall dramatically, but when the surface of the cooler star is hidden, the brightness will not fall as much. So you can look at the light curve and point to the deeper of the two eclipses and say, "That is where the hotter star is behind the cooler star."

Now change the argument to consider the diameters of the stars. **How could you look at the light curve of an eclipsing binary with total eclipses and find the ratio of the diameters?**

■ ■ ■

Connections: Perhaps the hardest thing to learn about a star is its mass, but stars orbiting around each other in binary systems can be analyzed to find their masses. Now you are ready to put the facts together.

9-5 A Survey of the Stars

YOU HAVE ACHIEVED the three goals set for you at the start of this chapter. You know how to find the luminosities, diameters, and masses of stars. Now you can put that data (**Window on Science 9-3**) together to paint a family portrait of the stars; like most family portraits, both similarities and differences are important clues to the history of the family. As you begin trying to understand how stars are born and how they die, you can ask a simple question: What is the average star like? Answering that question is both challenging and illuminating.

Mass, Luminosity, and Density

With enough data plotted in an H–R diagram, you can see the patterns that hint as to how stars are born, how they age, and how they die.

If you label an H–R diagram with the masses of the plotted stars, as in ■ Figure 9-23, you will discover that the main-sequence stars are ordered by mass. The most massive main-sequence stars are the hot stars. As you run your eye down the main sequence, you find lower-mass stars; and the lowest-mass stars are the coolest, faintest main-sequence stars.

Stars that do not lie on the main sequence are not in order according to mass. Giant stars are a jumble of different masses, and supergiants, although they tend to be more massive than giants, are in no particular order in the H–R diagram. In contrast, all white dwarfs have about the same mass, somewhere in the narrow range of 0.5 to about 1 solar mass.

Because of the systematic ordering of mass along the main sequence, these main-sequence stars obey a **mass–luminosity**

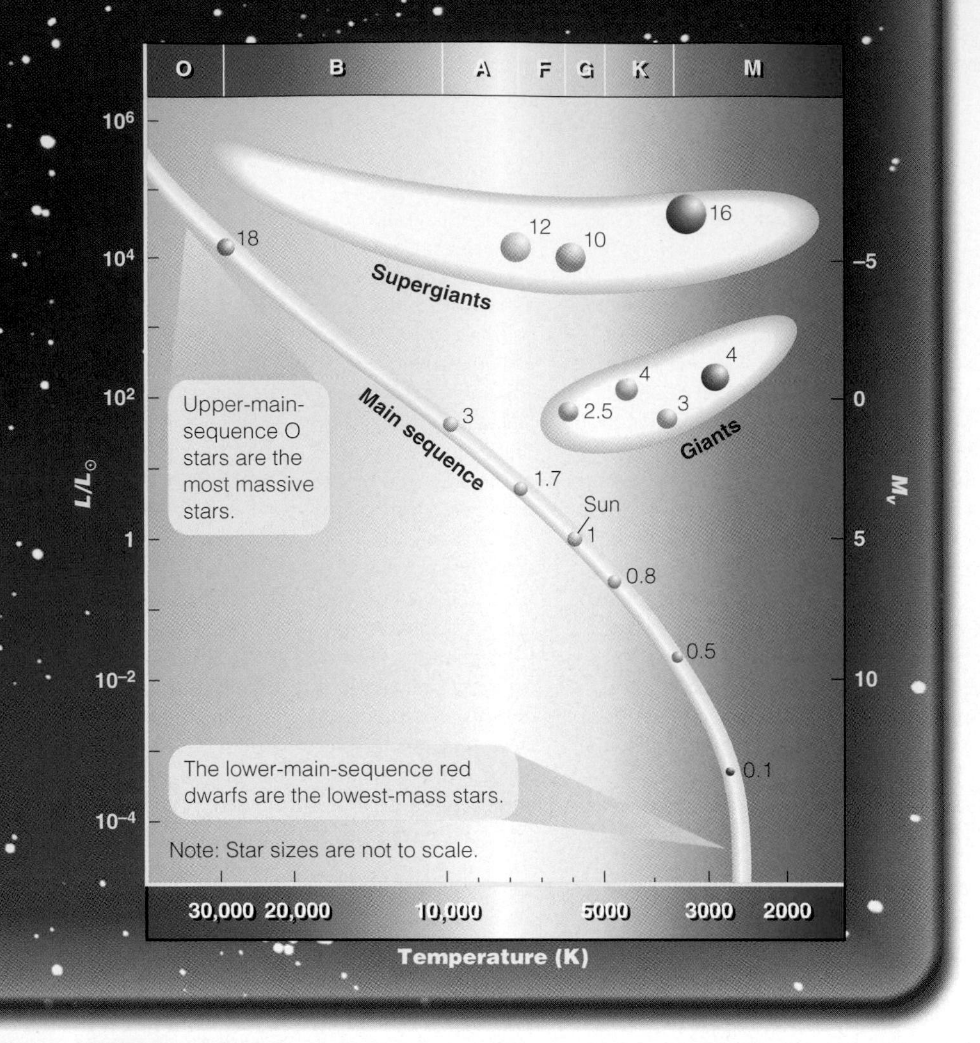

■ **Figure 9-23**

The masses of the plotted stars are labeled on this H–R diagram. Notice that the masses of main-sequence stars decrease from top to bottom but that masses of giants and supergiants are not arranged in any ordered pattern.

relation—the more massive a star is, the more luminous it is (■ Figure 9-24). In fact, the mass–luminosity relation can be expressed as a simple formula:

$$L = M^{3.5}$$

That is, a star's luminosity (in terms of the sun's luminosity) equals its mass (in solar masses) raised to the 3.5 power. For example, a star of 4 solar masses has a luminosity of approximately $4^{3.5}$, or $4 \times 4 \times 4 \times \sqrt{4}$. This equals 64×2, or 128. So a 4-solar-mass star will have a luminosity about 128 times the luminosity of the sun. This is only an approximate equation, as shown by the red line in Figure 9-24.

Notice how large the range in luminosity is. The observed range of masses extends from about 0.08 solar mass to about 83 solar masses—a factor of about 1000. But the range of luminosities extends from about 10^{-6} to about 10^6 solar luminosities—a factor of 10^{12}. Clearly, a small difference in mass causes a large difference in luminosity.

Giants and supergiants do not follow the mass–luminosity relation very closely, and white dwarfs not at all. In the next chapters, the mass–luminosity relation will help you understand how stars generate their energy.

Though mass alone does not reveal any pattern among giants, supergiants, and white dwarfs, density does. Once you know a star's mass and diameter, you can calculate its average density by dividing its mass by its volume. Stars are not uniform in density but are most dense at their centers and least dense near their surface. The center of the sun, for instance, is about 100 times as dense as water; its density near the visible surface is about 3400 times less dense than Earth's atmosphere at sea level. A star's average density is intermediate between its central and surface densities. The sun's average density is approximately 1 g/cm^3—about the density of water.

Main-sequence stars have average densities similar to the sun's, but giant stars, being large, have low average densities, ranging from 0.1 to 0.01 g/cm^3. The enormous supergiants have still lower densities, ranging from 0.001 to 0.000001 g/cm^3. These densities are thinner than the air you breathe; if you could insulate yourself from the heat, you could fly an airplane through these stars.

■ **Figure 9-24**

The mass–luminosity relation shows that the more massive a main-sequence star is, the more luminous it is. The open circles represent white dwarfs, which do not obey the relation. The red line represents the equation $L = M^{3.5}$.

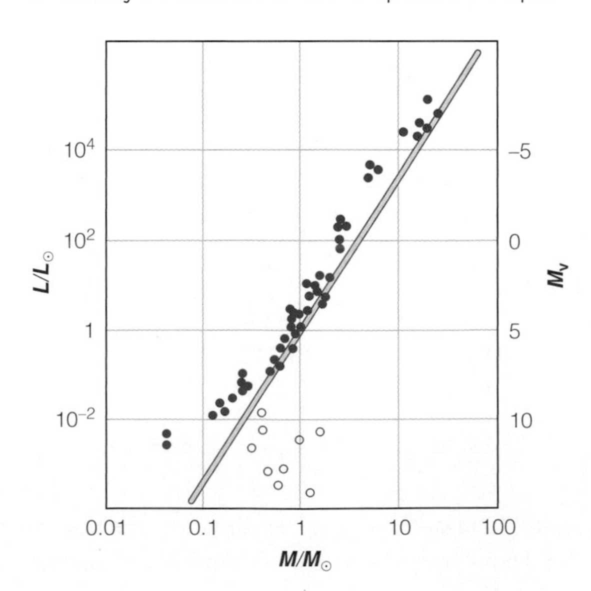

Only near the center would you be in any danger, for there the material is very dense—about 3,000,000 g/cm^3.

The white dwarfs have masses about equal to the sun's but are very small, only about the size of Earth. That means the matter is compressed to densities of 3,000,000 g/cm^3 or more. On Earth, a teaspoonful of this material would weigh about 15 tons.

Density divides stars into three groups. Most stars are main-sequence stars with densities like the sun's. Giants and supergiants are very-low-density stars, and white dwarfs are high-density stars. You will see in later chapters that these densities reflect different stages in the evolution of stars.

Ace Astronomy™ Log into AceAstronomy and select this chapter to see Astronomy Exercise "Mass-Luminosity Relation." You can discover your own relation from real stars.

Ace Astronomy™ Log into AceAstronomy and select this chapter to see Astronomy Exercise "H-R/Mass-Luminosity 3-D Graph." Look at the data in three dimentions and see how mass is related to position in the H-R diagram.

Surveying the Stars

If you want to know what the average person thinks about a certain subject, you take a survey. If you want to know what the average star is like, you must survey the stars. Such surveys reveal important relationships among the family of stars.

Not many decades ago, surveying large numbers of stars was an exhausting task, but modern computers have changed that. Specially designed telescopes controlled by computers can make millions of observations per night, and high-speed computers can compile and analyze these vast surveys and create easy-to-use databases. Chapter 6 mentioned the Sloan Digital Sky Survey and the 2Mass infrared survey. Those and other surveys produce mountains of data that astronomers can mine, searching for relationships within the family of stars.

What could you learn about stars from a survey of the stars near the sun? Because the sun is believed to be in a typical place in the universe, such a survey could reveal general characteristics of the stars and might reveal unexpected processes in the formation and evolution of stars. Study **The Family of Stars** on pages 210–211 and notice three important points:

1. First, taking a survey is difficult because you must be sure you get an honest sample. If you don't survey enough stars or if you don't notice some kinds of stars, you can get biased results.

2. The second important point is that most stars are faint, and the most luminous stars are rare. The most common kinds of stars are the lower-main-sequence red dwarfs and the white dwarfs.

3. Finally, notice that what you see in the sky is deceptive. Stars near the sun are quite faint, but luminous stars, although they are rare, are easily visible even at great distances. Many of the brighter stars in the sky are highly luminous stars that you see even though they lie far away.

The night sky is a beautiful carpet of stars, but they are not all the same. Some are giants and supergiants, and some are dwarfs. The family of the stars is rich in its diversity.

Building Scientific Arguments

What kind of stars do you see if you look at a few of the brightest stars in the sky?

This argument shows how careful you must be to interpret simple observations. When you look at the night sky, the brightest stars are mostly giants and supergiants. Most of the bright stars in Canis Major, for instance, are supergiants. Sirius, in Canis Major, is the brightest star in the sky, but it is a main-sequence star; it looks bright because it is very nearby, not because it is very luminous. In general, the supergiants and giants are so luminous that they stand out and look bright, even though they are not nearby. When you look at a bright star in the sky, you are probably looking at a highly luminous star—a supergiant or a giant. You can check the argument above by consulting the tables of the brightest and nearest stars in the Appendix.

Now revise your argument. **What kind of star do you see if you look at a few of the stars nearest to the sun?**

■ ■ ■

Connections: This chapter set out to find the basic properties of stars. Once you found the distance to the stars, you were able to find their luminosities, diameters, and masses—rather mundane data. But you have now discovered a puzzling situation. The largest and most luminous stars are so rare you might joke that they hardly exist, and the average stars are such dinky low-mass things they are hard to see even if they are near Earth in space. Why does nature make stars in this peculiar way? To answer that question, you must explore the birth, life, and death of stars. That quest begins in the next chapter.

The Family of Stars

1 What is the most common kind of star? Are some rare? Are some common? To answer those questions you must survey the stars. To do so you must know their spectral class, their luminosity class, and their distance. Your census of the family of stars produces some surprising demographic results.

1a You could survey the stars by observing every star within 62 pc of Earth. A sphere 62 pc in radius encloses a million cubic parsecs. Such a survey would tell you how many stars of each type are found within a volume of a million cubic parsecs.

62 pc

Earth

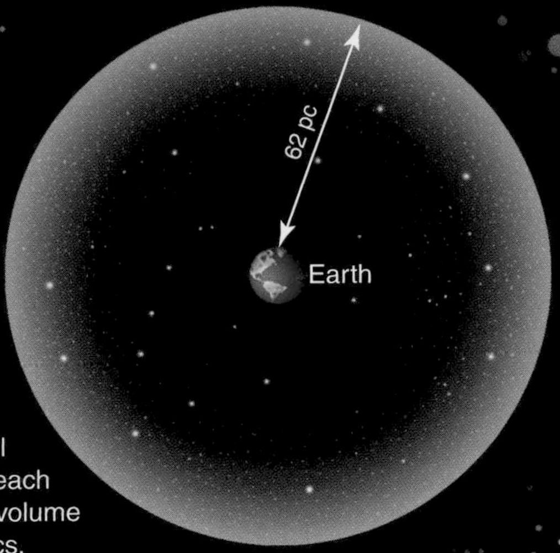

2 Your survey faces two problems.

1. The most luminous stars are so rare you find few in your survey region. There are no O stars at all within 62 pc of Earth.

2. Lower-main-sequence M stars, called red dwarfs, and white dwarfs are so faint they are hard to locate even when they are only a few parsecs from Earth. Finding every one of these stars in your survey sphere is a difficult task.

Spectral Class Color Key

- O and B
- A
- F
- G
- K
- M

The star chart in the background of these two pages shows most of the constellation Canis Major; stars are represented as dots with colors assigned according to spectral class. The brightest stars in the sky tend to be the rare, highly luminous stars, which look bright even though they are far away. Most stars are of very low luminosity, so nearby stars tend to be very faint red dwarfs.

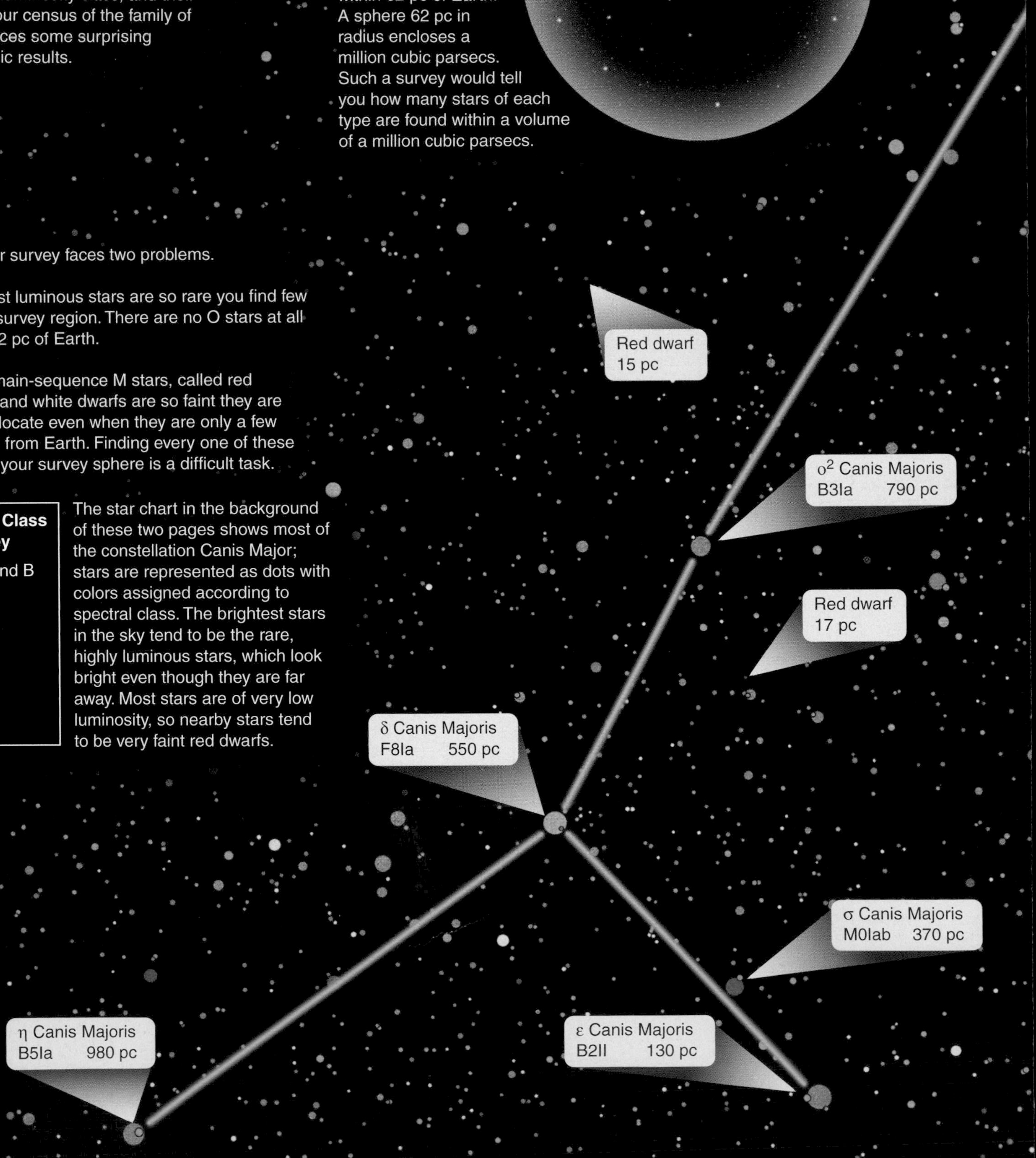

Red dwarf
15 pc

o^2 Canis Majoris
B3Ia 790 pc

Red dwarf
17 pc

δ Canis Majoris
F8Ia 550 pc

σ Canis Majoris
M0Iab 370 pc

η Canis Majoris
B5Ia 980 pc

ε Canis Majoris
B2II 130 pc

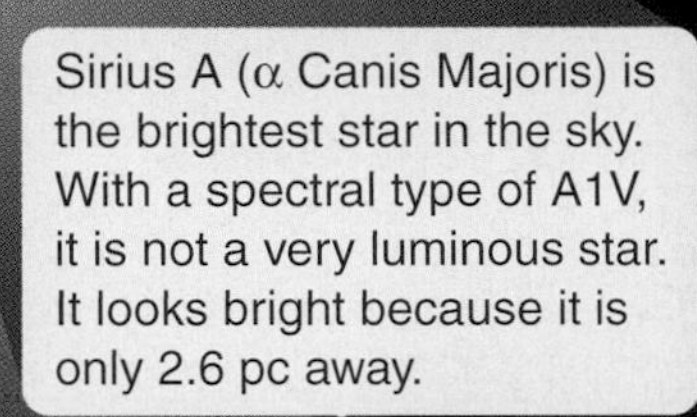

Sirius A (α Canis Majoris) is the brightest star in the sky. With a spectral type of A1V, it is not a very luminous star. It looks bright because it is only 2.6 pc away.

Sirius B is a white dwarf that orbits Sirius A. Although Sirius B is not very far away, it is much too faint to see with the unaided eye.

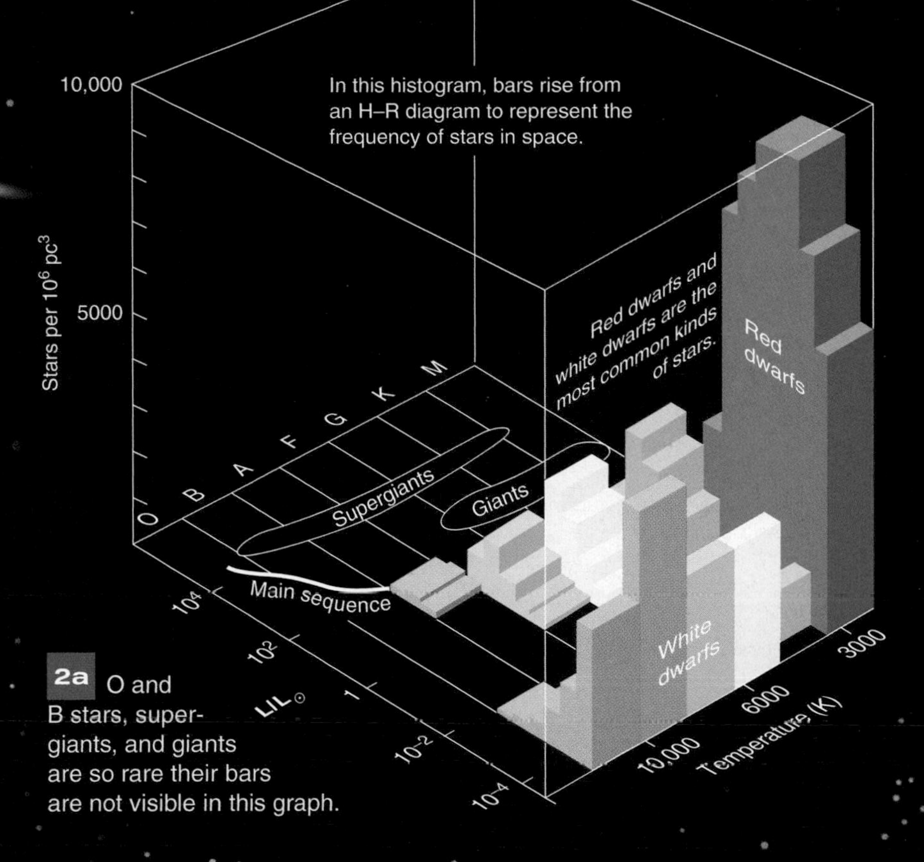

In this histogram, bars rise from an H–R diagram to represent the frequency of stars in space.
10,000
5000
Stars per 10^6 pc^3
O B A F G K M
Supergiants
Giants
Main sequence
Red dwarfs and white dwarfs are the most common kinds of stars.
Red dwarfs
White dwarfs
10^4
10^2
1
10^{-2}
10^{-4}
L/L$_\odot$
3000
6000
10,000
Temperature (K)

2a O and B stars, supergiants, and giants are so rare their bars are not visible in this graph.

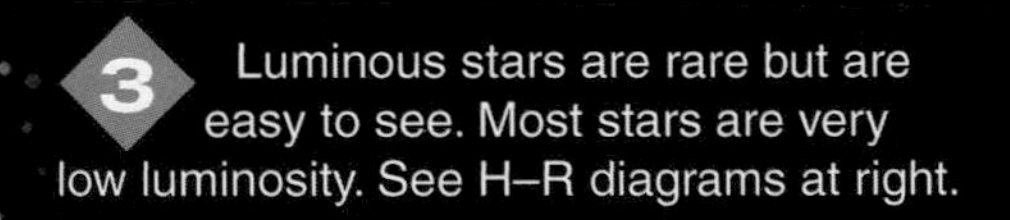

3 Luminous stars are rare but are easy to see. Most stars are very low luminosity. See H–R diagrams at right.

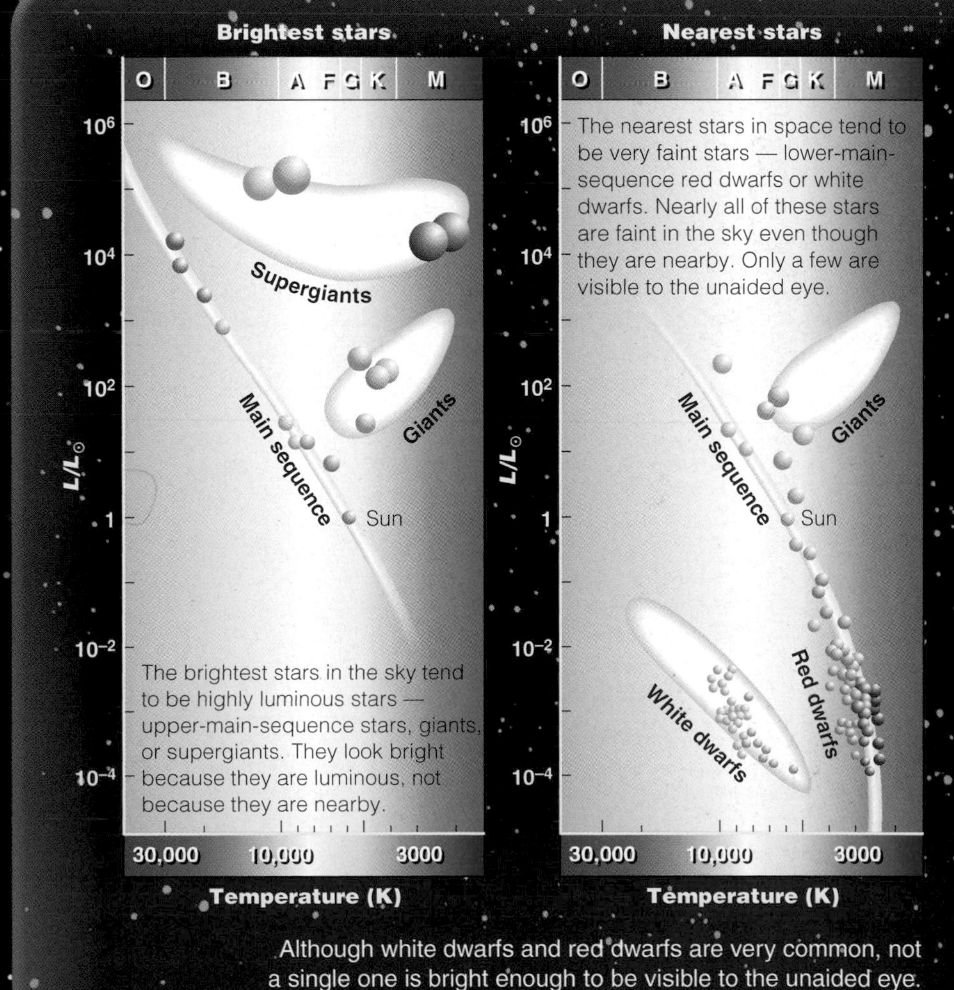

Brightest stars
O B A F G K M
10^6
10^4
10^2
1
10^{-2}
10^{-4}
L/L$_\odot$
Supergiants
Giants
Main sequence
Sun
The brightest stars in the sky tend to be highly luminous stars — upper-main-sequence stars, giants, or supergiants. They look bright because they are luminous, not because they are nearby.
30,000
10,000
3000
Temperature (K)
Nearest stars
O B A F G K M
The nearest stars in space tend to be very faint stars — lower-main-sequence red dwarfs or white dwarfs. Nearly all of these stars are faint in the sky even though they are nearby. Only a few are visible to the unaided eye.
Giants
Main sequence
Sun
Red dwarfs
White dwarfs
30,000
10,000
3000
Temperature (K)

Although white dwarfs and red dwarfs are very common, not a single one is bright enough to be visible to the unaided eye.

Study and Review Tools

Summary

9-1 | Measuring the Distances to Stars

How far away are the stars?

- Distance is critical in astronomy. Your goal in this chapter was to characterize the stars by finding their luminosities, diameters, and masses. Before you could begin, you needed to find their distances. Only by first knowing the distance to a star could you find its other properties.
- Astronomers can measure the distance to nearer stars by observing their parallaxes. The most distant stars are so far away that their parallaxes are unmeasurably small. To find the distances to these stars, astronomers must use spectroscopic parallax.
- Stellar distances are commonly expressed in parsecs. One parsec is 206,265 AU—the distance to an imaginary star whose parallax is 1 second of arc.

9-2 | Intrinsic Brightness

How much energy do stars make?

- Once you know the distance to a star, you can find its intrinsic brightness expressed as its absolute magnitude. The absolute magnitude of a star equals the apparent magnitude it would have if it were 10 pc away.
- The luminosity of a star, found from its absolute magnitude and its distance, is a measure of the total energy radiated by the star in one second. Luminosity is often expressed in terms of the luminosity of the sun.

9-3 | The Diameters of Stars

How big are stars?

- The H–R diagram is a plot of luminosity versus surface temperature. It is an important graph in astronomy because it sorts the stars into categories by size.
- Roughly 90 percent of normal stars, including the sun, fall on the main sequence, with the more massive stars being hotter and more luminous. The giants and supergiants, however, are much larger and lie above the main sequence. Some of the white dwarfs are very hot stars, but they fall below the main sequence because they are so small.
- The large size of the giants and supergiants means their atmospheres have low densities and their spectra have sharper spectral lines than the spectra of main-sequence stars. In fact, it is possible to assign stars to luminosity classes by the widths of their spectral lines. Class V stars are main-sequence stars. Giant stars, class III, have sharper lines; and supergiants, class I, have extremely sharp spectral lines.
- Spectroscopic parallax allows astronomers to estimate the distance to stars too far away for direct parallax measurements. By classifying a star according to spectral type and luminosity class, an astronomer can estimate its absolute magnitude from the H–R diagram and then find its distance by comparing absolute and apparent magnitudes.

9-4 | The Masses of Stars

How much matter do stars contain?

- The only direct way you can find the mass of a star is by studying binary stars. When two stars orbit a common center of mass, astronomers find their masses by observing the period and sizes of their orbits.
- Given the mass and diameter of a star, you can find its average density. On the main sequence, the stars are about as dense as the sun, but the giants and supergiants are very-low-density stars. Some are much thinner than air. The white dwarfs, lying below the main sequence, are tremendously dense.
- The mass–luminosity relation says that the more massive a star is, the more luminous it is. Main-sequence stars follow this rule closely, the most massive being the upper-main-sequence stars and the least massive the lower-main-sequence stars. Giants and supergiants do not follow the relation precisely, and white dwarfs not at all.

9-5 | A Survey of the Stars

What is the typical star like?

- A survey in the neighborhood of the sun shows that lower-main-sequence stars are the most common type. Giants and supergiants are rare, but white dwarfs are quite common, although they are faint and hard to find.

New Terms

stellar parallax (*p*) (p. 189)
parsec (pc) (p. 189)
proper motion (p. 190)
flux (p. 191)
absolute visual magnitude (M_V) (p. 192)
magnitude–distance formula (p. 192)
distance modulus ($m_V - M_V$) (p. 192)
luminosity (*L*) (p. 193)
absolute bolometric magnitude (p. 193)
H–R (Hertzsprung–Russell) diagram (p. 194)
main sequence (p. 196)
giant (p. 196)
supergiant (p. 197)
red dwarf (p. 197)
white dwarf (p. 197)
luminosity class (p. 198)
spectroscopic parallax (p. 199)
binary stars (p. 200)
visual binary system (p. 202)
spectroscopic binary system (p. 202)
eclipsing binary system (p. 204)
light curve (p. 204)
mass–luminosity relation (p. 207)

Review Questions

Ace Astronomy™ Assess your understanding of this chapter's topics with additional quizzing and animations at **http://ace.brookscole.com/sf9**

1. Why are parallax measurements made from Earth-based telescopes limited to the nearest stars?
2. How would having an observatory on Mars help astronomers measure parallax more accurately?
3. How did having an observatory in orbit above Earth's atmosphere help astronomers measure parallax more accurately?
4. What do the words *absolute* and *visual* mean in the term *absolute visual magnitude?*
5. For which stars does absolute visual magnitude differ least from absolute bolometric magnitude? Why?
6. What does luminosity measure that absolute visual magnitude does not?
7. How can a cool star be more luminous than a hot star? Give some examples.
8. How can you be certain that the giant stars are actually larger than the sun?
9. How can astronomers be sure that white dwarfs really are small?
10. What observations would you make to classify a star according to its luminosity class?

11. Describe the steps in using the method of spectroscopic parallax. Do you really measure a parallax?
12. Give the approximate radii and intrinsic brightnesses of stars in the following classes: G2 V, G2 III, G2 Ia.
13. Why does the orbital period of a binary star depend on the mass of the stars?
14. Why can astronomers find the masses of the stars in visual binary systems and in eclipsing binary systems but not in spectroscopic binary systems?
15. How do the ordered masses along the main sequence illustrate the mass–luminosity relation?
16. Why is it difficult to find out how common the most luminous stars are? the least luminous stars?
17. If all of the stars in the photo here are members of the same star cluster, then they all have about the same distance. Then why are three of the brightest much redder than the rest? What kind of star are they?

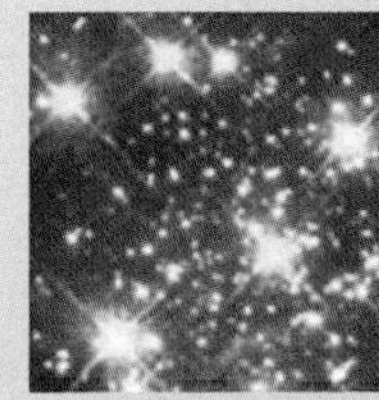
(NASA)

Discussion Questions

1. If someone asked you to compile a list of the stars nearest the sun based on your own observations, what kind of observations would you make, and how would you analyze them to detect nearby stars?
2. The sun is sometimes described as an average star. What is the average star really like?

Problems

1. If a star has a parallax of 0.050 second of arc, what is its distance in parsecs? in light-years? in AU?
2. If a star has a parallax of 0.016 second of arc and an apparent magnitude of 6, how far away is it, and what is its absolute magnitude?
3. Complete the following table:

m_V	M_V	d (pc)	P (seconds of arc)
___	7	10	___
11	___	1000	___
___	−2	___	0.025
4	___	___	0.040

4. If a main-sequence star has a luminosity of 400 $L_\odot$, what is its spectral type? (*Hint:* See Figure 9-13.)
5. If a star has an apparent magnitude equal to its absolute magnitude, how far away is it in parsecs? in light-years?
6. If a star has an absolute bolometric magnitude that is eight magnitudes brighter than the sun, what is the star's luminosity?
7. If a star has an absolute bolometric magnitude that is one magnitude fainter than the sun, what is the star's luminosity?
8. An O8 V star has an apparent magnitude of +1. Use the method of spectroscopic parallax to find the distance to the star.
9. Find the luminosity of the sun, given the radius of Earth's orbit and the solar constant (Chapter 8). Make your calculation in two steps. First, use $4\pi R^2$ to calculate the surface area (in square meters) of a sphere surrounding the sun with a radius of 1 AU. Second, multiply by the solar constant to find the total solar energy passing through the sphere in 1 second. That is the luminosity of the sun. Compare your result with that in Celestial Profile 1.
10. In the following table, which star is brightest in apparent magnitude? most luminous in absolute magnitude? largest? farthest away?

Star	Spectral Type	m_V
a	G2 V	5
b	B1 V	8
c	G2 Ib	10
d	M5 III	19
e	White dwarf	15

11. What is the total mass of a visual binary system if its average separation is 8 AU and its period is 20 years?
12. Measure the wavelengths of the iron lines in Figure 9-17 and plot a graph showing the radial velocities of the two stars in this system. What is the period? Assuming that the orbit is circular and edge-on, what is the total mass? What are the individual masses?
13. If the eclipsing binary in Figure 9-20 has a period of 32 days, an orbital velocity of 153 km/s, and an orbit that is nearly edge-on, what is the circumference of the orbit? the radius of the orbit? the mass of the system?
14. If the orbital velocity of the eclipsing binary in Figure 9-20 is 153 km/s and the smaller star becomes completely eclipsed in 2.5 hours, what is its diameter?
15. What is the luminosity of a 4-solar-mass main-sequence star? of a 9-solar-mass main-sequence star? of a 7-solar-mass main-sequence star?

Media Cluster

ACTIVE FIGURES

Ace Astronomy™ To access the resources in the Media Cluster, log into AceAstronomy at **http://ace.brookscole.com/sf9** and select Chapter 9.

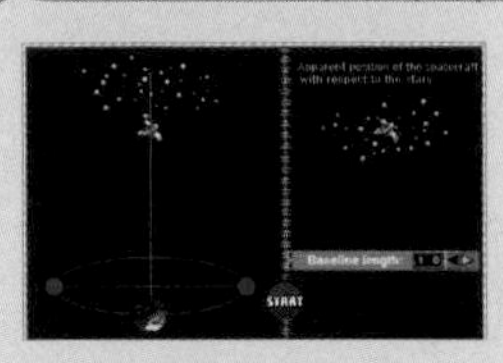

Brightness and Distance
In this animation, you can change the position of both the truck and the bicyclist in the figure and observe the effect on the intensity of light seen by the observer.

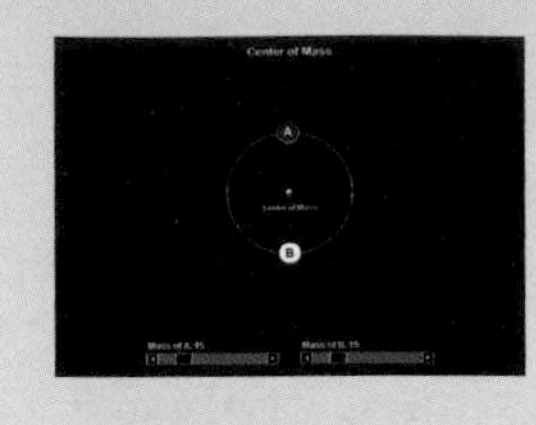

Center of Mass
You often think in terms of the moon orbiting Earth, but, more precisely, both bodies orbit each other around a common center of mass. Experiment with this animation to see how the ratio of the masses affects the relative size of the orbits.

ASTRONOMY EXERCISES

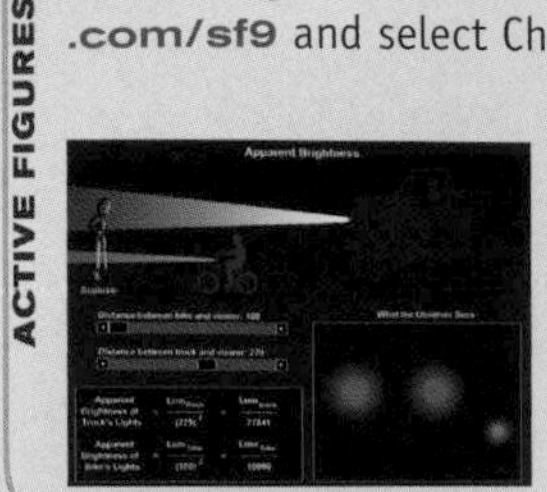

Parallax I and II
The apparent change in position of an object due to a change in the location of the observer is known as parallax. Use these two animations to study the concept of parallax in astronomy.

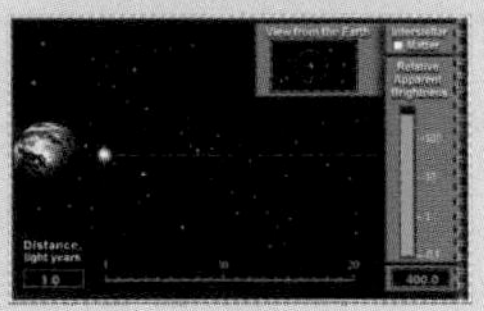

Apparent Brightness and Distance
The distance to stars affects how bright they look to you on Earth. You can change the distance to a star in this animation and see its apparent brightness change.

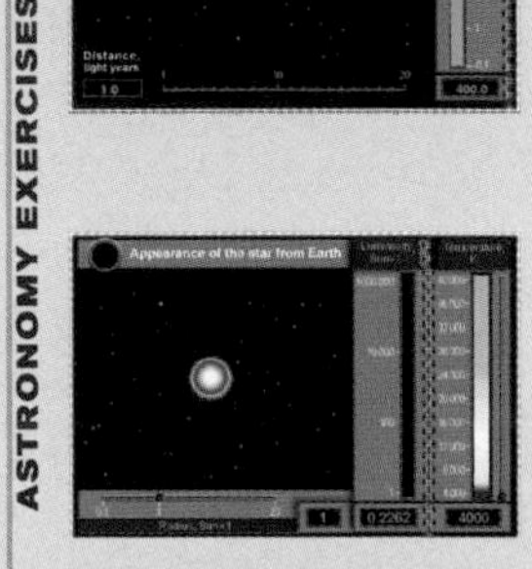

Stefan–Boltzmann Law II
In this exercise, you can vary the temperature of a star and its radius. As you change the temperature and its radius, notice what happens to the total energy output of the star as indicated by the luminosity meter.

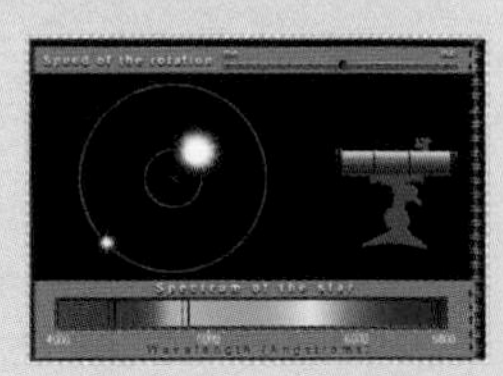

Spectroscopic Binaries
This animation shows a close double star system where each star orbits a common center of gravity. As the stars change position, notice what happens to the absorption lines in the spectrum of the star system.

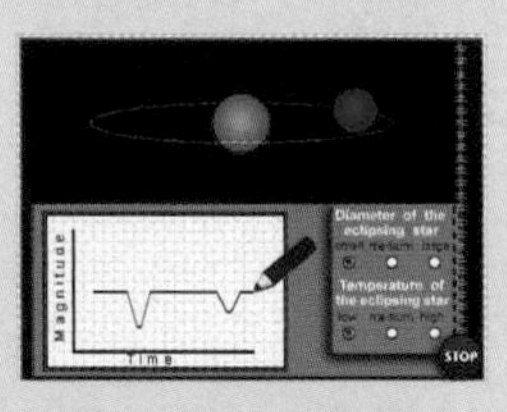

Eclipsing Binaries
Study an eclipsing binary system with this animation. As the stars move and periodically pass in front of one another, the total light output from the system is charted on a graph.

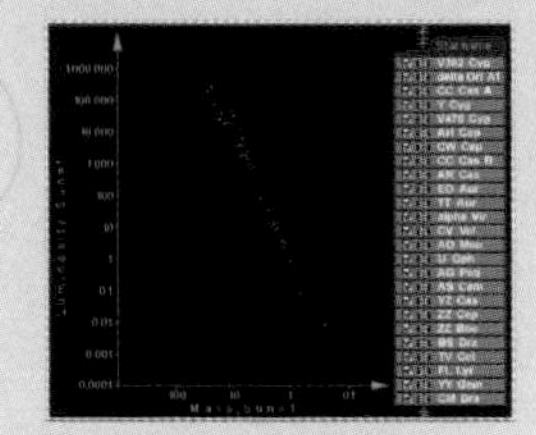

Mass–Luminosity Relation
This exercise provides a list of stars whose masses and luminosities have been measured. Click on each star to plot its point on the mass versus luminosity axes.

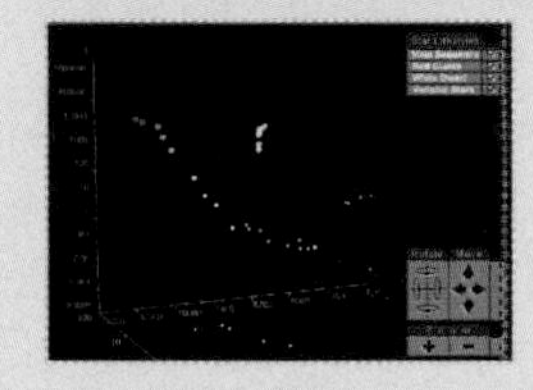

H–R/Mass–Luminosity 3-D Graph
Explore this 3-D graph and see how mass is related to position in the H–R diagram. You can select from main-sequence stars, red giants, white dwarfs, and variable stars.

VIRTUAL ASTRONOMY LABS

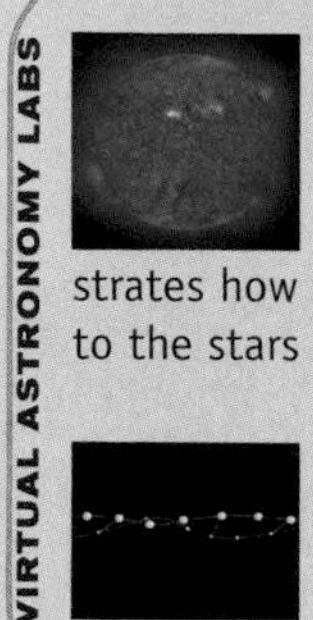

Lab 11: The Spectral Sequence and the H–R Diagram
This lab introduces the tools astronomers use to identify the stars of the main sequence and demonstrates how that information is used to estimate the distance to the stars and the age of a star cluster.

Lab 12: Binary Stars
This lab investigates some things that can be learned from different types of binary stars. Newton's form of Kepler's third law is used to determine stellar masses. You also see how the Doppler effect can be used in the study of binary systems.

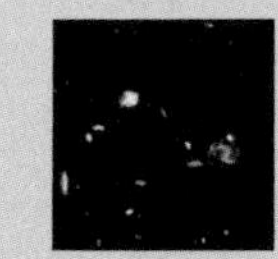

Lab 16: Astronomical Distance Scales
In this lab, you explore some methods for determining distances in astronomy.

Critical Inquiries for the Web

1. The Hertzsprung–Russell diagram was named for two famous astronomers. Who were they? What did they do to earn such an honor?
2. Algol is a famous binary star. What can you find out about the mythology of Algol?
3. An entire class of binary stars is called the Algol binary stars. How would you characterize such stars?

Exploring *TheSky*

1. Locate the following stars and determine their apparent magnitude, parallax, distance in parsecs and in light-years, and spectral classification. (*Hint:* To center on an object, use **Find** under the **Edit** menu and type the object's name followed by a period.)

   Sirius, Vega, Antares, Altair, Aldebaran, Deneb, Betelgeuse

2. Use the spectral type and parallax of the stars above to estimate their distance. Compare with the distances given in *TheSky*.
3. Take a survey of the stars. Center on Orion and adjust the field until it is about 100° wide. Click on the 10 brightest stars and record their spectral types. Now zoom in until only a few dozen stars are in the frame. Click on the 10 faintest stars and record their spectral types. Is there a difference between the brightest and faintest stars? Is the result what you expected? Are certain kinds of stars missing from this computer database?
4. Repeat Activity 3 for a region centered on the Big Dipper.

- Go to the Brooks/Cole Astronomy Resource Center **(http://astronomy.brookscole.com)** for critical thinking exercises, articles, and additional readings from InfoTrac College Edition, Brooks/Cole's online student library.

10 The Interstellar Medium

when he shall die,
Take him and cut him out
in little stars,
And he will make
the face of heaven so fine
That all the world
will be in love with night,
And pay no worship
to the garish sun.

SHAKESPEARE,
ROMEO AND JULIET III, ii, 21

Visual-wavelength image

JULIET LOVED ROMEO so much she compared him to the beauty of the stars (see the quotation above). Had she known what was between the stars, she might have compared him to that instead. True, the gas and dust between the stars are mostly dark and cold, but where it is illuminated by stars, that matter creates beautiful nebulae, and where it is densest it gives birth to beautiful stars. If there is beauty in vast extent and sweeping power, then the **interstellar medium,** the gas and dust between the stars, could steal worship from the garish stars. One reason you need to know about the interstellar medium is that it gives birth to stars and is enriched by the deaths of stars. The interstellar medium is an important part of the story of stars. Another reason is that it illustrates once again how astronomers use the interaction of light

Continued on page 218

From a glance at the night sky it may seem that the spaces between the stars are empty, but powerful cameras reveal great swirls of gas and dust.
(NASA, ESA, and The Hubble Heritage Team, STScI/AURA)

Guidepost

Looking Back

It is a thrill to photograph a faint galaxy a billion light-years from Earth; part of the fascination of astronomy is seeing the unseen. In previous chapters, you have studied the sun and stars in general, objects that are not too hard to detect. Now you are ready to study something that is nearly invisible.

This Chapter

Space is not empty. Great clouds of gas and dust drift between the stars, and this chapter will show you how astronomers use spectrographic analysis (Chapter 7) to understand that thinly scattered matter. In this chapter, you will answer three essential questions:

How do astronomers know that space isn't empty?

How do astronomers study the nearly invisible matter between the stars?

How does the matter between the stars interact with the stars?

Studying the matter between the stars is not an idle exercise. The gas and dust in space give birth to stars and are enriched by the deaths of stars.

Looking Ahead

As you begin studying the matter between the stars, you are beginning the story of stars. In the next few chapters you will see how stars are born, how they live, how they die, and what remains when they die. Are you surprised that stars die? That's just one of nature's little secrets.

AceAstronomy™ The AceAstronomy icon throughout the text indicates an opportunity for you to test yourself on key concepts and to explore animations and interactions on the AceAstronomy website at: http://ace.brookscole.com/sf9

and matter to understand the universe. But another reason to study the interstellar medium is that it is a beautiful part of the universe (■ Figure 10-1). Most people would use the phrase "the vacuum of space"; you are about to discover that space is not empty.

But be warned: In Chapter 1, you saw that we live in a disk-shaped galaxy, the Milky Way (Figure 1-11). The interstellar medium is confined mostly within a few hundred parsecs of the plane of the disk of our galaxy; if you journeyed out of the galaxy, out between the galaxies, you would find space a much better vacuum. In your study of the interstellar medium, you are learning about the disk of our galaxy in the neighborhood of the sun—probably a typical part of interstellar space.

Later you will need observations at radio, infrared, ultraviolet, and X-ray wavelengths to study the interstellar medium, but you can begin with one of your most sensitive instruments—eyeballs.

■ **Figure 10-1**

A swirl of dusty gas forms the Horse Head Nebula. The background nebula glows with the pink color of ionized hydrogen, and a bright blue star at lower left is just able to shine through the gas and dust. (Daniel Good)

10-1 Visible-Wavelength Observations

A QUICK GLANCE at the night sky does not reveal any matter between the stars, but evidence of its existence is visible to the naked eye in a familiar constellation.

Nebulae

On a cold, clear winter night, Orion hangs high in the southern sky, a large constellation composed of brilliant stars. If you look carefully at Orion's sword, you will see that one of the stars is a hazy cloud (Figure 2-4). A small telescope reveals even more such clouds of gas and dust. Astronomers refer to these clouds as **nebulae** (singular, nebula), from the Latin word for mist or cloud.

Study **Three Kinds of Nebulae** on pages 220–221 and notice three important points:

1. First, very hot stars can excite clouds of gas and dust to emit light, and that reveals that the clouds contain mostly hydrogen gas at very low densities.
2. Second, where slightly cooler stars illuminate clouds you see evidence that the dust in the clouds is made up of very small particles.
3. The third thing to notice is that some dense clouds of gas and dust are detectable only where they are silhouetted against background regions filled with stars or bright nebulae.

A detailed analysis of the spectra of nebulae can tell astronomers even more. Certain lines in the spectra of emission nebulae are called **forbidden lines** because they are almost never seen in excited gas on Earth. Two good examples are the strong green lines at 495.9 nm and 500.7 nm produced by oxygen atoms that have lost two electrons. (Following the convention for naming ions, twice-ionized oxygen is OIII.) The oxygen ions can become excited by collision with a high-energy photon or a rapidly moving ion or electron, and the atom can emit various-wavelength photons as its electron cascades back down to lower energy levels. However, transitions between certain energy levels are so unlikely they are called "forbidden." If an electron enters such an energy level, it will remain there for a relatively long time before it can decay further and emit the appropriate photon. For a normal transition, the electron might wait 10^{-8} to 10^{-7} second. If it enters one of these **metastable levels,** the electron may be stuck there for as long as an hour before it becomes likely to fall to a lower level and emit the proper photon.

Now you can understand why these forbidden lines aren't visible in laboratories on Earth. The atoms in a dense gas collide with each other so often that there isn't time for an electron in a metastable level to decay to a lower level and emit a photon. In a jar of gas in a laboratory, the gas is dense and the atoms collide so often that such electrons get excited back up to higher levels before they can drop downward and emit a photon.

And that tells you something important about emission nebulae. In those nebulae, the gas has a very low density, and an atom could go for an hour or more between collisions, giving an electron stuck in a metastable level time to decay to a lower level and emit a photon at a so-called forbidden wavelength. The forbidden lines are clear evidence that nebulae have very low densities. This is a dramatic example of how astronomers can use knowledge of atomic physics to understand astronomical objects.

The three kinds of nebulae are a good introduction to the interstellar medium, but there is more to learn. The next step is to think carefully about light passing through this gas and dust.

Extinction and Reddening

The dust in space is called **interstellar dust.** Its presence is dramatically evident in dark nebulae, but simple observations tell astronomers that the interstellar dust is spread throughout space, making up roughly 1 percent of the mass of the interstellar medium.

One way astronomers know that dust is present in the interstellar medium is that it makes distant stars appear fainter than they would if space were perfectly transparent. This phenomenon is called **interstellar extinction,** and in the neighborhood of the sun it amounts to about 2 magnitudes per thousand parsecs. That is, if a star is 1000 pc from Earth, it will look about 2 magnitudes fainter than it would if space were perfectly empty. If it were 2000 pc away, it would look about 4 magnitudes dimmer, and so on. This is a dramatic effect, and it shows that the interstellar medium is not confined to a few nebulae scattered here and there. The so-called vacuum of space is filled with a low-density, dusty gas. The spaces between the stars are far from empty.

Another way dust reveals its presence is through the effect it has on the colors of stars. An O star should be blue, but some stars with the spectrum of an O star look much redder than they should. Called **interstellar reddening,** this effect is produced by dust particles scattering light. As you saw in the case of the reflection nebulae, the dust particles are small, with diameters roughly equal to the wavelength of light, and they scatter shorter wavelengths better than longer wavelengths. That means they scatter blue photons more often than red photons. As light from a distant star travels toward Earth it loses some of its blue photons because of scattering, and consequently the star looks redder (■ Figure 10-2).

As was discussed earlier, this scattering of blue light is what makes the sky blue, but it is also what makes distant city lights look yellow. If you view the lights of a city at night from a high-flying aircraft or a distant mountaintop, the lights will look yellow. As you descend toward the city, the lights will look bluer.

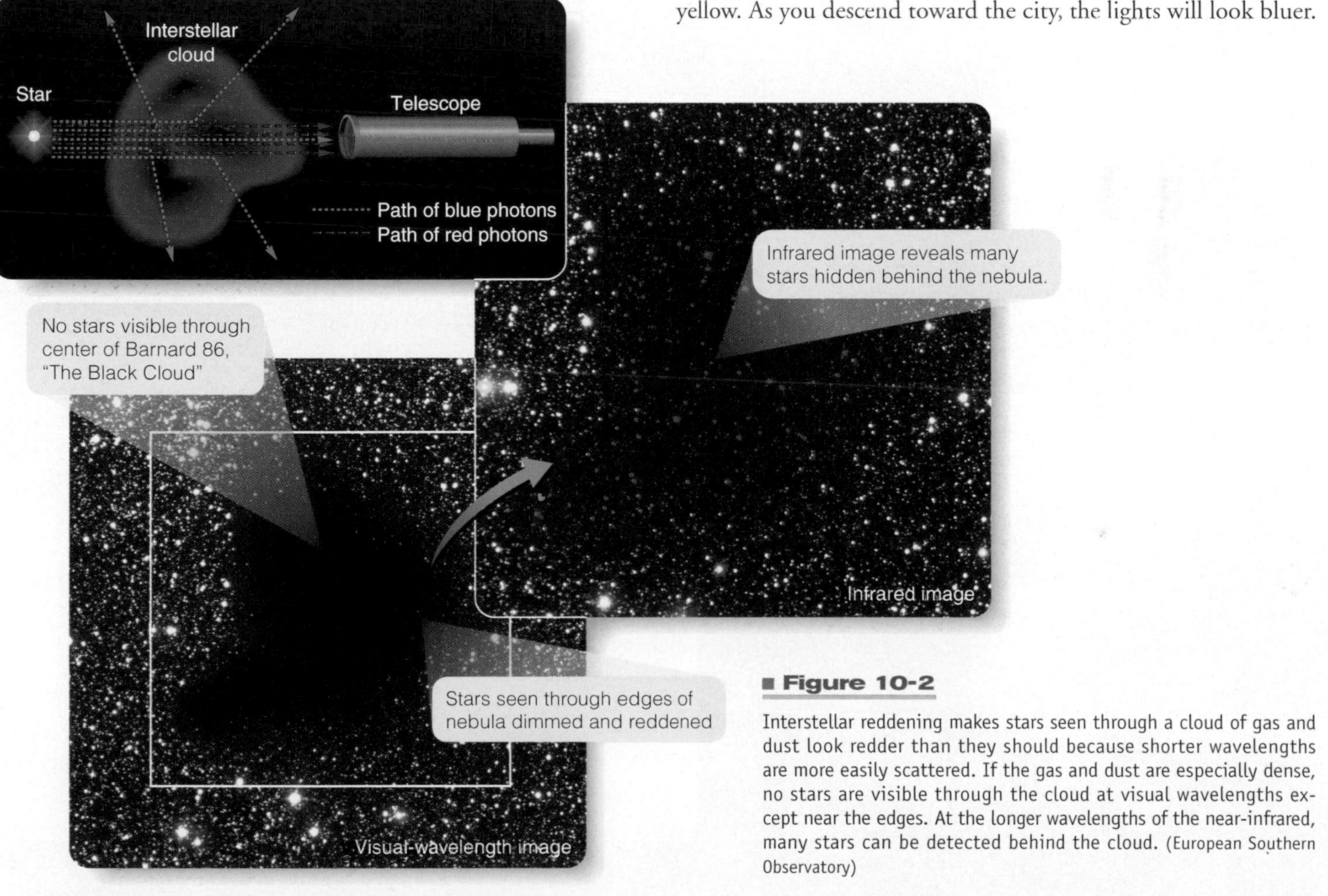

Interstellar cloud
Star
Telescope
Path of blue photons
Path of red photons
Infrared image reveals many stars hidden behind the nebula.
No stars visible through center of Barnard 86, "The Black Cloud"
Infrared image
Stars seen through edges of nebula dimmed and reddened
Visual-wavelength image

■ Figure 10-2

Interstellar reddening makes stars seen through a cloud of gas and dust look redder than they should because shorter wavelengths are more easily scattered. If the gas and dust are especially dense, no stars are visible through the cloud at visual wavelengths except near the edges. At the longer wavelengths of the near-infrared, many stars can be detected behind the cloud. (European Southern Observatory)

Three Kinds of Nebulae

1 **Emission nebulae** are produced when a hot star excites the gas near it to produce an emission spectrum. The star must be hotter than about B1 (25,000 K). Cooler stars do not emit enough ultraviolet radiation to ionize the gas. Emission nebulae have a distinctive pink color produced by the blending of the red, blue, and violet Balmer lines. Emission nebulae are also called **HII regions**, following the custom of naming gas with a roman numeral to show its state of ionization. HI is neutral hydrogen, and HII is ionized.

In an HII region, the ionized nuclei and free electrons are mixed. When a nucleus captures an electron, the electron falls down through the atomic energy levels, emitting photons at specific wavelengths. Spectra indicate that the nebulae have compositions much that like of the sun – mostly hydrogen. Emission nebulae have densities of 100 to 1000 atoms per cubic centimeter, better than the best vacuums produced in laboratories on Earth.

Visual-wavelength image

European Southern Observatory

Ace Astronomy™

Log into AceAstronomy and select this chapter to see Active Figure "Scattering." Watch photons scatter through Earth's atmosphere.

2 A **reflection nebula** is produced when starlight scatters from a dusty nebula. Consequently, the spectrum of a reflection nebula is just the reflected absorption spectrum of starlight. Gas is surely present in a reflection nebula, but it is not excited to emit photons. See image below.

2a Reflection nebulae look blue for the same reason the sky looks blue. Short wavelengths scatter more easily than long wavelengths. See image below.

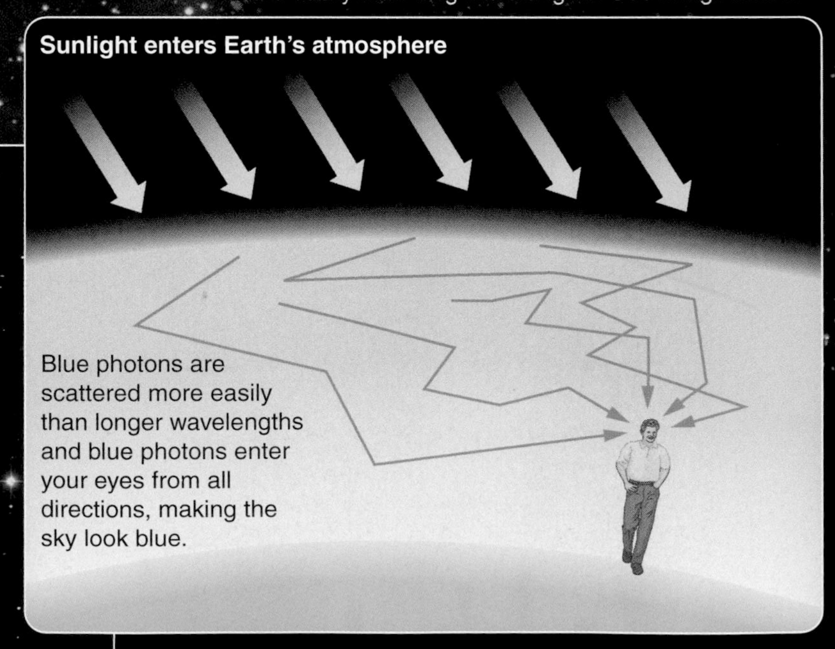

Sunlight enters Earth's atmosphere
Blue photons are scattered more easily than longer wavelengths and blue photons enter your eyes from all directions, making the sky look blue.

Reflection nebulae NGC 1973, 1975, and 1977 lie just north of the Orion nebula. The pink tints are produced by ionized gases deep in the nebulae.

Visual-wavelength image

AATB

2b The blue color of reflection nebulae at left shows that the dust particles must be very small in order to preferentially scatter the blue photons. Interstellar dust grains must have diameters ranging from 0.001 mm down to 1 nm or so.

The hottest star in the Pleiades star cluster is Merope, a B3 star. It is not hot enough to ionize the gas so you see a reflection nebula rather than an emission nebula.
Merope
Visual-wavelength image
Caltech
Reflection
Emission
Trifid Nebula
The Milky Way in Sagittarius contains two nebulae that dramatically demonstrate the difference between emission and reflection nebulae.
Emission
Merope
Visual
NASA
Visual
Lagoon Nebula
Daniel Good
Dark Nebula Barnard 86
Star Cluster NGC6520
Visual
AATB
Hubble Heritage Team, NASA
Visual
Northern Coalsack
Cygnus
Milky Way
Great Rift
Large dark nebulae obstruct the view of more distant stars and form holes and rifts along the Milky Way. The Great Rift extends from Cygnus to Sagittarius.
The Horsehead Nebula in Orion is a dark nebula silhouetted against a more distant emission nebula.

2c A dusty reflection nebula is located very close to the star Merope above. The glare from the star is caused by internal reflections in the telescope, but the wispy nature of the nebula is real. The intense light from the star is pushing the dust particles away and may destroy the little nebula over the next few thousand years.

3 **Dark nebulae** are dense clouds of gas and dust that obstruct the view of more distant stars. Some are generally round, but others are twisted and distorted, as shown at the left, suggesting that even when there are no nearby stars to ionize the gas or produce a reflection nebula, there are breezes and currents pushing through the interstellar medium.

The light from the city is reddened by microscopic particles in the air. If the particles are especially dense, people call them smog.

Astronomers can measure the amount of reddening by comparing two stars of the same spectral type, one of which is dimmed more than the other. The more obscured star will look redder. If you plot the difference in brightness between the two stars against wavelength, you get a curve that shows the reddening. That is, it shows how the starlight is dimmed at different wavelengths (■ Figure 10-3). In general, the light is dimmed in proportion to the reciprocal of the wavelength, a pattern typical of scattering from small dust particles. Laboratory measurements show that the high extinction at about 220 nm is caused by a form of carbon, evidence that some of the dust particles are carbon. Other evidence suggests that some grains contain silicates and metals and may have coatings of water ice, frozen ammonia, or carbon-based molecules.

Interstellar Absorption Lines

If you looked at the spectra of distant stars, you could see dramatic evidence of an interstellar medium. Of course, you would see spectral lines produced by the gas in the atmospheres of the stars, but you would also see sharp spectral lines produced by the gas in the interstellar medium. These **interstellar absorption lines** provide a new way to study the gas between the stars.

Astronomers can recognize interstellar absorption lines in three ways: by their ionization, by their widths, and by their multiple components. Some stellar spectra contain absorption lines that just don't belong because they represent the wrong ionization state. For example, if you looked at the spectrum of a very hot star such as an O star, you would expect to see no lines of once ionized calcium (CaII) because that ion cannot exist at the high temperatures in the atmosphere of such a hot star. But many O-star spectra contain lines of CaII. These lines must have been produced not in the star but in the interstellar medium.

In addition, the widths of the interstellar lines give away their identity. In the hot atmosphere of a star, the atoms move rapidly, and the random Doppler shifts broaden spectral lines. Also, the atoms collide with each other often enough to blur the electron energy levels, and that also broadens the spectral lines. These two effects were discussed as Doppler broadening and collisional broadening in Chapter 7. This blurring makes the lines in the spectrum of a main-sequence star quite broad, but even in the atmosphere of a giant or supergiant star, where the gas is less dense, the gas atoms move rapidly and collide often enough to broaden the spectral lines. Interstellar lines, on the other hand, are exceedingly sharp. This shows that the interstellar matter is extremely cold and has a low density. If it were hot, Doppler broadening would smear out the lines due to the motions of individual atoms. If the gas were dense, collisional broadening would produce wider lines. The exceedingly narrow widths of the interstellar absorption lines are typical of cold, low-density gas.

Another revealing characteristic of the interstellar lines is that they are often split into two or more components. The multiple components have slightly different wavelengths and appear to have been produced when the light from the star passed through different clouds of gas on its way to Earth. Because the clouds of gas have slightly different radial velocities, they produce absorption lines with slightly Doppler-shifted wavelengths. ■ Figure 10-4 illustrates

■ Figure 10-3

Interstellar extinction, the dimming of starlight by dust between the stars, depends strongly on wavelength. Infrared radiation is only slightly affected, but ultraviolet light is strongly scattered. The strong extinction at about 220 nm is caused by a certain form of carbon dust in the interstellar medium.

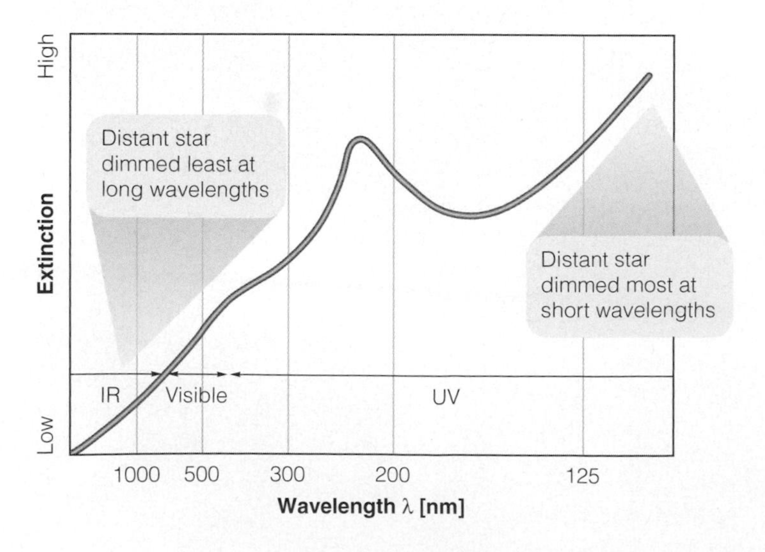

■ Figure 10-4

Interstellar absorption lines can be recognized in three ways. (a) The B0 supergiant ε Orionis is much too hot to show spectral lines of once-ionized calcium (CaII), yet this short segment of its spectrum reveals narrow, multiple lines of CaII (tick marks) that must have been produced in the interstellar medium. (The Observatories of the Carnegie Institution of Washington) (b) Spectral lines produced in the atmospheres of stars are much broader than the spectral lines produced in the interstellar medium. (Adapted from a diagram by Binnendijk) In both (a) and (b) the multiple interstellar lines are produced by separate interstellar clouds with slightly different radial velocities.

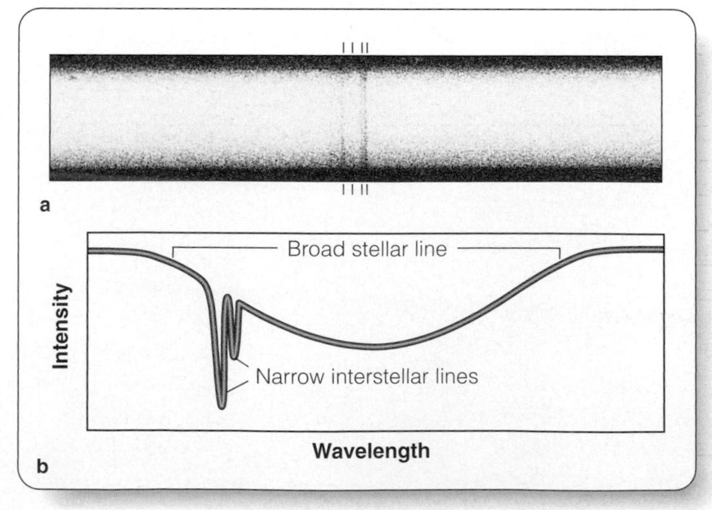

all three of these characteristics of interstellar absorption lines—the wrong ionization, narrow widths, and multiple components.

Clouds and What's in Between

Astronomers disagree as to the exact structure of the interstellar medium, and the boundaries and characteristics of clouds are ill defined. Nevertheless, you can get a general picture of the interstellar medium by classifying clouds into a few main varieties based on visual-wavelength observations.

Studies of interstellar absorption lines reveal clouds of neutral gas (and presumably dust) with densities of 10 to a few hundred atoms/cm^3. Because these clouds are not ionized, they are called **HI clouds.** These clouds must be 50 to 150 pc in diameter and have masses of a few solar masses. The gas temperature is only about 100 K. Starlight seems to pass through six to ten of these clouds for every 1000 pc near the plane of our galaxy. While it is easy to think of these clouds as more or less spherical blobs of gas, observations show that they are usually twisted into long filaments, flattened into thin sheets, or tangled into chaotic shapes (■ Figure 10-5)—further evidence that the interstellar medium is not static and motionless.

■ **Figure 10-5**

Like a flag in a strong wind, this dark nebula is being distorted and twisted by intense radiation from very luminous stars out of the picture to the left. (Anglo-Australian Observatory)

Between these HI clouds of neutral gas lies a hot **intercloud medium** with a temperature of a few thousand K and a density of only about 0.1 atom/cm^3. This intercloud medium consists of ionized hydrogen (HII). The gas is partially ionized by the ultraviolet photons in starlight. If it is to be in equilibrium with the HI clouds (if the HI clouds are not expanding or contracting), the pressure in both regions should be about the same.

The pressure in a gas depends on its density and its temperature (**Focus on Fundamentals 5**). In the interstellar medium, the HI clouds are cool but dense, and the HII gas of the intercloud medium is hot but low in density. That means the two regions could have similar pressures even though their temperatures are quite different. Where the pressures are equal, the HI clouds and the intercloud medium are stable—a bit like ice cubes floating in ice water. Nevertheless, observations reveal some regions where the pressures are not equal, and that is poorly understood.

By now you are probably wondering how the intercloud medium could be ionized when it is not close to hot stars. To understand why it is ionized, imagine that you are an atom floating in the intercloud medium. Ultraviolet photons from distant stars are not common, but they do whiz by now and then. Soon you become ionized by absorbing one of these photons and losing an electron. In a denser gas, you would quickly find another electron, capture it, and become a neutral atom again, but the interstellar medium has such a low density that the gas particles are spread far apart. You must wait a very long time before an electron comes close enough for you to capture. That is why the gas of the intercloud medium is ionized; the atoms spend almost all their time waiting for an electron to happen past.

Studies of interstellar absorption lines can tell you more about the composition of the interstellar gas. It is much like the composition of the sun. Hydrogen is most abundant, with helium second. Light elements such as carbon, nitrogen, and oxygen are as common as in the sun, but elements such as iron, calcium, and titanium are less abundant than they are in the sun. Most likely, these elements are missing from the gas because they have condensed to form the dust.

Building Scientific Arguments

What evidence can you cite that there is an interstellar medium?

Everything in science is based on evidence, so your argument should discuss observations. First, there is the simple fact that certain parts of the interstellar medium are visible—the nebulae. Emission, reflection, and dark nebulae show that interstellar space is not totally empty. Further, astronomers can detect interstellar extinction and reddening; the more distant stars look fainter than they should, and they also look redder. This not only shows that all of interstellar space is filled by some thin material, it also shows that some of the material is in the form of tiny dust specks. Gas atoms would not be very effective at reddening starlight, so the interstellar medium must contain dust as well as gas.

Focus on Fundamentals 5

Pressure

One of the most fundamental parameters in science is **pressure,** a measure of force per unit area. If you are going to understand astronomy, you must be sure you understand pressure. Doctors measure blood pressure, and astronomers measure gas pressure, but they are really measuring the same fundamental parameter.

Pressure is expressed in the units of force per unit area. If you inflate the tires on a car, you use the unit pounds per square inch. A typical pressure might be 34 lb/in^2. It is important to note that this is not the total force pushing out on the inside of the tire but only the force exerted on a single square inch. When you stand, your weight exerts a force on the floor, and the pressure under your shoes is your weight divided by the surface area of your shoes' soles. A typical pressure might be only 4 lb/in^2. If you step on someone's toe, that is the pressure you exert. Of course, if you were wearing ice skates, your weight would be spread over a much smaller area, the area of the bottom of the blade, and you might exert a pressure of 150 lb/in^2 or more. That's why skaters must be careful not to step on someone's toe. The pressure would be dangerously high.

Pressure pushes outward on the inside of a balloon.

Astronomers are most commonly interested in the behavior of matter when it is a gas, and pressure in a gas arises when atoms or molecules collide. Consider, for example, how the gas molecules colliding with the inside of a balloon exert an outward force on the rubber and keep the balloon inflated. If the gas is hot, the atoms or molecules move rapidly, and the resulting pressure is higher than for a cooler gas. If the gas is dense, there will be many gas particles colliding with the inside of the balloon, and the pressure will be higher than for a lower-density gas. In this way, pressure depends on both the temperature and the density of the gas.

Notice that pressure and density are related, but they are not at all the same thing. Density is a measure of the amount of matter in a given volume, and pressure is a measure of the force that matter exerts on its surroundings. A very-low-density gas and a very-high-density gas might have the same pressure if they had different temperatures.

In daily life, you think of pressure when you inflate an automobile tire, but pressure is common in nearly all of the sciences. Astronomers must consider pressure in thinking about the gas inside stars and the thin gas between the stars.

PRESSURE | MASS | ENERGY | TEMPERATURE AND HEAT | DENSITY

Perhaps this is enough evidence to convince someone that the spaces between the stars are not empty, but there is more. Expand your argument. **What do interstellar absorption lines reveal about the interstellar gas?**

■ ■ ■

Connections: Observations at visible wavelengths reveal important information about the interstellar medium, but to paint a complete picture of the matter between the stars, astronomers must observe at other wavelengths.

10-2 Long- and Short-Wavelength Observations

THE INTERSTELLAR MEDIUM is mostly invisible to human eyes, so astronomers must observe at the longer wavelengths of infrared and radio radiation and at the shorter wavelengths of ultraviolet and X-ray radiation to further explore the interstellar medium.

21-cm Observations

Cold, neutral hydrogen floating in space can emit electromagnetic radiation with a wavelength of 21 cm. This **21-cm radiation** allows radio astronomers to map the distribution of neutral hydrogen (HI) throughout our galaxy.

The existence of the 21-cm radiation was predicted theoretically in the 1940s by H. C. van de Hulst, but it was not detected until 1951. You can understand how a theoretical prediction could be made by thinking of the structure of a hydrogen atom. A hydrogen atom consists of a proton and an electron, and physicists know that both of these particles must spin like tiny tops. They can never stop. Because these particles have an electrostatic charge, their rapid spin is the same as the circulation of an electric current through a coil of wire; it creates a magnetic field. Because the charge on the proton is positive and the charge on the electron is negative, the magnetic fields are reversed when the particles spin in the same direction.

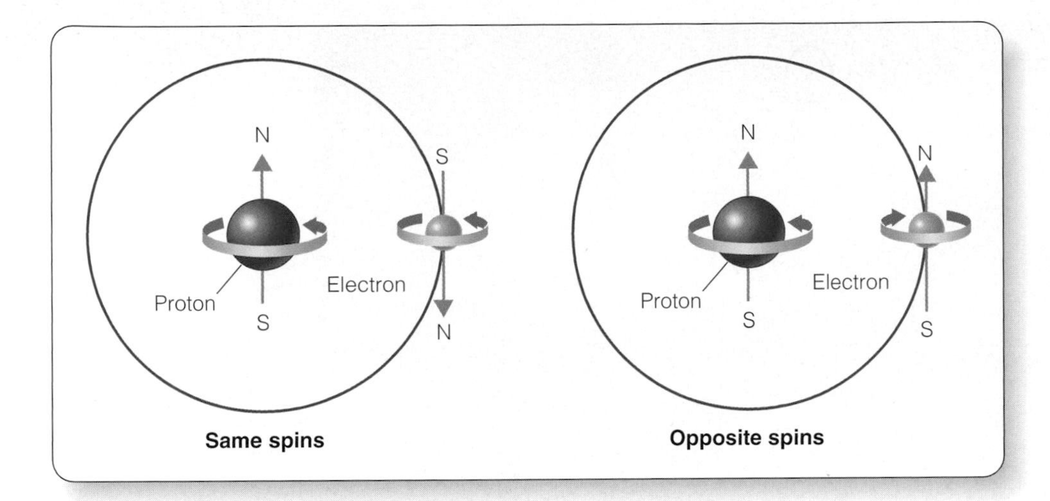

■ **Figure 10-6**

Both the proton and electron in a neutral hydrogen atom spin and consequently have small magnetic fields. When they spin in the same direction, their magnetic fields are reversed, and when they spin in opposite directions, their magnetic fields are aligned. As explained in the text, this allows cold, neutral hydrogen in space to emit radio photons with a wavelength of 21 cm.

The magnetic fields of the spinning proton and electron affect the binding energy that holds the electron to the proton in a hydrogen atom. You have probably played with small magnets and noticed that they repel each other in one orientation and attract each other if you turn one around. In the same way, the small magnetic fields produced by the spinning proton and electron can repel or attract each other. In one orientation the electron is slightly less tightly bound, and in the other orientation it is slightly more tightly bound (■ Figure 10-6). The energy difference is very small, but it makes a big difference in astronomy.

Now you can follow van de Hulst's logic and predict the existence of 21-cm radiation. Because an electron in the ground state of a hydrogen atom could spin in either of two directions—the same way as the proton or the opposite way—the ground state must really be two energy levels separated by the tiny amount of energy produced by the spinning particles. If an electron is spinning such that it is in the higher of the two states, it can spontaneously flip over and spin in the other direction. Then the atom must drop to the lower of the two energy levels, and the excess energy is radiated away as a photon. The energy difference is small, so the photon has a long wavelength—21 cm (■ Figure 10-7).

The 21-cm radiation was predicted theoretically, but it could not be detected in laboratory experiments. Of the two closely spaced energy levels, the upper one is metastable. Once an electron gets caught in the upper energy level, it will, on average, stay there for 11 million years before spontaneously dropping to the lower energy level and emitting a photon of 21-cm wavelength. The atoms in a gas in a laboratory jar will collide with each other millions of times a second, so none of those atoms can remain undisturbed with its electron in the upper energy level long enough to produce a photon of 21-cm radiation. Atoms in space, however, collide much less often, so a few do manage to emit 21-cm radiation. That's why the existence of 21-cm radiation had to be confirmed observationally by radio astronomers who detected it coming from clouds of neutral hydrogen in space.

The 21-cm observations give astronomers a way to map the cold, neutral hydrogen that fills much of our galaxy. Astronomers can locate individual clouds and measure their motion by observing the Doppler shifts in the 21-cm radiation (■ Figure 10-8).

The 21-cm radiation can map only neutral hydrogen. Ionized hydrogen lacks an electron, so it can't emit 21-cm radiation. Further, hydrogen atoms locked in molecules are also unable to

■ **Figure 10-7**

The lowest energy level of the hydrogen atom, the ground state, is actually two closely spaced energy levels that differ because the proton and electron spin. In this diagram, you would need a magnifying glass to distinguish the energy levels that make up the ground state. When an atom decays from the upper energy level to the lower energy level, it emits a photon with a wavelength of 21 cm.

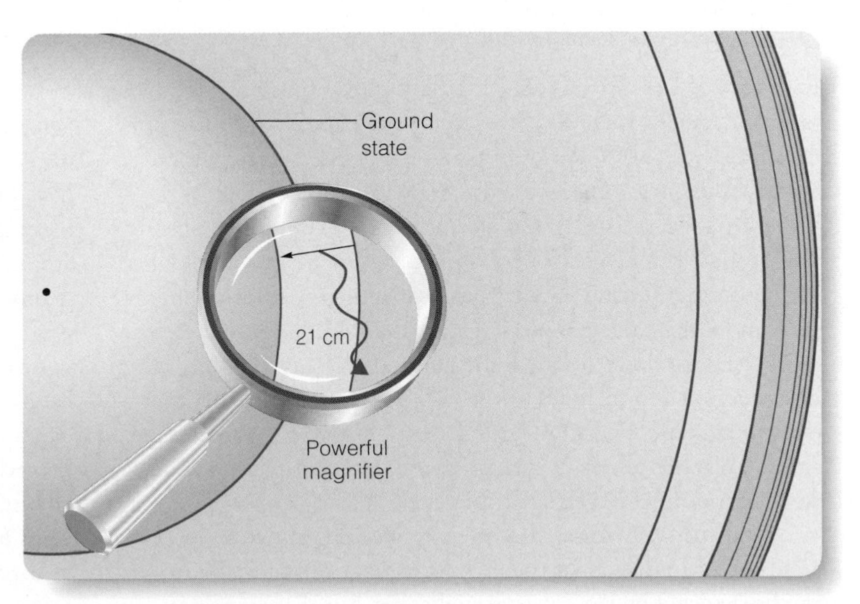

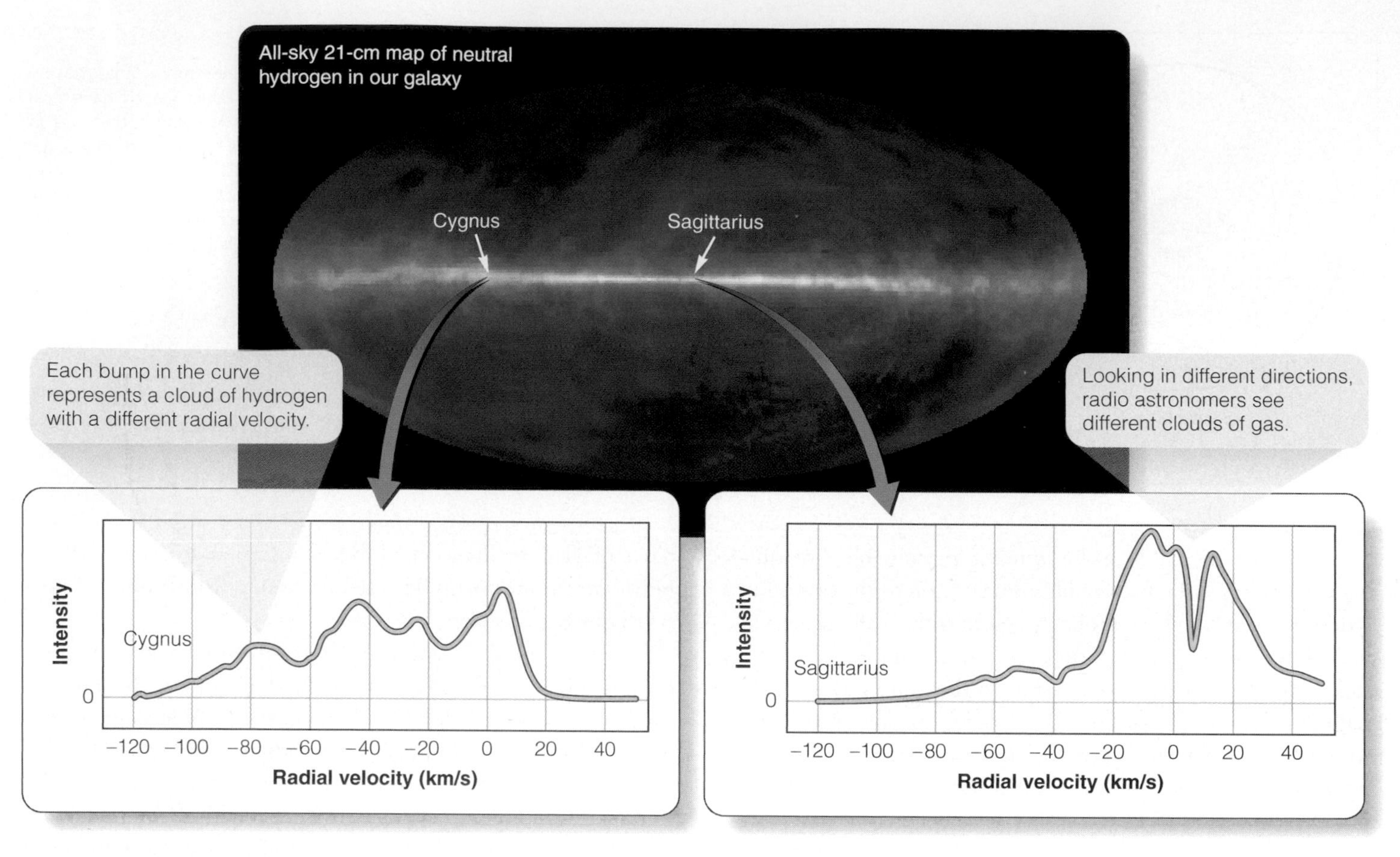

■ Figure 10-8

Most of the neutral hydrogen in the Milky Way Galaxy is in the plane of the disk-shaped galaxy—the bright band running from left to right. Notice how wispy and irregular the hydrogen is. (J. Dickey, UMn, F. Lockman, NRAO, SkyView) 21-cm radio observations reveal many clouds of neutral hydrogen orbiting the center of the galaxy. (Adapted from observations by Burton)

emit 21-cm wavelength photons. Astronomers must find other ways to study these parts of the interstellar medium.

Molecules in Space

Radio telescopes can also detect radiation from various molecules in the interstellar medium. A molecule can store energy in a number of different ways. For example, it can rotate at different rates, or the atoms in a molecule can vibrate as if they were linked together by small springs. If a molecule suffers a collision or absorbs a photon, it can be excited to vibrate and rotate in some higher energy state. Quickly, however, it will return to a lower energy state and radiate the excess energy as a photon. Because these energy levels are closely spaced, the emitted photons typically have low energies, and they fall in the radio or far-infrared part of the electromagnetic spectrum. Just as neutral hydrogen radiates at a specific wavelength of 21 cm, many natural molecules radiate at their own unique wavelengths.

Unfortunately for astronomers, molecules of hydrogen (H_2) are difficult to detect. In the far-ultraviolet, molecular hydrogen can be detected by the photons it absorbs, but these observations must be made by specialized telescopes in space. Mapping the location of molecular hydrogen would be easier if the molecules emitted radio-wavelength photons, but a molecule containing two identical atoms does not radiate in the radio part of the spectrum. However, clouds of gas dense enough to form molecular hydrogen also form tiny amounts of other molecules, and many of them are good emitters of radio energy. Nearly 100 different molecules have been detected (■ Table 10-1). Some are quite complex, and it is not clear how they form. Most astronomers believe that the atoms meet and bond to form molecules on the surfaces of dust grains. Some of these molecules have not yet been synthesized on Earth, but others are common, such as N_2O (nitrous oxide), also known as laughing gas. Ethyl alcohol, which some humans drink, has also been detected. Although this is a very rare molecule compared to molecular hydrogen, a single interstellar cloud can contain ethyl alcohol in amounts equivalent to 10^{28} fifths of whisky (about 100 Earth masses).

One of the most important of the interstellar molecules may be carbon monoxide (CO). It is one of the poisonous gases that come out of the tailpipes of cars, but it is also a very good emitter

■ **Table 10-1 | Selected Molecules Detected in the Interstellar Medium**

H_2	molecular hydrogen
H_2S	hydrogen sulfide
C_2	diatomic carbon
N_2O	nitrous oxide
CN	cyanogen
H_2CO	formaldehyde
CO	carbon monoxide
C_2H_2	acetylene
NO	nitric oxide
NH_3	ammonia
OH	hydroxyl
HCO_2H	formic acid
NaCl	common table salt
CH_4	methane
HCN	hydrogen cyanide
CH_3OH	methyl alcohol
H_2O	water
CH_3CH_2OH	ethyl alcohol

of radio energy at a wavelength of 2.6 mm. An interstellar cloud may contain only 1 CO molecule for every 10,000 molecules of hydrogen, but the CO can be detected while the molecular hydrogen cannot. Consequently, radio astronomers can map the interstellar clouds by searching for the CO radio emission; where they detect CO, they can be certain that molecular hydrogen is common.

Another important constituent of the interstellar medium is OH. Because this combination of an oxygen atom and a hydrogen atom is not electrically neutral, it is called a radical instead of a molecule. The OH radical forms in interstellar clouds and is a good emitter of radio energy.

These molecules are very fragile, and a high-energy photon such as an ultraviolet photon has enough energy to break the molecule into separate atoms. Consequently, the molecules cannot exist outside the dense clouds. Only deep inside the densest clouds, where dust absorbs and scatters the short-wavelength, high-energy photons, can the molecules survive. The very fact that these molecules are detected tells you that some of these **molecular clouds** must be very dense.

The molecules, especially CO, are such good radiators of energy that they cool the clouds to low temperatures. Thermal energy is present in the cloud as motion among the atoms and molecules. When a CO molecule collides with an atom or another molecule, some of the energy of motion can be stored in the rotation and vibration of the CO molecule. When the molecule emits that energy as a radio or infrared photon, the energy escapes from the cloud. In this way, molecular radiation can cool the interior of the cloud and keep it very cold.

The largest of these cool, dense clouds are called **giant molecular clouds.** They are 15 to 60 pc across and may contain 100 to 1,000,000 solar masses. The internal temperature is a frigid 10 K. Although these clouds are detected by their carbon monoxide molecular emission, remember that the gas is mostly hydrogen.

Giant molecular clouds are the nests of star birth. Deep inside these great clouds, gravity can pull the matter inward and create new stars. That is a story you will follow in detail in the next chapter. For now, you must meet the dirty part of the interstellar medium—the dust. To do this, you must trade your radio telescope for an infrared telescope.

Infrared Radiation from Dust

The dust in the interstellar medium makes up only about 1 percent of the mass, and at a temperature of 100 K (−143°C) or less it is very cold. Nevertheless, it is easy to detect at infrared wavelengths. To see how such cold dust can radiate a lot of energy, consider a simple experiment with the dust in an imaginary giant molecular cloud.

For the sake of quick calculation, assume that a giant molecular cloud has a mass of about 10^5 solar masses. Only 1 percent of that mass is dust, so all of the dust in the cloud amounts to a mass of 10^3 solar masses. Imagine that you could collect all of the dust into a single sphere. It would be only about 10 times the diameter of the sun, and its surface area would be about 100 times that of the sun. But suppose you left the dust as separate specks, each 0.0005 mm in diameter, a typical size. Then the cloud would contain about 10^{43} dust specks, and the total surface area of the dust would be about 10^{29} times that of the sun. When matter is finely divided into dust, it has a very large surface area, and even though it is much colder than the sun, its vast surface area can radiate tremendous amounts of infrared radiation.

In 1983, the Infrared Astronomy Satellite mapped the sky at far-infrared wavelengths and found the galaxy filled with infrared radiation from dust. Most of this dust is confined to the region near the plane of our galaxy, and, as you would expect, the dust is thickest where the gas is thickest. Much of the dust is distributed in wispy clouds that became known as the **infrared cirrus** because of their overall resemblance to cirrus clouds in Earth's atmosphere. The infrared cirrus consists of dusty clouds of interstellar matter with temperatures of about 30 K (■ Figure 10-9a). Studies of CO emission from molecular clouds show that at least some of the infrared cirrus is associated with molecular clouds within a few hundred parsecs of the sun. This suggests that the gas and dust in the molecular clouds is not uniform but rather very patchy.

As mentioned before, the interstellar dust appears to be composed of carbon, silicates, and iron, elements that are underabundant in the gas. Some grains appear to contain water ice contaminated with ammonia, methane, and other compounds.

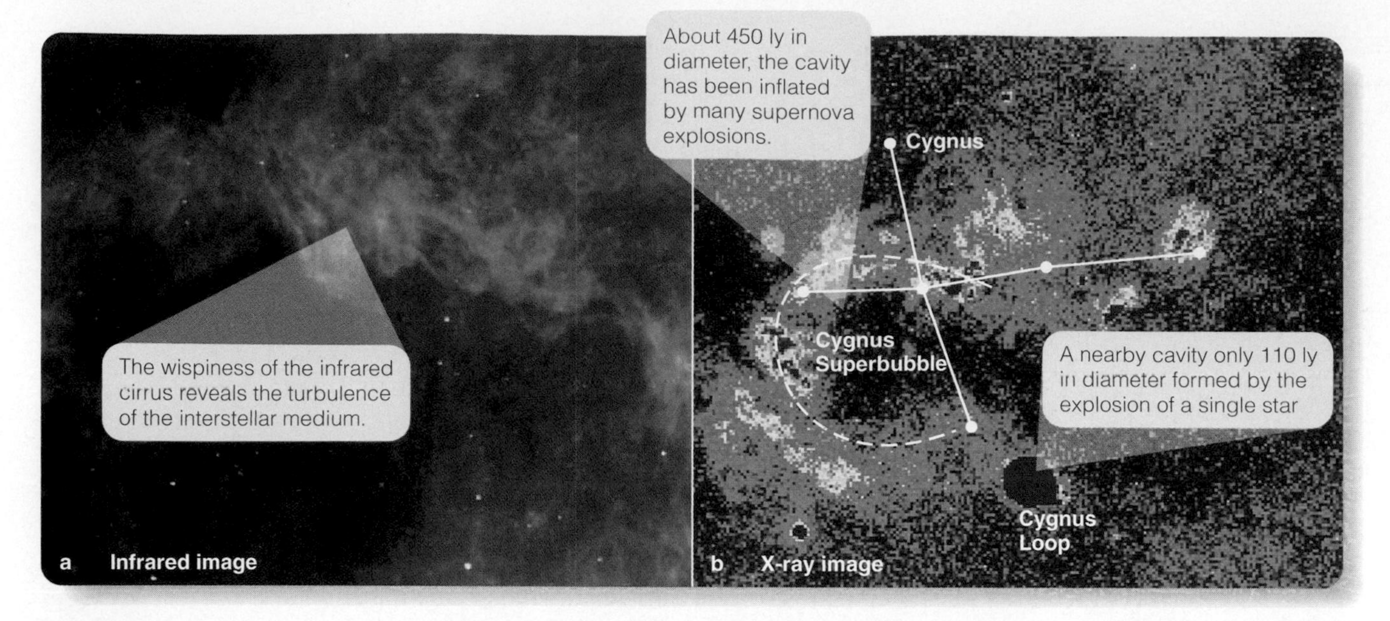
About 450 ly in diameter, the cavity has been inflated by many supernova explosions.
Cygnus
The wispiness of the infrared cirrus reveals the turbulence of the interstellar medium.
Cygnus Superbubble
A nearby cavity only 110 ly in diameter formed by the explosion of a single star
Cygnus Loop
a Infrared image
b X-ray image

Figure 10-9

Infrared and X-ray observations reveal different components of the interstellar medium. (a) This far-infrared image shows the infrared cirrus, wispy clouds of very cold dusty gas spread all across the sky. (NASA/IPAC, Courtesy Deborah Levine) (b) This ROSAT X-ray image shows the Cygnus Superbubble, a cavity filled with gas from exploding stars. The gas is heated and emits X rays where the expanding bubble pushes into the surrounding gas. (Courtesy Steve Snowden and Max-Planck Institute for Extraterrestrial Physics, Germany)

Infrared observations can help you understand the interstellar medium, but X-ray and ultraviolet observations complete the picture by revealing a part of the gas between the stars that is otherwise invisible.

X Rays from the Interstellar Medium

The interstellar medium seems very cold, so you would hardly expect to detect X rays. These high-energy photons are commonly produced by high temperatures. Nevertheless, X-ray telescopes can help you understand the interstellar medium.

X-ray telescopes above Earth's atmosphere have detected X rays from a part of the interstellar medium with a very high temperature. This gas has been called the **coronal gas** because it has temperatures of 10^6 K or higher, as does the sun's corona. Of course, the coronal gas in the interstellar medium is not related in any way to the actual corona of the sun.

You might expect the coronal gas to have a very high pressure because it is so hot, but it has a very low density. It contains only 0.0004 to 0.003 particles/cm^3. That is, you would have to search through a few hundred to a few thousand cubic centimeters of coronal gas to find a single particle—an ionized atom or an electron. Because of its very low density, the gas pressure in the coronal gas can be roughly the same as in the HI clouds and the intercloud medium. Nevertheless, exceptions are known. The cloud of interstellar matter near the sun has a pressure that is more than 20 times less than the pressure of neighboring coronal gas. Such differences in pressure are not well understood.

The coronal gas appears to originate when a massive star explodes violently in what astronomers call a supernova. Although rare, such explosions blast large amounts of very hot gas outward in expanding shells. Gas flowing away from very hot, young stars must add to the coronal gas. In some cases, neighboring shells of coronal gas may expand into each other and merge to form larger volumes, but an earlier theory that coronal gas fills a large network of tunnels and shells throughout interstellar space seems to be contradicted by evidence. Probably about 20 percent of the space between the stars is filled with isolated pockets of coronal gas.

The Cygnus superbubble appears to be related to the coronal gas. Located in Cygnus, it is a very large shell of hot gas about 450 pc in diameter (Figure 10-9b). The energy needed to create such a shell is equivalent to hundreds of supernova explosions. The bubble may have developed as a large cluster of stars was born and grew old and the most massive stars died in supernova explosions. Several of these large bubbles are known.

Ultraviolet Observations of the Interstellar Medium

You can divide the ultraviolet spectrum into the near-ultraviolet, with wavelengths only slightly shorter than visible light, and the far-ultraviolet, with much shorter wavelengths. Only a few decades ago, astronomers believed that far-ultraviolet photons could not travel far through the interstellar medium because they would be absorbed so easily by neutral hydrogen atoms. These atoms of neutral hydrogen are good absorbers of far-ultraviolet photons because the photons have enough energy to ionize the atoms. In fact, even one atom of HI per cubic centimeter would make the interstellar medium opaque to far-ultraviolet photons, and astronomers would be unable to see more than a light-year into space with a far-ultraviolet telescope.

When the Extreme Ultraviolet Explorer (EUVE) satellite was put into Earth orbit in 1992, it discovered that the interstellar medium was only partly cloudy. While some regions were filled with clouds of neutral hydrogen and so were opaque, other regions were filled with hot, ionized hydrogen that was transparent to far-ultraviolet photons. This confirms the description of the interstellar medium produced by X-ray observations.

The far-ultraviolet observations reveal that the sun is located just inside a large region of hot, ionized hydrogen, while only a few light-years from the sun lies a cool, neutral hydrogen cloud that is opaque to far-ultraviolet photons. This **local bubble or void** of gas in which the sun is located appears to be linked to other hot, transparent regions. In this way, ultraviolet observations, combined with X-ray observations, suggest that the apparently cold and empty regions between the stars are filled with a complex, evolving mixture of hot and cold gas.

Building Scientific Arguments

If hydrogen is the most common molecule, why do astronomers depend on the CO molecule to map molecular clouds?

This scientific argument links a bit of atomic physics with the chain of inference that leads from what can be observed to what astronomers need to know. Although hydrogen is the most common atom in the universe and molecular hydrogen the most common molecule, a molecule of hydrogen does not radiate in the radio part of the spectrum. Consequently, radio astronomers cannot detect it. But the much less common CO (carbon monoxide) molecule is a very efficient radiator of radio energy, so radio astronomers use it as a tracer of molecular clouds. Wherever radio telescopes reveal a great cloud of CO, you can be confident that most of the gas is molecular hydrogen.

You can also be confident that the molecular clouds contain dust, because it is the dust that protects the molecules in the cloud from the ultraviolet radiation that would otherwise break the molecules into atoms.

Dust doesn't radiate longer-wavelength radio energy, so you must study it in the infrared. Although the dust in a molecular cloud is very cold and makes up a small percentage of the total mass, it is a very good radiator of infrared radiation. Create a new argument using the physics of infrared emission from dust to answer a different question: **How can a small amount of cold dust radiate vast amounts of infrared radiation?**

■ ■ ■

Connections: From one end of the spectrum to the other, you have used every wavelength to study the interstellar medium. Now you can put your data in order and create a model of the gas and dust between the stars. That is the first step toward understanding how stars are born.

10-3 A Model of the Interstellar Medium

WHEN YOU LOOK at bright nebulae like the Great Nebula in Orion, you see a part of the interstellar medium, but what you see is only a very special region that happens to be close enough to hot, bright stars to become ionized and glow. Most of the interstellar medium is invisible to your eyes, so you must gather lots of evidence (**Window on Science 10-1**) and use it to develop a model of the interstellar medium.

For your model, you can divide the interstellar medium into four basic components (■ Table 10-2) and describe these components to explain how they interact and evolve. You will discover that their evolution is intimately connected to the process of star formation and star death.

Four Components of the Interstellar Medium

The interstellar medium is not at all uniform. Rather, it is lumpy, and the lumps differ dramatically in temperature and density.

HI clouds are cool, with temperatures of 50 to 150 K and densities of ten to a few hundreds of atoms per cubic centimeter. These clouds are only a few parsecs in diameter and contain a few solar masses.

Between the cool HI clouds lies the warm intercloud medium of HII, with temperatures of a few thousand Kelvin and densities of 0.01 atoms/cm^3. The intercloud medium is in approximate equilibrium with the HI clouds; they both have about the same pressure.

■ Table 10-2 I Four Components of the Interstellar Medium

Component	Temperature (K)	Density (atoms/cm^3)	Gas
HI clouds	50–150	1–1000	Neutral hydrogen; other atoms ionized
Intercloud medium	10^3–10^4	0.01	Partially ionized
Coronal gas	10^5–10^6	10^{-4}–10^{-3}	Highly ionized
Molecular clouds	20–50	10^3–10^5	Molecules

Window on Science 10-1

Understanding Science: Separating Facts from Theories

The fundamental work of science is testing theories by comparing them with facts. As you think about science, you need to distinguish clearly between facts and theories. The facts are the evidence against which scientists test theories.

Scientific facts are those observations or experimental results of which scientists are confident. An astronomer makes observations of stars, and a botanist collects samples of related plants. A fact could be a precise measurement, such as the mass of a star expressed as a specific number, or it could be a simple observation, such as that a certain butterfly no longer visits a certain mountain valley. In each case, the scientist is gathering facts.

A theory, however, is a conjecture as to how nature works. If you are uncertain of the theory, you might call it a hypothesis. In any case, these conjectures are not facts; they are attempts to explain how nature works. In a sense, a theory or hypothesis is a story that scientists have made up to explain how nature works in some specific case. These stories can be wonderfully detailed and ingenious, but without evidence they are nothing more than hunches.

In science, evidence is made up of facts, which could range from precise numerical measurements to the observation of the shape of a flower. (M. Seeds)

When one of these stories is tested against the facts and confirmed, scientists have more confidence that the story is more or less right. The more a story is tested successfully, the more confidence scientists have in it. The facts represent reality, and every theory or hypothesis must be repeatedly tested against reality.

You can't test one theory against another theory. Theories are not evidence; they are conjectures. If you were allowed to test one theory against another theory, you might fall into the trap of circular reasoning. "Elves make the flowers bloom. I know that elves exist because the flowers bloom." That is using two theories to confirm each other, and it leads to nonsense. Only facts can be evidence.

Nor can you use a theory to deduce facts. You can use a theory to make predictions, which you can then test against facts, but the predictions themselves can never be certainties, so they can't be facts. The only way to arrive at facts is to consult nature and make direct measurements or observations.

As you study different problems in astronomy, you must carefully distinguish between facts and theories. Facts are the basic building blocks of all science, and they can come only from the careful study of nature. As Galileo would say, you must read the book of nature.

Molecular clouds are especially dense. Molecules cannot survive if they are exposed to ultraviolet photons in starlight, so they can form only in the densest clouds, where dust absorbs and scatters ultraviolet photons. The molecular clouds can be very large, with diameters up to 60 pc and masses up to a million solar masses, but they are also very cold. Strong evidence suggests that stars are born when giant molecular clouds contract under the influence of their own gravity, a subject explored in detail in the next chapter.

Pushing through this stew of interstellar clouds are regions of coronal gas. Most of this very hot gas is probably produced in supernova explosions, although some may be gas flowing away from very hot stars. With temperatures up to a million degrees and densities as low as 10^{-4} atoms/cm^3, the coronal gas occupies a significant part of the interstellar medium.

Astronomers believe that the HI clouds make up about 25 percent of the interstellar mass, and the intercloud medium about 50 percent. The coronal gas contributes only 5 percent of the mass, although it seems to occupy about one-fifth of the volume. The giant molecular clouds amount to about 25 percent of the mass. If you add up these percentages, you get slightly more than 100 percent, which illustrates the uncertainty inherent in this model of the interstellar medium.

As the stars move through the interstellar medium, they meet no resistance, but they do have a dramatic effect on the gas and dust. The Pleiades, for example, is a relatively young star cluster that is moving rapidly through space. As it moves through the interstellar medium, it is leaving behind a trail like that left by a boat in water (■ Figure 10-10). This wake has been detected in the infrared and is apparently produced by the ultraviolet radiation from the stars in the cluster. None of the stars is hot enough to ionize the gas, but there are a number of relatively hot stars; the ultraviolet radiation from those stars heats, but does not quite ionize, the gas. The heated gas then expands and forms a wake showing the path of the cluster through the interstellar medium.

The Pleiades clearly illustrates the close relationship between the stars and the interstellar medium. In fact, you are now ready to outline a cycle that links the stars to the gas and dust between them.

The Interstellar Cycle

The story of the interstellar medium is closely linked to star formation. You will see in the next chapter that stars form in giant molecular clouds. As soon as a group of stars forms, the hottest stars begin ionizing the gas to produce emission nebulae. The pressure of the starlight and the gas flowing away from the hot,

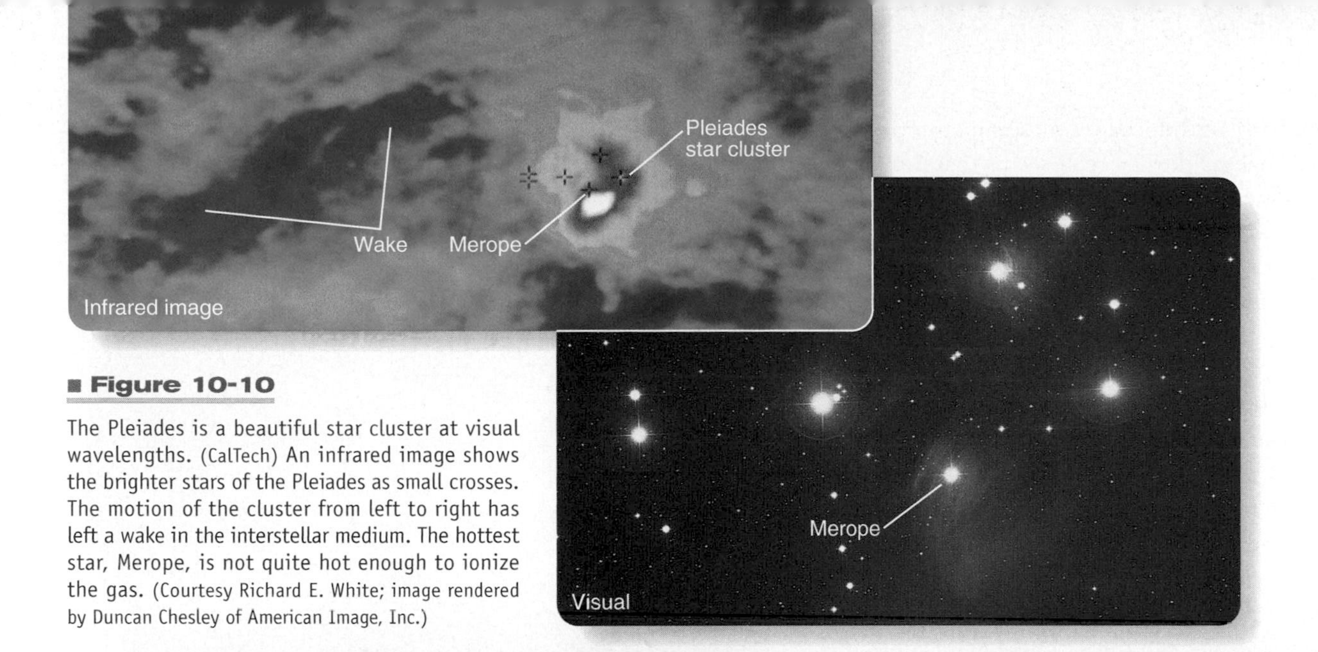

■ Figure 10-10

The Pleiades is a beautiful star cluster at visual wavelengths. (CalTech) An infrared image shows the brighter stars of the Pleiades as small crosses. The motion of the cluster from left to right has left a wake in the interstellar medium. The hottest star, Merope, is not quite hot enough to ionize the gas. (Courtesy Richard E. White; image rendered by Duncan Chesley of American Image, Inc.)

■ Figure 10-11

(a) The nebula N44C is part of a larger region where young, hot, massive stars have formed and where at least one supernova has blown a bubble filled with hot gas. (The Hubble Heritage Team, STScI/AURA/NASA) (b) This image of the entire N44 region combines a visual-wavelength image (white) with an X-ray image (red). X rays reveal regions of very-high-temperature gas that has been ejected from supernova explosions. (R. Chris Smith, CTIO and Y.-H. Chu, UICU)

young stars pushes the gas outward and may disrupt the cloud entirely.

The composition of the interstellar dust suggests that it is formed mostly in the atmospheres of cool stars. There the temperatures are low enough for some atoms to condense into specks of solid matter, much as soot can condense in a candle flame. The pressure of the starlight can push these dust specks out of the star and replenish the interstellar medium. Other stars that eject mass into space, such as supernovae, probably also add to the supply of interstellar dust.

The most massive stars die quickly in supernova explosions, and those tremendously violent events blast high-temperature gas outward that further disrupts the interstellar medium (■ Figure 10-11). Much of the motion in interstellar clouds and their twisted shapes are probably produced by these supernova explosions and the hot coronal gas they eject. It seems that our galaxy produces about two supernova explosions per century (although most are not visible from Earth because of obscuration by dust) and that these explosions keep the interstellar medium stirred and create the vast regions filled with coronal gas. In fact, the sun lies inside such a region of high-temperature, low-density gas called the local bubble or void. With a typical diameter of a few hundred parsecs, the local bubble may be a cavity in the interstellar medium inflated by a supernova explosion that occurred within the last million years or so.

While supernovae and hot stars keep the interstellar medium in motion, the natural gravitation of gas clouds and collisions between clouds may gradually build more massive clouds. In the most massive clouds, the dust protects the interior from ultraviolet photons, and molecules can form. These giant molecular clouds eventually give birth to new stars, and the cycle begins all over again.

The Trifid Nebula (■ Figure 10-12) is a dramatic illustration of this cycle. Measuring over 12 pc in diameter, the nebula is illuminated by a number of hot, young stars that have apparently formed recently from the gas. Near the stars, the gas is ionized and glows as a pink-red HII region; but farther from the stars, where the ultraviolet radiation is weaker, the gas is not ionized. Nevertheless, dust in the nebula scatters blue light, so this unionized part of the nebula is visible as a blue reflection nebula. Dark lanes of obscuring dust cross the face of the nebula as if to remind Earth's astronomers again of the importance of dust in the interstellar cycle.

■ Figure 10-12

The Trifid Nebula embodies an important part of the interstellar cycle. A dense cloud of gas has given birth to stars, and the hottest are ionizing the gas to produce the pink emission nebula, while dust scatters light to produce the blue reflection nebula. As the nebula is disrupted by the newborn stars, the gas and dust merge with the interstellar medium. (AURA/NOAO/NSF and Nigel Sharp)

Building Scientific Arguments

How can the coronal gas make up only a few percent of the mass but 20 percent of the volume of the interstellar medium?

Once again your argument needs to focus on some simple physics. The solution to this puzzle is the extremely low density of the coronal gas, roughly one particle for every thousand cubic centimeters. The coronal gas is the least dense part of the interstellar medium. It is a much better vacuum than any laboratory vacuum on Earth. It doesn't take much mass in coronal gas to occupy a large volume. In contrast, molecular clouds are the densest part, and they contain roughly a quarter of the mass, although they occupy only a small part of the volume.

The density and temperature of the different components of the interstellar medium determine how you could observe them. Create a new argument to consider observations. **What techniques would you use to observe the coronal gas and the molecular clouds?**

■ ■ ■

Connections: Your study of the interstellar medium is incomplete for two reasons. First, astronomers don't yet understand all of its components or how those components interact. Second, you can't fully understand the interstellar medium until you understand how stars are born and how they die. You begin that story in the next chapter.

Study and Review Tools

Summary

10-1 | Visible-Wavelength Observations

How do astronomers know that space isn't empty?

- The interstellar medium, the gas and dust between the stars, is mostly concentrated near the plane of our Milky Way Galaxy.
- Emission nebulae are clear evidence of an interstellar medium. The gas is ionized by ultraviolet radiation from nearby hot stars, and that makes the nebula glow like a giant neon sign. The red, blue, and violet Balmer lines blend together to produce the characteristic pink-red glow of ionized hydrogen.
- A reflection nebula is produced by gas and dust illuminated by a star that is not hot enough to ionize the gas. Rather, the dust scatters the starlight to produce a reflection of the stellar absorption spectrum. Because shorter-wavelength photons scatter more than longer-wavelength photons, reflection nebulae look blue. The daytime sky looks blue for the same reason.
- A dark nebula is a cloud of gas and dust that is visible because it blocks the light of distant stars. The irregular shapes of these dark nebulae reveal the turbulence in the interstellar medium.
- Further evidence of an interstellar medium is the extinction, or dimming, of the light of distant stars and interstellar reddening. Light from distant stars suffers scattering by dust particles in the interstellar medium, and blue light is scattered more than red light. This makes distant stars look redder than their spectral types suggest. The dependence of this extinction on wavelength shows that the scattering dust particles are very small. The dust is made of carbon, silicates, iron, and ice.
- Interstellar absorption lines in the spectra of distant stars are very narrow. The interstellar gas is cold and has a very low density, and this makes the interstellar lines much narrower than the spectral lines produced in stars. Multiple interstellar lines reveal that the light has passed through more than one interstellar cloud on its way to Earth.

10-2 | Long- and Short-Wavelength Observations

How do astronomers study the nearly invisible matter between the stars?

- Radio observations at a wavelength of 21 cm reveal multiple clouds of neutral hydrogen. These neutral clouds drift through a warmer but lower-density intercloud medium.
- Radio telescopes tuned to other wavelengths have detected nearly 100 different molecules in the interstellar medium, most of them found in giant molecular clouds.
- Infrared observations made from orbit have detected dust in the interstellar medium in the form of the infrared cirrus.
- X-ray and far-ultraviolet observations have detected very hot coronal gas produced by supernova explosions.

10-3 | A Model of the Interstellar Medium

How does the matter between the stars interact with the stars?

- The four main components of the interstellar medium are the small neutral HI clouds, the warm intercloud medium, coronal gas, and molecular clouds.
- Stars are born in the dense molecular clouds, and the energy from hot stars and supernova explosions causes currents in the interstellar medium and creates the coronal gas. The dust in the interstellar medium is formed in the atmospheres of cool stars and from gas ejected by supernova explosions.

New Terms

interstellar medium (p. 216)
nebula (p. 218)
forbidden line (p. 218)
metastable level (p. 218)
interstellar dust (p. 219)
interstellar extinction (p. 219)
interstellar reddening (p. 219)
emission nebula (p. 220)
HII region (p. 220)
reflection nebula (p. 220)
dark nebula (p. 221)
interstellar absorption lines (p. 222)
HI clouds (p. 223)
intercloud medium (p. 223)
pressure (p. 224)
21-cm radiation (p. 224)
molecular cloud (p. 227)
giant molecular cloud (p. 227)
infrared cirrus (p. 227)
coronal gas (p. 228)
local bubble or void (p. 229)

Review Questions

Ace Astronomy™ Assess your understanding of this chapter's topics with additional quizzing and animations at **http://ace.brookscole.com/sf9**

1. What evidence can you cite that the spaces between the stars are not totally empty?
2. What evidence can you cite that the interstellar medium contains both gas and dust?
3. How do the spectra of HII regions differ from the spectra of reflection nebulae? Why?
4. Why are interstellar lines so narrow? Why do some spectral lines forbidden in spectra on Earth appear in spectra of interstellar clouds and nebulae? What does that tell you?
5. How is the blue color of a reflection nebula related to the blue color of the daytime sky?
6. Why do distant stars look redder than their spectral types suggest?
7. If starlight on its way to Earth passed through a cloud of interstellar gas that was hot instead of very cold, would you expect the interstellar absorption lines to be broader or narrower than usual? Why?
8. How can the HI clouds and the intercloud medium have similar pressures when their temperatures are so different?
9. Why can the 21-cm radio emission line of neutral hydrogen be observed in the interstellar medium but not in the laboratory?
10. What does the shape of the 21-cm radio emission line of neutral hydrogen tell you about the interstellar medium?
11. What produces the coronal gas?
12. The image here shows two nebulae, one pink in the background and one black in the foreground. What kind of nebulae are these?

(NASA and the Hubble Heritage Team, STScI/AURA)

Discussion Questions

1. When you see distant streetlights through smog, they look dimmer and redder than they do normally. But when you see the same streetlights through fog or falling snow, they look dimmer but not redder. Use your knowledge of the interstellar medium to discuss the relative sizes of the particles in smog, fog, and snowstorms compared to the wavelength of light.
2. If you could see a few stars through a dark nebula, how would you expect their spectra and colors to differ from similar stars just in front of the dark nebula?

Problems

1. A small nebula has a diameter of 20 seconds of arc and a distance of 1000 pc from Earth. What is the diameter of the nebula in parsecs? in meters?
2. The dust in a molecular cloud has a temperature of about 50 K. At what wavelength does it emit the maximum energy? (*Hint:* Consider black body radiation, Chapter 7.)
3. Extinction dims starlight by about 1.9 magnitudes per 1000 pc. What fraction of photons survives a trip of 1000 pc? (*Hint:* Consider the definition of the magnitude scale in Chapter 2.)
4. If the total extinction through a dark nebula is 10 magnitudes, what fraction of photons makes it through the cloud? (*Hint:* See Problem 3.)
5. The density of air in a child's balloon 20 cm in diameter is roughly the same as the density of air at sea level, 10^{19} particles/cm^3. To how large a diameter would you have to expand the balloon to make the gas inside the same density as the interstellar medium, about 1 particle/cm^3? (*Hint:* The volume of a sphere is $\frac{4}{3}\pi R^3$.)
6. If a giant molecular cloud has a diameter of 30 pc and drifts relative to neighboring clouds at 20 km/s, how long will it take to travel its own diameter?
7. An HI cloud is 4 pc in diameter and has a density of 100 hydrogen atoms/cm^3. What is its total mass in kilograms? (*Hints:* The volume of a sphere is $\frac{4}{3}\pi R^3$, and the mass of a hydrogen atom is 1.67×10^{-27} kg.)
8. Find the mass in kilograms of a giant molecular cloud that is 30 pc in diameter and has a density of 300 hydrogen molecules/cm^3. (*Hint:* See Problem 7.)
9. At what wavelength does the coronal gas radiate most strongly? (*Hint:* Consider black body radiation, Chapter 7.)

Media Cluster

ACTIVE FIGURES

AceAstronomy™ To access the resources in the Media Cluster, log into AceAstronomy at **http://ace.brookscole.com/sf9** and select Chapter 10.

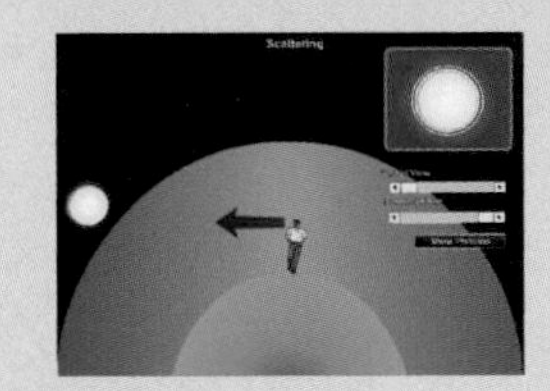

Scattering in Earth's Atmosphere
By absorption, reflection, and refraction, wavelengths of light are selected and filtered to produce the colorful world around you. In this animation, learn more about why you see blue skies and red sunsets and how the sun's position in the sky plays a part.

ASTRONOMY EXERCISES

Why Is the Sky Blue?
The sky is blue and the setting sun is red for the same reason that reflection nebulae are blue and distant stars are reddened. You can explore that process here in the topic begun in the Active Figure above.

Critical Inquiries for the Web

1. What causes the varied colors in images of gaseous nebulae that grace textbooks and websites? Search the Web for a color image of a nebula in the Messier catalog. Describe the structure and colors you see in terms of the concepts discussed in this chapter. (Be careful to distinguish real color from false color when answering this question. See Window on Science 6-2 for more information about false-color images.)
2. An interesting highlight in the history of astronomy is the discovery of "Nebulium." Search the Internet for information on this once-mysterious source of nebular spectral lines. How can astronomers today explain these lines in terms of known elements?

Exploring *TheSky*

1. The following nebulae are all star-formation regions. What kind of nebulae are they? (*Hint:* To center on an object, use **Find** under the **Edit** menu. Choose **Messier Objects** and pick from the list.)

   M42, M20, M8, M17
2. Locate M8 in *TheSky,* zoom in, and identify other nebulae in the region. Study the photo of NGC6559. What kind of nebula is it?
3. Locate M42, M20, and M8. Zoom in and identify the spectral type of the central star(s). What common characteristic do these stars share? Why?

- Go to the Brooks/Cole Astronomy Resource Center **(http://astronomy.brookscole.com)** for critical thinking exercises, articles, and additional readings from InfoTrac College Edition, Brooks/Cole's online student library.

11 The Formation of Stars

Jim he allowed [the stars] was made, but I allowed they happened. Jim said the moon could'a laid them; well, that looked kind of reasonable, so I didn't say nothing against it, because I've seen a frog lay most as many, so of course it could be done.

MARK TWAIN, *THE ADVENTURES OF HUCKLEBERRY FINN*

Visual + infrared image

THE STARS ARE NOT ETERNAL. When you look at the sky, you see hundreds of points of light, and each is an object like the sun held together by its gravity and generating tremendous energy in its core through nuclear reactions. The stars you see tonight are the same stars your parents, grandparents, and great grandparents saw. Stars change hardly at all in a human lifetime, but they are not eternal. Stars are born and stars die. This chapter begins that story. ▌ In this chapter, you will see how gravity creates stars from the thin gas of space and how nuclear reactions inside stars generate energy. You will see how the flow of that energy outward toward the surface of the star balances gravity and makes the stars stable. In the next two chapters, you will follow the life story of stars from the beginning of their stable lives to their ultimate deaths.

In this image of the Orion Nebula, every point of light is a star that was born recently from the gas and dust. At infrared wavelengths, you can see the warm glow of more stars taking form in the dark clouds behind the glowing nebula. (ESO)

Guidepost

Looking Back

Throughout the preceding chapters you have been building a general understanding of the sky and a facility with the tools of astronomy. The last chapter introduced you to the gas and dust between the stars. Now you are ready to make some stars.

This Chapter

This is the place where you will really begin to see how the universe works. Here you will begin putting together observations and theories to understand how nature makes stars. That will answer four essential questions:

How are stars born?

How do stars make energy?

How do stars maintain their stability?

What evidence do astronomers have that theories of star formation are correct?

Looking Ahead

The most important of the questions above may be the last. Testing theories against evidence is the basic skill required of all scientists, and you will use it over and over in the 15 chapters to come.

AceAstronomy™ The AceAstronomy icon throughout the text indicates an opportunity for you to test yourself on key concepts and to explore animations and interactions on the AceAstronomy website at: http://ace.brookscole.com/sf9

Nebula N44
Roughly 40 young stars are inflating a bubble of hot gas inside the nebula from which they formed.
100 ly
Visual-wavelength image
Small dark nebulae may form stars.
Aging supergiant has ejected ring of gas.
Newborn stars still in their birth nebulae
Visual-wavelength image

■ Figure 11-1

Nebula N44 has given birth to a large cluster of stars. (ESO) Various stages of star formation are evident in NGC3603, including a massive star ejecting gas as it approaches its end. (W. Brandner, JPL/IPAC, E. K. Grebel, Univ. of Washington, Y. Chu, Univ. of Illinois, and NASA)

11-1 Making Stars from the Interstellar Medium

STARS HAVE BEEN FORMING continuously since our galaxy took shape over 10 billion years ago. You can depend on this for two reasons. First, the sun is only about 5 billion years old, a relative newcomer compared to the older stars in our galaxy. Second, you can see hot, blue stars such as one of our Favorite Stars Spica (α Virginis), a B1 main-sequence star. As you will see in the next chapter, such massive stars have very short lives. In fact, a star like Spica can last only 10 million years and must have formed recently.

The key to understanding star formation is the correlation between young stars and clouds of gas. Where you find the youngest groups of stars, you also find large clouds of gas illuminated by the hottest and brightest of the new stars (■ Figure 11-1). This should make you suspect that stars form from such clouds, much as raindrops condense from the water vapor in a thundercloud. Indeed, the giant molecular clouds discussed in the preceding chapter can give birth to entire clusters of new stars.

The central problem for any discussion of star formation is how these large, low-density, cold clouds of gas become comparatively small, high-density, hot stars. Gravity is the key.

Star Birth in Giant Molecular Clouds

The giant molecular clouds are the sites of active star formation, yet they are very unlike stars. With a typical diameter of 50 pc and a typical mass exceeding 10^5 solar masses, a giant molecular cloud is vastly larger than a star. Also, the gas in a giant molecular cloud is about 10^{20} times less dense than a star and has temperatures of only a few degrees Kelvin. These clouds can form stars because gravity can force some small regions of the clouds to contract to high density and high temperature.

Radio observations show that at least some giant molecular clouds develop dense cores that are only 0.1 pc in radius and that contain roughly 1 solar mass. A single giant molecular cloud may contain many of these dense cores and can give birth to star clusters containing hundreds of stars. However, both theory and observations suggest that many giant molecular clouds cannot begin the formation of dense cores spontaneously. At least four factors resist the contraction of a gas cloud, and gravity must overcome those four factors before star formation can begin.

First, thermal energy in the gas is present as motion among the atoms and molecules. Even at the very low temperature of 10 K, the average hydrogen molecule moves at about 0.35 km/s (almost 800 mph). This thermal motion would make the cloud drift apart if gravity were too weak to hold it together.

The interstellar magnetic field is the second factor that gravity must overcome to make the cloud contract. Neutral atoms and molecules are unaffected by a magnetic field, but ions, having an electric charge, cannot move freely through a magnetic field. Although the gas in a molecular cloud is mostly neutral, there are some ions, and that means a magnetic field can exert a force on the gas. The magnetic field present all through our galaxy averages only about 10^{-4} times as strong as that on Earth, but it can act like an internal spring to resist the contraction of the gas cloud.

The third factor is rotation. Everything in the universe rotates to some extent. As the gas cloud begins to contract, it spins more and more rapidly as it conserves angular momentum, just as ice skaters spin faster as they pull in their arms (Figure 5-7). This rotation can become so rapid that it resists further contraction of the cloud.

The turbulence in the interstellar medium is the fourth thing that could prevent a cloud from contracting. In the previous chapter, you learned that nebulae are often twisted and distorted by moving currents in the interstellar medium. This turbulence could make it difficult for a large molecular cloud to contract.

Given these four resistive factors, it seems surprising that any giant molecular clouds can contract, form dense cores, and eventually form stars. In some cases, the gas in giant molecular clouds can gradually recombine with free electrons and become less ionized. Neutral gas is free to "slip past" the magnetic field and contract. This gradual process has been observed to form dense cores inside isolated molecular clouds.

In contrast, both theory and observation suggest that many giant molecular clouds are triggered to form stars by a passing **shock wave,** the astronomical equivalent of a sonic boom (■ Figure 11-2). During such a triggering event, a few regions of the large cloud can be compressed to such high densities that the resistive factors can no longer oppose gravity, and star formation begins.

At least four different processes can produce shock waves that trigger star formation. Supernova explosions (Chapter 13) can produce powerful shock waves that rush through the interstellar medium. Also, the ignition of very hot stars can ionize nearby gas and drive it away to produce a shock wave where it pushes into the colder, denser interstellar matter. A third trigger is the collision of molecular clouds. Because the clouds are large, they are likely to run into each other occasionally; and, because they contain magnetic fields, they cannot pass through each other. A collision between such clouds can compress parts of the clouds and trigger star formation. The fourth trigger is the spiral pattern of our Milky Way Galaxy (see Figure 1-11). One theory suggests that the spiral arms are shock waves that travel around the galaxy like the moving hands of a clock (Chapter 15). As a cloud passes through a spiral arm, the cloud could be compressed, and star formation could begin. Astronomers have found regions of star formation where these processes can be identified (■ Figure 11-3).

Of course, a single giant molecular cloud containing a million solar masses does not contract to form a single humongous

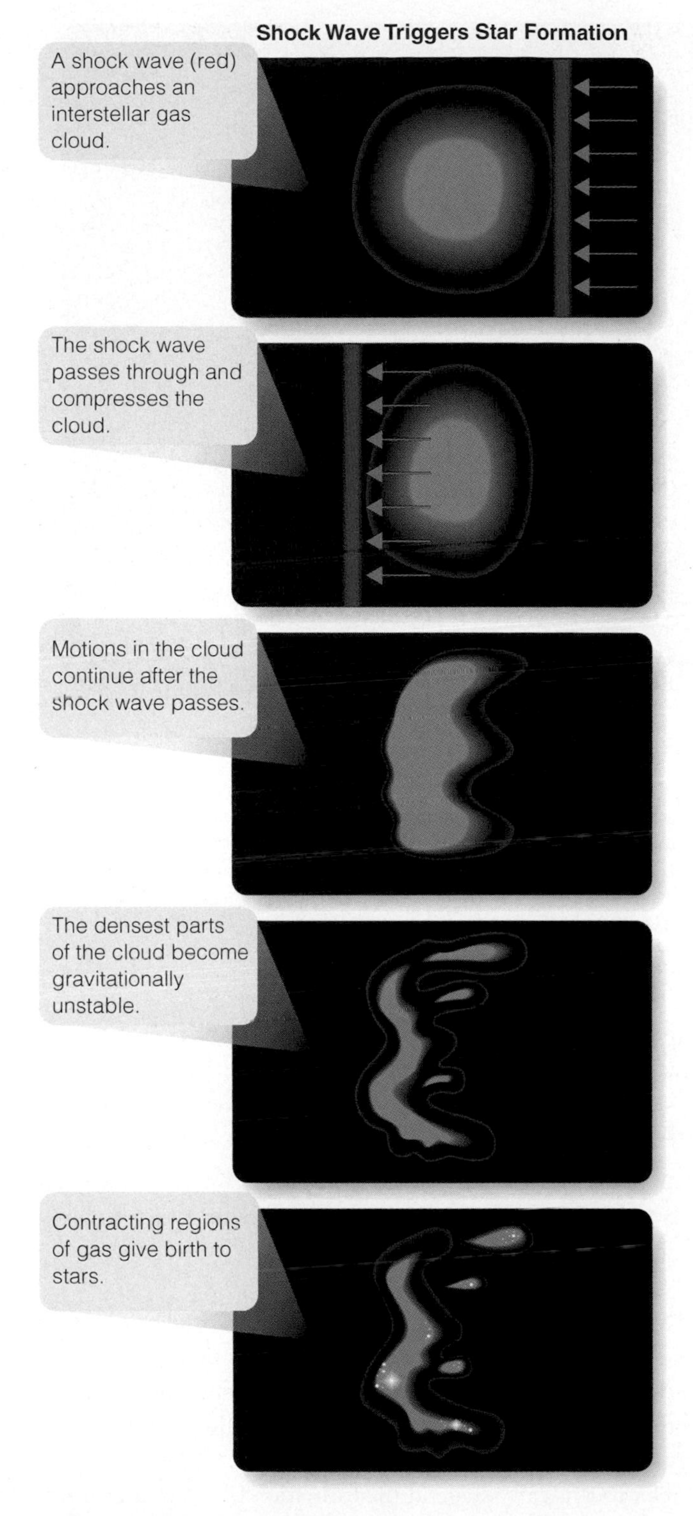

Shock Wave Triggers Star Formation
A shock wave (red) approaches an interstellar gas cloud.
The shock wave passes through and compresses the cloud.
Motions in the cloud continue after the shock wave passes.
The densest parts of the cloud become gravitationally unstable.
Contracting regions of gas give birth to stars.

■ **Figure 11-2**

In this summary of a computer model, an interstellar gas cloud is triggered into star formation by a passing shock wave. The events summarized here might span about 6 million years.

star. The cloud fragments and the densest parts form a number of dense cores. Exactly why a cloud fragments isn't fully understood, but the rotation, magnetic field, and turbulence probably

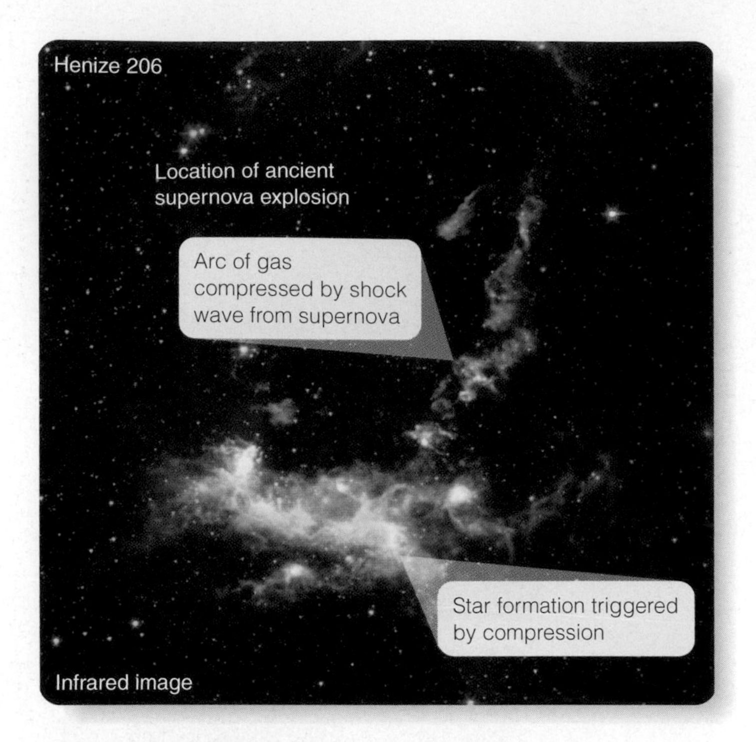

■ Figure 11-3

A few million years ago, a massive star exploded as a supernova in the upper left quarter of this image. The expanding shock wave compressed a nearby cloud of gas and triggered the birth of new stars. In this infrared image, you can see the arc of compressed gas left behind as the shock wave rushed through the gas clouds. Most of the bright, young stars in this nebula are hidden deep in dust clouds and are not yet visible at visual wavelengths. (NASA/JPL-Caltech/V. Gorjian, NOAO)

play important roles. In any case, a giant cloud of gas typically contracts to form a number of newborn stars.

Heating by Contraction

You can understand how low-density clouds of interstellar gas can become dense enough to make stars, but how can the horribly cold gas become hot enough to become a star? The answer, once again, is gravity.

To consider the formation of a star, shift your attention to a single dense core of gas destined to become a single star. Once a small cloud of gas begins to contract, gravity draws the atoms toward the center. That means the atoms are falling, and, like all things that fall, they gather speed as they fall. In fact, astronomers refer to this early stage in the formation of a star as **free-fall contraction.** Whereas the atoms may have had low velocities to start with, by the time they have fallen most of the way to the center of the cloud, they are traveling at high velocities. Thermal energy is the agitation of the particles in a gas, so this increase in velocity is a step toward heating the gas. But you can't say that the gas is hot simply because all of the atoms are moving rapidly. The air in the cabin of a jet airplane is traveling rapidly, but it isn't hot because all of the atoms are moving in generally the same direction with the plane. To convert the high velocity of the infalling atoms into thermal energy, the motion must become randomized, and that happens when the atoms begin to collide with one another as they fall into the central region of the cloud. The jumbled, random motion of the atoms represents thermal energy, and the temperature of the gas increases.

This is an important principle in astronomy. Whenever a cloud of gas contracts, the atoms move downward in the gravitational field and pick up speed, and the gas grows hotter. Astronomers express this by saying that gravitational energy is converted into thermal energy. Whenever a gas cloud expands, gas atoms move upward against gravity and lose speed, and the gas becomes cooler. Astronomers say that in this case thermal energy is converted into gravitational energy. This principle applies not only to clouds of interstellar gas but also to contracting and expanding stars, as you will see in the following chapters.

Your study of gas clouds has shown how nature can begin the contraction of dense cores in giant molecular clouds and how contraction can heat the gas. Now you are ready to construct a detailed story of the transformation from gas cloud to star.

Protostars

To follow the story of star formation farther, you must concentrate on a single fragment of a collapsing cloud as it contracts, heats up, and begins to behave like a star. Although the term ***protostar*** is used rather loosely by astronomers, it will be defined here to be a prestellar object that is hot enough to radiate infrared radiation but not hot enough to generate energy by nuclear fusion.

A protostar begins life as a contracting cloud of gas and dust falling inward under the influence of gravity (■ Figure 11-4). As the cloud contracts, it develops a higher-density region at the center and a low-density envelope. Mass continues to flow inward from the outer parts of the cloud. That is, the cloud contracts from the inside out, with the protostar taking shape deep inside an enveloping cloud of cold, dusty gas. These clouds have been called **cocoons** because they hide the forming protostar from view as it takes shape.

The contraction of the cloud cannot lead directly to the formation of a small protostar because the cloud has some net rotation. Everything rotates one way or another. The rotation of the cloud may be undetectable at first, but its rotation must grow more and more pronounced as it contracts and conserves its angular momentum. The rapidly spinning core of the cloud must flatten into a spinning disk like a blob of pizza dough spun into the air. Gas that has lost its angular momentum through collisions can sink directly to the center of the cloud, where the protostar begins to grow, surrounded by the disk. As more gas falls inward, it passes through the disk, giving up much of its angular momentum before it sinks into the protostar (■ Figure 11-5).

Basic physics predicts that protostars should form at the centers of spinning disks of gas, which are called **protostellar disks.** That is important because astronomers believe that planets form within these disks. Earth formed in a disk around the protosun

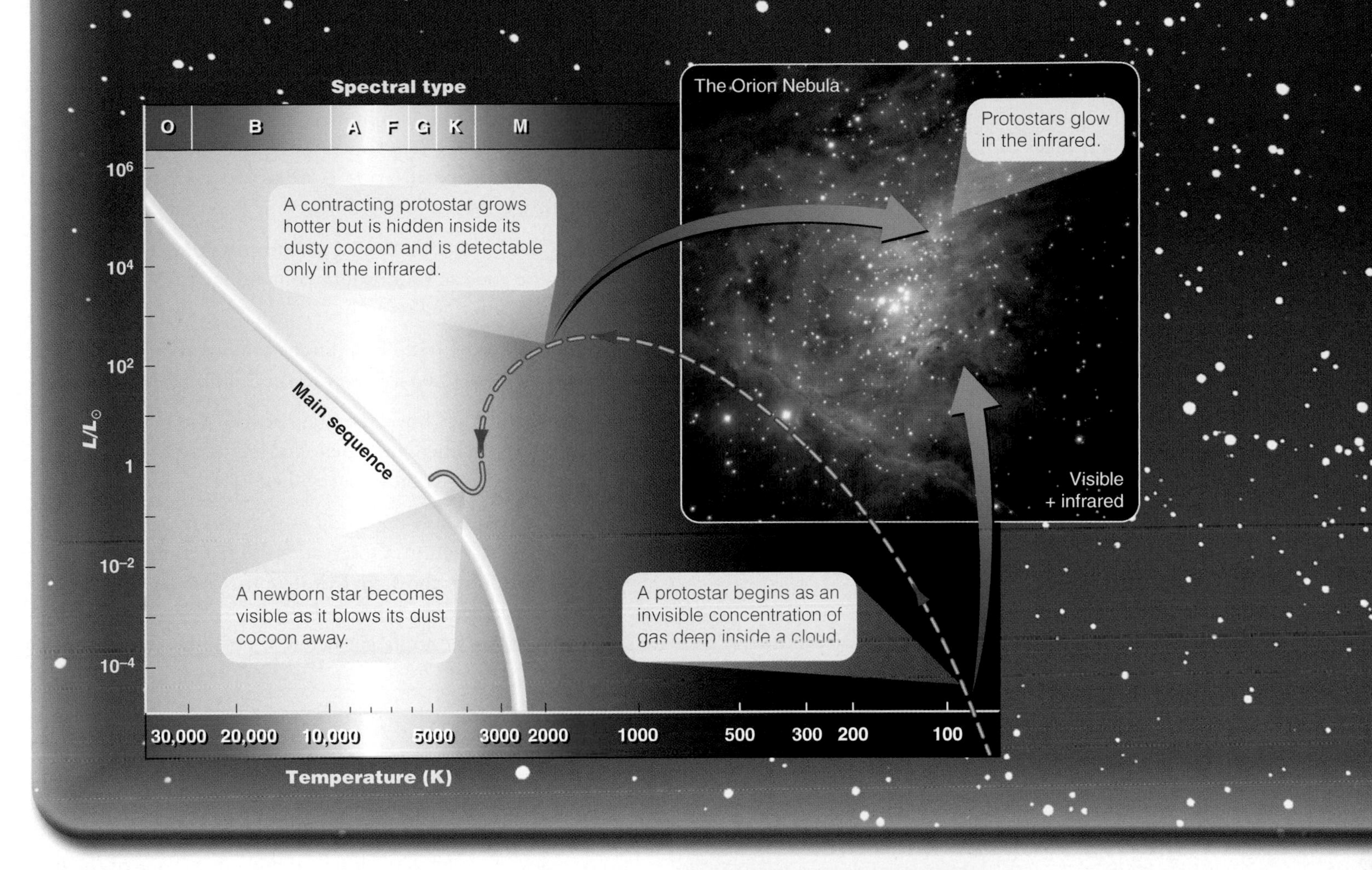

■ Figure 11-4

This H–R diagram has been extended to very low temperatures to show schematically the contraction of a dim, cool protostar. At visual wavelengths, protostars are invisible because they are deep inside dusty clouds of gas, but they are detectable at infrared wavelengths. The Orion Nebula contains both protostars and newborn stars that are just blowing their dust cocoons away. (ESO)

4.6 billion years ago. As you will see in Chapter 19, the evidence is very strong that planetary systems form in these protostellar disks.

If you could see a protostar without its cocoon, it would be a very luminous and cool red star in the upper right part of the H–R diagram. The dust cocoon, however, absorbs almost all of the visible radiation and, growing warm, reradiates the energy as infrared radiation. So you should not expect to see protostars at visible wavelengths, but they should be emitting infrared radiation.

When the protostar becomes hot enough, it can drive away the surrounding gas and dust and become visible. The location in the H–R diagram where protostars first emerge from their cocoons and become visible is called the **birth line.** Once a star crosses the birth line, it continues to contract toward the main sequence with a speed that depends on its mass. More massive stars have

Formation of a Protostellar Disk

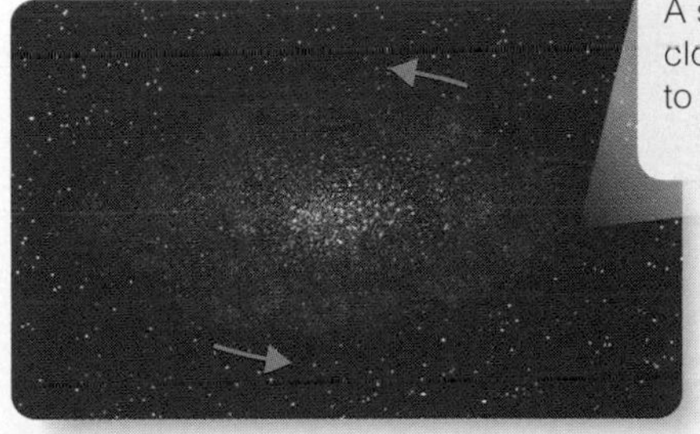

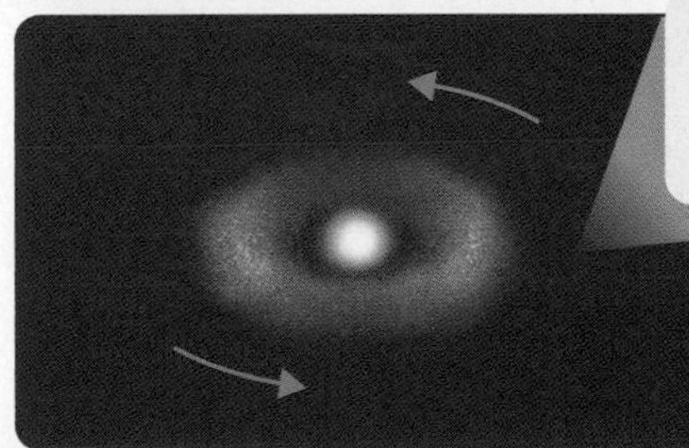

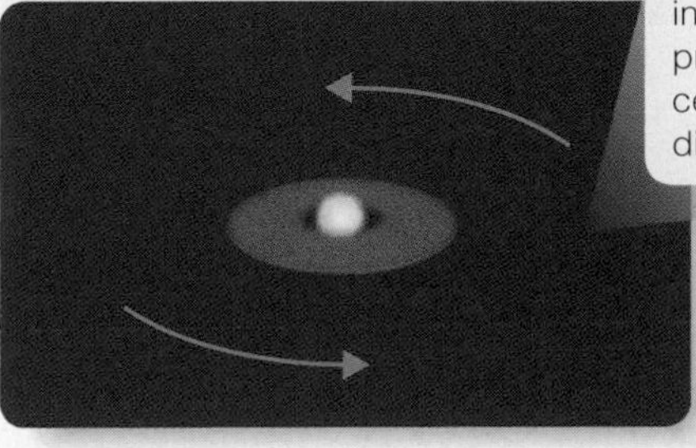

■ Figure 11-5

The rotation of a contracting gas cloud forces it to flatten into a disk, and the protostar grows at the center.

stronger gravity and contract more rapidly (■ Figure 11-6). The sun took about 30 million years to reach the main sequence, but a 30-solar-mass star takes only 30,000 years. A 0.2-solar-mass star needs about 1 billion years to contract from a gas cloud to the main sequence.

The theory of star formation takes you into an unearthly realm filled with unfamiliar processes and objects. You might think it was nothing more than a fairy tale if observations did not support the theory.

Evidence of Star Formation

In astronomy, evidence means observations. Consequently, astronomers must ask what observations confirm their theories of star formation. Unfortunately, a protostar is not easy to observe. The protostar stage is less than 0.1 percent of a star's lifetime, so although that is a long time in human terms, you cannot expect to find many stars in the protostar stage. Furthermore, protostars form deep inside clouds of dusty gas that absorb any light the protostar might emit. Astronomers must depend on observations at infrared wavelengths to search for hidden protostars.

■ Figure 11-6

The more massive a protostar is, the faster it contracts. A 1-$M_\odot$ star requires 30 million years to reach the main sequence. (Recall that $M_\odot$ means "solar mass.") The dashed line is the birth line, where contracting protostars first become visible as they dissipate their surrounding clouds of gas and dust. Compare with Figure 11-4, which shows the evolution of a protostar of about 1 $M_\odot$ as a dashed line up to the birth line and as a solid line from the birth line to the main sequence. (Illustration design by author)

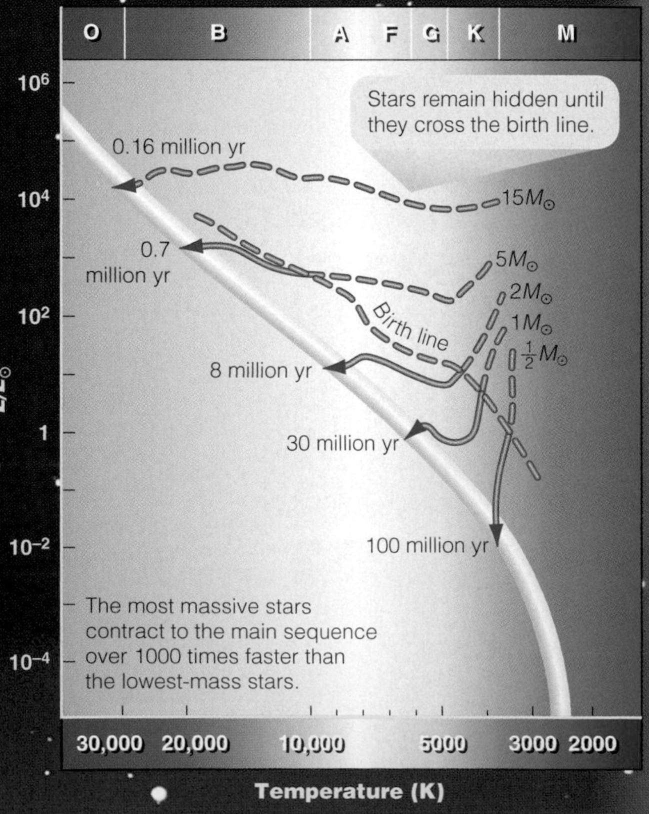

Study **Observational Evidence of Star Formation** on pages 244–245 and notice four important points:

1. You can be sure that star formation is going on right now because you can find regions containing stars so young they must have formed recently.
2. Visual and infrared observations can reveal dusty clouds of gas that seem to be in the process of forming stars.
3. Notice that newborn stars lie between the main sequence and the birth line—just where you would expect to find stars that have recently blown away their dust cocoons.
4. Finally, notice how observations provide clues to the process by which stars form. Observations of jets coming from hidden protostars show that protostars form surrounded by disks of gas and dust. Also, observations of small globules of gas and dust within larger nebulae show how star formation begins.

The origin of jets from spinning protostellar disks is not understood. Certainly the contracting, spinning disk contains tremendous energy, and theorists believe that magnetic fields become twisted tightly around the disk. Exactly how those fields squeeze hot gas out above and below the disk along the axis of rotation is not yet clear. But the detection of these jets was one of the first pieces of evidence that protostars are born at the centers of spinning disks.

In some cases, these dark disks of gas and dust are clearly visible around newborn stars (■ Figure 11-7). Such disks are exciting not only because they are evidence of star formation but also because they are the regions where planets form. You will meet such disks later in this chapter and again when you learn about the origin of our solar system in Chapter 19.

Some evidence of star formation is not at first obvious. An **association** is a widely distributed star cluster that is not held together by its own gravity—its stars wander away as the association ages. These associations consist of young stars, because the stars wander apart so quickly. The constellation Orion, a known region of star formation, is filled with T Tauri stars in a **T association.** T Tauri stars are relatively low-mass objects ranging from 0.75 to 3 solar masses, but the stars in **O associations,** extended groups of O stars, are more massive. It is not clear why some gas clouds give birth to compact star clusters held together by their own gravity and others give birth to larger associations not bound together by gravity. Associations are dramatic evidence of recent star formation because they drift apart so easily. They must have formed recently.

Not only can astronomers locate evidence of star formation, but they have found evidence that star formation can stimulate more star formation. If a gas cloud produces massive stars, those massive stars can ionize the gas nearby and drive it away. Where

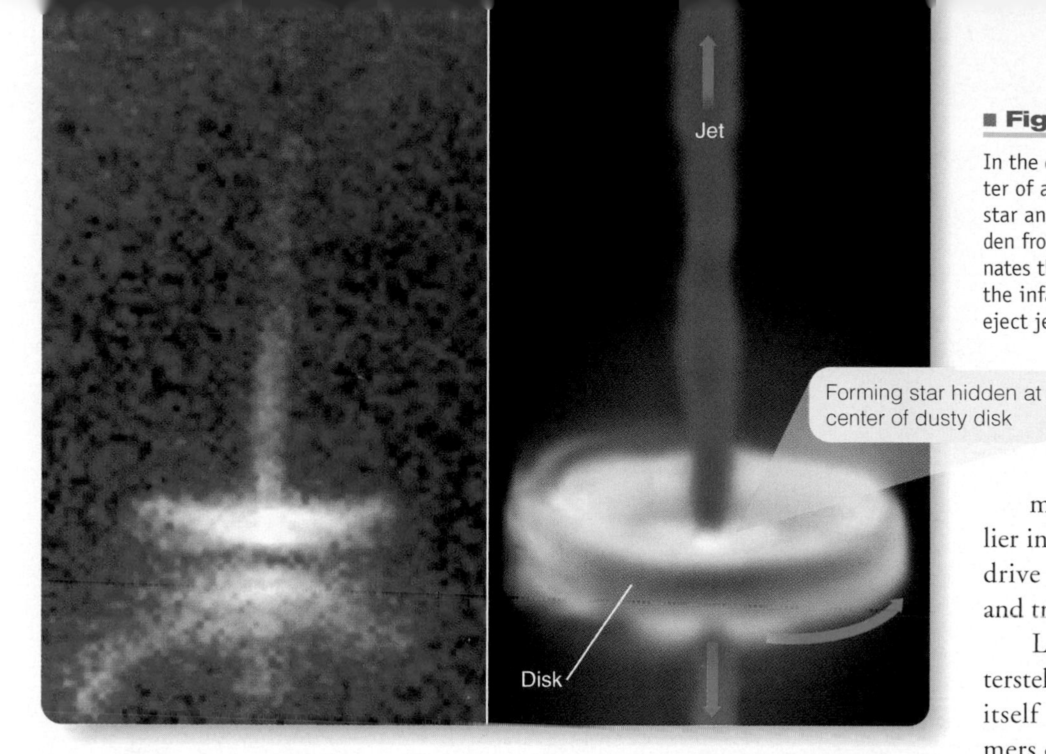

■ Figure 11-7

In the object HH30, a newly formed star lies at the center of a dense disk of dusty gas that is narrow near the star and thicker farther away. Although the star is hidden from you by the edge-on dusty disk, the star illuminates the inner surface of the disk. Interactions between the infalling material in the disk and the spinning star eject jets of gas along the axis of rotation. (C. Burrows, STScI & ESA, WFPC 2 Investigation Definition Team, NASA)

the intense radiation and hot gas pushes into surrounding gas, it can compress the gas and trigger more star formation. One sign of this process is the presence of **star-formation pillars,** columns of gas that point back at the source of the hot gas. Such pillars are produced by denser regions that protect the gas behind them as the blast of intense radiation and hot gas flows past (■ Figure 11-8). You can recognize this process in the overall shape of the Eagle Nebula on page 245 as well as in the smaller pillars within the nebula. Another way massive stars can trigger star formation is through the explosion of a massive star in a supernova. You read earlier in this chapter how such an explosion can drive a shock wave through surrounding gas and trigger more star formation (Figure 11-3).

Like a grass fire spreading through the interstellar medium, star formation can reignite itself if it creates massive stars, and astronomers can locate the remains of such episodes (■ Figure 11-9). Of course, lower-mass stars also form from such star formation, but they are not hot enough, nor do they explode as supernovae, so they alone cannot trigger further star formation.

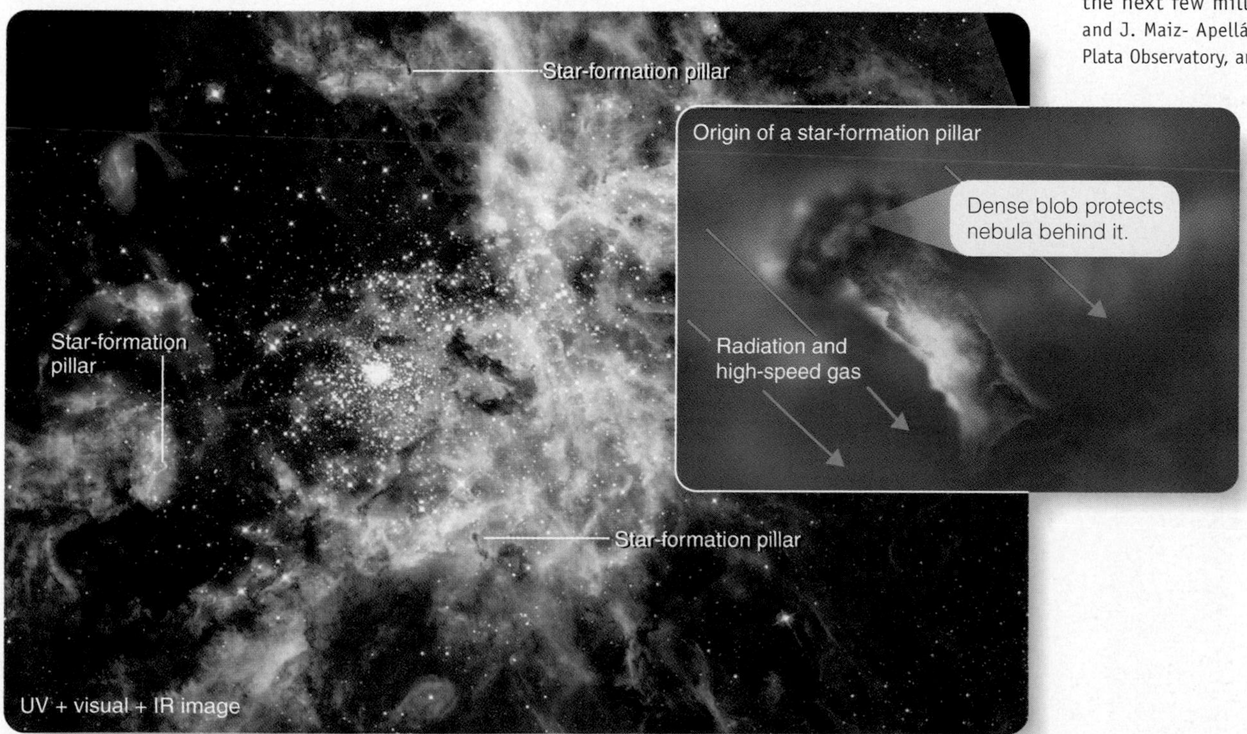

■ Figure 11-8

The hot, massive stars in the star cluster 30 Doradus are pouring out intense ultraviolet radiation and powerful winds of hot gas that are pushing back and compressing the surrounding nebula. Slightly denser regions in the nebula protect the gas behind them to form star-formation pillars a few light-years long that point back at the star cluster. The dense blobs are being compressed and may form more stars within the next few million years. (N. Walborn and J. Maiz- Apellániz, STScI, R. Barbá, La Plata Observatory, and NASA)

Observational Evidence of Star Formation

1 The nebula around the star S Monocerotis is bright with hot stars. Such stars live short lives of only a few million years, so they must have formed recently. Such regions of young stars are common. The entire constellation of Orion is filled with young stars and clouds of gas and dust.

Visual-wavelength image

1a Nebulae containing young stars usually contain T Tauri stars. These stars fluctuate irregularly in brightness, and many are bright in the infrared, suggesting they are surrounded by dust clouds and in some cases by dust disks. Doppler shifts show that gas is flowing away from many T Tauri stars. The T Tauri stars appear to be newborn stars just blowing away their dust cocoons. T Tauri stars appear to have ages ranging from 100,000 years to 100,000,000 years. Spectra of T Tauri stars show signs of an active chromosphere as we might expect from young, rapidly rotating stars with powerful dynamos and strong magnetic fields.

2 **Bok globules**, named after astronomer Bart Bok, are dense, dusty clouds seen silhouetted against glowing nebulae in the visual image below. Only a light-year or so in diameter, they contain from 10 to 1000 solar masses, and infrared observations show that most are very cold at their centers. Only a few have warm interiors and appear to be contracting to form protostars.

3 The star cluster NGC2264, imbedded in the nebula on this page, is only a few million years old. Lower-mass stars have not yet reached the main sequence, and the cluster contains many T Tauri stars (open circles), which are found above and to the right of the main sequence, near the birth line. The faintest stars in the cluster were too faint to be observed in this study.

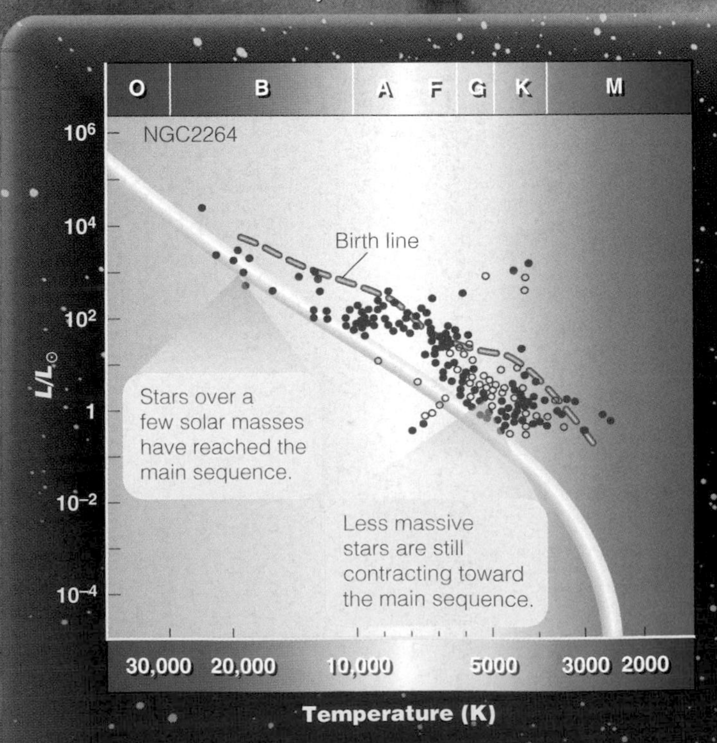

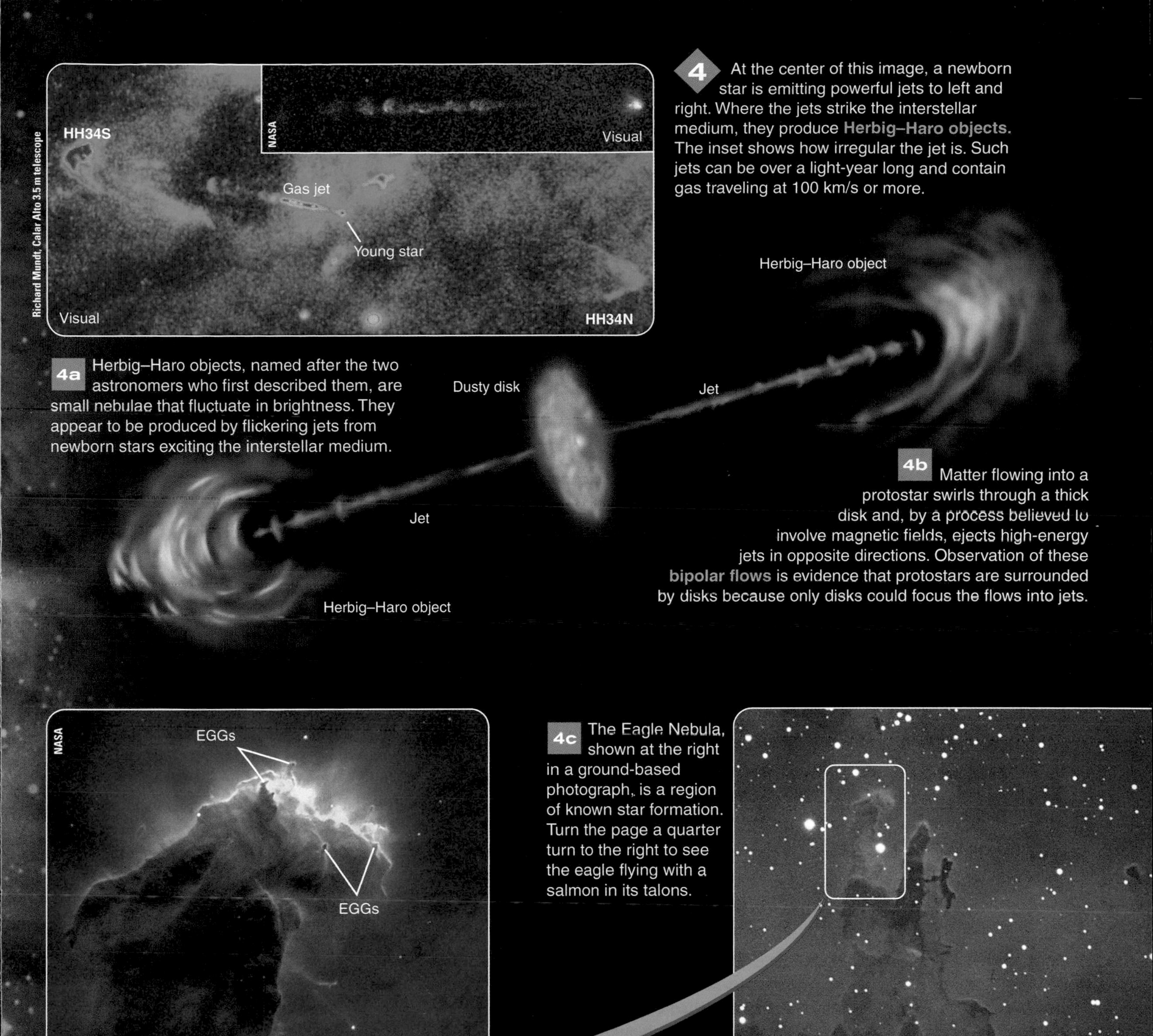

4 At the center of this image, a newborn star is emitting powerful jets to left and right. Where the jets strike the interstellar medium, they produce **Herbig–Haro objects.** The inset shows how irregular the jet is. Such jets can be over a light-year long and contain gas traveling at 100 km/s or more.

4a Herbig–Haro objects, named after the two astronomers who first described them, are small nebulae that fluctuate in brightness. They appear to be produced by flickering jets from newborn stars exciting the interstellar medium.

4b Matter flowing into a protostar swirls through a thick disk and, by a process believed to involve magnetic fields, ejects high-energy jets in opposite directions. Observation of these **bipolar flows** is evidence that protostars are surrounded by disks because only disks could focus the flows into jets.

4c The Eagle Nebula, shown at the right in a ground-based photograph, is a region of known star formation. Turn the page a quarter turn to the right to see the eagle flying with a salmon in its talons.

4d This Hubble Space Telescope image at left of part of the Eagle Nebula shows that radiation from bright stars out of the field to the upper right is evaporating the dust and driving away the gas to expose small globules of denser gas and dust. Infrared observations show that about 15 percent of these objects have formed protostars, and at least one seems to contain a newborn star (arrow). In part because these objects were first found in the Eagle Nebula, astronomers have enjoyed calling them EGGs — evaporating gaseous globules. They are evidently stars exposed in the act of forming.

Star Formation in the Orion Nebula

1 The visible Orion Nebula shown below is a pocket of ionized gas on the near side of a vast, dusty molecular cloud that fills much of the southern part of the constellation Orion. The molecular cloud can be mapped by radio telescopes. To scale, the cloud would be many times larger than this page. As the stars of the Trapezium were born in the cloud, their radiation has ionized the gas and pushed it away. Where the expanding nebula pushes into the larger molecular cloud, it is compressing the gas (see diagram at right) and may be triggering the formation of the protostars that can be detected at infrared wavelengths within the molecular cloud.

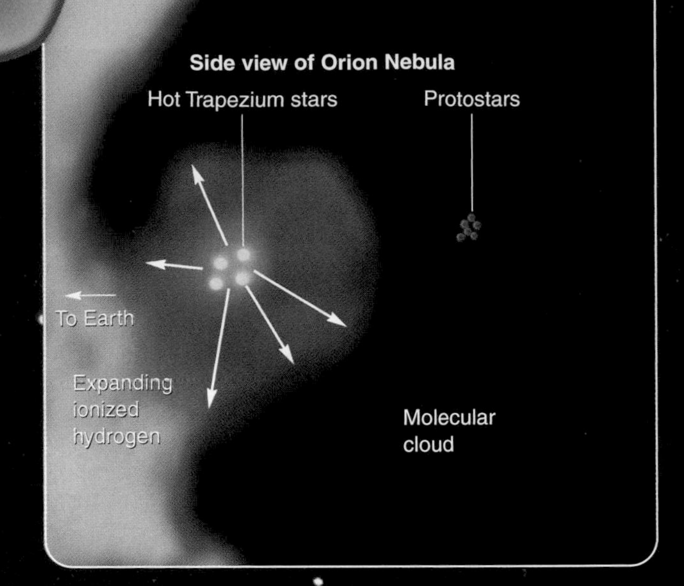

Hundreds of stars lie within the nebula, but only the four brightest, those in the Trapezium, are easy to see with a small telescope. A fifth star, at the narrow end of the Trapezium, may be visible on nights of good seeing.

The cluster of stars in the nebula is less than 2 million years old. This must mean the nebula is similarly young.

Trapezium

Visual-wavelength image
Daniel Good

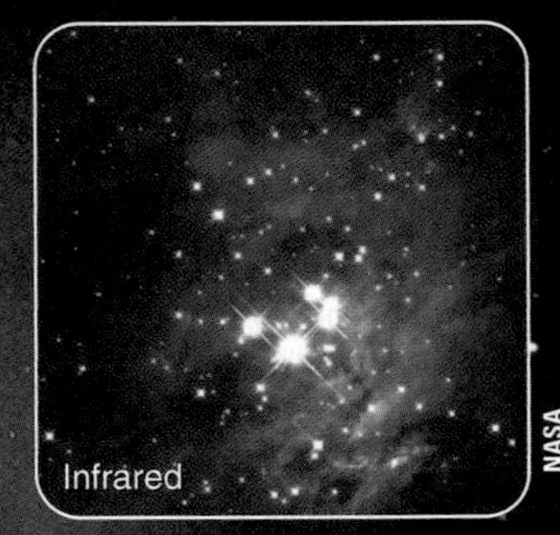

NASA

The near-infrared image above reveals about 50 low-mass, very cool stars that must have formed recently.

NASA/CXC/SAO

Roughly 1000 young stars with hot chromospheres appear in this X-ray image of the Orion Nebula.

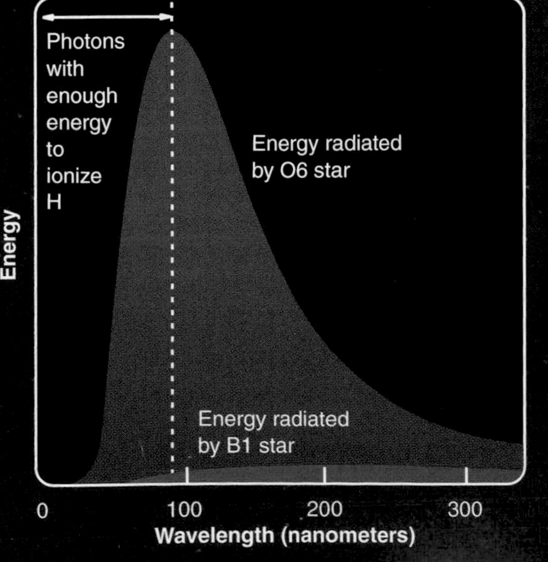

2 Of all the stars in the Orion Nebula, only one is hot enough to ionize the gas. Only photons with wavelengths shorter than 91.2 nm can ionize hydrogen. The second-hottest stars in the nebula are B1 stars, and they emit little of this ionizing radiation. The hottest star, however, is an O6 star 30 times the mass of the sun. At a temperature of 40,000 K, it emits plenty of photons with wavelengths short enough to ionize hydrogen. Remove that one star, and the nebula would turn off its emission.

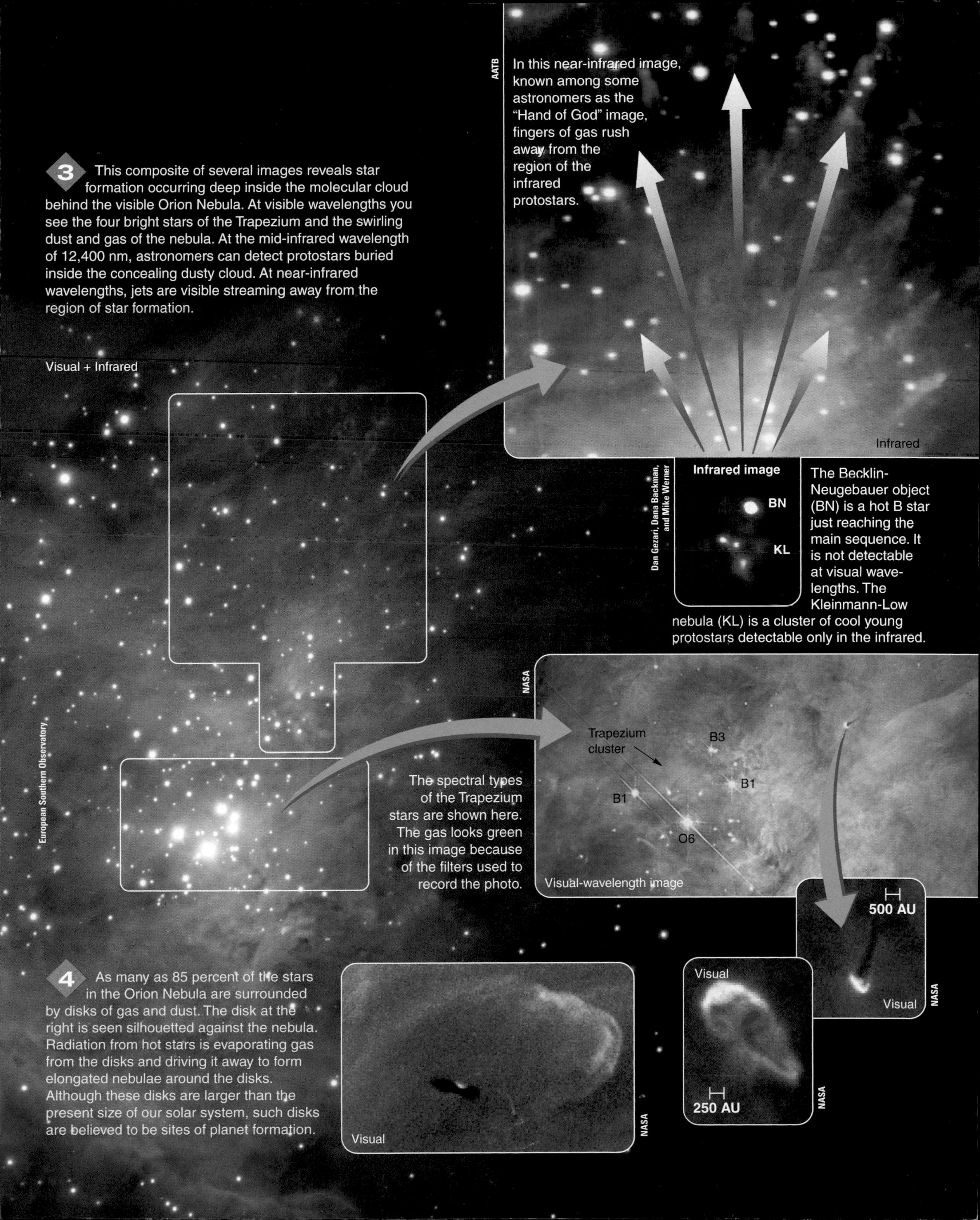

3 This composite of several images reveals star formation occurring deep inside the molecular cloud behind the visible Orion Nebula. At visible wavelengths you see the four bright stars of the Trapezium and the swirling dust and gas of the nebula. At the mid-infrared wavelength of 12,400 nm, astronomers can detect protostars buried inside the concealing dusty cloud. At near-infrared wavelengths, jets are visible streaming away from the region of star formation.

In this near-infrared image, known among some astronomers as the "Hand of God" image, fingers of gas rush away from the region of the infrared protostars.

The Becklin-Neugebauer object (BN) is a hot B star just reaching the main sequence. It is not detectable at visual wavelengths. The Kleinmann-Low nebula (KL) is a cluster of cool young protostars detectable only in the infrared.

The spectral types of the Trapezium stars are shown here. The gas looks green in this image because of the filters used to record the photo.

4 As many as 85 percent of the stars in the Orion Nebula are surrounded by disks of gas and dust. The disk at the right is seen silhouetted against the nebula. Radiation from hot stars is evaporating gas from the disks and driving it away to form elongated nebulae around the disks. Although these disks are larger than the present size of our solar system, such disks are believed to be sites of planet formation.

fusion. Now build a different argument. **How does the star manage to make exactly the right amount of energy to support its weight?**

■ ■ ■

Connections: You have traced the birth of stars from the first instability in the interstellar medium to the final equilibrium of nuclear fusion and gravity. You have constructed a complete story of star formation, but you expect evidence to support theories. One of the best places to search for evidence of star formation is in the Great Nebula in Orion. It is just part of a great storm of star birth sweeping through Orion.

11-4 The Orion Nebula

ON A CLEAR WINTER NIGHT, you can see with your naked eye the Great Nebula of Orion as a fuzzy wisp in Orion's sword. With binoculars or a small telescope it is striking, and through a large telescope it is breathtaking. At the center lie four brilliant blue-white stars known as the Trapezium, the brightest of a cluster of a few hundred stars. Surrounding the stars are the glowing filaments of a nebula more than 8 pc across. Like a great thundercloud illuminated from within, the churning currents of gas and dust suggest immense power. The significance of the Orion Nebula lies hidden, figuratively and literally, beyond the visible nebula. The region is ripe with star formation.

Evidence of Young Stars

You should not be surprised to find star formation in Orion. The constellation is a brilliant landmark in the winter sky because it is marked by hot, blue stars. These stars are bright in the sky, not because they are nearby, but because they are tremendously luminous. These O and B stars cannot live more than a few million years, so you can conclude that they must have been born recently. Furthermore, the constellation contains large numbers of T Tauri stars, which are known to be young. Orion is rich with young stars.

The history of star formation in the constellation of Orion is written in its stars. The stars at Orion's west shoulder are about 12 million years old, while the stars of Orion's belt are about 8 million years old. The stars of the Trapezium at the center of the Great Nebula are no older than 2 million years. Apparently, star formation began near the west shoulder, and the massive stars that formed there triggered the formation of the stars you see in Orion's belt. That star formation may have triggered the formation of the stars you see in the Great Nebula. Like a grass fire, star formation has swept across Orion from northwest to southeast.

Study **Star Formation in the Orion Nebula** on pages 252–253 and notice four points:

1. The nebula you see is only a small part of a vast, dusty molecular cloud. You see the nebula because the stars born within it have ionized the gas and driven it outward, breaking out of the molecular cloud.
2. Also notice that a single very hot star is almost entirely responsible for ionizing the gas.
3. Notice how infrared observations reveal clear evidence of active star formation deeper in the molecular cloud just to the northwest of the Trapezium.
4. Finally, notice that many stars visible in the Orion Nebula are surrounded by disks of gas and dust. Such disks do not last long and are clear evidence that the stars are very young.

In the next million years, the familiar outline of the Great Nebula will change, and a new nebula may begin to form as the protostars in the molecular cloud ionize the gas, drive it away, and become visible. Centers of star formation may develop and then dissipate as massive stars are born and force the gas to expand. If enough massive stars are born, they can blow the entire molecular cloud apart and bring the successive generations of star formation to a conclusion. The Great Nebula in Orion and its invisible molecular cloud are a beautiful and dramatic example of the continuing cycles of star formation.

Building Scientific Arguments

What did Orion look like to the ancient Egyptians, to the first humans, and to the dinosaurs?

Scientific arguments can do more than support a theory; they can change the way you think of the world around you. The Egyptian civilization had its beginning only a few thousand years ago, and that is not very long in terms of the history of Orion. The stars you see in the constellation are hot and young, but they are a few million years old, so the Egyptians saw the same constellation you see. (They called it Osiris.) Even the Orion Nebula hasn't changed very much in a few thousand years, and Egyptians may have admired it in the dark skies along the Nile.

Our oldest human ancestors lived about 3 million years ago, and that was about the time that the youngest stars in Orion were forming. Your earliest ancestors may have looked up and seen some of the stars you see, but some stars have formed since that time. Also, the Great Nebula is excited by the Trapezium stars, and they are not more than a few million years old, so your early ancestors probably didn't see the Great Nebula.

The dinosaurs saw something quite different. The last of the dinosaurs died about 65 million years ago, long before the birth of the brightest stars in the constellation you see. The dinosaurs, had they the brains to appreciate the view, might have seen bright stars along the Milky Way, but they didn't see Orion. All of the stars in the sky are moving through space, and the sun is orbiting the center of our galaxy. Over many millions of years, the stars move appreciable distances across the sky. The night sky above the dinosaurs contained totally different star patterns.

The Orion Nebula is the product of a giant molecular cloud, but such a cloud can't continue spawning new stars forever. Focus your argument to answer the following: **What processes limit star formation in a molecular cloud?**

■ ■ ■

Connections: The ancient Aztecs of central Mexico told the story of how the stars, known as the Four Hundred Southerners, were scattered across the sky when they lost a cosmic battle with their brother, the great war god Huitzilopochtli. Modern astronomy tells a less colorful story of the birth of stars, but the modern story is supported by evidence and leads you to further questions. If stars are born, then how do they die? You will begin that story in the next chapter.

Study and Review Tools

Summary

11-1 | Making Stars from the Interstellar Medium

How are stars born?

- Stars are born from the gas and dust of the interstellar medium.
- The existence of massive, hot stars such as Spica that cannot live very long and short-lived associations is strong evidence that star formation is a continuous process.
- The gravity of giant molecular clouds makes them contract, but that is resisted by thermal energy in the gas, magnetic fields, rotation, and turbulence. In at least some cases, clouds are compressed by passing shock waves, and star formation is triggered. The birth of massive stars can produce shock waves that trigger further star formation.
- The cold gas of interstellar space heats up as it contracts because the atoms fall inward and pick up speed. Astronomers say the gas is converting gravitational energy into thermal energy.
- Protostars form deep inside dusty cocoons and are not directly visible at visual wavelengths until they cross the birth line in the H–R diagram.
- Many, perhaps most, protostars form at the center of dusty disks, and jets of gas can be emitted along the axis of the spinning disk.

11-2 | The Source of Stellar Energy

How do stars make energy?

- Many stars make their energy the same way the sun does, using the proton–proton cycle.
- The CNO cycle is more efficient, but it requires a higher temperature than the proton–proton cycle. Both processes combine four hydrogen nuclei to make one helium nucleus plus energy.
- Stars more massive than 1.1 solar masses can make energy with the CNO cycle. Less massive stars can only use the proton–proton cycle.

11-3 | Stellar Structure

How do stars maintain their stability?

- Energy flows from the hot core to the cooler surface as radiation or as convection. Conduction is not important in most stars. Convection is important in stars because it stirs the material.
- The law of hydrostatic equilibrium says that the weight pressing down on a layer of gas in a star must be balanced by the pressure in the gas. That shows that the inner layers of stars must be hotter because they must support more weight.
- Upper-main-sequence stars, being more massive, must be hotter inside, and that allows them to use the CNO cycle. Because that cycle is so sensitive to temperature, the energy production occurs in a very small region near the center and the cores of the stars are convective zones. In their outer layers, these stars are radiative.
- Lower-main-sequence stars, being less massive, are not as hot inside and can fuse hydrogen using the proton–proton chain. Because that chain is not very sensitive to temperature, the energy generation is more widely spread through the star's core, and the deep interior is radiative. The outer layers of these stars are convective.
- The lowest-mass stars are so cool that their gases are rather opaque and they are convective throughout.

11-4 | The Orion Nebula

What evidence do astronomers have that theories of star formation are correct?

- The visible Orion Nebula is only a small part of a much larger dusty molecular cloud. Ionization by ultraviolet photons from the hottest star is lighting up the nebula and making it glow brightly.
- Infrared observations reveal clear evidence of active star formation deeper in the molecular cloud just to the northwest of the Trapezium.
- Many stars visible in the Orion Nebula are surrounded by disks of gas and dust. Such disks do not last long and are clear evidence that the stars are very young.

New Terms

shock wave (p. 239)
free-fall contraction (p. 240)
protostar (p. 240)
cocoon (p. 240)
protostellar disk (p. 240)
birth line (p. 241)
association (p. 242)
T association (p. 242)
O association (p. 242)
star-formation pillar (p. 243)
T Tauri star (p. 244)
Bok globule (p. 244)

12 Stellar Evolution

We should be unwise to trust scientific inference very far when it becomes divorced from opportunity for observational test.

SIR ARTHUR EDDINGTON, *THE INTERNAL CONSTITUTION OF THE STARS*

Visual-wavelength image

EVERY STAR HAD A beginning, and every star must have an ending, but in between they produce the light and energy that make our universe beautiful. The stars above you seem eternal, but the light and warmth they emit is produced by the fusion of nuclear fuels in their cores; and, even as you look at them, they are using up their fuels and drawing closer to their endings. The stars are, ever so slowly, going out. Although stars enjoy long, stable lives on the main sequence, they suffer the first pangs of stellar age as they gradually consume their hydrogen fuel. Once hydrogen is exhausted at their centers, they swell rapidly into giant stars 10 to 1000 times the size of the sun. For a short time, a few million to a billion years, they are majestic beacons visible across thousands of parsecs. Although astronomers say a star swells

Continued on page 260

Like lanterns, stars push back the darkness and illuminate the universe. These stars, located in the constellation Monoceros, the Unicorn, illuminate the Cone Nebula at left, and the Fox Fur Nebula at right. (T. A. Rector and B. A. Wolpa, NRAO/NOAO/AUI/AURA/NSF)

Guidepost

Looking Back

The life story of stars begins in the interstellar medium and continues through their birth in glowing nebulae. You have followed that story through the last two chapters, and now you are ready for the next step.

This Chapter

This chapter is about the long, stable middle age of stars on the main sequence and later as they swell to become giants. Here you will answer four essential questions:

How do astronomers know how stars evolve?

What are main-sequence stars like?

How do stars evolve after they exhaust hydrogen?

What evidence do astronomers have that stars really do evolve?

This discussion is a good example of how scientists use theory and evidence to understand how nature works.

Looking Ahead

This chapter is about how stars live. The next two chapters are about how stars die and the strange objects they leave behind.

Ace Astronomy™ The AceAstronomy icon throughout the text indicates an opportunity for you to test yourself on key concepts and to explore animations and interactions on the AceAstronomy website at: http://ace.brookscole.com/sf9

rapidly and spends a short time as a giant, these changes happen slowly compared with a human life. How can anyone be sure stars really do age? Stellar evolution and the formation of giant stars might be just an astronomer's daydream.

In this chapter, you will see the full interplay of theory and evidence. You will use the basic laws of physics, combined with your knowledge of the nature of stars, to create a theory to describe how main-sequence stars maintain their equilibrium and how such stars change when they exhaust their nuclear fuels. As you are warned in the quotation that opens this chapter, theory alone is never enough. At each step, astronomers must compare their theories with the evidence. In this chapter, you will discover that clusters of stars can make the slow evolution of stars visible even during our short lifetime, and you will find that some evolving stars can become unstable and pulsate like beating hearts. Those pulsating stars reveal more about how stars evolve.

From beginning to end, this chapter tells the story of the evolution of stars from main sequence to giant. You will be able to trace the passage of stars through this phase of their existence and see how they generate their energy and why they swell so large. The place to begin is with the main-sequence stars.

■ **Figure 12-1**

Stars shine because energy flows out of their cores, and they are held together by their own gravity. Here Merope, the brightest star in the Pleiades, is imbedded in wisps of gas through which the star cluster is drifting. The circular rings and spikes are a diffraction pattern produced by light inside the telescope. The actual star is an unresolved point of light. (NOAO/AURA/NSF)

12-1 Main-Sequence Stars

IF SHAKESPEARE WERE ALIVE today, he would probably have something sarcastic to say about modern astronomers. "How do these stargazers pretend to know the hearts of the eternal stars?" he might ask. In fact, one of the greatest triumphs of modern astronomy is the discovery that the stars are not eternal and that mere humans can indeed understand their hearts. Astronomers can know the conditions at the center of a star because stars are fundamentally very simple objects. You can begin by considering how astronomers use stellar models to understand the interior of main-sequence stars. Later you can use these same models to understand what happens to stars when they grow old.

Stellar Models

Every star is balanced between gravity trying to make it contract and internal pressure trying to make it expand (■ Figure 12-1). Astronomers describe the insides of stars by imagining that the stars are divided into concentric shells similar to the layers in an onion (Figure 11-12). Of course, stars are not made like onions, but defining these shells mathematically helps astronomers discuss the conditions at different levels inside the star—what astronomers call the "structure" of the star.

The structure of a star is described by four simple laws of physics, two of which you have already met. In Chapter 11, you met the law of hydrostatic equilibrium, which says that the weight pressing down on a shell in a star must be balanced by the pressure pushing outward in that shell. That is, at every level in a star, the weight of the overlying matter must be supported by the pressure. Of course, that means that there can be no empty shells; an empty gap could exert no pressure and would be unable to support the weight of the material above. Chapter 11 also introduced you to energy transport; the law of energy transport describes how energy flows from hot to cool regions by radiation, convection, or conduction.

To these two laws you can add two basic laws of nature. The **conservation of mass law** says that the total mass of the star must equal the sum of the masses in its shells. Of course, no shell can be empty, and there is no such thing as negative mass. The **conservation of energy law** says that the amount of energy flowing out the top of a shell in the star must be equal to the amount of energy coming in at the bottom of the shell plus whatever energy is generated within the shell. Of course, that's just logical. It means that the energy leaving the surface of the star, its luminosity, must equal the sum of the energy generated in all of the shells inside the star. This is like saying that the total number of new cars driving out of a factory must equal the sum of cars manufactured on each assembly line. No car can vanish into nothing or appear from nothing. Energy in a star may not vanish without a trace or appear out of nowhere.

The four laws of stellar structure, described in general terms in ■ Table 12-1, can be written as mathematical equations. By

■ Table 12-1 | The Four Laws of Stellar Structure

1. Hydrostatic equilibrium	The weight on each layer is balanced by the pressure in that layer.
2. Energy transport	Energy moves from hot to cool by radiation, convection, or conduction.
3. Conservation of mass	Total mass equals the sum of the shell masses. No gaps are allowed.
4. Conservation of energy	Total luminosity equals the sum of the energies generated in each shell.

solving those equations in a special way, astronomers can build a mathematical model of the inside of a star.

If you wanted to build a model of a star, you would have to divide the star into about 100 concentric shells and then write down the four equations of stellar structure for each shell. You would then have 400 equations that would have 400 unknowns, namely, the temperature, density, mass, and energy flow in each shell. Solving 400 equations simultaneously is not easy, and the first such solutions, done by hand before the invention of electronic computers, took months of work. Now a properly programmed computer can solve the equations in a few seconds and print a table of numbers that represent the conditions in each shell of the star. Such a table is a **stellar model.**

The table shown in ■ Figure 12-2 is a model of the sun. The bottom line, for radius equal to 0.00, represents the center of the sun, and the top line, for radius equal to 1.00, represents the surface. The other lines in the table show the temperature and density in each shell, the mass inside each shell, and the fraction of the sun's luminosity flowing outward through the shell. You can use the table to understand the sun. The bottom line tells you the temperature at the center of the sun is about 15 million Kelvin. At such a high temperature, the gas is highly transparent, and energy flows as radiation. Nearer the surface, the temperature is lower, the gas is more opaque, and the energy is carried by convection.

Notice that stellar models are quantitative; that is, properties have specific numerical values. Earlier in this book, you used models that were qualitative—the Babcock model of the sun's magnetic cycle, for instance. Both kinds of models are useful, but the quantitative model reveals deeper insights into how nature works, reflecting the power of mathematics as a precise way of thinking (**Window on Science 12-1**).

Stellar models let astronomers look into a star's past and future. In fact, they can use models as time machines to follow the evolution of stars over billions of years. To look into a star's future, for instance, you could use a stellar model to determine how

■ Figure 12-2

A stellar model is a table of numbers that represent conditions inside a star. Such tables can be computed using the four laws of stellar structure, shown here in mathematical form. The table in this figure describes the sun. (Illustration design by author)

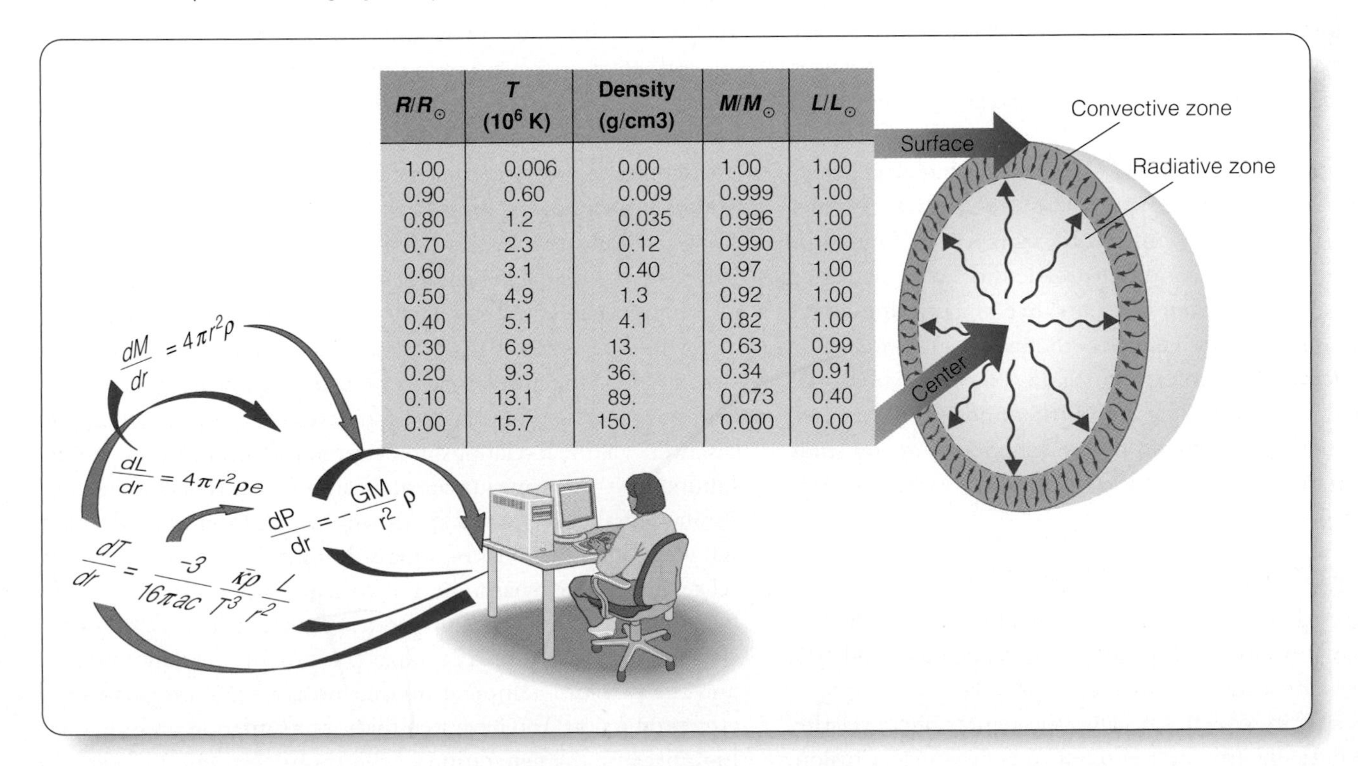

$R/R_\odot$	T (10^6 K)	Density (g/cm3)	$M/M_\odot$	$L/L_\odot$
1.00	0.006	0.00	1.00	1.00
0.90	0.60	0.009	0.999	1.00
0.80	1.2	0.035	0.996	1.00
0.70	2.3	0.12	0.990	1.00
0.60	3.1	0.40	0.97	1.00
0.50	4.9	1.3	0.92	1.00
0.40	5.1	4.1	0.82	1.00
0.30	6.9	13.	0.63	0.99
0.20	9.3	36.	0.34	0.91
0.10	13.1	89.	0.073	0.40
0.00	15.7	150.	0.000	0.00

Window on Science 12-1

Quantitative Thinking with Mathematical Models

One of the most powerful tools in science is the mathematical model, a group of equations carefully designed to mimic the behavior of the object scientists want to study. Astronomers build mathematical models of stars using only four equations, but other systems are much more complicated and may require many more equations.

Many sciences use mathematical models. Medical scientists have built mathematical models of the nerves that control the human heart, and physicists have built mathematical models of the inside of an atomic nucleus. Economists have built mathematical models of certain aspects of economic systems, such as the municipal bond market, and Earth scientists have built mathematical models of Earth's atmosphere. In each case, the mathematical model allows the scientists to study something that is too difficult to study in the real world. The model can reveal regions that cannot be observed, speed up a slow process, slow down a fast process, and allow scientists to perform experiments that would be impossible in reality. Astronomers, for example, can change the abundance of certain chemical elements in a model star to see how its structure depends on its composition.

Many of these mathematical models require very large and fast computers, in some cases supercomputers. A modern computer takes only seconds to compute a stellar model, but models such as the motions of the planets in our solar system millions of years into the future require the largest and fastest computers in the world and highly sophisticated mathematical methods. Most astronomers are apt computer programmers, and some are experts. A few astronomers have built their own specialized computers to calculate specific kinds of models. The mathematical model is one of the most important tools in astronomy.

As is true for any scientific model, a mathematical model is only as reliable as the assumptions that go into its creation. In Window on Science 2-1, you saw that the celestial sphere was an adequate model of the sky for some purposes but breaks down if extended too far. So, too, with mathematical models. You can think of a mathematical model as a numerical expression of one or more theories. While such models can be very helpful, they are always based on theory and so must be compared with the real world at every opportunity. Models must always be tested against experiment and observation. Otherwise, they might lead you astray.

Before any new airplane flies, engineers build mathematical models to test its stability. (The Boeing Company)

fast the star uses its fuel in each shell. As the fuel is consumed, the chemical composition of the gas changes, and the amount of energy generated declines. By calculating the rate of these changes, you could predict what the star will look like at any point in the future.

Although this sounds simple, it is actually a highly challenging problem involving nuclear and atomic physics, thermodynamics, and sophisticated computational methods. Only since the 1950s have electronic computers made the rapid calculation of stellar models possible, and the advance of astronomy since then has been heavily influenced by the use of such models to study the structure and evolution of stars. The summary of star formation in this chapter is based on thousands of stellar models. You will continue to rely on theoretical models as you study main-sequence stars in the next section and the deaths of stars in the next chapter.

Why Is There a Main Sequence?

You can now understand why there is a main sequence. The clue is the law of hydrostatic equilibrium, and you can use models of stars to analyze this clue.

There is a main sequence because the centers of contracting protostars eventually grow hot enough to begin nuclear fusion. Deuterium, the heavy isotope of hydrogen, is the first nuclear fuel to fuse, but it is so rare and produces so little energy it has no real effect on a contracting star. Hydrogen fusion is the big powerhouse; when it begins, it stops the contraction. Stars reach equilibrium somewhere along a line in the H–R diagram that astronomers call the main sequence. Hot stars are more luminous, and cool stars are less luminous, as you would expect. There is, however, a mystery about the main sequence that you can now solve. Why does the luminosity of a star depend on its mass?

In Chapter 9, you used binary stars to find the masses of stars, and you discovered that the masses of main-sequence stars are ordered along the main sequence. The least massive stars are at the bottom, and the most massive stars are at the top. Further, you discovered a direct relationship between the mass of a star and its luminosity—the mass–luminosity relation. This is one of the most fundamental observations in astronomy, and stellar models can tell you why there must be a mass–luminosity relation, and that tells you why there has to be a main sequence.

The keys to the mass–luminosity relation are the law of hydrostatic equilibrium, which says that pressure must balance weight, and the pressure–temperature thermostat, which regulates energy production. You have seen that a star's internal pressure stays high because the generation of thermonuclear energy keeps its interior hot. Because more massive stars have more weight pressing down on the inner layers, their interiors must have high pressures and thus must be hot. For example, the temperature at the center of a 15-solar-mass star is about 34,000,000 K, more than twice the central temperature of the sun.

Because massive stars have hotter cores, their nuclear reactions burn more fiercely. That is, their pressure–temperature thermostat is set higher. The nuclear fuel at the center of a 15-solar-mass star fuses over 3000 times more rapidly than the fuel at the center of the sun. The rapid reactions in the cores of massive stars produce more energy, but that energy cannot remain in the core. The energy must flow outward toward the cooler surface, and, in doing so, the flowing energy heats each level in the star and enables it to support the weight pressing inward. When all that energy reaches the surface, it radiates into space and makes the star highly luminous. That means there must be a mass–luminosity relation because each star must support its weight by generating nuclear energy, and more massive stars have more weight to support.

The main sequence is elegant in its simplicity. It exists because stars balance their weight by fusing hydrogen in their cores. To understand the main-sequence stars even better, you should next look at its top and bottom ends.

The Upper End of the Main Sequence

Stellar structure gives astronomers a way to think about the extreme ends of the main sequence, the most massive and least massive stars. The first are rare, but the latter are common. Nevertheless, both are very difficult to study.

There are two reasons why there is an upper limit to the mass of stars. First, it appears that gas clouds contracting to form a star can fragment and form two or more stars. If a gas cloud contains more mass, it is more likely to fragment, so there aren't many very massive stars because those gas clouds broke into smaller fragments and formed multiple-star systems.

But, second, stellar models reveal that stars of roughly 100 solar masses are unstable at formation. To support the tremendous weight in such stars, the internal gas must be very hot, and that means it must emit floods of radiation that flow outward through the star and blow gas away from the star's surface in powerful stellar winds. This mass loss from very massive stars could reduce a 60-solar-mass star to less than 30 solar masses in only a million years.

This process sets an upper limit on the mass of stars, but it is difficult to test the theory because it is hard to find truly massive stars. Most of the O and B stars in the sky have masses of 10 to 25 solar masses. The survey of stars at the end of Chapter 9 revealed that the stars at the upper end of the main sequence are very rare, so astronomers must search to great distances to find just a few. Nevertheless, a few stars are known that are thought to be very massive, and their spectra contain blueshifted emission lines. Kirchhoff's laws tell you that emission lines come from excited low-density gas, and the blueshift must be caused by a Doppler shift in gas coming toward Earth. These stars are losing mass.

■ Figure 12-3 shows a famous star, Eta Carinae, that is believed to be a massive binary in which the stars contain 60 solar masses and 70 solar masses. They may have formed with about 100 solar masses each, but mass loss is reducing their mass. An eruption 150 years ago made Eta Carinae the second brightest star in the sky and ejected the two expanding lobes of dusty gas. Although the star has faded, it is still very active, and more recent eruptions have ejected jets and an equatorial disk of gas and dust. Whatever their status, these massive stars are clearly unstable.

The Lower End of the Main Sequence

The lower end of the main sequence is difficult to study, not because the stars are rare but because they are dim. If a red dwarf from the lower end of the main sequence replaced the sun, it would shine only a few times brighter than the full moon. Such stars are difficult to find even when they are only a few light-years away.

Stellar models predict that stars less massive than 0.08 solar mass cannot get hot enough to ignite hydrogen fusion. These **brown dwarfs** should cool as they convert their gravitational energy into heat and radiate it away. A star of 0.08 solar mass fusing hydrogen has a surface temperature of about 2500 K, but brown dwarfs should have temperatures of 1000 K or so, giving them a color even ruddier than red dwarfs—thus the term "brown dwarf." Brown dwarfs were difficult to find because they are so faint, but large surveys and infrared studies have turned up lots. Some are located in binary systems with normal stars (■ Figure 12-4a), but large numbers are free-floating objects without stellar companions (Figure 12-4b).

Brown dwarfs are clearly different from normal stars. Some brown dwarfs have methane bands in their spectra, and that means they must be quite cool. Methane molecules would be broken up at the temperatures of true stars. Color variations suggest that some brown dwarfs may be cool enough to have weather patterns.

The discovery of brown dwarfs has created a controversy. Are they failed stars, or are they planets? To discuss this, astronomers use a unit of mass related to Jupiter, the giant planet in our solar system. Jupiter is one thousand times less massive than the sun, and astronomers refer to the mass of brown dwarfs in Jupiter masses. A brown dwarf must be less than about 80 Jupiter masses. Generally, astronomers think of stars as bodies that generate energy by nuclear fusion, and a star less than 80 Jupiter masses can't heat its center to the 2.7 million Kelvin temperature needed to start hydrogen fusion. Then brown dwarfs aren't stars. But astronomers tend to think of a planet as a nonluminous body that orbits a star, so can a brown dwarf floating free in space be called a planet?

Brown dwarfs more massive than 13 Jupiter masses can fuse deuterium, a heavy isotope of hydrogen, and some astronomers draw the line between stars and planets at 13 Jupiter masses.

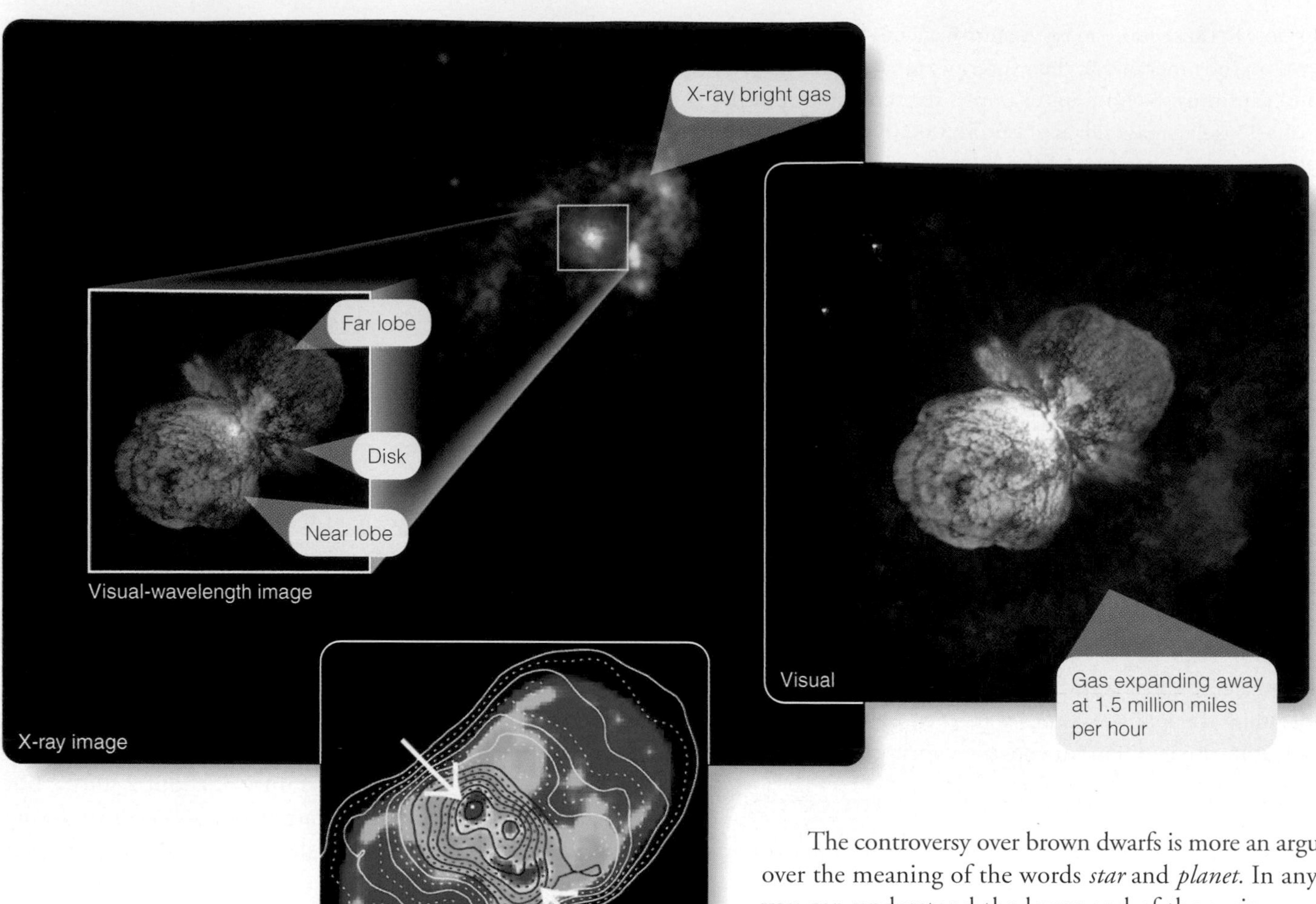

Figure 12-3

The star Eta Carinae is a binary containing two stars that are so massive they are rapidly losing mass. At visual wavelengths, two inflating lobes are visible with a disk of ejected material between like a plate pressed between two basketballs. Each lobe is about half a light-year in diameter. At X-ray wavelengths, very hot gas is excited by collision with high-speed gas ejected from the stars. An infrared image reveals a 15-solar-mass torus (doughnut shape) of gas and dust squeezing the outflowing gas into the two lobes. (NASA/CXC/SAO/HST; Jon Morse and J. Hester; IR image: ISO, Courtesy ESA)

However, deuterium fusion does not last very long, does not generate much energy, and does not stop the star from cooling. Not everyone agrees that this is a good dividing line.

Another approach to the controversy is to note how these objects form. A star forms from the contraction of a gas cloud, but astronomers believe planets form from the accumulation of solid bits of matter in disk-shaped nebulae around stars (a subject discussed in Chapter 19). In that case, free-floating brown dwarfs can't be planets.

The controversy over brown dwarfs is more an argument over the meaning of the words *star* and *planet*. In any case, you can understand the lower end of the main sequence. Stars less massive than 80 Jupiter masses cannot ignite hydrogen fusion; and, even if they are more massive than 13 Jupiter masses and deuterium fusion generates a little energy, they must continue contracting until the internal gas becomes so dense it cannot contract further. At that point, the object radiates its thermal energy away and slowly cools.

The Life of a Main-Sequence Star

A normal main-sequence star supports its weight by fusing hydrogen into helium, but its supply of hydrogen is limited. As it consumes its hydrogen, the chemical composition in its core changes, and the star evolves. Mathematical models of stars allow astronomers to follow that evolution.

Hydrogen fusion combines four nuclei into one. Consequently, as a main-sequence star consumes its hydrogen, the total number of nuclei in its interior decreases. Each newly made helium nucleus can exert the same pressure as a hydrogen nucleus; but, because the gas has fewer nuclei, its total pressure is less. This unbalances the gravity–pressure stability, and gravity squeezes the core of the star more tightly. As the core contracts, its temperature and density increase, and the nuclear reactions burn faster, releasing more energy and making the star more luminous. This additional energy flowing outward through the envelope

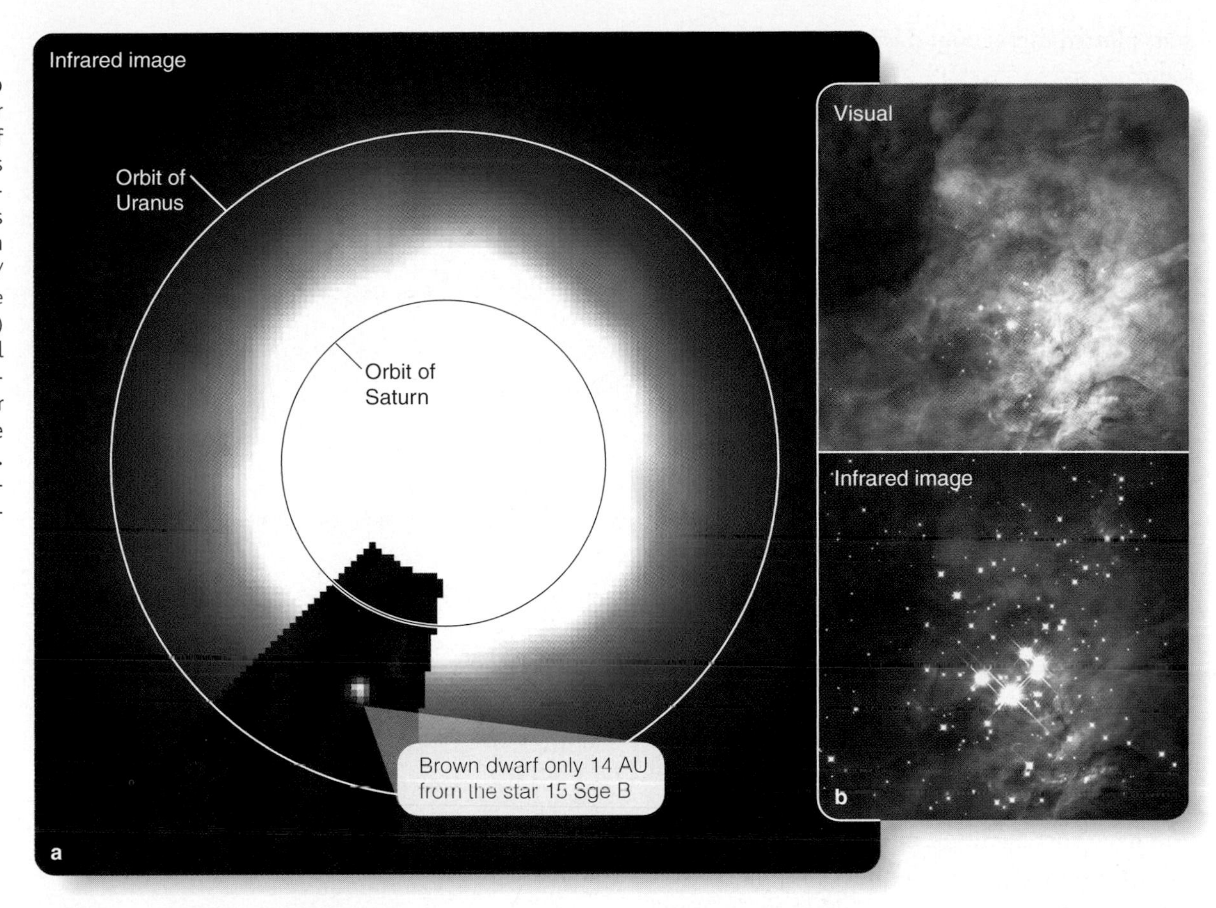

■ **Figure 12-4**

(a) Computer processing to remove glare from the star 15 Sge B reveals a brown dwarf only 14 AU from its star. It is about as far from its companion star as the giant planets Saturn and Uranus are from the sun. (Gemini Observatory/University of Hawaii Institute for Astronomy/Michael Liu/NSF) (b) Too dim to see at visual wavelengths, about 50 free-floating brown dwarfs appear in an infrared image of the center of the Orion nebula. The bright stars of the Trapezium are located at the center. (STScI and NASA)

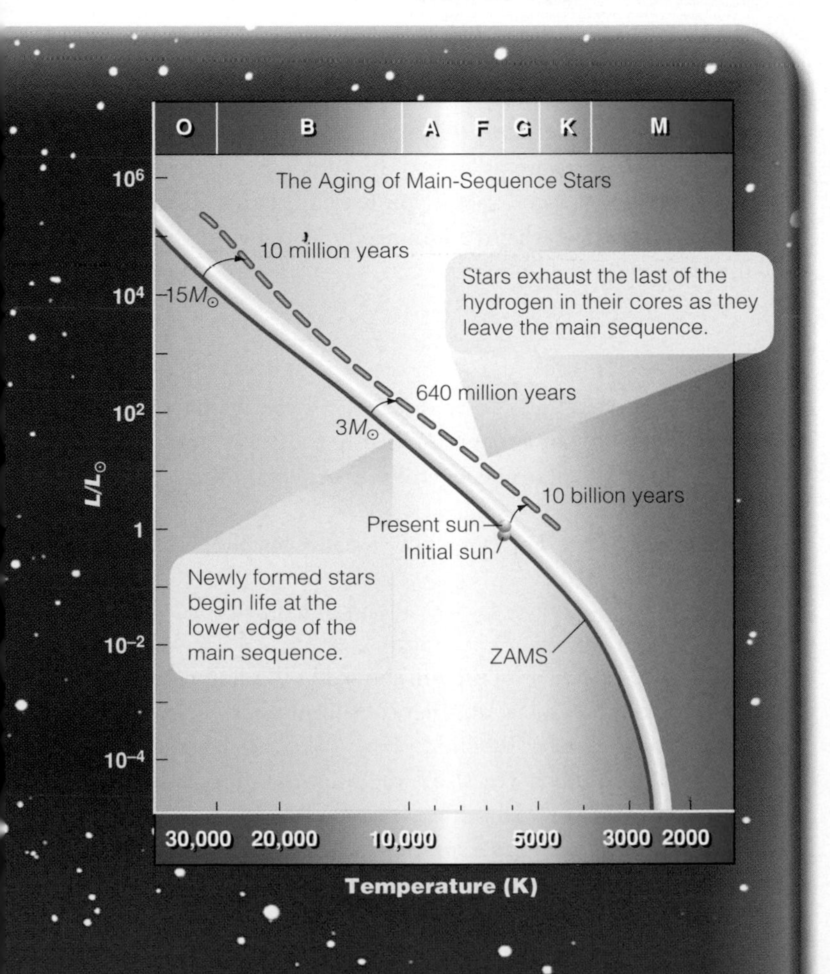

forces the outer layers to expand and cool, so the star becomes slightly larger, brighter, and cooler.

As a result of these gradual changes in main-sequence stars, the main sequence is not a sharp line across the H–R diagram but rather a band (■ Figure 12-5). A star begins its stable life fusing hydrogen and settles on the lower edge of this band, the **zero-age main sequence (ZAMS).** As it combines hydrogen nuclei to make helium nuclei, the star slowly changes. In the H–R diagram, the point that represents the star's luminosity and surface temperature moves upward and to the right, eventually reaching the upper edge of the main sequence, just as the star exhausts nearly all of the hydrogen in its center. Astronomers find main-sequence

■ **Active Figure 12-5**

Contracting protostars reach stability at the lower edge of the main sequence, the zero-age main sequence (ZAMS). As a star converts hydrogen in its core into helium, it moves slowly across the main sequence, becoming slightly more luminous and slightly cooler. Once a star consumes all of the hydrogen in its core, it can no longer remain a stable main-sequence star. More massive stars age rapidly, but less-massive stars use up the hydrogen in their cores more slowly and live longer main-sequence lives.

Ace Astronomy™ Log into AceAstronomy and select this chapter to see Active Figure "Future of the Sun." See how the future evolution of the sun will affect Earth.

stars plotted throughout this band at various stages of their main-sequence lives.

These gradual changes in the sun will spell trouble for Earth. When the sun began its main-sequence life about 5 billion years ago, it was only about 75 percent of its present luminosity. This, by the way, makes it difficult to explain how Earth has remained at roughly its present temperature for at least 3 billion years. Some experts suggest that Earth's atmosphere has gradually changed and compensated for the increasing luminosity of the sun.

By the time the sun leaves the main sequence in a few billion years, it will have twice its present luminosity. This will raise the average temperature on Earth by at least 19°C (34°F). As this happens over the next few billion years, the polar caps will melt, the oceans will evaporate, and much of the atmosphere will vanish into space. Clearly, the future of Earth as the home of life is limited by the evolution of the sun.

Once a star leaves the main sequence, it evolves rapidly and dies. The average star spends 90 percent of its life fusing hydrogen on the main sequence. This explains why 90 percent of all normal stars are main-sequence stars. You are most likely to see a star during that long, stable period when it is on the main sequence.

The number of years a star spends on the main sequence depends on its mass (■ Table 12-2). Massive stars use fuel rapidly and live short lives, but low-mass stars conserve their fuel and shine for billions of years. For example, a 25-solar-mass star will exhaust its hydrogen and die in only about 7 million years. This means, for one thing, that life is very unlikely to develop on planets orbiting massive stars. These stars do not live long enough for life to get started and evolve into complex creatures. You will learn about this problem in detail in Chapter 26.

Very-low-mass stars, the red dwarfs, use their fuel so slowly they should survive for 200 to 300 billion years. Because the universe seems to be only about 14 billion years old, red dwarfs must still be in their infancy. None of them should have exhausted their hydrogen fuel yet.

Nature makes more low-mass stars than massive stars, but this fact is not sufficient to explain the vast numbers of low-mass stars that fill the sky. An additional factor is the stellar lifetimes. Because low-mass stars live long lives, there are more of them in the sky than massive stars. Look at page 211 and notice how much more common the lower-main-sequence stars are than the massive O and B stars. The main-sequence K and M stars are so faint they are difficult to locate, but they are very common. The O and B stars are luminous and easy to locate; but, because of their fleeting lives, there are never more than a few on the main sequence at any one time.

The Life Expectancies of Stars

To understand how nature makes stars and how stars evolve, you must be able to estimate how long they can survive, and that turns out to be easy. In general, massive stars live short lives and lower-mass stars live long lives, but you can make more accurate estimates by using simple stellar models. In fact, you can calculate the approximate life expectancy of a star from its mass.

■ **Table 12-2 | Main-Sequence Stars**

Spectral Type	Mass (sun = 1)	Luminosity (sun = 1)	Approximate Years on Main Sequence
O5	40	405,000	1×10^6
B0	15	13,000	11×10^6
A0	3.5	80	440×10^6
F0	1.7	6.4	3×10^9
G0	1.1	1.4	8×10^9
K0	0.8	0.46	17×10^9
M0	0.5	0.08	56×10^9

Because main-sequence stars consume their fuel at an approximately constant rate, you can estimate the amount of time a star spends on the main sequence—its life expectancy T—by dividing the amount of fuel by the rate of fuel consumption. This is a common calculation. If you drive a truck that carries 20 gallons of fuel and uses 5 gallons of fuel per hour, you know the truck can run for 4 hours.

The amount of fuel a star has is proportional to its mass, and the rate at which it burns its fuel is proportional to its luminosity; that means you could make a first estimate of the star's life expectancy by dividing mass by luminosity. A 2-solar-mass star is about 11 times more luminous than the sun and should live about 2/11, or 18 percent, as long as the sun. You can, however, make the calculation even easier if you remember that the mass–luminosity relation says that the luminosity of a star equals $M^{3.5}$. The life expectancy then is:

$$T = \frac{M}{L} = \frac{M}{M^{3.5}} \quad \text{or} \quad T = \frac{1}{M^{2.5}}$$

This means that you can estimate the life expectancy of a star by dividing 1 by the star's mass raised to the 2.5 power. If you express the mass in solar masses, the life expectancy will be in solar lifetimes.

For example, how long can a 4-solar-mass star live?

$$T = \frac{1}{4^{2.5}} = \frac{1}{4 \cdot 4 \cdot \sqrt{4}} = \frac{1}{32} \text{ solar lifetimes}$$

Detailed studies of models of the sun show that the sun, presently 5 billion years old, can last another 5 billion years. So, a solar lifetime is approximately 10 billion years, and a 4-solar-mass star will last about (10 billion years)/32, or about 310 million years.

This estimation of stellar life expectancies is very approximate. For example, the model ignores mass loss, which may affect the life expectancies of very luminous and very faint stars.

Nevertheless, it serves to illustrate an important point. Stars that are only slightly more massive than the sun have dramatically shorter lifetimes on the main sequence.

Building Scientific Arguments

Why is there a main sequence?

Most scientific arguments are simple chains of ideas that begin from a well-understood idea and lead to a less obvious conclusion. You can begin with the simple observation that weight must be balanced by pressure. That is just the principle of hydrostatic equilibrium. A contracting protostar is not quite in equilibrium, and gravity squeezes it tighter and tighter. As it contracts, its interior heats up; and, when it gets hot enough to fuse hydrogen into helium, the pressure–temperature thermostat takes over and regulates energy production so that the star makes just enough energy to support its own weight. That means that massive stars, having more weight to support, must have higher internal pressures and therefore must make more energy. Energy must flow from hot to cool, so that energy flows from the hot core of the star out to the cooler surface and is radiated into space. Massive stars, having more weight to support, must have their pressure–temperature thermostats set higher, and that makes them more luminous. They reach stability along the upper main sequence. Less-massive stars support less weight and reach stability along the lower main sequence. There is a main sequence because stars fuse hydrogen to support their own weight.

Now revise your argument. **Why do massive stars have such short life expectancies?**

■ ■ ■

Connections: When a star finally exhausts its hydrogen fuel, it can no longer resist the pull of its own gravity. The contraction that began when it was a protostar resumes and begins the process that leads to the star's death. It can delay its end by fusing other fuels, but, as you will discover in Chapter 13, nothing can steal gravity's final victory.

12-2 Post-Main-Sequence Evolution

IN EARLIER CHAPTERS, you probably had questions about giant stars: Why are giant stars so large? Why are they so uncommon? And why do they have such low densities? Now you are ready to answer those questions by discussing the evolution of stars after they leave the main sequence.

Expansion into a Giant

To understand how stars evolve, you must remember that they are not well mixed; that is, their interiors are not stirred. The centers of lower-mass stars like the sun are radiative, meaning the energy moves as radiation and not as circulating currents of heated gas. The gas does not move deep inside such stars, and that means they are not mixed at all. More massive stars have convective cores that mix the central regions (Figure 11-14), but these regions are not very large, and so, for the most part, these stars, too, are not mixed. (The lowest-mass stars are an exception that you will examine in the next chapter.)

In this respect, stars are like a campfire that is not stirred; the ashes accumulate at the center, and the fuel in the outer parts never gets used. Nuclear fusion consumes hydrogen nuclei and produces helium nuclei, the "ashes," at the star's center. Nothing mixes the interior of the star, so the helium nuclei remain where they are in the center of the star, and the hydrogen in the outer parts of the star is not mixed down to the center where it can be fused.

The helium ashes that accumulate in the star's core cannot fuse into heavier elements because the temperature is too low. As a result, the helium accumulates, the hydrogen is used up, and the core becomes an inert ball of helium. As this happens, the energy production in the core falls, and the weight of the outer layers forces the core to contract.

Although the contracting helium core cannot generate nuclear energy, it does grow hotter because it converts gravitational energy into thermal energy (see Chapter 11). The rising temperature heats the unprocessed hydrogen just outside the core, hydrogen that was never before hot enough to fuse. When the surrounding hydrogen becomes hot enough, it ignites in a shell of fusing hydrogen. Like a grass fire burning outward from an exhausted campfire, the hydrogen-fusing shell burns outward, leaving helium ash behind and increasing the mass of the helium core.

At this stage in its evolution, the star overproduces energy; that is, it produces more energy than it needs to balance its own gravity. The helium core, having no nuclear energy sources, must contract, and that contraction converts gravitational energy into thermal energy—the contraction heats the helium core. Some of that heat leaks outward through the star. At the same time, the hydrogen-fusing shell produces energy as the contracting core brings fresh hydrogen closer to the center of the star and heats it to high temperature. The result is a flood of energy flowing outward through the outer layers of the star, forcing them to puff up and swelling the star into a giant (■ Figure 12-6).

The expansion of the envelope dramatically changes the star's location in the H–R diagram. As the outer layers of gas expand, energy is absorbed in lifting and expanding the gas. The loss of that energy lowers the temperature of the gas. Consequently, the point that represents the star in the H–R diagram moves quickly to the right (in less than a million years for a star of 5 solar masses). A medium-mass star like the sun expands and cools to become a red giant (■ Figure 12-7). As the radius of a giant star continues to increase, the enlarging surface area makes the star more luminous, moving its point upward in the H–R diagram. One of our

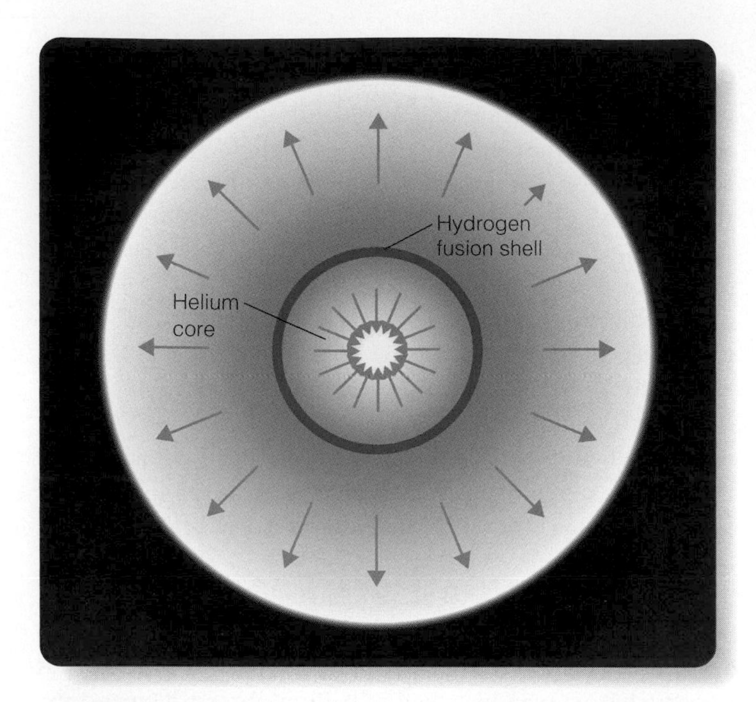

■ Figure 12-6

When a star runs out of hydrogen at its center, it ignites a hydrogen-fusing shell. The helium core contracts and heats while the envelope expands and cools. (For a scale drawing, see Figure 12-9.)

Now you can understand a few things you noticed in earlier chapters. Giants and supergiants are large because they have expanded, and they have very low densities for the same reason. Also, the giant and supergiant region of the H–R diagram contains a jumble of different mass stars because all main-sequence stars funnel through this part of the diagram as they die. (See Figure 9-23.)

Degenerate Matter

Although the hydrogen-fusing shell can force the envelope of the star to expand, it cannot stop the contraction of the helium core. Because the core has no energy source, gravity squeezes it tighter, and it becomes very small. If you represent the helium core of a 5-solar-mass star with a quarter, the outer envelope of the star would be about the size of a baseball diamond. Yet the core would contain about 12 percent of the star's mass compressed to very high density. When gas is compressed to such extreme densities, it begins to behave in astonishing ways that can alter the evolution of the star. To follow the story of stellar evolution, you must consider the behavior of gas at extremely high densities.

Normally, the pressure in a gas depends on its temperature. The hotter the gas is, the faster its particles move, and the more pressure it exerts. But the gas inside a star is ionized, so there are

Favorite Stars, Aldebaran, the glowing red eye of Taurus the bull, is such a red giant, with a diameter 25 times that of the sun but with a surface temperature only half that of the sun.

More massive stars cross higher in the H–R diagram and become supergiants, stars even larger than giants. Consider two more Favorite Stars. Betelgeuse (α Orionis) is a very cool, red supergiant over 800 times the diameter of the sun. Rigel (β Orionis) is a supergiant 50 times larger than the sun. It may seem odd to say that Rigel, a blue star, has expanded and cooled. At a temperature of 12,000 K, Rigel looks quite blue to your eyes. When it was on the main sequence, Rigel had a much hotter surface, but it would not have looked much bluer because you cannot see the ultraviolet radiation that the hottest stars emit. Any star as hot or hotter than Rigel will look blue to your eyes, as shown in Figure 7-2. So, blue Rigel is indeed a star that has expanded and cooled.

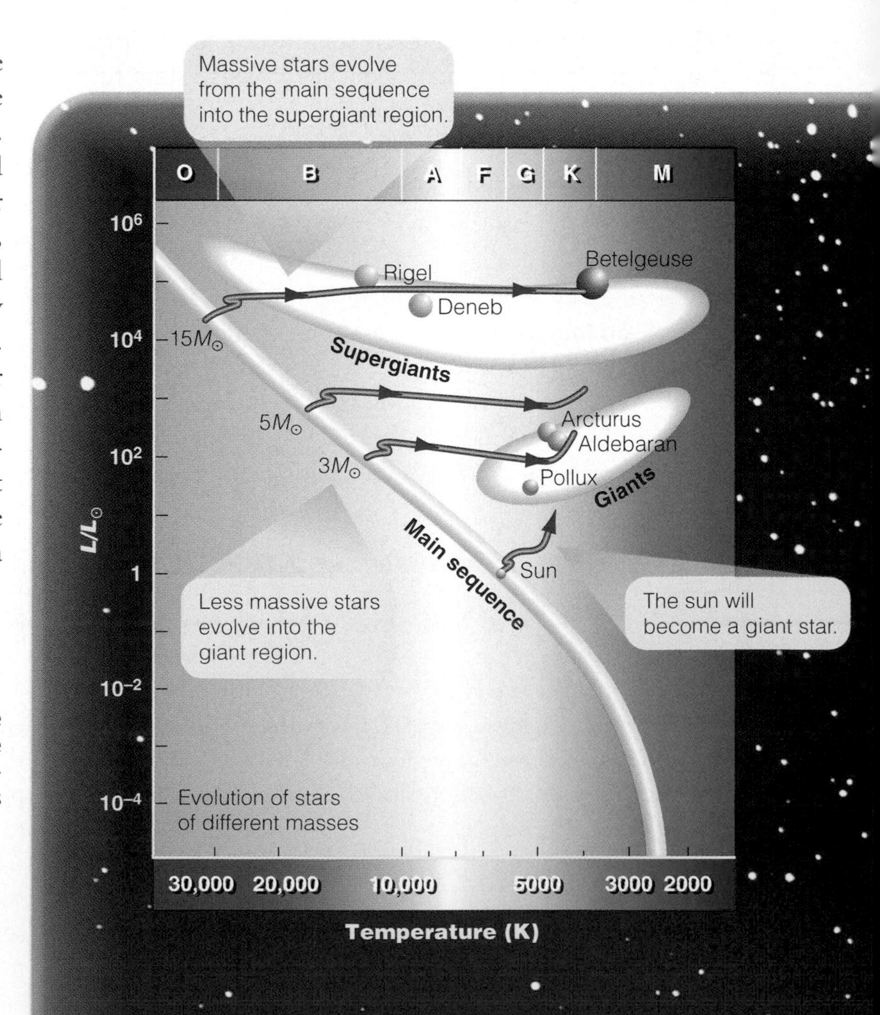

■ Active Figure 12-7

The evolution of a massive star moves the point that represents it in the H–R diagram to the right of the main sequence into the region of the supergiants such as Rigel and Betelgeuse. The evolution of a medium-mass star moves its point in the H–R diagram into the region of the giants such as those shown here.

AceAstronomy™ Log into AceAstronomy and select this chapter to see Active Figure "Evolution of Stars" and compare the evolution of a massive star to that of a low-mass star.

two kinds of particles, atomic nuclei and free electrons. If the gas is compressed to very high densities, in the core of a giant star, for example, the difference between these two kinds of particles is important.

If the density is very high, the particles of the gas are forced close together, and two laws of quantum mechanics become important. First, quantum mechanics says that the moving electrons confined in the star's core can have only certain amounts of energy, just as the electron in an atom can occupy only certain energy levels (see Chapter 7). You can think of these permitted energies as the rungs of an energy ladder. An electron can occupy any rung but not the spaces between.

The second quantum-mechanical law (called the Pauli Exclusion Principle) says that two identical electrons cannot occupy the same energy level. Because electrons spin in one direction or the other, two electrons can occupy an energy level if they spin in opposite directions. That level is then completely filled, and a third electron cannot enter because, whichever way it spins, it will be identical to one or the other of the two electrons already in the level. Thus, no more than two electrons can occupy the same energy level.

A low-density gas has few electrons per cubic centimeter, so there are plenty of energy levels available. If a gas becomes very dense, however, nearly all of the lower energy levels may be occupied (■ Figure 12-8). In such a gas, a moving electron cannot slow down, because slowing down would decrease its energy, and

■ Figure 12-8

Electron energy levels are arranged like rungs on a ladder. In a low-density gas, many levels are open, but in a degenerate gas, all lower energy levels are filled.

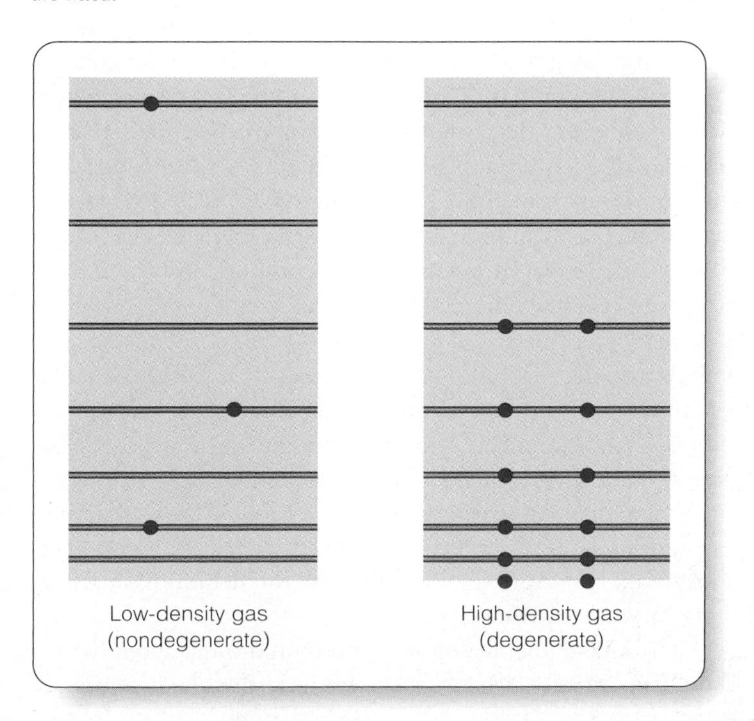

there are no open energy levels for it to drop down to. It can speed up only if it can absorb enough energy to leap to the top of the energy ladder, where there are empty energy levels.

When a gas is so dense that the electrons are not free to change their energy, astronomers call it **degenerate matter.** Although it is a gas, it has two peculiar properties that can affect the star. First, the degenerate gas resists compression. To compress the gas, you would have to push against the moving electrons, and changing their motion means changing their energy. That requires tremendous effort, because you would have to boost them to the top of the energy ladder. That makes degenerate matter, though still a gas, harder to compress than the hardest steel.

Second, the pressure of degenerate gas does not depend on temperature but rather on the speed of the electrons, which cannot be changed without tremendous effort. The temperature, however, depends on the motion of all the particles in the gas, both electrons and nuclei. If you were to add thermal energy to the gas, most of that energy would go to speed up the motions of the nuclei, and only a few electrons would be able to absorb enough energy to reach the empty energy levels at the top of the energy ladder. This means that changing the temperature of the gas has almost no effect on the pressure.

These two properties of degenerate matter become important when stars end their main-sequence lives and approach their final collapse (**Window on Science 12-2**). Eventually, many stars collapse into white dwarfs, and you will discover that these tiny stars are made of degenerate matter. But long before that, the cores of many giant stars become so dense that they are degenerate, a situation that can produce a cosmic bomb when helium begins to fuse.

Helium Fusion

Hydrogen fusion leaves behind helium ash, which cannot fuse at the relatively low temperatures of hydrogen fusion. Helium nuclei have a positive charge twice that of a hydrogen nucleus (a proton), which means the Coulomb barrier is higher. At the temperature of hydrogen fusion, the particles move too slowly and collide too gently to produce helium fusion. Consequently, the helium that accumulates is an ash that cannot fuse until the temperature rises significantly.

As the star becomes a giant star, fusing hydrogen in a shell, the inner core of helium ash contracts and grows hotter. It may even become degenerate. Finally, as the temperature approaches 100,000,000 K, helium nuclei begin to fuse together to make carbon.

You can summarize the helium-fusing process in two steps:

$$^{4}\text{He} + {}^{4}\text{He} \rightarrow {}^{8}\text{Be} + \gamma$$
$$^{8}\text{Be} + {}^{4}\text{He} \rightarrow {}^{12}\text{C} + \gamma$$

This process is complicated by the fact that beryllium-8 is very unstable and may break up into two helium nuclei before it can

Window on Science | 12-2

Causes in Nature: The Very Small and the Very Large

One of the most interesting lessons of science is that the behavior of very small things often determines the structure and behavior of very large things. In degenerate matter, the quantum mechanical behavior of electrons helps determine the evolution of giant stars. Such links between the very small and the very large are common in astronomy, and you will see in later chapters how the nature of certain subatomic particles may determine the fate of the entire universe.

This is a second reason why astronomers study atoms. Recall that the first reason is that atoms interact with light, and astronomers must understand that interaction in order to analyze the light. The second reason is the link between the very small and the very large. It would be impossible to understand the evolution of the largest stars if astronomers failed to understand how degenerate electrons behave. Similarly, astronomers studying galaxies must understand how microscopic dust in space reddens starlight, and astronomers studying planets must know how molecules form crystals.

As you study any science, look for the way very small things affect very large things. Sociologists and psychologists know that the mass behavior of very large groups of people can depend on the behavior of a few key individuals. Biologists study the visible consequences of atomic bonds in molecules called genes, and meteorologists know that tiny changes in temperature in one part of the world can affect worldwide climate weeks or months later.

Nature is the sum of many tiny parts, and science is the study of nature. One way to do science is to search out the tiny causes that determine the way nature behaves on the grandest scales.

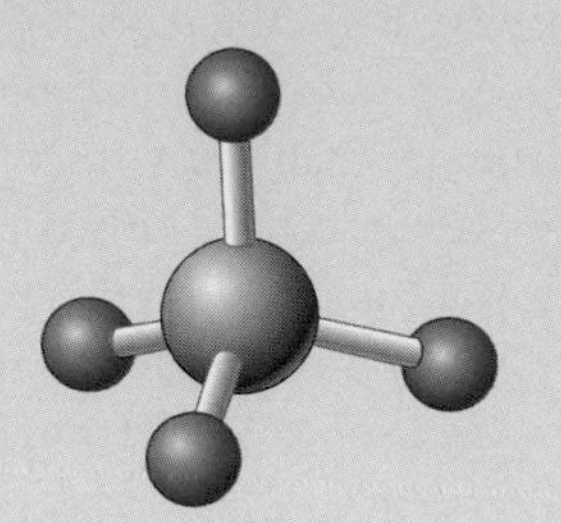

The properties of atoms and molecules determine the structure of the largest objects in the universe.

absorb another helium nucleus. Three helium nuclei can also form carbon directly, but such a triple collision is unlikely. Because a helium nucleus is called an alpha particle, astronomers often refer to helium fusion as the **triple alpha process.**

Some stars begin helium fusion gradually, but stars in a certain mass range begin helium fusion with an explosion called the **helium flash.** This explosion is caused by the density of the helium, which can reach 1,000,000 g/cm^3. On Earth, a teaspoon of this material would weigh as much as a large truck. At these densities, the gas is degenerate, and its pressure no longer depends on temperature. That means the pressure–temperature thermostat that controls the nuclear fusion reactions no longer works.

When the helium ignites, it generates energy, which raises the temperature. Because the pressure–temperature thermostat is not working in the degenerate gas, the core does not respond to the higher temperature by expanding. Rather, the higher temperature forces the reactions to go faster, which makes more energy, which raises the temperature, which makes the reactions go faster, and so on. The ignition of helium fusion in a degenerate gas results in a runaway explosion so violent that for a few minutes the helium core generates more energy than an entire galaxy. At its peak, the core generates 10^{14} times more energy per second than the sun.

Although the helium flash is sudden and powerful, it does not destroy the star. In fact, if you were observing a giant star as it experienced a helium flash, you would see no outward evidence of an eruption. The helium core is quite small (■ Figure 12-9), and all of the energy of the explosion is absorbed by the distended envelope. Also, the helium flash is a very short-lived event. In a matter of minutes, the core of the star becomes so hot it is no longer degenerate, the pressure–temperature thermostat brings the helium fusion under control, and the star proceeds to fuse helium steadily in its core. You will learn about this post-helium-flash evolution later.

Not all stars experience a helium flash. Stars less massive than about 0.4 solar mass can never get hot enough to ignite helium, and stars more massive than about 3 solar masses ignite helium before their cores become degenerate (■ Figure 12-10). In such stars, pressure depends on temperature, so the pressure–temperature thermostat keeps the helium fusion under control.

If the helium flash occurs only in some stars and is a very short-lived event that is not visible from outside the star, why should you worry about it? The answer is that it limits the reliability of the mathematical models astronomers use to study stellar evolution. The medium-mass stars, which experience the helium flash, are those that astronomers must consider most carefully in studying stellar evolution. Massive stars are not very common; low-mass stars evolve so slowly that it is hard to find evidence of low-mass stellar evolution. For those reasons, studies of stellar evolution must concentrate on medium-mass stars, which do experience the helium flash. But the helium flash occurs so rapidly and so violently that computer programs cannot follow the changes in the star's internal structure in detail. To follow the evolution of medium-mass stars like the sun past the helium flash, astronomers must make assumptions about the way the helium flash affects stars' internal structures.

There is another reason why you should think about the helium flash. By seeing what happens when the pressure–temperature

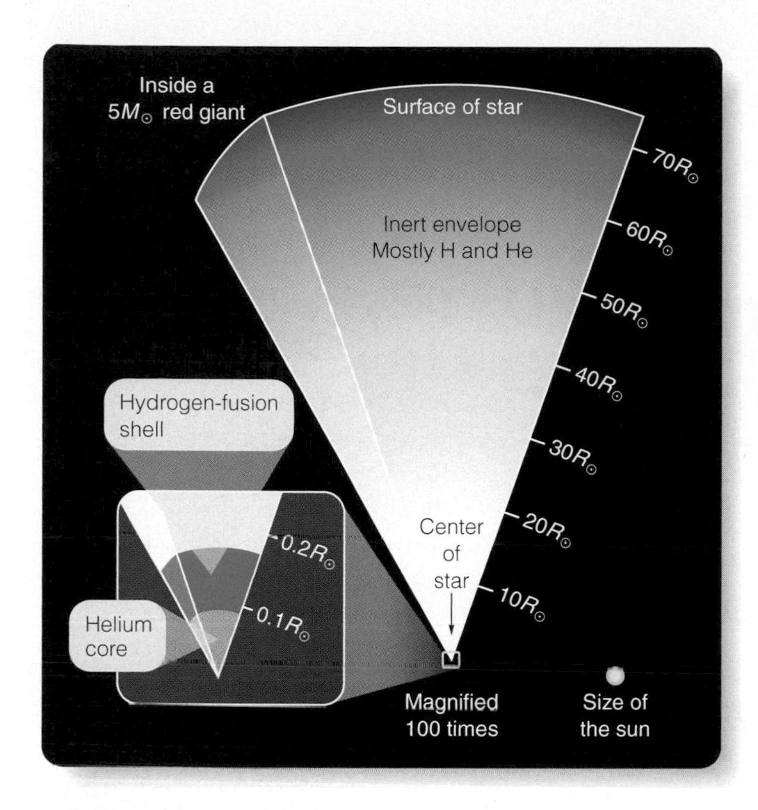

■ Active Figure 12-9

When a star runs out of hydrogen at its center, the core of helium contracts to a small size, becomes very hot, and begins nuclear fusion in a shell (blue). The outer layers of the star expand and cool. The red giant star shown here has an average density much lower than the air at Earth's surface. Here $M_\odot$ stands for the mass of the sun, and $R_\odot$ stands for the radius of the sun. (Illustration design by the author)

Ace Astronomy™ Log into AceAstronomy and select this chapter to see Active Figure "Inside Stars." Compare the interior of stars with that of the sun.

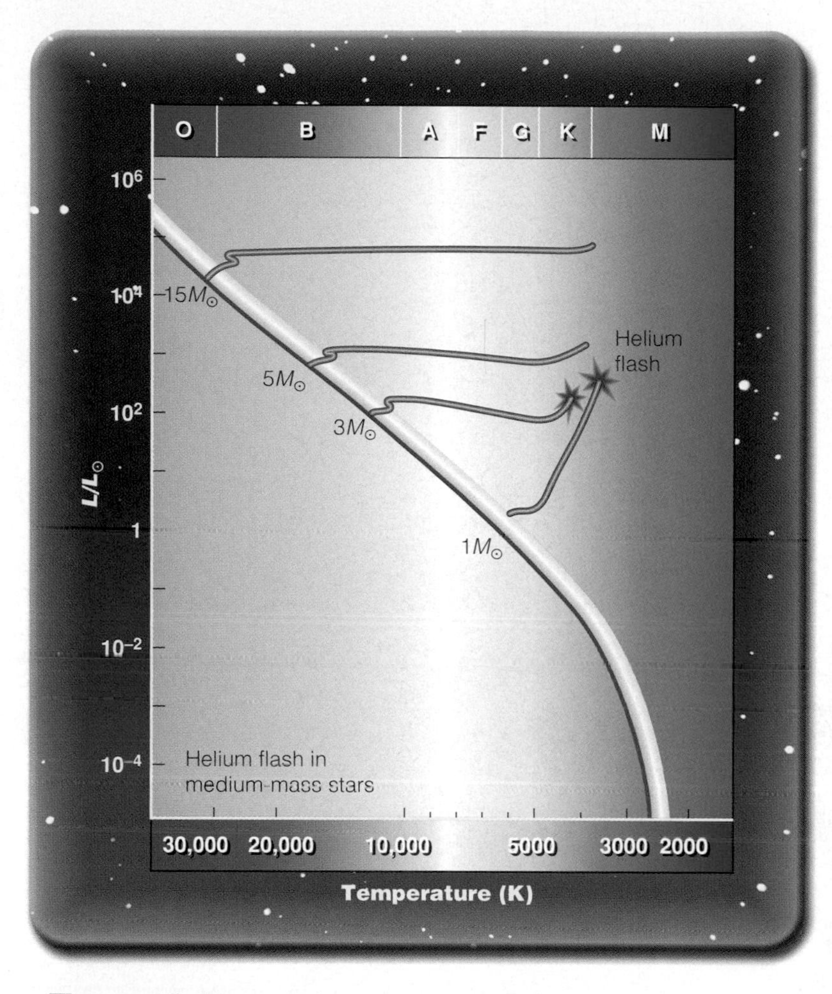

■ Figure 12-10

Stars like the sun suffer a helium flash, but more massive stars begin helium fusion without a helium flash. Stars less than 0.4 times the mass of the sun cannot get hot enough to ignite helium. The zero-age main sequence is shown here in red at the lower edge of the main sequence.

thermostat is turned off, you can appreciate its importance in stars like the sun.

Post-helium-flash evolution is generally understood. As the core temperature rises, the gas ceases to be degenerate, and helium fusion proceeds under the control of the pressure–temperature thermostat. Throughout these events, the hydrogen-fusion shell continues to produce energy, but the new helium-fusion energy produced in the core makes the core expand. That expansion absorbs energy previously used to support the outer layers of the star. As a result, the outer layers contract, and the surface of the star grows slightly hotter. In the H–R diagram, the point that represents the star initially moves down toward lower luminosity and to the left toward higher temperature. Later, as the star stabilizes fusing helium in its core, its luminosity recovers, but its surface grows hotter.

Helium fusion produces carbon and oxygen as "ash," and these accumulate at the center of the star. They do not fuse because the core is not hot enough. Thus, the helium in the core is gradually converted into carbon–oxygen ashes. As this happens, the core contracts, grows hotter, and ignites a helium-fusion shell. At this stage, the star has two shells producing energy. A hydrogen-fusion shell eats its way outward, leaving behind helium ash; a helium-fusion shell eats its way outward into the helium, leaving behind carbon–oxygen ash. Unable to generate energy in its carbon–oxygen core, the star expands its outer layers, and its point in the H–R diagram moves back toward the right, completing a loop. You can see that small loop in the giant region of the H–R diagram as shown in ■ Figure 12-11. Expanded drawings show how the interior of the star changes as it uses up its fuels.

Fusing Elements Heavier Than Helium

Modeling the evolution of stars after helium exhaustion requires great sophistication. The inert core of carbon–oxygen ash contracts and becomes hotter. Stars more massive than about 4 solar masses

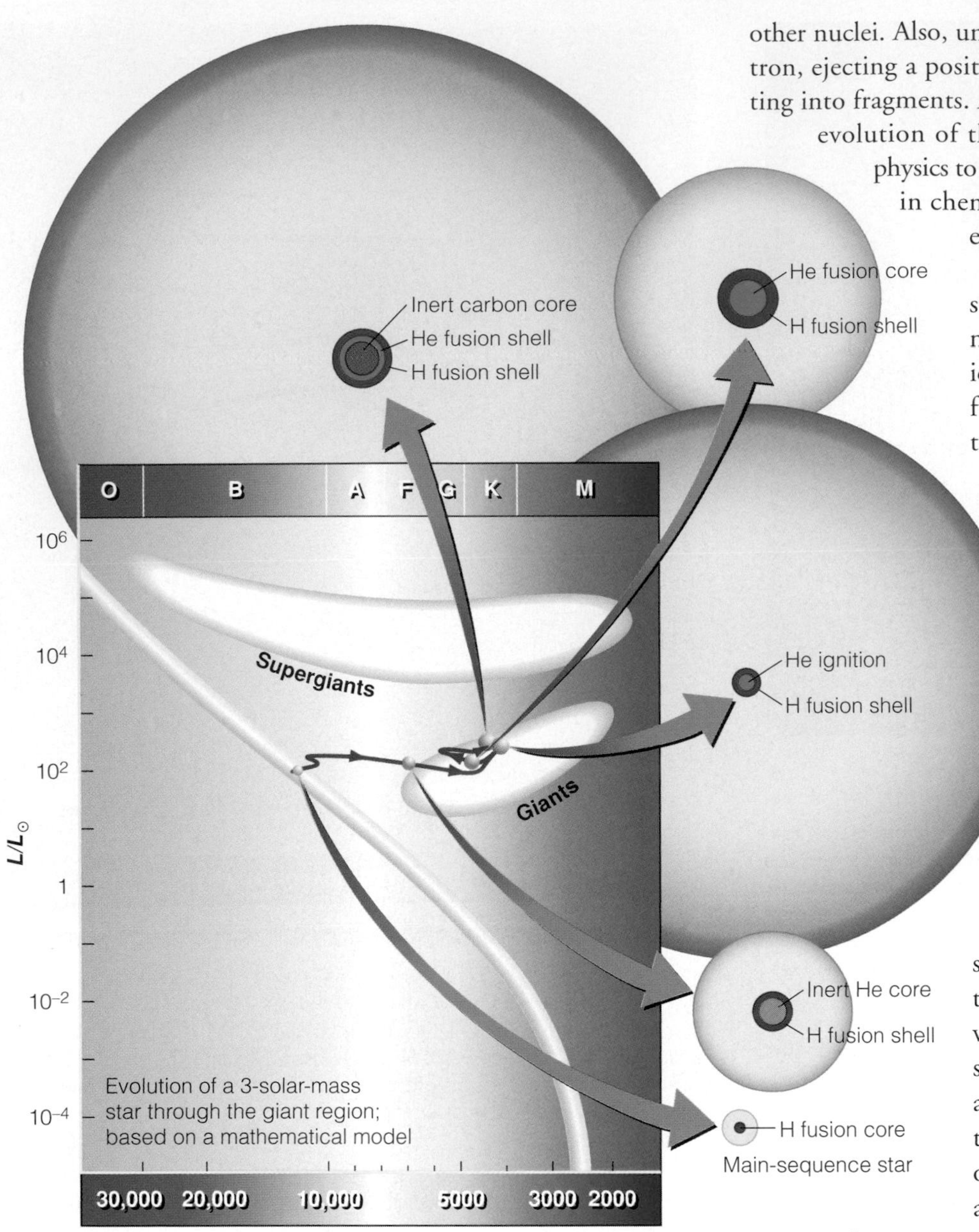

■ **Figure 12-11**

When a main-sequence star exhausts the hydrogen in its core, it evolves rapidly to the right in the H–R diagram as it expands to become a cool giant. These changes are caused at first by the ignition of a hydrogen fusion shell. When helium ignites in the core, the star contracts and heats up slightly, but when helium begins to fuse in a shell, the star again expands and cools.

can heat their cores to temperatures as high as 600,000,000 K, hot enough to fuse carbon. Subsequently, the stars can fuse oxygen, silicon, and other heavy elements.

Carbon fusion in a star begins a complex network of reactions illustrated in ■ Figure 12-12, where each circle represents a possible nucleus and each arrow represents a different nuclear reaction. Nuclei can react by capturing a proton, capturing a neutron, capturing a helium nucleus, or by combining directly with other nuclei. Also, unstable nuclei can decay by ejecting an electron, ejecting a positron, ejecting a helium nucleus or by splitting into fragments. Astronomers who model the structure and evolution of these stars must use sophisticated nuclear physics to compute the energy production and changes in chemical composition as the stars fuse heavier elements.

These stars develop complex internal structure as they fuse heavier and heavier elements in concentric shells. Like great spherical cakes with many layers, the stars may be fusing a number of different elements simultaneously in concentric shells.

Eventually, all stars face collapse. Less massive stars cannot get hot enough to ignite the heavier nuclei. The more massive stars can ignite heavy nuclei, but these stars consume their fuels at a tremendous rate, so they run through their available fuel quickly (■ Table 12-3). No matter what the star's mass, it will eventually run out of usable fuels, and gravity will win the struggle with pressure. You will explore the ultimate deaths of stars in the next chapter.

The time a star spends as a giant or supergiant is small compared with its life on the main sequence. The sun, for example, will spend about 10 billion years as a main-sequence star but only about a billion years as a giant. The more massive stars pass through the giant stage even more rapidly. Because of the short time a star spends as a giant, you are unlikely to see many such stars. This illustrates an important principle in astronomy: The shorter the time a given evolutionary stage takes, the less likely you are to see stars in that particular stage. That explains why you see a great many main-sequence stars but few giants (see page 211).

Building Scientific Arguments

What uncertainties limit the accuracy of our story of stellar evolution?

A scientific argument is not meant to convince others but to test your own understanding, and that means it should include a discussion of any uncertainties in the analysis. In the case of stellar evolution, there are two things that limit the accuracy of the models. First, some events in the history of a star happen so fast that computers can't predict the results in a reasonable amount of time. In fact, some astronomers estimate that a computer following the evolution of a model

■ Figure 12-12

Carbon fusion involves many possible reactions that build numerous heavy nuclei.

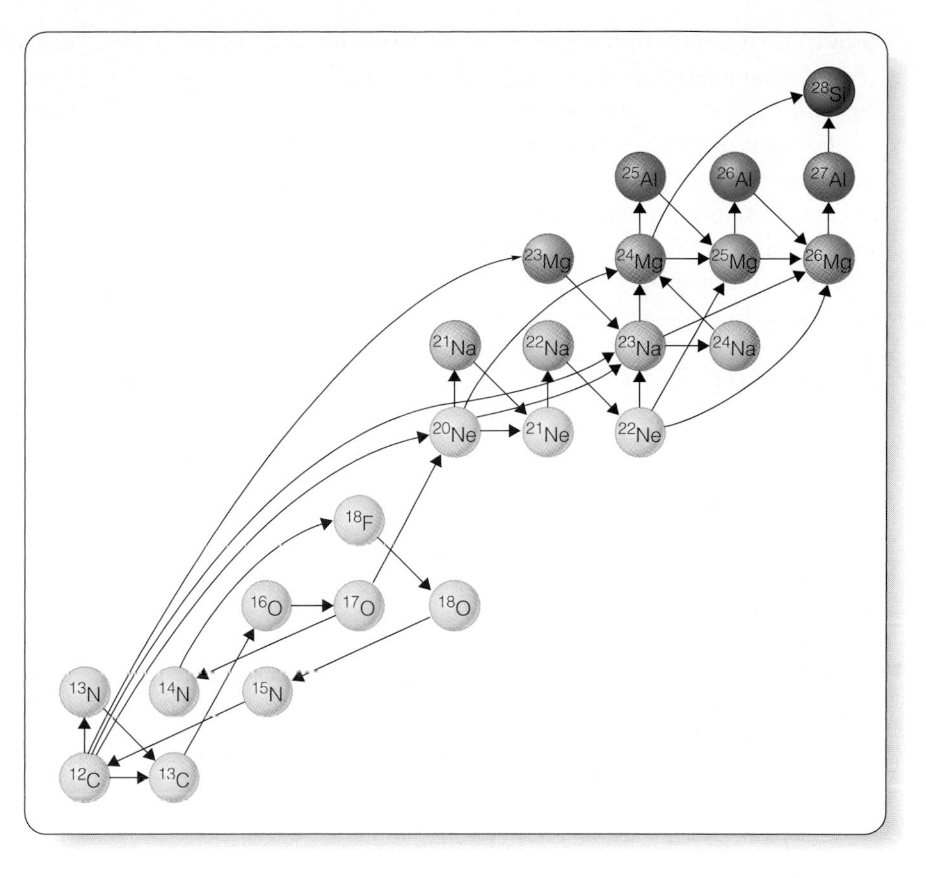

star undergoing the helium flash would have to recompute the model in such short time steps that the computer program would take millions of years to predict events that occur inside the star in just seconds. Truly rapid changes such as explosions in stars make the models uncertain.

Second, you have seen how the general properties of a giant star can depend on the subatomic properties of matter. Not all of those subatomic properties are well understood. For example, neutrinos oscillate, and scientists don't entirely understand how that happens. If some subatomic particle behaves differently than expected, then stellar models may be slightly different from the real stars they are supposed to represent.

Can you add to your argument? **What other physical processes have been omitted from stellar models? What features of the sun's surface suggest motions and forces that are ignored in most stellar models?**

■ ■ ■

Connections: Humans do not live long enough to see the stars evolving, so astronomers must test their theoretical story of stellar evolution against observation. To paraphrase the quotation at the beginning of this chapter, you shouldn't believe any theory until you compare it with reality—seeing is believing. Consequently, in the rest of this chapter, you will search for evidence to support the theory of stellar evolution. As you search, you will discover some interesting consequences of the fact that the stars are not eternal.

12-3 Evidence of Evolution: Star Clusters

THE THEORY OF STELLAR EVOLUTION is so complex and involves so many assumptions that astronomers would have little confidence in it were it not for H–R diagrams of clusters of stars. By observing the properties of star clusters of different ages, you can see clear evidence of the evolution of stars.

To grasp the difficulty of understanding stellar evolution, consider an analogy. Suppose a visitor to Earth who had never seen a tree wandered through a forest for an hour looking at mature trees, fallen seeds, young saplings, fallen logs, and rising

■ Table 12-3 | Nuclear Reactions in Massive Stars

Nuclear Fuel	Nuclear Products	Minimum Ignition Temperature	Main-Sequence Mass Needed to Ignite Fusion	Duration of Fusion in a 25-$M_\odot$ Star
H	He	4×10^6 K	0.1 $M_\odot$	7×10^6 yr
He	C, O	120×10^6 K	0.4 $M_\odot$	0.5×10^6 yr
C	Ne, Na, Mg, O	0.6×10^9 K	4 $M_\odot$	600 yr
Ne	O, Mg	1.2×10^9 K	~8 $M_\odot$	1 yr
O	Si, S, P	1.5×10^9 K	~8 $M_\odot$	~ 0.5 yr
Si	Ni to Fe	2.7×10^9 K	~8 $M_\odot$	~ 1 day

sprouts. Could such an observer understand the life cycle of the trees? Astronomers face the same problem when they try to understand the life story of the stars.

Humans do not live long enough to see stars evolve. You see only a momentary glimpse of the universe as it is during your lifetime, a snapshot in which all the stages in the life cycle of the stars are represented. Unscrambling these stages and putting them in order is a difficult task. Only by looking at selected groups of stars—star clusters—can you see the pattern.

Observing Star Clusters

Just as Sherlock Holmes studies peculiar dust on a lamp shade as a clue to a crime, astronomers look at star clusters and say, "Aha!" Each star cluster freezes a moment in the evolution of the cluster and makes the evolution of the stars visible.

The stars in a cluster all formed at about the same time and from the same cloud of gas, so they must be about the same age and composition. The differences you see among stars in a cluster must arise from differences in mass, and that makes stellar evolution visible. Study **Star Cluster H–R Diagrams** on pages 276–277 and notice three important points:

1. There are two kinds of star clusters, but they are similar in the way their stars evolve. You will learn more about these clusters in a later chapter.

2. Notice that you can estimate the age of a star cluster by observing the distribution of the points that represent its stars in the H–R diagram.

3. Finally, notice that the shape of a star cluster's H–R diagram is governed by the evolutionary path the stars take. By comparing clusters of different ages, you can visualize how stars evolve almost as if you were watching a film of a star cluster evolving over billions of years.

Were it not for star clusters, astronomers would have little confidence in the theories of stellar evolution. Star clusters make that evolution visible and assure astronomers that they really do understand how stars are born, live, and die.

The Evolution of Star Clusters

What you know about star formation and stellar evolution can help you understand star clusters and how they evolve.

A star cluster is formed when a cloud of gas contracts, fragments, and forms a group of stars. As you recall from Chapter 11, some groups of stars, called associations, are so widely scattered the group is not held together by its own gravity. The stars in an association wander away from each other rather quickly. Some groups of stars are more compact; even after the stars become luminous and the gas and dust are blown away, gravity holds the group together as a star cluster.

Open clusters are not as old or as crowded as the globular clusters, and that helps explain their appearance. Close encounters between stars are rare in an open cluster where the stars are further apart, and such clusters have an irregular appearance. The globular clusters appear to be nearly perfect globes because the stars are much closer together and encounters between stars are more common. The globular clusters have had time to evenly distribute the energy of motion among all of the stars so they have settled into a more uniform, spherical shape.

As a star cluster ages, some stars traveling a bit faster than others can escape. Globular clusters are compact and massive and have survived for 11 billion years or more. A star cluster with only a few stars widely distributed may evaporate completely as its stars escape one by one. Our sun probably formed in a star cluster 5 billion years ago, but there is no way to trace it back to its family home even if that star cluster still exists.

Building Scientific Arguments

Why do open clusters contain only a small number of giant stars but many main-sequence stars?

Sometimes the critical factor in a scientific argument is timing. When you look at a star cluster, you see a snapshot of the cluster that freezes it in time. In the H–R diagram of an open cluster, you see a main sequence containing many stars, but typically you see significantly fewer giant stars. Those giant stars were once main-sequence stars in the cluster, but they exhausted the hydrogen fuel in their cores and expanded to become giants. A star like the sun can fuse hydrogen on the main sequence for many billions of years but can remain a giant for only a billion years or less. More massive stars evolve even faster. Because of this, at any given moment, you see most of the stars in a star cluster on the main sequence and only a few, caught in the snapshot, as giant stars.

Now create a new argument. **How do H–R diagrams reveal the age of a star cluster?**

■ ■ ■

Connections: Star clusters provide dramatic confirmation that stars evolve and also reveal some of the details of stellar structure and evolution. Certain pulsating giant stars provide even further information about how stars work.

12-4 Evidence of Evolution: Variable Stars

FOR MILLENNIA, poets and astronomers believed that the stars were constant and unchanging. Modern theory shows that the stars are slowly evolving as they consume their fuels. But can you really be sure the stars are changing and evolving? Certain stars that pulsate and change their brightness provide evidence of stellar evolution and illustrate once more the importance of energy flowing outward through the layers of the stars.

A **variable star** is any star that changes its brightness in a periodic way. Variable stars include many different kinds of stars. Some variable stars are eclipsing binaries (Chapter 9), but many others are stars that expand and contract, heat and cool, and grow more luminous and less luminous because of internal characteristics. These stars are often called **intrinsic variables** to distinguish them from eclipsing binaries. Of these intrinsic variables, one particular kind is centrally important to modern astronomy even though the first star of that class was discovered centuries ago.

Cepheid and RR Lyrae Variable Stars

In 1784, the deaf and mute English astronomer John Goodricke, then 19 years old, discovered that the star δ Cephei is variable, changing its brightness by almost a magnitude over a period of 5.37 days (■ Figure 12-13a). Goodricke died at the age of 21, but his discovery ensures his place in the history of astronomy. Since 1784, hundreds of stars like δ Cephei have been found, and they are now known as **Cepheid variable stars.**

Cepheid variables are supergiant or bright giant stars of spectral type F or G. The fastest complete a cycle—bright to faint to bright—in about 2 days, whereas the slowest take as long as 60 days. A plot of a variable's magnitude versus time, a light curve, shows a typical rapid rise to maximum brightness and a slower decline (Figure 12-13b). Some Cepheids change their brightness by only 0.5 magnitude (about 10 percent), while others change by more than a magnitude.

Supergiants and bright giants are rare, and not all are Cepheids. Nevertheless, some familiar stars are Cepheids. One of our Favorite Stars, Polaris, the North Star, is a Cepheid with a period of 3.9696 days and a very low amplitude.

The **RR Lyrae variable stars** are related to the Cepheids. The RR Lyrae stars pulsate with periods of less than a day and are fainter than the Cepheids.

Studies of Cepheids and RR Lyrae stars reveal two related facts of great importance. First, there is a **period–luminosity relation** that connects the period of pulsation of a Cepheid to its luminosity. The longer-period Cepheids are about 40,000 times more luminous than the sun, while the shortest-period Cepheids are only a few hundred times more luminous than the sun. Second, there are two types of Cepheids. Type I Cepheids, including δ Cephei itself, have chemical compositions like that of the sun, but type II Cepheids and the RR Lyrae stars are poor in elements heavier than helium. These two facts are clues that will help you understand variable stars as evidence of stellar evolution.

Pulsating Stars

Why should a star pulsate? Why should there be a period–luminosity relation? Why are there two types of Cepheids? The

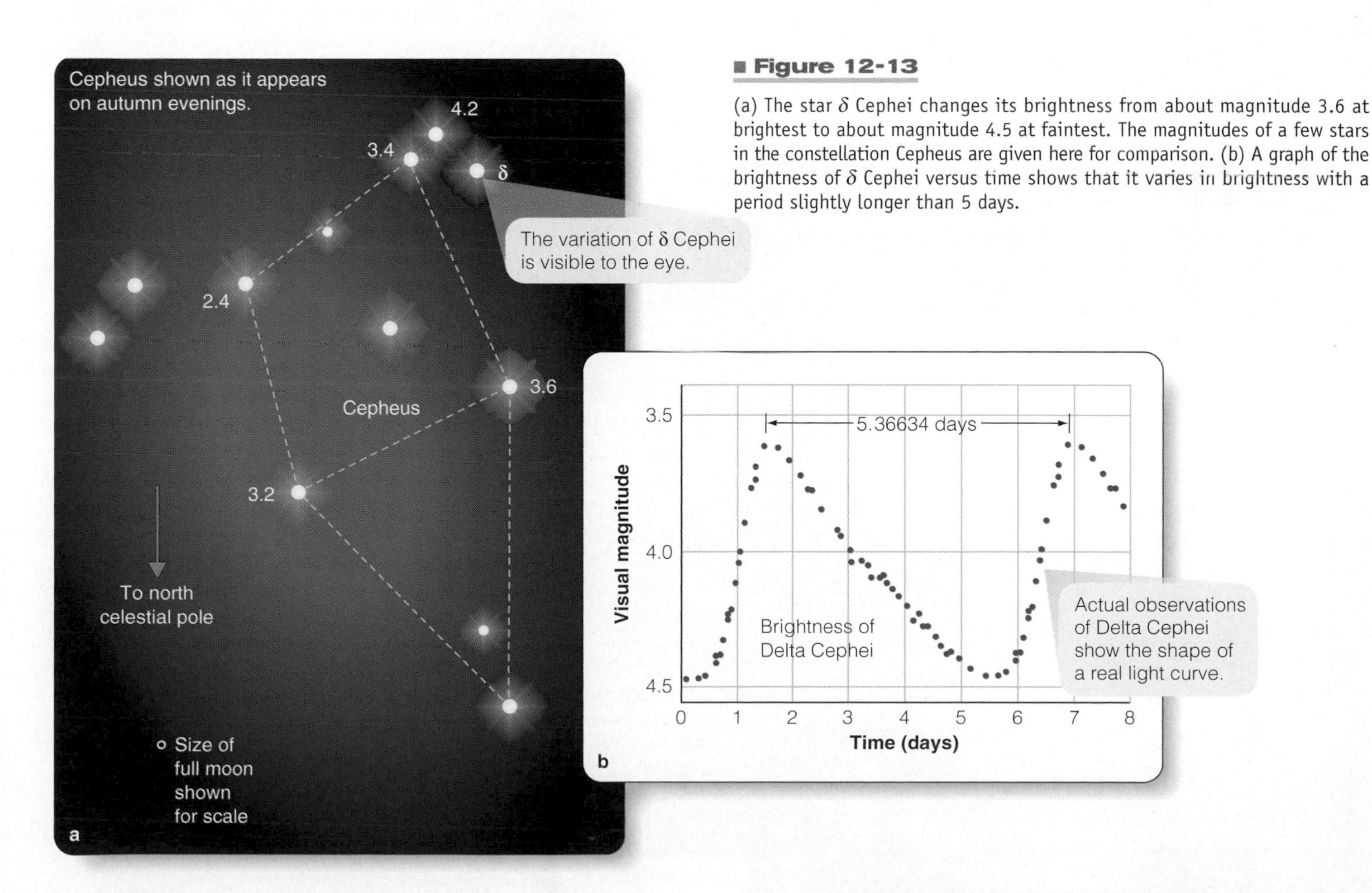

■ **Figure 12-13**

(a) The star δ Cephei changes its brightness from about magnitude 3.6 at brightest to about magnitude 4.5 at faintest. The magnitudes of a few stars in the constellation Cepheus are given here for comparison. (b) A graph of the brightness of δ Cephei versus time shows that it varies in brightness with a period slightly longer than 5 days.

Star Cluster H–R Diagrams

1 An **open cluster** is a collection of 10 to 1000 stars in a region about 25 pc in diameter. Some open clusters are quite small and some are large, but they all have an open, transparent appearance because the stars are not crowded together.

In a star cluster each star follows its orbit around the center of mass of the cluster.

AURA/NOAO/NSF

Visual-wavelength image

Open Cluster
The Jewel Box

1a A **globular cluster** can contain 10^5 to 10^6 stars in a region only 10 to 30 pc in diameter. The term "globular cluster" comes from the word "globe," although globular cluster is pronounced like "glob of butter." These clusters are nearly spherical, and the stars are much closer together than the stars in an open cluster.

Globular Cluster
47 Tucanae

Visual-wavelength image

ATFB

Astronomers can construct an H–R diagram for a star cluster by plotting a point to represent the luminosity and temperature of each star.

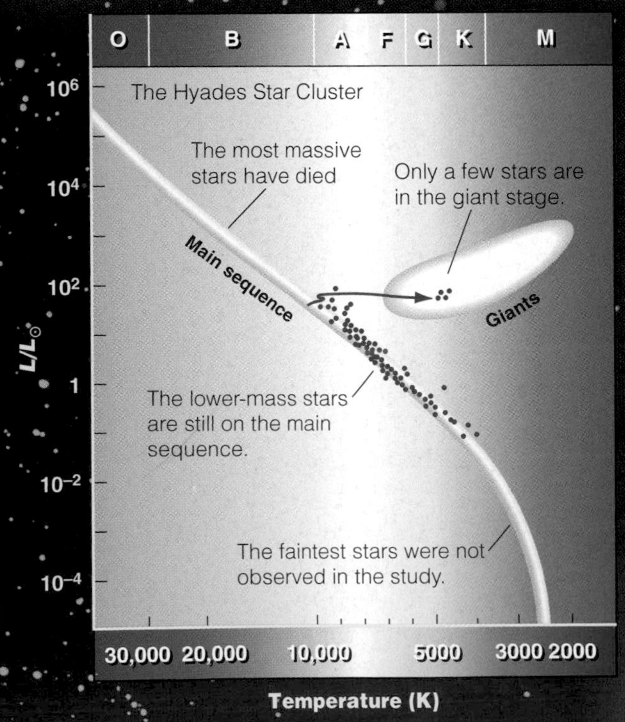

O B A F G K M
The Hyades Star Cluster
The most massive stars have died
Only a few stars are in the giant stage.
Main sequence
Giants
The lower-mass stars are still on the main sequence.
The faintest stars were not observed in the study.
10^6 10^4 10^2 1 10^{-2} 10^{-4}
$L/L_{\odot}$
30,000 20,000 10,000 5000 3000 2000
Temperature (K)

2 The H–R diagram of a star cluster can make the evolution of stars visible. The key is to remember that all of the stars in the star cluster have the same age but differ in mass. The H–R diagram of a star cluster provides a snapshot of the evolutionary state of the stars at the time you happen to be alive. The diagram here shows the 650- million-year-old star cluster called the Hyades. The upper main sequence is missing because the more massive stars have died, and our snapshot catches a few medium-mass stars leaving the main sequence to become giants.

As a star cluster ages, its main sequence grows shorter like a candle burning down. You can judge the age of a star cluster by looking at the turnoff point, the point on the main sequence where stars evolve to the right to become giants. Stars at the **turnoff point** have lived out their lives and are about to die. Consequently, the life expectancy of the stars at the turnoff point equals the age of the cluster.

Ace Astronomy™ Log into AceAstronomy and select this chapter to see Active Figure "Cluster Turnoff" and notice how the shape of a cluster's H–R diagram changes with time.

3 From theoretical models of stars, you could construct a film to show how the H–R diagram of a star cluster changes as it ages. You can then compare theory (left) with observation (right) to understand how stars evolve. Note that the time step for each frame in this film increases by a factor of 10.

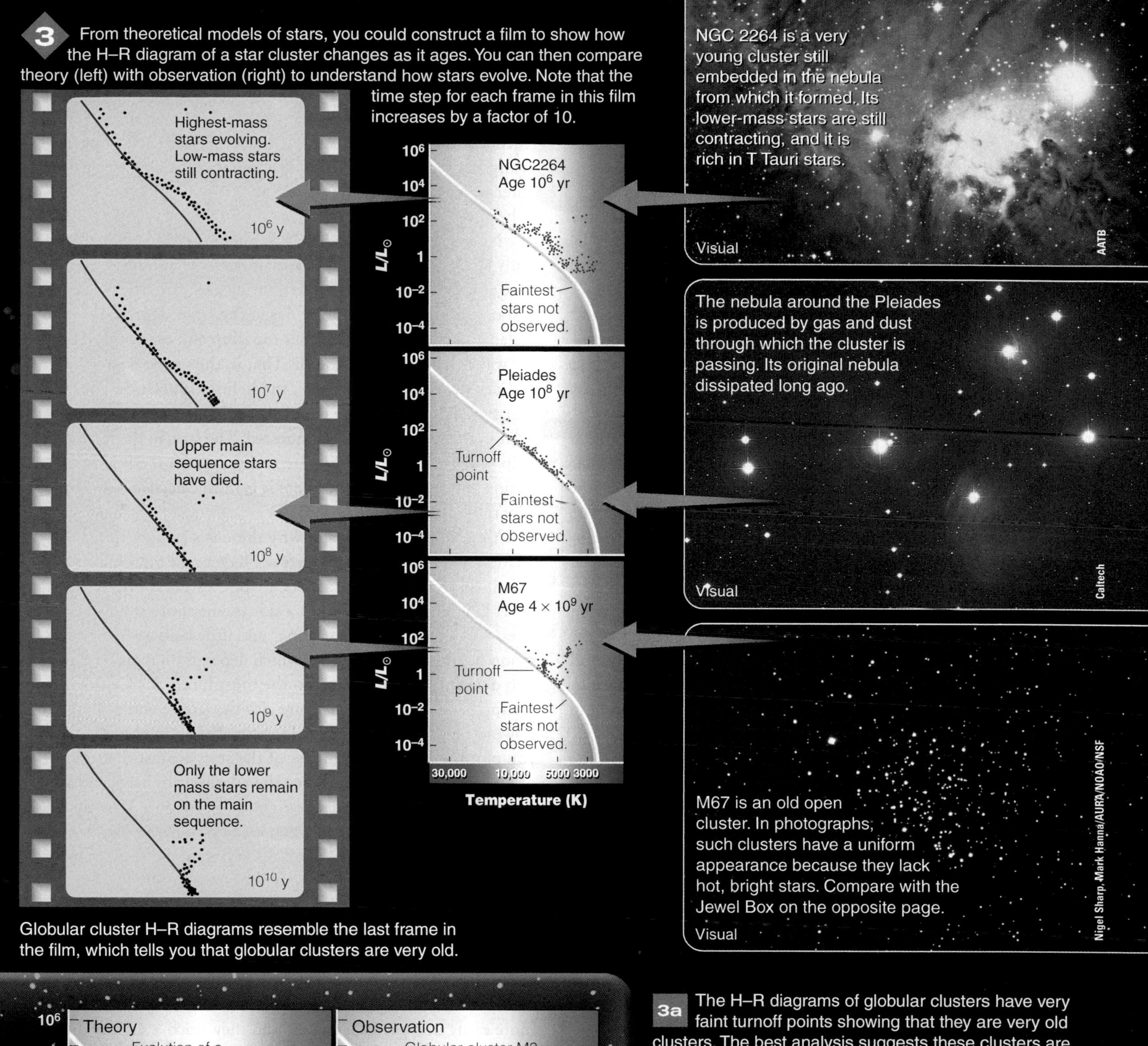

NGC 2264 is a very young cluster still embedded in the nebula from which it formed. Its lower-mass stars are still contracting, and it is rich in T Tauri stars.

The nebula around the Pleiades is produced by gas and dust through which the cluster is passing. Its original nebula dissipated long ago.

M67 is an old open cluster. In photographs, such clusters have a uniform appearance because they lack hot, bright stars. Compare with the Jewel Box on the opposite page.

Globular cluster H–R diagrams resemble the last frame in the film, which tells you that globular clusters are very old.

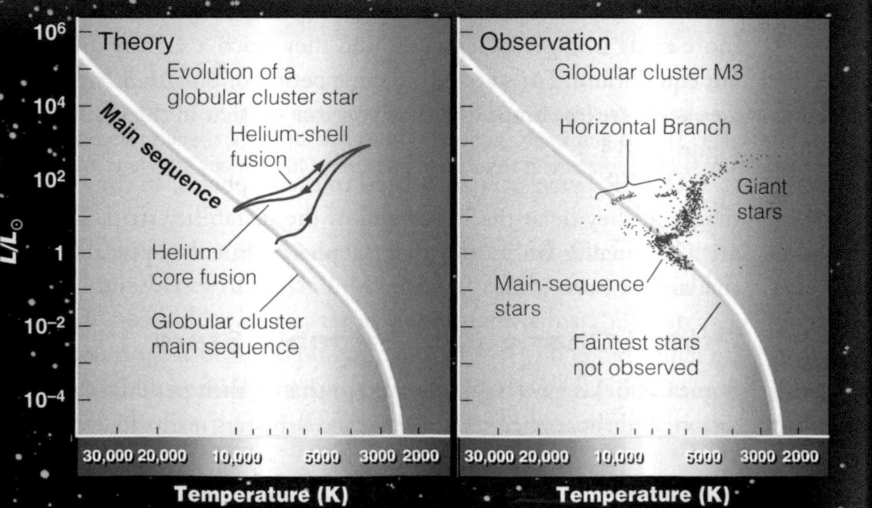

3a The H–R diagrams of globular clusters have very faint turnoff points showing that they are very old clusters. The best analysis suggests these clusters are about 11 billion years old.

The **horizontal branch** stars are giants fusing helium in their cores and then in shells. The shape of the horizontal branch outlines the evolution of these stars.

The main-sequence stars in globular clusters are fainter and bluer than the zero-age main sequence. Spectra reveal that globular cluster stars are poor in elements heavier than helium, and that means their gases are less opaque. That means energy can flow outward more easily, which makes the stars slightly smaller and hotter. Again the shape of star cluster H–R diagrams illustrates principles of stellar evolution.

The Formation of Planetary Nebulae

1 Simple observations tell astronomers what planetary nebulae are like. Their angular size and their distance indicate that their radii range from 0.2 to 3 light-years. The presence of emission lines in their spectra assures astronomers they are excited, low-density gas. Doppler shifts show they are expanding at 10 to 20 km/s. If you divide radius by velocity, you find planetary nebulae are no more than 10,000 years old. Older planetary nebulae evidently become mixed into the interstellar medium.

NGC6369 Visual

Hubble Heritage Team, STScI/AURA/NASA

This nearly spherical planetary nebula has a low-luminosity outer envelope and a highly excited inner region.

The Ring Nebula in the constellation Lyra is visible even in small telescopes. Note the hot blue star at its center and the radial texture in the gas, suggesting outward motion.

Visual

Hubble Heritage Team, AURA/STScI/NASA

1a Astronomers find about 1500 planetary nebulae in the sky. Because planetary nebulae are short-lived formations, you can conclude that they must be a common part of stellar evolution. Medium-mass stars up to a mass of about 8 solar masses are destined to die by forming planetary nebulae.

2 The process that produces planetary nebulae involves two stellar winds. First, as an aging giant, the star gradually blows away its outer layers in a slow breeze of low-excitation gas that is not easily visible. Once the hot interior of the star is exposed, it ejects a high-speed wind that overtakes and compresses the gas of the slow wind like a snowplow, while ultraviolet radiation from the hot remains of the central star excites the gases to glow like a giant neon sign.

Slow stellar wind from a red giant

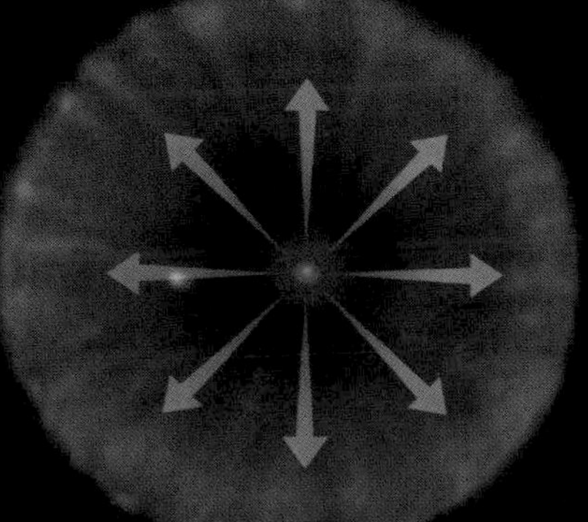

The gases of the slow wind are not easily detectable.

Fast wind from exposed interior

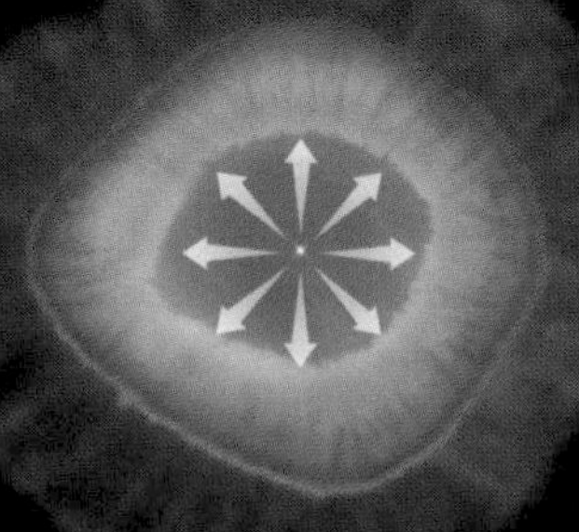

You see a planetary nebula where the fast wind compresses the slow wind.

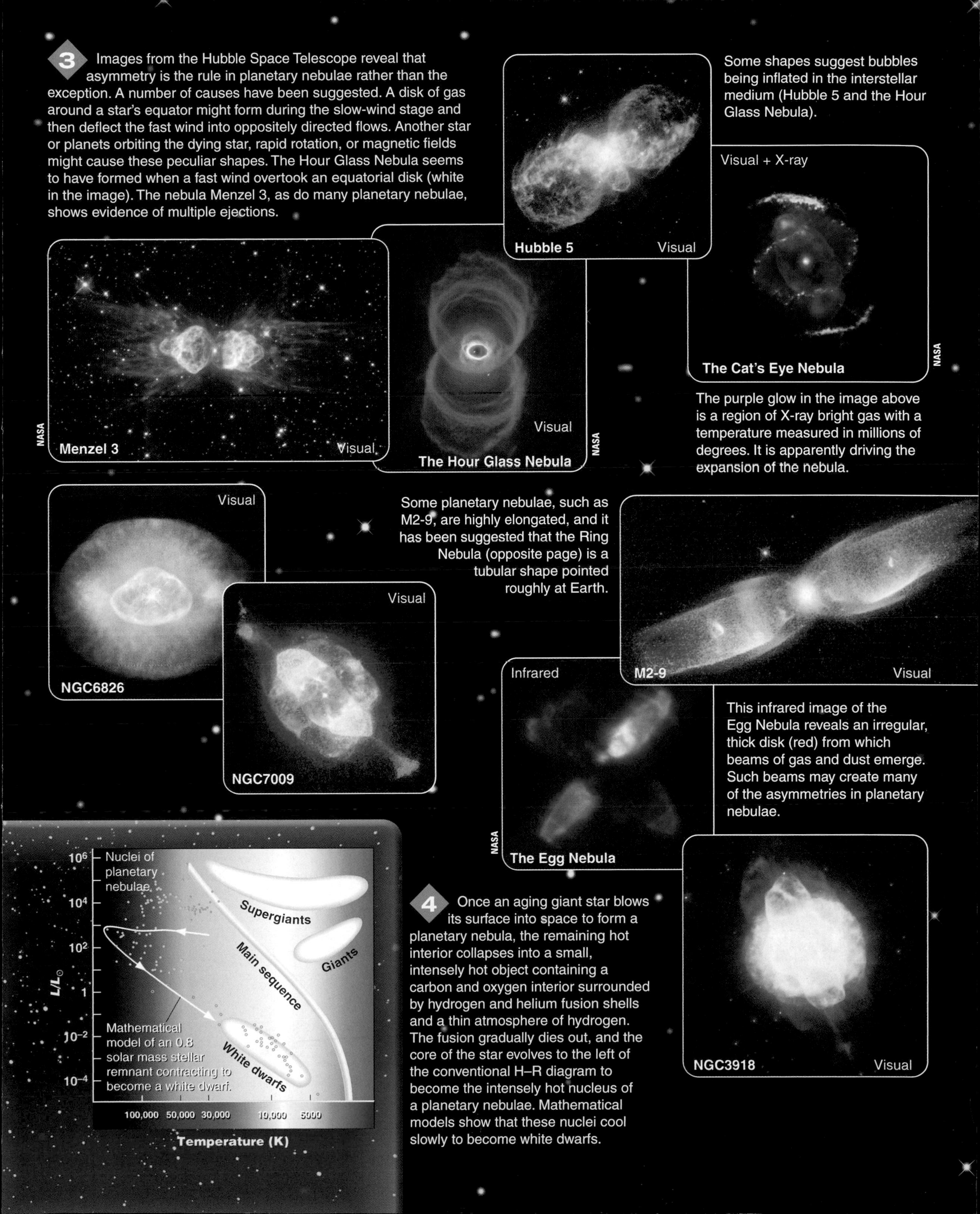

3 Images from the Hubble Space Telescope reveal that asymmetry is the rule in planetary nebulae rather than the exception. A number of causes have been suggested. A disk of gas around a star's equator might form during the slow-wind stage and then deflect the fast wind into oppositely directed flows. Another star or planets orbiting the dying star, rapid rotation, or magnetic fields might cause these peculiar shapes. The Hour Glass Nebula seems to have formed when a fast wind overtook an equatorial disk (white in the image). The nebula Menzel 3, as do many planetary nebulae, shows evidence of multiple ejections.

Some shapes suggest bubbles being inflated in the interstellar medium (Hubble 5 and the Hour Glass Nebula).

The purple glow in the image above is a region of X-ray bright gas with a temperature measured in millions of degrees. It is apparently driving the expansion of the nebula.

Some planetary nebulae, such as M2-9, are highly elongated, and it has been suggested that the Ring Nebula (opposite page) is a tubular shape pointed roughly at Earth.

This infrared image of the Egg Nebula reveals an irregular, thick disk (red) from which beams of gas and dust emerge. Such beams may create many of the asymmetries in planetary nebulae.

4 Once an aging giant star blows its surface into space to form a planetary nebula, the remaining hot interior collapses into a small, intensely hot object containing a carbon and oxygen interior surrounded by hydrogen and helium fusion shells and a thin atmosphere of hydrogen. The fusion gradually dies out, and the core of the star evolves to the left of the conventional H–R diagram to become the intensely hot nucleus of a planetary nebulae. Mathematical models show that these nuclei cool slowly to become white dwarfs.

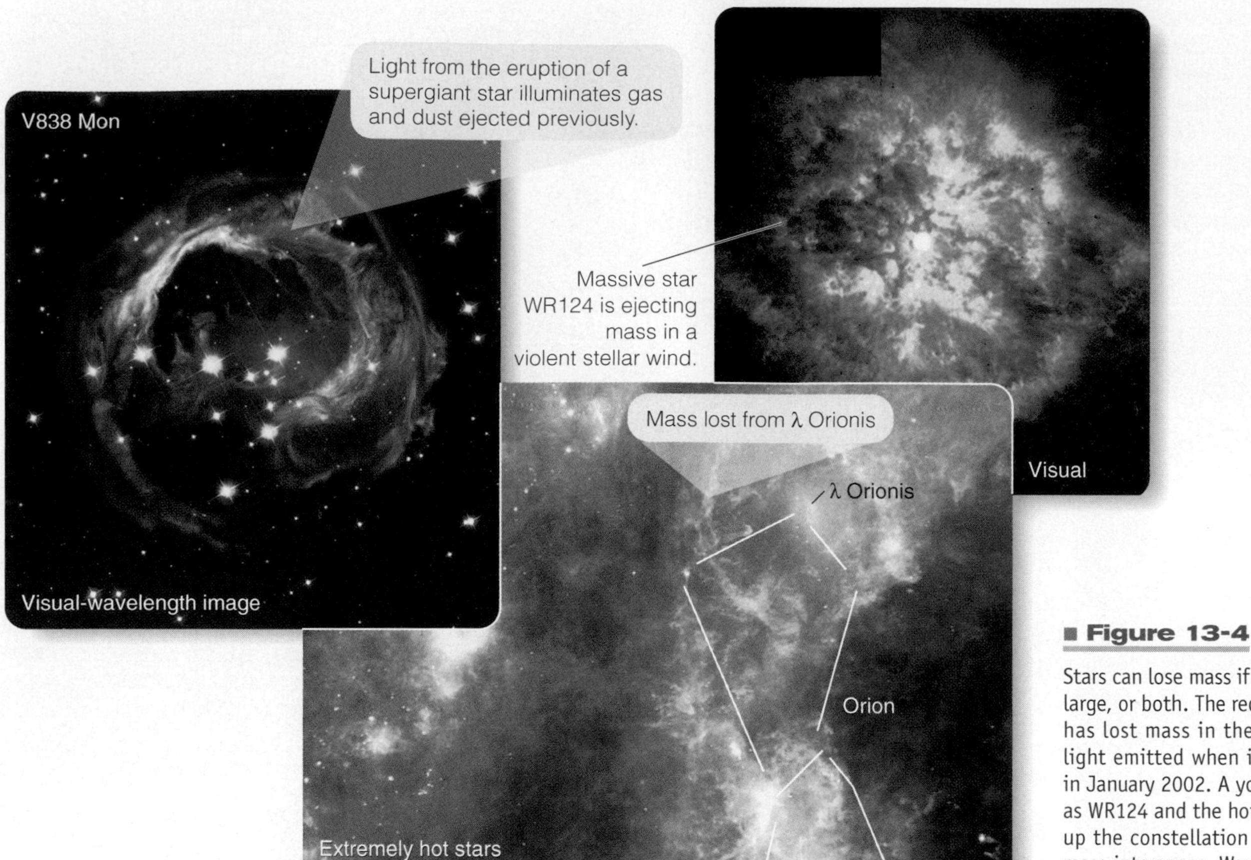
V838 Mon
Light from the eruption of a supergiant star illuminates gas and dust ejected previously.
Visual-wavelength image
Massive star WR124 is ejecting mass in a violent stellar wind.
Visual
Mass lost from λ Orionis
λ Orionis
Orion
Extremely hot stars such as those in Orion can drive gas away.
Orion Nebula
Infrared image

Figure 13-4

Stars can lose mass if they are very hot, very large, or both. The red supergiant V838 Mon has lost mass in the past, as revealed by light emitted when it erupted temporarily in January 2002. A young massive star such as WR124 and the hot, blue stars that make up the constellation Orion constantly lose mass into space. Warmed dust in these gas clouds can make them glow in the infrared. (V838 Mon: NASA and The Hubble Heritage Team AURA/STScI; WR124: NASA; Orion: NASA/IPAC courtesy Deborah Levine)

wind carrying gas into space and pushing back the interstellar medium. Eventually, the loss of mass will expose deeper layers in the sun. These extremely hot layers will heat the gas around the sun and propel it outward to scoop up previously expelled gases. If the sun becomes hot enough, it will ionize the gas and produce a planetary nebula. In any case, the last remains of the sun will contract and slowly cool to form a small, dense white dwarf. The surface will be very hot but so small that the sun will be about 100 times fainter than it is now.

The story of stellar evolution predicts that our sun will someday become a giant star and then a white dwarf. Focus your scientific argument on a slightly different part of the story. **As the sun becomes a white dwarf, what will have become of all the hydrogen the sun once contained?**

■ ■ ■

Connections: Lower-main-sequence stars like the sun die by producing beautiful planetary nebulae and shrinking to become white dwarfs. Now you are ready to see what happens when one of these dying stars has a stellar companion.

13-2 The Evolution of Binary Stars

SO FAR YOU HAVE BEEN THINKING about the deaths of stars as if they were all single objects that never interact. But more than half of all stars are members of binary star systems. Most such binaries are far apart, and one of the stars can swell into a giant and eventually collapse without affecting the companion star. In some binary systems, however, the two stars orbit close together. When the more massive star begins to expand, it interacts with its companion star in peculiar ways. These interacting binary stars are fascinating objects themselves, but they are also important because they help explain observed phenomena such as nova explosions. In the next chapter, you will see how they can help astronomers find black holes.

Mass Transfer

Binary stars can sometimes interact by transferring mass from one star to the other. To understand this process, you need to think about how the gravity of the two stars controls the matter in the binary system.

Each star in a binary system is held together by its own gravity: but, because the system is rotating, there are unexpected forces acting on loose matter between the stars. If you could gently release a pebble into such a system, it might fall into one star, fall into the other star, or be ejected from the system. What happened to your pebble would depend on where you dropped it. Each star controls a region of space near it, and if you could make these regions visible they would look like two teardrop-shaped lobes enclosing the stars and meeting tip to tip between the stars. These two volumes are called the **Roche lobes,** and the dumbbell-shaped surface of the two lobes is called the **Roche surface.** If you released your pebble inside one of the Roche lobes, it would belong gravitationally to the star that controls that lobe.

Because of the interaction of the gravity of the two stars and their rotation around their center of mass, there are five points of stability in the orbital plane called **Lagrangian points.*** ■ Figure 13-5 illustrates the arrangement of the Roche surface around the stars and the Lagrangian points.

The Lagrangian points are important because they are locations where matter is stable in the revolving binary system. If you could carefully release your pebble at a Lagrangian point, it would remain at that point and circle the center of mass along with the stars. Rather than a pebble, think of a bit of naturally occurring gas in the binary system. Gas at the L_2 or L_3 point is critically stable, meaning that any small disturbance will make it leak away. In contrast, gas at the L_4 or L_5 point can be trapped. (You will meet the L_4 and L_5 points again when you study planetary satellites and the orbits of asteroids in later chapters.) As you think about binary stars, the inner Lagrangian point L_1 located between the two stars is the most important. If matter from one star can reach the inner Lagrangian point, it can flow onto the other star. That is, the stars can transfer mass through the inner Lagrangian point.

In general, there are only two ways matter can escape from a star and reach the inner Lagrangian point. First, if a star has a strong stellar wind, some of the gas blowing away from the star can pass through the inner Lagrangian point and be captured by the other star. Second, if an evolving star expands so far that it fills its Roche surface, it will be forced to take on the teardrop shape of the Roche surface. You have seen that effect in your study of binary stars (Figure 9-21). If the star expands farther, matter will flow through the inner Lagrangian point (like water in a pond flowing over a dam) and will fall into the other star. Mass transfer driven by a stellar wind tends to be slow, but mass transfer driven by an expanding star can occur rapidly.

*The Lagrangian points are named after French mathematician Joseph Louis Lagrange, who solved this famous mathematical problem around the time of the French Revolution.

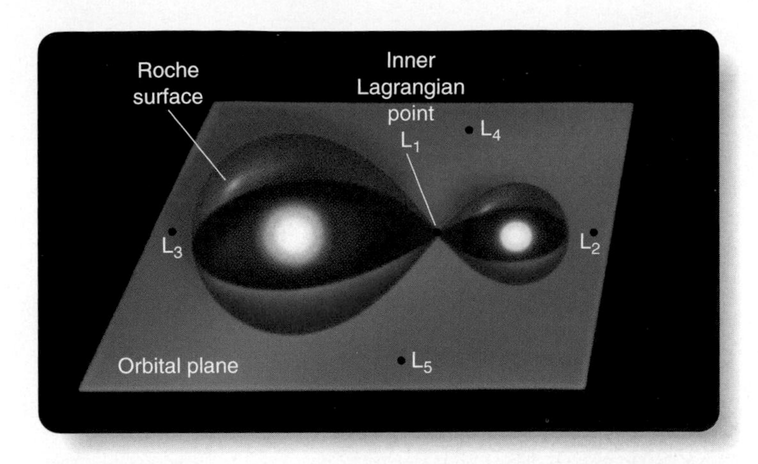

■ **Figure 13-5**

A pair of binary stars control the region of space located inside the Roche surface. The Lagrangian points are locations of stability, with the inner Lagrangian point making a connection through which the two stars can transfer matter.

Evolution with Mass Transfer

Mass transfer between stars can affect their evolution in surprising ways. In fact, it provides the explanation for a problem that puzzled astronomers for many years.

In some binary systems, the less massive star has become a giant, while the more massive star is still on the main sequence. That seems backward. If more massive stars evolve faster than lower-mass stars, how does the low-mass star in such binaries manage to leave the main sequence first? This is called the Algol paradox, after the binary system Algol (see Figure 9-22).

Mass transfer explains how this could happen. Imagine a binary system that contains a 5-solar-mass star and a 1-solar-mass companion (■ Figure 13-6). The two stars formed at the same time, so the more massive star must evolve faster and leave the main sequence first. When it expands into a giant, it can fill its Roche surface and transfer matter to the low-mass companion. In that way, the massive star will shrink into a lower-mass star, and the companion will gain mass and become a massive star still on the main sequence. If you observed such a system after the mass transfer ended, you might see a binary system such as Algol containing a 5-solar-mass main-sequence star and a 1-solar-mass giant.

The evolution of close binary stars could result in one of the stars having its outer layers "peeled" away. A massive star expanding to become a giant could lose its outer layers to its companion and then collapse to form a lower-mass peculiar star. A few such peculiar stars are known.

Another exotic result of the evolution of close binary systems is the merging of the stars. Astronomers see many binaries in which both stars have expanded to fill their Roche surfaces and spill mass out into space. If the stars are close enough together and expand rapidly enough, theorists believe, the two stars could merge into a single, rapidly rotating giant star. Most giants rotate slowly because they conserved angular momentum as they expanded and consequently slowed their rotation. Examples of rapidly rotating giant stars are known, however, and they are believed to be the results of merged binary stars. Inside the distended envelope of such a star, the cores of the two stars could continue to orbit each other until friction slowed them down and they sank to the center.

■ Figure 13-6

A pair of stars orbiting close to each other can exchange mass and modify their evolution.

The Evolution of a Binary System

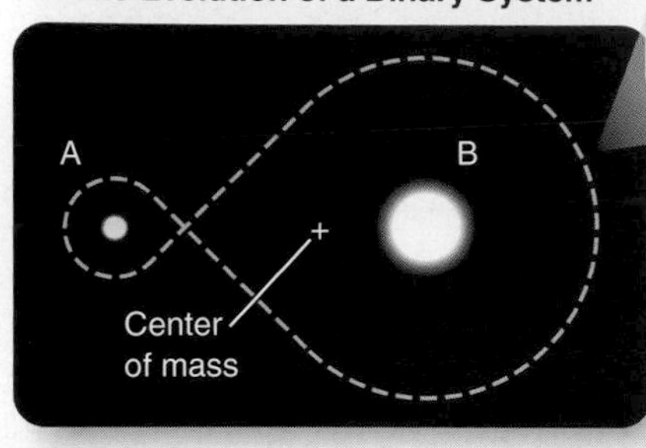

A
B
Center of mass
Star B is more massive than Star A.

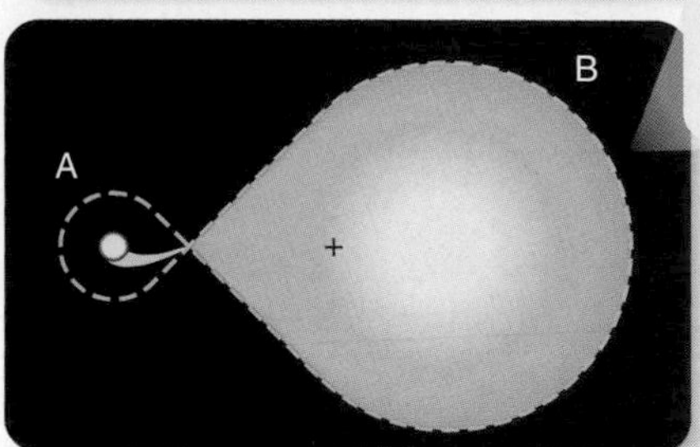

A
B
Star B becomes a giant and loses mass to Star A.

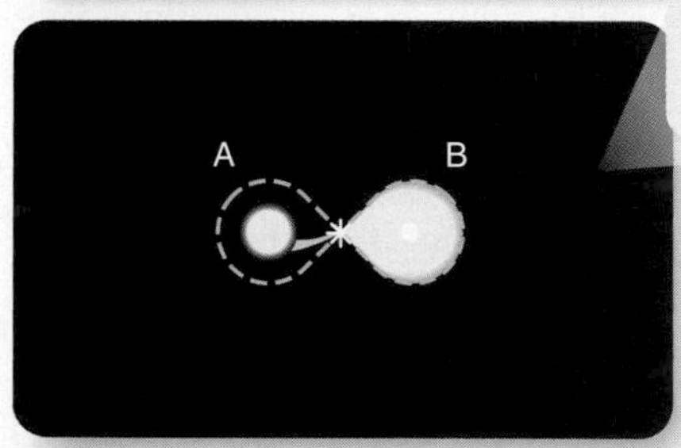

A
B
Star B loses mass, and Star A gains mass.

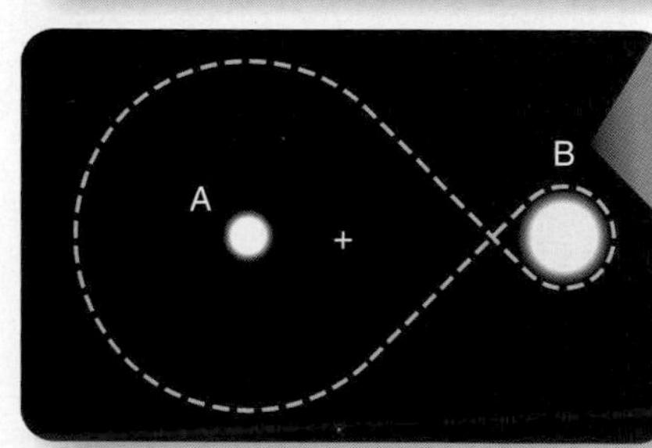

A
B
Star A is a massive main-sequence star with a lower-mass giant companion—an Algol system.

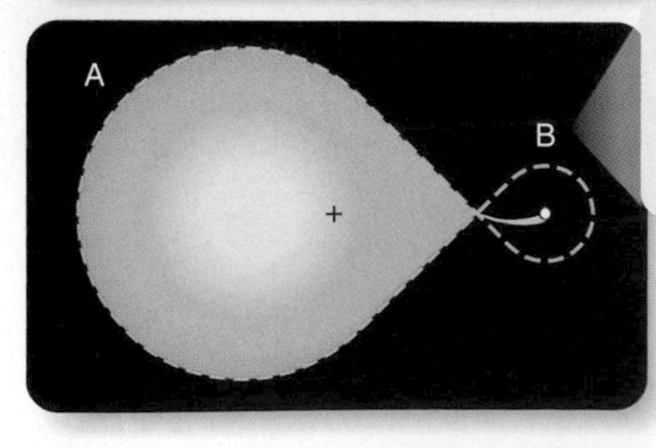

A
B
Star A has now become a giant and loses mass back to the white dwarf that remains of Star B.

Mass transfer can lead to dramatic violence. The first four frames of Figure 13-6 show mass transfer producing a system like Algol. The last frame shows an additional stage in which the giant star has expelled its outer layers and has collapsed to form a white dwarf. The more massive companion has expanded, transferring matter back onto the white dwarf. Such systems can become the site of tremendous explosions. To see how this can happen, you need to consider in detail how mass falls into a star.

Accretion Disks

Matter flowing through the inner Lagrangian point cannot fall directly into the white dwarf because of the conservation of angular momentum (as described in Chapter 5). Instead, it falls into a whirlpool around the white dwarf. For a common example, consider a bathtub full of water. Gentle currents in the water give it some angular momentum, but you can't see its slow circulation until you pull the stopper. Then, as the water rushes toward the drain, conservation of angular momentum forces it to form a whirlpool. This same effect forces gas falling into a white dwarf to form a whirling disk of gas called an **accretion disk** (■ Figure 13-7).

Two important things happen in an accretion disk. First, the gas in the disk grows very hot due to friction and tidal forces. The disk also acts as a brake, ridding the gas of its angular momentum and allowing it to fall into the white dwarf. The temperature of the gas in the inner parts of an accretion disk can become very high, in some cases exceeding 1,000,000 K and emitting X rays.

Now you are ready to put the pieces together and explain one of the ways a star can explode. All you need is mass transfer onto a white dwarf.

Nova Explosions

At the beginning of this chapter you saw that the word *nova* refers to a new star that appears in the sky for a while and then fades

■ Figure 13-7

Matter falling into a compact object forms a whirling accretion disk. Friction and tidal forces can make the disk very hot.

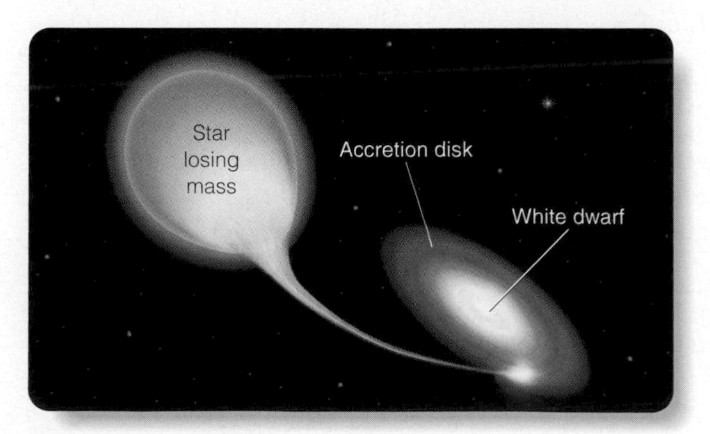

Star losing mass
Accretion disk
White dwarf

■ Figure 13-8

Nova Cygni 1975 near maximum at about second magnitude (a) and later when it had declined to about eleventh magnitude (b). (Photo © UC Regents)

away (■ Figure 13-8). The evidence shows that it is not a new star at all. After a nova fades, astronomers can photograph the spectrum of the object, and they invariably find a closely spaced binary star containing a normal star and a white dwarf. A nova is evidently an explosion involving a white dwarf in a binary system.

Observational evidence can tell you how nova explosions occur. As the explosion begins, spectra show blueshifted absorption lines, which tells you the gas is dense and coming toward Earth at a few thousand kilometers per second. After a few days, the spectral lines change to emission lines, which tells you the gas has thinned, but the blueshifts remain, showing that a cloud of debris has been ejected into space.

You can understand nova explosions as the result of mass transfer from a normal star through the inner Lagrangian point into an accretion disk around the white dwarf. As the matter loses its angular momentum in the accretion disk, it drifts inward and eventually settles onto the surface of the white dwarf. Because the matter came from the surface of a normal star, it is rich in unused fuel, mostly hydrogen, and when it accumulates on the surface of the white dwarf, it forms a layer of unprocessed fuel.

As the layer of fuel grows deeper, it becomes hotter and denser and eventually becomes degenerate, much like the core of a sun-like star approaching the helium flash. In such a gas, the pressure–temperature thermostat does not work, so the layer of unfused hydrogen is a thermonuclear bomb waiting to explode. By the time the white dwarf has accumulated about 100 Earth masses of hydrogen, the temperature at the base of the hydrogen layer reaches millions of degrees, and the density is 10,000 times the density of water. Suddenly, the hydrogen begins to fuse on the proton–proton chain, and the energy released drives the temperature so high that the CNO cycle begins. With no pressure–temperature thermostat to control the fusion, the temperature shoots to 100 million degrees in seconds. Quickly the temperature rises high enough to force the gas to expand, and it stops being degenerate. But it is too late. By the time the gas begins to expand, the amount of energy released is enough to blow the surface layers of the star into space as a violently expanding shell of hot gas, which is visible from Earth as a nova.

The explosion of its surface hardly disturbs the white dwarf or its companion star. Mass transfer quickly resumes, and a new layer of fuel begins to accumulate. How fast the fuel builds up depends on the rate of mass transfer. Accordingly, you can expect novae to repeat each time an explosive layer accumulates. Many novae take thousands of years to build an explosive layer, but some take only a few years (■ Figure 13-9).

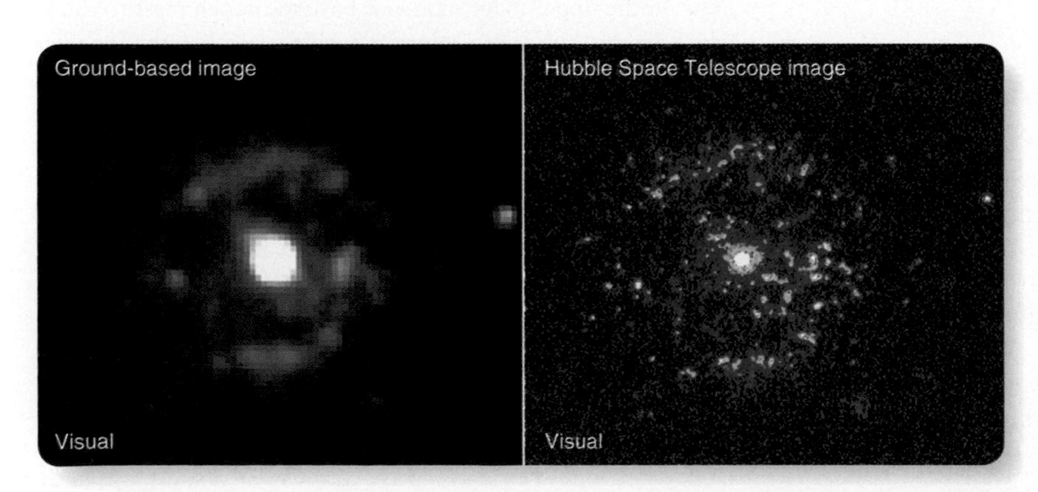

■ Figure 13-9

Nova T Pyxidis erupts about every two decades, expelling shells of gas into space. The shells of gas are visible from ground-based telescopes, but the Hubble Space Telescope reveals much more detail. The shell consists of knots of excited gas that presumably form when a new shell collides with a previous shell. (M. Shara and R. Williams, STScI; R. Gilmozzi, ESO; and NASA)

The End of Earth

Astronomy is about us. Although you have been reading about the deaths of medium-mass stars, white dwarfs, and novae explosions, you have also been reading about the future of our planet. The sun is a medium-mass star and must eventually die by becoming a giant, possibly producing a planetary nebula, and collapsing into a white dwarf. That will spell the end of Earth.

Mathematical models of the sun suggest that it may survive for 5 billion years or so, but it is already growing more luminous as it fuses hydrogen into helium. In a few billion years, it will exhaust hydrogen in its core and swell into a giant star about 100 times its present radius. That giant sun will be about as large as the orbit of Earth, so that will mark the end of our world. Whether or not the expanding sun becomes large enough to totally engulf Earth, its growing luminosity will certainly evaporate our oceans, drive away our atmosphere, and even vaporize most of Earth's crust.

While it is a giant star, the sun will lose mass into space. This mass loss is a relatively gentle process, so any cinder that might remain of Earth would not be disturbed. Of course, when the sun finally collapses into a white dwarf, the solar system will become a much colder place, but not much of Earth will remain by then. If the collapsing sun becomes hot enough, it will ionize the expelled gas to form a planetary nebula. Some models suggest that the sun is not quite massive enough to light up a planetary nebula. It will collapse into a hot object that will cool to become a white dwarf, but it may not become hot enough to ionize the expelled gas and "turn on" a planetary nebula. An astronomer recently expressed disappointment that the dying sun might not leave behind a beautiful planetary nebula, but that embarrassment lies a few billion years in the future.

There is no danger that the sun will explode as a nova; it has no binary companion. Also, as you will see, the sun is not massive enough to die the violent death of the massive stars.

The most important lesson of astronomy is that we are part of the universe and not just observers. The atoms you are made of are destined to return to the interstellar medium in just a few billion years. That's a long time, and it is possible that the human race will migrate to other planetary systems. That might save the human race, but our planet is star dust.

Building Scientific Arguments

How does modern astronomy explain the Algol paradox? Scientific arguments combine evidence with theory to explain nature, and some of the most interesting arguments explain what seem to be impossible situations. When you encounter a paradox in nature, it is usually a warning that you don't understand things as well as you thought you did. The Algol paradox is a good example. The binary star Algol contains a lower-mass giant star and a more massive main-sequence star. Because the two stars must have formed together, they must be the same age. Then the more massive star should have evolved first and left the main sequence. But in binary systems such as Algol, the lower-mass star has left the main sequence first, or so it seems.

You can understand this paradox if you add mass transfer to your argument. The first star to leave the main sequence must be the more massive star; but if the stars are close together, the star that is initially more massive can fill its Roche surface and transfer mass back to its companion. The companion can grow more massive, and the giant can grow less massive. In this way, the lower-mass giant star may have originally been more massive and may have evolved away from the main sequence, leaving its companion behind to increase in mass as the giant decreased.

Mass transfer can explain the Algol paradox, and it can also explain the violent explosions called novae. Develop a new argument. **How does mass transfer explain why novae can explode over and over in the same binary system?**

■ ■ ■

Connections: Your study of the deaths of low- and medium-mass stars has led you to planetary nebulae, white dwarfs, and novae explosions. But there are more violent fish in the cosmic sea. What causes supernovae? To find out, you need to consider how massive stars die.

13-3 The Deaths of Massive Stars

YOU HAVE SEEN that low- and medium-mass stars die relatively quietly as they exhaust their hydrogen and helium and then eject their surface layers to form planetary nebulae. In contrast, massive stars live spectacular lives and destroy themselves in violent explosions.

Nuclear Fusion in Massive Stars

Stars on the upper main sequence have too much mass to die as white dwarfs, but their evolution begins much like that of their lower-mass cousins. They consume the hydrogen in their cores and ignite hydrogen shells; as a result, they expand and become giants or, for the most massive stars, supergiants. Their cores contract and fuse helium first in the core and then in a shell, producing a carbon–oxygen core.

A massive star can lose significant mass as it ages; but, if it still has a mass over 4 solar masses when its carbon–oxygen core contracts, it can reach a temperature of 600 million Kelvin and ignite carbon fusion. The fusion of carbon produces heavier nuclei such as oxygen and neon. As soon as the carbon is exhausted in the core, the core contracts, and carbon ignites in a shell. This

■ Active Figure 13-10

Massive stars live fast and die young. The three shown here are among the most massive stars known, containing 100 solar masses or more. They are rapidly ejecting gas into space. The centers of these massive stars develop Earth-size cores (magnified 100,000 times in this figure) composed of concentric layers of gases undergoing nuclear fusion. The iron core at the center leads eventually to a star-destroying explosion. (AFGL2591: Gemini Observatory/NSF/C. Aspin; Eta Carinae and the Pistol Star: NASA)

AceAstronomy™ Log into AceAstronomy and select this chapter to see the Active Figure "Inside Stars." Compare the interior of a sunlike star with that of a massive star.

pattern of core ignition followed by shell ignition continues with fuel after fuel, and the star develops a layered structure at its center (■ Figure 13-10), with a hydrogen-fusion shell above a helium-fusion shell above a carbon- fusion shell and so on. After carbon fusion, oxygen, neon, magnesium, and heavier elements fuse, right up to iron.

The fusion of these nuclear fuels goes faster and faster as the massive star evolves rapidly. Recall that massive stars must consume their fuels rapidly to support their great weight, but other factors also cause the heavier fuels like carbon, oxygen, and silicon to fuse at increasing speed. For one thing, the amount of energy released per fusion reaction decreases as the mass of the fusing atom increases. To support its weight, a star must fuse oxygen much faster than it fused hydrogen. Also, there are fewer atoms in the core of the star by the time heavy atoms begin to fuse. Four hydrogens made a helium atom, and three heliums made a carbon, so there are 12 times fewer atoms of carbon available for fusion than there were of hydrogen. This means that heavy-element fusion goes very quickly in massive stars (■ Table 13-1). Hydrogen fusion can last 7 million years in a 25-solar-mass star, but that same star will fuse its oxygen in six months and its silicon in a day.

The Iron Core

Heavy-element fusion ends with iron, because nuclear reactions that use iron as a fuel cannot produce energy. Nuclear reactions can produce energy if they proceed from less tightly bound nuclei to more tightly bound nuclei. As shown in Figure 8-8, both nuclear fission and nuclear fusion produce nuclei that are more tightly bound than the starting fuel. Look again at Figure 8-8 and notice that iron is the most tightly bound nucleus of all. No nuclear reaction, fission or fusion, that starts with iron can produce a more tightly bound nucleus, and that means that iron is a dead end.

When a massive star develops an iron core, nuclear fusion cannot produce energy, and the core contracts and grows hotter. The shells around the core burn outward, fusing lighter elements and leaving behind more iron, which further increases the mass of the core. When the mass of the iron core exceeds 1.3 to 2 solar masses, the core must collapse.

As the core begins to collapse, two processes can make it contract even faster. Heavy nuclei in the core can capture high-energy electrons, thus removing thermal energy from the gas. This robs the gas of some of the pressure it needs to support the crushing weight of the outer layers. Also, in more massive stars, temperatures are so high that many photons have gamma-ray wavelengths and can break more massive nuclei into less massive nuclei. In the process, the gamma rays are absorbed. This reversal of nuclear fusion absorbs energy and allows the core to collapse even faster.

Although a massive star may live for millions of years, its iron core—about 500 km in diameter—collapses in only a few thousandths of a second. This collapse happens so rapidly that the most powerful computers are not capable of following the details. One thing is clear, however. The collapse of the iron core in a massive star triggers a star-destroying explosion—a supernova.

■ Table 13-1 | Heavy-Element Fusion in a 25-$M_\odot$ Star

Fuel	Time	Percentage of Lifetime
H	7,000,000 years	93.3
He	500,000 years	6.7
C	600 years	0.008
O	0.5 years	0.000007
Si	1 day	0.00000004

The Supernova Deaths of Massive Stars

Modern theory predicts that the collapse of a massive star can eject the outer layers of the star to produce a supernova explosion while the core of the star collapses to form a neutron star or a black hole. You will learn more about neutron stars and black holes in the next chapter; here you can concentrate on the process that triggers the supernova explosion.

A supernova explosion is rare, remote, rapid, and violent. It is an event in nature that is extremely difficult to study in person, which is why astronomers have used powerful mathematical modeling techniques and high-speed supercomputers to explore the inside of a star exploding as a supernova. Such models allow astronomers to experiment on an exploding star as if the star were in a laboratory beaker.

Mathematical models reveal that the key to the supernova explosion is the collapse of the iron core, which allows the rest of the interior of the star to fall inward, creating a tremendous "traffic jam" as all the nuclei try to fall toward the center. It is as if all the residents of Indiana suddenly tried to drive their cars as fast as possible into the center of Indianapolis. There would be a tremendous traffic jam downtown. As more cars rushed in, the traffic jam would spread outward into the suburbs. Similarly, as the inner core of the star falls inward, a shock wave (a traffic jam) develops and begins to move outward. Containing about 100 times more energy than necessary to destroy the star, such a shock wave was first thought to be the cause of the supernova explosion.

Computer models revealed, however, that the shock wave spreading outward through the collapsing star stalls within a few hundredths of a second. Matter flowing inward smothers the shock wave and pushes it back into the star. If this happened in a real star, it presumably would collapse without any visible explosion.

However, theory predicts that 99 percent of the energy released in the collapse will appear as neutrinos. In the sun, neutrinos zip outward unimpeded by the gas of the solar layers, but in a collapsing star the density is as high as 10^{12} g/cm^3, nearly as dense as an atomic nucleus. This gas is opaque to neutrinos, and they are partially absorbed in the gas. Not only does the tremendous burst of neutrinos remove energy from the core and allow the core to collapse even faster, but the neutrinos are absorbed outside the core and heat those layers. For some years, astronomers thought that these neutrinos could spread energy across the shock wave and, within a quarter of second or so, reaccelerate the stalled shock wave. The computer models, however, refused to explode.

The final ingredient that the models needed to make them explode was turbulent convection. When the collapse begins, the very center of the star forms a highly dense core, the beginnings of a neutron star, and the infalling material bounces off that core. With the temperature shooting up, the bouncing material produces highly turbulent convection currents that give the stalled shock wave an outward boost (■ Figure 13-11). Within a second or so, the shock wave begins to push outward, and after just a few hours it bursts out through the surface, blasting the star apart in a supernova explosion.

The supernova visible from Earth is the brightening of the star as its outer layers are blasted outward. As the months pass, the cloud of gas expands, thins, and begins to fade. But the way it fades in some cases tells astronomers about the death throes of the star. Essentially all of the iron in the core of the star is destroyed when the core collapses, but the violence in the outer layers can produce densities and temperatures high enough to trigger nuclear fusion reactions that produce as much as half a solar mass of radioactive nickel-56. The nickel gradually decays to form radioactive cobalt, which decays to form normal iron. The rate at which the supernova fades matches the rate at which these radioactive elements decay. Thus the destruction of iron in the core of the star is matched by the production of iron through nuclear fusion in the expanding outer layers.

The presence of nuclear fusion in the outer layers of the supernova testifies to the violence of the explosion. A typical supernova is equivalent to the explosion of 10^{28} megatons of TNT—about 3 million solar masses of high explosive.

Collapsing massive stars can trigger cosmic violence, but astronomers have observed more than one kind of supernova.

The Exploding Core of a Supernova

The core of a massive supergiant has begun to collapse at the lower left corner of this model.

Matter continues to fall inward (blue and green) as the core expands outward (yellow) creating a shock wave.

To show the entire star at this scale, this page would have to be 30 kilometers in diameter.

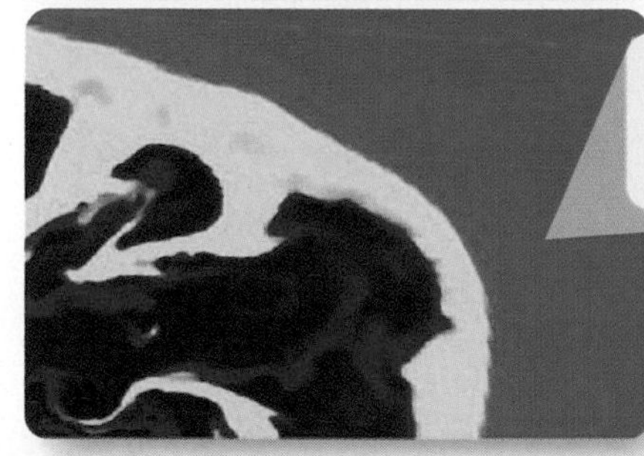

Only 0.4 s after beginning, violent convection in the expanding core (red) pushes outward.

The shock wave will blow the star apart as a neutron star forms at the extreme lower left corner.

■ Figure 13-11

As the iron core of a massive star begins to collapse, intensely hot gas triggers violent convection. Even as the outer parts of the core continue to fall inward, the turbulence blasts outward and reaches the surface of the star within hours, creating a supernova eruption. This diagram is based on mathematical models and shows only the exploding core of the star. (Courtesy Adam Burrows, John Hayes, and Bruce Fryxell)

Types of Supernovae

Supernovae are rare, and only a few have been seen in our galaxy, but astronomers have been able to observe supernovae occurring in other galaxies (■ Figure 13-12). From these data accumulated over decades, astronomers have noticed that there are two main types. **Type I supernovae** have spectra that contain no hydrogen lines. They reach a maximum brightness about 4 billions times more luminous than the sun and decline rapidly at first and then more slowly. **Type II supernovae** have spectra containing hydrogen lines. They reach a maximum brightness up to about 0.6 billion times more luminous than the sun, decline to a standstill, and then fade rapidly. The light curves in ■ Figure 13-13 summarize the behavior of the two types of supernovae.

The evidence is clear that type II supernovae occur when a massive star develops an iron core and collapses. Such supernovae occur in regions of active star formation where there is plenty of

■ **Figure 13-12**

This supernova (arrow) was seen in the outskirts of the central disk of galaxy NGC4526. From its spectrum and from the way its brightness faded with time, astronomers recognize it as a type Ia supernova produced by the collapse and total destruction of a white dwarf. (High-Z Supernova Search Team, HST, NASA)

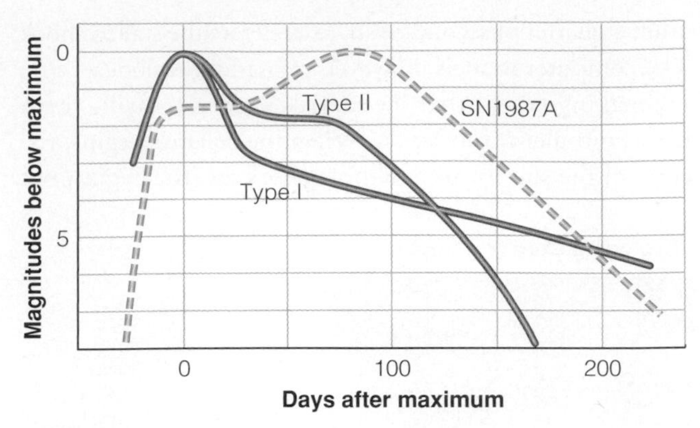

■ **Figure 13-13**

Type I supernovae decline rapidly at first and then more slowly, but type II supernovae pause for about 100 days before beginning a steep decline. Supernova 1987A was odd in that it did not rise directly to maximum brightness. These light curves have been adjusted to the same maximum brightness. Generally, type II supernovae are about two magnitudes fainter than type I.

gas and dust. These are the regions where you would expect to find massive stars. The spectra of type II supernovae contain hydrogen lines, as you would expect from the explosion of a massive star that contains large amounts of hydrogen in its outer layers.

Type I supernovae show no hydrogen in their spectra, but there are two kinds of type I supernovae that have dramatically different causes. Both involve binary stars. Type Ia supernovae are often found in regions where star formation ended long ago. That means they can't be caused by massive stars, which don't live very long. Also, type Ia supernovae lack hydrogen lines in their spectra, so they can't be caused by the deaths of massive stars. Rather, the evidence shows that a type Ia occurs when a white dwarf gaining mass in a binary star system exceeds the Chandrasekhar limit and collapses.

The collapse of a white dwarf is different from the collapse of a massive star because the core of the white dwarf contains usable fuel. As the collapse begins, the temperature shoots up, but the gas cannot halt the collapse because it is degenerate, and the pressure–temperature thermostat is turned off in the core. Even as carbon fusion begins, the increased temperature cannot increase the pressure and make the gas expand and slow the reactions. The core of the white dwarf is a bomb. The carbon–oxygen core fuses suddenly in violent nuclear reactions called **carbon deflagration.** The word *deflagration* means to be totally destroyed by fire, in this case by nuclear fusion. In a flicker of a stellar lifetime the entire star is consumed, and the outermost layers are blasted away in a violent explosion that at its brightest is three to six times more luminous than a type II supernova. The white dwarf is entirely destroyed. No neutron star or black hole is left behind. Of course, you would see no hydrogen lines in the spectrum of a type Ia supernova because white dwarfs contain very little hydrogen.

Type Ib supernovae are less common. Their spectra contain no hydrogen lines, but they do occur in regions where you would expect to find young stars. They are believed to occur when a massive star in a binary system loses its hydrogen-rich outer layers to its companion star. The remains of the massive star could develop an iron core and collapse, producing a supernova explosion that lacks hydrogen lines in the spectrum. Some astronomers have referred to these as "peeled" supernovae, meaning that the massive star has had its hydrogen-rich outer layers peeled away by its binary companion.

To summarize, a type II supernova is caused by the collapse of a massive star. A type Ia is caused by the collapse of a white dwarf. A type Ib is caused by the collapse of a massive star that has lost its outer envelope of hydrogen.

Much of what you have learned so far about supernovae has been based on theory, so it is time to compare theory with observations of real supernova explosions. These frequent reality checks are a distinguishing characteristic of science.

Observations of Supernovae

In 1054, Chinese astronomers saw a "guest star" appear in the constellation now known as Taurus, the bull. The star quickly became so bright it was visible in the daytime. After a month's time, it slowly faded, taking almost two years to vanish from sight. When modern astronomers turned their telescopes to the location of the guest star, they found a cloud of gas about 1.35 pc in radius, expanding at 1400 km/s. Projecting the expansion back in time, they concluded that the expansion must have begun about nine centuries ago, just when the guest star made its visit. From this and other evidence, astronomers conclude that the nebula, now called the Crab Nebula because of its shape (■ Figure 13-14), marks the site of the 1054 supernova.

The Crab Nebula contains a critical clue. The glowing filaments appear to be excited gas flung outward by the explosion, but the hazy glow in the inner nebula is something else. Radio observations show that the gas in the nebula is emitting **synchrotron radiation**—electromagnetic energy radiated by high-speed electrons spiraling through a magnetic field. Low-speed electrons radiate at longer wavelengths, and high-speed electrons radiate at shorter wavelengths, so synchrotron radiation is spread over a wide range of wavelengths. The foggy glow of light in the Crab Nebula is synchrotron radiation at the very short wavelengths of visible light, and that must mean the electrons are traveling at tremendous speeds. In the nine centuries since the Crab supernova explosion, the electrons should have radiated their energy away and slowed down. Evidently they have not, so there must be an energy source in the Crab Nebula that is producing very-high-speed electrons. You will follow this clue in the next chapter and discover a neutron star at the center of the Crab Nebula.

The Crab Nebula is important because it is linked to a supernova explosion that was actually seen to occur. Supernovae are rare. Only a few have been seen with the naked eye in recorded history. Arab astronomers saw one in 1006, and the Chinese saw the Crab supernova in 1054. European astronomers observed two—one in 1572 (Tycho's supernova) and one in 1604 (Kepler's supernova). Also, the guest stars of 185, 386, 393, and 1181 may have been supernovae. For 383 years following 1604, no naked-eye supernova appeared, and then in February 1987 a star in the southern sky exploded. You will meet the great supernova of 1987 later in this section, but for now let's continue to explore supernovae in general.

Most supernovae are discovered in distant galaxies, and searching for these supernovae was once a tedious job. Robotic telescopes now observe long lists of galaxies each night searching for the appearance of a new star. Detecting supernovae early is important because astronomers need to study the early stages of the explosion in order to understand the complex processes that occur in the rapidly expanding cloud of gas.

The supernova explosion fades to obscurity in a year or two, but an expanding shell of gas marks the explosion site. The gas, originally expelled at 10,000 to 20,000 km/s, may carry away one-fifth of the mass of the star. The collision of that expanding gas with the surrounding interstellar medium can sweep up even more gas and excite it to produce a **supernova remnant,** the nebulous remains of a supernova explosion.

Supernova remnants look quite delicate and do not survive very long—a few tens of thousands of years—before they gradually mix with the interstellar medium and vanish. The Crab Nebula is a young remnant, only about 950 years old, and it isn't very large, only a few parsecs in diameter. Older remnants can be larger. Some supernova remnants are detectable only at radio and X-ray wavelengths. They have become too tenuous to emit detectable light, but the collision of the expanding hot gas with the interstellar medium can generate radio and X-ray radiation and allows astronomers to create images of them at these nonvisible wavelengths. In general, supernova remnants are tenuous spheres of gas expanding into the interstellar medium (■ Figure 13-15). You saw in Chapter 11 that the compression of the interstellar medium by expanding supernova remnants can trigger star formation.

The Great Supernova of 1987

Until 1987, astronomers had never looked at a bright supernova through a telescope. Nearly all supernovae are in distant galaxies and thus are very faint. The last supernova visible to the naked eye, Kepler's supernova of 1604, predated the invention of the telescope. Then, in late February 1987, the news raced around the world: Astronomers in Chile had discovered a naked-eye supernova in the Large

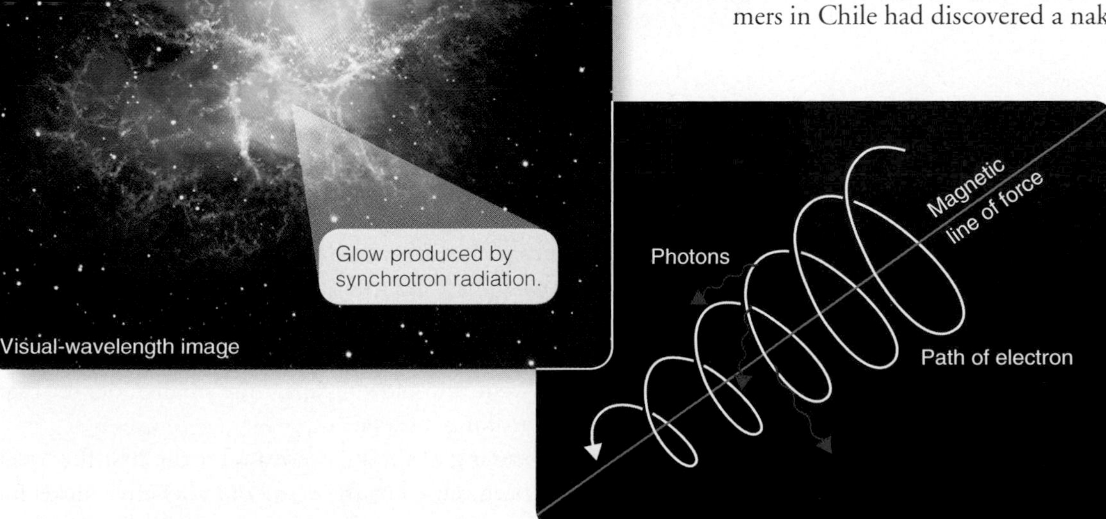

■ Figure 13-14

The Crab Nebula is located in the constellation Taurus the bull, just where Chinese astronomers saw a brilliant guest star in AD 1054. Over tens of years, astronomers can measure the motions of the filaments as they expand away from the center. Doppler shifts confirm that the near side of the nebula is moving toward us. The foggy glow is synchrotron radiation produced by high-speed electrons moving through a magnetic field. (ESO)

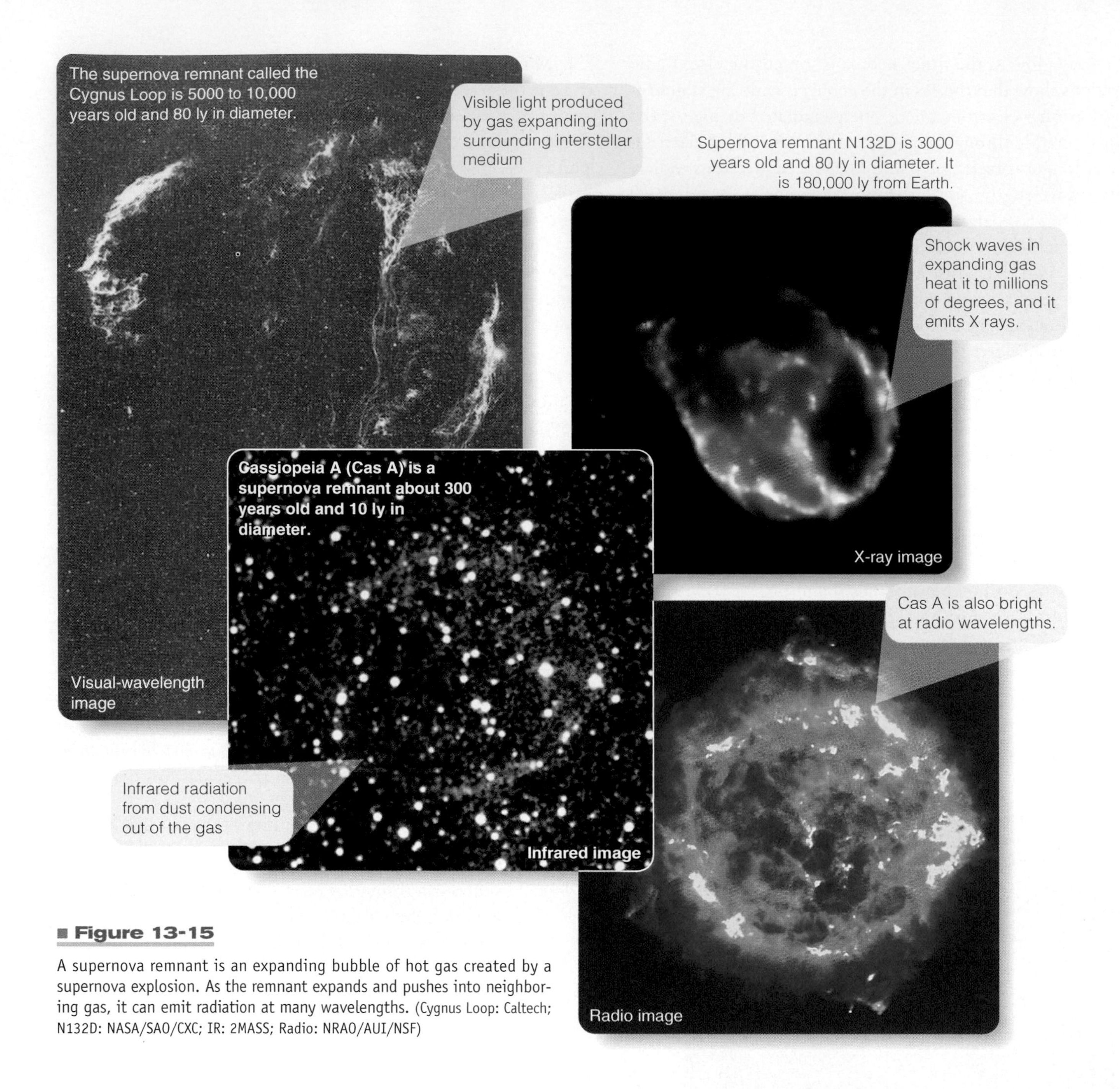

■ Figure 13-15

A supernova remnant is an expanding bubble of hot gas created by a supernova explosion. As the remnant expands and pushes into neighboring gas, it can emit radiation at many wavelengths. (Cygnus Loop: Caltech; N132D: NASA/SAO/CXC; IR: 2MASS; Radio: NRAO/AUI/NSF)

Magellanic Cloud, a small galaxy very near our Milky Way Galaxy (■ Figure 13-16). Because the supernova was only 20 degrees from the south celestial pole, it could be studied only from southern latitudes. It was named SN1987A to denote the first supernova discovered in 1987.

The hydrogen-rich spectrum suggested that the supernova was a type II, caused by the collapse of the core of a massive star. As the months passed, however, the light curve proved to be odd (Figure 13-13) in that it paused for a few weeks before rising to its final maximum. From photographs of the area made some years before, astronomers were able to determine that the star that exploded, cataloged as Sanduleak −69°202, was not the expected red supergiant but rather a hot, blue supergiant of only 20 solar masses and 50 solar radii, not extreme for a supergiant. Theorists now believe that the star was chemically poor in elements heavier than helium and had consequently contracted and heated up after a phase as a cool, red supergiant, during which it lost mass into space. The relatively small size of the supergiant may explain the pause in the light curve. Much of the energy of the explosion went into blowing apart the smaller, denser-than-usual star and making it expand.

The brightening of the supernova after the first few weeks seems to have been caused by the decay of radioactive nickel into cobalt. Theory predicts the production of such nickel atoms in the

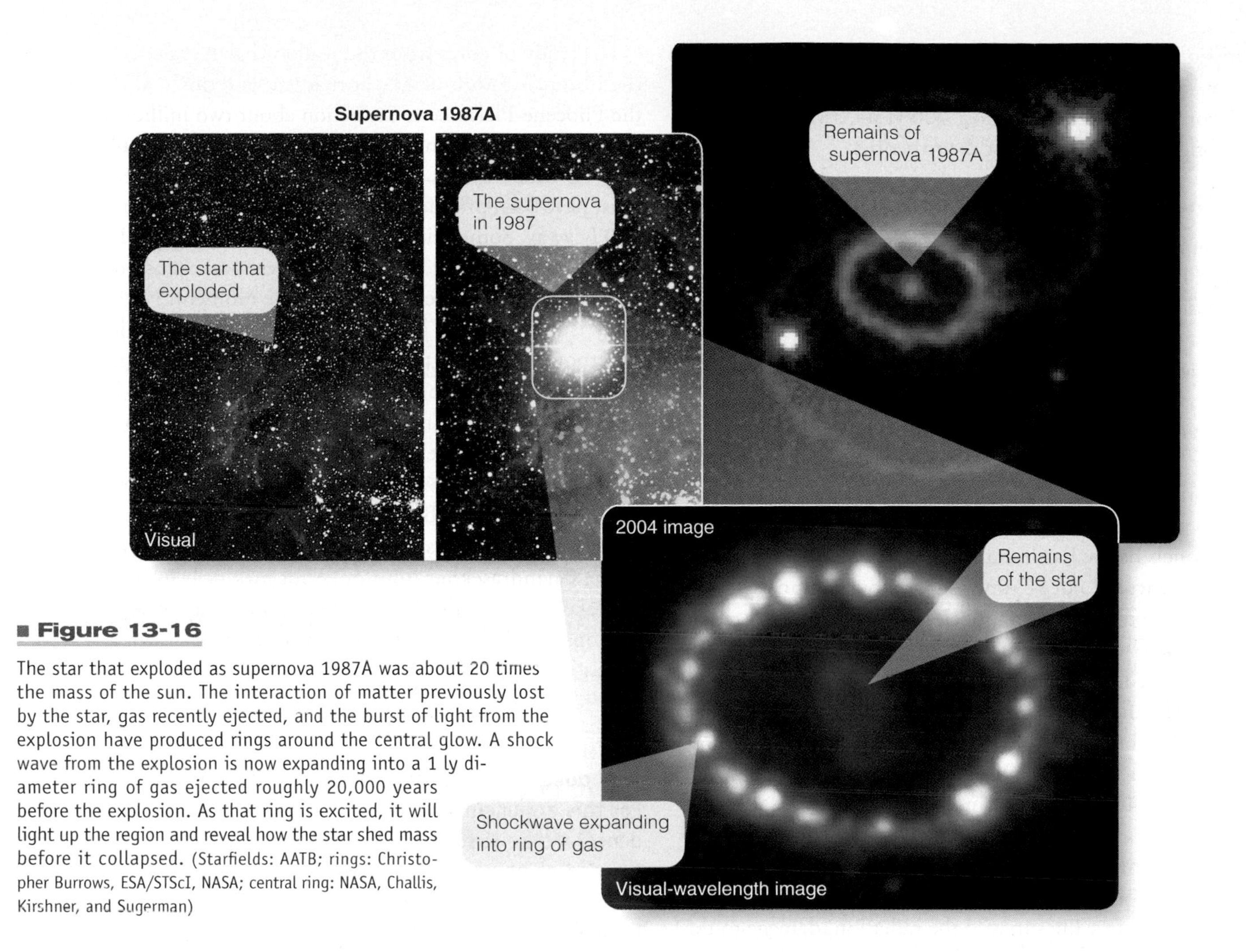

■ **Figure 13-16**

The star that exploded as supernova 1987A was about 20 times the mass of the sun. The interaction of matter previously lost by the star, gas recently ejected, and the burst of light from the explosion have produced rings around the central glow. A shock wave from the explosion is now expanding into a 1 ly diameter ring of gas ejected roughly 20,000 years before the explosion. As that ring is excited, it will light up the region and reveal how the star shed mass before it collapsed. (Starfields: AATB; rings: Christopher Burrows, ESA/STScI, NASA; central ring: NASA, Challis, Kirshner, and Sugerman)

explosion, and their decay into cobalt would release gamma rays that would heat the expanding shell of gas and make it brighter. About 0.07 solar mass of nickel was produced, about 20,000 times the mass of Earth.

The cobalt atoms are also unstable, but they decay more slowly, so it was not until sometime later, after much of the nickel had decayed, that the decay of cobalt into iron began providing energy to keep the expanding gas hot and luminous. Although these processes had been predicted, they were clearly observed in SN1987A. Gamma rays from the decay of cobalt to iron were detected, and cobalt and iron are clearly visible in the infrared spectra of the supernova.

As the supernova dimmed, photographs revealed bright rings of gas (Figure 13-16). Comparison with mathematical models shows that the rings were produced by two stellar winds. When the star was a red supergiant, it expelled a slow stellar wind. Later, when it became a blue supergiant, it expelled a fast stellar wind. The interaction of these two winds shaped by a magnetic field and illuminated by the supernova explosion produces the rings. Compare this with the rings seen in some of the planetary nebulae shown on page 291. In more recent years, a shock wave from the supernova explosion has begun having an impact on the bright ring of gas and exciting it to glow brightly, as shown at the bottom of Figure 13-16.

Two independent observations confirm that SN1987A probably gave birth to a neutron star. Theory predicts that the collapse of a massive star's core should liberate a tremendous blast of neutrinos that leave the star hours before the shock wave from the interior blows the star apart. Two independent neutrino detectors, one in Ohio and one in Japan, recorded a burst of neutrinos passing through Earth at 2:35:41 AM EST on February 23, 1987, about 18 hours before the supernova was seen. The data show that the neutrinos rushed toward Earth from the direction of the supernova. The detectors caught only 19 neutrinos during a 12-second interval, but recall that neutrinos hardly ever react with normal matter. Only 19 were detected, but the full flood of neutrinos must have been immense. Within a few seconds of that time, roughly 20 trillion neutrinos passed harmlessly through each human body on Earth. The detection of the neutrino blast confirms that the collapsing core gave birth to a neutron star.

The expanding gas shell of the supernova will continue to thin and cool, and eventually astronomers on Earth will be able

14 Neutron Stars and Black Holes

Almost anything is easier to get into than out of.

AGNES ALLEN

X-ray image

GRAVITY ALWAYS WINS. In a star's struggle to withstand its own gravity, the star must eventually exhaust its fuel, and gravity must win. Gravity ensures that the star's last remains must eventually reach one of three final states—white dwarf, neutron star, or black hole. You studied the first of these compact objects in the previous chapter, and now you are ready to complete the story of the stars. ▌ Theory predicts that neutron stars and black holes exist, but science depends on evidence. Can astronomers find objects in the sky that are real neutron stars and black holes? Scientists always fall back on evidence—the final reality check on their theories.

▌ Continued on page 310 ▌

Yielding finally to its own gravity, a massive star collapsed and died, leaving behind this supernova remnant 40 ly in diameter. A neutron star located at the center is about the size of a mountain on Earth and emits 30,000 times more energy than the sun. (NASA/CXC/SAO)

Guidepost

Looking Back

Through the last three chapters you have traced the story of stars from their birth as clouds of gas in the interstellar medium to their final collapse. By now you are asking a simple question, "What is left?" The answer, of course, depends on the mass of the star. Stars like the sun leave behind white dwarfs, but more massive stars leave behind the strangest beasts in the cosmic zoo.

This Chapter

Now you are ready to meet neutron stars and black holes, and your exploration will answer five essential questions:

How did scientists predict the existence of neutron stars?

What is the evidence that neutron stars really exist?

How did scientists predict the existence of black holes?

What is the evidence that black holes really exist?

What happens when matter falls into a neutron star or black hole?

You can tell from the first four questions above that this chapter will show you clear examples of how astronomers combine theory and evidence to understand nature.

Looking Ahead

This chapter ends the story of individual stars, but it does not end the story of stars. In the next chapter, you will begin exploring the giant communities in which stars live—the galaxies.

AceAstronomy™ The AceAstronomy icon throughout the text indicates an opportunity for you to test yourself on key concepts and to explore animations and interactions on the AceAstronomy website at: http://ace.brookscole.com/sf9

14-1 Neutron Stars

A **NEUTRON STAR** is a star of a little over 1 solar mass compressed into a radius of about 10 km. Its density is so high that the matter is stable only as a fluid of neutrons. Theory predicts that such an object would spin a number of times a second, be nearly as hot at its surface as the inside of the sun, and have a magnetic field a trillion times stronger than Earth's (■ Figure 14-1). Two questions should occur to you immediately. First, how could any theory predict such a wondrously unbelievable star? And second, do such neutron stars really exist?

Theoretical Prediction of Neutron Stars

The neutron was discovered in the laboratory in 1932, and its properties suggested something fantastic. Neutrons spin in much the way that electrons do, which means that neutrons must obey the Pauli exclusion principle. In that case, if neutrons are packed together tightly enough, they can become degenerate just as electrons do. White dwarfs are supported by degenerate electrons; and, in 1932, the Russian physicist Lev Landau predicted that neutron stars might exist supported by degenerate neutrons. Of course, the inside of a neutron star would have to be much denser than the inside of a white dwarf.

Only two years later, in 1934, Walter Baade and Fritz Zwicky provided a pedigree for neutron stars. Not knowing of Landau's prediction, they suggested that some of the most luminous novae in the historical record were not true novae but were caused by the collapse of a massive star in an even larger explosion they called a supernova. What was left of the core of the star, they proposed, was a small, high-density neutron star.

■ Figure 14-1

A supernova explosion seen in AD 1181 left behind an expanding supernova remnant. The Chandra X-Ray Observatory has imaged the nebula in X rays and finds a tiny hot object within—a neutron star. (NASA/SAO/CXC/P. Slane et al.)

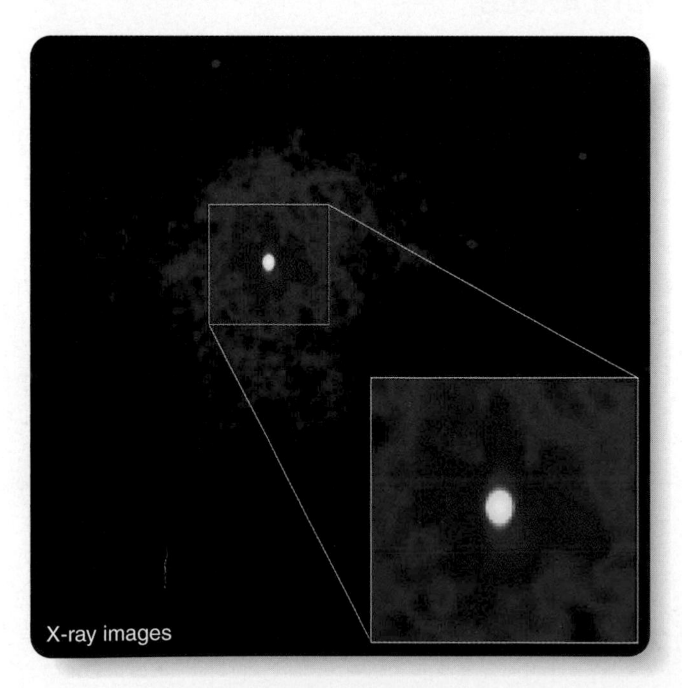

Atomic physics provides an explanation of how the collapsing core of a massive star could form a neutron star. If the collapsing core is more massive than the Chandrasekhar limit of 1.4 solar masses, then it cannot reach stability as a white dwarf. The weight is too great to be supported by degenerate electrons. The collapse of the core continues, and the atomic nuclei are broken apart by gamma rays. Almost instantly, the increasing density forces the freed protons to combine with electrons and become neutrons by the emission of neutrinos:

$$e + p \rightarrow n + \nu$$

The burst of neutrinos, as you saw in the previous chapter, helps blow the star apart; and the core of the star, in a fraction of a second, is transformed into a dense ball of neutrons left behind by the supernova explosion as a neutron star.

As you saw in the previous chapter, a star of 8 solar masses or less could lose enough mass to die as a planetary nebula leaving behind a white dwarf. More massive stars, up to a limit of about 15 solar masses, will lose mass rapidly, but they cannot reduce their mass fast enough and will apparently face death as a supernova explosion leaving behind a neutron star (■ Figure 14-2). (You will see later in this chapter that the most massive stars probably leave behind black holes.)

Theoretical calculations predict that a neutron star will be only 10 or so kilometers in radius (■ Figure 14-3) and will have a density of about 10^{14} g/cm^3. On Earth, a sugar-cube-sized lump of this material would weigh 100 million tons. This is roughly the density of the atomic nucleus, and you can think of a neutron star as matter with all of the empty space squeezed out of it.

How massive can a neutron star be? That is a critical question, and it's a difficult one to answer because physicists don't know the strength of pure neutron material. Such matter can't be made in the laboratory, so its properties must be predicted theoretically. The most widely accepted calculations suggest that a neutron star cannot be more massive than 2 to 3 solar masses. If a neutron star were more massive than that, the degenerate neutrons would not be able to support the weight, and the object would collapse (presumably into a black hole).

Simple physics, the physics you have used in previous chapters to discuss normal stars, predicts that neutron stars should be hot, spin rapidly, and have strong magnetic fields. You have seen that contraction heats the gas in a star. As the gas particles fall inward, they pick up speed; and, when they collide, their high speeds become thermal energy. The sudden collapse of the core of a massive star to a radius of 10 km should heat it to millions of degrees. Furthermore, neutron stars should cool slowly because the heat can escape only from the surface, and neutron stars are so small they have little surface from which to radiate.

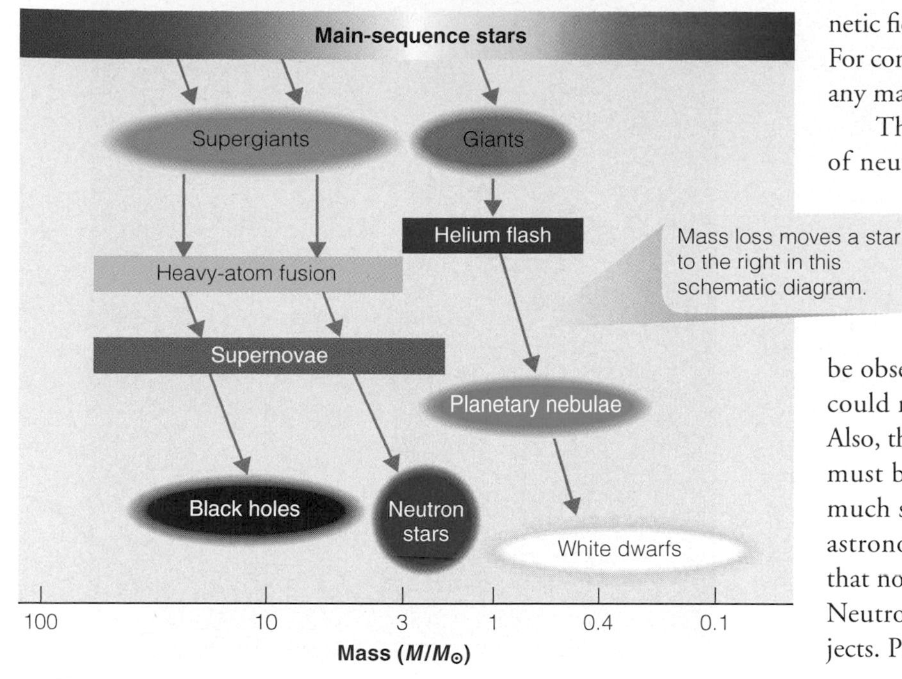

■ Figure 14-2

How a star evolves depends on its mass. Mass loss can change a star's fate by reducing its mass as it evolves. The mass limits of the categories shown here are not well known and are given only for purposes of illustration. (Figure design by author)

In this way, basic theory predicts that neutron stars should be very hot.

The conservation of angular momentum predicts that neutron stars should spin rapidly. All stars rotate to some extent because they form from swirling clouds of interstellar matter. As such a star collapses into a neutron star, it must rotate faster because it conserves angular momentum. Recall that you see this happen when ice skaters spin slowly with their arms extended and then speed up as they pull their arms closer to their bodies (see Figure 5-7). In the same way, a collapsing star must spin faster as it pulls its matter closer to its axis of rotation. If the sun collapsed to a radius of 10 km, its period of rotation would decrease from 25 days to about 0.001 second. You might expect the collapsed core of a massive star to rotate 10 or 100 times a second.

Basic theory also predicts that a neutron star should have a powerful magnetic field. Remember from your study of the sun's atmosphere in Chapter 8 that a magnetic field passing through an ionized gas is "frozen in." Whatever magnetic field a star has is frozen into the gas of the star. When the star collapses, the magnetic field is carried along and squeezed into a smaller area, which could make the field a billion times stronger. Some stars have magnetic fields over 1000 times stronger than the sun's, so you might expect a neutron star to have a magnetic field as much as a trillion times stronger than the sun's. For comparison, that is about 10 million times stronger than any magnetic field ever produced in the laboratory.

Theory allowed astronomers to predict the properties of neutron stars, but it also predicted that such objects should be difficult to observe. Neutron stars are very hot, so Wien's law of black body radiation (Chapter 7) told astronomers that neutron stars would radiate most of their energy in the X-ray part of the spectrum, radiation that could not be observed in the 1940s and 1950s because astronomers could not put their telescopes above Earth's atmosphere. Also, the small surface areas of neutron stars mean that they must be faint objects. They are hot, but they don't have much surface area from which to radiate. Consequently, astronomers of the mid-20th century were not surprised that none of the newly predicted neutron stars were found. Neutron stars were, at that point, entirely theoretical objects. Progress came not from theory but from observation.

The Discovery of Pulsars

In November 1967, Jocelyn Bell, a graduate student at Cambridge University in England, found a peculiar pattern on the paper chart from a radio telescope. Unlike other radio signals from celestial bodies, this was a series of regular pulses (■ Figure 14-4). At first she and the leader of the project, Anthony Hewish, thought the signal was interference, but they found it day after day in the same place in the sky. Clearly, it was celestial in origin.

■ Figure 14-3

A tennis ball and a road map illustrate the relative size of a neutron star. Such an object, containing slightly more than the mass of the sun, would fit with room to spare inside the beltway around Washington, DC. (M. Seeds)

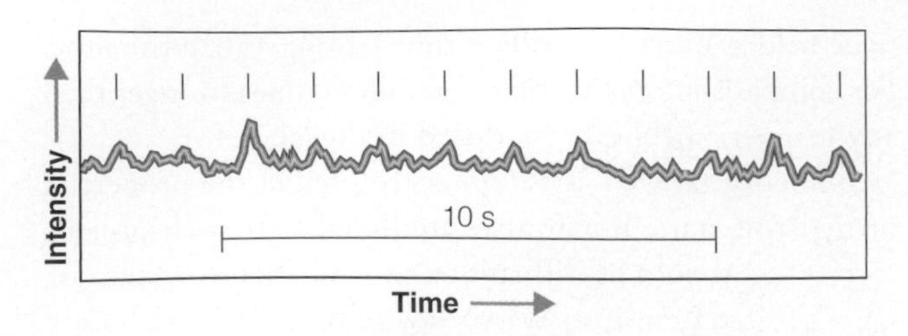

■ **Figure 14-4**

The 1967 detection of regularly spaced pulses in the output of a radio telescope led to the discovery of pulsars. This record of the radio signal from the first pulsar, CP1919, contains regularly spaced pulses (marked by ticks). The period is 1.33730119 seconds.

Another possibility, that it was a radio signal from a distant civilization, led them to consider naming it LGM for Little Green Men. But within a few weeks, the team found three more objects in other parts of the sky, pulsing with different periods. The objects were clearly natural, and the team dropped the name LGM in favor of **pulsar**—a contraction of *pulsing star.* The pulsing radio source Bell had observed with her radio telescope was the first known pulsar.

As more pulsars were found, astronomers argued over their nature. Periods ranged from 0.033 to 3.75 seconds and were nearly as exact as an atomic clock. Months of observation showed that many of the periods were slowly growing longer by a few billionths of a second per day. Whatever produced the regular pulses had to be highly precise, nearly as exact as an atomic clock, but it had to gradually slow down.

It was easy to eliminate possibilities. Pulsars could not be stars. A normal star, even a small white dwarf, is too big to pulse that fast. Nor could a star with a hot spot on its surface spin fast enough to produce the pulses. Even a small white dwarf would fly apart if it spun 30 times a second.

The pulses themselves gave the astronomers a clue. The pulses last only about 0.001 second. This places an upper limit on the size of the object producing the pulse. If a white dwarf blinked on and then off in that interval, astronomers would not see a 0.001-second pulse. The near side of the white dwarf would be about 6000 km closer to Earth, and light from the near side would arrive 0.022 seconds before the light from the bulk of the white dwarf. In that way, its short blink would be smeared out into a longer pulse. This is an important principle in astronomy—an object cannot change its brightness appreciably in an interval shorter than the time light takes to cross its diameter. If pulses from pulsars are no longer than 0.001 second, then the objects cannot be larger than about 300 km (190 miles) in diameter and could be smaller.

Only a neutron star is small enough to be a pulsar. In fact, a neutron star is so small, it can't vibrate slowly enough, but it can spin as fast as 1000 times a second without flying apart. The missing link between pulsars and neutron stars was found in late 1968, when astronomers discovered a pulsar at the heart of the Crab Nebula (■ Figure 14-5). The Crab Nebula is a supernova remnant, and theory predicts that some supernovae leave behind a neutron star.

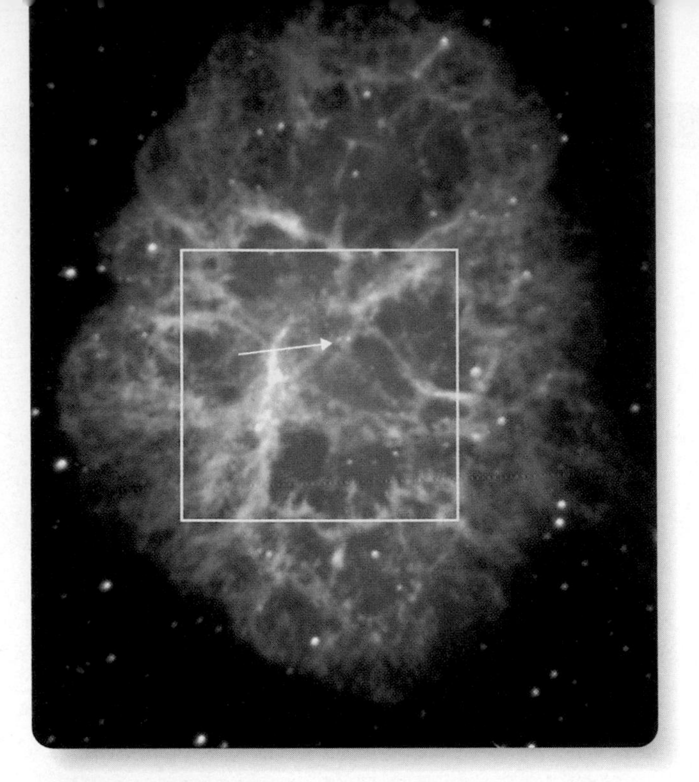

■ **Figure 14-5**

The pulsar at the center of the Crab Nebula (arrow) is detectable in visual-wavelength photographs. The star to the right of the pulsar lies much closer to Earth and is not in the Crab Nebula. The white box outlines the area imaged on page 315. (Caltech)

The short pulses and the discovery of the pulsar in the Crab Nebula are strong evidence that pulsars are neutron stars. If you combine theory and observation, you can devise a model of a pulsar.

A Model Pulsar

Scientists often work by building a model of a natural phenomenon—not a physical model made of plastic and glue but an intellectual conception of how nature works in a specific instance. The model may be limited and incomplete, but it helps them organize their theories and observations. A model of a pulsar will help you draw together a lot of different ideas.

The modern model of a pulsar has been called the **lighthouse model** and is shown in **The Lighthouse Model of a Pulsar** on pages 314–315. Notice three important points:

1. A pulsar does not pulse but rather emits beams of radiation that sweep around the sky as the neutron star rotates. If the beams do not sweep over Earth, the pulses will not be detectable by Earth's radio telescopes.

2. Notice that the mechanism that produces the beams involves extremely high energies and is not fully understood.

3. Also notice how modern space telescopes observing at nonvisual wavelengths can help confirm and refine the model.

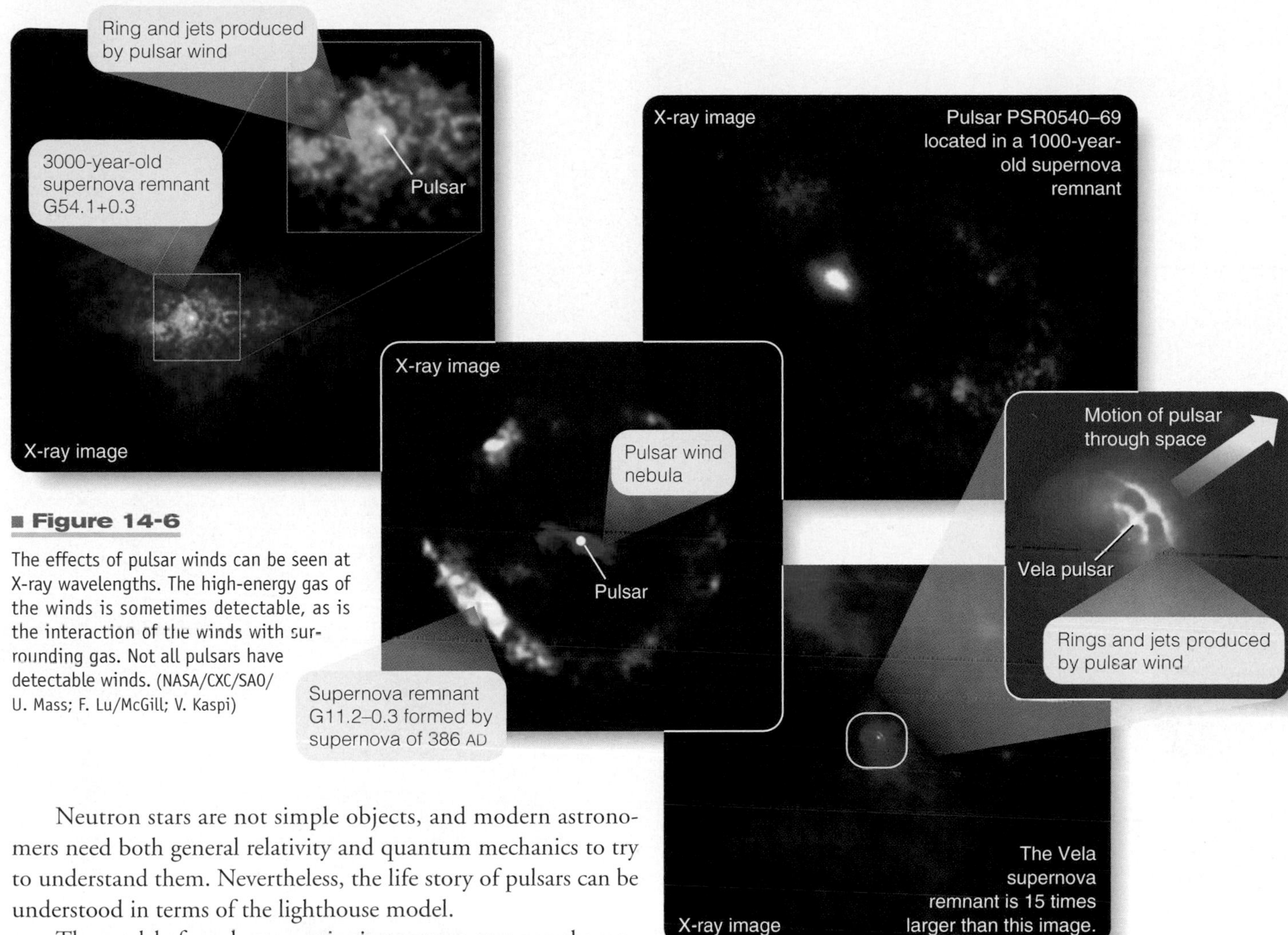

■ **Figure 14-6**

The effects of pulsar winds can be seen at X-ray wavelengths. The high-energy gas of the winds is sometimes detectable, as is the interaction of the winds with surrounding gas. Not all pulsars have detectable winds. (NASA/CXC/SAO/U. Mass; F. Lu/McGill; V. Kaspi)

Neutron stars are not simple objects, and modern astronomers need both general relativity and quantum mechanics to try to understand them. Nevertheless, the life story of pulsars can be understood in terms of the lighthouse model.

The model of a pulsar as a spinning neutron star won the support of many astronomers because of two properties of pulsars. First, many pulsars are slowing down. Their periods are increasing by a few billionths of a second each day—a change radio astronomers can measure using atomic clocks. Evidently, the spinning neutron star is converting some of its energy of rotation into various kinds of electromagnetic energy and a powerful outflow of high-speed particles called a **pulsar wind** (■ Figure 14-6). About 99.9 percent of the energy released by the slowing of the neutron star is carried away by the pulsar wind, and only 0.1 percent goes into producing the radio beams. The energy that keeps the Crab Nebula glowing nearly 1000 years after the explosion is coming from the rotational energy of the neutron star and the pulsar wind that the neutron star produces.

The second property of pulsars that supports the neutron-star model of a pulsar is the **glitch**—a sudden increase in the pulse rate seen in some pulsars (■ Figure 14-7). Two theories have been

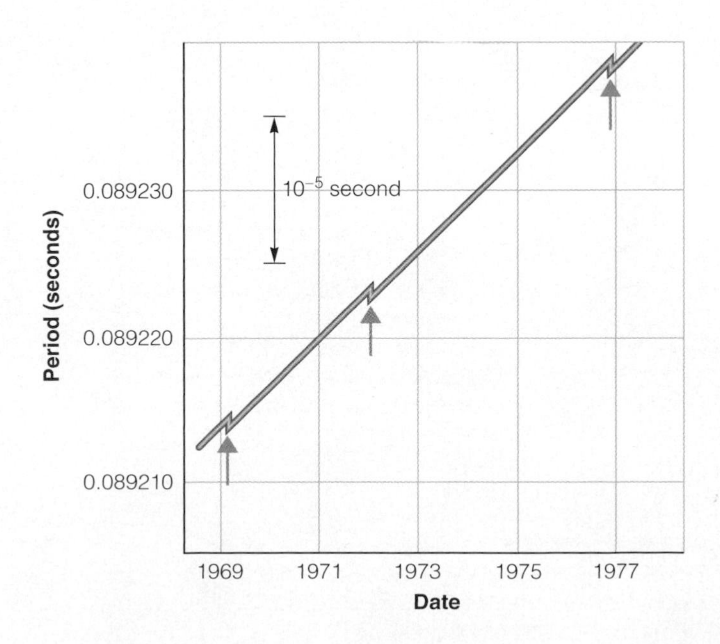

■ **Figure 14-7**

Soon after pulsars were discovered, radio astronomers accumulated enough data to show that pulsars were very gradually slowing down. That is, their pulses were growing longer. Some pulsars, such as the Vela pulsar whose data are shown here, experience glitches in which the pulsar suddenly speeds up only to resume its more leisurely decline.

The Lighthouse Model of a Pulsar

1 Astronomers think of pulsars not as pulsing objects, but rather as objects emitting beams. As they spin, the beams sweep around the sky; when a beam sweeps over Earth, observers detect a pulse of radiation. Understanding the details of this lighthouse model is a challenge, but the implications are clear. Although a neutron star is only a few kilometers in radius, it can produce powerful beams. Also, observers tend to notice only those pulsars whose beams happen to sweep over Earth.

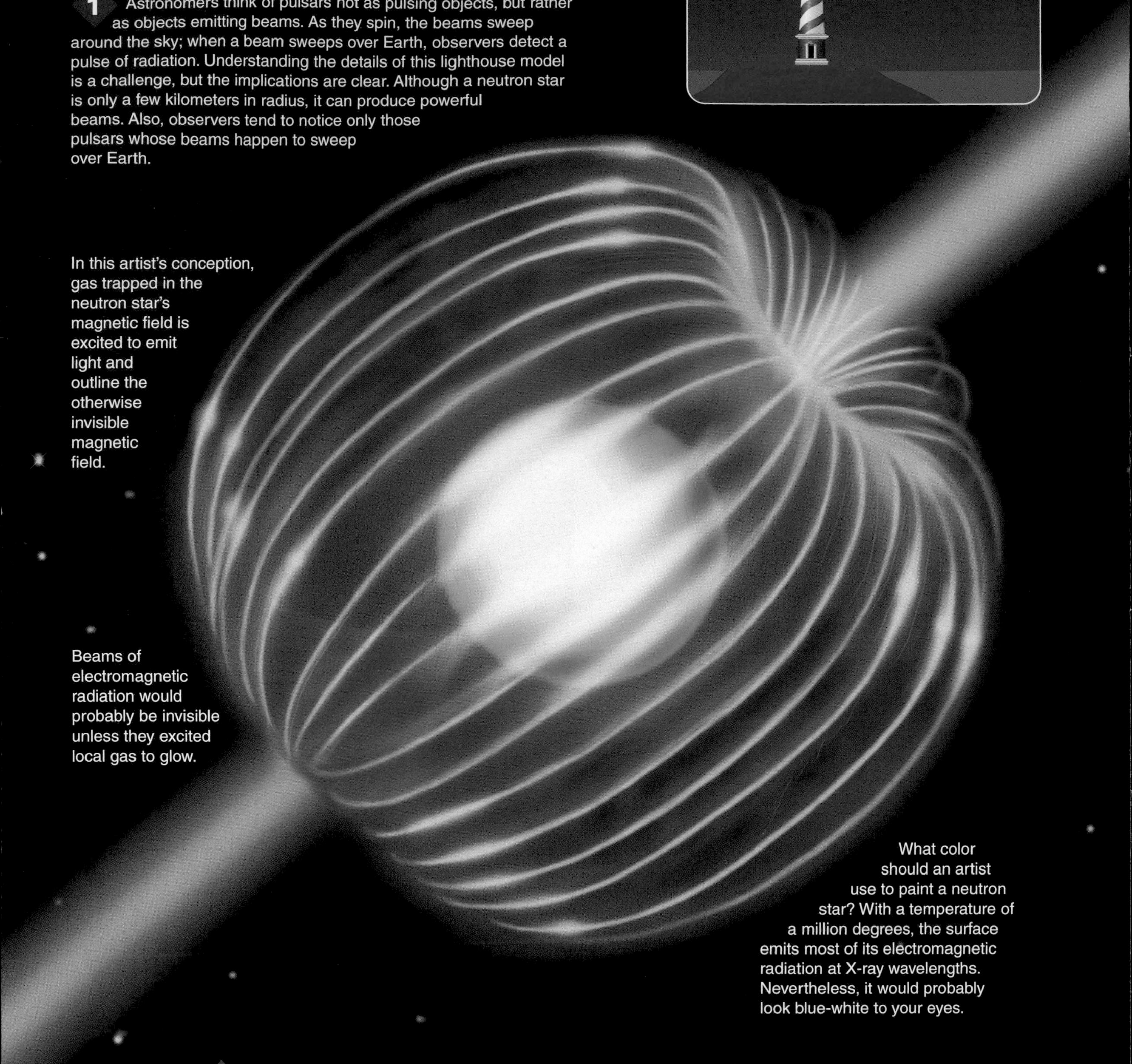

In this artist's conception, gas trapped in the neutron star's magnetic field is excited to emit light and outline the otherwise invisible magnetic field.

Beams of electromagnetic radiation would probably be invisible unless they excited local gas to glow.

What color should an artist use to paint a neutron star? With a temperature of a million degrees, the surface emits most of its electromagnetic radiation at X-ray wavelengths. Nevertheless, it would probably look blue-white to your eyes.

2 How a neutron star can emit beams is one of the challenging problems of modern astronomy, but astronomers have a general idea. A neutron star contains a powerful magnetic field and spins very rapidly. The spinning magnetic field generates a tremendously powerful electric field, and the field causes the production of electron–positron pairs. As these charged particles are accelerated through the magnetic field, they emit photons in the direction of their motion, which produce powerful beams of electromagnetic radiation emerging from the magnetic poles.

Neutron Star Rotation with Beams

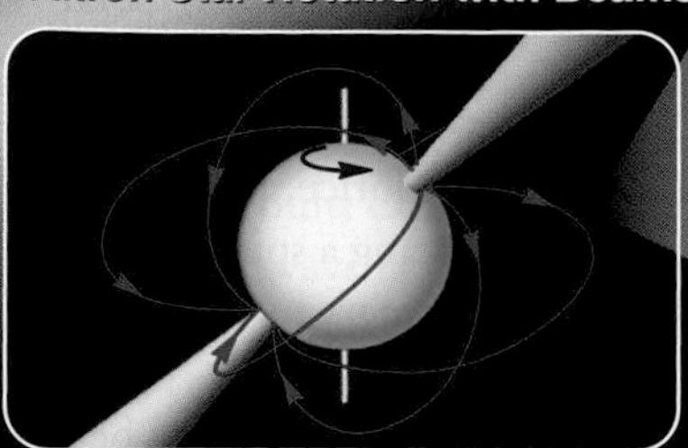

As in the case of Earth, the magnetic axis of a neutron star could be inclined to its rotational axis.

The rotation of the neutron star will sweep its beams around like beams from a lighthouse.

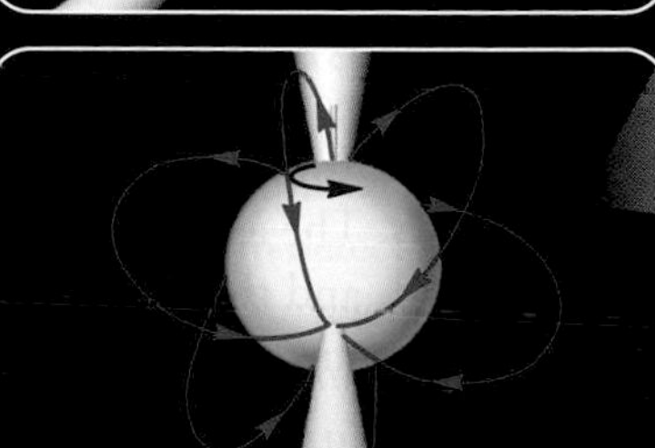

While a beam points roughly toward Earth, observers detect a pulse.

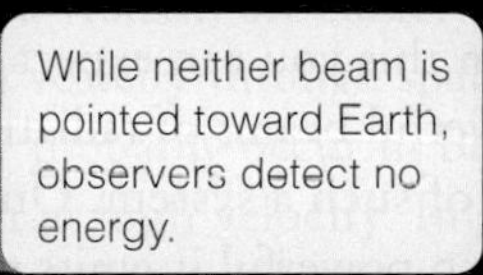

While neither beam is pointed toward Earth, observers detect no energy.

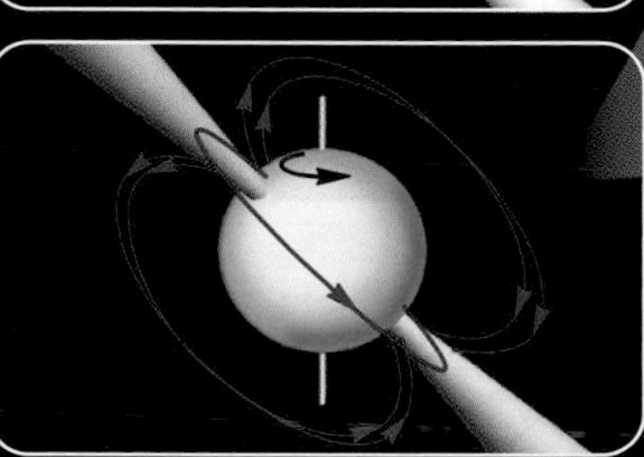

Beams may not be as exactly symmetric as in this model.

Log into AceAstronomy and select this chapter to see the Active Figure called "Neutron Star." Adjust the inclination of the neutron star's magnetic field to produce pulses.

3 Observations at many different wavelengths help astronomers understand pulsars. The hazy glow of the Crab Nebula is produced by synchrotron radiation. In the nearly 10 centuries since the supernova, the high-speed electrons should have radiated their energy away, and the synchrotron radiation should have faded. Evidently, the Crab pulsar powers the nebula. The Hubble Space Telescope image at right shows the region boxed in Figure 14-5. Circular wisps excited by the pulsar change and flicker from day to day.

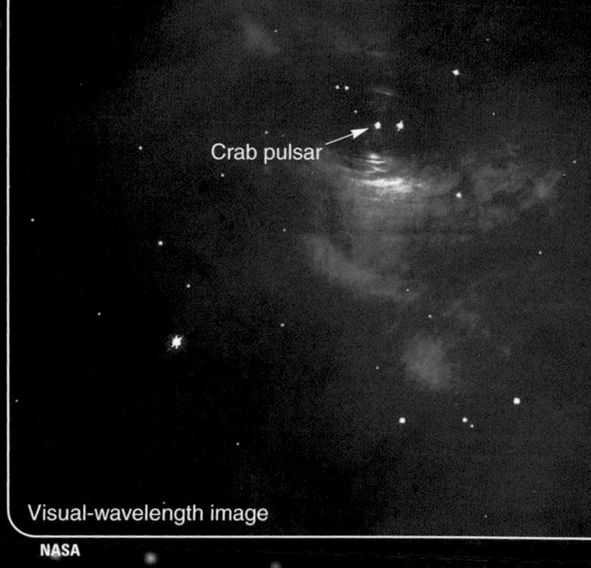

Visual-wavelength image

NASA

3a Red and yellow in this image show the radio and visual extent of the Crab Nebula. Blue, an X-ray image, reveals that the central pulsar has flung off rings of highly ionized gas in a disk up to a light-year in radius. The pulsar is ejecting jets of high-energy particles perpendicular to the disk.

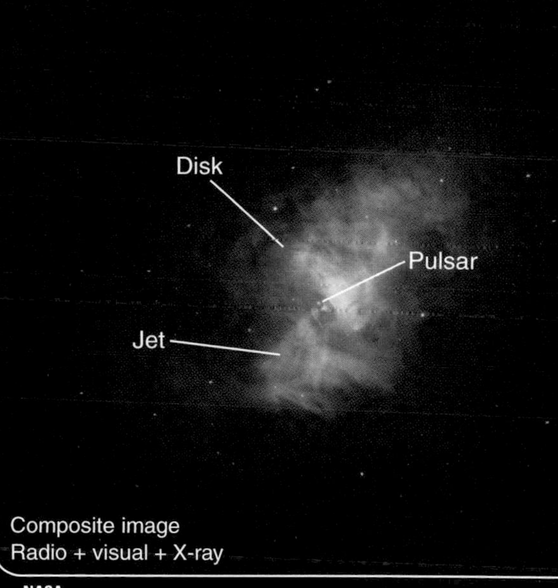

Composite image
Radio + visual + X-ray

NASA

X-ray image of Puppis supernova remnant

Neutron star

NASA

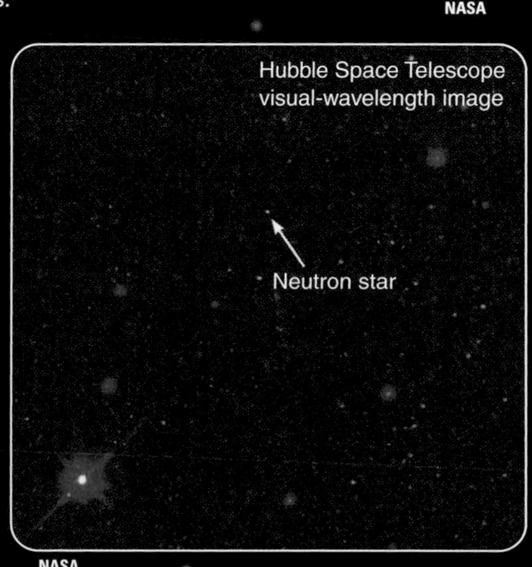

Hubble Space Telescope visual-wavelength image

NASA

3b If a pulsar's beams do not sweep over Earth, observers detect no pulses, and the neutron star is difficult to find. A few such objects are known, however. The Puppis A supernova remnant is about 4000 years old and contains a point source of X rays believed to be a neutron star. The isolated neutron star in the right-hand image has a temperature of 700,000 K.

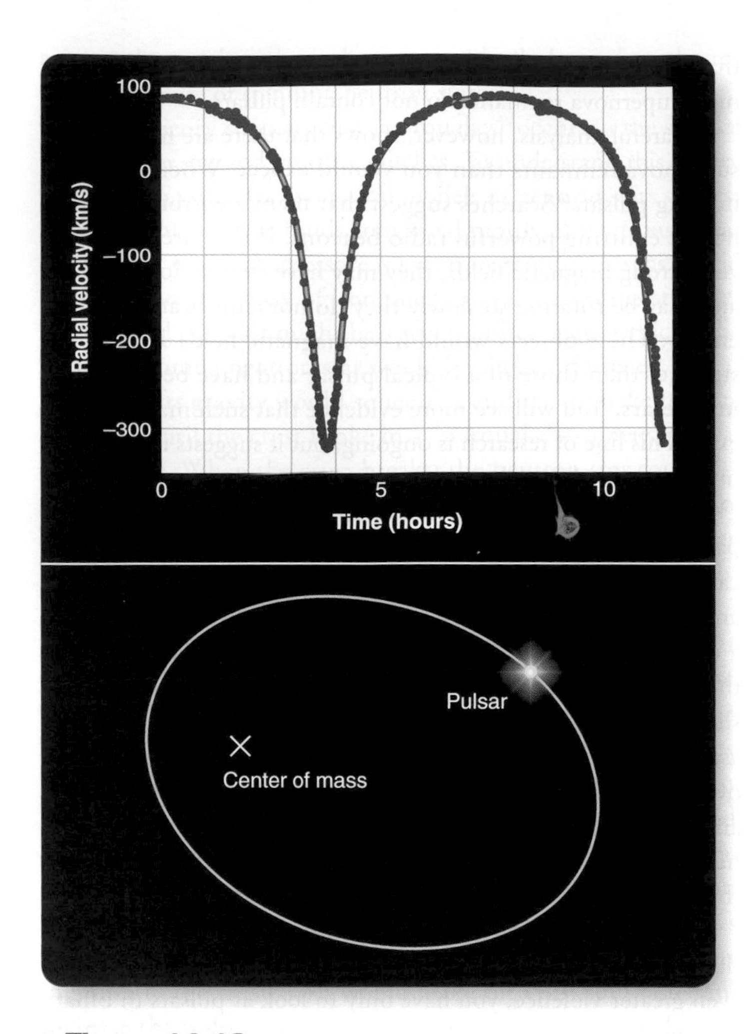

■ Figure 14-10

The radial velocity of pulsar PSR 1913+16 can be found from the Doppler shifts in its pulsation. Analysis of the radial velocity curve allows astronomers to determine the pulsar's orbit. Here, the center of mass does not appear to be at a focus of the elliptical orbit because the orbit is inclined. (Adapted from data by Joseph Taylor and Russell Hulse)

could be analyzed to find the shape of the pulsar's orbit (■ Figure 14-10). When Taylor and Hulse analyzed PSR 1913+16, they discovered that the binary system consisted of two neutron stars separated by a distance roughly equal to the radius of our sun.

Yet another surprise was hidden in the motion of PSR 1913+16. In 1916, Einstein's general theory of relativity described gravity as a curvature of space-time. Einstein realized that any rapid change in a gravitational field should spread outward at the speed of light as **gravitational radiation.** Gravity waves have not been detected, but Taylor and Hulse were able to show that the orbital period of the binary pulsar was slowly growing shorter because the stars are gradually spiraling toward each other. They are radiating orbital energy away as gravitational radiation. Taylor and Hulse won the Nobel prize in 1993 for their work with binary pulsars.

Dozens of binary pulsars have been found; by analyzing the Doppler shifts in their pulse periods, astronomers can estimate the mass of the neutron stars. Typical masses are about 1.35 solar masses, in good agreement with models of neutron stars.

In 2004, radio astronomers announced the discovery of a double pulsar. The two pulsars orbit each other in only 2.4 hours, and their spinning beams sweep over Earth (■ Figure 14-11). One spins with a period of 0.023 seconds (44 times a second), and the other spins in 2.8 seconds. This system is a pulsar jackpot because the orbits are nearly edge-on to Earth and the powerful magnetic fields eclipse each other, giving astronomers a chance to study their size and structure. Not only that, but the theory of general relativity predicts that they are emitting gravitational radiation and that their separation is decreasing by 7 mm per year. The two neutron stars will merge in 85 million years, presumably to trigger a supernova explosion. In the meantime, the steady decrease in orbital period can be measured and gives astronomers a further test of general relativity and gravitational radiation.

Binary pulsars can emit strong gravitational waves because the neutron stars contain large amounts of mass in a small volume. This also means that binary pulsars can be sites of tremendous violence because of the strength of gravity at the surface of a neutron star. An astronaut stepping onto the surface of a neutron star would be instantly smushed into a layer of matter only 1 atom thick. Matter falling onto a neutron star can release titanic amounts of energy. If you dropped a single marshmallow onto the surface of a neutron star from a distance of 1 AU, it would hit with an impact equivalent to a 3-megaton nuclear warhead. In general, a particle falling from a large distance to the surface of a neutron star will release energy equivalent to 0.2 mc^2, where m is the par-

■ Figure 14-11

Artist's impression of the double pulsar. One star must have exploded to form a pulsar, and later the other star did the same. Gravitational radiation causes the neutron stars to drift toward each other, and they will merge in 85 million years, presumably to trigger another supernova explosion. (John Rowe Animations)

ticle's mass at rest. Even a small amount of matter flowing from a companion star to a neutron star can generate high temperatures and release X rays and gamma rays.

As an example of such an active system, consider Hercules X-1. It emits pulses of X rays with a period of about 1.2 seconds, but every 1.7 days the pulses vanish for a few hours (■ Figure 14-12). You can understand this system by comparing it to an eclipsing binary star. Hercules X-1 seems to contain a 2-solar-mass star with a temperature of 7000 K and a neutron star that orbit each other with a period of 1.7 days. Matter flowing from the normal star into an accretion disk around the neutron star can reach temperatures of millions of degrees and emit a powerful X-ray glow. Interactions with the neutron star's magnetic field can produce beams of X rays that sweep around with the rotating neuron star. Earth receives a pulse of X rays every time a beam points our way. The X rays shut off every 1.7 days when the neutron star is eclipsed behind the normal star. The X rays from the neutron star and its accretion disk heat the near side of the normal star to about 20,000 K. As the system rotates, astronomers on Earth alternately see the hot side of the star and then the cool side, and its brightness at visible wavelengths varies. Hercules X-1 is a complex system and is still not well understood, but this quick analysis will show you how complex and powerful such binary systems are during mass transfer.

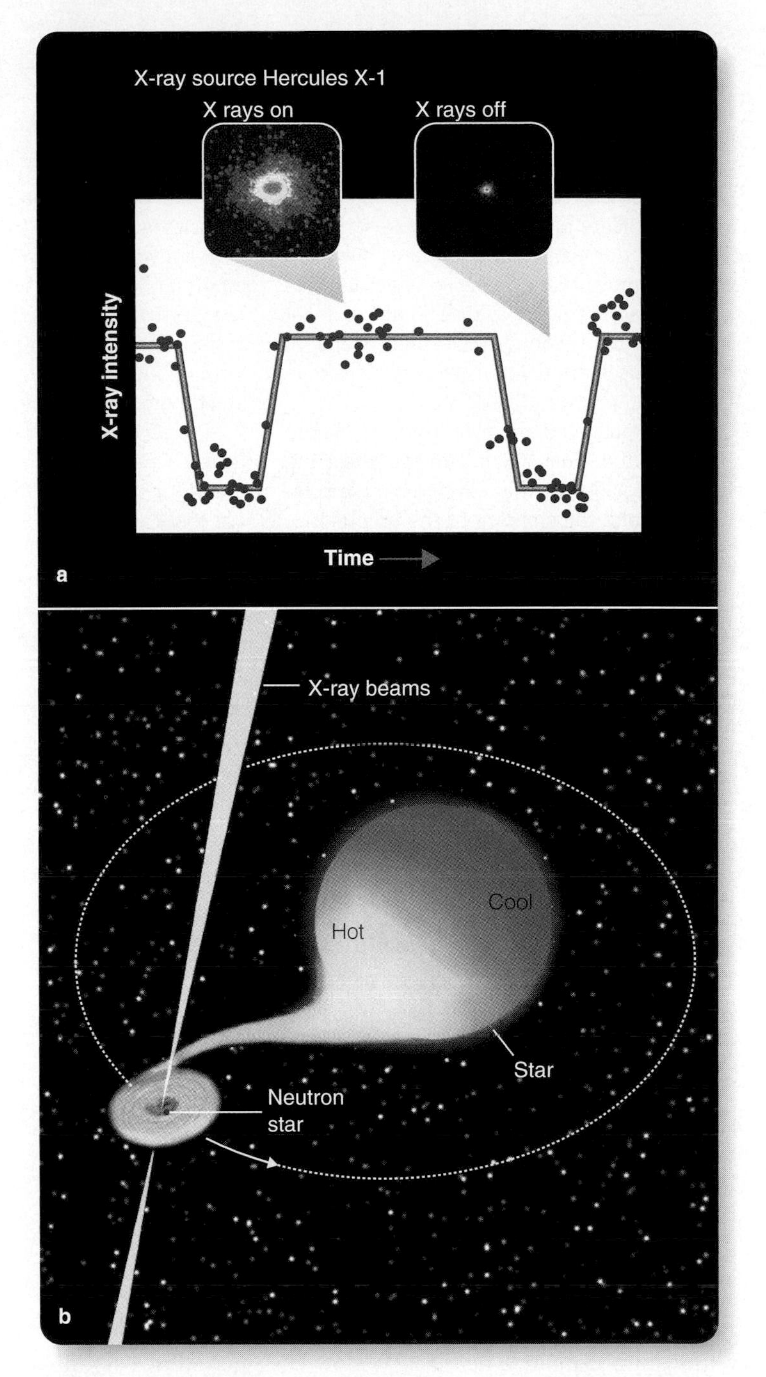

■ Figure 14-12

Sometimes the X-ray pulses from Hercules X-1 are on, and sometimes they are off. A graph of X-ray intensity versus time looks like the light curve of an eclipsing binary. (Insets: J. Trümper, Max-Planck Institute) (b) In Hercules X-1, matter flows from a star into an accretion disk around a neutron star producing X rays, which heat the near side of the star to 20,000 K compared with only 7000 K on the far side. X rays turn off when the neutron star is eclipsed behind the star.

The Fastest Pulsars

This discussion of pulsars suggests that newborn pulsars should blink rapidly and old pulsars should blink slowly, but the handful that blink the fastest may be quite old. One of the fastest known pulsars is cataloged as PSR 1937+21 in the constellation Vulpecula. It pulses 642 times a second and is slowing down only slightly. The energy stored in the rotation of a neutron star at this rate is equal to the total energy of a supernova explosion, so it seemed difficult at first to explain this pulsar. It now appears that PSR 1937+21 is an old neutron star that has gained mass and rotational energy from its companion in the binary system. Like water hitting a mill wheel, the matter falling on the neutron star has spun it up to 642 rotations per second. With its old, weak magnetic field, it slows down very slowly and will continue to spin for a very long time.

A number of other very fast pulsars have been found. They are known generally as **millisecond pulsars** because their pulse periods are almost as short as a millisecond (0.001 s). This produces some fascinating physics because the pulse period of a pulsar equals the rotation period of the neutron star. If a neutron star 10 km in radius spins 642 times a second, as does PSR 1937+21, then the period is 0.0016 second, and the equator of the neutron star must be traveling about 40,000 km/s. That is fast enough to flatten the neutron star into an ellipsoidal shape and is nearly fast enough to break it up.

All scientists should be made honorary citizens of Missouri, the "Show Me" state, because scientists demand evidence. The hypothesis that the millisecond pulsars are spun up by mass transfer from a companion star is quite reasonable, but astronomers demand reality checks, and evidence has been found. For example, the pulsar PSR J1740−5340 has a period of 42 milliseconds and

Window on Science 14-1

The Impossibility of Proof in Science

No scientific theory or hypothesis can be proved correct. You can test a theory over and over by performing experiments or making observations, but you can never prove that the theory is absolutely true. It is always possible that you have misunderstood the theory or the evidence, and the next observation you make might disprove the theory.

For example, you might propose the theory that the sun is mostly iron. You could test the theory by looking at the iron lines in the solar spectrum, and the strength of the iron lines would suggest that your theory is right. Although your observation has confirmed your theory, it has not proven the theory is right. You might confirm the theory many times before you realized that iron absorbs photons much more efficiently than hydrogen. Although the hydrogen lines are weak in the sun's spectrum, they show that most of the atoms in the sun are hydrogen and not iron.

The nature of scientific thinking can lead to two kinds of misconceptions. Sometimes nonscientists will say, "You scientists just want to tear everything down—you don't believe in anything." Scientists test a theory over and over to test its worth. If a theory survives many tests, scientists begin to have confidence it is true, but they continue to test their understanding of nature at every opportunity.

The second misconception nonscientists have is revealed when they say, "You scientists are never sure of anything." Again, the scientist knows that no theory can be proven correct. That the sun will rise tomorrow is very likely, and scientists have great confidence in that theory. But in the end it is still a theory.

People will say of an idea they dislike, "That is only a theory," as if a theory were simply a random guess. In fact, a theory can be a well-tested truth in which all scientists have great confidence. Yet you can never prove that any theory is absolutely true.

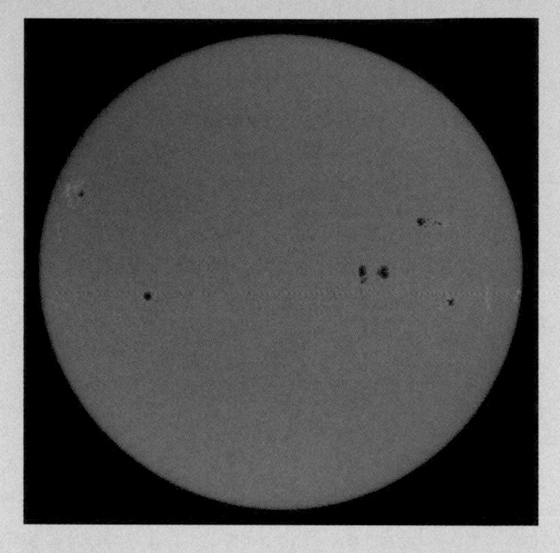

It is only a theory, but astronomers have tremendous confidence that the sun is made almost entirely of hydrogen and helium. (SOHO/MDI)

is orbiting with a bloated red star that is losing mass to the neutron star. This appears to be a pulsar in the act of being spun up to high speed. For another example, consider the X-ray source XTE J1751−305, a pulsar with a period of only 2.3 milliseconds. X-ray observations show that it is in the act of gaining mass from a companion star. The orbital period is only 42 minutes, and the mass of the companion star is only 0.014 solar masses. The evidence suggests this neutron star has devoured all but the last morsel of its binary partner.

Although some millisecond pulsars have binary companions, some are solitary neutron stars. How did they get spun up if they don't have a companion star? A pulsar known as the Black Widow may explain. The Black Widow has a period of 1.6 milliseconds, meaning it is spinning 622 times per second, and it orbits with a low-mass companion. Presumably the neutron star was spun up by mass flowing from the companion, but spectra show that the blast of radiation and high-energy particles from the neutron star is now boiling away the surface of the companion. The Black Widow has eaten its fill and is now evaporating the remains of its companion. It will soon be a solitary millisecond pulsar (■ Figure 14-13).

"Show me," say scientists; and, in the case of neutron stars, the evidence seems very strong. Of course, a theory can never be proven absolutely true (**Window on Science 14-1**), but the evidence for neutron stars is so strong that astronomers have great confidence that such objects really do exist. Other theories

■ Figure 14-13

The Black Widow pulsar and its companion star are moving rapidly through space, creating a shock wave like the bow wave of a speedboat. The shock wave confines high-energy particles shed by the pulsar into an elongated cocoon (red). (X-ray: NASA/CXC/ASTRON/B. Stappers et al.; Optical: AAO/J. Bland-Hawthorn & H. Jones)

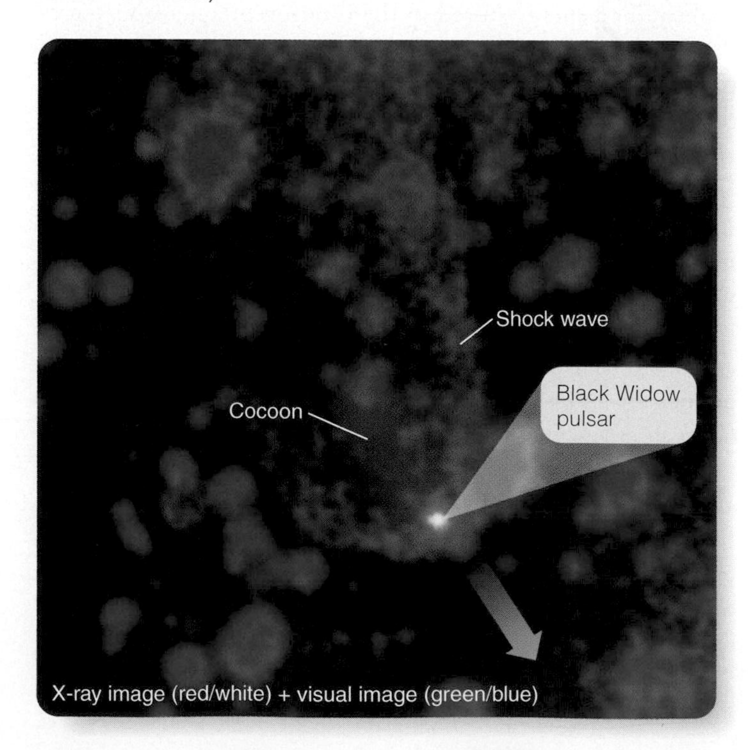

that describe how they emit beams of radiation and how they form and evolve are less certain, but continuing observations at many wavelengths are revealing more about these last embers of massive stars. In fact, observations have turned up objects no one predicted.

Pulsar Planets

Finding planets orbiting stars other than the sun is very difficult, and hardly two hundred are known. Oddly, the first such planets were found orbiting a neutron star.

Because a pulsar's period is so precise, astronomers can detect tiny variations by comparison with atomic clocks. When astronomers checked pulsar PSR 1257+12, they found variations in the period of pulsation much like those caused by the orbital motion of the binary pulsar (■ Figure 14-14a). However, in the case of PSR 1257+12, the variations were much smaller; and, when they were interpreted as Doppler shifts, it became evident that the pulsar was being orbited by at least two objects with planetlike masses of 4.3 and 3.9 Earth masses. The gravitational tugs of the planets make the pulsar wobble about the center of mass of the system by no more than 800 km, and that produces the tiny changes in period.

Astronomers greeted this discovery with both enthusiasm and skepticism. As usual, they looked for ways to test the hypothesis. Simple gravitational theory predicts that the planets should interact and slightly modify each other's orbit. When the data were analyzed, that interaction was found, further confirming the hypothesis that the variations in the period of the pulsar are caused by planets. In fact, further data revealed the presence of a third planet of about the mass of Earth's moon, and a fourth planet with a mass of about 100 Earth masses is now believed to follow a much larger orbit. This illustrates the astonishing precision of studies based on pulsar timing.

Astronomers wonder how a neutron star can have planets. The planets that orbit PSR 1257+12 are very close to the pulsar; the inner three orbit at 0.19 AU, 0.36 AU, and 0.47 AU—closer to the pulsar than Venus is to the sun. Any planets that orbit a star so closely would be lost or vaporized when the star exploded. Furthermore, a star about to explode as a supernova would be a large giant or a supergiant, and planets only a few astronomical units distant would be inside such a large star and could not survive. It seems more likely that these planets are the remains of a stellar companion that was devoured by the neutron star. In fact, the pulsar is very fast (162 pulses per second), suggesting that it was spun up in a binary system.

Another pulsar planet has been found orbiting in a binary system containing a neutron star and a white dwarf. Because this system is located in a very old star cluster and contains a white dwarf, astronomers suspect that the planet may be very old. Planets probably orbit other neutron stars, and small shifts in the timing of the pulses may eventually reveal their presence.

You can imagine what these worlds might be like. Planets formed from the remains of dying stars might be rich in heavy elements, but truly ancient planets might be poor in such elements. You can imagine visiting these worlds, landing on their surfaces, and hiking across their valleys and mountains. Above you, the neutron star would glitter in the sky, a tiny point of light.

■ Figure 14-14

(a) The dots in this graph are observations showing that the period of pulsar PSR 1257+12 varies from its average value by a fraction of a billionth of a second. The blue line shows the variation that would be produced by planets orbiting the pulsar. (b) As the planets orbit the pulsar, they cause it to wobble by less than 800 km, a distance that is invisibly small in this diagram. (Adapted from data by Alexander Wolszczan)

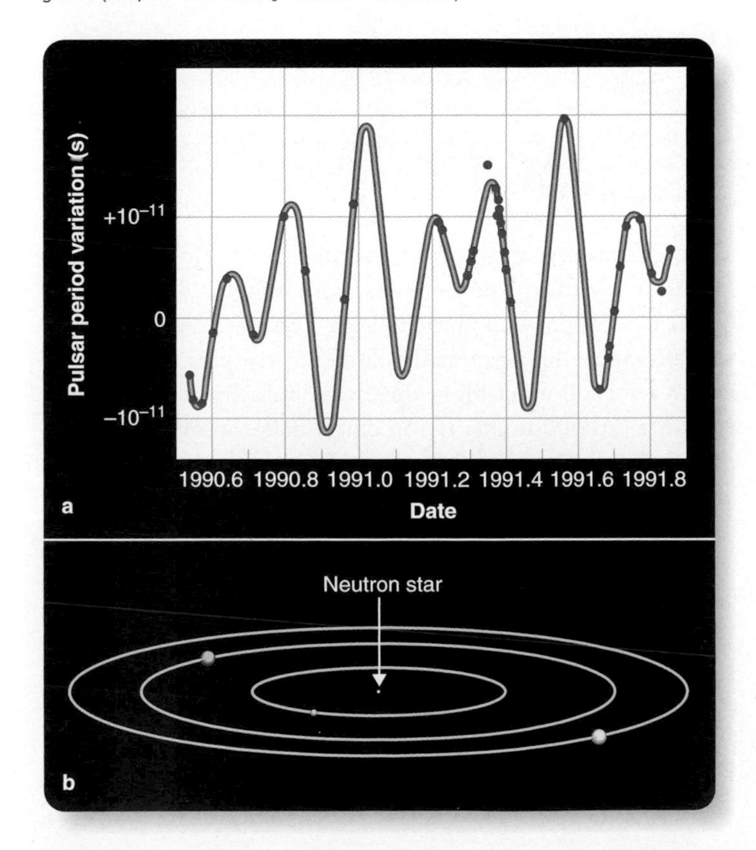

Building Scientific Arguments

Why are neutron stars detectable at X-ray wavelengths? This argument draws together a number of ideas you know from previous chapters. First, you should remember that a neutron star is very hot because of the heat released when it contracts to a radius of 10 km. It could easily have a surface temperature of 1,000,000 K, and Wien's law (Chapter 7) tells you that such an object will radiate most intensely at a very short wavelength, typical of X rays. However, you know that the total luminosity of a star depends on its surface temperature and its surface area, and a neutron star is so small it can't radiate much energy. X-ray telescopes have found such neutron stars, but they are not easy to locate.

There is, however, a second way a neutron star can radiate X rays. If a normal star in a binary system loses mass to a neutron star companion, the inflowing matter will form a very hot accretion disk that can radiate intense X rays easily detectable by X-ray telescopes orbiting above Earth's atmosphere. Further, any matter that hits the surface of the neutron star will impact with so much energy that it will be heated to very high temperatures and will radiate X rays.

Now build a new argument as if you were seeking funds for a research project. **What observations would you make to determine whether a newly discovered pulsar was young or old, single or a member of a binary system, alone or accompanied by planets?**

■ ■ ■

Connections: Perhaps the strangest planets in the universe are those orbiting pulsars. But however strange a pulsar planet may be, you can imagine going to one. The next topic, in contrast, seems beyond the reach of even your imagination.

14-2 Black Holes

YOU HAVE NOW STUDIED white dwarfs and neutron stars, two of the three end states of dying stars. Now you are ready to turn to the third—black holes.

Although the physics of black holes is difficult to discuss without sophisticated mathematics, simple logic is sufficient to predict that they should exist. The problem is to consider their predicted properties and try to find objects in the heavens that could be real black holes. More difficult than the search for neutron stars, the quest for black holes has nevertheless met with success.

To begin your study of black holes, consider a simple question. How fast must an object travel to escape from the surface of a celestial body?

Escape Velocity

Suppose you threw a baseball straight up. How fast must you throw it if it is not to come down? Of course, gravity would pull back on the ball, slowing it, but if the ball were traveling fast enough to start with, it would never come to a stop and fall back. Such a ball would escape from Earth. The escape velocity is the initial velocity an object needs to escape from a celestial body (■ Figure 14-15). (See Chapter 5.)

■ Figure 14-15

Escape velocity, the velocity needed to escape from a celestial body, depends on mass. The escape velocity at the surface of a very small, low-mass body would be so low you could jump into space. Earth's escape velocity is much larger, about 11 km/s (25,000 mph).

Whether you are discussing a baseball leaving Earth or a photon leaving a collapsing star, the escape velocity depends on two things, the mass of the celestial body and the distance from the center of mass to the escaping object. If the celestial body has a large mass, its gravity is strong, and you need a high velocity to escape; but if you begin your journey farther from the center of mass, the velocity needed is less. For example, to escape from Earth, a spaceship would have to leave Earth's surface at 11 km/s (25,000 mph), but if you could launch spaceships from the top of a tower 1000 miles high, the escape velocity would be only 8.8 km/s (20,000 mph). If you could make an object massive enough or small enough, its escape velocity could be greater than the speed of light. Relativity shows that nothing can travel faster than the speed of light, so even photons, which have no mass, would be unable to escape. Such a small, massive object could never be seen because light could not leave it.

Long before Einstein and relativity, the Rev. John Mitchell, a British gentleman astronomer, realized that Newton's laws of gravity and motion had peculiar consequences. In 1783, Mitchell pointed out that an object 500 times the radius of the sun but of the same density would have an escape velocity greater than the speed of light. Then, "all light emitted from such a body would be made to return towards it." Mitchell didn't know it, but he was talking about a black hole.

Ace Astronomy™ Log into AceAstronomy and select this chapter to see Astronomy Exercise "Escape Velocity." Try to escape from the planet.

Schwarzschild Black Holes

If the core of a star collapses and contains more than about three solar masses, no known force can stop it. The object cannot stop collapsing when it reaches the size of a white dwarf because degenerate electrons cannot support the weight. It cannot stop when it reaches the size of a neutron star because degenerate neutrons cannot support the weight. No force remains to stop the object from collapsing to zero radius.

As an object collapses, its density and the strength of its surface gravity increase; and, if an object collapses to zero radius, its density and gravity become infinite. Mathematicians call such a point a **singularity;** but in physical terms it is difficult to imagine an object of zero radius. Some theorists believe that a singularity is impossible and that when the laws of physics are better understood, they will show that the collapse halts at some subatomic radius and does not go to zero. Astronomically, it makes little difference.

If the contracting core of a star becomes small enough, the escape velocity in the region around it is so large that no light can escape. You can receive no information about the object or about the region of space near it, and such a region is called a **black hole.** If the core of an exploding star collapsed into a black hole, the core would vanish without a trace. Consequently, a supernova remnant that lacks a central pulsar may harbor a black hole, but you must be cautious. There are other reasons a supernova might not leave behind a compact object (■ Figure 14-16).

The boundary of the black hole is called the **event horizon,** because any event that takes place inside the event horizon is invisible to an outside observer. To see how a black hole can exist, you need to consider general relativity.

In 1916, Albert Einstein published a mathematical theory of space and time that became known as the general theory of relativity. Einstein treated space and time as a single entity—space-time. His equations showed that gravity could be described as a curvature of space-time, and almost immediately the astronomer Karl Schwarzschild found a way to solve the equations to describe the gravitational field around a single, nonrotating, electrically neutral lump of matter. That solution contained the first general relativistic description of a black hole; nonrotating, electrically neutral black holes are now known as Schwarzschild black holes. In most cases in astronomy, astronomers can use the Schwarzschild solution to think about black holes. You will see later in this chapter what difference rotation makes.

Schwarzschild's solution shows that if matter is packed into a small enough volume, then space-time curves back on itself. Objects can still follow paths that lead into the black hole, but no path leads out, so nothing can escape, not even light. For that reason, the inside of the black hole is totally beyond the view of an outside observer.

The event horizon is the boundary between the isolated volume of space-time and the rest of the universe, and the radius of the event horizon is called the **Schwarzschild radius, R_S**—the radius within which an object must shrink to become a black hole (■ Figure 14-17).

Although Schwarzschild's work was highly mathematical, his conclusion is quite simple. The Schwarzschild radius (in meters) depends only on the mass of the object (in kilograms):

$$R_S = \frac{2GM}{c^2}$$

In this simple formula, G is the gravitational constant, M is the mass, and c is the speed of light. A bit of arithmetic shows that a 1 solar-mass black hole will have a Schwarzschild radius of 3 km, a 10-solar-mass black hole will have a Schwarzschild radius of 30 km, and so on (■ Table 14-1). Even a very massive black hole would not be very large.

■ Figure 14-16

Some supernova remnants contain no neutron star, perhaps because the supernova formed a black hole instead. However, this supernova remnant, formed by Tycho Brahe's supernova of 1572, has the chemical composition typical of the remains of a type Ia supernova. Such supernovae are believed to destroy the white dwarf entirely and do not leave behind a neutron star or a black hole. (John P. Hughes, Rutgers University)

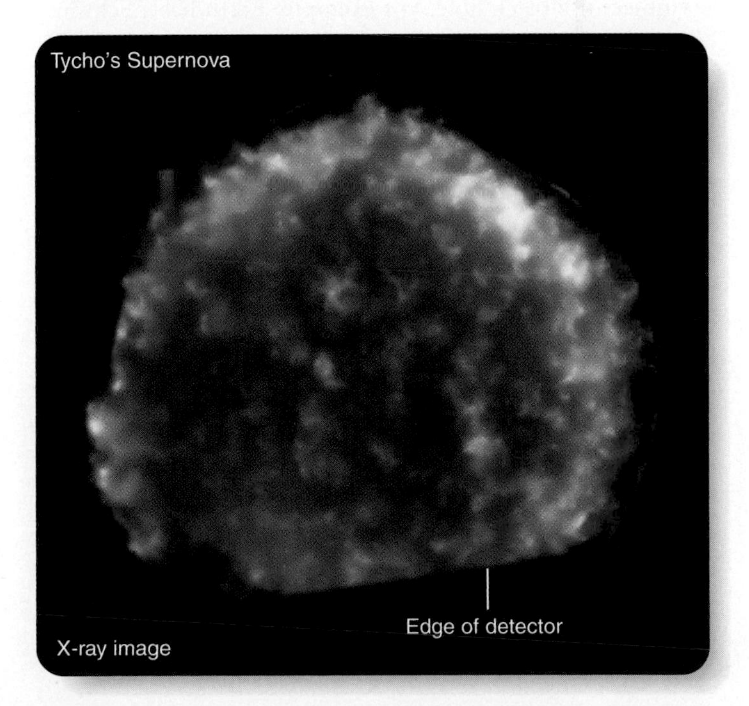

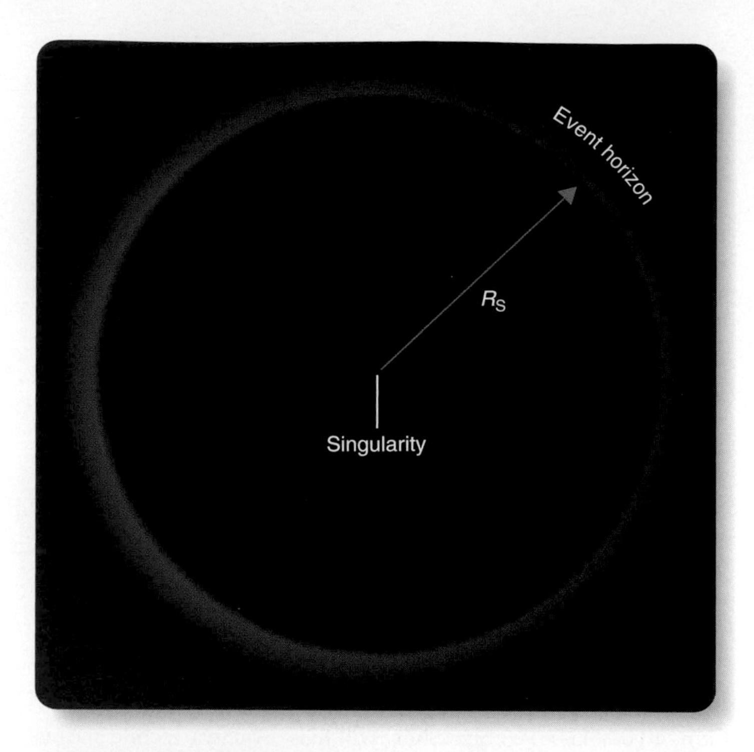

■ Active Figure 14-17

A black hole forms when an object collapses to a small size (perhaps to a singularity) and the escape velocity in its neighborhood is so great that light cannot escape. The boundary of this region is called the event horizon because any event that occurs inside is invisible to outside observers. The radius of the region is R_S, the Schwarzschild radius.

Ace Astronomy™ Log into AceAstronomy and select this chapter to see the Active Figure "Schwarzschild Radius." Take control of this diagram.

Every object with mass has a Schwarzschild radius, but not every object is a black hole. For example, Earth has a Schwarzschild radius of about 1 cm, but it could become a black hole only if you squeezed it inside that radius. Fortunately, Earth will not collapse spontaneously into a black hole because it has a small mass and is supported by the mechanical strength of the iron and rock in its interior. Only exhausted stellar cores more massive than three solar masses can form black holes under the sole influence of their own gravity. In this chapter, you are interested in black holes that might originate from the deaths of stars. These black holes would have masses slightly larger than 3 solar masses. In later chapters, you will encounter black holes whose masses might exceed a million solar masses.

Do not think of black holes as giant vacuum cleaners that will pull in everything in the universe. A black hole is just a gravitational field, and at a reasonably large distance its gravity is no greater than that of a normal object of similar mass. If the sun were replaced by a 1-solar-mass black hole, the orbits of the planets would not change at all. The gravity of a black hole becomes extreme only if you approach close to it (■ Figure 14-18). The universe contains many black holes. So long as you and other objects stay at a safe distance from the black holes, they have no catastrophic effects.

Ace Astronomy™ Log into AceAstronomy and select this chapter to see Astronomy Exercise "Black Hole." You can change the mass of an object and watch it become a black hole.

■ Table 14-1 | The Schwarzschild Radius

Object	Mass ($M_\odot$)	Radius
Star	10	30 km
Star	3	9 km
Star	2	6 km
Sun	1	3 km
Earth	0.000003	0.9 cm

Black Holes Have No Hair

Theorists who study black holes are fond of saying, "Black holes have no hair." By that they mean that once matter forms a black hole, it loses almost all of its normal properties. A black hole made of a collapsed star will be indistinguishable from a black hole of the same mass made from peanut butter and fake-fur mittens. Once the matter is inside the event horizon, it retains only three properties—mass, angular momentum, and electrical charge.

The Schwarzschild black hole is represented by a solution to Einstein's equations for the special case where the object has only mass. Schwarzschild black holes do not rotate or have charge. The solutions for rotating or charged black holes (or for rotating, charged black holes) are more difficult and have been found in only the last few decades. Generally, rotating, charged black holes are similar to Schwarzschild black holes.

It seems that astronomers need not worry about charged black holes because stars, whose collapse presumably forms black holes, cannot have large electrostatic charges. Suppose that you could give the sun a large positive charge. It would begin to repel protons in its corona and attract electrons and would soon return to neutral charge. For this reason, you can expect stars and black holes to be electrically neutral.

But everything in the universe seems to rotate, and collapsing stars spin rapidly as they conserve angular momentum. Consequently, you should probably expect black holes to have angular momentum. In 1963, New Zealand mathematician Roy P. Kerr found a solution to Einstein's equations that describes a rotating black hole. This is now known as the **Kerr black hole.**

The mass of a black hole curves neighboring space-time, and the Kerr solution predicts that the rotation of a black hole drags space-time around with it. The **ergosphere** is a region outside the

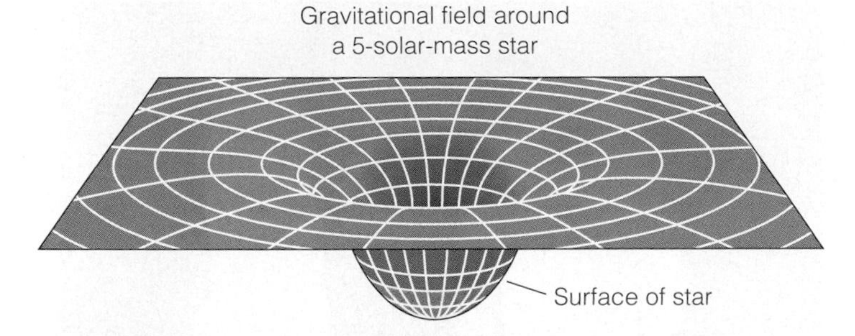

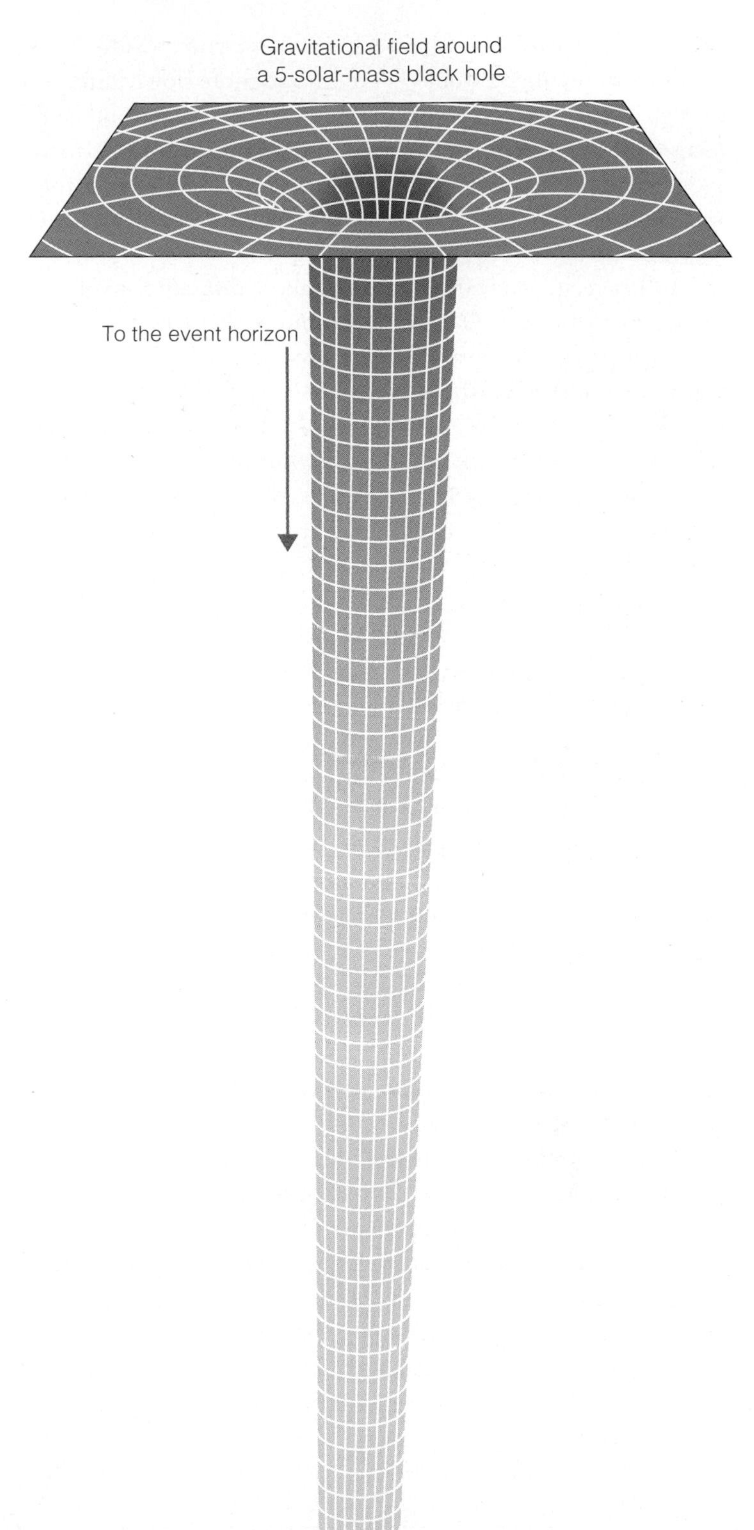

Figure 14-18

If you fell into the gravitational field of a star, you would hit the star's surface before you fell very far. Because a black hole is so small, you could fall much deeper into its gravitational field and eventually cross the event horizon. At a distance, the two gravitational fields are the same.

event horizon in which space-time rotates with the rotating black hole so powerfully that nothing could avoid being dragged along. No one has ever approached a rotating black hole, so no one has any idea what it might feel like to enter a region where space-time was whirling past.

The word *ergosphere* comes from the Greek word *ergo,* meaning "work," because the rotating space-time in the ergosphere can do work on a particle; that is, the particle can gain energy. In particular, the Kerr solution shows that a particle that enters the ergosphere can break into two pieces, one falling into the black hole and the other escaping with more energy than it had when it entered. In this way, energy can be extracted from a rotating black hole, and, as a result, the black hole slows its rotation very slightly.

The Kerr solution is a fascinating bit of theoretical physics, but it has an important application in astronomy. Almost certainly, black holes rotate, and matter falling into black holes must pass through the ergosphere. This suggests you should expect to find situations where energy is extracted from rotating black holes.

A Leap Into a Black Hole

Before you can search for real black holes, you need to understand what theory predicts about the appearance of a black hole. To explore that idea, imagine that you leap, feet first, into a Schwarzschild black hole.

If you were to leap into a black hole of a few solar masses from a distance of an astronomical unit, the gravitational pull would not be very large, and you would fall slowly at first. Of course, the longer you fell and the closer you came to the center, the faster you would travel. Your wristwatch would tell you that you fell for about 65 days before reaching the event horizon.

Your friends who stayed behind would see something different. They would see you falling more slowly as you came closer to the event horizon because, as explained by general relativity, clocks slow down in curved space-time. This is known as **time dilation.**

In fact, your friends would never actually see you cross the event horizon. To them you would fall more and more slowly until you seemed hardly to move. Generations later, your descendants could focus their telescopes on you and see you still inching closer to the event horizon. You, however, would have sensed no slowdown and would conclude that you had crossed the event horizon after only about 65 days.

Other relativistic effects would make it difficult for your descendents to see you. As light travels out of a gravitational field, it loses energy, and its wavelength grows longer. This is known as a **gravitational redshift** (■ Table 14-2). Light leaving you and traveling away from the black hole would suffer a larger and larger gravitational redshift. In addition, as your inward velocity grew higher and higher, a relativistic effect would cause more and more of the light leaving you to be emitted in the forward direction into the black hole. This would reduce the amount of light leaving you and traveling outward, making you even more difficult to see. Although you would notice none of these effects as you fell toward the black hole, your friends would need to observe at longer wavelengths and with larger telescopes to detect you.

While these relativistic effects seem merely peculiar, other effects would be quite unpleasant. Imagine again that you are falling feet first toward the event horizon of a black hole. You would feel your feet, which would be closer to the black hole, being pulled in more strongly than your head. This is a tidal force, and at first it would be minor. But as you fell closer, the tidal force would become very large. Another tidal force would compress you as your left side and your right side both fell toward the center of the black hole. For any black hole with a mass like that of a star, the tidal forces would crush you laterally and stretch you longitudinally long before you reached the event horizon (■ Figure 14-19). The friction from such severe distortions of your body would heat you to millions of degrees, and you would emit X rays and gamma rays. (Needless to say, this would render you inoperative as a thoughtful observer.)

Some years ago a popular book suggested that you could travel through the universe by jumping into a black hole in one place and popping out of another somewhere far across space. That might make for good science fiction, but tidal forces would make it an unpopular form of transportation even if it worked. You would certainly lose your luggage.

■ Table 14-2 | The Gravitational Redshift

Object	Redshift (percent)
Sun	0.0002
White dwarf	0.01
Neutron star	20
Black hole event horizon	Infinite

■ Figure 14-19

Leaping feet first into a black hole, a person of normal proportions (left) would be distorted by tidal forces (right) long before reaching the event horizon around a typical black hole of stellar mass. Tidal forces would stretch the body lengthwise while compressing it laterally. Friction from this distortion would heat the body to high temperatures.

Your imaginary leap into a black hole is not frivolous. You now know how to find a black hole: Look for a strong source of X rays. It may be a black hole into which matter is falling.

The Search for Black Holes

Do black holes really exist? Beginning in the 1970s, astronomers searched for observational evidence that their theories were correct. They tried to find one or more objects that were obviously black holes. That very difficult search is a good illustration of how the unwritten rules of science help you understand nature (**Window on Science 14-2**).

A black hole alone is totally invisible because nothing can escape from the event horizon. But a black hole into which matter is flowing would be a source of X rays. Of course, X rays can't escape from inside the event horizon, but X rays emitted by the heated matter flowing into the black hole could escape if the

Window on Science 14-2

Natural Checks on Fraud in Science

Fraud is actually quite rare in science. The nature of science makes fraud difficult, and the way scientists publish their research makes it almost impossible. In fact, you can think of science as a set of unwritten rules of behavior that have evolved to prevent scientists from lying to one another or to themselves, even by accident.

Suppose for a moment that you wanted to commit scientific fraud. You would have to invent data supposedly obtained from experiment or observation. You might invent X-ray data supposedly obtained by observing an X-ray binary star. Or, if you were interested in theory, you might invent a fraudulent mathematical calculation of the physics going on in an X-ray binary. You might get away with it for a short time, but one of the most important rules in science is that good results must be reproducible.

Other people must be able to repeat your observations, experiments, and calculations. In fact, most scientists routinely repeat other scientists' work as a way of getting started on a new research topic. As soon as someone tries to repeat your fraudulent research, you will be caught. The more important a scientific result is, the sooner other scientists will repeat it, so you don't have much of a chance of getting away with scientific fraud. In this way, science is self-correcting.

Even if you could invent some convincing scientific research, you would probably have difficulty publishing it. When a scientist submits an article to a scientific journal, it is subject to peer review. That is, the editor of the journal sends the article to one or two other experts in the field for comment and suggestions. These reviewers often make helpful suggestions, but they may also point out errors that have to be fixed before the journal can publish the article. In some cases, an article may be so flawed the editor will refuse to publish it at all. If you submitted your fraudulent research on X-ray binaries, the reviewers would almost certainly notice things wrong with it, and it would never get into print.

Scientists know the rules, and they use them. If someone makes a big discovery and is interviewed by the press, scientists will begin asking each other, "Has this work been published in a peer-reviewed journal yet?" That is, they want to know if other experts have checked the work. Until research is peer reviewed and published, it isn't quite official, and most scientists would treat the results with care.

Fraud isn't impossible in science. Some cases have even been in the national news. Big grant money is a terrible temptation. But, in science, "the truth will out." Because of the way scientists reproduce research and because of the way research is published, fraud is quite rare among scientists.

X rays were emitted before the matter crossed the event horizon. An isolated black hole will not have much matter flowing into it, but a black hole in a binary system might receive a steady flow of matter transferred from the companion star. This suggests that you can search for black holes by searching among X-ray binaries.

Some X-ray binaries, such as Hercules X-1, contain a neutron star, and they will emit X rays much as would a binary containing a black hole. You can tell the difference between a neutron star and a black hole in an X-ray binary in two ways. If the compact object emits pulses, you know it is a neutron star. Otherwise, you must depend on the mass of the object. If the compact object has a mass greater than 3 solar masses, the object can't be a neutron star, and you can conclude that it must be a black hole.

The first X-ray binary suspected of harboring a black hole was Cygnus X-1, the first X-ray object discovered in Cygnus. It contains a supergiant B0 star and a compact object orbiting each other with a period of 5.6 days. Matter flows from the B0 star as a strong stellar wind, and some of that matter enters a hot accretion disk around the compact object (■ Figure 14-20). The accretion disk is about five times larger in diameter than the orbit of Earth's moon, and the inner few hundred kilometers of the disk have a temperature of about 2 million Kelvin—hot enough to radiate X rays. The compact object is invisible, but Doppler shifts in the spectrum reveal the motion of the B0 star around the center of mass of the binary. From the geometry of the orbit, astronomers can calculate the mass of the compact

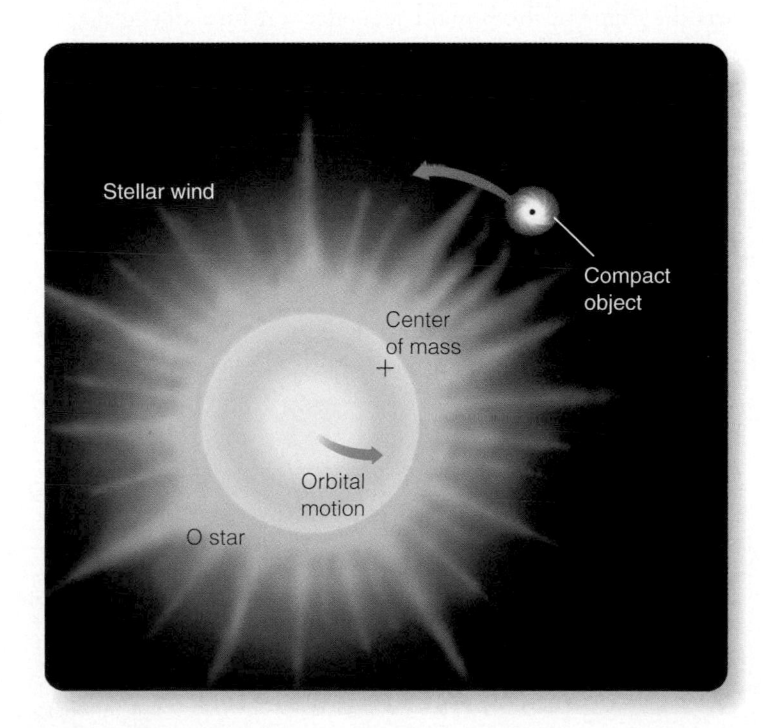

■ **Figure 14-20**

The X-ray source Cygnus X-1 is a supergiant B0 star and a compact object orbiting each other. Gas from the B0 star's stellar wind flows into the hot accretion disk, and the X rays detected come from the disk.

object—at least 3.8 solar masses, well above the maximum for a neutron star.

To confirm that black holes exist, astronomers needed to find a conclusive example, an object that couldn't be anything else. Cygnus X-1 didn't quite pass that test when it was first discovered. Perhaps the B0 star was not a normal star, said some astronomers, and that would make the mass of the compact object uncertain. Also, some astronomers suspected there might be a third star in the system, and that would distort the analysis. At the time, astronomers could not conclusively show that Cygnus X-1 contained a compact object with a mass greater than 3 solar masses. More recent research has given astronomers more confidence that the compact object is indeed a black hole with a mass slightly less than 10 solar masses.

Astronomers needed a more conclusive example than Cygnus X-1, and, as X-ray telescopes found more X-ray objects, the list of black hole candidates grew to a few dozen. A few of these objects, such as the first two in ■ Table 14-3, contain massive stellar companions, either giants, supergiants, or massive main-sequence stars. This makes such systems difficult to analyze, because such massive companions dominate the system. A good example of this sort is LMC X-3 (LMC refers to the Large Magellanic Cloud, a small galaxy near our own). The compact object in LMC X-3 has a mass of about 10 solar masses. One reason astronomers think the compact object is so massive is that it distorts the shape of the B main-sequence star into an egg shape; and, as the system rotates, the light from the B star varies because telescopes on Earth detect light from the side of the egg and then from the end. By analyzing the light change, astronomers can determine the shape of the B star and from that find the mass of the compact object.

Many of the black hole candidates are binary systems in which the normal star is a lower-mass main-sequence star. Such systems do not remain X-ray sources continuously but suffer X-ray nova outbursts as matter flows rapidly into the accretion disk. A year or so after an outburst, the flow has stopped, the accretion disk has dimmed, and astronomers can detect the spectrum of the main-sequence star. The spectral type and Doppler motions of the main-sequence companion reveal the mass of the compact object with only limited uncertainty. The lower-mass stellar companion makes these X-ray nova systems easier to analyze than are systems with massive companions.

A number of examples of these X-ray nova systems are shown in Table 14-3. A0620-00 is an old nova that erupted again in 1975. It contains an ordinary main-sequence K star and a compact object that orbit each other with a period of 7.75 hours. From the orbital motion and the distortion of the K star, astronomers conclude the compact object must have a mass between 5 and 15 solar masses. Three of the best understood examples are V404 Cygni; J1655-40, also known as Nova Scorpii 1994; and QZ Vul. The compact objects in these systems seem much too massive to be anything but black holes.

The growing list of X-ray binaries with compact objects exceeding 3 solar masses has convinced astronomers that black holes really do exist. The problem now is to understand how these objects interact with matter flowing into them through accretion disks to produce X rays, gamma rays, and jets of matter.

Building Scientific Arguments

How can a black hole emit X rays?

A common misconception about black holes is that they can emit nothing. By constructing a careful argument, you can show that a black hole can be a powerful source of energy. Of course, once a bit of matter falling into a black hole crosses the event horizon, no light or other electromagnetic radiation it emits can escape from the black hole. The matter becomes lost to view. But, if it emitted radiation before it crossed the event horizon, that radiation could escape, and you could detect it. Furthermore, the powerful gravitational field near a black hole stretches and distorts infalling matter, and internal friction heats the matter to millions of degrees. Wien's law

■ **Table 14-3 | Nine Black-Hole Candidates**

Object	Location	Companion Star	Orbital Period	Mass of Compact Object
Cygnus X-1	Cygnus	B0 supergiant	5.6 days	$>3.8\ M_\odot$
LMC X-3	Dorado	B3 main-sequence	1.7 days	$\sim10\ M_\odot$
A0620-00	Monocerotis	K main-sequence	7.75 hours	$10 \pm 5\ M_\odot$
V404 Cygni	Cygnus	K main-sequence	6.47 days	$12 \pm 2\ M_\odot$
J1655-40	Scorpius	F main-sequence	2.61 days	$6.9 \pm 1\ M_\odot$
QZ Vul	Vulpecula	K main-sequence	8 hours	$10 \pm 4\ M_\odot$
4U 1543-47	Lupus	A main-sequence	1.123 days	$2.7–7.5\ M_\odot$
V4641 Sgr	Sagittarius	B supergiant	2.81678 days	$8.7–11.7\ M_\odot$
XTE J1118+480	Ursa Major	K main-sequence	0.170113 days	$>6\ M_\odot$

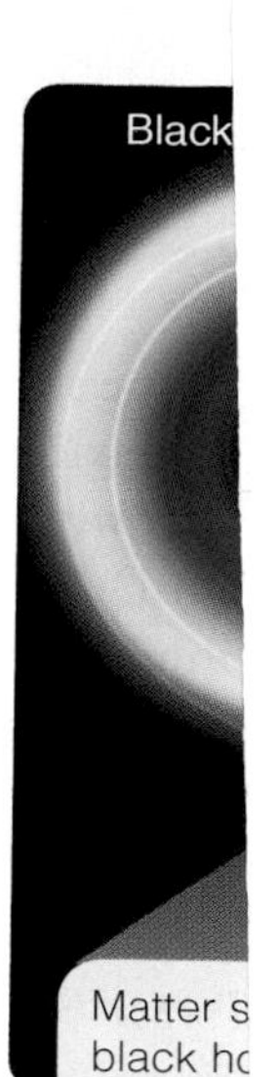

(Chapter 7) says that matter at such a high temperature should emit X rays. Any X rays emitted before the matter crosses the event horizon will escape, and you can look for black holes by looking for X-ray sources. Of course, an isolated black hole will probably not have much matter falling in, but black holes in binary systems may have large amounts of matter flowing in from the companion star. Consequently, the best place to search for black holes is in X-ray binaries.

A good scientific argument includes both theory and evidence. Expand your argument to include observations. **What observations would you make of an X-ray binary system to distinguish between a black hole and a neutron star?**

■ ■ ■

Connections: Today's astronomers are confident that black holes really exist. Modern research is revealing that both neutron stars and black holes can produce powerful phenomena as matter plunges in.

14-3 Compact Objects with Disks and Jets

NEUTRON STARS AND BLACK HOLES seem to be exotic objects, and they generate equally exotic phenomena. By studying those phenomena, you can learn more about the strange objects.

X-Ray Bursters

Beginning in the 1970s, X-ray telescopes revealed that some objects emit irregularly spaced bursts of X rays. Typically, bursts that follow a long quiet period are especially large (■ Figure 14-21), and this suggests that some mechanism is accumulating energy that is released by the bursts. The longer the quiet phase, the more energy accumulates.

These **X-ray bursters** are thought to be binaries containing neutron stars. The X-ray burster 4U 1820−30 in the globular cluster NGC6624, for example, appears to be a neutron star pulling mass away from a white dwarf (■ Figure 14-22). Matter from the white dwarf first flows into an accretion disk and then falls to the surface of the neutron star where the impact heats the gas. Hydrogen fuses on the surface of the star, leaving helium ash to accumulate. When the helium reaches a depth of about a meter, it fuses explosively into carbon and produces an X-ray burst. The rapid increase in brightness (in a few seconds) and the total amount of energy produced (up to 100,000 times the luminosity of the sun) fit well with theoretical models of an explosion on the surface of an object as small as a neutron star. Of course, mass continues to accumulate, producing burst after burst. Notice the similarity with the mechanism that produces nova explosions on the surfaces of white dwarfs.

Dozens of X-ray bursters are known, and astronomers suspect that they occur in binary systems where matter flows through accretion disks and falls into neutron stars.

Accretion Disk Observations

When material falls into a neutron star or black hole it forms an accretion disk, and high-speed observations have been able to reveal some of the processes that go on in such disks.

When a blob of matter gets caught in an accretion disk, it can orbit so fast its orbital period is measured in thousandths of a second. Because the blob of matter looses energy by friction, it spirals inward, and its orbital period grows shorter. This process can be detected as a quick flicker as the disk emits a short series of pulses of electromagnetic radiation. Some accretion disks produce pulses with a period as short as 0.00075 second, as blobs of material orbit only a dozen kilometers from the center of the compact object. Because the pulses last only a few cycles and because the period of the pulses grows rapidly shorter, they are called **quasi-periodic oscillations (QPOs).** Do not confuse these rapid flickering pulses with the regular click of pulses emitted by a pulsar.

QPOs can be observed in J1550−564, a star shedding mass into an accretion disk around a black hole. The flow of mass is irregular and causes X-ray flares. Short flickering strings of pulses are seen with periods as short as 0.003 second, and the period decreases rapidly. This suggests the energy is being emitted by

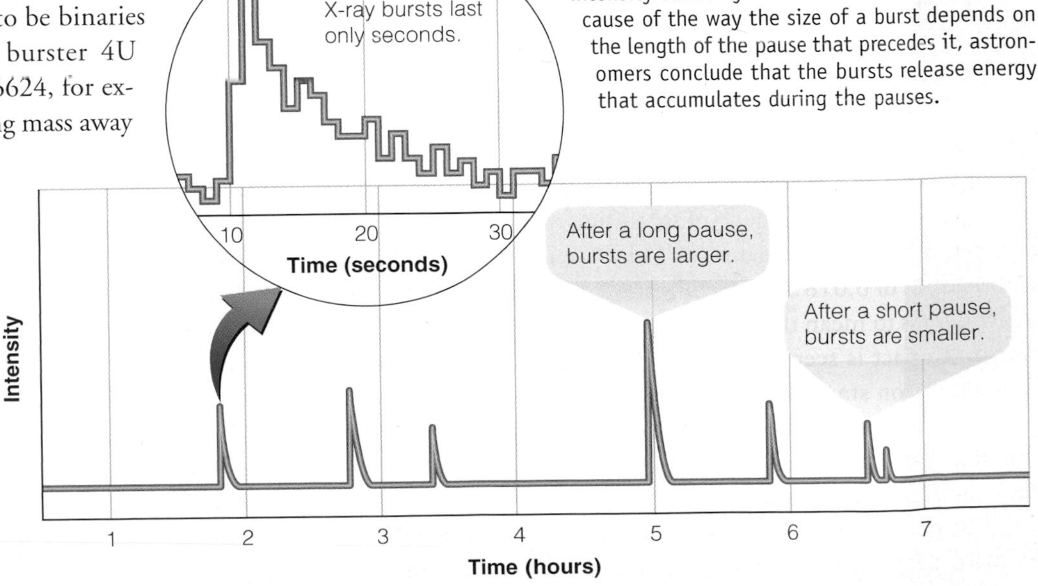

■ **Figure 14-21**

X-ray bursters emit bursts of X rays that rise to full intensity suddenly and then fade in seconds. Because of the way the size of a burst depends on the length of the pause that precedes it, astronomers conclude that the bursts release energy that accumulates during the pauses.

Sagittarius A*

1 The constellation of Sagittarius is so filled with stars and with gas and dust you can see nothing at visual wavelengths of the center of our galaxy.

The image below is a wide-field radio image of the center of our galaxy. Many of the features are supernova remnants (SNR), and a few are clouds of star formation. Peculiar features such as threads, the Arc, and the Snake may be gas trapped in magnetic fields. At the center lies Sagittarius A, the center of our galaxy.

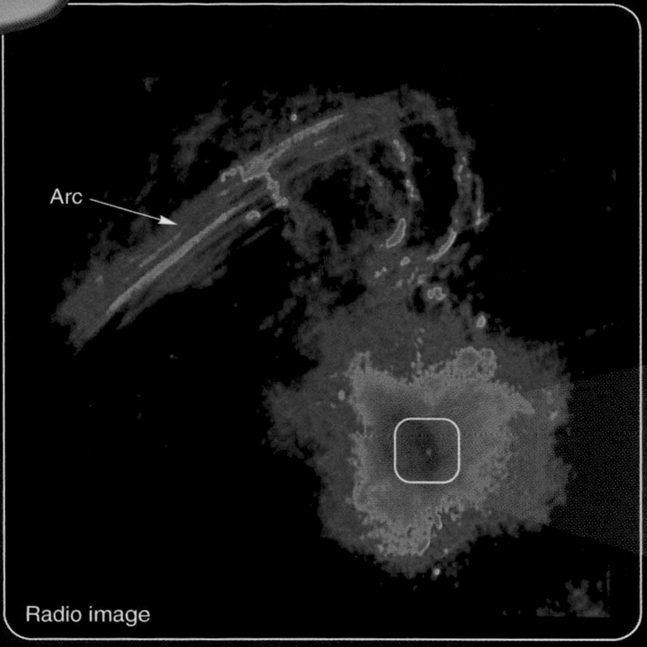

The radio map above shows Sgr A and the Arc filaments, 50 parsecs long. The image was made with the VLA radio telescope. The contents of the white box are shown on the opposite page.

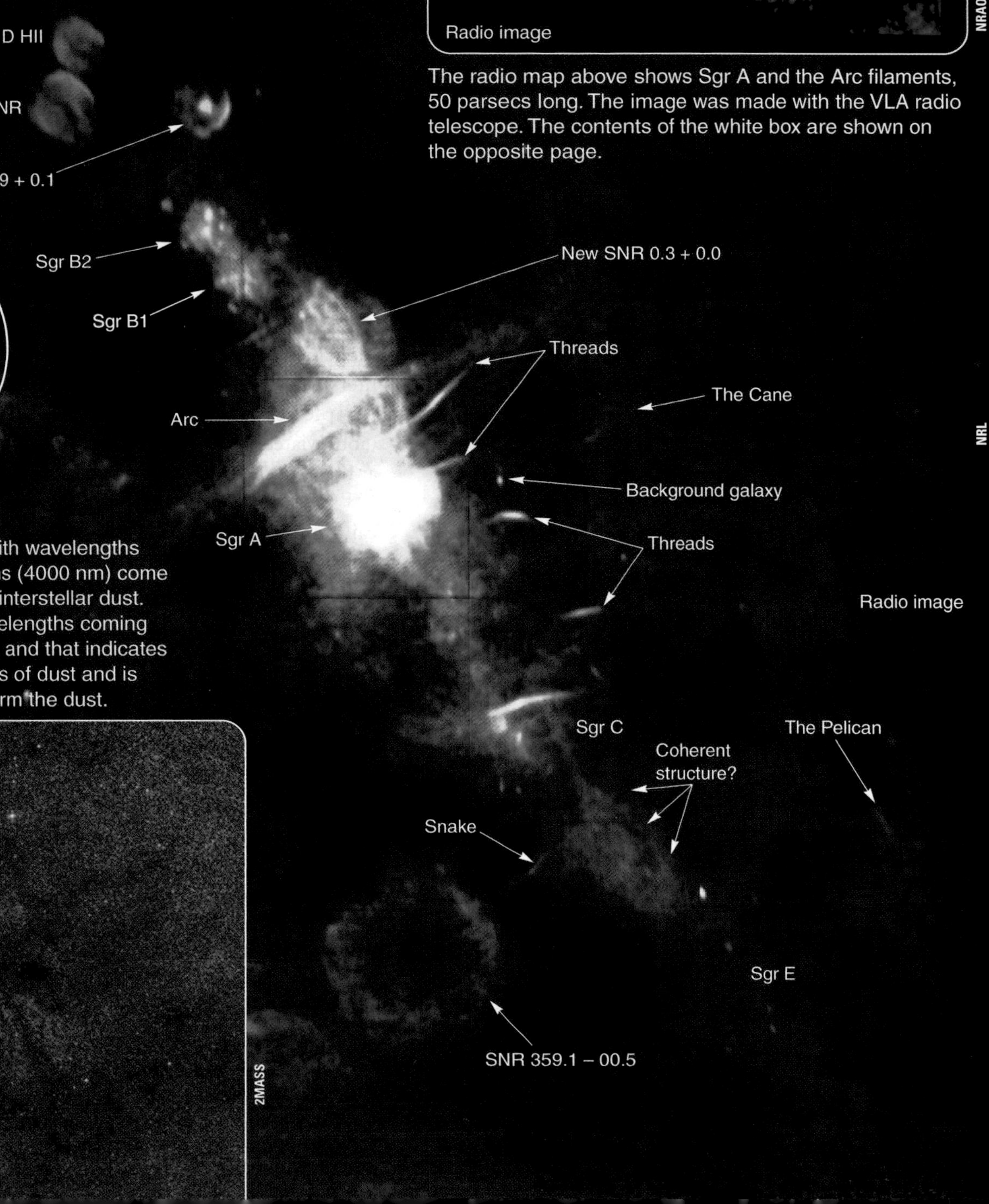

2 Infrared photons with wavelengths longer than 4 microns (4000 nm) come almost entirely from warm interstellar dust. The radiation at these wavelengths coming from Sagittarius is intense, and that indicates that the region contains lots of dust and is crowded with stars that warm the dust.

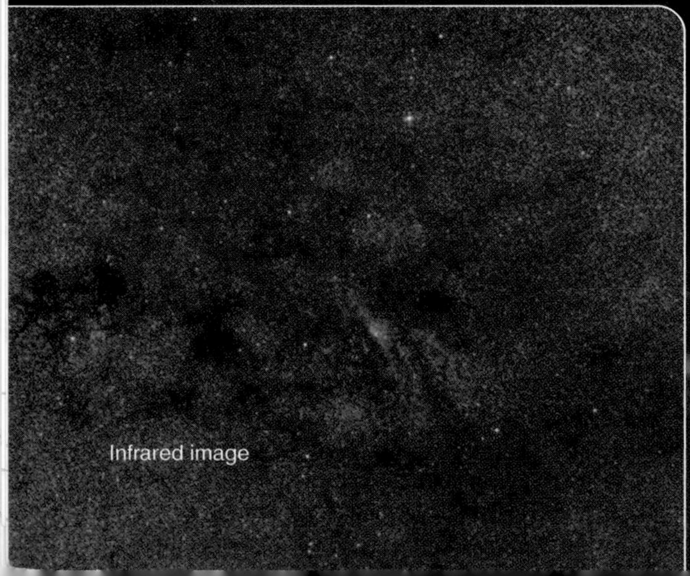

Visual-

a

c

■ Figure

(a) At visible
stars. (b) In
1820–30, co
conception s
around the n

material in
bits of shor
Cyg X-
between 3.8
with periods
which seem
But no imp
into a neutr
material ap
tional red s
they cannot
Further
study of 12

1a This high-resolution radio image of Sgr A (the white boxed area on the opposite page) reveals a spiral swirl of gas around an intense radio source known as Sgr A*, the presumed central object in our galaxy. About 3 pc across, this spiral lies in a low-density cavity inside a larger disk of neutral gas. The arms of the spiral are thought to be streams of matter flowing into Sgr A* from the inner edge of the larger disk (drawing at right).

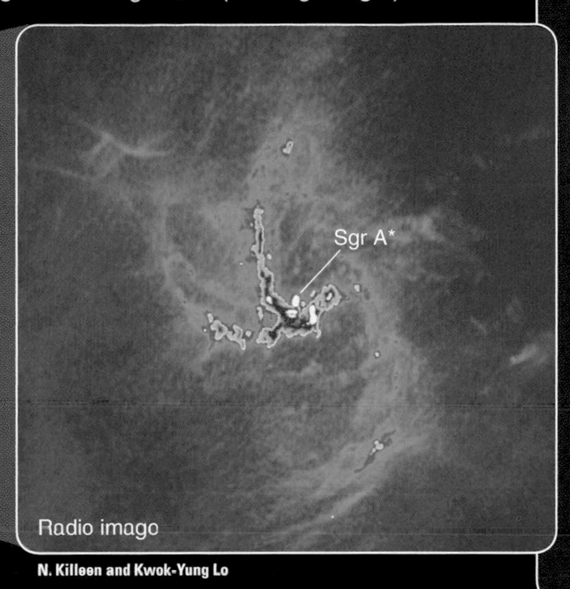

N. Killeen and Kwok-Yung Lo

Evidence of a Black Hole at the Center of Our Galaxy

3 Since the middle 1990s, astronomers have been able to use large infrared telescopes and active optics to follow the motions of stars orbiting around Sgr A*. A few of those orbits are shown here. The size and period of the orbit allows astronomers to calculate the mass of Sgr A* using Kepler's third law. The orbital period of the star SO-2, for example, is 15.2 years and the semimajor axis of its orbit is 950 AU. The combined motions of the observed stars suggest that Sgr A* has a mass of 2.6 million solar masses.

The Chandra X-ray Observatory has imaged Sgr A* and detected over 2000 other X ray sources in the area.

NASA/CXC/MIT/F.K. Baganoff et al.

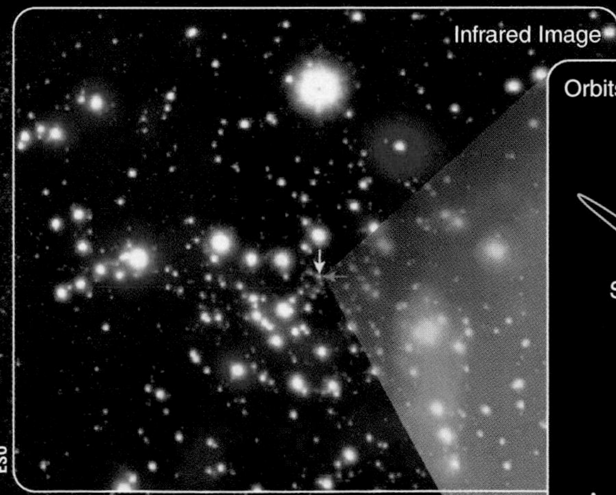

ESO

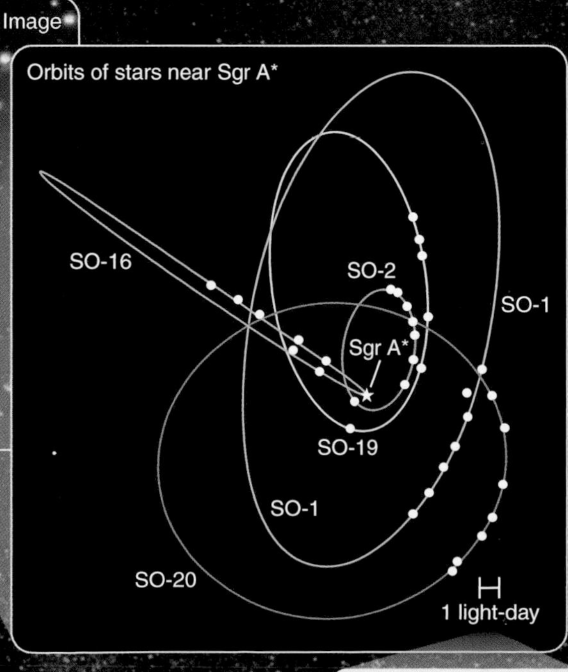

At its closest, SO-2 comes within 17 light-hours of Sgr A*. Alternative theories that Sgr A* is a cluster of stars, of neutron stars, or of stellar black holes are eliminated. Only a single black hole could contain so much mass in so small a region.

Our solar system is half a light-day in diameter.

3a A black hole with a mass of 2.6 million solar masses would have an event horizon smaller than the smallest dot in this diagram. A slow dribble of only 0.0002 solar masses of gas per year flowing into the black hole could produce the observed energy. A sudden increase as when a star falls in could produce a violent eruption.

The evidence of a massive black hole at the center of our galaxy seems conclusive. It is much too massive to be the remains of a dead star, however, and astronomers conclude that it probably formed as the galaxy first took shape.

Galaxy Classification

1 **Elliptical galaxies** are round or elliptical, contain no visible gas and dust, and lack hot, bright stars. They are classified with a numerical index ranging from 1 to 7; E0s are round, and E7s are highly elliptical. The index is calculated from the largest and smallest diameter of the galaxy used in the following formula and rounded to the nearest integer.

$$\frac{10(a - b)}{a}$$

a

b

Outline of an E6 galaxy

AURA/NOAO/NSF

Visual-wavelength image

M87 is a giant elliptical galaxy classified E1. It is a number of times larger in diameter than our own galaxy and is surrounded by a swarm of over 500 globular clusters.

Anglo-Australian Telescope Board

Visual

The Leo 1 dwarf elliptical galaxy is not many times bigger than a globular cluster.

2 **Spiral galaxies** contain a disk and spiral arms. Their halo stars are not visible, but presumably all spiral galaxies have halos. Spirals contain gas and dust and hot, bright O and B stars, as shown at right. The presence of short-lived O and B stars alerts us that star formation is occurring in these galaxies. Sa galaxies have larger nuclei, less gas and dust, and fewer hot, bright stars. Sc galaxies have small nuclei, lots of gas and dust, and many hot, bright stars. Sb galaxies are intermediate.

Anglo-Australian Telescope Board

Sb

NGC 3627 Visual

2a Roughly 2/3 of all spiral galaxies are **barred spiral galaxies** classified SBa, SBb, and SBc. They have an elongated nucleus with spiral arms springing from the ends of the bar, as shown at left. Our own galaxy is a barred spiral.

AceAstronomy™

Log into AceAstronomy and select this chapter to see Active Figure "Galaxy Types" and review the classification of galaxies.

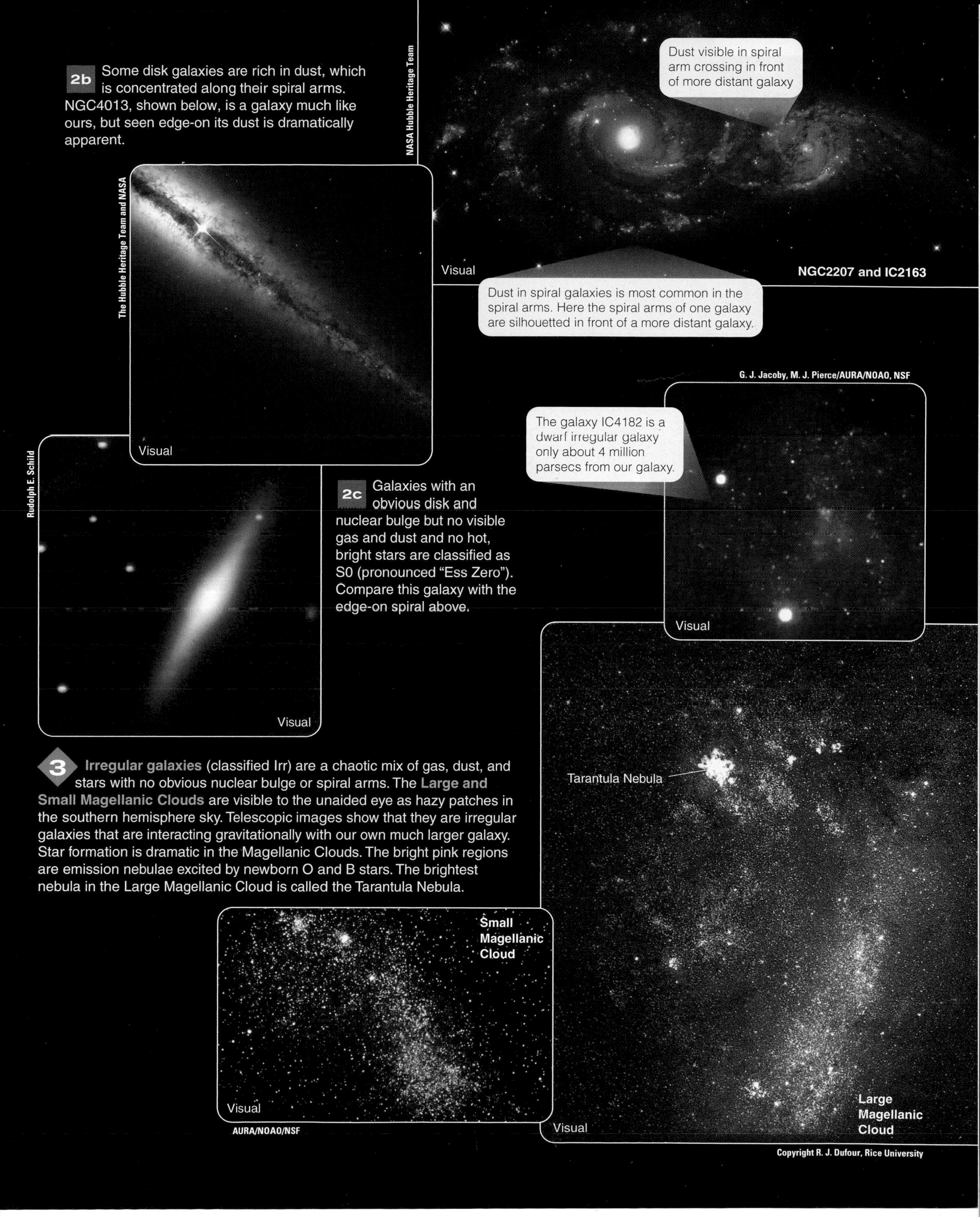

2b Some disk galaxies are rich in dust, which is concentrated along their spiral arms. NGC4013, shown below, is a galaxy much like ours, but seen edge-on its dust is dramatically apparent.

2c Galaxies with an obvious disk and nuclear bulge but no visible gas and dust and no hot, bright stars are classified as S0 (pronounced "Ess Zero"). Compare this galaxy with the edge-on spiral above.

3 **Irregular galaxies** (classified Irr) are a chaotic mix of gas, dust, and stars with no obvious nuclear bulge or spiral arms. The **Large and Small Magellanic Clouds** are visible to the unaided eye as hazy patches in the southern hemisphere sky. Telescopic images show that they are irregular galaxies that are interacting gravitationally with our own much larger galaxy. Star formation is dramatic in the Magellanic Clouds. The bright pink regions are emission nebulae excited by newborn O and B stars. The brightest nebula in the Large Magellanic Cloud is called the Tarantula Nebula.

Interacting Galaxies

1 When two galaxies collide, they can pass through each other with the stars so small and so far apart that they never collide. Gas clouds and magnetic fields do collide, but the biggest effects may be tidal. Even when two galaxies just pass near each other, tides can cause dramatic effects, such as long streamers called **tidal tails.** In some cases, two galaxies can merge and form a single galaxy.

Tidal Distortion

Small galaxy passing near a massive galaxy.

Gravity of a second galaxy represented as a single massive object

1a When a galaxy swings past a massive object such as another galaxy, tides are severe. Stars near the massive object try to move in smaller, faster orbits while stars further from the massive object follow larger, slower orbits. Such tides can distort a galaxy or even rip it apart.

Galaxy interactions can stimulate the formation of spiral arms

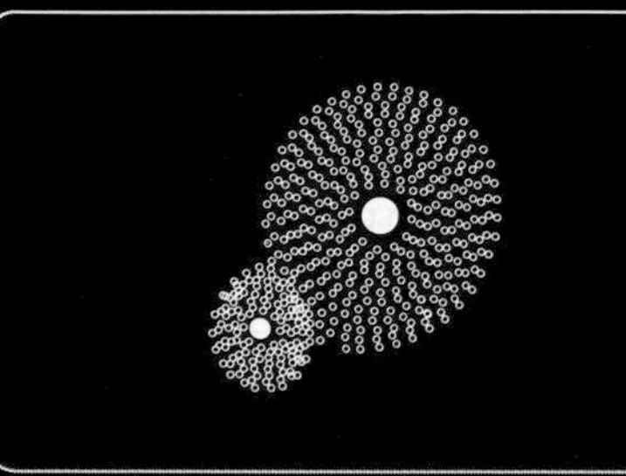

In this computer model, two uniform disk galaxies pass near each other.

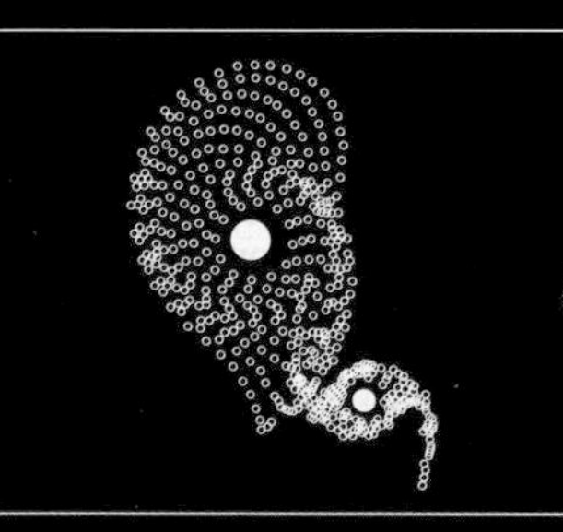

The small galaxy passes behind the larger galaxy so they do not actually collide.

Tidal forces deform the galaxies and trigger the formation of spiral arms.

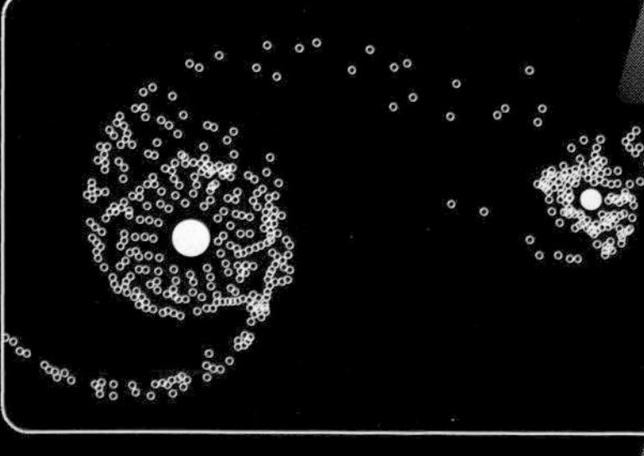

The upper arm of the large galaxy passes in front of the small galaxy.

Allen Beechel

A photo of the well-known Whirlpool Galaxy resembles the computer model.

NOAO

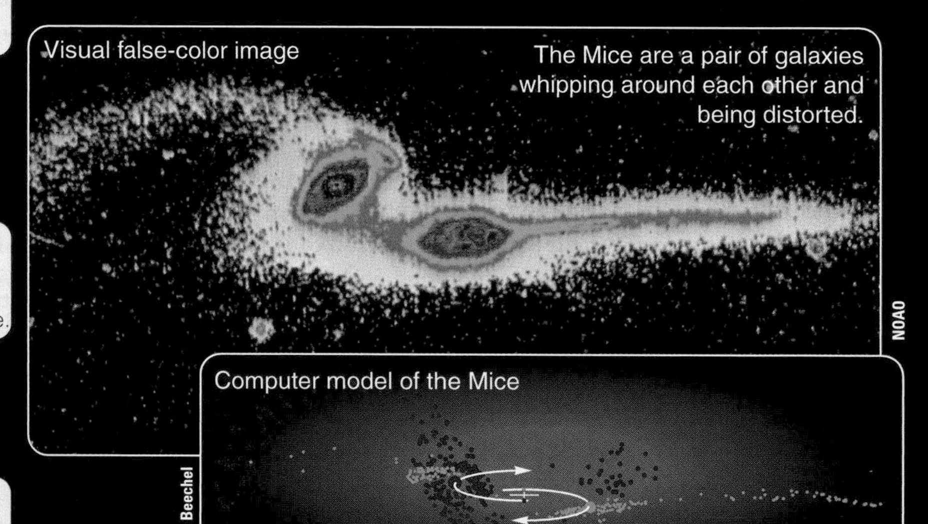

The Mice are a pair of galaxies whipping around each other and being distorted.

NOAO

Allen Beechel

1b The merger of galaxies is called **galactic cannibalism.** Models show that merging galaxies spiral around their common center of mass while tides rip stars away and form shells.

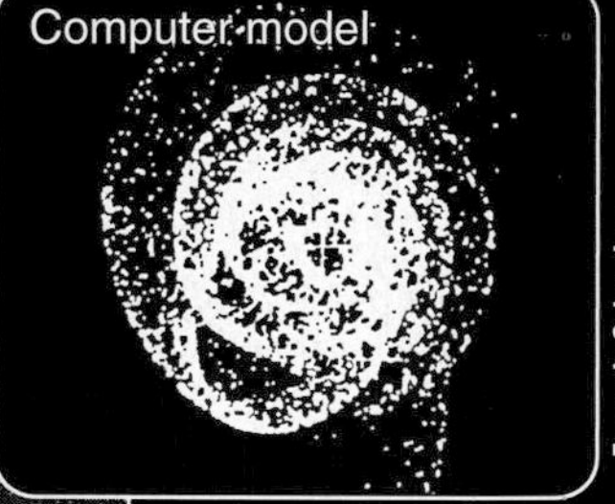

François Schweizer and Alan Toomre

Shells of stars

Visual enhanced image

NOAO

Such shells have been found around elliptical galaxies such as NGC5128. It is peculiar in many ways and even has a belt of dusty gas. The shells revealed in this enhanced image are evidence that the giant galaxy has cannibalized at least one smaller galaxy. The giant galaxy itself may be the result of the merger of two large galaxies.

2 The collision of two galaxies can trigger firestorms of star formation as gas clouds are compressed. Galaxies NGC4038 and 4039 have been known for years as the Antennae because the long tails visible in Earth-based photos resemble the antennae of an insect. Hubble Space Telescope images reveal that the two galaxies are blazing with star formation. Roughly a thousand massive star clusters have been born.

Spectra show that the galaxy is 10 to 20 times richer in elements like magnesium and silicon. Such metals are produced by massive stars and spread by supernova explosions.

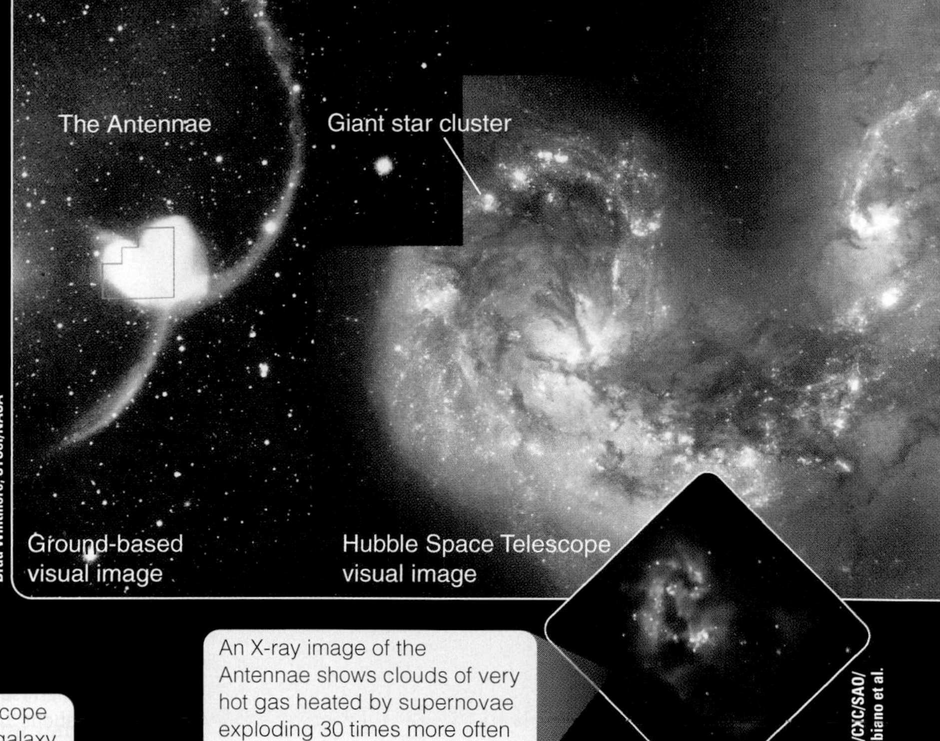

Brad Whitmore, STScI/NASA

An X-ray image of the Antennae shows clouds of very hot gas heated by supernovae exploding 30 times more often than in our own galaxy.

X-ray image

NASA/CXC/SAO/ G. Fabbiano et al.

3 Evidence of past galaxy mergers shows up in the motions inside some galaxies. NCG7251 is a highly distorted galaxy with tidal tails in this ground-based image.

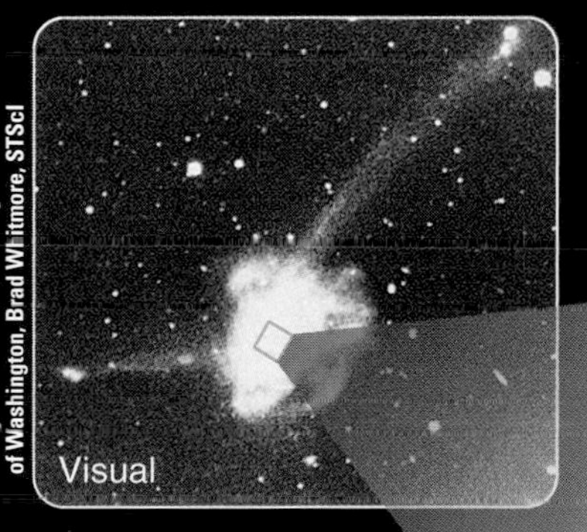

François Schweizer, Carnegie Inst. of Washington, Brad Whitmore, STScI

This Hubble Space Telescope image of the core of the galaxy reveals a small spiral spinning backward in the heart of the larger galaxy.

This counter rotation suggests that NCG7251 is the remains of two oppositely rotating galaxies that merged about a billion years ago.

3a Radio evidence of past mergers: Doppler shifts reveal the rotation of the spiral galaxy M64. The upper part of the galaxy has a redshift and is moving away from Earth, and the bottom part of the galaxy has a blueshift and is approaching. A radio map of the core of the galaxy reveals that it is rotating backward. This suggests a merger long ago between two galaxies that rotate in opposite directions.

Rotation of galaxy M64

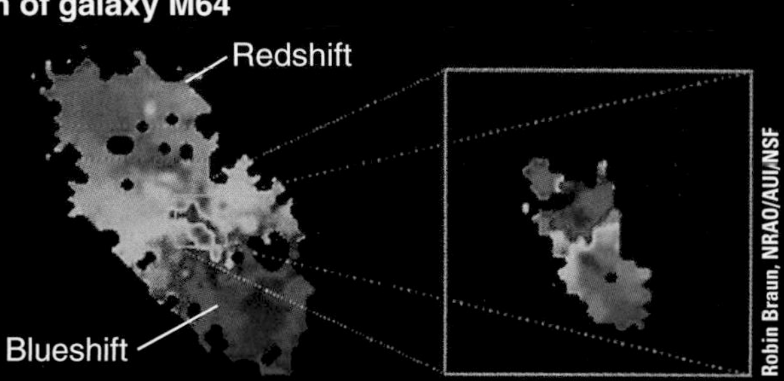

Robin Braun, NRAO/AUI/NSF

Evidence of galactic cannibalism: Giant elliptical galaxies in rich clusters sometimes have multiple nuclei, thought to be the densest parts of smaller galaxies that have been absorbed and only partly digested.

Multiple nuclei

Visual false-color image

Michael J. West

4 Galaxy AM 0644-741 below was once a normal spiral galaxy, but now it is a **ring galaxy**. Mathematical models show that such galaxies are produced by high-speed collisions in which a smaller galaxy passes through another galaxy almost perpendicular to the disk. Ring galaxies tend to have nearby companions, suggestive evidence of past encounters.

Visual-wavelength image

Hubble Heritage Team, NASA/ESA/AURA/STScI

Cosmic Jets and Radio Lobes

Size of Milky Way Galaxy

Hot spot

1 Many radio sources consist of two bright lobes — double-lobed radio sources — with a galaxy, often a peculiar or distorted galaxy, located between them. Evidence suggests these active galaxies are emitting jets of high-speed gas that inflate the lobes as cavities in the intergalactic medium. Where the jets impact the far side of the cavities, they create **hot spots**.

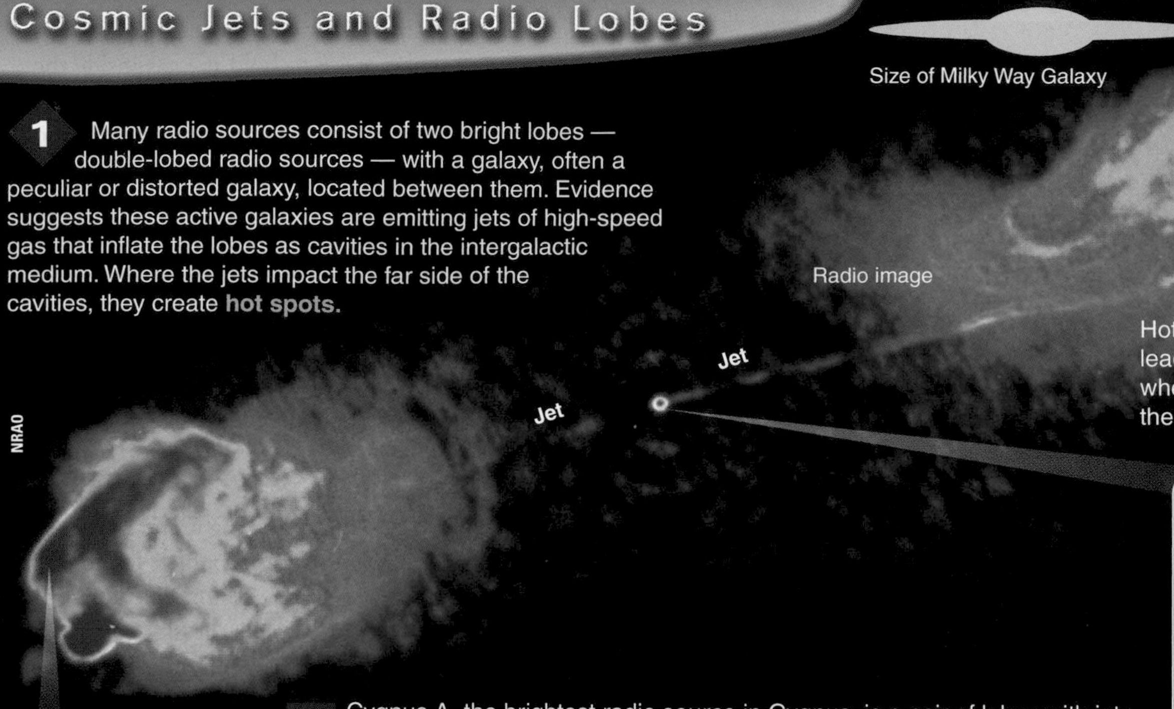

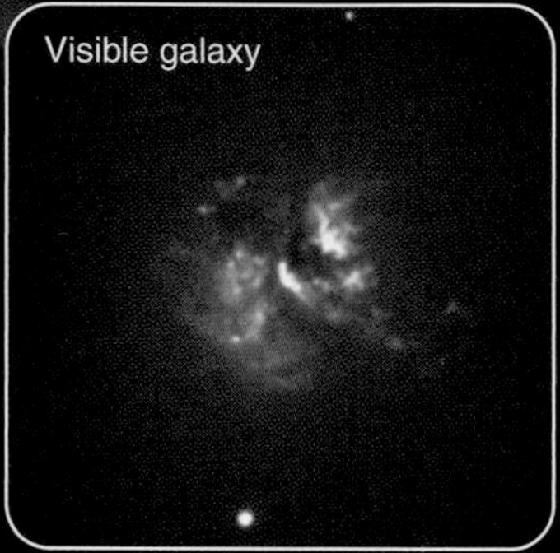

Radio image

Jet

Jet

NRAO

Hot spots lie on the leading edge of a lobe where the jet pushes into the surrounding gas.

Visible galaxy

Hot spot

1a Cygnus A, the brightest radio source in Cygnus, is a pair of lobes with jets leading from the nucleus of a highly disturbed galaxy. In this false-color image, the areas of strongest radio signals are shown in red and the weakest in blue. Because the radio energy detected is synchrotron radiation, astronomers conclude that the jets and lobes contain very-high-speed electrons, usually called relativistic electrons, spiraling through magnetic fields about 1000 times weaker than Earth's field. The total energy in a radio lobe is about 10^{53} J — what you would get if you turned the mass of a million suns directly into energy.

Used with permission, Fosbury R. A. E., Vernet, J., Villar-Martin, M., Cohen, M. H., Ogle, P. M., Tran, H. D. & Hook, R. N. 1998, Optical continuum structure of Cygnus A. "KNAW colloquium on: The most distant radio galaxies," Amsterdam, 15–17 October 1997, Roettgering H, Best P and Lehnert M eds, Reidel, astro-ph/9803310

2 Radio galaxy NGC5128 lies between two radio lobes and appears to be a giant elliptical galaxy experiencing a collision with a dusty spiral galaxy. It is known to radio astronomers as Centaurus A, an inner pair of radio lobes and a larger outer pair.

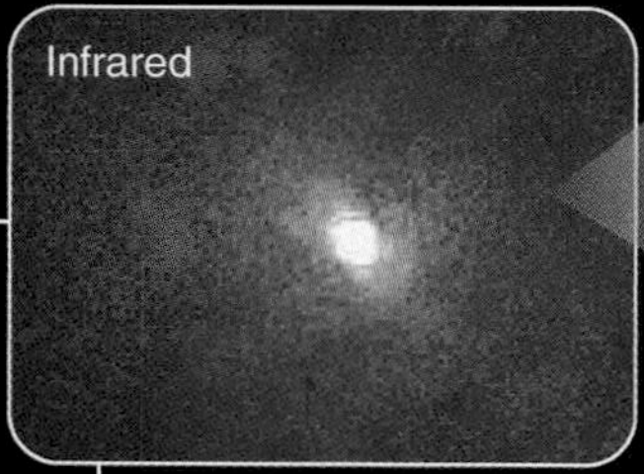

Infrared

An infrared image of the center of the galaxy NGC5128 reveals a small, bright disk of hot gas surrounding the nucleus.

NASA/CXC/NRAO/VLA

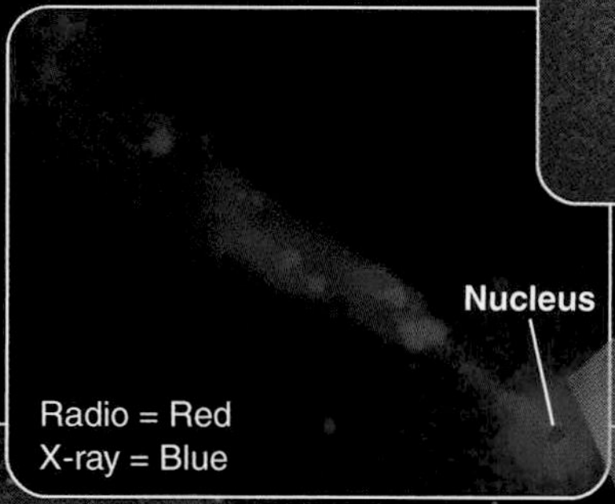

E. Schreier, STScI, NASA

Nucleus

Radio = Red
X-ray = Blue

The combined radio and X-ray image at the left shows a high-energy jet at the very center of the galaxy pointing to the upper left into the northern radio lobe.

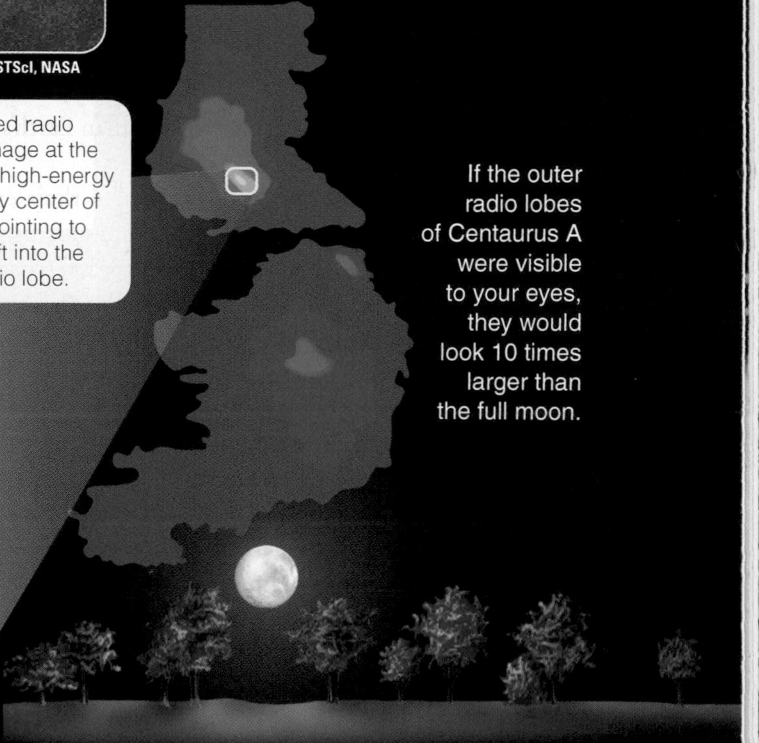

If the outer radio lobes of Centaurus A were visible to your eyes, they would look 10 times larger than the full moon.

Radio (green)

X-ray (blue)

Visual + radio + X-ray

X-ray: NASA/CXC/M. Karovska et al., Radio: NRAO/VLA/Schiminovich, et al., Radio: NRAO/VLA/J. Condon et al., Optical: Digitized Sky Survey U.K. Schmidt Image/STScI.

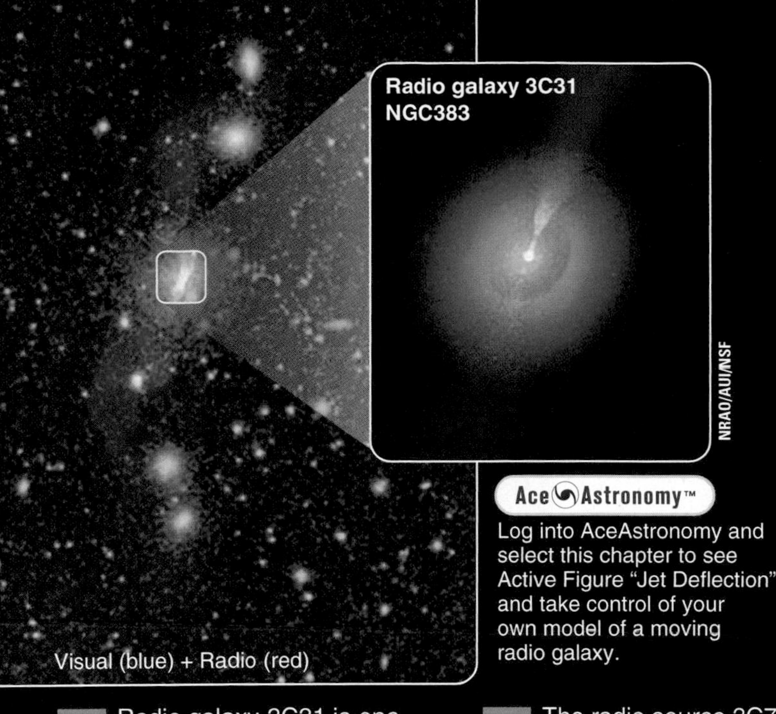

Ace Astronomy™

Log into AceAstronomy and select this chapter to see Active Figure "Jet Deflection" and take control of your own model of a moving radio galaxy.

3 The radio jets from NGC1265 are being left behind as the galaxy moves rapidly through the gas of the intergalactic medium. Twists in the tails are presumably caused by motions of the active nucleus.

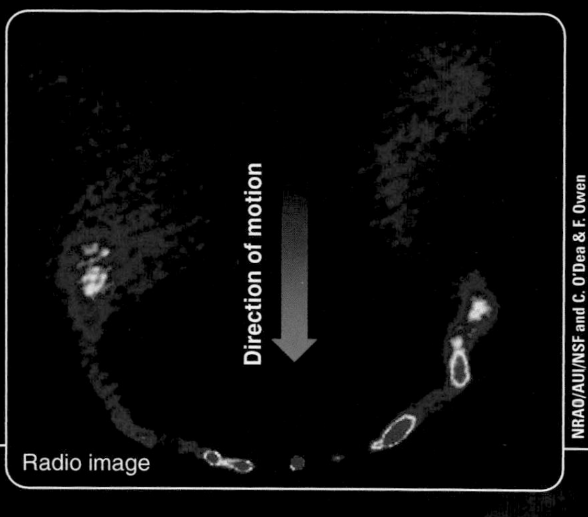

3a Radio galaxy 3C31 is one of a chain of galaxies. It has ejected jets from its core that twist, presumably because the active nucleus is orbiting another object such as the nucleus of a recently absorbed galaxy.

3b The radio source 3C75 is produced by two galaxies experiencing a close encounter. As the active nuclei whip around each other, their jets twist and turn. The size of the visible galaxies would be about the size of cherries at the scale of this image.

Nucleus

Nucleus

Radio image

NRAO/AUI/NSF and F.N. Owen, C.P. O'Dea, M. Inoue, & J. Eilek

4 High-energy jets appear to be caused by matter flowing into a supermassive black hole in the core of an active galaxy. Conservation of angular momentum forces the matter to form a whirling accretion disk around the black hole. How that produces a jet is not entirely understood, but it appears to involve magnetic fields that are drawn into the accretion disk and tightly wrapped to eject high-temperature gas. The twisted magnetic field confines the jets in a narrow beam and causes synchrotron radiation.

4a Jets from active galaxies may have velocities from thousands of kilometers per second up to a large fraction of the speed of light. Compare this with the jets in bipolar flows, where the velocities are only a few hundred kilometers per second. Active-galaxy jets can be millions of light-years long. Bipolar-flow jets are typically a few light-years long. The energy is different, but the geometry is the same.

Black hole

Magnetic field lines

Accretion disk

4b Excited matter traveling at very high speeds tends to emit photons in the direction of travel. Consequently, a jet pointed roughly toward Earth will look brighter than a jet pointed more or less away. This may explain why some radio galaxies have one jet brighter than the other, as in Cygnus A shown at the top of the opposite page. It may also explain why some radio galaxies appear to have only one jet. The other jet may point generally away from Earth and be too faint to detect, as in the case of NGC5128, also shown on the opposite page.

Adapted from a diagram by Ann Field, NASA, STScI.

18 Cosmology in the 21st Century

The Universe, as has been observed before, is an unsettlingly big place, a fact which for the sake of a quiet life most people tend to ignore.

DOUGLAS ADAMS, *THE RESTAURANT AT THE END OF THE UNIVERSE*

LOOK AT YOUR THUMB. The matter in your thumb was present in the fiery beginning of the universe. **Cosmology,** the study of the universe as a whole, can tell you where your matter came from, and it can tell you where your matter is going. One possibility suggested by the newest theories of cosmology is that the expansion of the universe will eventually rip the galaxies away from each other, rip stars away from galaxies, rip planets away from their stars, rip stars apart, rip molecules apart, and eventually rip every atom in the universe to pieces. Cosmologists like to call that the big rip, but you can relax; even if it happens, it won't be for 30 billion years or more. ❙ Cosmology is a mind-bendingly weird subject, and you can enjoy it for its strange ideas and theories. After all, it is fun to think about space stretching like a rubber sheet and

❙ Continued on page 418 ❙

All of the energy and matter in the universe, including the matter in your body, began in the big bang. In this computer model of the evolution of the universe soon after the big bang, matter is being drawn together to form great clouds of galaxies. (Courtesy MPA and Joerg Colberg)

Guidepost

Looking Back

Since Chapter 1 you have been on an outward journey through the universe. You have studied the appearance of the night sky seen from our planet, the birth and death of stars, and the interaction of galaxies. Now you have reached the limit of your journey in space and in time—the study of the universe as a whole.

This Chapter

The ideas in this chapter are the biggest and the most difficult in all of science. Can you imagine an edge to the universe or the first instant of time? Perhaps you have heard that the universe began in an event called the big bang. Now it is time to try to imagine what it was like.

As you explore cosmology, you will find answers to four essential questions:

Does the universe have an edge in space or time?

What evidence exists that the universe began with a big bang?

How can the universe expand if it has no edge?

Why is the universe the way it is?

These questions are deceptively simple, but they will lead you to a new understanding of what you are and where you are.

Looking Ahead

Once you have finished this chapter, you will have modern insight into the nature of the universe, and it will be time to focus on your place in that universe—the subject of the rest of this book.

Ace Astronomy™ The AceAstronomy icon throughout the text indicates an opportunity for you to test yourself on key concepts and to explore animations and interactions on the AceAstronomy website at: http://ace.brookscole.com/sf9

invisible energy pushing the universe to expand faster and faster. As you wonder about the origin of vast walls of galaxy clusters, notice that it is all supported by evidence. Cosmology, however strange it may seem, is a serious and logical attempt to understand how the universe works.

The ideas of cosmology may seem weird at first glance, but this chapter will help you climb the cosmology pyramid one easy step at a time (■ Figure 18-1). You can begin with your common expectations; you have some ideas about what the universe is like. Start with those and test them against observations, compare them with theories, and step-by-step you can build a modern understanding of cosmology. Each step in the pyramid is small, but it leads to some astonishing insights into how the universe works and how you came to be a part of it.

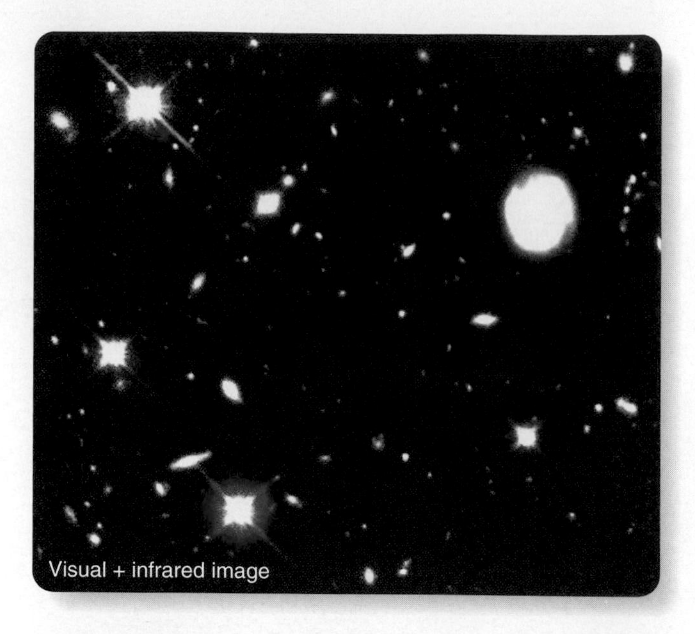

■ Figure 18-2

The entire sky is filled with galaxies. Some lie in clusters of thousands, and others are isolated in nearly empty voids between the clusters. In this image of a typical spot on the sky, bright objects with spikes caused by diffraction in the telescope are nearby stars. All other objects are galaxies ranging from the nearby face-on spiral at upper right to the most distant galaxies visible only in the infrared, shown as red in this composite image. (R. Williams, STScI HDF-South Team, NASA)

18-1 Introduction to the Universe

EVERYONE KNOWS what a gold mine looks like, but few people have actually visited a gold mine. If you explored a real gold mine, you might be surprised at what you find. Similarly, everyone has an impression of the universe as a vast depth filled with galaxies (■ Figure 18-2); but, as you begin exploring the universe, you must sort out your expectations so they do not mislead you. The first step is to deal with an expectation so obvious that most people, for the sake of a quiet life, don't even think about it.

■ Figure 18-1

Climbing the cosmology pyramid step-by-step isn't very difficult, and it leads to some fascinating ideas about the origin and evolution of the universe.

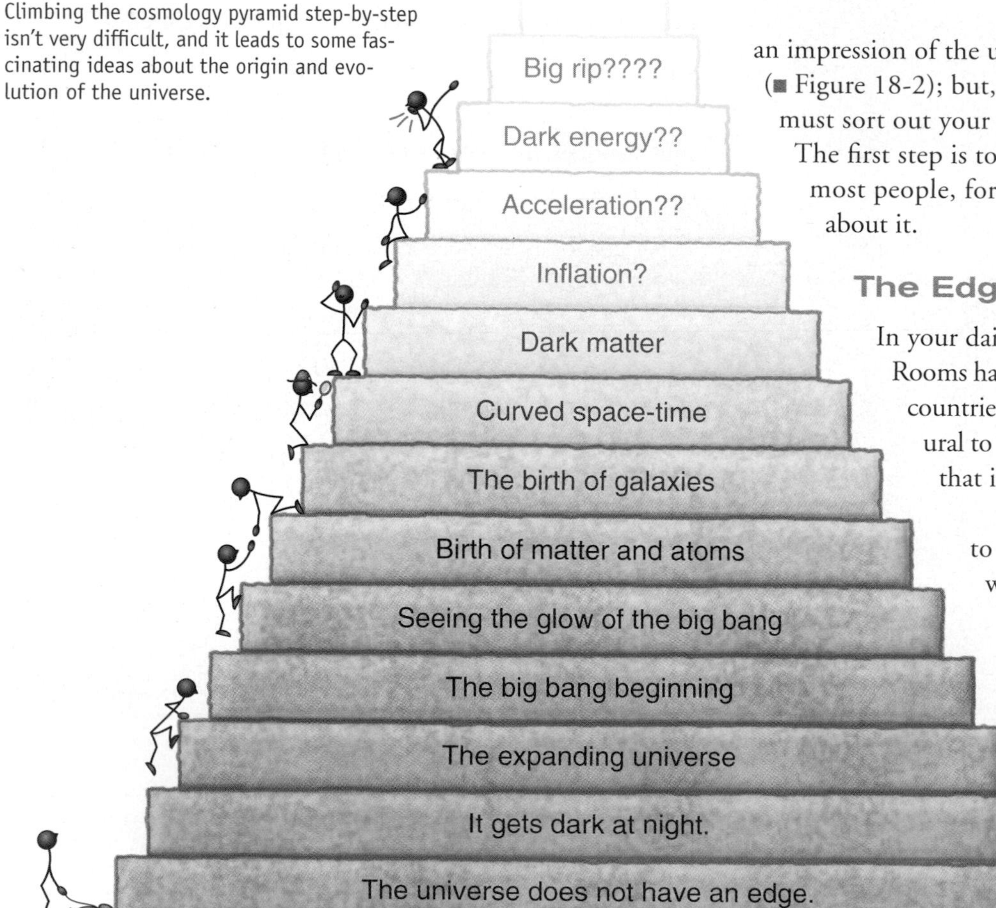

The Edge–Center Problem

In your daily life, you are accustomed to boundaries. Rooms have walls, athletic fields have boundary lines, countries have borders, oceans have shores. It is natural to think of the universe as having an edge, but that idea can't be right.

If the universe had an edge, imagine going to that edge. What would you find there: a wall of cardboard? a great empty space? nothing? This is not an edge to the distribution of matter, but an edge to space itself, so presumably you could not reach beyond the edge and feel around.

An edge to the universe seems to violate common sense, and modern cosmologists assume that the universe cannot have an edge. That is, the universe must be unbounded.

If the universe has no edge, then it cannot have a center. You find the centers of things—galaxies, globular clusters, oceans, pizzas—by referring to their edges. If the universe cannot have an edge, then it cannot have a center.

You must think carefully about the universe and avoid using the ideas of an edge or a center. Those ideas are misleading. Of course, if the universe is infinite, if it extends without limit in all directions, then it has no edge or center. There is no edge–center problem if you think of the universe as infinite. Could the universe be finite—could it contain a limited volume? You will see later in this chapter how the universe could be finite but have no edge or center. But before you go further with that idea, you must deal with another common expectation, the beginning of the universe.

The Necessity of a Beginning

Of course you have noticed that the night sky is dark. However, reasonable assumptions about the geometry of the universe can lead to the conclusion that the night sky should glow as brightly as a star's surface. This conflict between observation and theory is called **Olbers's paradox** after Heinrich Olbers, a Viennese physician and astronomer, who discussed the problem in 1826.

However, Olbers's paradox is not Olbers's, and it isn't a paradox. The problem of the dark night sky was first discussed by Thomas Digges in 1576 and was further analyzed by astronomers such as Johannes Kepler in 1610 and Edmund Halley in 1721. Olbers gets the credit through an accident of scholarship on the part of modern cosmologists who did not know of previous discussions. What's more, Olbers's paradox is not a paradox. You will be able to understand why the night sky is dark by revising your assumptions about the nature of the universe.

To begin, let's state the so-called paradox. Suppose you assume that the universe is infinite and uniformly filled with stars. (The aggregation of stars into galaxies makes no difference to your argument.) If you look in any direction, your line of sight must eventually reach the surface of a star. Look at ■ Figure 18-3, which uses the analogy of lines of sight in a forest (**Window on Science 18-1**). When you are deep in a forest, every line of sight terminates on a tree trunk, and you cannot see out of the forest. By analogy, every line of sight from Earth out into space should eventually terminate on the surface of a star, and the entire sky should be as bright as the surface of an average star. It should not get dark at night.

Of course, the more distant stars would be fainter than nearby stars because of the inverse square law. However, the farther you look into space, the larger the volume you see and the more stars you see; the two effects cancel out. Then given the assumptions, every spot on the sky must be occupied by the surface of a star, and the night sky should not be dark.

Can you imagine the entire sky glowing with the brightness of the surface of the sun? The glare would be overpowering. In fact, the radiation would rapidly heat Earth and all other celestial

■ Figure 18-3

(a) Every direction you look in a forest eventually reaches a tree trunk, and you cannot see out of the forest. (Photo courtesy Janet Seeds) (b) If the universe is infinite and uniformly filled with stars, then any line from Earth should eventually reach the surface of a star. This assumption predicts that the night sky should glow as brightly as the surface of the average star, a puzzle commonly referred to as Olbers's paradox.

1 In 1929, Edwin Hubble discovered that the redshifts of the galaxies are proportional to distance — a relationship now known as the Hubble law. It was taken as dramatic evidence that the universe is expanding.

Distance is the separation between two points in space, and time is the separation between two events. Albert Einstein showed how to relate space to time and treat the whole as space-time. You can think of space-time as a canvas on which the universe is painted, a canvas that can stretch.

Hubble Heritage Team/STScI/AURA/NASA

1a Astronomers often express galaxy redshifts as apparent velocities of recession, but these redshifts are not Doppler shifts. They are caused by the expansion of space-time.

For decades textbooks have described the cosmological redshifts using Einstein's relativisitic Doppler formula, but that formula applies to motion through space and not to the behavior of space itself. The true relation between redshift and apparent velocity of recession is not accurately known.

Notice that this description of the expansion of the universe means the galaxies are not really moving any more than the raisins are swimming through the bread dough. Except for orbital motion among neighboring galaxies, the galaxies are motionless in space-time, and it is space-time that is stretching like a rubber sheet and carrying the galaxies away from each other.

1b The stretching of space-time not only moves the galaxies away from each other, but it lengthens the wavelength of photons traveling through space-time. See illustration below.

What are the cosmological redshifts?

A distant galaxy emits a short-wavelength photon toward our galaxy.

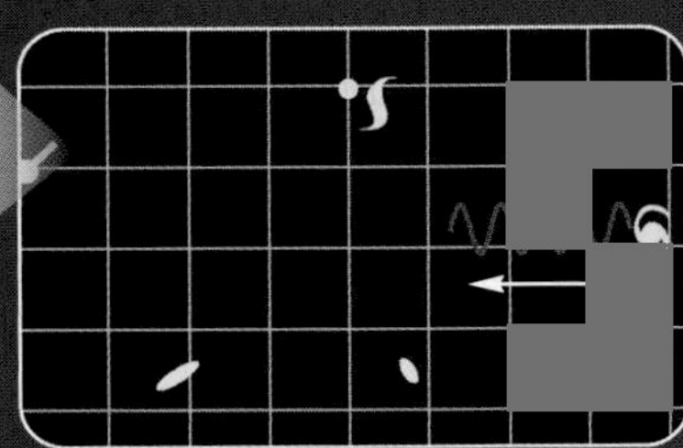

Grid shows expansion of space-time.

The expansion of space-time stretches the photon to longer wavelength as it travels.

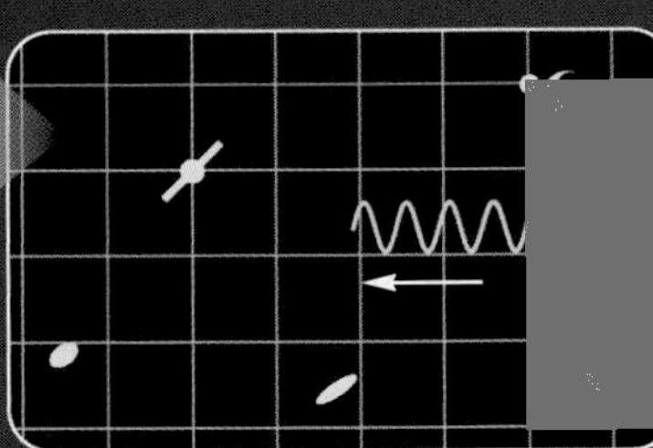

The farther the photon has to travel, the more it is stretched.

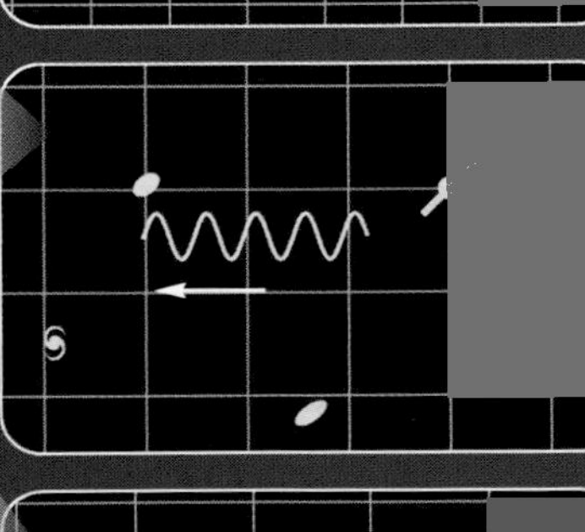

When the photon arrives at our galaxy, you see it with a longer wavelength — a redshift that is proportional to distance.

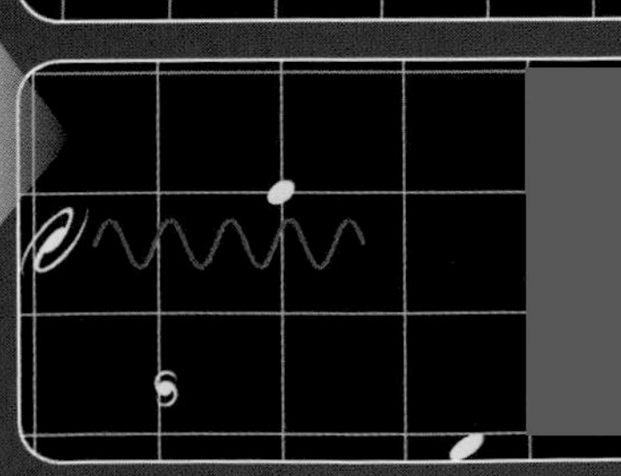

2 To answer these questions you must use Einstein's theory of general relativity, which predicts that the presence of mass can curve space-time, a curvature you experience as gravity. Furthermore, the theory predicts that even empty space-time can have a curvature, so what you see in the universe depends on the overall curvature of space-time. Finally, the theory shows that space-time could have the property of expansion. That is, the expansion of space-time is not caused by external influences, but is part of space-time itself.

Could the universe be finite and still have no edge?

How can it expand if it has no edge?

How can the universe expand if there is no space outside for it to expand into?

2a To think about space-time curvature, you can use an analogy of a two-dimensional ant on an orange. If he is truly two-dimensional, he will not be able to understand up and down. He can only travel forward and back, right and left. Then as he walks over the surface of his spherical universe, he will eventually realize he has been everywhere because his universe is covered by his footprints. He will conclude that he lives in a finite universe that has no edge and no center.

Similarly, our universe could be curved back on itself so that it could be finite but you would never find an edge. To find out how our universe is really curved, astronomers must study very distant objects.

Notice that the center of the orange cannot be the center of the ant's two-dimensional universe because the center of the orange does not lie on the surface. That is, the center of the orange is not part of the ant's universe.

Analogy Ahead

Don't forget: These analogies are only two-dimensional. When you think about our universe, you must think of a three-dimensional universe. It is as difficult for you to imagine curvature in our three-dimensional universe as it is for ants to imagine curvature in their two-dimensional universes.

2b There are three possible ways our universe might be curved. It could have positive curvature, analogous to the surface of a sphere. Models of such universes are called **closed universes** because they are finite. Of course, ours could be a **flat universe** with zero curvature. Another possibility is that our universe has negative curvature, analogous to a two-dimensional universe shaped like a saddle or a potato chip. Models of these universes are called **open universes.**

r

Aha!

A positively curved universe, also called a closed universe, is finite and has no edge. In the two-dimensional analogy, ants on an orange would notice that the areas of large circles were less than πr^2.

A flat universe must be infinite or it will have an edge. In a two-dimensional analogy, ants on a flat sheet of paper would find that all circles have areas of πr^2.

Hey!
C'mere!

r

3 Although our ant might be unable to sense a third dimension, it could still measure the curvature of its universe by drawing circles. On a flat universe, the area of a circle would always be πr^2 no matter how big the circle was. But on a spherical surface, large circles would contain less than πr^2. On a saddle-shaped surface, large circles would contain more than πr^2. In the same way, you could detect the curvature of our three-dimensional universe, but you would have to make measurements over very large distances.

Whether our universe is closed, flat, or open, it cannot have an edge or a center.

What?

r

A negatively curved universe, also called an open universe, must be infinite or it will have an edge. In a two-dimensional analogy, ants on a saddle shape will discover that large circles have areas greater than πr^2.

This is not a small issue. Judging by their gravitational fields, galaxies and clusters of galaxies contain as much as 10 times more matter than you would expect from what you see. To correct for this dark matter, you would not just add a small percentage. You have to multiply by a *factor* as large as 10.

Looking at the universe of visible galaxies is like looking at a ham sandwich and seeing only the mayonnaise. Most of the universe is invisible, and most astronomers now believe that the invisible matter is not the normal matter of which you and the stars are made. To follow that line of evidence, you must think again about the atoms made during the big bang.

During the first few minutes of the big bang, nuclear reactions converted some protons into helium and a small amount into other elements. How much of these elements was created depends critically on the density of the material. Deuterium, for example, is an isotope of hydrogen in which the nucleus contains a proton and a neutron. The amount of deuterium produced in the big bang depends strongly on the density of normal matter. If there were a lot of normal particles such as protons and neutrons, then they would have collided with the deuterium nuclei and converted them into helium. If the density of normal particles was less, more deuterium would have survived.

Lithium is another nucleus that could have been made in small amounts during the big bang. Figure 18-9 shows that there is a gap between helium and lithium; there is no stable nucleus with atomic mass 5, so regular nuclear reactions during the big bang could not convert helium into lithium. If, however, the density of normal matter such as protons and neutrons was high enough, a few nuclear reactions could have leaped the gap and produced a few atoms of the isotope lithium-7.

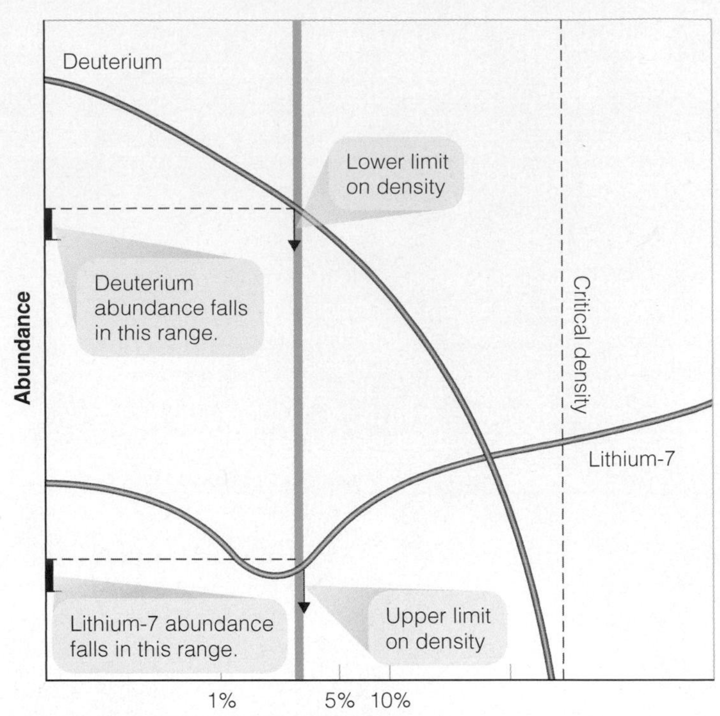

■ Figure 18-15

This diagram compares observation with theory. Theory predicts how much deuterium and lithium-7 you would observe for different densities of normal matter (red and blue curves). The observed density of deuterium falls in a narrow range shown at upper left and sets a lower limit on the possible density of normal matter. The observed density of lithium-7 sets an upper limit. This means the true density of normal matter must fall in a narrow range represented by the green column. Certainly, the density of normal matter is much less than the critical density.

Deuterium is so easily converted into helium that none can be made in stars. In fact, stars destroy what deuterium they have by converting it into helium. Lithium too is destroyed in stars. Using the largest telescopes, astronomers have been able to measure the chemical abundance of gas clouds near quasars. The look-back times to these gas clouds is so great that they appear as they were before stars could have altered the abundance of the elements. As shown in ■ Figure 18-15, the observed amount of deuterium sets a lower limit on the density of the universe, and the observed abundance of lithium-7 sets an upper limit. The normal matter that you and the stars are made of cannot make up much more than 4 percent of the critical density. Yet observations show that galaxies and galaxy clusters contain large amounts of dark matter. The protons and neutrons that make up normal matter belong to a family of subatomic particles called *baryons,* so cosmologists believe that the dark matter cannot be baryons. Only a small amount of the mass in the universe can be baryonic; the dark matter must be **nonbaryonic matter.**

Theorists have suggested that the dark matter is made of weakly interacting massive particles (WIMPS), which could be hundreds of times more massive than a proton. If WIMPs almost never interact with other particles, they would be almost undetectable except for their gravitational influence. There is, however, no conclusive evidence that WIMPs actually exist. A controversial experiment has hinted that a WIMP called a neutralino may have been detected in a laboratory. If physicists can establish that WIMPs actually exist and measure their density in the universe, it may help cosmologists explain the dark matter.

For some years, astronomers thought that neutrinos could be an important part of the dark matter. You saw in Chapter 8 how observations made by underground detectors suggested that neutrinos coming from the sun oscillate, and that is important because, according to quantum mechanics, if neutrinos oscillate, they cannot have zero mass. Although the mass of neutrinos has not been measured, it is clearly very small. Even a small mass might be important because there are so many neutrinos in the universe—10^8 neutrinos for every normal particle.

Theoretical models of galaxy formation in the early universe show that the dark matter can't be mostly neutrinos. Dark matter composed of neutrinos and similar particles that travel at or near the speed of light is called **hot dark matter.** Such fast-moving particles do not clump together easily and could not have stimulated the formation of objects as small as galaxies and clusters of galaxies. The most successful models of galaxy formation require that the dark matter be made up of **cold dark matter,** meaning the particles move slowly and can clump into smaller structures. WIMPs, for example, are massive and slow moving and, if they exist, could clump together in the early universe and help pull matter into galaxy-size clouds.

Nonbaryonic dark matter does not interact significantly with normal matter or with photons, which is why you can't see it. But that means that dark matter was not affected by the intense radiation that dominated the universe when it was very young. That radiation prevented normal matter from contracting to begin forming galaxies. But the dark matter was immune, and it could contract to form clouds. Once the density of radiation fell low enough, normal matter could begin falling into the clouds of dark matter to form the first galaxies. Dark matter could have given galaxy formation a head start soon after the big bang. Models with cold, nonbaryonic dark matter are most successful at forming galaxies and clusters of galaxies early in the history of the universe.

Now that you understand more about dark matter, you can evaluate the possibility that the halos of galaxies contain large numbers of very-low-mass stars, brown dwarfs, or planets too faint to see. These **MACHOs** (for Massive Compact Halo Objects) can be detected if they pass between telescopes on Earth and a distant star. The gravitational field of the MACHO acts as a small gravitational lens, focusing the light of the star and making it grow brighter for a period of a few weeks as the MACHO passes between the star and Earth. Extensive searches have detected such events (■ Figure 18-16) but not in large numbers. Furthermore, other searches have turned up lots of white dwarfs in the halo of our galaxy. However common these halo objects are, they are made of baryonic matter, and the abundance of deuterium and lithium shows that the dark matter must be nonbaryonic. It does not appear that MACHOs and white dwarfs can make up a significant part of the dark matter.

Although there is good evidence that dark matter exists, it has proven difficult to identify. No form of dark matter has been found that is abundant enough to provide the critical density needed to make the universe flat. In fact, all forms of dark matter appear to add up to less than 30 percent of the critical density. As you will see later in this chapter, there is more to the universe than meets the eye, and more even than the dark matter.

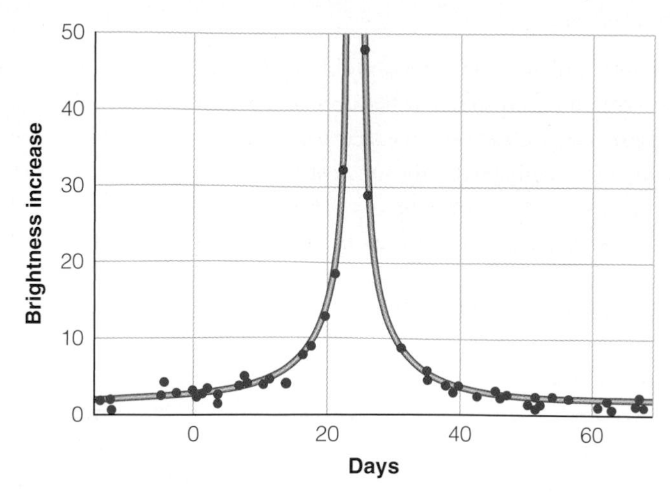

■ **Figure 18-16**

Measurements of the brightness of a distant star (dots) increased and then decreased exactly as predicted by general relativity (curve) for a star's passage behind a gravitational lens caused by a MACHO. In this example, the star brightened by a factor of over 40.

Building Scientific Arguments

Why do astronomers think that dark matter can't be baryonic?

Good scientific arguments always fall back on evidence. In this case the evidence is very strong. Small amounts of isotopes like deuterium and lithium-7 were produced in the first minutes of the big bang, and the abundance of those elements depends strongly on the density of protons and neutrons. Because these particles belong to the family of particles called baryons, astronomers refer to normal matter as baryonic. Measurements of the abundance of deuterium and lithium-7 show that the universe cannot contain more baryons than about 4 percent of the critical density. Yet observations of galaxies and galaxy clusters show that dark matter must make up almost 30 percent of the critical density. Consequently, astronomers conclude that the dark matter must be made up of nonbaryonic particles.

Finding the dark matter is important because the density of matter in the universe determines its curvature. Expand your argument to discuss theory. **How does curvature allow you to avoid an edge or center in a finite model of the universe?**

■ ■ ■

Connections: By adding the curvature of space-time and effects of dark matter to your cosmology, you have made your theories much more sophisticated. You are now ready to explore the most recent and most exciting advances in cosmology.

18-3 21st-Century Cosmology

IF YOU ARE A LITTLE DIZZY from the weirdness of curved space-time and dark matter, make sure you are sitting down before you read much further. As the 21st century began, cosmologists made a startling discovery, and all around the world astronomers looked at each other and said, "What? What!" The most amazing thing about these amazing discoveries is that they fit so well with some of the things you have been learning in this chapter. To get a running start on these new discoveries, you'll have to go back a couple of decades.

Inflation

In 1980, astronomers faced a problem. The big bang models could not explain two important features of the universe. To solve those two problems, a new theory of the big bang was created, and that theory has been startlingly successful. To introduce the new theory, you can begin with the two problems.

One of the problems is called the **flatness problem.** The universe seems to be balanced near the boundary between an open and a closed universe. That is, it seems nearly flat. Given the vast range of possibilities, from zero to infinite, it seems peculiar that the density of the universe is within a factor of 10 of the critical density that would make it flat. If dark matter is as common as it seems, the density may be even closer than a factor of three to being perfectly flat.

Even a small departure from critical density when the universe was young would be magnified by subsequent expansion. To be so near critical density now, the density of the universe during its first moments must have been within 1 part in 10^{49} of the critical density. So the flatness problem is: Why is the universe so nearly flat?

The second problem with the big bang theory is the isotropy of the primordial microwave background radiation. When astronomers correct for the motion of Earth, they see the same background radiation in all directions to at least 1 part in 1000. Yet when you look at background radiation coming from two points in the sky separated by more than a degree, you look at two parts of the big bang that were not causally connected when the radiation was emitted. That is, when recombination occurred and the gas of the big bang became transparent to the radiation, the universe was not old enough for any signal to have traveled from one of these regions to the other. Thus, the two spots you look at did not have time to exchange heat and even out their temperatures. Then how did every part of the entire big bang universe get to be so nearly the same temperature by the time of recombination? This is called the **horizon problem** because the two spots are said to lie beyond their respective light-travel horizons.

The key to these two problems and to others involving subatomic physics may lie with the theory called the **inflationary universe** because it predicts a sudden inflation when the universe was very young, a violent expansion even more extreme than that predicted by the big bang theory.

To understand the inflationary universe, you need to recall that physicists know of only four forces—gravity, the electromagnetic force, the strong force, and the weak force (Chapter 8). You are familiar with gravity, and the electromagnetic force is responsible for making magnets stick to refrigerator doors and cat hair stick to wool sweaters charged with static electricity. The strong force holds atomic nuclei together, and the weak force is involved in certain kinds of radioactive decay.

For many years, theorists have tried to unify these forces; that is, they have tried to describe the forces with a single mathematical law. A century ago, James Clerk Maxwell showed that the electric force and the magnetic force were really the same effect, and physicists now count them as a single electromagnetic force. In the 1960s, theorists succeeded in unifying the electromagnetic force and the weak force in what they called the electroweak force, effective only for processes at very high energy. At lower energies, the electromagnetic force and the weak force behave differently. Now theorists have found ways of unifying the electroweak force and the strong force at even higher energies. These new theories are called **grand unified theories,** or **GUTs.**

According to the inflationary universe, the universe expanded and cooled until about 10^{-35} second after the big bang, when it became so cool that the electroweak force and the strong force began to disconnect from each other and behave in different ways. This released tremendous energy, which suddenly inflated the universe by a factor between 10^{20} and 10^{30} (■ Figure 18-17). At that time the part of the universe that is now visible from Earth, the entire observable universe, was no larger than the volume of an atom, but it suddenly inflated to the volume of a cherry pit and then continued its slower expansion to its present extent.

That sudden inflation can solve the flatness problem and the horizon problem. The sudden inflation of the universe would have forced whatever curvature it had toward zero, just as inflating a balloon makes a small spot on its surface flatter. Consequently you now live in a universe that is almost perfectly flat because of

■ Figure 18-17

When the universe was very young and hot (top), the four forces of nature were indistinguishable. As the universe began to expand and cool, the forces separated and triggered a sudden inflation in the size of the universe.

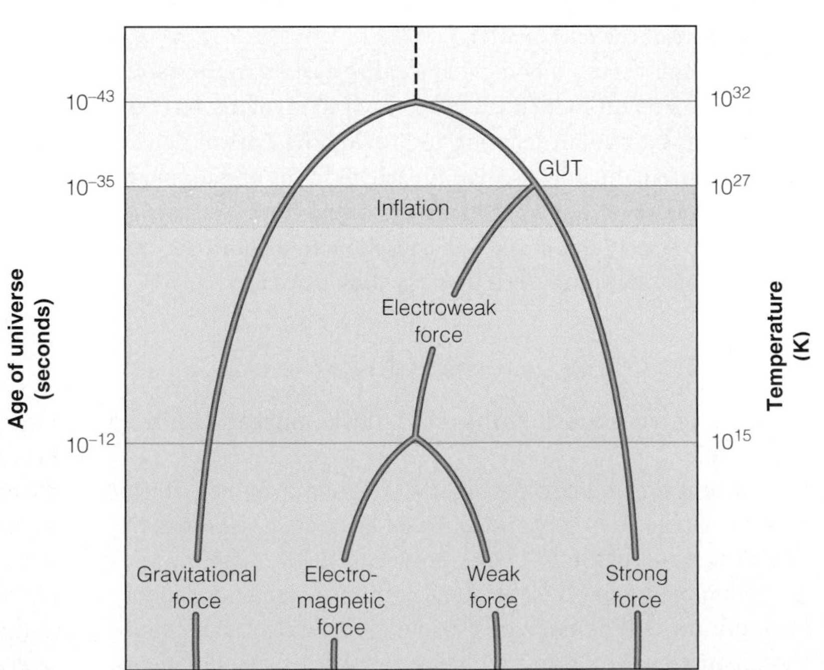

that sudden inflation long ago. In addition, because the observable part of the universe was once no larger in volume than an atom, it had plenty of time to equalize its temperature before inflation occurred. Now you live in a universe with the same temperature for the background radiation in all directions.

The inflationary universe is based, in part, on quantum mechanics, and a slightly different aspect of quantum mechanics may explain why there was a big bang at all. Theorists believe that a universe totally empty of matter could be unstable and decay spontaneously by creating pairs of particles until it was filled with the hot, dense state called the big bang. This theoretical discovery has led some cosmologists to believe that the universe could have been created by a chance fluctuation in space-time. In the words of physicist Frank Wilczyk, "The reason there is something instead of nothing is that 'nothing' is unstable."

The inflationary theory predicts that the universe is almost perfectly flat. That is, the true density must equal the critical density. A theory can never be used as evidence, but the beauty of the inflationary theory has given many cosmologists confidence that the universe must be flat. Observations, however, seem to say that the universe does not contain enough matter (baryonic plus dark) to be flat. Can there be more to the universe than baryonic matter and dark matter? What could be weirder than dark matter? Read on.

The Acceleration of the Universe

Ever since Hubble discovered the expansion of the universe, astronomers have known what to expect. They have also known that expectations are not reliable; and, as the 21st century approached, astronomers made a mind-boggling discovery that contradicted everyone's expectations.

Both common sense and mathematical models suggest that as the galaxies recede from each other, the expansion should be slowed by gravity trying to pull the galaxies toward each other. How much the expansion is slowed should depend on the amount of matter in the universe. If the density of matter is less than the critical density, the expansion should be slowed only slightly, and the universe should expand forever. If the density of matter in the universe is greater than the critical density, the expansion should be slowing down dramatically, and the universe should eventually stop expanding and begin contracting. Notice that this is the same as saying a low-density universe should be open and a high-enough-density universe should be closed.

For decades, astronomers struggled to measure the Hubble constant and detect the slowing of the expansion. A direct measurement of the rate of slowing would reveal the true curvature of the universe. This was one of the key projects for the Hubble Space Telescope, and two teams of astronomers spent years making the measurements.

Detecting a change in the rate of expansion is a difficult project because it requires accurate measurements of the distances to very remote galaxies, and both teams used the same technique. They calibrated type Ia supernovae as distance indicators. A type Ia supernova occurs when a white dwarf gains matter from a companion star, exceeds the Chandrasekhar limit, and collapses in a supernova explosion. Because all such white dwarfs should collapse at the same mass limit, they should all produce explosions of the same luminosity, and that makes them good distance indicators.

In Chapter 16, you read that astronomers refer to distance indicators such as Cepheid variable stars as *standard candles,* objects of known luminosity. Type Ia supernovae have been described as *standard bombs.* They are titanic explosions, but they all reach the same peak brightness, and that makes them good distance indicators.

The two teams calibrated type Ia supernovae by locating such supernovae occurring in nearby galaxies whose distance was known from Cepheid variables and other reliable distance indicators. Once the peak luminosity of type Ia supernovae had been determined, they could be used to find the distance to much more distant galaxies.

Both teams announced their results in 1998. The expansion of the universe is not slowing down. It is speeding up! That is, the expansion of the universe is accelerating (■ Figure 18-18).

The announcement that the expansion of the universe is accelerating made astronomers stop and stare at each other. It was totally unexpected. It was simultaneously an exciting, puzzling, and revealing discovery, but astronomers immediately began testing it. It depends critically on the calibration of type Ia supernovae as distance indicators, and some astronomers suggested that the calibration might be wrong (Window on Science 15-1). Some unknown kind of dust might make very distant supernovae too faint without making them too red. No reddening had been detected, so normal dust seemed to be ruled out. Or perhaps the very distant supernovae were too faint because astronomers see them at great look-back times, and some unknown process made those early supernovae a bit less luminous than more recent supernovae.

Those problems with the calibration are ruled out by more recently discovered supernovae. For example, SN1997ff, recognized in 2001 as a type Ia supernova, is too faint for its redshift of 1.7 and confirms acceleration (■ Figure 18-19). More type Ia supernovae discovered by the Hubble Space Telescope rule out problems with the calibration. The universe really does seem to be expanding faster and faster.

Dark Energy and Acceleration

If the expansion of the universe is accelerating, then there must be a force of repulsion in the universe, and astronomers are struggling to understand what it could be. One possibility leads back to 1916.

When Albert Einstein published his theory of general relativity in 1916, he recognized that his equations describing space and time implied that space had to contract or expand. The galaxies could not float unmoving in space because their gravity would pull them toward each other. The only solutions seemed to be a

19 The Origin of the Solar System

What place is this?

Where are we now?

CARL SANDBURG, *GRASS*

MICROSCOPIC CREATURES LIVE in the roots of your eyelashes. Don't worry. Everyone has them, and they are harmless.* They hatch, fight for survival, mate, lay eggs, and die in the tiny spaces around the roots of your eyelashes without doing any harm. Some live in renowned places—the eyelashes of a glamorous movie star—but the tiny beasts are not self-aware; they never stop to say, "Where are we?" Humans are more intelligent; humans have the ability to wonder where we are in the universe and how we came here. ▌ You should study the solar system for many reasons. You need to understand Earth as a planet because six billion of us are living on Earth and changing it in ways we don't understand. You should also study

***Demodex folliculorum* has been found in 97 percent of individuals and is a characteristic of healthy skin.

Continued on page 450

The human race lives on a planet in a planetary system that appears to have formed in a nebula around the protostar that became the sun. This artist's impression shows the formation of the giant planet Jupiter. (NASA)

Guidepost

Looking Back

You have become an expert on the universe. You have studied the appearance, origin, structure, and evolution of stars, galaxies, and the universe itself. But so far, your studies have left out one important class of objects—planets. Now it is time for you to correct that omission.

This Chapter

In this chapter, you will look back on what you have learned and find your place. You are a planetwalker. What does that mean? Where do you fit into the cosmos? Most of all, you need to know how the solar system formed. That will tell you how your home planet was produced by the processes you have been studying. As you explore our solar system in space and time, you will find answers to four essential questions:

What theories account for the origin of the solar system?

What properties must these theories explain?

How do planets form?

Is our solar system unique?

Looking Ahead

One reason you should learn about the origin of the solar system is that you live here, but there is another reason. In the next three chapters you will explore in more detail each of the planets, plus the comets and asteroids. By studying the origin of the solar system first, you give yourself a framework for understanding these fascinating worlds.

AceAstronomy™ The AceAstronomy icon throughout the text indicates an opportunity for you to test yourself on key concepts and to explore animations and interactions on the AceAstronomy website at: http://ace.brookscole.com/sf9

the solar system because, as you are about to discover, there are more planets in the universe than stars. Above all, you should study the solar system because it is your home in the universe. Because humans are an intelligent species, we have the right and the responsibility to wonder what we are. Our kind have inhabited this solar system for at least a million years, but only within the last hundred years have we begun to understand what a solar system is. Like sleeping passengers on a train, we waken, look out at the passing scenery, and mutter, "What place is this? Where are we now?"

19-1 Theories of Earth's Origin

YOU ARE LINKED through a great chain of origins that leads backward through time to the first instant when the universe began 13.7 billion years ago. The gradual discovery of the links in that chain is one of the most exciting adventures of the human intellect. In earlier chapters, you studied some of that story: the origin of the universe in the big bang, the formation of galaxies, the origin of stars, and the growth of the chemical elements. Here you will explore further to consider the formation of planets.

Early Hypotheses

The earliest theories for Earth's origin are myths and folktales that go back beyond the beginning of recorded history. In addition, almost all religions contain an account of the origin of the world. It was not until about the time of Galileo that philosophers began searching for rational explanations for natural phenomena. While people like Copernicus, Kepler, and Galileo tried to find logical explanations for the motions of Earth and the other planets, other philosophers began thinking about the origin of the planets.

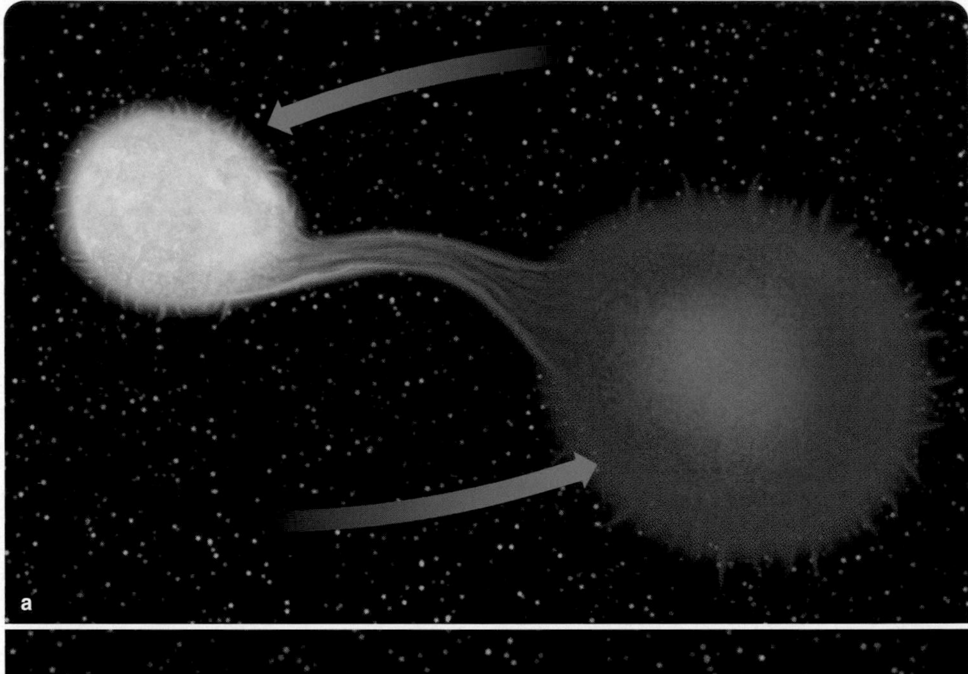

■ **Figure 19-1**

(a) The passing star hypothesis was catastrophic. It proposed that the sun was hit by or had a very close encounter with a passing star and that matter torn from the sun and the star formed planets orbiting the sun. The theory is no longer accepted. (b) Laplace's nebular hypothesis was evolutionary. It suggested that a contracting disk of matter conserved angular momentum, spun faster, and shed rings of matter that formed planets.

The first rational theory for Earth's origin was proposed by the French philosopher and mathematician René Descartes (1596–1650). Because he lived and wrote before the time of Newton, Descartes did not have the concept of gravitation as the dominant force in the universe. Rather, he believed that force was communicated by contact between bodies and that the entire universe was filled with vortices of whirling invisible particles. In 1644, he proposed that the sun and planets formed when a large vortex contracted and condensed. In this way, his hypothesis explained the general properties of the solar system known at the time.

A century later, in 1745, the French naturalist Georges-Louis de Buffon (1707–1788) proposed an alternative hypothesis that the planets were formed when a passing star collided with or passed close to the sun and pulled matter out of the sun and the star. The matter condensed to form the planets, and they fell into orbit around the sun (■ Figure 19-1a). This **passing star hypothesis** was popular off and on for two centuries, but it con-

Window on Science 19-1

Two Kinds of Theories: Evolution and Catastrophe

Many theories in science can be classified as either evolutionary, in that they involve gradual processes, or catastrophic, in that they depend on specific, unlikely events. Scientists have generally preferred evolutionary theories. Nevertheless, catastrophic events do occur.

Something in people prefers catastrophic theories, perhaps because everyone likes to see spectacular violence from a safe distance, which may explain the success of movies that include lots of car crashes and explosions. Also, cataclysmic theories resonate with Old Testament accounts of catastrophic events and special acts of creation. Thus, people have an understandable interest in catastrophic theories.

Nevertheless, most scientific theories are evolutionary. Such theories do not depend on unlikely events or special acts of biblical creation. For example, geologists study theories of mountain building that are evolutionary, with the mountains being pushed up slowly as centuries pass. All the evidence of erosion and the folding of rock layers shows that the process is gradual. Because most such natural processes are evolutionary, scientists sometimes find it difficult to accept any theory that depends on catastrophic events.

You will see in this and later chapters that catastrophes do occur. The planets, for example, are bombarded by debris from space, and some of those impacts are very large. As you study astronomy or any other natural science, notice that most theories are evolutionary but that you must allow for the possibility of unpredictable catastrophic events.

Mountains evolve to great heights by rising slowly, not catastrophically. (Janet Seeds)

tains serious flaws. First, stars are very small compared to the distances between them, and thus they collide very infrequently. In the entire history of our galaxy, stars have probably collided only a few times. More important, the gas pulled from the sun and the star would be much too hot to condense to make planets. Furthermore, even if planets formed, they would not go into stable orbits.

The hypotheses of Descartes and Buffon fall into two broad categories. Descartes's hypothesis is **evolutionary** in that it calls upon common, gradual events to produce the sun and planets. If it is correct, stars with planets are very common. Buffon's hypothesis, on the other hand, is **catastrophic.** It calls on unlikely, sudden events to produce the solar system, and thus it implies that solar systems are very rare. While your imagination may be tempted by colliding stars, modern hypotheses for the origins of the planets are evolutionary, with, as you will see, a few astonishing catastrophes thrown in (**Window on Science 19-1**).

The modern theory of the origin of the solar system had its true beginning with Pierre-Simon de Laplace (1749–1827), the brilliant French astronomer and mathematician. In 1796, he combined Descartes's vortex with Newton's gravity to produce a model of a rotating cloud of matter contracting under its own gravitation and flattening into a disk—the **nebular hypothesis.** As the disk grew smaller, it had to conserve angular momentum and spin faster and faster. Laplace reasoned that, when it was spinning as fast as it could, the disk would shed its outer edge to leave behind a ring of matter. Then the disk could contract further, speed up again, and leave another ring. In this way, he imagined, the contracting disk would leave behind a series of rings, each to become a planet circling the newborn sun at the center of the disk (Figure 19-1b).

According to the nebular hypothesis, the sun should be spinning very rapidly, or, to put it another way, the sun should have most of the angular momentum of the solar system. (Recall from Chapter 5 that angular momentum is the tendency of a rotating object to continue rotating.) As astronomers studied the planets and the sun, however, they found that the sun rotated slowly and that the planets moving in their orbits had most of the angular momentum in the solar system. In fact, the rotation of the sun contains only about 0.3 percent of the angular momentum of the solar system. Because the nebular hypothesis could not explain this **angular momentum problem,** it was never fully successful, and astronomers toyed with various versions of the passing star hypothesis for over a century.

In contrast to the astronomy of earlier centuries, 20th-century astronomy applied modern physics to the stars and galaxies, and that illuminated the origins of the stars and the elements. You can now trace that great chain of origins that began with the big bang and led to the matter of which you are made. First you should review the story of the origin of matter, and then you can add the story of how that matter formed your home world.

A Review of the Origin of Matter

The matter in your body came into existence within minutes of the beginning of the universe. Astronomers have strong evidence that the universe began in an event called the big bang

(Chapter 18); and, by the time the universe was three minutes old, the protons, neutrons, and electrons in your body had come into existence. You are made of very old matter.

Although those particles formed quickly, they were not linked together to form the atoms that are common today. Most of the matter was hydrogen, and about 25 percent was helium. Very few heavier atoms were made in the big bang. Although your body does not contain helium, it does contain many of those ancient hydrogen atoms unchanged since the universe began.

Within a few hundred million years after the big bang, matter began to collect to form galaxies containing billions of stars. You have learned how nuclear reactions inside stars combine low-mass atoms such as hydrogen to make heavier atoms (Chapter 8). Generation after generation of stars cooked the original particles, linking them together to build atoms such as carbon, nitrogen, and oxygen (Chapter 15). Those are common atoms in your body. Even the calcium atoms in your bones were assembled inside stars.

Massive stars produce iron in their cores, but much of that iron core is destroyed when the star collapses and explodes as a supernova. Most of the iron in your body was produced by carbon fusion in type Ia supernovae and by the expanding matter ejected by type II supernovae. Atoms heavier than iron were created by rapid nuclear reactions that can occur only during supernova explosions (Chapter 13). Gold and silver are rare in your body, but iodine is critical in your thyroid gland, and those heavy atoms were produced during the violent deaths of massive stars.

Our galaxy contains at least 100 billion stars, of which the sun is one. It formed from a cloud of gas and dust about 5 billion years ago, and the atoms in your body were part of that cloud. How the sun took shape, how the cloud gave birth to the planets, how the atoms in your body found their way onto Earth and into you is the story of this chapter. As you explore the origin of the solar system, keep in mind the great chain of origins that created the atoms. As the geologist Preston Cloud remarked, "Stars have died that we might live."

The Solar Nebula Hypothesis

A rotating cloud of gas contracts and flattens...

to form a thin disk of gas and dust around the forming sun at the center.

Planets grow from gas and dust in the disk and are left behind when the disk clears.

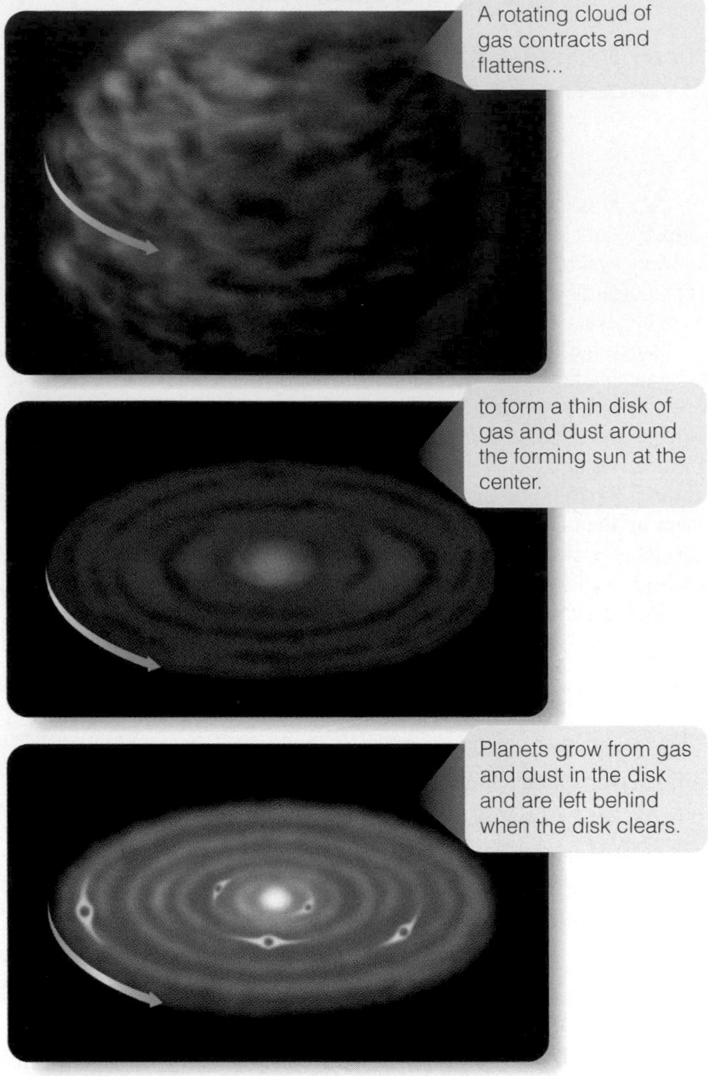

■ Figure 19-2

The solar nebula hypothesis proposes that the planets formed along with the sun.

The Solar Nebula Hypothesis

The stars gave birth to the heavy atoms of which our world is made, and the modern theory of the origin of the planets is based on star formation. The **solar nebula hypothesis** proposes that the planets were formed from the disk of gas and dust that surrounded the sun as it formed (■ Figure 19-2).

As stars form in contracting clouds, they remain surrounded by cocoons of dust and gas, and the rotation of the cloud causes that dust and gas to form a spinning disk around the protostar. When the center of the star grows hot enough to ignite nuclear reactions, its surface quickly heats up, becomes more luminous, and blows away the gas and dust cocoon.

The solar nebula hypothesis supposes that planets form in the rotating disks of gas and dust around young stars. You have seen clear evidence that disks of gas and dust are common around young stars. Infrared observations of T Tauri stars, for instance, show that some are surrounded by gas clouds rich in dust, and spectra show that these stars are blowing away their nebulae at speeds up to 200 km/s. Bipolar flows from protostars (Chapter 11) were the first evidence of such disks, but modern techniques can image the disks directly (see Figure 11-7 and page 253).

Our own planetary system probably formed in such a disk-shaped cloud around the sun. When the sun became luminous enough, the remaining gas and dust were blown away into space, leaving the planets orbiting the sun. Notice that the solar nebula hypothesis also suffers from the angular momentum problem. Before modern astronomers could take the theory seriously, they had to understand why the sun now rotates so slowly, a puzzle you will be able to solve later in this chapter.

If the solar nebula hypothesis is correct, then our Earth and the other planets of the solar system formed billions of years ago as the sun condensed from the interstellar medium. If that is true, then planets form as a by-product of star formation, and most stars should have planets.

Building Scientific Arguments

Why does the solar nebula theory imply planets are common? Often, the implications of a theory are more important in building a scientific argument than the theory's own conjecture about nature. The solar nebula theory is an evolutionary theory; and, if it is correct, the planets of our solar system formed from the disk of gas and dust that surrounded the sun as it condensed from the interstellar medium. That suggests it is a common process. Most stars form with disks of gas and dust around them, and planets should form in such disks. Planets should be very common in the universe.

Now build a new scientific argument to consider the old catastrophic theory. **Why did the passing star hypothesis suggest that planets are very rare?**

■ ■ ■

Connections: The solar nebula hypothesis seems to fit with what you know about star formation, so it is time to construct a detailed theory to describe planet formation. The first step is to survey the solar system to discover the distinguishing characteristics of the planetary system. Those are the characteristics that a successful theory must explain.

19-2 A Survey of the Solar System

TO TEST THEIR THEORIES, astronomers must search the present solar system for evidence of its past. In this section, you will survey the solar system and compile a list of its most significant characteristics, potential clues to how it formed.

You should begin with the most general view of the solar system. It is, in fact, almost entirely empty space (Figure 1-7). Imagine that you reduce the solar system until Earth is the size of a grain of table salt, about 0.3 mm (0.01 in.) in diameter. The moon is a speck of pepper about 1 cm (0.4 in.) away, and the sun is the size of a small plum 4 m (13 ft) from Earth. Mercury, Venus, and Mars are grains of salt. Jupiter is an apple seed 20 m (66 ft) from the sun, and Saturn is a smaller seed over 36 m (120 ft) away. Uranus and Neptune are slightly larger than average salt grains, and Pluto, at the edge of the solar system, is a speck of pepper over 150 m (500 ft) from the central plum. Although your model solar system would be larger than a football field, you would need a powerful microscope to detect the asteroids. The planets are tiny specks of matter scattered around the sun—the last remains of the solar nebula.

Revolution and Rotation

The planets revolve* around the sun in orbits that lie close to a common plane. The orbit of Mercury, the closest planet to the sun, is tipped 7° to Earth's orbit, and Pluto's orbit is tipped 17.2°. The rest of the planets' orbital planes are inclined by no more than 3.4°. As you can see, the solar system is basically disk shaped.

The rotation of the sun and planets on their axes also seems related to this disk shape. The sun rotates with its equator inclined only 7.25° to Earth's orbit, and most of the other planets' equators are tipped less than 30°. The rotations of Venus, Uranus, and Pluto are peculiar, however. Venus rotates backward compared with the other planets, and both Uranus and Pluto rotate on their sides (with their equators almost perpendicular to their orbits). You will explore these planets in detail in Chapters 22 and 24, but later in this chapter you will be able to understand how they could have acquired their peculiar rotations.

Apparently, the preferred direction of motion in the solar system—counterclockwise as seen from the north—is also related to its disk shape. All the planets revolve counterclockwise around the sun; and, with the exception of Venus, Uranus, and Pluto, they rotate counterclockwise on their axes. Furthermore, nearly all of the moons in the solar system, including Earth's moon, orbit around their planets counterclockwise. With only a few exceptions, most of which are understood, revolution and rotation in the solar system follow a disk theme.

Two Kinds of Planets

Perhaps the most striking clue to the origin of the solar system comes from the division of the planets into two categories: the small Earthlike worlds and the giant Jupiterlike worlds. The difference is so dramatic that it is hard to keep from shouting, "Aha, this must mean something!" Study **Terrestrial and Jovian Planets** on pages 454–455 and notice three important points:

1. Notice how the two kinds of planets are distinguished by their location. The four inner planets are quite different from the next four outward.
2. Also notice how common craters are. Almost every solid surface in the solar system is covered with craters.
3. Finally, notice how the planets are distinguished by individual properties such as rings, clouds, and moons. Any theory of the origin of the planets needs to explain these properties.

The division of the planets into two families is a clue to how our solar system formed, but you can learn little from the

*Recall from Chapter 2 that the words *revolve* and *rotate* refer to different motions. A planet revolves around the sun but rotates on its axis. Cowboys in the old west didn't carry revolvers. They carried rotators.

Terrestrial and Jovian Planets

1 The distinction between the terrestrial planets and the Jovian planets is dramatic. The inner four planets, Mercury, Venus, Earth, and Mars, are **terrestrial planets,** meaning they are small, dense, rocky worlds with little or no atmosphere. The outer four planets, Jupiter, Saturn, Uranus, and Neptune, are **Jovian planets,** meaning they are large, low-density worlds with thick atmospheres and liquid interiors. Pluto does not fit this scheme, being small but low density; you will see in a later chapter that it is a very special world and might not be considered a planet at all.

Mercury

Venus

Earth

Moon

Mars

Jupiter

Saturn

Uranus

Neptune

Pluto

Sun

The planets and the sun to scale. Saturn's rings would just reach from Earth to the moon.

Planetary orbits to scale. The terrestrial planets lie quite close to the sun, whereas the Jovian planets are spread far from the sun outside the asteroid belt. The elliptical shape of Pluto's orbit is visible here.

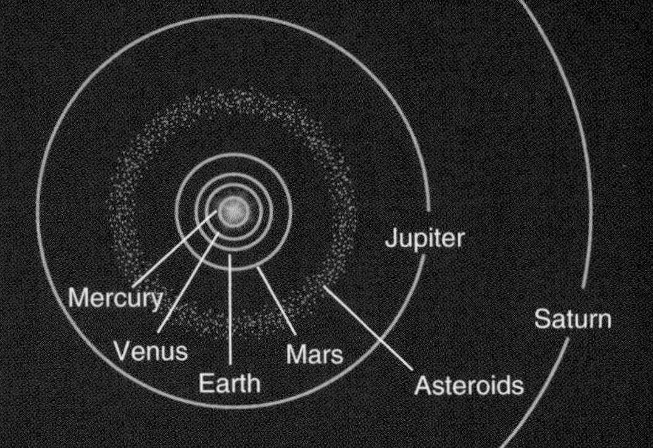

1a Of the terrestrial planets, Earth is most massive, but the Jovian planets are much more massive. Jupiter is over 300 Earth masses, and Saturn is nearly 100 Earth masses. Uranus and Neptune are 15 and 17 Earth masses.

Mercury is only 40 percent larger than Earth's moon, and its weak gravity cannot retain a permanent atmosphere. Like the moon, it is covered with craters from meteorite impacts.

Mercury

Earth's moon

NASA

UCO/Lick Observatory

2 Craters are common on all of the surfaces in the solar system that are strong enough to retain them. Earth has about 150 impact craters, but many more have been erased by erosion. Besides the planets, the asteroids and nearly all of the moons in the solar system are scarred by craters. Ranging from microscopic to hundreds of kilometers in diameter, these craters have been produced over the ages by meteorite impacts. When astronomers see a rocky or icy surface that contains few craters, they know that the surface is young.

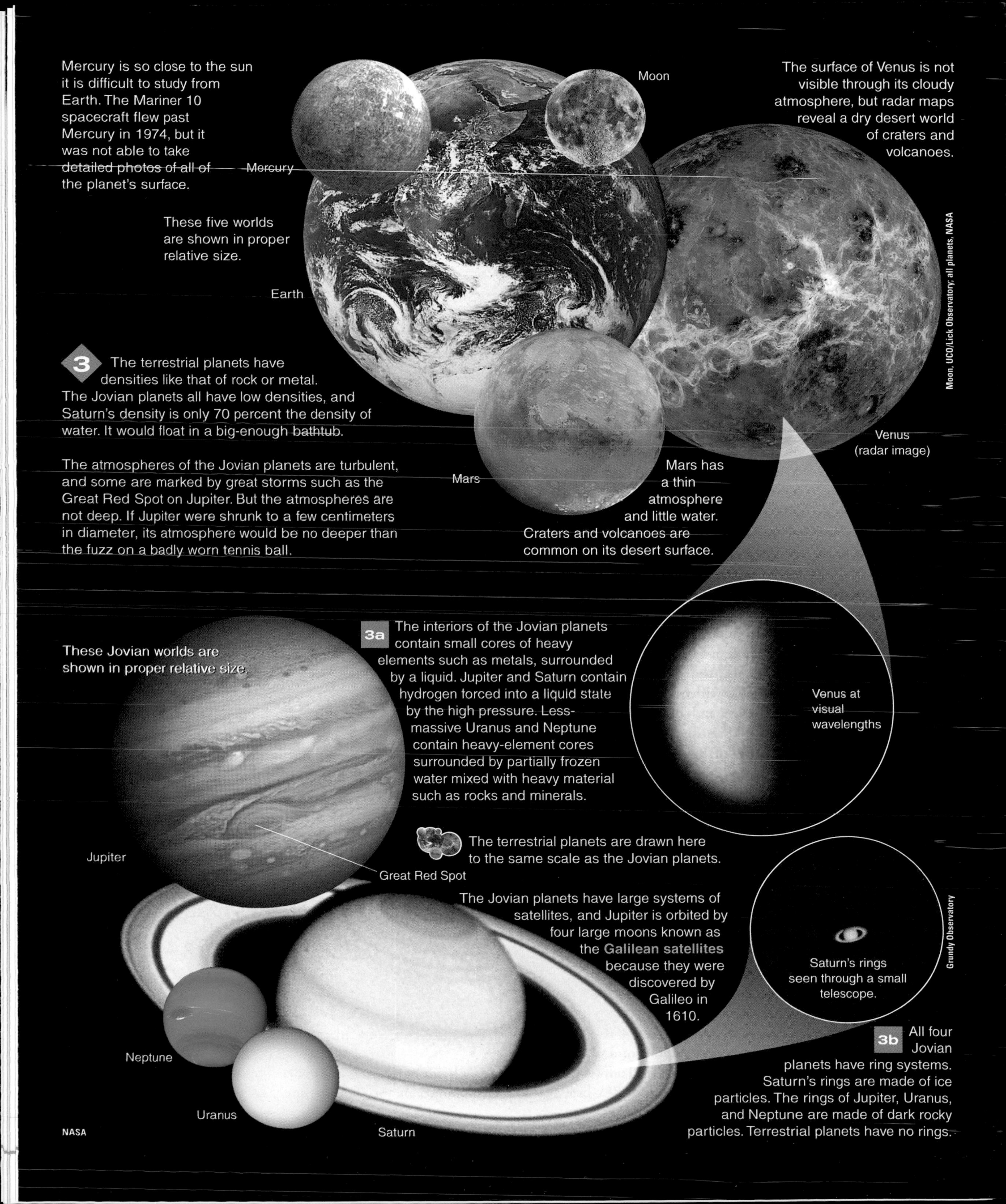
Mercury is so close to the sun it is difficult to study from Earth. The Mariner 10 spacecraft flew past Mercury in 1974, but it was not able to take detailed photos of all of the planet's surface.
Mercury
Moon
The surface of Venus is not visible through its cloudy atmosphere, but radar maps reveal a dry desert world of craters and volcanoes.
These five worlds are shown in proper relative size.
Earth
Moon, UCO/Lick Observatory; all planets, NASA
3 The terrestrial planets have densities like that of rock or metal. The Jovian planets all have low densities, and Saturn's density is only 70 percent the density of water. It would float in a big-enough bathtub.
The atmospheres of the Jovian planets are turbulent, and some are marked by great storms such as the Great Red Spot on Jupiter. But the atmospheres are not deep. If Jupiter were shrunk to a few centimeters in diameter, its atmosphere would be no deeper than the fuzz on a badly worn tennis ball.
Mars
Mars has a thin atmosphere and little water. Craters and volcanoes are common on its desert surface.
Venus (radar image)
These Jovian worlds are shown in proper relative size.
3a The interiors of the Jovian planets contain small cores of heavy elements such as metals, surrounded by a liquid. Jupiter and Saturn contain hydrogen forced into a liquid state by the high pressure. Less-massive Uranus and Neptune contain heavy-element cores surrounded by partially frozen water mixed with heavy material such as rocks and minerals.
Venus at visual wavelengths
The terrestrial planets are drawn here to the same scale as the Jovian planets.
Jupiter
Great Red Spot
The Jovian planets have large systems of satellites, and Jupiter is orbited by four large moons known as the Galilean satellites because they were discovered by Galileo in 1610.
Saturn's rings seen through a small telescope.
Grundy Observatory
3b All four Jovian planets have ring systems. Saturn's rings are made of ice particles. The rings of Jupiter, Uranus, and Neptune are made of dark rocky particles. Terrestrial planets have no rings.
Neptune
Uranus
Saturn
NASA

Window on Science 20-1

Scientific Imagination: Seeing the Unseen

One of the most fascinating aspects of science is its power to reveal the unseen. That is, it gives you an opportunity to learn about regions you can never visit. You saw this in earlier chapters when you studied the inside of the sun and stars, the surface of neutron stars, the event horizon around black holes, the cores of active galaxies, and more. In this chapter, you have seen Earth's core.

Not only does science take you to places you can never visit, but it takes you to scales you can never explore. For example, physicists can explore the inside of an atom, and biologists can study the structure of a virus. Not only can you stretch the scale of space, but you can also stretch the scale of time. Geologists can watch mountain ranges rising slowly from the plains and eroding away to grit. Astronomers can watch the 200-million-year-long whirl of two galaxies colliding.

In this respect, astronomy, geology, and psychology are particularly similar. All three sciences study realms that cannot be explored in person. The insides of stars, Earth's ancient history, and the workings of the human mind lie beyond immediate experience. Only by careful observation and careful thought can scientists learn to understand these aspects of the natural world.

Science can reveal the unseen because it can combine human imagination with the predictive power of the laws of nature and the precision of mathematics. This calls on scientists to be as creative as artists and as precise as surgeons to produce reliable theories and models, but the reward is one of the great thrills of science—exploring beyond the limits of normal human experience.

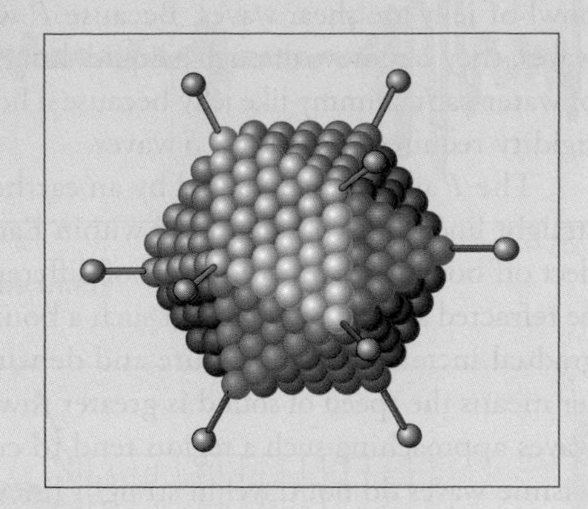

The electron microscope allows biologists to study the elegant structure of the virus that causes the common infection called pinkeye. (From *Virus Ultrastructure: Electron Micrograph Images*. Copyright 1995 by Linda M. Stannard. Reprinted with permission from Linda M. Stannard. Found at: http://www.web.uct.ac.za/depts/mmi/stannard/adeno.html)

up to 60 km thick, and thinnest under the oceans, where it is only about 10 km thick. Unlike the mantle, the crust is brittle and breaks much more easily than the plastic mantle.

Perhaps no region is more immediate yet more inaccessible than Earth's core. Only a few thousand kilometers from where you sit, Earth's interior is beyond your immediate reach. Yet you can send your imagination where your body can never go (**Window on Science 20-1**). Earth's seismic activity reveals some of Earth's innermost secrets. But there is another source of evidence about Earth's interior—its magnetic field.

The Magnetic Field

Apparently, Earth's magnetic field is a direct result of its rapid rotation and its molten metallic core. The internal heat forces the liquid core to churn with convection while Earth's rotation turns the core about an axis. This core is a highly conductive iron–nickel alloy, a better electrical conductor than copper, the material commonly used for electrical wiring. The rotation of this convecting, conducting liquid generates Earth's magnetic field in a process called the dynamo effect (■ Figure 20-7). This is believed to be the same process that generates the solar magnetic field in the convective layers of the sun, and you will meet it again when you explore other planets.

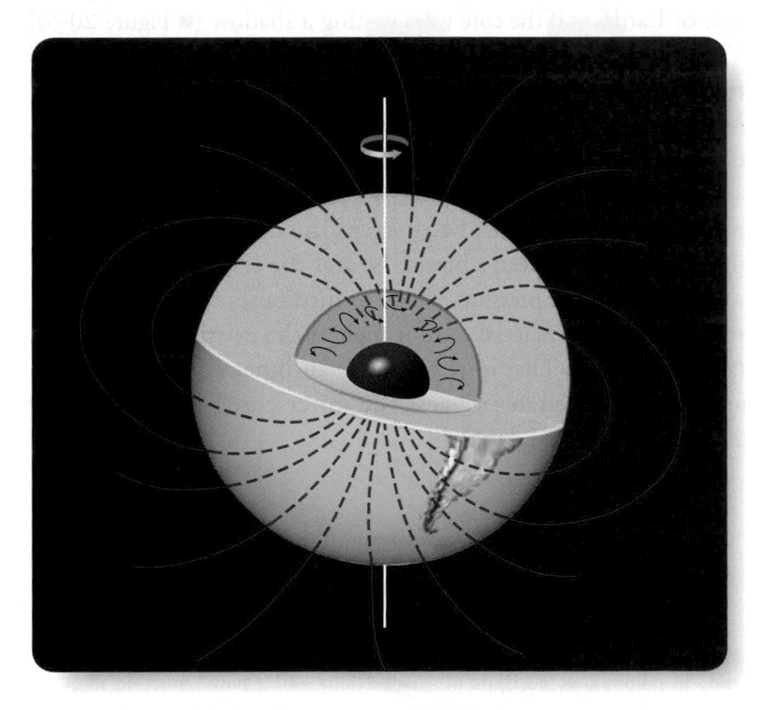

■ Figure 20-7

The dynamo effect couples convection in the liquid core with Earth's rotation to produce electric currents that are believed to be responsible for Earth's magnetic field.

Earth's magnetic field protects it from the solar wind. Blowing outward from the sun at about 400 km/s, the solar wind consists of ionized gases carrying a small part of the sun's magnetic field. When the solar wind encounters Earth's magnetic field, it is deflected like water flowing around a boulder in a stream. The surface where the solar wind is first deflected is called the **bow shock,** and the cavity dominated by Earth's magnetic field is called the **magnetosphere** (■ Figure 20-8a). High-energy particles from the solar wind leak into the magnetosphere and become trapped

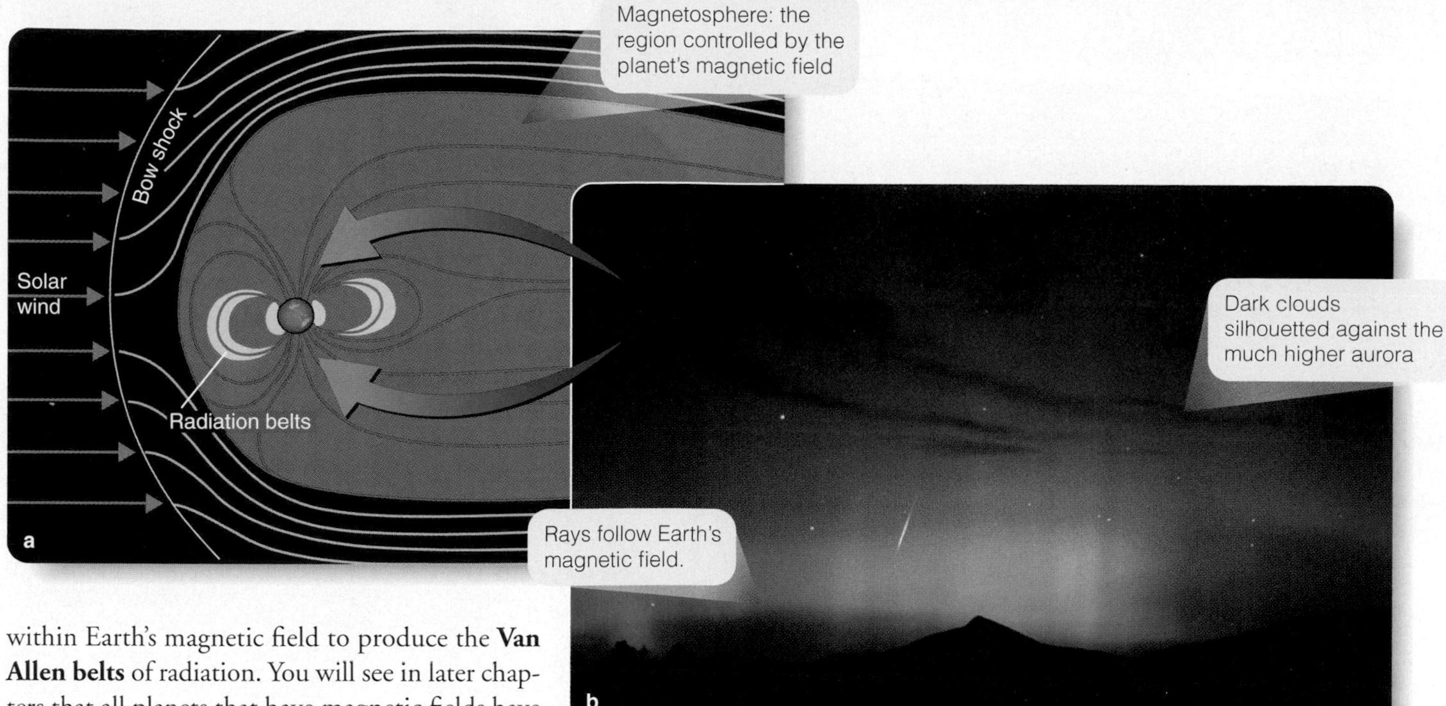

■ **Figure 20-8**

Earth's magnetic field dominates space around Earth by deflecting the solar wind and trapping high-energy particles in radiation belts. Around the north and south magnetic poles, where the magnetic field enters Earth's atmosphere, powerful currents can flow down and excite gas atoms to emit photons, which produces auroras. Colors are produced as different atoms are excited. Note the meteor (shooting star). (Jimmy Westlake)

within Earth's magnetic field to produce the **Van Allen belts** of radiation. You will see in later chapters that all planets that have magnetic fields have bow shocks, magnetospheres, and radiation belts.

One dramatic result of Earth's magnetic field are the auroras, glowing rays and curtains of light in the upper atmosphere (Figure 20-8b). The solar wind carries charged particles past Earth's extended magnetic field, and this generates tremendous electrical currents that flow into Earth's atmosphere near the north and south magnetic poles. The currents ionize gas atoms in Earth's atmosphere, and when the ionized atoms capture electrons and recombine, they emit light as if they were part of a vast "neon" sign. That is why the spectrum of an aurora is an emission spectrum.

Although you can be confident that Earth's magnetic field is generated within its molten core, many mysteries remain. For example, rocks retain traces of the magnetic field in which they solidify, and some contain fields that point backward. That is, they imply that Earth's magnetic field was reversed at the time they solidified. Careful analysis of such rocks indicates that Earth's field has reversed itself every million years or so, with the north magnetic pole becoming the south magnetic pole and vice versa. These reversals are poorly understood, but they may be related to changes in the core convection.

Convection in Earth's core is important because it generates the magnetic field. As you will see in the next section, convection in the mantle constantly remakes Earth's surface.

AceAstronomy™ Log into AceAstronomy and select this chapter to see Astronomy Exercise "Convection and Magnetic Fields." Take control of a planetary interior.

Earth's Active Crust

Earth's crust is composed of lower-density rock that floats on the mantle. The image of a rock floating may seem odd, but recall that the rock of the mantle is very dense. Also, just below the crust, the mantle rock tends to be highly plastic, so great sections of low-density crust do indeed float on the semiliquid mantle like great lily pads floating on a pond.

The motion of the crust and the erosive action of water make Earth's crust highly active. Look at **The Active Earth** on pages 486–487 and notice three important points:

1. The motion of crustal plates produces much of the geological activity on Earth. Earthquakes, volcanism, and mountain building are all linked to motions in the crust and the location of plate boundaries.

2. Notice how the continents on Earth's surface have moved and changed over periods of hundreds of millions of years. A hundred million years is only 0.1 billion years, so sections of Earth's crust are in rapid motion.

3. Notice that most of the geological features you know—mountain ranges, the Grand Canyon, and even the outline of the continents—are recent products of Earth's active surface. Earth's surface is constantly renewed. The oldest rocks on Earth, small crystals called zircons from western Australia, are 4.3 billion years old. Most of the crust is much younger than that. Most of the mountains and valleys you see around you are no more than a few tens of millions of years old.

The Active Earth

1 Our world is an astonishingly active planet. Not only is it rich in water and therefore subject to rapid erosion, but its crust is divided into moving sections called plates. Where plates spread apart, lava wells up to form new crust; where plates push against each other, they crumple the crust to form mountains. Where one plate slides over another, you see volcanism. This process is called **plate tectonics**, referring to the Greek word for "builder." (An architect is literally an arch builder.)

A **rift valley** forms where continental plates begin to pull apart. The Red Sea has formed where Africa has begun to pull away from the Arabian peninsula.

1a Evidence of plate tectonics was first found in ocean floors, where plates spread apart and magma rises to form **midocean rises** made of rock called **basalt**, a rock typical of solidified lava. Radioactive dating shows that the basalt is younger near the midocean rise. Also, the ocean floor carries less sediment near the midocean rise. As Earth's magnetic field reverses back and forth, it is recorded in the magnetic fields frozen into the basalt. This produces a magnetic pattern in the basalt that shows that the seafloor is spreading away from the midocean rise.

A typical view of planet Earth

Mountains are common on Earth, but they erode away rapidly because of the abundant water.

1b A **subduction zone** is a deep trench where one plate slides under another. Melting releases low-density magma that rises to form volcanoes such as those along the northwest coast of North America, including Mt. St. Helens.

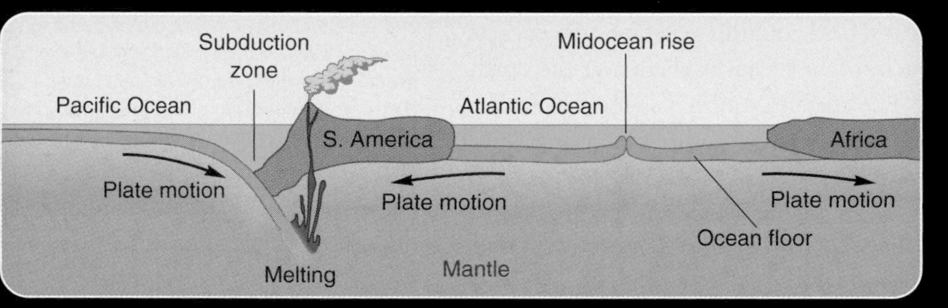

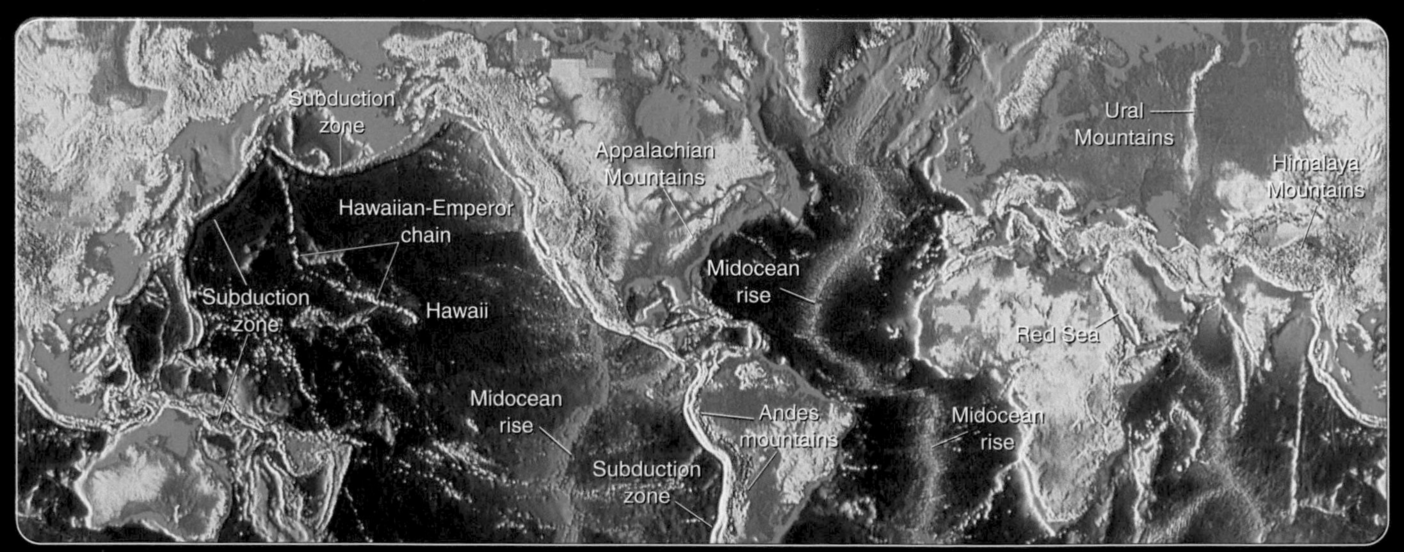

1c Hot spots caused by rising magma in the mantle can poke through a plate and cause volcanism such as that in Hawaii. As the Pacific plate has moved northwestward, the hot spot has punched through to form a chain of volcanic islands, now mostly worn below sea level. **Folded mountain ranges** can form where plates push against each other. For example, the Ural Mountains lie between Europe and Asia, and the Himalaya Mountains are formed by India pushing north into Asia. The Appalachian Mountains are the remains of a mountain range pushed up when North America was pushed against Africa.

Ace Astronomy™ Log into AceAstronomy and select this chapter to see Active Figure "Hot Spot Volcanoes." Notice how the moving plate can produce a chain of volcanic peaks, mostly under water in the case of Earth.

1d The floor of the Pacific Ocean is sliding into subduction zones in many places around its perimeter. This pushes up mountains such as the Andes and triggers earthquakes and active volcanism all around the Pacific in what is called the Ring of Fire. In places such as southern California, the plates slide past each other, causing frequent earthquakes.

Yellow lines on this globe mark plate boundaries. Red dots mark earthquakes since 1980. Earthquakes within the plate, such as those at Hawaii, are related to volcanism over hot spots in the mantle.

Continental Drift

Not long ago, Earth's continents came together to form one continent.

Pangaea

200 million years ago

Pangaea broke into a northern and a southern continent.

Notice India moving north toward Asia.

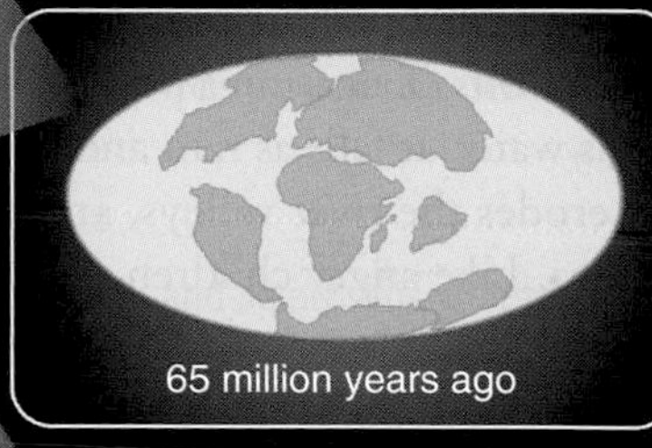

The continents are still drifting on the highly plastic upper mantle.

2 The floor of the Atlantic Ocean is not being subducted. It is locked to the continents and is pushing North and South America away from Europe and Africa at about 3 cm per year, a motion called *continental drift*. Radio astronomers can measure this motion by timing pulsars from European and from American radio telescopes. Roughly 200 million years ago, North and South America were joined to Europe and Africa. Evidence of that lies in similar fossils and similar rocks and minerals found in the matching parts of the continents. Notice how North and South America fit against Europe and Africa like a puzzle.

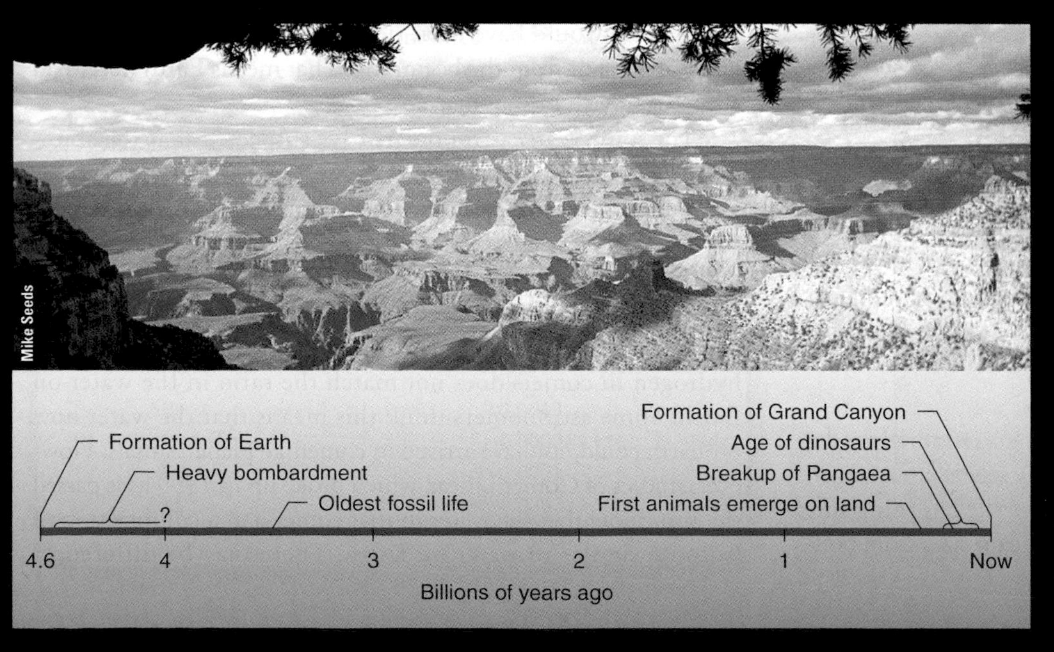

3 Plate tectonics pushes up mountain ranges and causes bulges in the crust, and water erosion wears the rock away. The Colorado River began cutting the Grand Canyon only about 10 million years ago when the Colorado plateau warped upward under the pressure of moving plates. That sounds like a long time ago, but it is only 0.01 billion years. A mile down, at the bottom of the canyon, lie rocks 0.57 billion years old, the roots of an earlier mountain range that stood as high as the Himalayas. It was pushed up, worn away to nothing, and covered with sediment long ago. Many of the geological features we know on Earth have been produced by very recent events.

Impact Cratering

1 The craters that cover the moon and many other bodies in the solar system were produced by the high-speed impact of eorites of all sizes. Meteorites striking the moon travel 10 to 60 km/s and can hit with the energy of many nuclear bombs.

A meteorite striking the moon's surface can deliver tremendous energy and can produce an impact crater 10 or more times larger in diameter than the meteorite. The vertical scale is exaggerated at right for clarity.

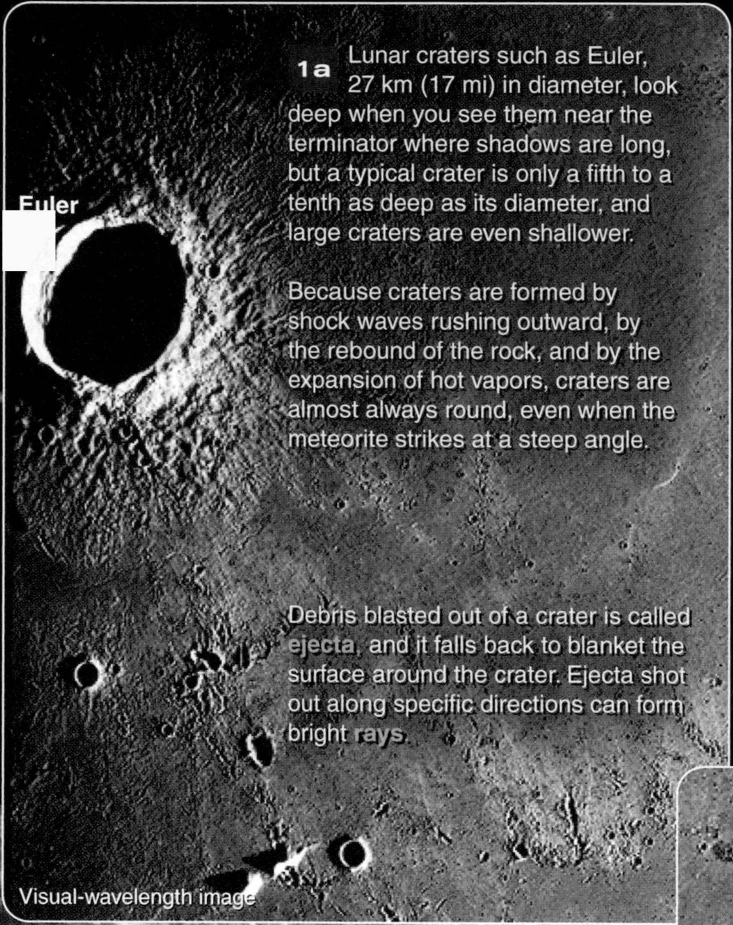

1a Lunar craters such as Euler, 27 km (17 mi) in diameter, look deep when you see them near the terminator where shadows are long, but a typical crater is only a fifth to a tenth as deep as its diameter, and large craters are even shallower.

Because craters are formed by shock waves rushing outward, by the rebound of the rock, and by the expansion of hot vapors, craters are almost always round, even when the meteorite strikes at a steep angle.

Debris blasted out of a crater is called **ejecta**, and it falls back to blanket the surface around the crater. Ejecta shot out along specific directions can form bright **rays**.

Euler

Visual-wavelength image

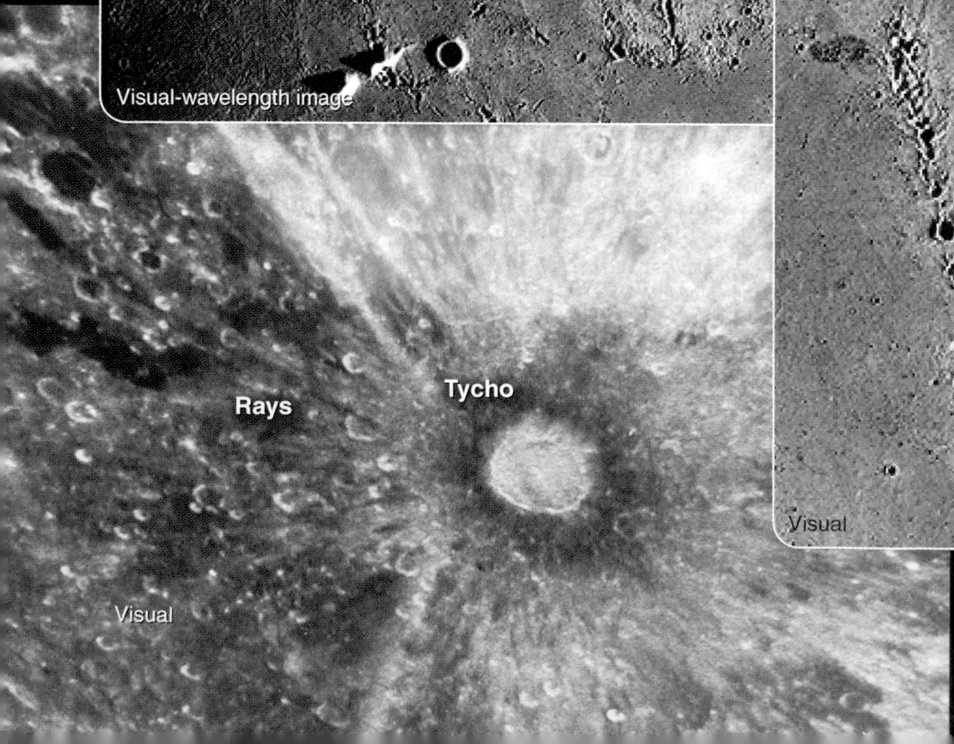

NASA

Impact Cratering

A meteorite approaches the lunar surface at high velocity.

On impact, the meteorite is deformed, heated, and vaporized.

The resulting explosion blasts out a round crater.

Slumping produces terraces in crater walls, and rebound can raise a central peak.

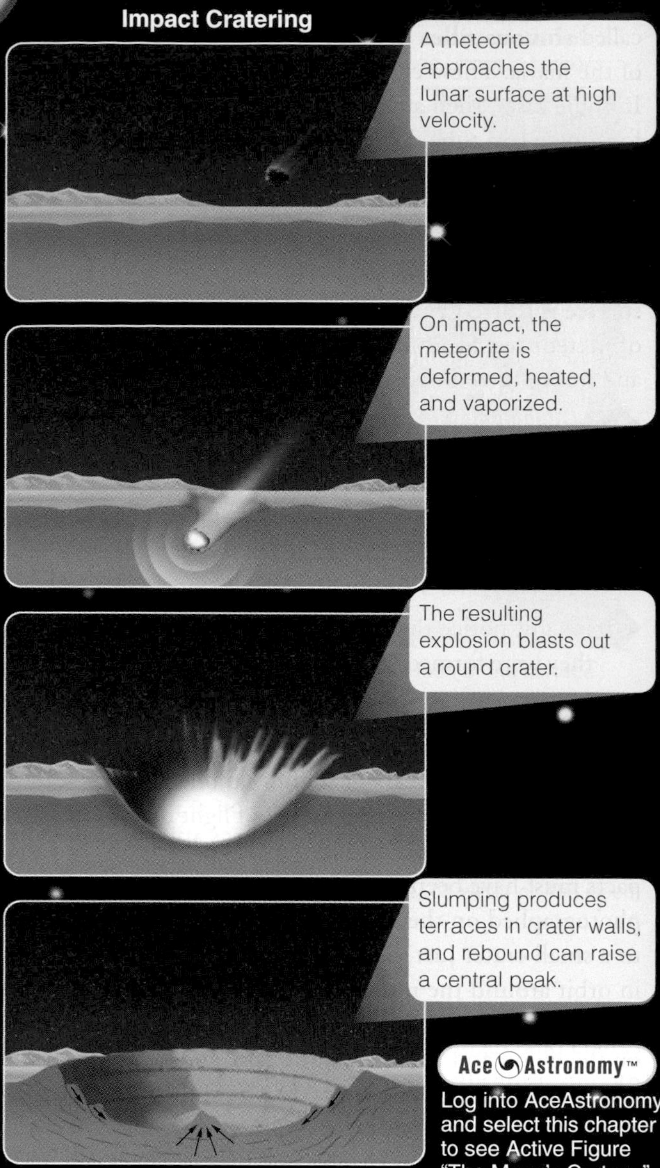

AceAstronomy™
Log into AceAstronomy and select this chapter to see Active Figure "The Moon's Craters." Notice that the structure of the craters depends on their size.

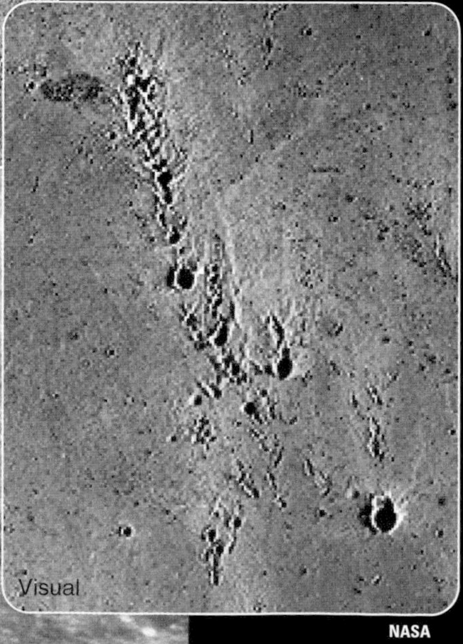

1b Rock ejected from distant impacts can fall back to the surface and form smaller craters called **secondary craters.** The chain of craters here is a 45-km-long chain of secondary craters produced by ejecta from the large crater Copernicus 200 km out of the frame to the lower right.

Bright ejecta blankets and rays gradually darken as sunlight darkens minerals and small meteorites stir the dusty surface. Bright rays are signs of youth. Rays from the crater Tycho, perhaps only 100 million years old, extend halfway around the moon.

Visual

NASA

Rays

Tycho

Visual

NASA

2 Plum Crater, 40 m (130 ft) in diameter, was visited by Apollo 16 astronauts. Note the many smaller craters visible. Lunar craters range from giant impact basins to tiny pits in rocks struck by **micrometeorites,** meteorites of microscopic size.

Visual-wavelength images

Mare Orientale

Solidified lava

NASA

2a In larger craters, the deformation of the rock can form one or more inner rings concentric with the outer rim. The largest of these craters are called **multiringed basins.** In Mare Orientale on the west edge of the visible moon, the outermost ring is almost 900 km (550 mi) in diameter.

2b The energy of an impact can melt rock, some of which falls back into the crater and solidifies. When the moon was young, craters could also be flooded by lava welling up from below the crust.

A few meteorites found on Earth have been identified chemically as fragments of the moon's surface blasted into space by cratering impacts. The fragmented nature of these meteorites indicates that the moon's surface has been battered by impact craters.

Meteorite from moon

3 Most of the craters on the moon were produced long ago when the solar system was filled with debris from planet building. As that debris was swept up, the cratering rate fell rapidly, as shown below.

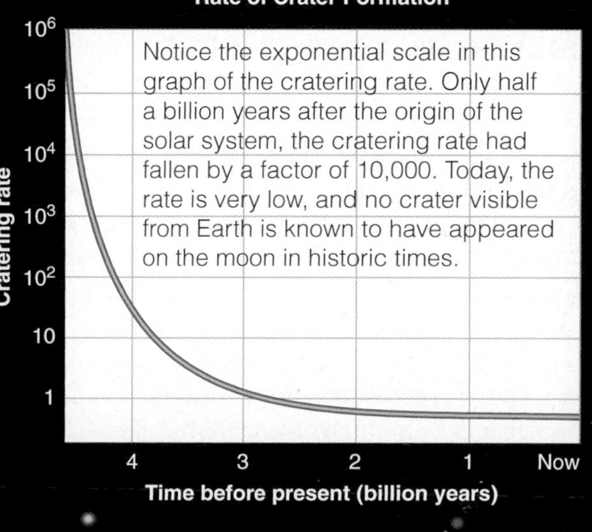

Notice the exponential scale in this graph of the cratering rate. Only half a billion years after the origin of the solar system, the cratering rate had fallen by a factor of 10,000. Today, the rate is very low, and no crater visible from Earth is known to have appeared on the moon in historic times.

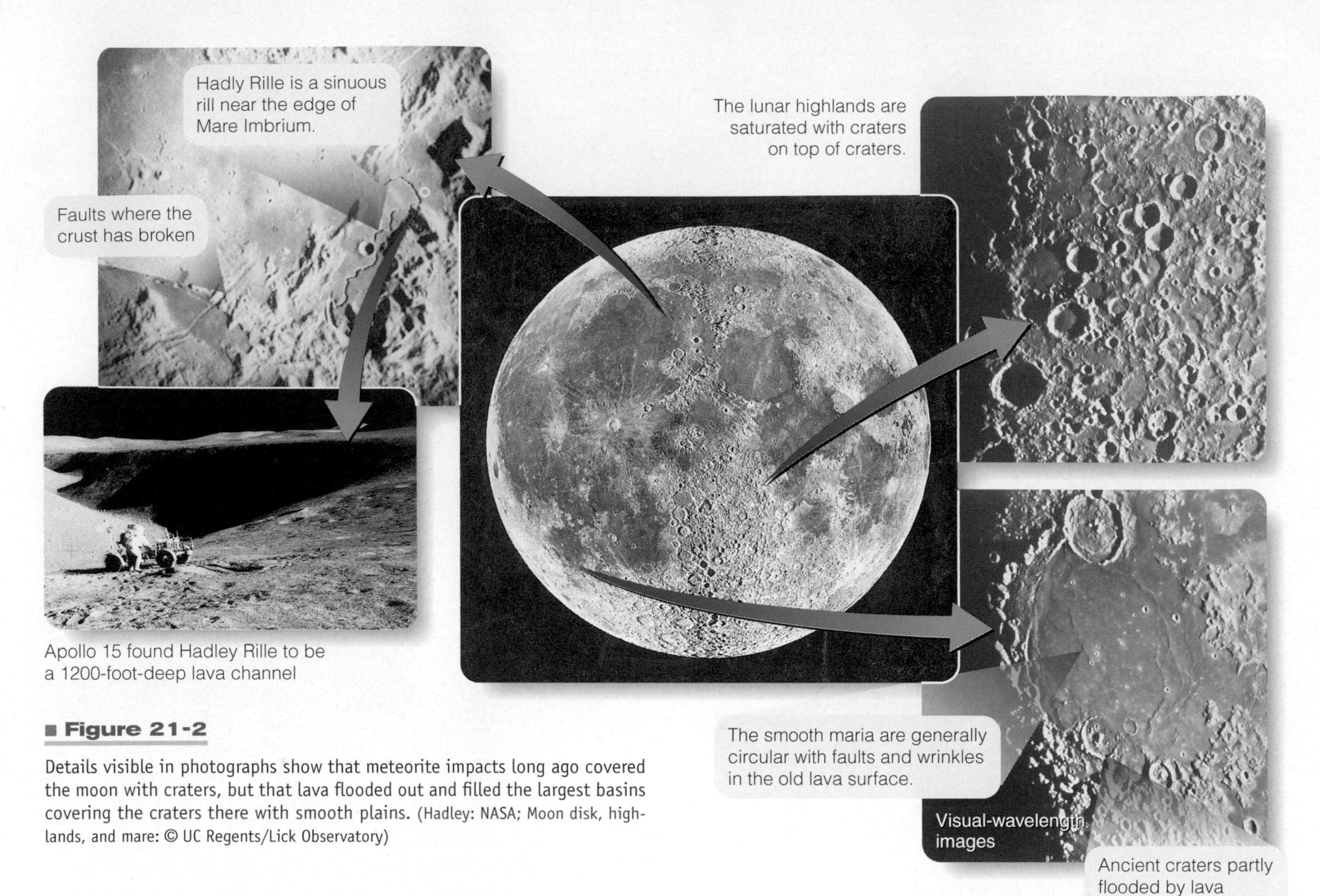

Hadly Rille is a sinuous rill near the edge of Mare Imbrium.
Faults where the crust has broken
Apollo 15 found Hadley Rille to be a 1200-foot-deep lava channel
The lunar highlands are saturated with craters on top of craters.
The smooth maria are generally circular with faults and wrinkles in the old lava surface.
Visual-wavelength images
Ancient craters partly flooded by lava

■ Figure 21-2

Details visible in photographs show that meteorite impacts long ago covered the moon with craters, but that lava flooded out and filled the largest basins covering the craters there with smooth plains. (Hadley: NASA; Moon disk, highlands, and mare: © UC Regents/Lick Observatory)

The maria fill the lunar lowlands, and they contain few craters, as you can see in Figure 21-2. Among the few craters that do scar the maria, many have bright rays, suggesting that they are younger craters. For examples, locate the craters Kepler and Copernicus in Figure 21-1.

You can conclude from this that the highlands are old regions where you see craters from early in the history of the solar system. Most of those craters were formed during the heavy bombardment over 4 billion years ago. The maria that fill the lowlands must have formed later, after the decline of the heavy bombardment. That means the maria are younger than the highlands. These are called **relative ages** because you can tell which features formed first, but you can't give ages in years.

The cratering curve shown on page 501 can reveal **absolute ages,** meaning ages in years. Studies of cratered surfaces in the solar system have allowed planetary scientists to calibrate the cratering curve. The lava flows that filled the lowlands and covered up the craters there could not have formed earlier than 3 to 4 billion years ago. Had they occurred earlier, the surfaces of the maria would show more craters.

Absolute ages derived from the cratering curve cannot be precise because planetary scientists suspect that the cratering rate may not have fallen as smoothly as the cratering curve suggests. Although the debris in the early solar system was gradually removed, bursts of cratering may have occurred. A major impact between large planetesimals, for example, might shower the solar system with fragments and produce a storm of crater impacts. Mathematical models suggest that the final formation of Uranus and Neptune might have flung icy planetesimals through the solar system like giant shrapnel. The true cratering curve might have rises and dips superimposed on a gradual decline.

Whatever the exact timing of the heavy bombardment, it is clear that the craters in the highlands are old and that the later lava flows that formed the maria covered over any craters that had been formed in the lowlands. Using such clues, astronomers were able to begin telling the story of the moon's surface (**Window on Science 21-1**). But the view from Earth does not provide enough evidence. To really understand the lunar surface, humans had to go there.

The Apollo Missions

On May 25, 1961, President John Kennedy addressed Congress and committed the United States to landing a human being on the moon by 1970. The reasons for that decision are complicated

Window on Science 21-1

How Hypotheses and Theories Unify the Details

Like any technical subject, science includes a mass of details, facts, figures, measurements, and observations. It is easy to be overwhelmed by the flood of details, but one of the most important characteristics of science comes to your rescue. The goal of science is not to discover more details but to explain the details with a unifying hypothesis or theory. A good theory is like a basket that makes it easier for you to carry a large assortment of details.

This is true of all the sciences. When a psychologist begins studying the way the human eye and brain respond to moving points of light, the data are a sea of detailed measurements and observations. Once the psychologist forms a hypothesis about the way the eye and brain interact, the details fall into place as parts of a logical story. If you understand the hypothesis, the details all fit together and make sense, and you can remember the details without blindly memorizing tables of facts and figures. Remember that the goal of science is understanding nature, not memorizing details.

Scientists are in the storytelling business. The stories are often called hypotheses or theories, but they are really just stories to explain how nature works. The difference between scientific stories and works of fiction lies in the use of facts. Scientific stories are constructed to fit all known facts and are then tested over and over against new facts obtained by observation and experiment.

When you try to tell the story of each planet in our solar system, you pull together all the hypotheses, theories, and observations and try to make them into a logical history of how the planet got to be the way it is. Of course, your stories will be incomplete because astronomers don't understand all the factors in planetary astronomy. Nevertheless, your story of each planet will draw together the known facts and details and attempt to make them a logical whole.

When scientists create a hypothesis, it draws together a great many observations and measurements. (Phyllis Leber)

Memorizing a list of facts can give you a false feeling of security, just as when someone memorizes the names of things without understanding them (Window on Science 2-2). Rather than memorizing facts, you should search for the unifying hypothesis that pulls the details together into a single story. That is the goal of science.

and related more to economics, international politics, and the stimulation of technology than to science, but the Apollo program became a fantastic adventure in science, a flight to the moon that changed how we all think about Earth.

Flying to the moon is not particularly difficult; with powerful enough rockets and enough food, water, and air, it is a straightforward trip. Landing on the moon is more difficult but not impossible. The moon's gravity is only one-sixth that of Earth, and there is no atmosphere to disturb the trajectory of the spaceship. Getting to the moon is simple and landing is possible; the difficulty is doing both on one trip. The spaceship must carry food, water, and air for a number of days in space plus fuel and rockets for midcourse corrections and for a return to Earth. All of this adds up to a ship that is too massive to make a safe landing on the lunar surface. The solution was to take two spaceships to the moon, one to ride in and one to land in (■ Figure 21-3).

The command module was the long-term home and command center for the trip. Three astronauts had to live in it for a week, and it had to carry all the life-support equipment, navigation instruments, computers, power packs, and so on for a week's jaunt in space. The lunar landing module (LM for short) was tacked to the front of the command module like a bicycle strapped to the front of the family camper. It carried only enough fuel and supplies for the short trip to the lunar surface, and it was built to minimize weight and maximize maneuverability.

The weaker gravity of the moon made the design of the LM simpler. Landing on Earth requires reclining couches for the astronauts, but the trip to the lunar surface involved smaller accelerations. In an early version of the LM, the astronauts sat on what looked like bicycle seats, but these were later scrapped to save weight. The astronauts had no seats at all in the LM, and once they began their descent and acquired weight, they stood at the controls held by straps, riding the LM like two daredevils riding a rocket surfboard.

Lifting off from the lunar surface, the LM saved weight by leaving the larger descent rocket and support stage behind. Only the astronauts, their instruments, and their cargo of rocks returned to the command module orbiting above. Again, the astronauts in the LM blasted up from the lunar surface standing at the controls. The rocket engine that lifted them back into lunar orbit was not much bigger than a dishwasher.

The most complicated parts of the trip were the rendezvous and docking between the tiny remains of the LM and the command module, aided by radar systems and computers. The astronauts rejoined their colleague in the command module, transferred their moon rocks, and jettisoned the LM. Only the command module returned to Earth.

The first manned lunar landing was made on July 20, 1969. While Michael Collins waited in orbit around the moon, Neil Armstrong and Edwin Aldrin, Jr., took the LM down to the

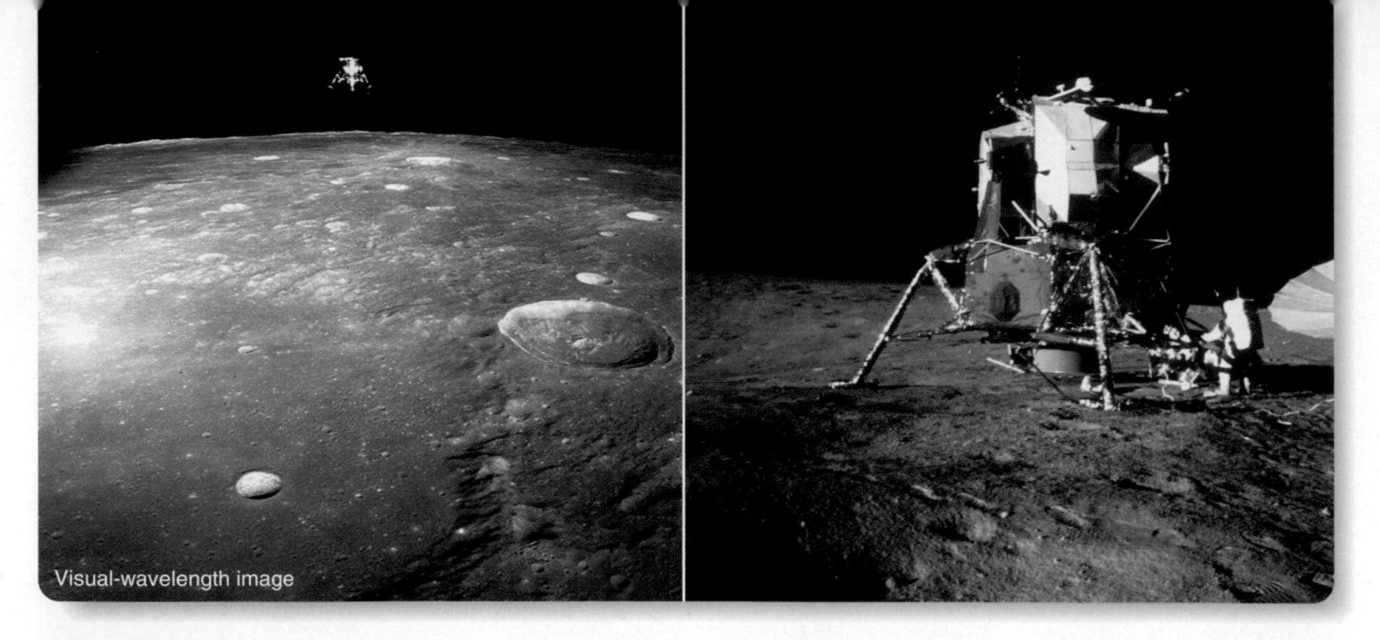

Figure 21-3

The Apollo 12 lunar module. The two astronauts stood in a space hardly bigger than two telephone booths. The metal skin was so thin it was easily flexible, like metal foil, and the legs of the module, designed specifically for the moon's weak gravity, could not support the lander's weight on Earth. Only the upper half of the lander blasted off from the surface to return the astronauts to the command module in orbit around the moon. (NASA)

surface. Although much of the descent was controlled by computers, the astronauts had to override a number of computer alarms and take control of the LM to avoid a boulder-strewn crater bigger than a football field. Climbing down the ladder, Armstrong and then Aldrin stepped onto the lunar surface.

Between July 1969 and December 1972, 12 people reached the lunar surface and collected 380 kg (840 lb) of rocks and soil (■ Table 21-1). The flights were carefully planned to visit different regions and develop a comprehensive history of the lunar surface.

Table 21-1 | Apollo Lunar Landings

Apollo Mission*	Astronauts: Commander LM Pilot CM Pilot	Date	Mission Goals	Sample Weight (kg)	Typical Samples	Ages (10^9 y)
11	Armstrong Aldrin Collins	July 1969	First manned landing; Mare Tranquillitatis	21.7	Mare basalts	3.48–3.72
12	Conrad Bean Gordon	Nov. 1969	Visit Surveyor 3; sample Oceanus Procellarum (mare)	34.4	Mare basalts	3.15–3.37
14	Shepard Mitchell Roosa	Feb. 1971	Fra Mauro, Imbrium ejecta sheet	42.9	Breccia	3.85–3.96
15	Scott Irwin Worden	July 1971	Edge of Mare Imbrium and Apennine Mountains, Hadley Rille	76.8	Mare basalts; highland anorthosite	3.28–3.44 4.09
16	Young Duke Mattingley	April 1972	Sample highland crust; Cayley formation (ejecta); Descartes	94.7	Highland basalt; breccia	3.84 3.92
17	Cernan Schmitt Evans	Dec. 1972	Sample highland crust; dark halo craters; Taurus–Littrow	110.5	Mare basalt; highland breccia; fractured dunite	3.77 3.86 4.48

*The Apollo 13 mission suffered an explosion on the way to the moon and did not land.

Apollo 17, the last Apollo mission to the moon, landed in the highlands in December 1972.

Apollo 11 landed in the lunar lowlands in July 1969.

■ Figure 21-4

Apollo 11, the first mission to the moon, landed on the smooth surface of Mare Tranquillitatis in the lunar lowlands and the horizon was straight and level. When Apollo 17 landed at Taurus–Littrow in the lunar highlands, the astronauts found the horizon mountainous and the terrain rugged. Landing sites for the other Apollo missions are shown. (Moon: © UC Regents/Lick Observatory; Astronauts: NASA)

The first flights went into relatively safe landing sites (■ Figure 21-4)—Marc Tranquillitatis for Apollo 11 and Oceanus Procellarum for Apollo 12. Apollo 13 was aimed at a more complicated site, but an explosion in an oxygen tank on the way to the moon ended all chances of a landing and nearly cost the astronauts their lives. They succeeded in using the life support in the LM to survive the trip around the moon, and they eventually landed safely in the crippled command module.

The last four Apollo missions, 14 through 17, sampled geologically important places on the moon. Apollo 14 visited the Fra Mauro region, which is covered by ejecta from the impact that dug the basin now filled by Mare Imbrium. Apollo 15 visited the edge of Mare Imbrium at the foot of the Apennine Mountains and examined Hadley Rille (Figure 21-2). Apollo missions 16 and 17 visited highland regions to sample the older parts of the lunar crust (Figure 21-4). Almost all of the lunar samples from these six landings are now held at the Planetary Materials Laboratory at the Johnson Space Center in Houston. They are a national treasure containing clues to the beginnings of our solar system.

Moon Rocks

If you visited the moon, where the gravity is one-sixth that on Earth, you would discover that you weighed only one-sixth what you weigh on Earth. The Apollo astronauts found it easy to dance across the lunar surface conducting experiments and collecting geological specimens.

Of all of the rock samples that the Apollo astronauts carried back to Earth, every one is igneous. That is, they formed by the cooling and solidification of molten rock. No sedimentary rocks were found, which is consistent with the moon never having had liquid water on its surface. In addition, the rocks were extremely dry. Almost all Earth rocks contain 1 to 2 percent water, either as free water trapped in the rock or as water molecules chemically bonded with certain minerals. But moon rocks contain no water at all.

Rocks from the lunar maria are dark-colored, dense basalts much like the solidified lava produced by the Hawaiian volcanoes (■ Figure 21-5). These rocks are rich in heavy elements such as iron, manganese, and titanium, which give them their dark color. Some of the basalts are **vesicular,** meaning that they contain holes caused by bubbles of gas in the molten rock. Like bubbles in a carbonated beverage, these bubbles do not form while the magma is under pressure. Only when the molten rock flows out onto the surface, where the pressure is low, do bubbles appear. The presence of vesicular basalts assures that these rocks

The Apollo astronauts found that all moon rocks are igneous, meaning they solidified from molten rock.
Rocks exposed on the lunar surface become pitted by micrometeorites.
Vesicular basalt contains bubbles frozen into the rock when it was molten.
A breccia is formed by rock fragments bonded together by heat and pressure.

Figure 21-5

Rocks returned from the moon show that the moon formed in a molten state, that it was heavily fractured by cratering when it was young, and that it is now affected mainly by micrometeorites grinding away at surface rock. (NASA)

formed in lava flows that reached the surface and did not solidify underground.

Absolute ages of the mare basalts can be found from the radioactive atoms they contain (Chapter 19), and these ages range from about 3.1 to 3.8 billion years. That means the lava flows that formed these rocks must have happened some time after the end of the heavy bombardment.

The highlands are composed of low-density rock containing calcium-, aluminum- and oxygen-rich minerals that would be among the first to solidify and float to the top of molten rock. The crustal rocks range in age from 4.0 to 4.5 billion years old, significantly older than the mare basalts. The highlands contain **anorthosite,** a light-colored rock that contributes to the highlands' bright contrast with the dark, iron-rich basalts of the lowlands. The rocks of the highlands, although badly shattered by impacts, represent the moon's original low-density crust, whereas the mare basalts rose as molten rock from the deep crust and upper mantle.

A large fraction of the lunar rocks are **breccias,** rocks that are made up of fragments of earlier rocks cemented together by heat and pressure. Evidently, meteorite impacts have broken up the rocks and fused them together time after time.

If you went to the moon, you would get your spacesuit dirty. Both the highlands and the lowlands of the moon are covered by a layer of powdered rock and crushed fragments called the **regolith.** It is about 10 m deep on the maria but over 100 m deep in certain places in the highlands. About 1 percent of the regolith is meteoric fragments; the rest is the smashed remains of moon rocks that have been pulverized by the constant rain of meteorites. The smallest meteorites, the micrometeorites, do the most damage by their constant sandblasting of the lunar surface, which grinds the rock down to fine dust. The Apollo astronauts found that the dust coated their spacesuits and equipment and made a mess where it got tracked into the Lunar Landing Module. From all this you can see that impact cratering, a theme of this chapter, dominates the lunar surface and is even responsible for the lunar regolith.

Window on Science 21-2

Understanding Natural Processes: Follow the Energy

One of the best ways to think about a scientific problem is to follow the energy.

According to the principle of cause and effect, every effect must have a cause, and every cause must involve energy (Focus on Fundamentals 2). Energy moves from regions of high concentration to regions of low concentration and in doing so produces changes. For example, coal burns to make steam in a power plant, and the steam passes through a turbine and then escapes into the air. In flowing from the burning coal to the atmosphere, the heat spins the turbine and makes electricity.

It is surprising how commonly scientists use energy as a key to understanding nature. A biologist might ask where certain birds get the energy to fly thousands of miles, and an economist might ask where the economic energy to support the creation of new investments comes from. Energy is everywhere; and when it moves, whether it is birds, money, or molten magma, it causes change. Energy is the cause in cause and effect.

In earlier chapters, the flow of energy from the inside of a star to its surface helped you understand how stars, including the sun, work. You saw that the outward flow of energy supports the star against its own weight, drives convection currents that produce magnetic fields, and causes surface activity such as spots, prominences, and flares. You were able to understand stars because you could follow the flow of energy outward from their interiors.

You can think of a planet by following the energy. The heat in the interior of a planet may be left over from the formation of the planet, or it may be heat generated by radioactive decay, but it must flow outward toward the cooler surface, where it is radiated into space. In flowing outward, the heat can cause convection currents in the mantle, magnetic fields, plate motions, quakes, faults, volcanism, mountain building, and more.

When you think about any world, be it a small asteroid or a giant planet, think of it as a source of heat that flows outward through the planet's surface into space. If you can follow that energy flow, you can understand a great deal about the world. A planetary astronomer once said, "The most interesting thing about any planet is how its heat gets out."

Heat flows out of Earth's interior and generates geological activity such as that at Yellowstone National Park. (M. Seeds)

The moon rocks are old, dry, igneous, and badly shattered by impacts. You can use these facts, combined with what you know about lunar features, to tell the story of the moon.

The History of the Moon

The four-stage history of the moon is dominated by a single fact—it is small, only one-fourth the diameter of Earth. The escape velocity is low, so it has been unable to hold any atmosphere, and it cooled rapidly as its internal heat flowed outward into space. As you will see when you visit other worlds, a planet's geology is largely driven by energy in the form of heat flowing outward from its interior (**Window on Science 21-2**). Small worlds have less heat and lose it more rapidly, so the moon's small size has been critical in its history.

The Apollo moon rocks show that the moon must have formed in a molten state. Planetary geologists now refer to the newborn moon as a sea of magma. Denser materials sank to form a small core; and, as the magma ocean cooled, low-density minerals floated to the top to form a low-density crust. In this way the moon must have differentiated into core, mantle, and crust. The radioactive ages of the moon rocks show that the surface solidified from 4.6 to 4.1 billion years ago. The moon has a low density and no magnetic field, so its dense core must be small. The core may still retain enough heat to be partially molten, but it can't contain much iron or the dynamo effect would produce a magnetic field.

The second stage, cratering, began as soon as the crust solidified, and the older highlands show that the cratering was intense during the first 0.5 billion years—during the heavy bombardment at the end of planet building.

The moon's crust was shattered to a depth of 10 kilometers or so, and the largest impacts formed giant multiringed crater basins hundreds of kilometers in diameter, such as Mare Orientale. This led to the third stage—flooding. Although Earth's moon cooled rapidly after its formation, some process such as radioactive decay heated material deep in the crust, and part of it melted. Molten rock followed the cracks up to the surface and flooded the giant basins with successive lava flows of dark basalts from 3.8 to 3.2 billion years ago. This formed the maria (■ Figure 21-6).

Make a special note that the lava that floods out on the surfaces of planets does not come from the molten core. That is a common misconception. The lava comes from the lower crust and possibly from the top of the mantle. The pressure is lower there, and that lowers the melting point of the rock. Radioactive decay can generate enough heat to melt portions of the rock, and if there is enough heat and if there are faults and cracks, the magma can reach the surface and form volcanoes and lava flows. Whenever you see lava flows on a planet, you can be sure heat is flowing

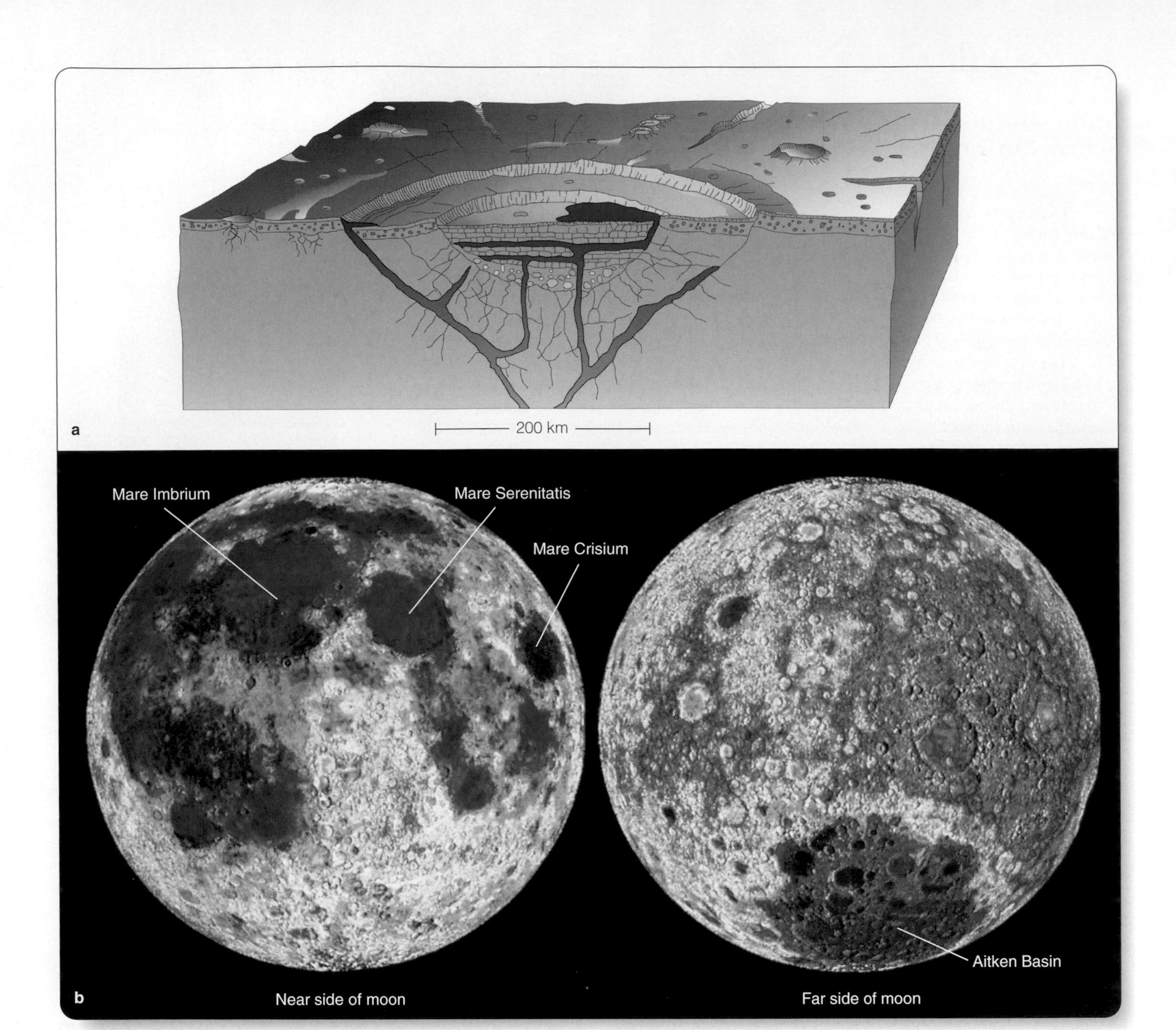

■ Figure 21-6

(a) Major impacts broke the lunar crust and produced large basins. Lava flooded up through the fractures and flooded the basins to produce maria. (Adapted from a diagram by William Hartmann) (b) In these maps, color indicates elevation, with red the highest regions and purple the lowest. Maria on the near side are generally circular, low, lava-filled basins. Note the circular outline of the labeled maria. The crust on the far side is thicker, and there is less flooding. Even the Aitken Basin contains little lava flooding. (NASA/Clementine)

out of the interior (one of the themes of this chapter), but the lava itself did not come all the way from the core.

Some maria on the moon, such as Mare Imbrium, Mare Serenitatis, Mare Humorum, and Mare Crisium, retain their round shapes, but others are irregular because the lava overflowed the edges of the basin or because the shape of the basin was modified by further cratering. The moving lava formed channels that are seen from Earth as sinuous rills. The weight of the maria pressed the crater basins downward, and the solidified lava was compressed and formed wrinkle ridges visible even in small telescopes. The tension at the edges of the maria broke the hard lava to produce straight fractures and faults. All of these features are visible in

■ **Figure 21-7** ▶

Apollo 14 landed on rugged terrain suspected of being ejecta from the Imbrium impact. The large boulder here is ejecta that, at some time in the past, fell here from an impact beyond the horizon. (NASA)

■ **Figure 21-8**

Lava flooding after the end of the heavy bombardment filled a giant multiringed basin and formed Mare Imbrium. (Don Davis)

Origin of Mare Imbrium

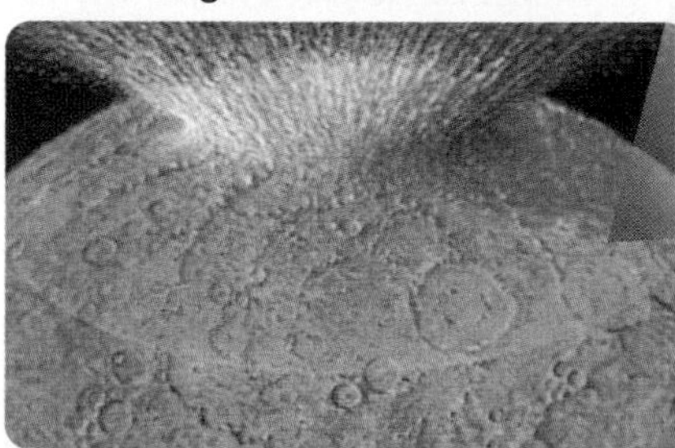

Four billion years ago, an impact forms a multiringed basin over 1000 km (700 mi) in diameter.

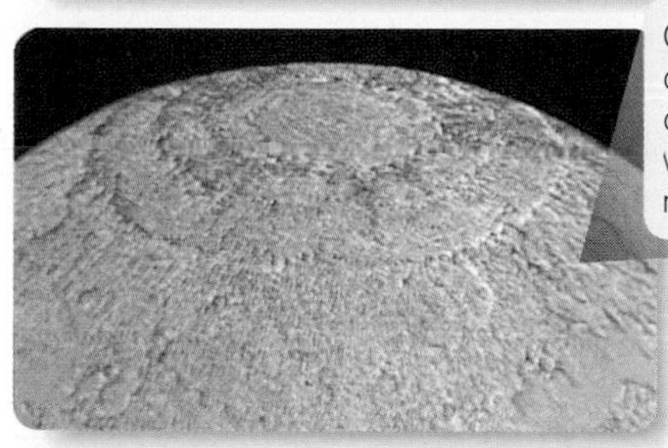

Continuing impacts crater the surface but do not erase the high walls of the multiringed basin.

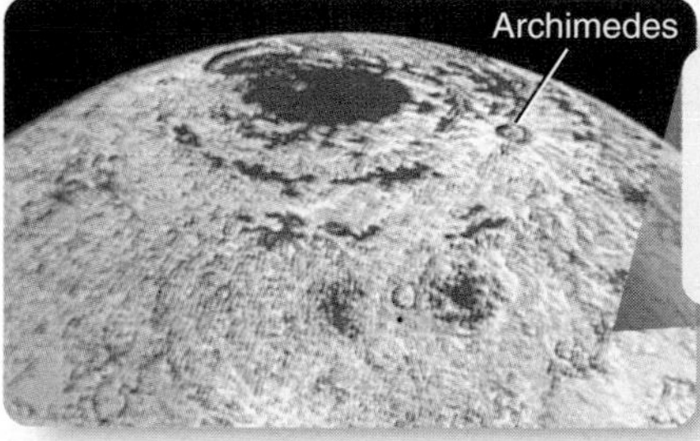

Archimedes
Impacts form a few large craters, and, starting about 3.8 billion years ago, lava floods low regions.

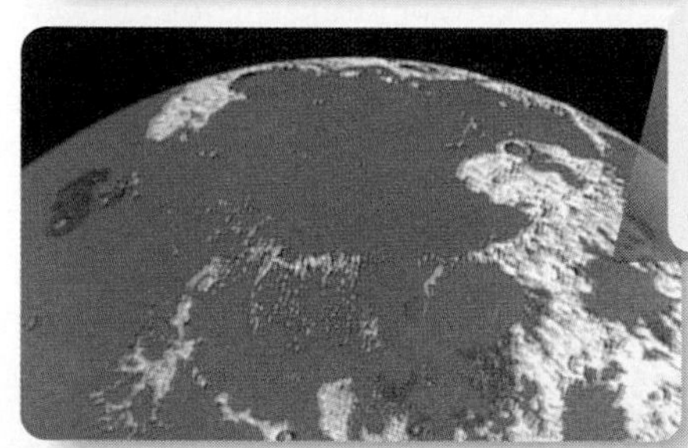

Repeated lava flows cover most of the inner rings and overflow the basin to merge with other flows.

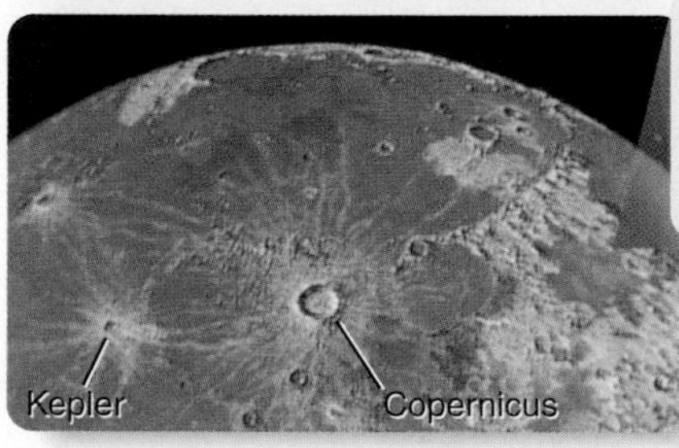

Impacts continue, including those that formed the relatively young craters Copernicus and Kepler.
Kepler
Copernicus

Figure 21-2. Later cratering and overlapping lava floods further modified the maria. Consequently you should think of the maria as accumulations of features reflecting the moon's complex history.

Mare Imbrium is a dramatic example of how the great basins became the maria. Near the end of the heavy bombardment, roughly 4 billion years ago, a planetesimal the size of Rhode Island struck the moon and blasted out a giant multiringed basin. The impact was so violent the ejecta blanketed 16 percent of the moon's surface. Apollo 14 landed at Fra Mauro to sample the ejecta (■ Figure 21-7). After the cratering rate fell at the end of the heavy bombardment, lava flows welled up time after time and flooded the Imbrium Basin, burying all but the highest parts of the giant multiringed basin. The Imbrium Basin is now a large, generally round mare marked by only a few craters that have formed since the last of the lava flows (■ Figure 21-8).

This story of the moon might suggest that it was a violent place during the cratering phase, but large impacts were in fact rare; and, except for a continuous rain of smaller meteorites, the moon was, for the most part, a peaceful place even during the heavy bombardment. Had you stood on the moon at that time you would have experienced a continuous rain of micrometeorites and much less common pebble-size impacts.

Centuries might pass between major impacts. Of course, a large impact far beyond the horizon might have threatened to bury you under ejecta or jolted you by seismic shocks. You could have felt the Imbrium impact anywhere on the moon, but had you been standing on the side of the moon opposite that impact, you would have been at the focus of seismic waves traveling around the moon from different directions. When the waves met under your feet, the surface would have jerked up and down by as much as 10 m. You would not have liked that. The place on the moon opposite the Imbrium Basin is a strangely disturbed landscape called **jumbled terrain.** You will see similar effects of large impacts on other worlds.

Studies of our moon show that its crust is thinner on the side toward Earth, perhaps due to tidal effects. Consequently, while lava flooded the basins on the Earthward side, it was unable to rise through the thicker crust to flood the lowlands on the far side. The largest impact basin in the solar system is the south pole Aitkin Basin on the moon (Figure 21-6b). It is about 2500 kilometers (1500 miles) in diameter and as deep as 13 kilometers (8 miles) in places, but flooding has never filled it with smooth lava flows.

The fourth stage, slow surface evolution, is limited because our moon lacks water and has cooled rapidly. Flooding on Earth included water, but the moon has never had an atmosphere and thus has never had liquid water. With no air and no water, erosion is limited to the constant bombardment of micrometeorites and rare larger impacts.

The moon is small, and small worlds cool rapidly because they have a large ratio of surface area to volume. You may have noticed that a small cupcake fresh from the oven cools more rapidly than a large cake. The moon has lost much of its internal heat, and it is the outward flow of heat that produces geological activity. That means the moon is mostly inactive. The crust of the moon rapidly grew thick and never divided into moving plates. There are no rift valleys or folded mountain chains on the moon. The last lava flows on the moon ended one or two billion years ago when the moon's temperature fell too low to maintain subsurface lava.

The overall terrain on the moon is almost fixed. On Earth a billion years from now, plate tectonics will have totally changed the shapes of the continents, and erosion will have long ago worn away the Rocky Mountains. On the moon, impacts will have formed only a few more large craters, and nearly all of the lunar scenery will be unchanged. Micrometeorites, however, will have blasted the soil, erasing the footprints left by the Apollo astronauts and reducing the equipment they left behind to peculiar chemical contamination in the regolith at the six Apollo landing sites.

You have studied the story of the moon's evolution in detail for later comparison with other planets and moons in our solar system, but the story has skipped one important question: Where did Earth get such a large satellite?

The Origin of Earth's Moon

Over the last two centuries, astronomers developed three different hypotheses for the origin of Earth's moon, but these traditional ideas have failed to survive comparison with the evidence. A theory proposed in the 1980s may hold the answer. You can begin by testing the three unsuccessful theories against the evidence to see why they failed.

The first of the three traditional theories, the **fission hypothesis,** supposes that the moon formed by the fission of Earth. If the young Earth spun fast enough, tides raised by the sun might make it break into two parts (■ Figure 21-9a). If this separation occurred after Earth differentiated, the moon would have formed from crust material. In that way, the theory explains the moon's low density.

But the fission theory has problems. No one knows why the young Earth should have spun so fast, nearly ten times faster than today, nor where all that angular momentum went after the fission. In addition, the moon's orbit is not in the plane of Earth's equator, as it would be if it had formed by fission.

The second traditional theory is the **condensation** (or double-planet) **hypothesis.** It supposes that Earth and the moon condensed as a double planet from the same cloud of material (Figure 21-9b). However, if they formed from the same material, they should have the same chemical composition and density, which they don't. The moon is very poor in iron and related elements such as titanium and in volatiles such as water vapor and sodium. Yet the moon contains almost exactly the same ratios of oxygen isotopes as does Earth's mantle. The condensation theory cannot explain these compositional differences.

The third theory is the **capture hypothesis.** It supposes that the moon formed somewhere else and was later captured by Earth (Figure 21-9c). You might imagine the moon forming inside the orbit of Mercury, where the heat would prevent the condensation of solid metallic grains and only high-melting-point metal oxides could have solidified. A later encounter with Mercury could "kick" the moon out to Earth.

The capture theory was not popular because it requires highly unlikely events involving interactions with Mercury and Earth to move the moon from place to place. Scientists are always suspicious of explanations that require a chain of unlikely coincidences. Also, upon encountering Earth, the moon would have been moving so rapidly that Earth's gravity would have been unable to capture it without ripping the moon to fragments through tidal forces.

Until recently, astronomers were left with no acceptable theory to explain the origin of the moon, and they occasionally joked that the moon could not exist. But during the 1980s, plan-

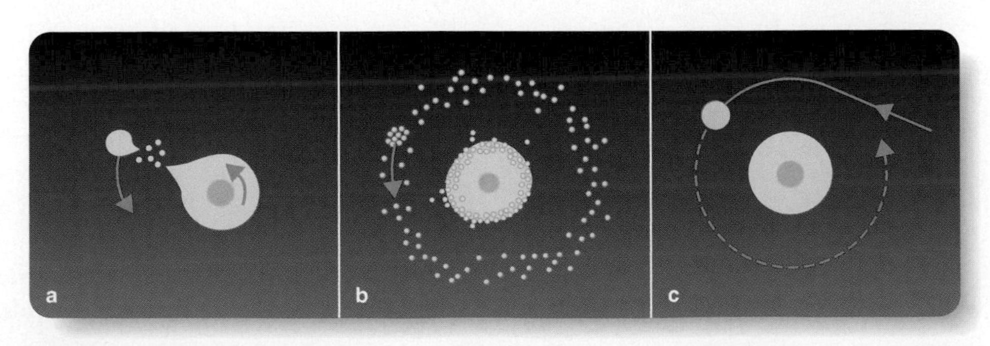

■ Figure 21-9

Three traditional theories for the moon's origin. (a) Fission theories suppose that Earth and the moon were once one body and broke apart. (b) Condensation theories suppose that the moon formed at the same time and from the same material as Earth. (c) Capture theories suggest that the moon formed elsewhere and was captured by Earth. None of these theories explains all the facts.

etary astronomers developed a new theory that combines the best aspects of the fission hypothesis and the capture hypothesis.

The **large-impact hypothesis** supposes that the moon formed from debris ejected into a disk around Earth by the impact of a large body. The impacting body may have been twice as large as Mars. Instead of saying that Earth was hit by a large body, it may be more nearly correct to say that Earth and the moon resulted from the collision and merger of two very large planetesimals. The resulting large body became Earth, and the ejected debris formed the moon (■ Figure 21-10). Such an impact would have melted the proto-Earth, and the material falling together to form the moon would have been heated hot enough to melt. This fits the evidence from moon rocks that the moon formed as a sea of magma.

This hypothesis would explain a number of things. The collision had to occur at a steep angle to eject enough matter to make the moon. That is, the objects did not collide head-on. Such a glancing collision would have spun the resulting material rapidly enough to explain the observed angular momentum in the Earth–moon system. If the two colliding planetesimals had already differentiated, the ejected material would be mostly iron-poor mantle and crust. Calculations show that the iron core of the impacting body could have fallen into the larger body that became Earth. This would explain why the moon is so poor in iron and why the abundances of other elements are so similar to rocks from Earth's mantle. Also, the material that eventually became the moon would have remained in a disk long enough for volatile elements, which the moon lacks, to be lost to space.

The moon may be the remains of a giant impact. Until recently, astronomers have been reluctant to consider such catastrophic events, but a number of lines of evidence suggest that planets have been affected by giant impacts. Consequently, the third theme identified in the introduction to this chapter, giant impacts, has the potential to help you understand other worlds. Catastrophic events are rare, but they can occur.

The Large-Impact Hypothesis

A protoplanet nearly the size of Earth differentiates to form an iron core.

Another body that has also formed an iron core strikes the larger body and merges, trapping most of the iron inside.

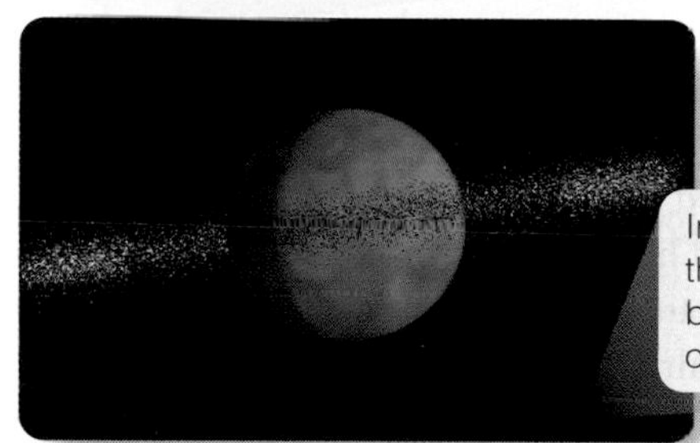

Iron-poor rock from the mantles of the two bodies forms a ring of debris.

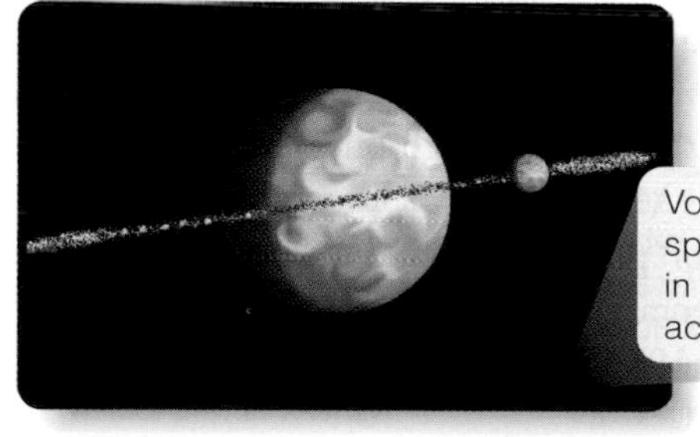

Volatiles are lost to space as the particles in the ring begin to accrete larger bodies.

Eventually the moon forms from the iron-poor and volatile-poor matter in the disk.

■ Figure 21-10

Sometime before the solar system was 50 million years old, a collision produced Earth and the moon in its inclined orbit.

Building Scientific Arguments

If the moon was intensely cratered by the heavy bombardment and then formed great lava plains, why didn't the same thing happen on Earth?

Is this argument obvious? It is still worth reviewing as a way to test your understanding. In fact, the same thing did happen on Earth. Although the moon has more craters than Earth, the moon and Earth are the same age, and both were battered by meteorites during the heavy bombardment. Some of those impacts on Earth must have been large and dug giant multiringed basins. Lava flows must have welled up through Earth's crust and flooded the lowlands to form great lava plains much like the lunar maria.

Earth, however, is a larger world and has more internal heat, which escapes more slowly than the moon's heat did. The moon is now geologically dead, but Earth is very active, with heat flowing outward from the interior to drive plate tectonics. The moving plates long ago erased all evidence of the cratering and lava flows dating from Earth's youth.

Volcanoes

1 Molten rock (magma) is less dense than its surroundings and tends to rise. Where it bursts through Earth's crust, you see volcanism. The two main types of volcanoes on Earth provide good examples for comparison with those on Venus and Mars.

On Earth, **composite volcanoes** form above subduction zones where the descending crust melts and the magma rises to the surface. This forms chains of volcanoes along the subduction zone, such as the Andes along the west coast of South America.

Magma rising above subduction zones is not very fluid, and it produces explosive volcanoes with sides as steep as 30°.

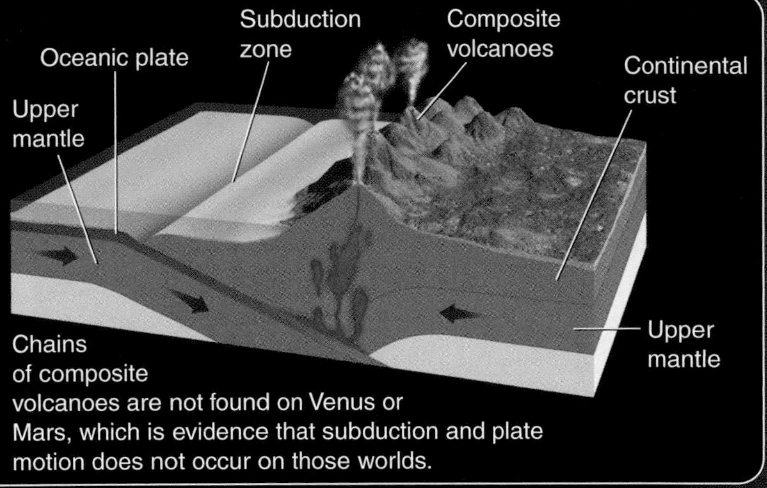

Chains of composite volcanoes are not found on Venus or Mars, which is evidence that subduction and plate motion does not occur on those worlds.

Based on *Physical Geology,* 4th edition, James S. Monroe and Reed Wicander, Wadsworth Publishing Company. Used with permission.

Mount St. Helens exploded northward on May 18, 1980, killing 63 people and destroying 600 km^2 (230 mi^2) of forest with a blast of winds and suspended rock fragments that moved as fast as 480 km/hr (300 mph) and had temperatures as hot as 350°C (660°F). Note the steep slope of this composite volcano.

USGS

Shield volcano

Lava flow

Magma chamber

Oceanic crust

1a A **shield volcano** is formed by highly fluid lava (basalt) that flows easily and creates low-profile volcanic peaks with slopes of 3° to 10°. The volcanoes of Hawaii are shield volcanoes that occur over a hot spot in the middle of the Pacific plate.

Magma collects in a chamber in the crust and finds its way to the surface through cracks.

Magma forces its way upward through cracks in the upper mantle and causes small, deep earthquakes.

A hot spot is formed by a rising convection current of magma moving upward through the hot, deformable (plastic) rock of the mantle.

The Cascade Range composite volcanoes are produced by an oceanic plate being subducted below North America and partially melting.

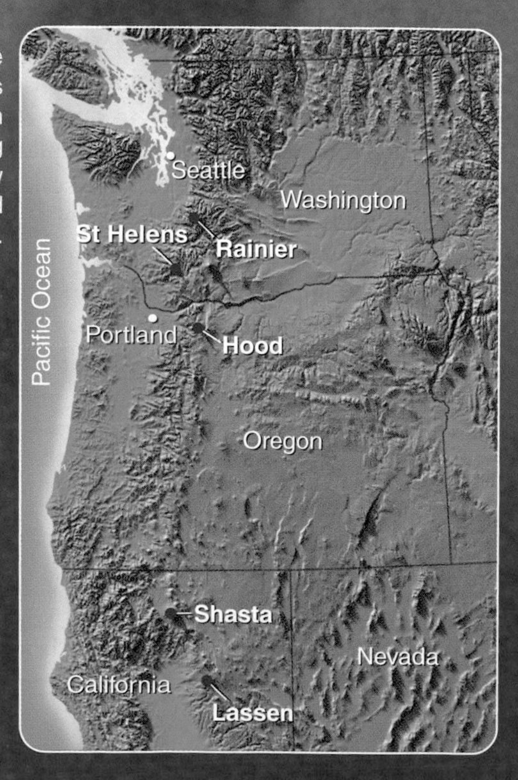

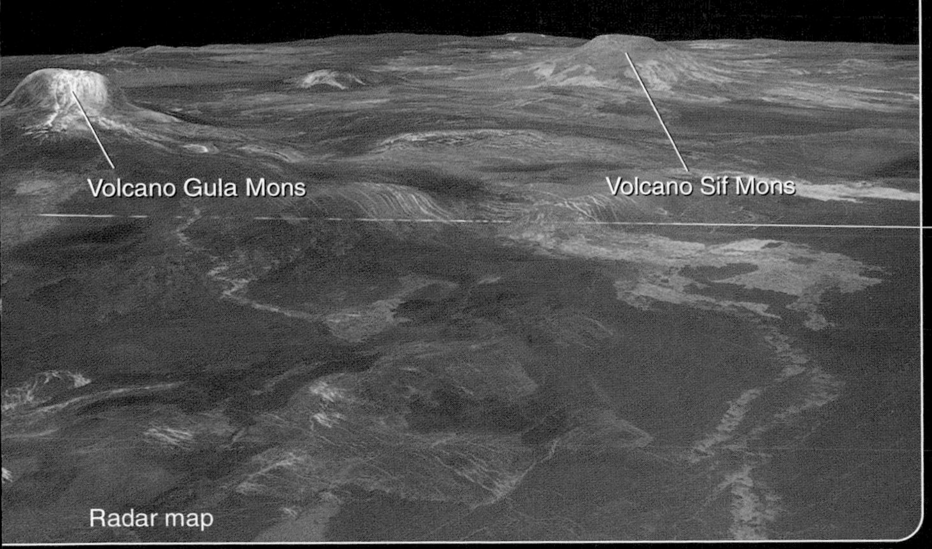

NASA

2 Volcanoes on Venus are shield volcanoes. They appear to be steep sided in some images created from Magellan radar maps, but that is because the vertical scale has been exaggerated to enhance detail. The volcanoes of Venus are actually shallow-sloped shield volcanoes.

2a This computer model of a mountain with the vertical scale magnified 10 times appears to have steep slopes such as those of a composite volcano.

A true profile of the computer model shows the mountain has very shallow slopes typical of shield volcanoes.

Mike Seeds

3 Volcanism over a hot spot results in repeated eruptions that build up a shield volcano of many layers. Such volcanoes can grow very large.

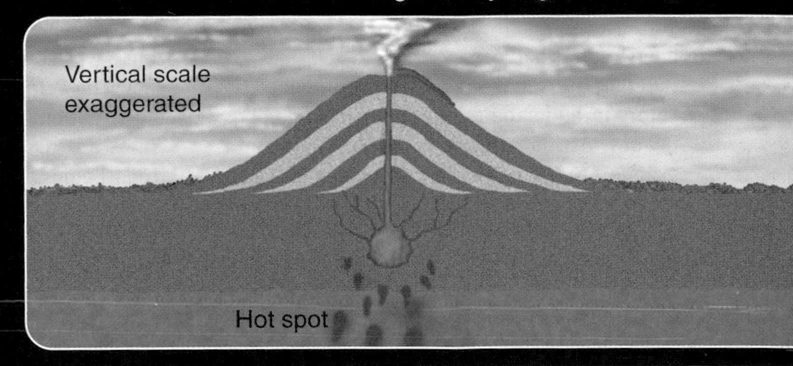

Old volcanic island eroded below sea level

Plate motion

Hot spot

If the crustal plate is moving, magma generated by the hot spot can repeatedly penetrate the crust to build a chain of volcanoes. Only the volcanoes over the hot spot are active. Older volcanoes slowly erode away. Such volcanoes cannot grow large because the moving plate carries them away from the hot spot.

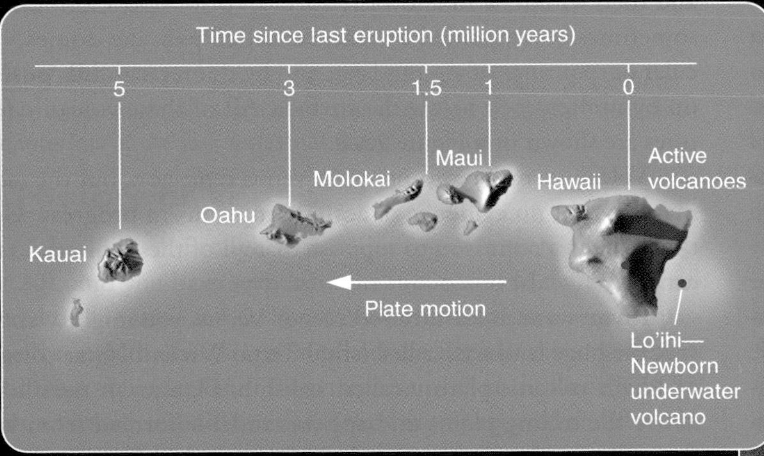

Ace Astronomy™

Log into AceAstronomy and select this chapter to see the Active Figure "Hot Spot Volcanoes" and compare volcanism on Earth with that on Venus.

3a The volcanoes that make up the Hawaiian Islands as shown at left have been produced by a hot spot poking upward through the middle of the moving Pacific plate.

NASA

3b The plate moves about 9 cm/yr and carries older volcanic islands northwest, away from the hot spot. The volcanoes cannot grow extremely large because they are carried away from the hot spot. New islands form to the southeast over the hot spot.

Olympus Mons at right is the largest volcano on Mars. It is a shield volcano 25 km (16 mi) high and 700 km (440 mi) in diameter at its base. Its vast size is evidence that the crustal plate must have remained stationary over the hot spot. This is evidence that Mars has not had plate tectonics.

Olympus Mons contains 95 times more volume than the largest volcano on Earth, Mauna Loa in Hawaii.

Caldera from repeated eruptions

Digital elevation map

Jupiter's Atmosphere

1 You probably won't ever visit Jupiter's atmosphere. Its cloud layers are deathly cold, and the deeper layers that are warmer have a crushingly high pressure. There is no free oxygen to breathe; the gases are roughly three-quarters hydrogen and a quarter helium, plus traces of water vapor, methane, ammonia, and similar molecules. Traces of sulfur and molecules containing sulfur probably make it smell bad. Of course, Jupiter has no surface, so there isn't even a place to stand. Jupiter is a nice planet to look at, but it's not a place to visit.

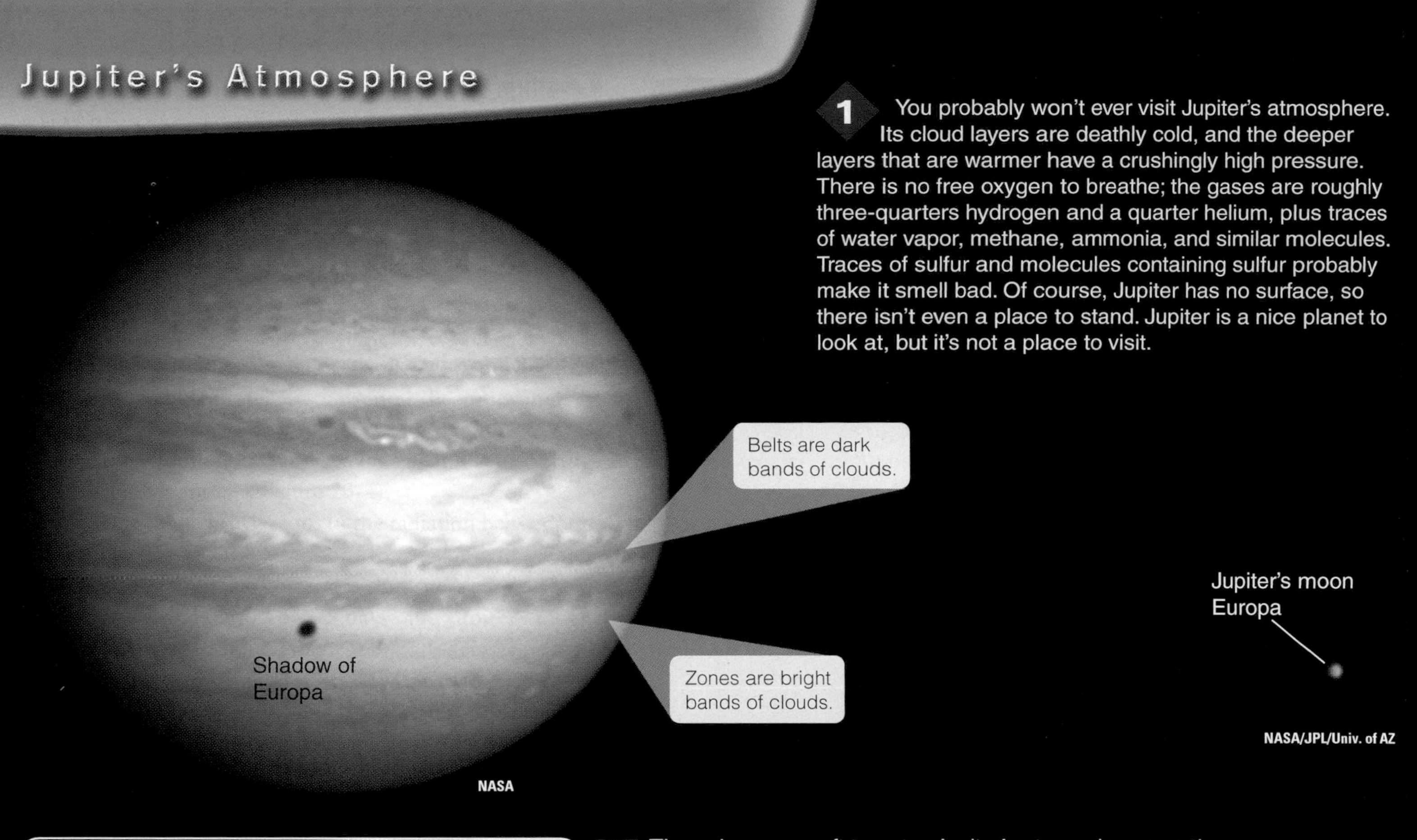

1a The only spacecraft to enter Jupiter's atmosphere was the Galileo probe. Released from the Galileo spacecraft, the probe entered Jupiter's atmosphere in December 1995. It parachuted through the upper atmosphere of clear hydrogen, released its heat shield, and then fell through Jupiter's stormy atmosphere until it was crushed by the increasing pressure.

Lightning bolts are common in Jupiter's turbulent clouds.

Hughes Aircraft Co

Jupiter's atmosphere is a very thin layer of turbulent gas above the liquid interior. It makes up only about 1 percent of the radius of the planet.

NASA/JPL

The Great Red Spot at right is a giant circulating storm in one of the southern zones. It has lasted at least 300 years since astronomers first noticed it after the invention of the telescope. Smaller spots are also circulating storms.

2 The visible clouds on Jupiter are composed of ammonia crystals, but models predict that deeper layers of clouds contain ammonia hydrosulfide crystals, and deeper still lies a cloud layer of water droplets. These compounds are normally white, so planetary scientists think the colors arise from small amounts of other molecules formed by lightning or by sunlight.

If you could put thermometers in Jupiter's atmosphere at different levels, you would discover that the temperature rises below the uppermost clouds.

Far below the clouds, the temperature and pressure climb so high the gaseous atmosphere merges gradually with the liquid hydrogen interior and there is no surface.

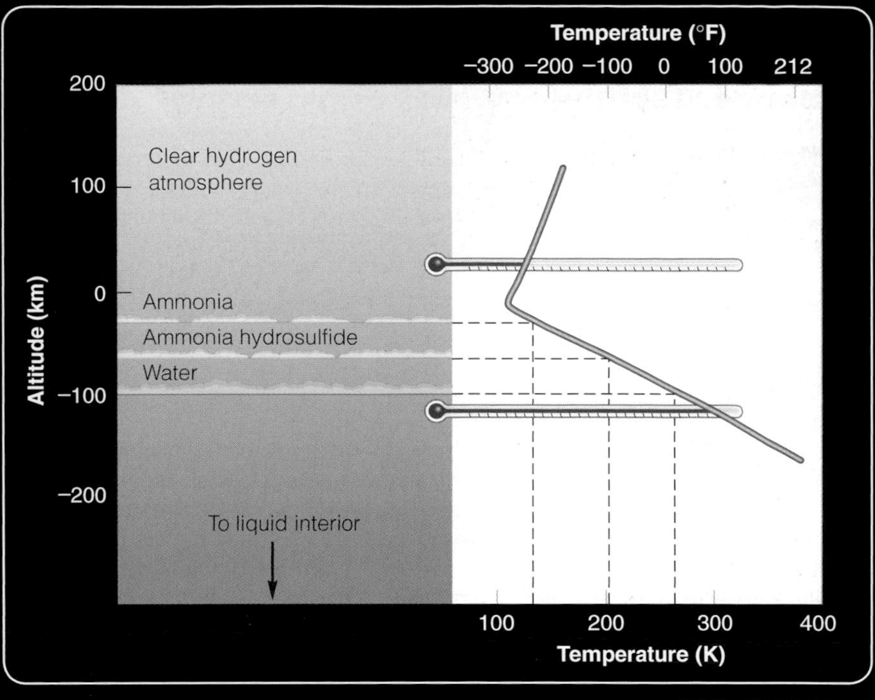

Log into Ace Astronomy and select this chapter to see the Active Figure called "Planetary Atmospheres." Notice the temperatures at which the cloud layers form.

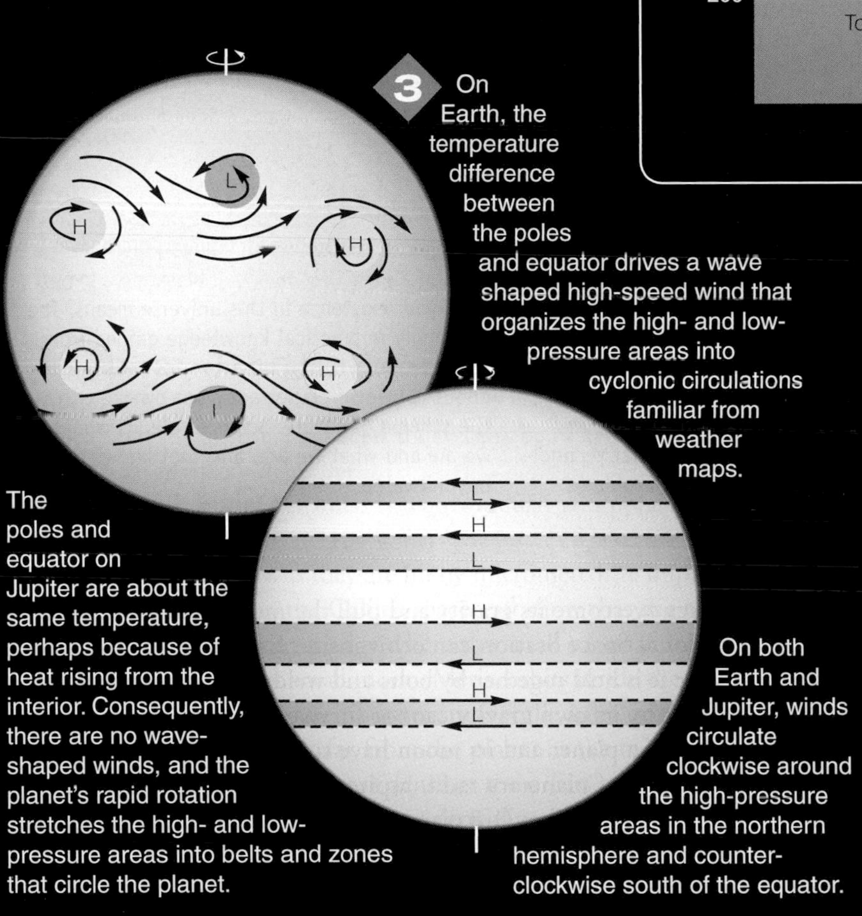

3 On Earth, the temperature difference between the poles and equator drives a wave shaped high-speed wind that organizes the high- and low-pressure areas into cyclonic circulations familiar from weather maps.

The poles and equator on Jupiter are about the same temperature, perhaps because of heat rising from the interior. Consequently, there are no wave-shaped winds, and the planet's rapid rotation stretches the high- and low-pressure areas into belts and zones that circle the planet.

On both Earth and Jupiter, winds circulate clockwise around the high-pressure areas in the northern hemisphere and counter-clockwise south of the equator.

Zones are brighter than belts because rising gas forms clouds high in the atmosphere, where sunlight is strong.

Belt

Zone

Altitude

North

Equator

4 The two white ovals here are counterclockwise, high-pressure weather systems that have been visible in Jupiter's southern hemisphere since they formed in the 1930s. These two huge storms merged into one spot in February 1998. The pear-shaped circulation visible between the two storms vanished during the merger.

Features in Jupiter's atmosphere may be stable for decades or centuries, but even the Giant Red Spot may someday vanish.

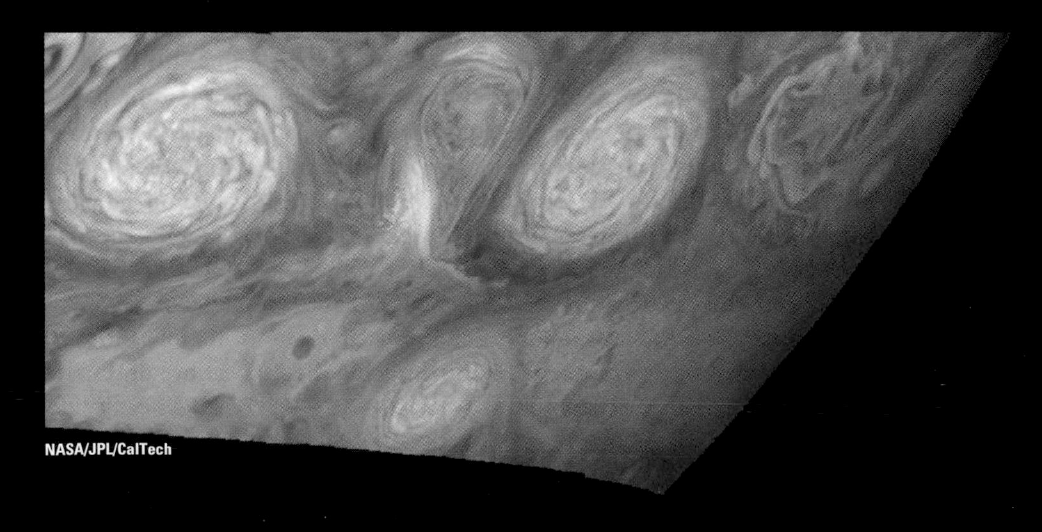

NASA/JPL/CalTech

The Ice Rings of Saturn

1 The brilliant rings of Saturn are made up of billions of ice particles ranging from microscopic specks to chunks bigger than a house. Each particle orbits Saturn in its own circular orbit. Much of what astronomers know about the rings was learned when the Voyager 1 spacecraft flew past Saturn in 1980, followed by the Voyager 2 spacecraft in 1981. The Cassini Spacecraft reached orbit around Saturn in 2004. From Earth, astronomers see three rings labeled A, B, and C. Voyager and Cassini images reveal over a thousand ringlets within the rings.

Saturn's rings can't be leftover material from the formation of Saturn. The rings are made of ice particles, and the planet would have been so hot when it formed that it would have vaporized and driven away any icy material. Rather, the rings must be debris from collisions between passing comets and Saturn's icy moons. Such impacts should occur every 10 million years or so, and they would scatter ice throughout Saturn's system of moons. The ice would quickly settle into the equatorial plane, and some would become trapped in rings. Although the ice may waste away due to meteorite impacts and damage from radiation in Saturn's magnetosphere, new impacts could replenish the rings with fresh ice. The bright, beautiful rings you see today may be only a temporary enhancement caused by an impact that occurred since the extinction of the dinosaurs.

Encke's division

Cassini's division

A ring

B ring

C ring
The Crepe ring

As in the case of Jupiter's ring, Saturn's rings lie inside the planet's Roché limit where the ring particles cannot pull themselves together to form a moon.

Because it is so dark, the C ring has been called the Crepe ring, referring to the black, semitransparent cloth associated with funerals.

Earth to scale

Visual-wavelength image

NASA

1a An astronaut could swim through the rings. Although the particles orbit Saturn at high velocity, all particles at the same distance from the planet orbit at about the same speed, so they collide gently at low velocities. If you could visit the rings, you could push your way from one icy particle to the next. This artwork is based on a model of particle sizes in the A ring.

The C ring contains boulder-size chunks of ice, whereas most particles in the A and B rings are more like golf balls, down to dust-size ice crystals. Further, C ring particles are less than half as bright as particles in the A and B rings. Cassini observations show that the C ring particles contain less ice and more minera

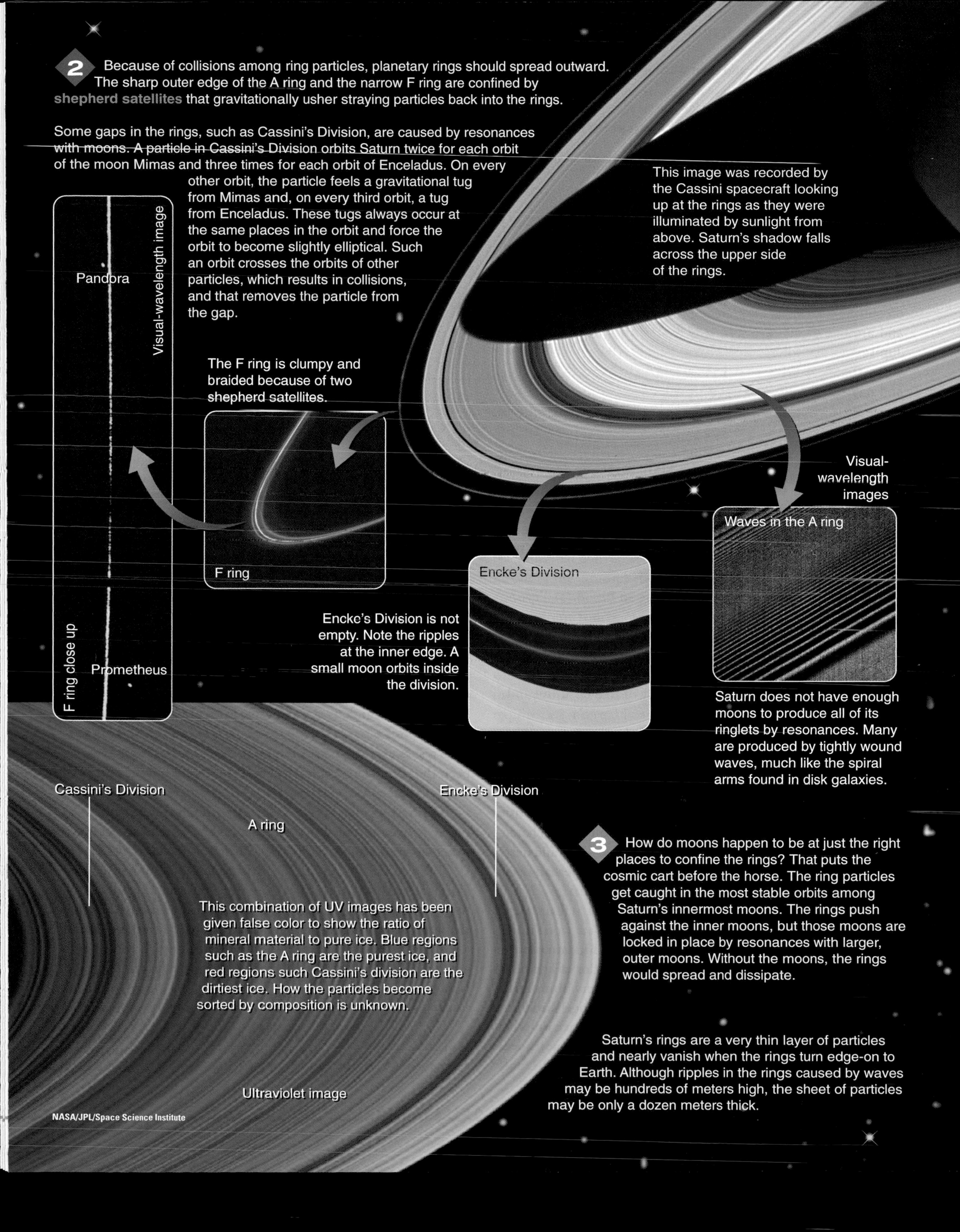

2 Because of collisions among ring particles, planetary rings should spread outward. The sharp outer edge of the A ring and the narrow F ring are confined by **shepherd satellites** that gravitationally usher straying particles back into the rings.

Some gaps in the rings, such as Cassini's Division, are caused by resonances with moons. A particle in Cassini's Division orbits Saturn twice for each orbit of the moon Mimas and three times for each orbit of Enceladus. On every other orbit, the particle feels a gravitational tug from Mimas and, on every third orbit, a tug from Enceladus. These tugs always occur at the same places in the orbit and force the orbit to become slightly elliptical. Such an orbit crosses the orbits of other particles, which results in collisions, and that removes the particle from the gap.

This image was recorded by the Cassini spacecraft looking up at the rings as they were illuminated by sunlight from above. Saturn's shadow falls across the upper side of the rings.

The F ring is clumpy and braided because of two shepherd satellites.

Encke's Division is not empty. Note the ripples at the inner edge. A small moon orbits inside the division.

Saturn does not have enough moons to produce all of its ringlets by resonances. Many are produced by tightly wound waves, much like the spiral arms found in disk galaxies.

This combination of UV images has been given false color to show the ratio of mineral material to pure ice. Blue regions such as the A ring are the purest ice, and red regions such Cassini's division are the dirtiest ice. How the particles become sorted by composition is unknown.

3 How do moons happen to be at just the right places to confine the rings? That puts the cosmic cart before the horse. The ring particles get caught in the most stable orbits among Saturn's innermost moons. The rings push against the inner moons, but those moons are locked in place by resonances with larger, outer moons. Without the moons, the rings would spread and dissipate.

Saturn's rings are a very thin layer of particles and nearly vanish when the rings turn edge-on to Earth. Although ripples in the rings caused by waves may be hundreds of meters high, the sheet of particles may be only a dozen meters thick.

NASA/JPL/Space Science Institute

The Rings of Uranus and Neptune

1 The rings of Uranus were discovered in 1977, when Uranus crossed in front of a star. During this occultation, astronomers saw the star dim a number of times before and again after the planet crossed over the star. The dips in brightness were caused by rings circling Uranus.

More rings were discovered by Voyager 2. The rings are identified in different ways depending on when and how they were discovered.

AceAstronomy™

Log into AceAstronomy and select this chapter to see the Active Figure "Uranus's Ring Detection" and animate this diagram.

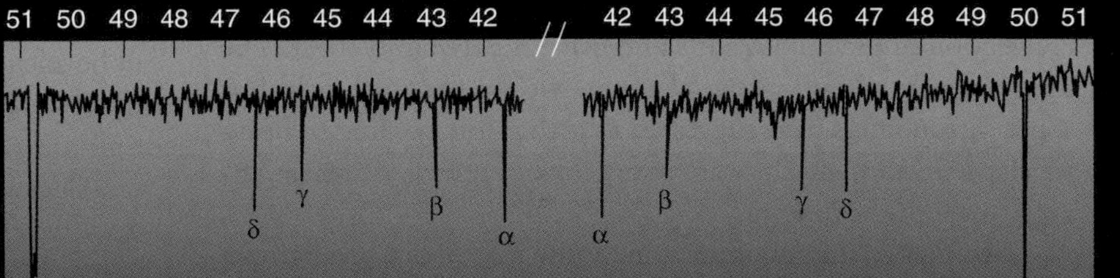

Notice the eccentricity of the ε ring. It lies at different distances on opposite sides of the planet.

ε
λ
η
γ
δ
β
α
6
5
4
1986 U2R

2 The albedo of the ring particles is only about 0.015, darker than lumps of coal. If the ring particles are made of methane-rich ices, radiation from the planet's radiation belts could break the methane down to release carbon and darken the ices. The same process may darken the icy surface of Uranian moons.

The narrowness of the rings suggests they are shepherded by small moons. Voyager 2 found Ophelia and Cordelia shepherding the ε ring. Other small moons must be shepherding the other narrow rings. Such moons must be structurally strong to hold themselves together inside the planet's Roche limit.

Ophelia

Cordelia

The eccentricity of the ε ring is apparently caused by the eccentric orbits of Ophelia and Cordelia.

2a When the Voyager 2 spacecraft looked back at the rings illuminated from behind by the sun, the rings were not bright. That is, the rings are not bright in forward-scattered light. That means they must not contain much dust. The nine main rings contain particles no smaller than meter-sized boulders.

Uranus

3 Ring particles don't last forever as they collide with each other and are exposed to radiation. The rings of Uranus may need to be resupplied with fresh particles occasionally as impacts on icy moons scatter icy debris.

Collisions among ring particles produce dust, which is thinly scattered inward from the λ ring, but the high, tenuous atmosphere of Uranus is slowing the dust particles and making them fall into the planet. The rings of Uranus actually contain very little dust.

The rings of Neptune are bright in forward-scattered light, as in the image above, and that indicates that the rings contain significant amounts of dust. The ring particles are as dark as those that circle Uranus, however, so they probably contain methane-rich ice darkened by radiation.

4 The brightness of Neptune is hidden behind the black bar in this Voyager 2 image. Two narrow rings are visible, and a wider, fainter ring lies closer to the planet. More ring material is visible between the two narrow rings.

5 When Neptune occulted stars, astronomers sometimes detected rings and sometimes did not. From that they concluded that Neptune might have ring arcs. Computer enhancement of this Voyager 2 visual-wavelength image shows arcs, regions of higher density, in the outer ring.

The ring arcs visible in the outer ring appear to be generated by the gravitational influence of the moon Galatea, but other moons must also be present to confine the rings.

4a Neptune's rings lie in the plane of the planet's equator and inside the Roche limit. The narrowness of the rings suggests that shepherd moons must confine them, and a few such moons have been found among the rings. There must be more small moons to confine the rings.

Naiad

Galetea

Thalassa

Adams

LeVerrier

Despina

Galle

4c Like the rings of Uranus, the ring particles that orbit Neptune cannot have survived since the formation of the planet. Occasional impacts on Neptune's moons must scatter debris and resupply the rings with fresh particles.

4b Neptune's rings have been given names associated with the planet's history. English astronomer Adams and French astronomer LeVerrier predicted the existence of Neptune from the motion of Uranus. The German astronomer Galle discovered the planet in 1846 based on LeVerrier's prediction.

Neptune

25 Meteorites, Asteroids, and Comets

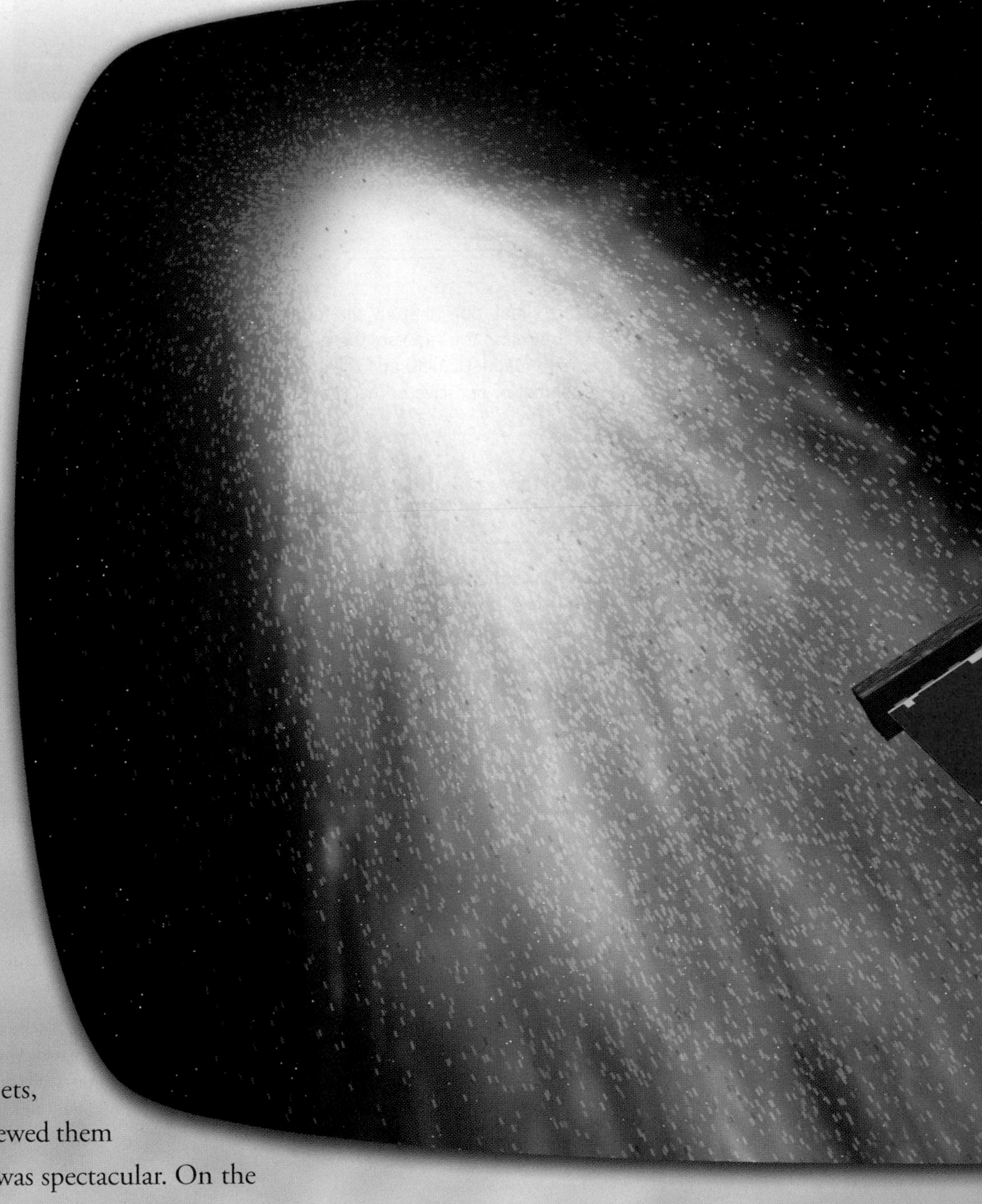

When they shall cry "PEACE, PEACE" then cometh sudden destruction!

COMET'S CHAOS?— What Terrible events will the Comet bring?

FROM A RELIGIOUS PAMPHLET PREDICTING THE END OF THE WORLD BECAUSE OF THE APPEARANCE OF COMET KOHOUTEK, 1973

YOU ARE NOT AFRAID of comets, of course; but not long ago, people viewed them with terror. In 1910, Comet Halley was spectacular. On the night of May 19, Earth actually passed through the tail of the comet—and millions of people panicked. The spectrographic discovery of cyanide gas in the tails of comets led many to believe that life on Earth would end. Householders in Chicago stuffed rags around doors and windows to keep out the gas, and bottled oxygen was sold out. Con artists in Texas sold comet pills and inhalers to ward off the noxious fumes. An Oklahoma newspaper reported (in what was apparently a hoax) that a religious sect tried to sacrifice a virgin to the comet. Throughout history, bright comets have been seen as portents of doom. Even in our own time, the appearance of bright comets has generated predictions

Continued on page 618

Comets can be terrifying to the superstitious, but they are dramatically beautiful and carry clues to the origin of the solar system. The Stardust spacecraft flew through the dust and gas spewing from comet Wild 2 in 2004. (NASA/JPL)

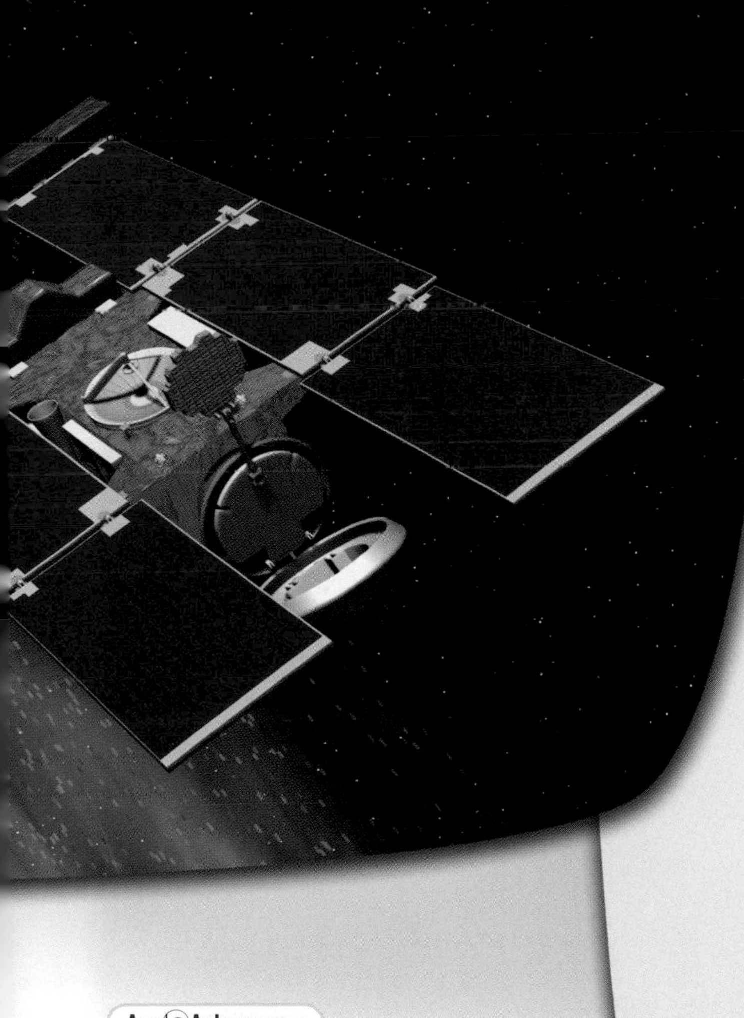

Guidepost

Looking Back

Everything has to have come from somewhere, and you began wondering where the solar system came from back in Chapter 19. You have now visited all the planets and many of the moons in the solar system, gathering evidence and making observations. But you found that the planets and moons did not offer many clues to the formation of the solar system because they have been heavily altered by cratering and by heat flowing out of their interiors. You will find the smaller objects in the solar system less altered.

This Chapter

Compared with planets, the comets and asteroids are unevolved objects. You will find them much as they were when they formed 4.6 billion years ago. The fragments of these objects that fall to Earth, the meteors and meteorites, will give you a close look at these ancient planetesimals. As you explore, you will find answers to four essential questions:

Where do meteors and meteorites come from?

What are the asteroids?

Where do comets come from?

What happens when an asteroid or comet hits Earth?

Looking Ahead

As you finish this chapter, you will have an astronomer's insight into your place in nature. You live on the surface of a planet. There are other planets. Are they inhabited too? That is the subject of the next chapter.

Ace Astronomy™ The AceAstronomy icon throughout the text indicates an opportunity for you to test yourself on key concepts and to explore animations and interactions on the AceAstronomy website at: http://ace.brookscole.com/sf9

Observations of Asteroids

1 Seen from Earth, asteroids look like faint points of light moving in front of distant stars. Not many years ago they were known mostly for drifting slowly through the field of view and spoiling long time exposures. Some astronomers referred to them as "the vermin of the sky." Spacecraft have now visited asteroids, and the images radioed back to Earth show that the asteroids are mostly small, gray, irregular worlds heavily cratered by impacts.

The Near Earth Asteroid Rendezvous (NEAR) spacecraft visited the asteroid Eros in 2000 and found it to be heavily cratered by collisions and covered by a layer of crushed rock ranging from dust to large boulders. The NEAR spacecraft eventually landed on Eros.

Visual-wavelength image

Eros appears to be a solid fragment of rock.

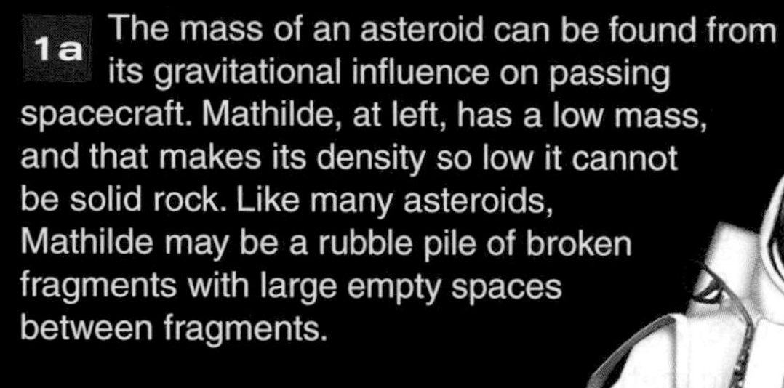

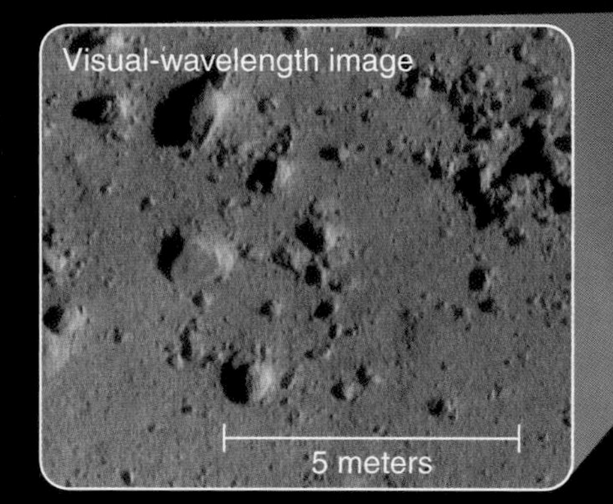

Most asteroids are too small for their gravity to pull them into a spherical shape. Impacts break them into irregularly shaped fragments.

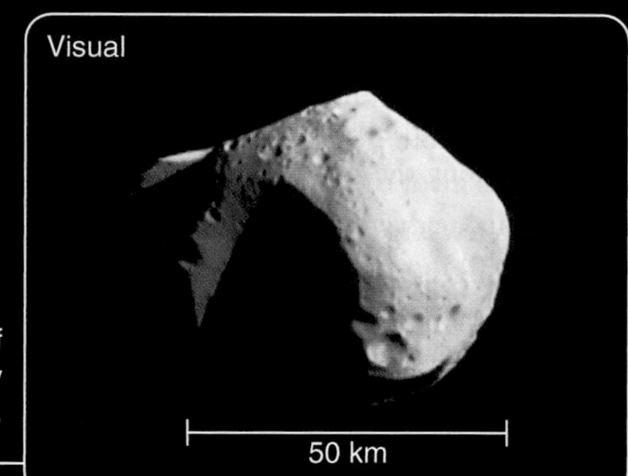

The surface of Mathilde is very dark rock.

1a The mass of an asteroid can be found from its gravitational influence on passing spacecraft. Mathilde, at left, has a low mass, and that makes its density so low it cannot be solid rock. Like many asteroids, Mathilde may be a rubble pile of broken fragments with large empty spaces between fragments.

If you walked across the surface of an irregularly shaped asteroid such as Eros, you would find gravity very weak; and in many places, it would not be perpendicular to the surface.

Enhanced visual image

5 km

Like most asteroids, Gaspra would look gray to your eyes; but, in this enhanced image at left, color differences probably indicate difference in mineralogy.

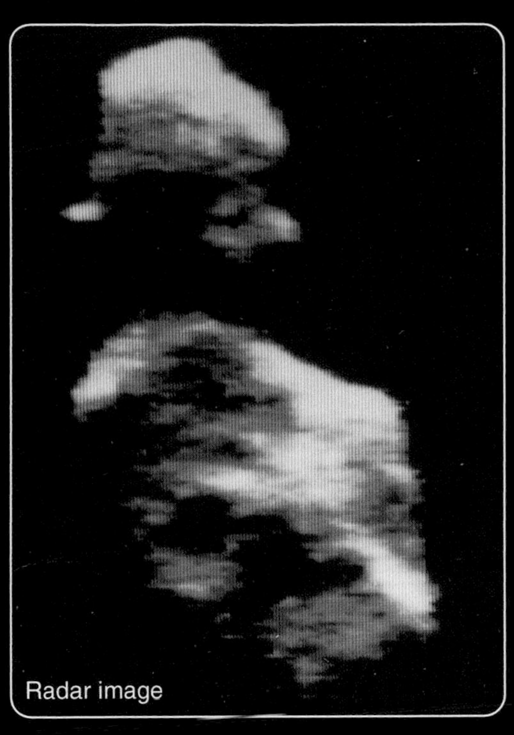

2 Asteroids that pass near Earth can be imaged by radar. The asteroid Toutatis is revealed to be a double object—two objects orbiting close to each other or actually in contact.

Ida

Dactyl

30 km

Enhanced visual + infrared

Double asteroids are more common than was once thought, reflecting a history of collisions and fragmentation. The asteroid Ida is orbited by a moon Dactyl only about 1.5 km in diameter.

Occasional collisions among the asteroids release fragments, and Jupiter's gravity scatters them into the inner solar system as a continuous supply of meteorites.

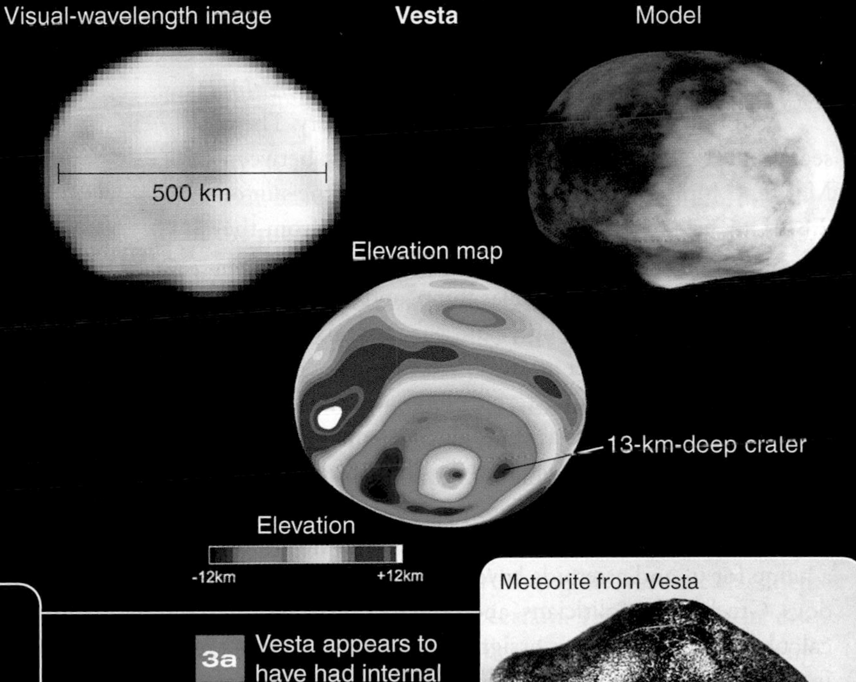

3 The large asteroid Vesta, as shown at right, provides evidence that some have suffered geological activity. No spacecraft has visited it, but its spectrum resembles that of solidified lava. Images made by the Hubble Space Telescope allow the creation of a model of its shape. It has a huge crater at its south pole. A family of small asteroids is evidently composed of fragments from Vesta, and a certain class of meteorites, spectroscopically identical to Vesta, are believed to be fragments from the asteroid. The meteorites appear to be solidified basalt.

3a Vesta appears to have had internal heat at some point in its history, perhaps due to the decay of radioactive minerals. Lava flows have covered at least some of its surface.

Meteorite from Vesta

5 cm

2 in.

Bright

Dark

Albedo (reflected brightness)

0.4, 0.3, 0.2, 0.1, 0.06, 0.04

M

S

C

Common in the inner asteroid belt

Common in the outer asteroid belt

Grayer

Redder

0.8, 1.0, 1.2, 1.4, 1.6

Ultraviolet minus visual color index

4 Although asteroids would look gray to your eyes, they can be classified according to their albedos (reflected brightness) and spectroscopic colors. As shown at left, S-types are brighter and tend to be reddish. They are the most common kind of asteroid and appear to be the source of the most common chondrites.

M-type asteroids are not too dark but are also not very red. They may be mostly iron-nickel alloys.

C-type asteroids are as dark as lumps of sooty coal and appear to be carbonaceous.

Comet Observations

1 A **type I** or **gas tail** is produced by ionized gas carried away from the nucleus by the solar wind. The spectrum of a gas tail is an emission spectrum. The atoms are ionized by the ultraviolet light in sunlight. The wisps and kinks in gas tails are produced by the magnetic field embedded in the solar wind.

Spectra of gas tails reveal atoms and ions such as H_2O, CO_2, CO, H, OH, O, S, C, and so on. These are released by the vaporizing ices or produced by the breakdown of those molecules. Some gases, such as hydrogen cyanide (HCN), must be formed by chemical reactions.

Gas tail (Type I)

Dust tail (Type II)

1a A **type II** or **dust tail** is produced by dust from the vaporizing ices of the nucleus. The dust is pushed gently outward by the pressure of sunlight, and it reflects an absorption spectrum, the spectrum of sunlight. The dust is not affected by the magnetic field of the solar wind, so dust tails are more uniform than gas tails. Dust tails are often curved because the dust particles follow their individual orbits around the sun once they leave the nucleus.

Nucleus

1b The nucleus of a comet (not visible here) is a small, fragile lump of porous rock containing ices of water, carbon dioxide, ammonia, and so on. Comet nuclei can be 10 to 100 km in diameter.

When a spacecraft named ICE passed through the gas tail of a comet, it found a magnetic field from the solar wind draped over the nucleus like seaweed draped over a fishhook.

Coma

The **coma** of a comet is the cloud of gas and dust that surrounds the nucleus. It can be over 1,000,000 km in diameter, bigger than the sun.

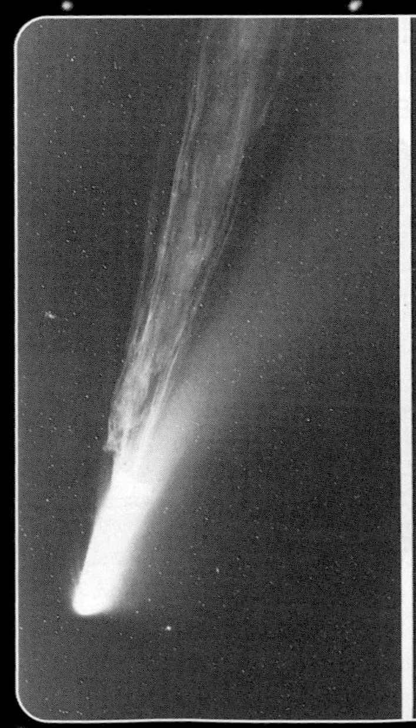

1c Comet Mrkos in 1957 shows how the gas tail can change from night to night due to changes in the magnetic field in the solar wind.

Visual-wavelength images

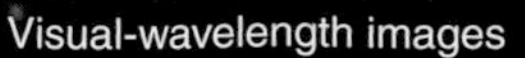

Ace Astronomy™

Log into AceAstronomy and select this chapter to see Active Figure "Build a Comet." See how energy from the sun shapes a comet.

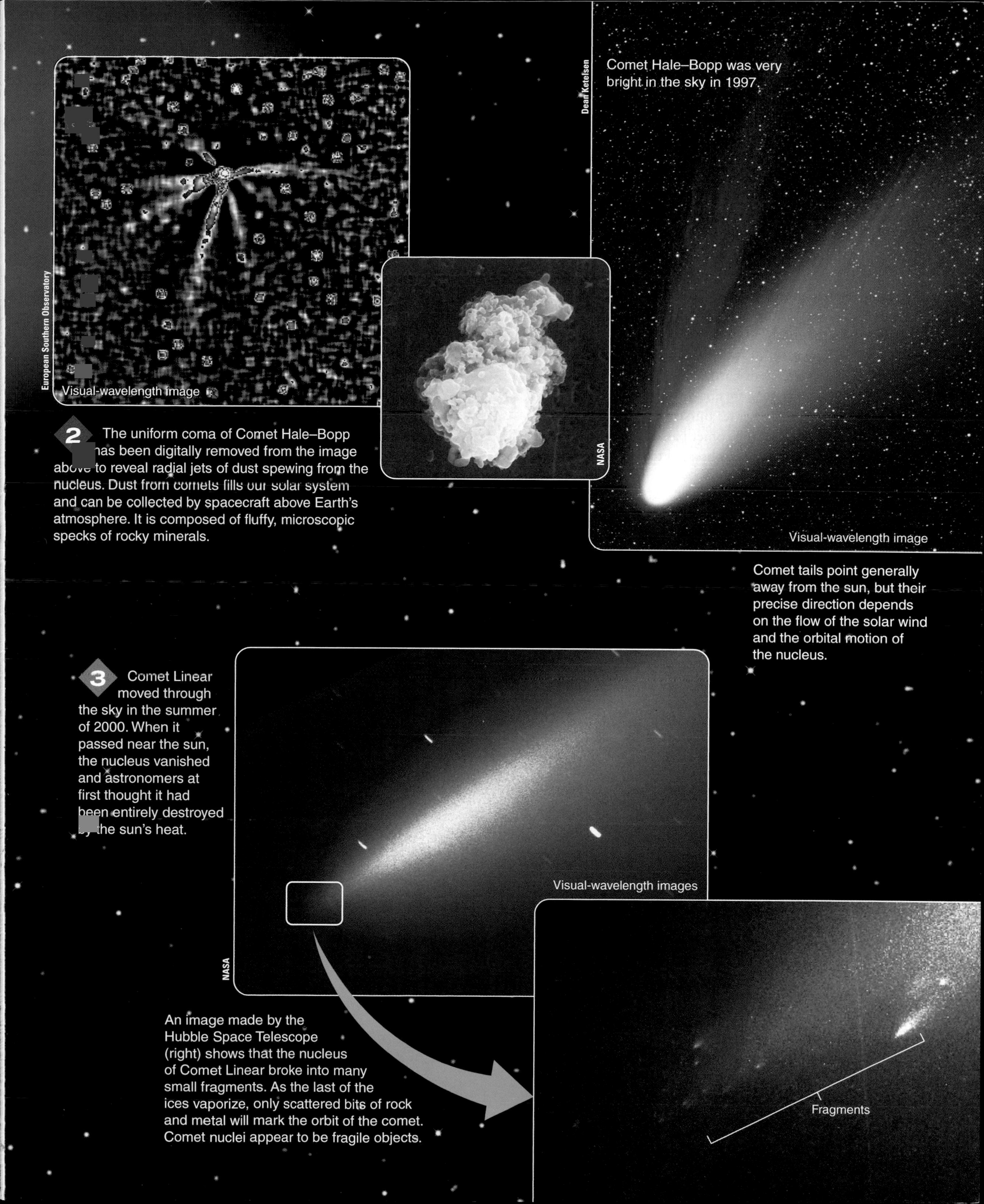

2 The uniform coma of Comet Hale–Bopp has been digitally removed from the image above to reveal radial jets of dust spewing from the nucleus. Dust from comets fills our solar system and can be collected by spacecraft above Earth's atmosphere. It is composed of fluffy, microscopic specks of rocky minerals.

Comet Hale–Bopp was very bright in the sky in 1997.

Comet tails point generally away from the sun, but their precise direction depends on the flow of the solar wind and the orbital motion of the nucleus.

3 Comet Linear moved through the sky in the summer of 2000. When it passed near the sun, the nucleus vanished and astronomers at first thought it had been entirely destroyed by the sun's heat.

An image made by the Hubble Space Telescope (right) shows that the nucleus of Comet Linear broke into many small fragments. As the last of the ices vaporize, only scattered bits of rock and metal will mark the orbit of the comet. Comet nuclei appear to be fragile objects.

26 Life on Other Worlds

Did I solicit thee from
Darkness to promote me?

JOHN MILTON, *PARADISE LOST*

As LIVING THINGS, we have been promoted from darkness. We are made of heavy atoms that could not have formed at the beginning of the universe. Successive generations of stars fusing light elements into heavier elements have built the atoms so important to our existence. When a dark cloud of interstellar gas enriched with these heavy atoms fell together to form our sun, a small part of the cloud gave birth to the planet you inhabit. Your atoms were in that cloud and have been part of Earth ever since it condensed from the solar nebula. ▌ If life can originate on other worlds, and if intelligence is a natural result of the evolution of life forms, then you might expect that other intelligent races inhabit other worlds. Visits between worlds seem impossible, but perhaps distant civilizations can be detected by radio. ▌ Alien life forms could

▌ Continued on page 652 ▌

Every life form we know of has evolved to live somewhere on Earth. The Wekiu bug lives with the astronomers at 13,600 feet atop Hawaiian volcano Mauna Kea. It lives in the icy cinders and eats insects carried up by ocean breezes. (Kris Koenig/Coast Learning Systems)

Guidepost

Looking Back

This chapter is either unnecessary or critical, depending on your point of view. If you believe that astronomy is the study of the physical universe above the clouds, then you are done; the last 25 chapters completed your study of astronomy. But if you believe that astronomy is the study of your position in the universe, not just your physical location but also your role as a living being in the evolution of the universe, then everything you have done so far was just preparation for this chapter.

This Chapter

You will begin by tackling a question that has troubled scientists and philosophers for centuries: What is life? You won't get an answer to the question, but just asking it will illuminate the problem and prepare you to search for life on other worlds. In fact, as you read this chapter, you will ask four essential questions:

What is life?

How did life originate on Earth?

Could life begin on other worlds?

Could Earthlings communicate with civilizations on other worlds?

You won't be able to answer these questions yet, but often in science asking a good question is more important than getting an answer.

Looking Ahead

You are different now. You have explored the universe from the phases of the moon to the big bang, from the origin of Earth to the death of the sun. Astronomy is important, not because it is about stars and galaxies but because it is about us. It tells us what we are, and once you know astronomy, you see yourself and your world in a different way. Astronomy changes us. You are different now.

Ace Astronomy™ The AceAstronomy icon throughout the text indicates an opportunity for you to test yourself on key concepts and to explore animations and interactions on the AceAstronomy website at: http://ace.brookscole.com/sf9

Window on Science | 26-1

The Nature of Scientific Explanation

Science is a way of understanding the world around you, and the heart of that understanding is the explanations that science provides for natural phenomena. Whether you call these explanations stories, histories, theories, or hypotheses, they are attempts to describe how nature works based on fundamental rules of evidence and intellectual honesty. While you may take these explanations as factual truth, you should understand that they are not the only explanations that satisfy the rules of logic.

A separate class of explanations involves religion, and those explanations can be quite logical. The Old Testament description of the creation of the world, for instance, does not fit scientific observations, but if you accept the existence of an omnipotent being, then the biblical explanation is internally logical and acceptable. Of course, it is not a scientific explanation, but religion is a matter of faith and not subject to the rules of evidence. Religious explanations follow their own logic, and it is wrong to demand that they follow the rules of evidence that govern scientific explanations, just as it is wrong to demand that scientific explanations accept certain religious beliefs on faith. Both kinds of explanations are logical, but the rules are different.

If scientific explanations are not the only logical explanations, then why do people give them such weight? First, you should notice the tremendous success of scientific explanations in producing technological advances in your daily lives. Diseases such as chickenpox are more childhood irritations than life-threatening illnesses thanks to the application of scientific explanations to modern medicine. The power of science to shape your world can lead you to think that its explanations are unique. But, second, the process called science depends on the use of evidence to test and perfect explanations, and the logical rigor of this process gives scientists great confidence in their conclusions.

Scientific explanations have given us tremendous insight into the workings of nature, and consequently both scientists and nonscientists tend to forget that there can be other logical explanations. The so-called conflict between science and religion has been symbolized for centuries by the trial of Galileo. That conflict is easier to understand when you consider the nature of scientific explanations and the role of evidence in testing scientific understanding.

Galileo's telescope gave him a new way to know about the world.

be quite different from Earthlings, but if they are alive, then they must share certain characteristics. Your goal in this chapter is to use your knowledge of science to explain the greatest of mysteries—the origin and evolution of life on Earth and on other worlds (**Window on Science 26-1**).

26-1 The Nature of Life

WHAT IS LIFE? Philosophers have struggled with that question for thousands of years, so it is unlikely that you will find an answer here. But you must have a working model of life before you can speculate on its occurrence on other worlds. To that end, you can identify two important aspects in living things, a physical basis and a unit of controlling information.

The Physical Basis of Life

On Earth, the physical basis of life is the carbon atom (■ Figure 26-1). Because of the way this atom bonds to other atoms, it can form long, complex, stable chains that are capable of extracting, storing, and utilizing energy. Other chemical bases of life may exist. Science fiction stories and movies abound with silicon creatures, living things whose body chemistry is based on silicon rather than carbon. However, silicon forms weaker bonds than carbon does, and it cannot form double bonds as easily. Consequently, it cannot form the long, complex, stable chains that carbon can. Silicon is 135 times more common on Earth than carbon is, yet there are no silicon creatures among us. All Earth life is carbon based. The likelihood that distant planets are inhabited by silicon people seems small, but you should not rule out life based on noncarbon chemistry.

In fact, nonchemical life might be possible. What is required is some mechanism capable of supporting the extraction and utilization of energy that has been identified as life. One could at least imagine life based on electromagnetic fields and ionized gas. No one has ever met such a creature, but science fiction writers conjure up all sorts.

Clearly, you could range far in space and time, theorizing about different bases for alien life, but to make progress you need to discuss what is known best—carbon-based life on Earth. How can a lump of carbon-rich matter live? The answer lies in the information that guides its life processes.

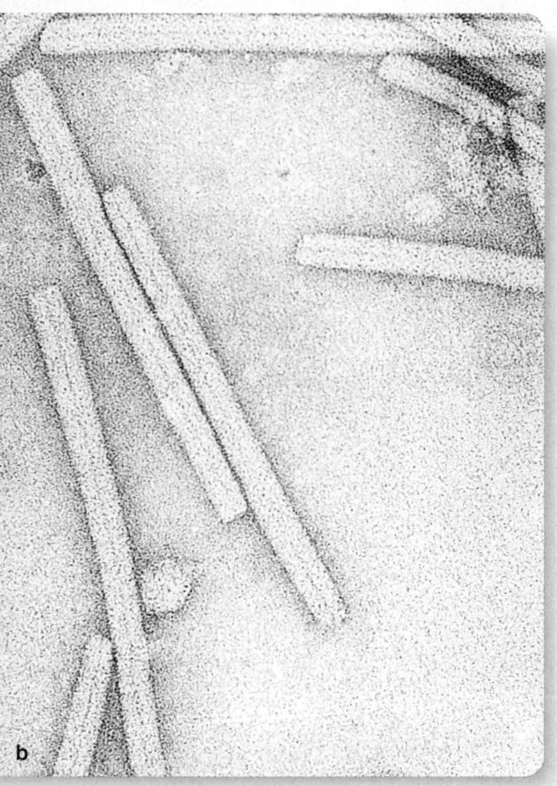

■ **Figure 26-1**

All living things on Earth are based on carbon chemistry. Even the long molecules that carry genetic information, DNA and RNA, have a framework defined by chains of carbon atoms. (a) Katie, a complex mammal, contains about 30 AU of DNA. (Michael Seeds) (b) Each rod of the tobacco mosaic virus contains a single spiral strand of RNA about 0.01 mm long. (L. D. Simon) All life on Earth stores its genetic information in such carbon-chain molecules.

Information Storage and Duplication

Living cells are tiny chemical factories. They must store all of the recipes for those chemicals in a safe place, use them to fulfill the cell's task, and hand down duplicates of the recipes to offspring. That information is encoded on long molecules.

Study **DNA: The Code of Life** on pages 654–655 and notice three important points:

1. The chemical recipes of life are stored as templates on DNA molecules. The templates automatically guide specific chemical reactions within the cell.
2. The instructions stored in DNA are the genetic information handed down to offspring. When people say, "You have your mother's eyes," they are talking about DNA codes.
3. Notice how the DNA molecule reproduces itself when a cell divides so that each new cell contains a copy of the original information.

Although the DNA molecule must preserve its coded information from damage and make accurate copies, it must also be capable of making mistakes. To see why, you must consider how new DNA recipes are created.

Modifying the Information

You are probably a little surprised to hear that DNA needs to make mistakes, but it is true. If living things are to survive for many generations, then the information stored in their DNA must be able to change as the environment changes. Without change in DNA, a slight warming of the climate, for example, might kill a species of plant, in turn starving the rabbits, deer, and other plant eaters, and leaving the hawks, wolves, and mountain lions with no prey. If the information stored in DNA could never change, then environmental changes would quickly drive life forms to extinction. If life is to survive in a changing world, then the information in DNA must be changeable. Living things must evolve.

Species evolve by **natural selection.** Each time an organism reproduces, its offspring receives copies of the data stored in the DNA, but some variation is possible. For example, most of the rabbits in a litter may be normal, but it is possible for one to get a DNA recipe that gives it stronger teeth. If it has stronger teeth, it may be able to eat something other than the plant the others depend on; and, if that plant is becoming scarce, the rabbit with stronger teeth has a survival advantage. It can eat other plants and so will be healthier than its littermates and have more offspring. Some of these offspring will also have stronger teeth, as the altered DNA recipes are handed down to the new generation. In this way, nature selects and preserves those attributes that contribute to the survival of the species. Those creatures that are unfit die. Natural selection is merciless to the individual, but it gives the species the best possible chance to survive in a changing environment.

The only way nature can obtain new DNA patterns from which to select the best is from DNA molecules that have changed. This can happen through chance mismatching of base pairs—mistakes—in the reproduction of the DNA molecule. Another way

DNA: The Code of Life

1 The key to understanding life is information — the information that guides all of the processes in an organism. In most living things on Earth, that information is stored on a long spiral molecule called **DNA (deoxyribonucleic acid).**

1a The DNA molecule looks like a spiral ladder with rails made of phosphates and sugars. The rungs of the ladder are made of four chemical bases arranged in pairs. The bases always pair the same way. That is, base A always pairs with base T, and base G always pairs with base C.

1b Information is coded on the DNA molecule by the order in which the base pairs occur. To read that code, molecular biologists have to "sequence the DNA." That is, they must determine the order in which the base pairs occur along the DNA ladder.

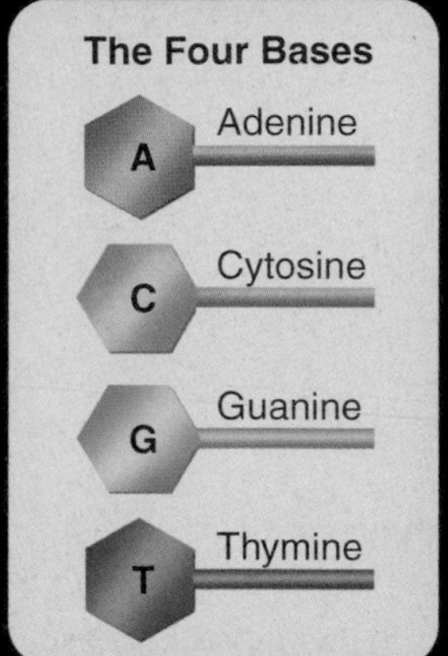

2 DNA automatically combines raw materials to form important chemical compounds. The building blocks of these compounds are relatively simple **amino acids.** Segments of DNA act as templates that guide the amino acids to join together in the correct order to build specific **proteins,** chemical compounds important to the structure and function of organisms. Some proteins called **enzymes** regulate other processes. In this way, DNA recipes regulate the production of the compounds of life.

The traits you inherit from your parents, the chemical processes that animate you, and the structure of your body are all encoded in your DNA.

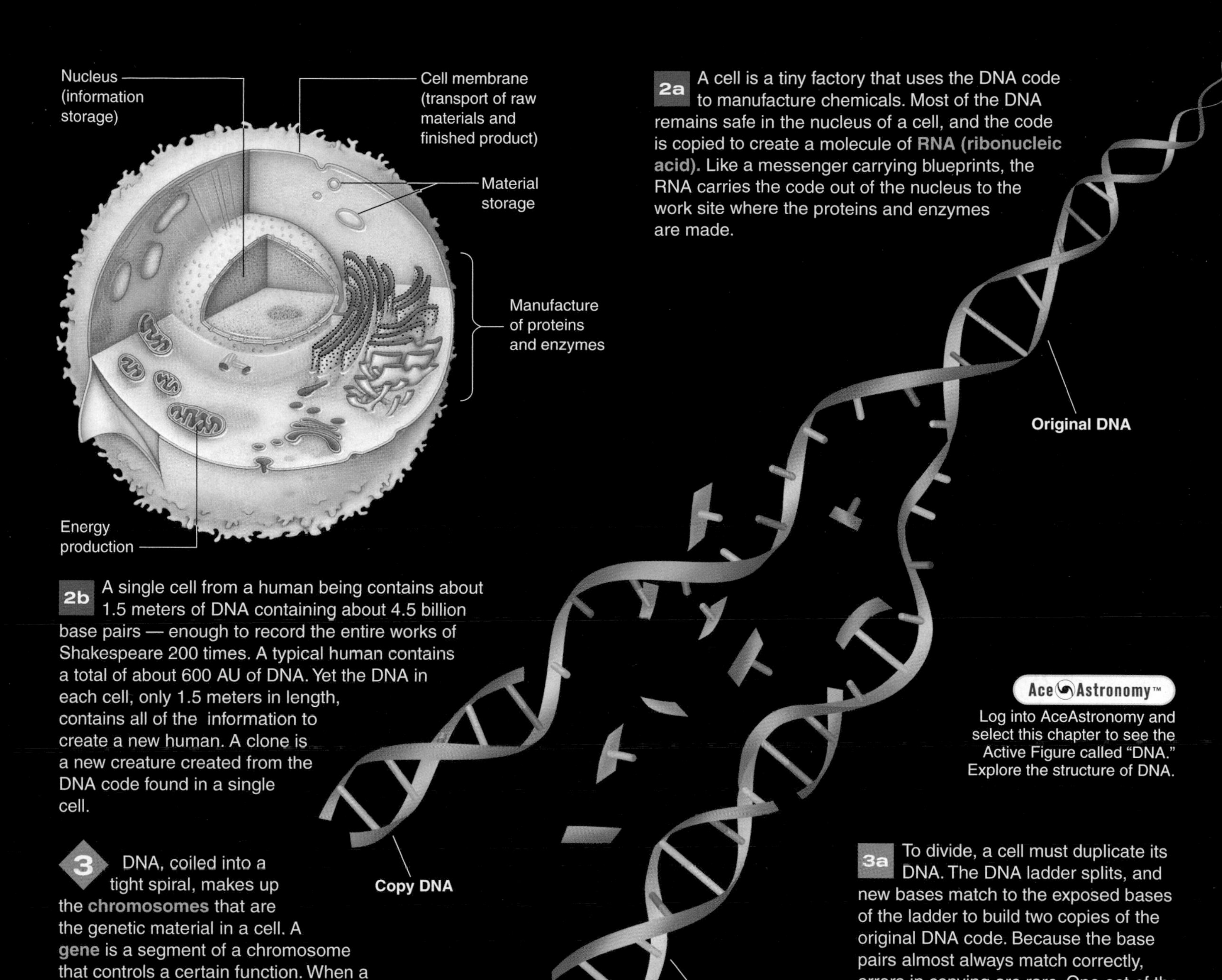

2a A cell is a tiny factory that uses the DNA code to manufacture chemicals. Most of the DNA remains safe in the nucleus of a cell, and the code is copied to create a molecule of **RNA (ribonucleic acid).** Like a messenger carrying blueprints, the RNA carries the code out of the nucleus to the work site where the proteins and enzymes are made.

2b A single cell from a human being contains about 1.5 meters of DNA containing about 4.5 billion base pairs — enough to record the entire works of Shakespeare 200 times. A typical human contains a total of about 600 AU of DNA. Yet the DNA in each cell, only 1.5 meters in length, contains all of the information to create a new human. A clone is a new creature created from the DNA code found in a single cell.

Ace Astronomy™

Log into AceAstronomy and select this chapter to see the Active Figure called "DNA." Explore the structure of DNA.

3 DNA, coiled into a tight spiral, makes up the **chromosomes** that are the genetic material in a cell. A **gene** is a segment of a chromosome that controls a certain function. When a cell divides, each of the new cells receives a copy of the chromosomes, as genetic information is handed down to new generations.

3a To divide, a cell must duplicate its DNA. The DNA ladder splits, and new bases match to the exposed bases of the ladder to build two copies of the original DNA code. Because the base pairs almost always match correctly, errors in copying are rare. One set of the DNA code goes to each of the two new cells.

Cell Reproduction by Division

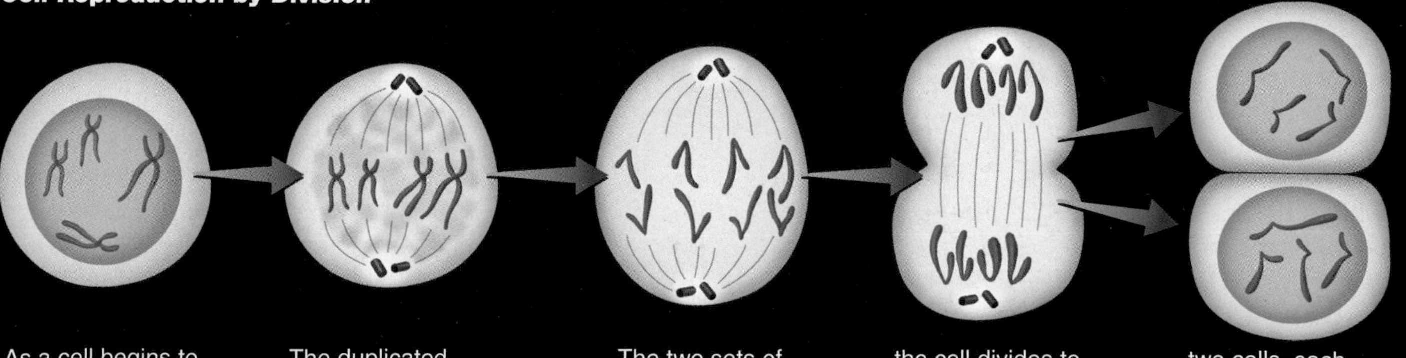

As a cell begins to divide, its DNA duplicates itself.

The duplicated chromosomes move to the middle.

The two sets of chromosomes separate, and . . .

the cell divides to produce . . .

two cells, each containing a full set of the DNA code.

this can occur is through damage to reproductive cells from exposure to radioactivity such as cosmic rays or natural radioactivity in the soil. In any case, an offspring born with altered DNA is called a **mutant.** Most mutations make no difference at all because they change segments of DNA that are not being used. Many mutations are fatal, and the individual dies long before it can have offspring of its own. But in rare cases, a mutation may give a species a new survival advantage. Then natural selection makes it likely that the new DNA message will survive and be handed down, making the species more capable of surviving.

Evolution is not random. Of course, the errors that occur in the DNA code are indeed random, but natural selection is not random. Those changes in the DNA that help a species survive are selected and preserved in future generations. With each passing generation, the species becomes more fit to survive in its environment.

Evolution is natural and automatic, but humans can control it. If you drink milk or eat cheese you are benefiting from the careful breeding of milk cows that has extended over centuries. Do you own a dog or a cat? Controlled evolution has created miniature poodles, huge Doberman pinschers, regal Siamese cats, and tailless Manx cats. But evolution works against us too. Some insecticides don't work anymore because the insects that were susceptible were killed off, and those that survived have a different gene that makes them immune. You have probably heard that some bacteria have built up a resistance to antibiotics. Again, the germs are evolving and handing on the genes—the bits of information—that help them survive.

Building Scientific Arguments

Why can't the information in DNA be permanent? Sometimes the most valuable scientific arguments are those that challenge common misconceptions. It seems so obvious that DNA codes must not change, but that's a common misconception. The information stored in a creature's DNA provides all of the recipes that make the creature what it is. For example, the DNA in a starfish must contain all the recipes for making the various kinds of proteins needed to consume and digest food. That information must be passed on to offspring starfish, or they will be unable to survive. But the information must be changeable because the environment is changeable. Ice ages come and go, mountains rise, lakes dry up, and ocean currents shift. If the environment changes in some way, one or more of the recipes may no longer work. In this example, a change in the temperature of the ocean water may kill off the specific shellfish the starfish eat. If they can't digest other shellfish, the entire species will become extinct. Natural variation in DNA means that among all the infant starfish in any generation, a few of the recipes are different; if the environment changes, all of the old-style starfish may die, but a few—those with the different DNA—can carry on.

The survival of life depends on this delicate balance between reliable reproduction and the introduction of small variations in DNA information. Now go the next step in your argument. **What are some of the ways these small changes in DNA can arise?**

■ ■ ■

Connections: Life is based not only on information but also on the duplication of information. Today, that process seems so complex that it is hard to imagine how it could have begun.

26-2 The Origin of Life

IF LIFE ON EARTH is based on the storage of information in these long, complex, carbon-chain molecules, how could it have ever gotten started? Obviously, 4.5 billion chemical bases didn't just happen to drift together to form the DNA formula for a human being. The key is evolution. Once a life form begins to reproduce itself, natural selection preserves the most advantageous traits. Over long periods of time spanning thousands, perhaps millions, of generations, the life form becomes more fit to survive. This often means the life form becomes more complex. That means life could have begun as a very simple process that gradually became more sophisticated as it was modified by evolution.

This is a critical point, so let's repeat it. Life is now very complex, but it didn't have to begin by some astonishingly unlikely combination of atoms to make a complicated creature. Life could have begun as a simple organism that automatically made copies of itself. Once reproduction began, natural selection made living things into the complex creatures in the world today. Of course, the key is evidence; what evidence exists to confirm this hypothesis?

The Origin of Life on Earth

The oldest fossils hint that life began in the sea. The first living things on Earth left behind a very poor fossil record. They were at first single-celled creatures; and, even when they became more complex, multicellular creatures, they contained no hard parts such as bones or shells that form good fossils. Yet a few traces of these earliest living things can be found, and they are all ocean creatures.

Although the oldest rocks contain no obvious fossils, microscopes reveal traces of ancient life. Rock from western Australia that is nearly 3.5 billion years old contains microscopic features that may be fossils of living things (■ Figure 26-2), and spectroscopic analysis reveals the presence of organic matter. While the experts debate the details in this controversial field, you can conclude that life was present on Earth as simple organisms at least 3.5 billion years ago. That's roughly one billion years after Earth formed.

A little over a half-billion years ago something happened, perhaps a change in Earth's climate, and life exploded into a wide di-

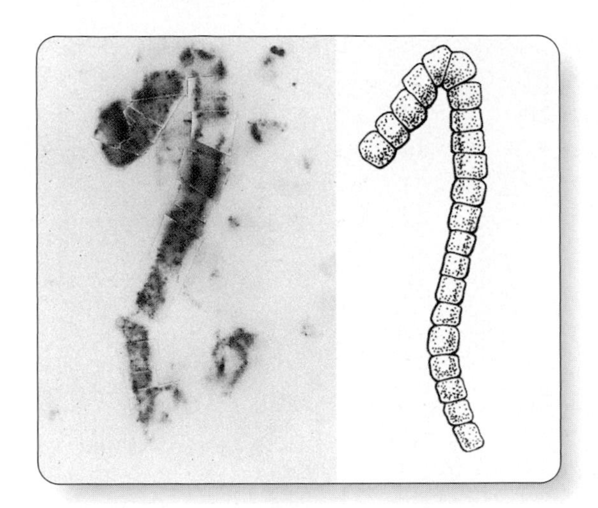

■ **Figure 26-2**

Among the oldest fossils known, this microscopic filament resembles modern bacterial forms (artist's reconstruction at right). This fossil was found in the 3.5-billion-year-old chert of the Pribara Block in northwestern Australia. (Courtesy J. William Schopf)

■ **Figure 26-3**

Trilobites made their first appearance in the Cambrian oceans. The smallest were almost microscopic, and the largest were bigger than dinner plates. This example, about the size of a human hand, lived 400 million years ago in an ocean floor that is now a limestone deposit in Pennsylvania. (Grundy Observatory photograph)

versity of complex forms. This marks the beginning of the **Cambrian period** and is sometimes called the *Cambrian explosion.* The best known of the Cambrian creatures may be the trilobites (■ Figure 26-3). Although the Cambrian creatures were diverse, there are no Cambrian fossils of land plants or animals. Evidently, land surfaces were totally devoid of life until only 400 million years ago.

The fossil record shows that life began as simple organisms in the sea soon after Earth formed. Only recently, in geological terms, has life become complex. It isn't difficult to understand how simple creatures evolve to become complex. The problem is trying to understand how nonliving atoms became simple creatures. That's the hard bit right at the beginning.

The key to the origin of this life may lie in an experiment performed by Stanley Miller and Harold Urey in 1952. This **Miller experiment** sought to reproduce the conditions on Earth under which life began. In a closed glass container, the experimenters placed water (to represent the oceans); the gases hydrogen, ammonia, and methane (to represent the primitive atmosphere); and an electric arc (to represent lightning bolts). The apparatus was sterilized, sealed, and set in operation (■ Figure 26-4).

After a week, Miller and Urey stopped the experiment and analyzed the material in the flask. Among the many compounds the experiment produced, they found four amino acids (building blocks of protein), various fatty acids, and urea, a molecule common to many life processes. Evidently, the energy from the electric arc had molded the atmospheric gases into some of the basic components of living matter. Other energy sources, such as hot silica (to simulate hot lava spilling into the sea) and ultraviolet radiation (to simulate sunlight), produce similar results.

More recent studies of the composition of meteorites and models of planet formation suggest that Earth's first atmosphere did not resemble the gases used in the Miller experiment. Earth's first atmosphere was probably composed of carbon dioxide, nitrogen, and water vapor. This finding, however, does not invalidate the Miller experiment. When such gases are processed in a Miller apparatus, some organic molecules appear, although not as many as in the first trials.

The Miller experiment did not create life, of course, nor did it necessarily imitate the exact conditions on the young Earth. That is not the lesson to take from this famous experiment. Rather, it is important because it shows that complex organic molecules form naturally in a wide variety of circumstances. The chemical deck is stacked to deal nature a hand of complex molecules. If you could travel back in time, you would probably find Earth's first oceans filled with a rich mixture of organic compounds in what some have called the **primordial soup.**

The next step on the journey toward life is for the compounds dissolved in the oceans to link up and form larger molecules. Amino acids, for example, can link together to form proteins. This linkage occurs when amino acids join together end-to-end and release a water molecule (■ Figure 26-5). For many years, experts

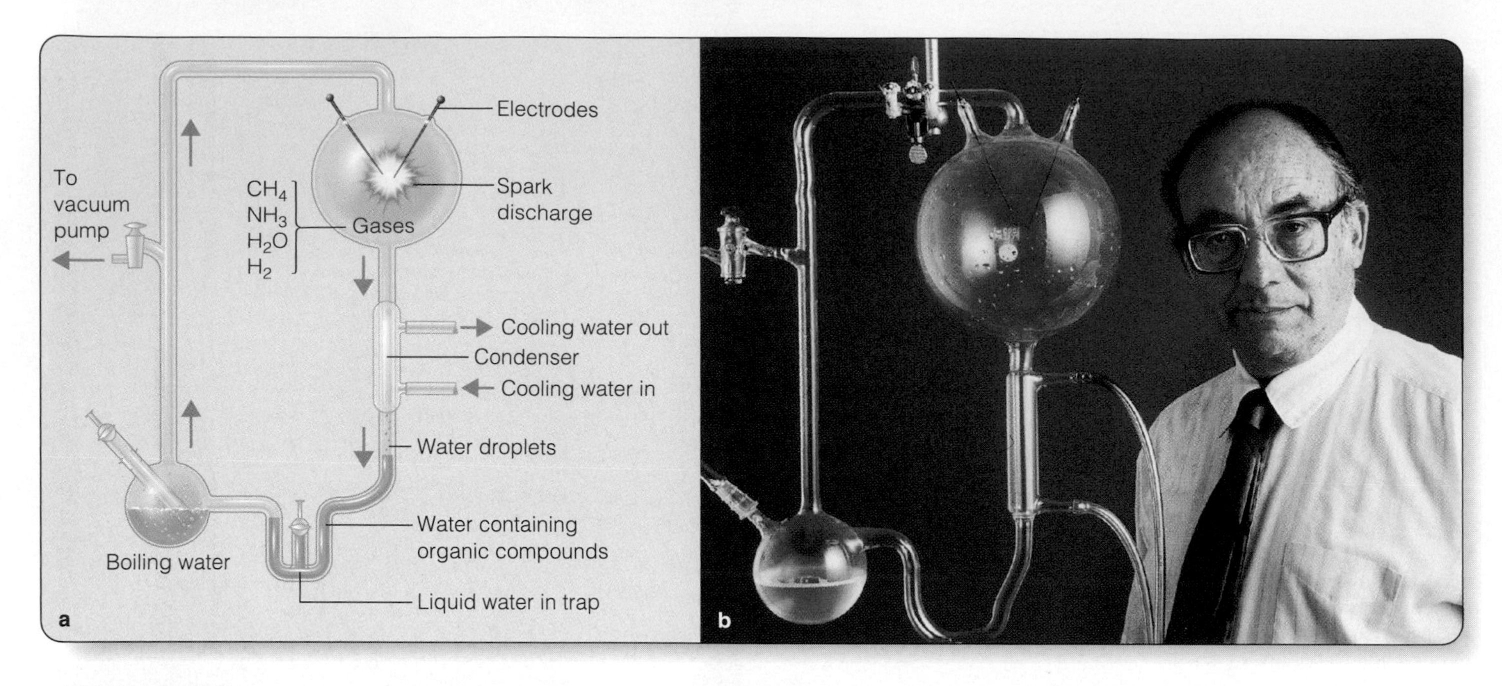

Figure 26-4

(a) The Miller experiment circulated gases through water in the presence of an electric arc. This simulation of primitive conditions on Earth produced amino acids, the building blocks of proteins. (b) Stanley Miller with a Miller apparatus. (Courtesy Stanley Miller)

have assumed that this process must have happened in sun-warmed tidal pools where evaporation concentrated the broth. But more recent studies suggest that the young Earth was subject to extensive volcanism and large meteorite impacts that periodically modified the climate enough to destroy any life forms exposed on the surface. Rather it seems likely that the early growth of complex molecules took place among the hot springs along the midocean ridges. These complex molecules would not have been photosynthetic—taking energy from sunlight. Rather they would have taken energy from the heat and chemicals emerging from the hot springs of the midocean ridge. That energy could have powered the growth of long protein chains. Deep in the oceans, they would have been safe from climate changes.

Although these proteins might have contained hundreds of amino acids, they would not have been alive. Not yet. Such molecules would not have reproduced but would have merely linked together and broken apart at random. Because some molecules are more stable than others, however, and because some molecules bond together more readily than others, this automatic **chemical evolution** would have led to the concentration of the varied

Figure 26-5

Amino acids can link together through the release of a water molecule to form long carbon-chain molecules. The amino acid in this hypothetical example is alanine, one of the simplest.

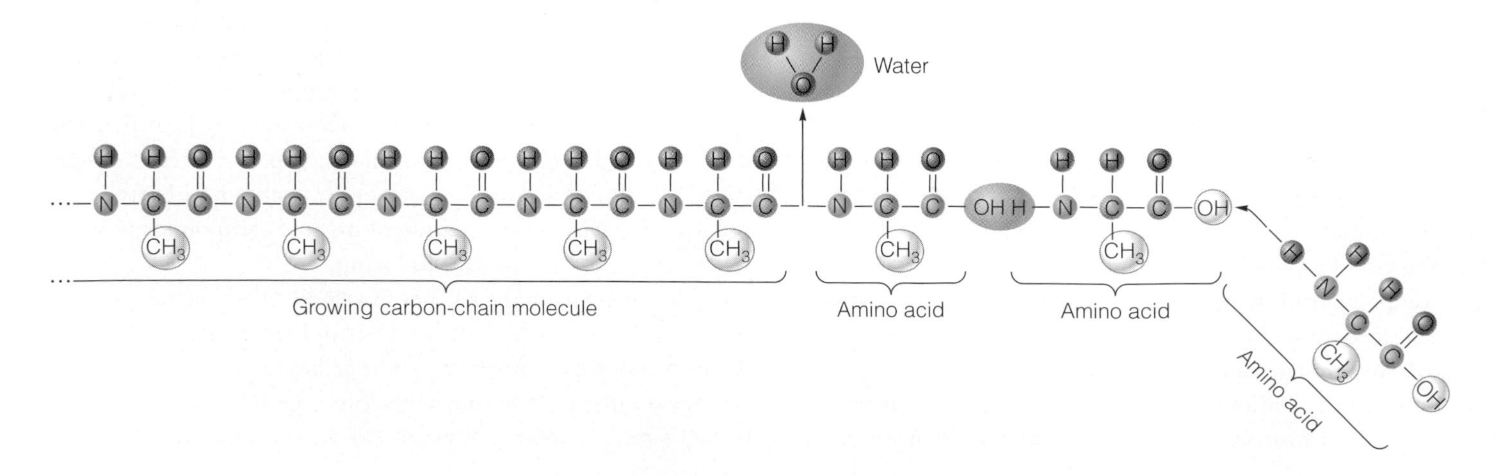

smaller molecules into the most stable larger forms. Eventually, somewhere in the oceans, a molecule took shape that automatically made copies of itself. At that point, the chemical evolution of molecules became the biological evolution of living things.

You might enjoy speculating about an alternative theory that proposes that primitive living things such as reproducing molecules did not originate on Earth but came here in meteorites or comets. Radio astronomers have found a wide variety of organic molecules in the interstellar medium, and some studies have found similar compounds inside meteorites (■ Figure 26-6). Such molecules form so readily that you should be surprised if they were not present in space. A few investigators, however, have speculated that living, reproducing molecules originated in space and came to Earth as a cosmic contamination. If this were true, every planet in the universe would be contaminated with the seeds of life. You may find this theory fun to think about, but it is presently untestable, and an untestable theory is of little use in science.

Whether life originated in the oceans or in space, you still face the problem of how life began—that hard bit at the beginning in which nonliving atoms become reproducing molecules. The exact details are not yet understood, but the important point is that natural processes could have done it. Just because you don't understand something doesn't mean it is impossible. The evidence seems conclusive that life began as reproducing molecules in Earth's oceans.

Which came first, reproducing molecules or the cell? Because you probably think of the cell as the basic unit of life, this question seems to make no sense, but in fact the cell may have originated during chemical evolution. If a dry mixture of amino acids is heated, the acids form long, proteinlike molecules that, when poured into water, collect to form microscopic spheres that function in ways similar to cells (■ Figure 26-7). They have a thin membrane surface, they can absorb material from their surroundings, they grow in size, and they can divide and bud just as cells do. They contain no large molecule that copies itself, however. So it is possible that the structure of the cell originated first and the reproducing molecules later.

An alternative theory proposes that the replicating molecule developed first. Such a molecule would have been exposed to damage if it had been bare, so the first to manufacture or attract a protective coating of protein would have had a significant survival advantage. If this was the case, the protective cell membrane was a later development of biological evolution.

The first living things must have been single-celled organisms much like modern bacteria. Some of the oldest fossils known are **stromatolites,** structures produced by communities of photosynthesizing bacteria that grew in mats and, year by year, deposited layers of minerals that were later fossilized. One of the

■ Figure 26-6

A sample of the Murchison meteorite, a carbonaceous chondrite that fell in 1969 near Murchison, Australia. Analysis of the interior of the meteorite revealed evidence of amino acids. Whether the first building blocks of life originated in space is unknown, but the amino acids found in meteorites illustrate how commonly amino acids and other complex molecules occur even in the absence of living things. (Courtesy Chip Clark, National Museum of Natural History)

■ Figure 26-7

Single amino acids can be assembled into long proteinlike molecules. When such material cools in water, it can form microspheres, microscopic spheres with double-layered boundaries similar to cell membranes. Microspheres may have been an intermediate stage in the evolution of life between complex molecules and cells holding molecules reproducing genetic information. (Courtesy Sidney Fox and Randall Grubbs)

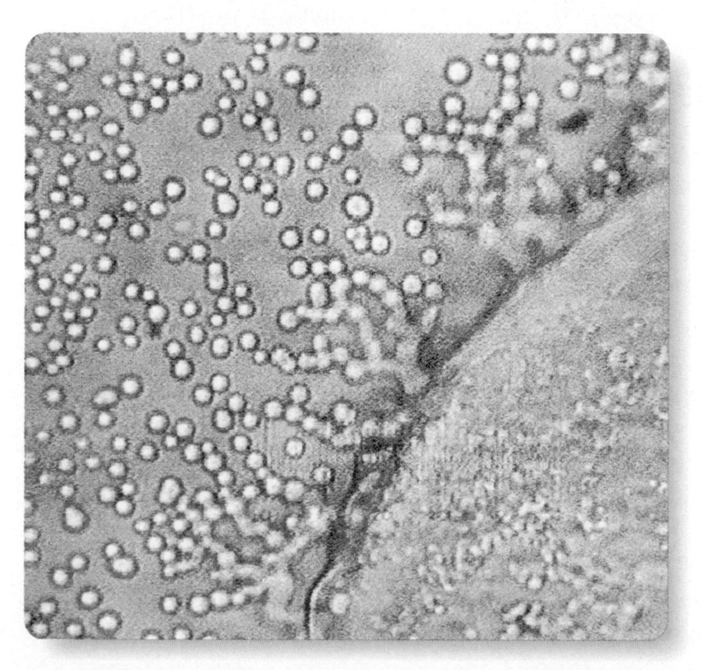

■ **Active Figure 26-8**

A 3.5-billion-year-old fossil stromatolite from western Australia is one of the oldest known fossils (inset). Stromatolites were formed, layer by layer, by mats of bacteria living in shallow water. Such life may have been common in shallow seas when Earth was young. Stromatolites are still being formed today in similar environments. (Mural by Peter Sawyer; photo courtesy Chip Clark, National Museum of Natural History)

Ace Astronomy™ **Log into AceAstronomy and select this chapter to see the Active Figure "Future of the Sun." Watch Earth's temperature change as the sun ages.**

oldest such fossils known is believed to be 3.5 billion years old (■ Figure 26-8). If such bacteria were common when Earth was young, the early atmosphere may have contained a small amount of oxygen produced by the photosynthesis. Atmospheric models suggest that an oxygen abundance of only 0.1 percent would have been sufficient to provide an ozone screen that would protect organisms from the sun's ultraviolet radiation.

How evolution shaped creatures to live in the ancient oceans, to photosynthesize and respire, to become multicellular, and to reproduce sexually is a fascinating story, but this chapter cannot explore those details. You can see that the hard bit in the story is how life began, and there appear to be natural chemical processes that could have led to molecules automatically duplicating themselves. Once some DNA-like molecule formed, evolution was automatic; and, over billions of years, the genetic information stored in living things accumulated those qualities that favored survival, and unsuccessful variations did not get passed down to new generations. As Samuel Butler said, "The chicken is the egg's way of making another egg." In that sense, all living matter on Earth is the physical expression of DNA's automatic tendency to continue its existence.

Perhaps this seems harsh. Human experience goes far beyond mere reproduction. *Homo sapiens* has art, poetry, music, philosophy, religion, science. Perhaps all of the great accomplishments of our intelligence represent more than mere reproduction of DNA. Nevertheless, intelligence, the ability to analyze complex situations and respond with appropriate action, must have begun as a survival mechanism. For example, a fixed escape strategy stored in the DNA is a disadvantage for a creature that frequently moves from one environment to another. A rodent that always escapes from predators by automatically climbing the nearest tree would be in serious jeopardy if it met a hungry fox in a treeless clearing. Even a faint glimmer of intelligence might allow the rodent to analyze the situation and, finding no trees, to choose running over climbing. Intelligence, of which *Homo sapiens* is so proud, may have developed in ancient creatures as a way of making them more versatile without adding tremendous amounts of information to their DNA library.

Any discussion of the evolution of life seems to involve highly improbable coincidences until you consider how many years have passed in the history of Earth. You can read the words—4.6 billion years—so easily, but it is in truth hard to grasp the meaning of such a long period of time.

Geologic Time

Humanity is a very new experiment on planet Earth. You can fit the history of Earth into a single chart such as ■ Figure 26-9. You can take comfort in thinking that creatures like us have walked on Earth for roughly 3 million years, but when you add our history to the chart, you will discover that the entire history of humanity makes up no more than a thin line at the top. In fact, if you tried to represent the entire 4.6-billion-year history of Earth on the chart, the portion describing the rise of complex life after the Cambrian explosion would be an unreadably small segment. It is expanded in Figure 26-9 just to make it legible.

One way to represent the evolution of life is to compress the 4.6-billion-year history of Earth into a one-year-long video program. In such a program, Earth forms as the video begins on January 1; through all of January and February it cools and is cratered, and the first oceans form. Search as you might, you would find no trace of life in these oceans until sometime in March or early April, when the first living things develop. You would need a microscope to see them. The slow development of these simplest of living forms grinds on through the spring and summer of the video. The entire 4-billion-year history of Precambrian evolution lasts until the video reaches mid-November, when the primitive ocean life explodes into the more complex Cambrian organisms such as trilobites and many other specialized creatures (■ Figure 26-10).

While the yearlong video plays on and on, you might amuse yourself by looking at the land instead of the oceans, but you would be disappointed. The land is a lifeless waste with no plants or animals of any kind. Not until November 28 in the video does life appear on the land; but, once it does, it evolves rapidly into a wide range of plants and animals. Dinosaurs, for example, appear about December 12 and vanish by Christmas evening as mammals and birds flourish.

Throughout the one-year run of the video there have been no humans; and even during the last days of the year, as the mammals rise to dominate the landscape, there are no people. In the early evening of December 31, vaguely human forms move through the grasslands, and by late evening they begin making stone tools. The Stone Age lasts till about 11:45 PM, and the first signs of civilization, towns and cities, do not appear until 11:54 PM. The Christian era begins only 14 seconds before the New Year, and the Declaration of Independence is signed with but 1 second to spare.

By imagining the history of Earth as a yearlong video, you can place the rise of life in perspective. Tremendous amounts of time were needed for the first simple living things to evolve in the oceans, and even more time was needed for the evolution of complex creatures that could colonize the land. As life became more complex, it evolved and diversified faster and faster, as if evolution were drawing on a growing library of solutions that had been previously invented with great effort to solve earlier problems. The burst of diversity on land led eventually to the rise of intelligent creatures like us, a process that has taken 4.6 billion years.

If life could originate on Earth and develop into intelligent creatures, perhaps the same thing could have happened on other planets. This raises three questions. First, could life originate on another world if conditions were suitable? No one knows for sure, but you have found natural, chemical processes that could lead to living things, so your answer to this question should probably be yes. The second question is, will life always evolve toward intelligence? Perhaps intelligence arose on Earth under unusual conditions, but you should recall that intelligence appears to be a way to make a species more versatile and consequently more likely to survive. If that is true, then intelligence may develop under a wide range of conditions if enough time is available. But what of the third question: Are suitable conditions so rare that life almost never gets started? The only way to answer that is to search for life on other planets. You can begin in the next section by searching for life on the other planets in our solar system.

Life in Our Solar System

Although science fiction writers imagine life based on something other than carbon chemistry, no one knows of any real examples to illustrate how such life might arise and survive. To make progress in your search for life, you need to limit your discussion to life as it is known on Earth and the conditions it requires. The most important requirement is the presence of liquid water, not only as part of the chemical reactions of life but also as a medium to transport nutrients and wastes within the organism. Also, it seems that life on Earth began in the oceans and developed there for nearly 4 billion years before it was able to emerge onto the land. Certainly, any world where you hope to find life must have liquid water, and that means it must have moderate temperatures.

The liquid water requirement automatically eliminates many worlds in our solar system. The moon is airless, and although some data suggest ice frozen in the soil at its poles, it has never had liquid water on its surface. In the vacuum of the lunar surface, liquid water would boil away rapidly. Mercury too is airless and cannot have had liquid water on its surface for long periods of time. Venus has some traces of water vapor in its atmosphere, but it is much too hot for liquid water to survive. If there were any lakes or oceans of water on its surface when it was young, they must have evaporated quickly. Even if life began there, no traces would be left now.

The inner solar system seems too hot, and the outer solar system seems too cold. The Jovian planets have deep atmospheres; and, at a certain level, they have moderate temperatures where water might condense into liquid droplets. But it seems unlikely that life could begin there. The Jovian planets have no surfaces where oceans could nurture the beginning of life, and currents in

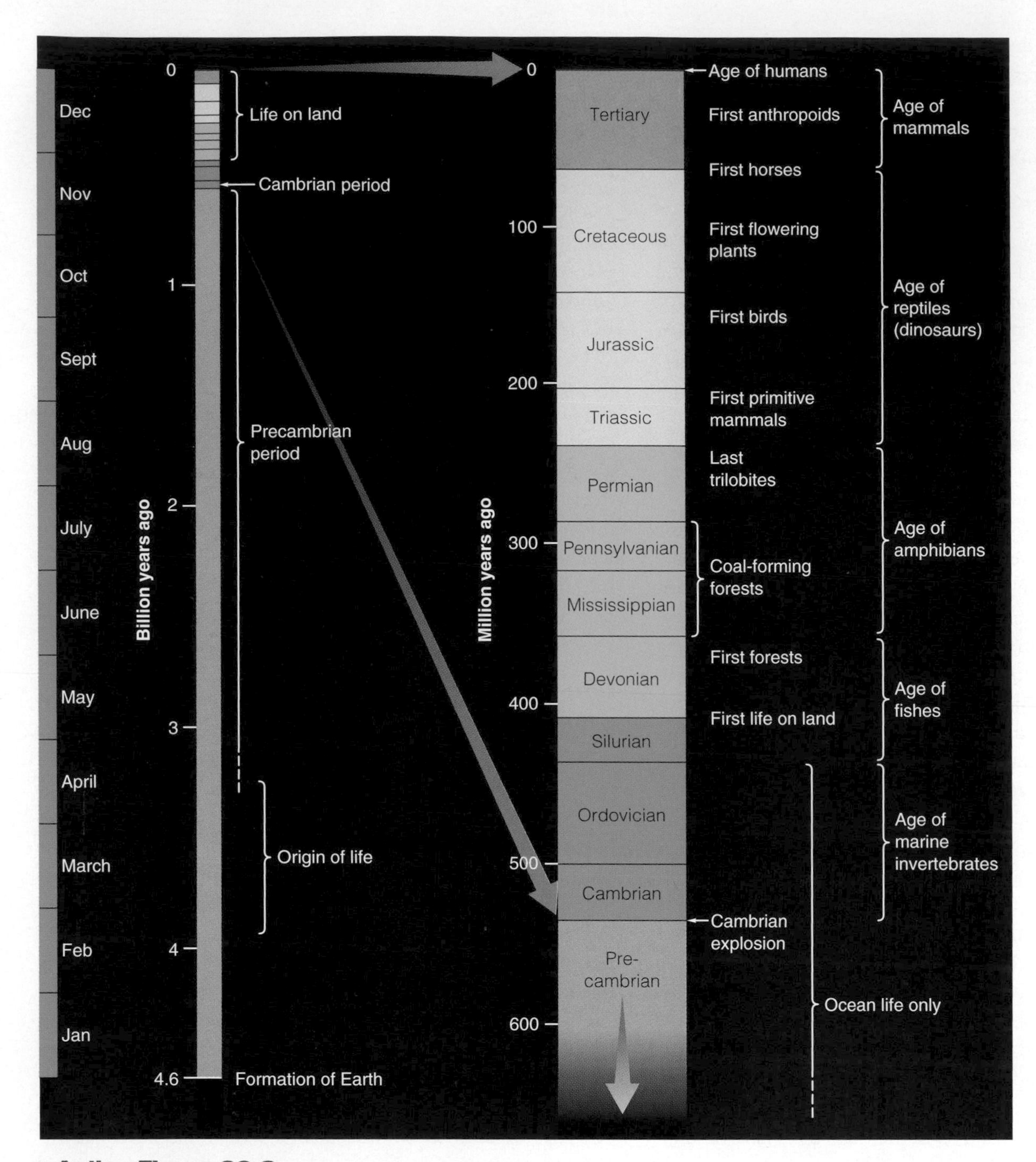

■ Active Figure 26-9

Complex life has developed on Earth only recently. If the entire history of Earth were represented in a time line (left), you would have to magnify the end of the line to see details such as life leaving the oceans and dinosaurs appearing. The age of humans would still be only a thin line at the top of your diagram. If the history of Earth were a yearlong videotape, humans would not appear until the last hours of December 31.

Ace Astronomy™ Log into AceAstronomy and select this chapter to see the Active Figure "Earth Calendar." Animate this diagram.

the atmosphere seem destined to circulate gas and water droplets from regions of moderate temperature to other levels that are much too hot or too cold for life to survive.

A few of the satellites of the Jovian planets might have suitable conditions for life. Jupiter's moon Europa seems to have a liquid-water ocean below its icy crust (see Figure 23-12), and minerals dissolved in that water would provide a rich broth of possibilities for chemical evolution. Nevertheless, Europa is not a promising site to search for life because conditions may not have remained stable for the billions of years needed for life to evolve beyond the microscopic stage. The subsurface ocean is kept from freezing by tidal heating. If Jupiter's moons interact gravitationally and modify their orbits, Europa may have been frozen solid at some points in history. Such periods of freezing would proba-

■ **Figure 26-10**

Before the Cambrian explosion, life consisted mostly of single-celled creatures living alone or in organized colonies. During the Cambrian period, life became complex. Anomalocaris (upper left) was about the size of your hand and had specialized organs including eyes, coordinated fins for active swimming, gripping mandibles capable of raking through the seafloor and capturing prey, and a powerful, toothed maw located below its head (upper right). Notice Opabinia at center with its long snout. (D. W. Miller)

bly prevent life from developing. Drilling through the icy crust of Europa to search for life in its ocean will be a wonderful adventure for future generations, but Europa does not seem to be a good bet to harbor any form of complex life.

Saturn's moon Titan has an atmosphere of nitrogen, argon, and methane and may have lakes or small oceans of liquid methane on its surface. You saw in Chapter 23 how sunlight can convert the methane in the atmosphere to organic smog particles that settle to the surface, and the images radioed back to Earth by the Huygens probe as it parachuted down to the surface of Titan showed that dark material filled drainage channels and lowlands. The chemistry of life that might crawl or swim on such a world is unknown, but life there may be unlikely because of the temperature. The surface of Titan is a deadly −179°C (−290°F). Chemical reactions occur slowly or not at all at such low temperatures, so the chemical evolution needed to begin life may never have occurred on Titan.

Mars is the most likely place for life in our solar system. The evidence, however, is not encouraging. In 1976, two robotic spacecraft, Viking 1 and Viking 2, landed at two different places on the Martian surface. The spacecraft scooped up soil samples and subjected them to tests for the presence of living organisms. For example, they gave some soil samples a dose of nutrient-rich water and watched for signs the nutrients were taken up by biological processes. The results seem negative. Although some peculiar chemical processes were detected, no evidence was found that clearly indicated the presence of any living things in the soil.

Approximate true-color image

■ **Figure 26-11**

Mars rover Opportunity descended into the crater Endurance and then looked back up the crater wall. In places, Opportunity found layering in the rocks that showed they had formed as layers of sand and silt deposited by flowing water. The rover also found minerals that showed that water was once present at the site. Signs of past life may lie hidden in these rocks. (NASA/JPL/Cornell)

Rovers on Mars and orbiting spacecraft have found dramatic evidence that liquid water once flowed over the surface (■ Figure 26-11), but that does not mean that life did originate there. Only detailed study of the rocky soil can answer that question, and that means Earth must send a geologist to Mars or bring rocks from Mars back to Earth. Nature has performed the latter task for us.

Meteorite ALH84001 (■ Figure 26-12a) was found on the Antarctic ice in 1984. Years later an analysis of its chemical composition showed that it had originated on Mars. Being a good, skeptical scientist, you are probably wondering how anyone could know that the meteorite came from Mars. The evidence is solid.

■ Figure 26-12

(a) Meteorite ALH84001 is one of many meteorites known to have originated on Mars. It was claimed that the meteorite contained chemical and physical traces of ancient life on Mars, including what appear to be fossils of microscopic organisms (b). The evidence has not been confirmed, and the validity of the claim is highly questionable. (NASA)

The Viking landers measured the abundance of oxygen isotopes on Mars, and the pattern of isotopes was different from Earth and from meteorites. But the isotopes in ALH84001 are a precise match with the Viking results. The meteorite must have come from Mars.

ALH84001 was probably part of debris ejected into space by a large impact on Mars. Some of that debris fell to Earth, and a small number of these meteorites have been found, mostly in Antarctica because that's the easiest place to find meteorites. You might have heard of ALH84001; it was in the news because a team of scientists studied it and announced in 1996 that it contained chemical and physical traces of ancient life on Mars (Figure 26-12b).

The discovery was received with great excitement by the press and the public. It would, indeed, be dramatic evidence if true, because it would show that life could begin on another world. Scientists were excited too, but being professionally skeptical (see Window on Science 19-3), they began testing the results immediately. In many cases, the tests did not confirm the conclusion that life once existed on Mars. Some chemical contamination from water on Earth has occurred, and some chemicals in the meteorite may have originated without the presence of life. The physical features that look like fossil bacteria may be mineral formations in the rock. Although studies of ALH84001 continue, it does not provide unchallenged evidence that life once existed on Mars. Carl Sagan once said, "Extraordinary claims require extraordinary proof," and in this case the evidence is at best inconclusive.

Spacecraft now visiting Mars are revealing the past history of water there and painting a more detailed picture of present conditions. Spacecraft are planned that will eventually land on Mars, collect rocks, and return them to Earth. Nevertheless, conclusive evidence may have to wait until a geologist in a spacesuit can scramble down the dry streambeds of Mars cracking open rocks and searching for fossils.

Your search for life on the other worlds in our solar system has been inconclusive. You didn't find life, but you didn't find evidence to completely rule it out. So far as anyone knows at present, our solar system is bare of life except for Earth. Consequently the search for life in the universe takes you to other planetary systems.

Life in Other Planetary Systems

Might life exist in other solar systems? To consider this question, you need to consider how common planets are and what conditions a planet must fulfill for life to originate and evolve to intelligence. The first question is astronomical; the second is biological. The ability of scientists to discuss the problem of life outside our solar system is severely limited by lack of experience. They know only one planet well—Earth.

In Chapter 19 you learned that planets form as a natural byproduct of star formation and that a number of extrasolar planets have been found circling nearby stars. From this you can conclude that planetary systems are very common.

If a planet is to become a suitable home for life, it must have a stable orbit around its sun. This is simple in a solar system like our own, but in a binary system most planetary orbits are unstable. Most planets in such systems would not last long before they were swallowed up by one of the stars or ejected from the system. Consequently, single stars are the most likely to have planets suitable for life. Because our galaxy contains at least 10^{11} stars, half of which are single, there could be roughly 5×10^{10} planetary systems in which life might exist.

A few million years of suitable conditions does not seem to be enough time to originate life. On our planet it took about a billion years for the first living things to appear and 4.6 billion years for intelligence to evolve. Clearly, conditions on a planet must remain acceptable over a long time. This eliminates massive stars that remain stable on the main sequence for only a few million years. If it takes a few billion years for life to originate and evolve to intelligence, no star hotter than about F5 will do. This is not really a serious restriction, because upper-main-sequence stars are rare anyway.

In previous sections, you saw how life on Earth depends on water. In the past, astronomers have used that to define a **life zone** (or ecosphere) around a star, a region within which planets have temperatures that permit the existence of liquid water. A cool star has a small life zone, and a hot star has a large life zone. The life zone around the sun extends from about the orbit of Venus to the orbit of Mars.

For a number of decades astronomers have used the life zone to eliminate lower main-sequence stars as candidates for life. The M dwarfs at the bottom of the main sequence are so faint that a planet would have to orbit very close to stay warm. Some astronomers have argued that such a planet would become tidally locked to its star, and one side of the planet would be in perpetual darkness. Water would tend to freeze out on the dark side and end chances for life to develop. Whether the atmosphere could distribute heat to the dark side is unknown. Also, M stars sometimes suffer violent flares that might make nearby planets unsuitable for life. Most astronomers thought the dinky red dwarfs were not good candidates for life. However, extrasolar planets have been discovered orbiting these M stars, and the large numbers of these extrasolar planets have lead some astronomers to suggest that at least some might be stable enough to develop life.

Recent discoveries, however, are making the whole idea of a life zone seem just a little bit of an oversimplification. For one thing, scientists on Earth are finding living things in astonishing places, such as the bottom of icy lakes in Antarctica and far underground in solid rock. Living things have been found in boiling springs where the water is highly acidic. It appears that life is much tougher than some scientists thought. Also, spacecraft visiting other worlds have found conditions that might support life outside the so-called life zone of the sun. The biggest water ocean in our solar system is under the ice on Jupiter's moon Europa. Liquid has flooded parts of Neptune's moon Triton, and Saturn's moon Titan has a complex surface chemistry of carbon-rich organic molecules. It isn't hard to imagine life in these environments, and they all lie outside the conventional life zone of the sun. Perhaps the life zone is most useful as a warning that life is proving to be more varied than many people expected.

One restriction on life is the slow evolution of a planet's star. You learned in Chapter 9 that main-sequence stars gradually grow more luminous as they convert their hydrogen to helium. The sun was only about 70 percent as luminous when it formed as it is today. As a star grows more luminous, it should slowly warm its planets. A planet might have liquid water on its surface long enough for life to begin but then be sterilized as its star slowly grows more luminous and drives away the planet's atmosphere and oceans. Perhaps a planet must form at just the right distance from its star to remain hospitable to life for billions of years.

The fundamental question is simple. If conditions are right, will life begin? Early in this chapter you found natural chemical processes that seem capable of forming molecules that make copies of themselves. You saw that evolution would automatically shape these molecules to be most likely to survive. Perhaps the fundamental question should be, "What could prevent life from beginning?" Given what you know about life, it seems that it should arise whenever conditions permit, and our galaxy should be filled with planets that are inhabited with living creatures.

Building Scientific Arguments

What evidence can you cite that life is at least possible on other worlds?

A good scientific argument is based on evidence, but it also includes careful analysis. The evidence is limited almost entirely to Earth, but it is promising. Fossils show that life originated in Earth's oceans almost 4 billion years ago, and biologists have proposed relatively simple chemical processes that could have formed these first reproducing molecules. Fossils show that life developed from a very slow beginning into more and more complex creatures that filled the oceans. The pace of evolution quickened dramatically about half a billion years ago, the beginning of the Cambrian period, when life began taking on complex forms; later, when life emerged onto the surface of the land, it evolved rapidly to produce the tremendous diversity of today. Human intelligence has been a very recent development; it is only a few million years old.

If this process occurred naturally on Earth, then it seems reasonable that it could have occurred on other worlds as well. This conclusion hinges mostly on a reasonable belief that the laws of nature apply on other worlds as they do on Earth. Now expand your argument: **What conditions should you expect of other worlds that host life?**

■ ■ ■

Connections: It is both easy and fun to speculate about life on other worlds, but it leads to a simple question: If there really is life on other worlds, why can't we detect it? Why haven't we heard from alien civilizations?

26-3 Communication with Distant Civilizations

IF OTHER CIVILIZATIONS EXIST, perhaps we Earthlings can communicate with them in some way. Sadly, travel between the stars is more difficult in real life than in science fiction—and may

Window on Science 26-2

UFOs and Space Aliens

If you discuss life on other worlds, then you might be tempted to use UFO sightings and supposed visits by aliens from outer space as evidence to test your hypotheses. Scientists don't do so for two reasons, both related to the reliability of these observations.

First, the reputation of the sources of UFO sightings and alien encounters does not inspire confidence that these data are reliable. Most people hear of such events via grocery store tabloids, daytime talk shows, or sensational "specials" on viewer-hungry cable networks. You must consider the low reputation of the media that report UFOs and space aliens. Most of these reports are simply made up for the sake of sensation, and you should not try to use them as reliable evidence.

Second, the remaining UFO sightings, those not simply made up, do not survive careful examination. Most are mistakes or unconscious misinterpretations of natural events made by honest people. A number of unbiased studies have found no grounds for believing in UFOs. In short, there is no conclusive evidence that Earth has been visited by aliens from space.

That's too bad. A confirmed visit by intelligent creatures from beyond our solar system would answer many questions. It would be exciting, enlightening, and, like any real adventure, a bit scary. But none of the UFO sightings is dependable, and we are left with no direct evidence of intelligent life on other worlds.

UFOs from space are fun to think about, but there is no evidence that they are real.

in fact be impossible. If physical visits are impossible, perhaps radio communication is the way to say hello. Again, nature places restrictions on such conversations, but the restrictions are not too severe. As you will see, the real problem lies with the life expectancy of civilizations.

Travel between the Stars

Practically speaking, roaming among the stars is tremendously difficult because of three limitations: distance, speed, and fuel. The distances between stars are almost beyond comprehension. It does little good to explain that if you use a golf ball in New York City to represent the sun, the nearest star would be another golf ball in Chicago. It is only slightly better to note that the fastest commercial jet would take about 4 million years to reach the nearest star.

The second limitation is a speed limit—you cannot travel faster than the speed of light. Though science fiction writers invent hyperspace drives so their heroes can zip from star to star, the speed of light is a natural and unavoidable limit that cannot be exceeded. This, combined with the large distances between stars, makes interstellar travel very time consuming.

The third limitation is that you can't even approach the speed of light without using a fantastic amount of fuel. Even if you ignore the problem of escaping from Earth's gravity, you must still use energy stored in fuel to accelerate to high speed and to decelerate to a stop when you reach our destination. To return to Earth, assuming you wish to, you have to repeat the process, and that takes more fuel.

These changes in velocity require a tremendous amount of fuel. If you flew a spaceship as big as a large yacht to a star 5 light-years (1.5 pc) away and wanted to get there in only 10 years, you would use 40,000 times as much energy as the United States consumes in a year.

Travel for a few individuals might be possible if they accept very long travel times. That would require some form of suspended animation (currently unknown) or colony ships that carry a complete, though small, society in which people are born, live, and die generation after generation as the ship travels through space. Whether the occupants of such a ship would retain the social and emotional characteristics of humans over a long voyage is questionable.

These three limitations not only make it difficult for us to leave our solar system but would also make it difficult for aliens to visit Earth. Reputable scientists have studied "unidentified flying objects" (UFOs) and related phenomena and have never found any evidence that Earth is being visited or has ever been visited by aliens from other worlds (**Window on Science 26-2**). Consequently, humans are unlikely ever to meet an alien face-to-face. The only way to communicate with other civilizations is via radio.

Radio Communication

Nature places two restrictions on radio communication with distant societies. One has to do with simple physics, is well understood, and merely makes the communication difficult. The second has to do with the fate of technological civilizations, is still unresolved, and may severely limit the number of societies detectable by radio.

Radio signals are electromagnetic waves that travel at the speed of light. Because even the nearest civilizations must be a few light-years away, this limits conversations with distant be-

An anticoded message

1 0 1 0 0 1 1 1 1 1
0 0 1 0 1 0 0 1 0 0
0 1 0 1 0 0 1 0 1 0
0 1 0 1 0

5 rows of 7

1	0	1	0	0	1	1
1	1	1	0	0	1	0
1	0	0	1	0	0	0
1	0	1	0	0	1	0
1	0	0	1	0	1	0

7 rows of 5

1	0	1	0	0
1	1	1	1	1
0	0	1	0	1
0	0	1	0	0
0	1	0	1	0
0	1	0	1	0
0	1	0	1	0

■ **Figure 26-13**

An anticoded message is designed for easy decoding. Here a string of 35 radio pulses, represented as 1s and 0s, can be arranged in only two ways, as 5 rows of 7 or 7 rows of 5. The second way produces a friendly message. Any number of pulses can be used so long as it is the product of two prime numbers. Then the pulses can be arranged in only two ways.

ings. If you ask a question of a creature 10 light-years away, you will have to wait 20 years for a reply. Clearly, the give-and-take of normal conversation will be impossible.

Instead, you could simply broadcast a radio beacon of friendship to announce your presence as an intelligent being. In fact, the human race is already broadcasting a recognizable beacon. Short-wavelength radio signals, such as TV and FM, have been leaking into space for the last 50 years at least. Any civilization within 50 light-years might already have detected Earth.

If you intentionally broadcast a signal, you can anticode it. That is, you could arrange it to make it easy to decode. You could transmit pulses to represent 1s and gaps to represent 0s. A message counting through the first few prime numbers would distinguish your signal from natural sources of radio noise. You could even transmit a picture by sending a string of 1s and 0s that can be arranged in only two ways. One way produces nonsense, but the other way produces a meaningful picture (■ Figure 26-13).

In 1974, at the dedication of the 1000-ft radio telescope at Arecibo, radio astronomers transmitted such a signal toward the globular cluster M13, which is located 26,000 ly from Earth. When the signal finally arrives, any aliens who detect it will be able to arrange its 1679 pulses in only two ways, as 23 rows of 73 or 73 rows of 23. The second arrangement will form a picture that describes life on Earth (■ Figure 26-14).

It took only minutes to transmit the Arecibo message. If more time were taken, a more detailed picture could be sent; and, if you were sure your radio telescope was pointed at a listening civilization, you could send a long series of pictures. With pictures you could teach aliens your language and tell them all about human life, its difficulties, and its accomplishments.

If you can think of sending such signals, aliens could think of it too. If you pointed your radio telescope in the right direction and listened at the right wavelength, you might hear other intelligent races calling out to one another. This raises two questions: Which stars are the best candidates, and what wavelengths are most likely? You have already answered the first question. Main-sequence G and K stars have the most favorable characteristics,

■ **Active Figure 26-14**

The Arecibo message to M13 begins by counting from 1 to 10 and goes on to describe our solar system and the biochemistry of life on Earth (color added for clarity). Binary numbers give the height of the human figure (1110) and the diameter of the telescope dish (100101111110) in units of the wavelength of the signal, 12.3 cm. (NASA)

AceAstronomy™ Log into AceAstronomy and select this chapter to see the Active Figure "Interstellar Communication." Send messages and probes to nearby stars.

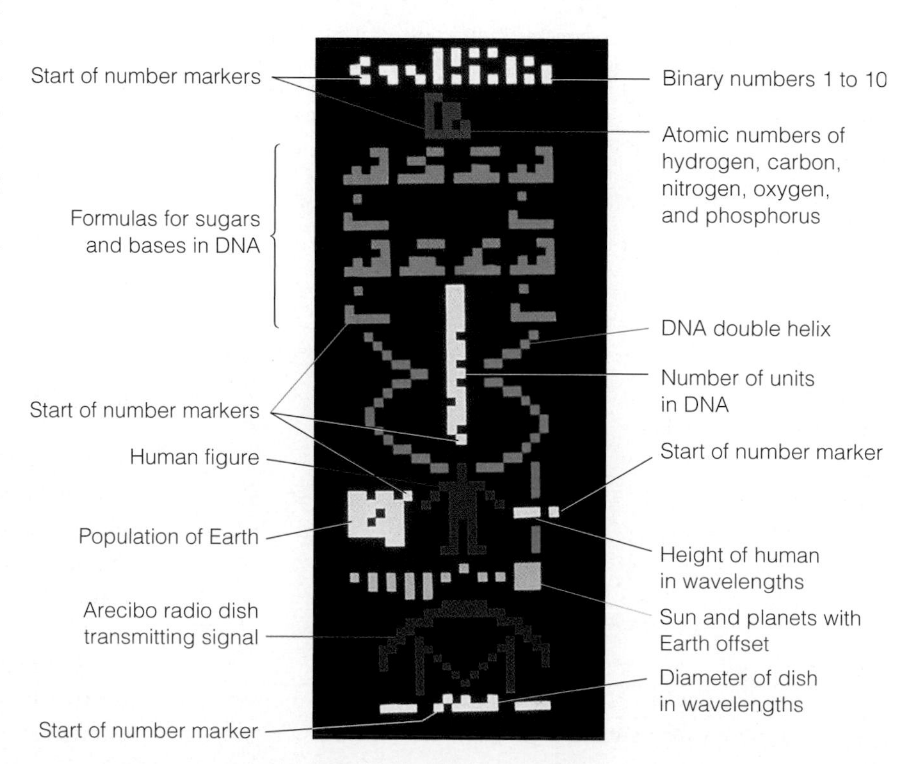

and maybe M stars are worth a look. But the second question is more complex.

Only certain wavelengths are useful for communication. Wavelengths longer than about 30 cm can't be used because the signal would be lost in the background radio noise from our galaxy. Nor can wavelengths much shorter than 1 cm be used because of absorption within our atmosphere. So only a certain range of wavelengths, a radio window, is open for communication.

This communications window is very wide, so a radio telescope would take a long time to tune over all the wavelengths in the window searching for intelligent signals. Nature may have provided a way to narrow the search, however. Within the communications window lie the 21-cm line of neutral hydrogen and the 18-cm line of OH. The interval between these two lines has been dubbed the **water hole** because the combination of H and OH yields water (H_2O). Water is the fundamental solvent in our life form, so it might seem natural for similar water creatures to call out to each other at wavelengths in the water hole (■ Figure 26-15). But even silicon creatures, if they were competent to build radio telescopes, would be familiar with the 21-cm line of hydrogen and the 18-cm line of OH.

This is not idle speculation. A number of searches for extraterrestrial radio signals have been made, and some major searches are now under way. The field has become known as **SETI,** Search for Extra-Terrestrial Intelligence, and it has generated heated debate. Some scientists and philosophers argue that life on other worlds can't possibly exist, so it is a waste of money to search. Others argue that life on other worlds is common and could be detected with present technology. Congress funded a NASA search for a short time but then ended support in the early 1990s because political leaders feared public reaction. In fact, the annual cost of a major search is only about as much as a single Air Force attack helicopter. The controversy may spring in part from the theological and philosophical controversy that would result from the discovery of intelligent life on another world.

In spite of the controversy, major searches have been made, and others are under way. The NASA SETI project canceled by Congress was completed with private funds as Project Phoenix. The SETI Institute was founded in 1984 and has pursued a number of important searches using radio telescopes at major radio astronomy observatories to listen to the radio emissions of a long list of candidate stars. About two billion radio frequency bands must be examined for each star, so the project depends heavily on computers to search for signs of artificial radio transmissions in the recorded radio emissions of each star.

Astonishing amounts of computer power are needed to search for weak or unusual signals. Consequently, the Berkeley SETI team, with the support of the Planetary Society, has recruited owners of personal computers linked to the Internet to participate in a project called seti@home. Participants download a screen saver that searches data files from the Arecibo radio telescope for meaningful signals whenever the owner is not using the computer. About 4 million people have signed up and are providing 1000 years of computer time per day for the project. For information, locate the seti@home project at http://setiathome.ssl.berkeley.edu/.

META II is a major search for radio signals in the southern sky, and the ingenious search called SERENDIP (Search for Extraterrestrial Radio Emission from Nearby Developed Intelligent Populations) rides piggyback on the 305-m Arecibo Telescope searching for signals wherever the radio astronomers point the telescope. Another major search is scanning the sky for intelligent signals in the form of pulsating light flashes.

The SETI Institute, working with the University of California, Berkeley, is building the Allen Telescope Array, which will eventually include 350 dish antennas (■ Figure 26-16). It will be a powerful radio telescope, but it will also be used to search for radio transmissions from other worlds.

At radio wavelengths, noise is a problem. Poorly designed radio transmitters generate noise at unexpected frequencies, and other electronic devices, such as computers, emit radio frequency signals. Furthermore, society, industry, and governments press to use wider and wider sections of the electromagnetic spectrum. Radio astronomers struggle to hear through the radio babble, and SETI searches suffer because many of the noise signals mimic the patterns expected from other civilizations. It would be ironic if we fail to detect signals from another world because our own world has become too noisy.

Ultimately, the chances of success depend on the number of inhabited worlds in our galaxy, and that number is difficult to estimate.

■ Active Figure 26-15

Radio noise from various sources makes it difficult to detect distant signals at wavelengths longer than 30 cm or shorter than 1 cm. In this range, radio emission from H atoms and from OH molecules marks a small wavelength range dubbed the water hole, which may be a likely place for communication.

AceAstronomy™ Log into AceAstronomy and select this chapter to see the Active Figure "Drake Equation." How many civilizations might be ready for communication?

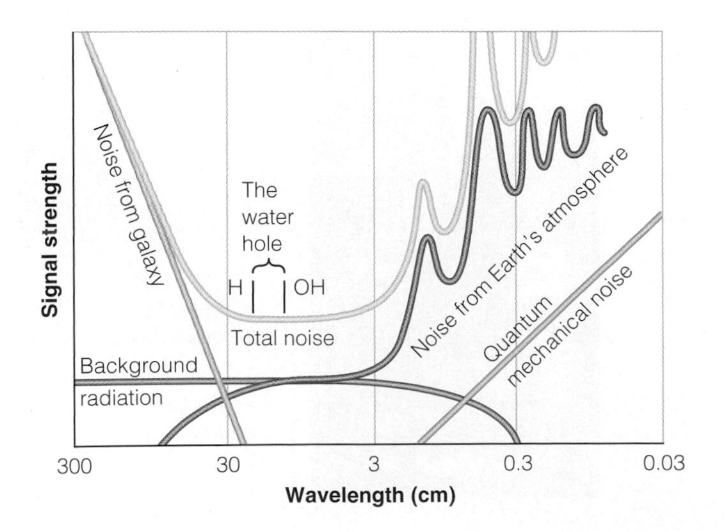

■ **Figure 26-16**

The Allen Telescope Array now being built in California will eventually grow to include 350 radio dishes, each 6 meters in diameter, in an arrangement precisely designed to maximize resolution. As radio astronomers aim the telescope at galaxies and nebulae of interest, state-of-the-art computer systems will analyze stars in the field of view searching for signals from distant civilizations. (Isaac Gary/SETI Institute)

How Many Inhabited Worlds?

The technology exists, and given enough time the searches will find other inhabited worlds, assuming there are at least a few out there. If intelligence is common, then the signals should be found soon—in the next few decades—but if intelligence is rare in the universe, it may be a very long time before SETI confirms that we are not alone.

Simple arithmetic can provide an estimate of the number of technological civilizations available for communication, N_c. The first proposed formula for N_c is now known as the **Drake equation,** named after radio astronomer Frank Drake, a pioneer in the search for extraterrestrial intelligence. It is better to use a version of the Drake equation modified slightly to make it a bit easier to understand. The formula gives N_c, the number of communicative civilizations in a galaxy, as

$$N_c = N^* \cdot f_P \cdot n_{LZ} \cdot f_L \cdot f_I \cdot F_S$$

N^* is the number of stars in a galaxy, and f_P represents the fraction of all stars that have planets. If all single stars have planets, f_P is about 0.5. The factor n_{LZ} is the average number of planets in a solar system suitably placed in the life zone, f_L is the fraction of suitable planets on which life begins, and f_I is the fraction of planets on which life forms evolve to intelligence. These factors can be roughly estimated, but the remaining factor is much more uncertain.

F_S is the fraction of a star's life during which the life form is communicative. Here you can assume that a star lives about 10 billion years. If a society survives at a technological level for only 100 years, the chances of communicating with it are small. But a society that stabilizes and remains technological for a long time is much more likely to be in the communicative phase at the proper time to signal to Earth. If you assume that technological societies destroy themselves in about 100 years, F_S is 100 divided by 10 billion, or 10^{-8}. But if societies can remain technological for a million years, then F_S is 10^{-4}. The influence of the factors in the formula is shown in ■ Table 26-1.

If the optimistic estimates are true, there may be a communicative civilization within a few dozen light-years of Earth, and it could be found by searching through only a few thousand stars. On the other hand, if the pessimistic estimates are correct, Earth may be the only planet in our galaxy capable of communication. It all depends on how technological societies function and survive.

Building Scientific Arguments

Why does the number of inhabited worlds that could be detected depend on how long civilizations survive at a technological level?

This scientific argument depends on timing and requires careful analysis. If you turned a radio telescope toward the sky and scanned millions of frequency bands for lots of stars, you

■ **Table 26-1 | The Number of Technological Civilizations per Galaxy**

		Estimates	
	Variables	**Pessimistic**	**Optimistic**
N^*	Number of stars per galaxy	2×10^{11}	2×10^{11}
f_P	Fraction of stars with planets	0.01	0.5
n_{LZ}	Planets per star in life zone for over 4 billion years	0.01	1
f_L	Fraction of suitable planets on which life begins	0.01	1
f_I	Fraction of life forms that evolve to intelligence	0.01	1
F_S	Fraction of star's life during which a technological society survives	10^{-8}	10^{-4}
N_C	Number of communicative civilizations per galaxy	2×10^{-5}	10×10^{6}

Afterword

The aggregate of all our joys and sufferings, thousands of confident religions, ideologies and economic doctrines, every hunter and forager, every hero and coward, every creator and destroyer of civilizations, every king and peasant, every young couple in love, every hopeful child, every mother and father, every inventor and explorer, every teacher of morals, every corrupt politician, every superstar, every supreme leader, every saint and sinner in the history of our species, lived there on a mote of dust, suspended in a sunbeam.

CARL SAGAN (1934–1996)

Earth photographed by Voyager 1 from the edge of the solar system. (NASA)

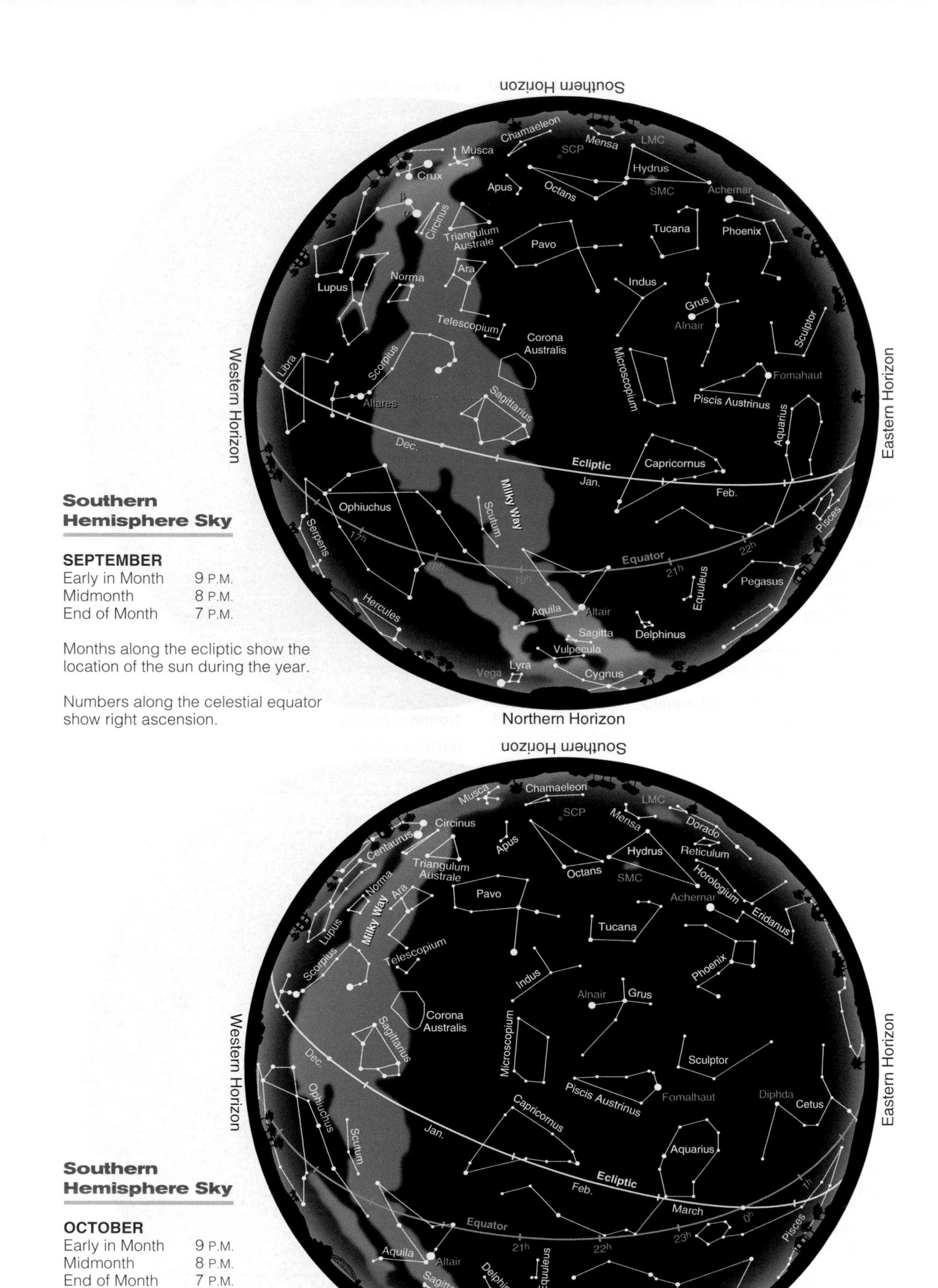

Southern Hemisphere Sky

SEPTEMBER

Early in Month	9 P.M.
Midmonth	8 P.M.
End of Month	7 P.M.

Months along the ecliptic show the location of the sun during the year.

Numbers along the celestial equator show right ascension.

Southern Hemisphere Sky

OCTOBER

Early in Month	9 P.M.
Midmonth	8 P.M.
End of Month	7 P.M.

Months along the ecliptic show the location of the sun during the year.

Numbers along the celestial equator show right ascension.

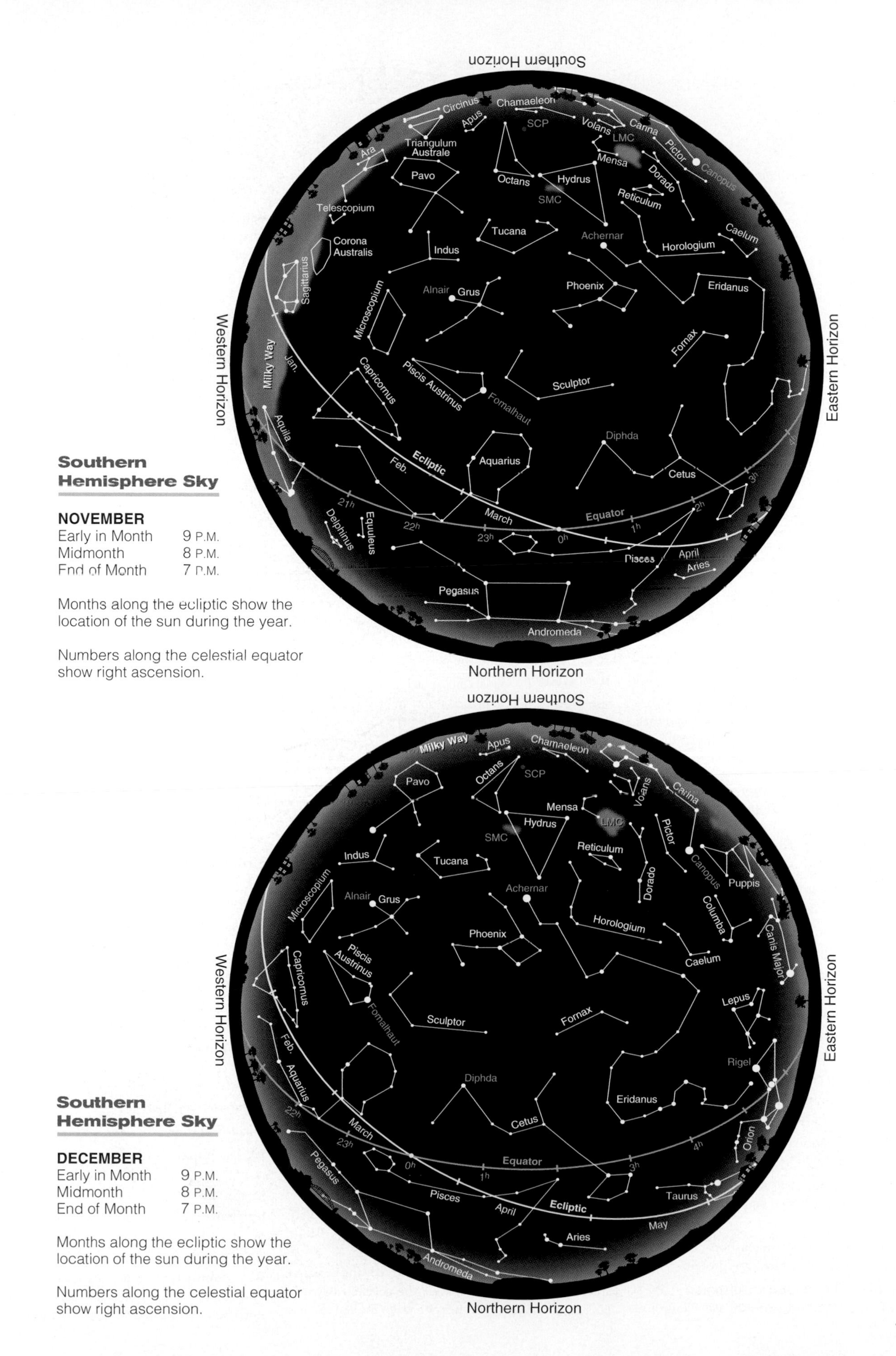

Southern Hemisphere Sky

NOVEMBER

Early in Month	9 P.M.
Midmonth	8 P.M.
End of Month	7 P.M.

Months along the ecliptic show the location of the sun during the year.

Numbers along the celestial equator show right ascension.

Southern Hemisphere Sky

DECEMBER

Early in Month	9 P.M.
Midmonth	8 P.M.
End of Month	7 P.M.

Months along the ecliptic show the location of the sun during the year.

Numbers along the celestial equator show right ascension.

Glossary

Numbers in parentheses refer to the page where the term is first discussed in the text.

absolute age (502) An age determined in years, as from radioactive dating (see also *relative age*).

absolute bolometric magnitude (193) The absolute magnitude you would observe if you could detect all wavelengths.

absolute visual magnitude (M_V) (192) Intrinsic brightness of a star; the apparent visual magnitude the star would have if it were 10 pc away.

absolute zero (138) The lowest possible temperature; the temperature at which the particles in a material, atoms or molecules, contain no energy of motion that can be extracted from the body.

absorption line (146) A dark line in a spectrum; produced by the absence of photons absorbed by atoms or molecules.

absorption spectrum (dark-line spectrum) (146) A spectrum that contains absorption lines.

acceleration (89) A change in a velocity; a change in either speed or direction. (See *velocity.*)

acceleration of gravity (87) A measure of the strength of gravity at a planet's surface.

accretion (461) The sticking together of solid particles to produce a larger particle.

accretion disk (294) The whirling disk of gas that forms around a compact object such as a white dwarf, neutron star, or black hole as matter is drawn in.

achondrite (624) Stony meteorite containing no chondrules or volatiles.

achromatic lens (114) A telescope lens composed of two lenses ground from different kinds of glass and designed to bring two selected colors to the same focus and correct for chromatic aberration.

active galactic nucleus (AGN) (396) The central energy source of an active galaxy.

active galaxy (396) A galaxy that is a source of excess radiation, usually radio waves, X rays, gamma rays, or some combination.

active optics (121) Optical elements whose position or shape is continuously controlled by computers.

active region (175) An area on the sun where sunspots, prominences, flares, and the like occur.

adaptive optics (121) Computer-controlled telescope mirrors that can at least partially compensate for seeing.

albedo (489) The fraction of the light hitting an object that is reflected.

alt-azimuth mounting (121) A telescope mounting capable of motion parallel to and perpendicular to the horizon.

amino acid (654) One of the carbon-chain molecules that are the building blocks of protein.

Angstrom (Å) (111) A unit of distance; 1 Å = 10^{-10} m; often used to measure the wavelength of light.

angular diameter (21) A measure of the size of an object in the sky; numerically equal to the angle in degrees between two lines extending from the observer's eye to opposite edges of the object.

angular distance (21) A measure of the separation between two objects in the sky; numerically equal to the angle in degrees between two lines extending from the observer's eye to the two objects.

angular momentum (93) The tendency of a rotating body to continue rotating; mathematically, the product of mass, velocity, and radius.

angular momentum problem (451) An objection to Laplace's nebular hypothesis that cited the slow rotation of the sun.

annular eclipse (48) A solar eclipse in which the solar photosphere appears around the edge of the moon in a bright ring, or annulus. The corona, chromosphere, and prominences cannot be seen.

anorthosite (506) Rock of aluminum and calcium silicates found in the lunar highlands.

antimatter (425) Matter composed of antiparticles, which upon colliding with a matching particle of normal matter annihilate and convert the mass of both particles into energy. The antiproton is the antiparticle of the proton, and the positron is the antiparticle of the electron.

aphelion (27) The orbital point of greatest distance from the sun.

apogee (48) The orbital point of greatest distance from Earth.

Apollo–Amor object (631) Asteroid whose orbit crosses that of Earth (Apollo) and Mars (Amor).

apparent visual magnitude (m_V) (16) The brightness of a star as seen by human eyes on Earth.

archaeoastronomy (58) The study of the astronomy of ancient cultures.

association (242) Group of widely scattered stars (10 to 1000) moving together through space; not gravitationally bound into a cluster.

asterism (14) A named group of stars not identified as a constellation, e.g., the Big Dipper.

asteroid (456) Small rocky world; most asteroids lie between Mars and Jupiter in the asteroid belt.

astronomical unit (AU) (6) Average distance from Earth to the sun; 1.5×10^8 km, or 93×10^6 miles.

atmospheric window (112) Wavelength regions in which Earth's atmosphere is transparent—at visual, infrared, and radio wavelengths.

aurora (181) The glowing light display that results when a planet's magnetic field guides charged particles toward the north and south magnetic poles, where they strike the upper atmosphere and excite atoms to emit photons.

autumnal equinox (26) The point on the celestial sphere where the sun crosses the celestial equator going southward. Also, the time when the sun reaches this point and autumn begins in the northern hemisphere—about September 22.

Babcock model (177) A model of the sun's magnetic cycle in which the differential rotation of the sun winds up and tangles the solar magnetic field in a 22-year cycle. This is thought to be responsible for the 11-year sunspot cycle.

Balmer series (147) Spectral lines in the visible and near-ultraviolet spectrum of hydrogen produced by transitions whose lowest orbit is the second.

barred spiral galaxy (374) A spiral galaxy with an elongated nucleus resembling a bar from which the arms originate.

basalt (486) Dark, igneous rock characteristic of solidified lava.

belt (561) One of the dark bands of clouds that circle Jupiter parallel to its equator; generally red, brown, or blue-green; believed to be regions of descending gas.

belt–zone circulation (585) The atmospheric circulation typical of Jovian planets. Dark belts and bright zones encircle the planet parallel to its equator.

big bang (422) The theory that the universe began with a violent explosion from which the expanding universe of galaxies eventually formed.

big rip (439) The possible fate of the universe if dark energy increases rapidly and the expansion of spacetime pulls galaxies, stars, and ultimately atoms apart.

binary star (200) One of a pair of stars that orbit around their common center of mass.

binding energy (143) The energy needed to pull an electron away from its atom.

bipolar flow (245) Oppositely directed jets of gas ejected by some protostellar objects.

birth line (241) In the H–R diagram, the line above the main sequence where protostars first become visible.

black body radiation (139) Radiation emitted by a hypothetical perfect radiator; the spectrum is continuous, and the wavelength of maximum emission depends only on the body's temperature.

black dwarf (289) The end state of a white dwarf that has cooled to low temperature.

black hole (323) A mass that has collapsed to such a small volume that its gravity prevents the escape of all radiation; also, the volume of space from which radiation may not escape.

blazar (403) See *BL Lac object.*

BL Lac object (403) Object that resembles a quasar; thought to be the highly luminous core of a distant galaxy emitting a jet almost directly toward Earth.

blueshift (153) The shortening of the wavelengths of light observed when the source and observer are approaching each other.

Bok globule (244) Small, dark cloud only about 1 ly in diameter that contains 10 to 1000 $M_\odot$ of gas and dust; believed related to star formation.

bow shock (484) The boundary between the undisturbed solar wind and the region being deflected around a planet or comet.

breccia (506) A rock composed of fragments of earlier rocks bonded together.

bright-line spectrum (146) See *emission spectrum.*

brown dwarf (263) A very cool, low-luminosity star whose mass is not sufficient to ignite nuclear fusion.

butterfly diagram (174) See *Maunder butterfly diagram.*

CAI (624) Calcium–aluminum-rich inclusions found in some meteorites; believed to be very old.

Cambrian period (657) A geological period 0.6 to 0.5 billion years ago during which life on Earth became diverse and complex. Cambrian rocks contain the oldest easily identifiable fossils.

capture hypothesis (510) The theory that the moon formed elsewhere in the solar system and was later captured by Earth.

carbonaceous chondrite (623) Stony meteorite that contains both chondrules and volatiles. These may be the least altered remains of the solar nebula still present in the solar system.

carbon deflagration (300) The process in which the carbon in a white dwarf is completely consumed by nuclear fusion, producing a type Ia supernova explosion.

carbon–nitrogen–oxygen (CNO) cycle (247) A series of nuclear reactions that use carbon as a catalyst to combine four hydrogen atoms to make one helium atom plus energy; effective in stars more massive than the sun.

Cassegrain focus (120) The optical design of a reflecting telescope in which the secondary mirror reflects light back down the tube through a hole in the center of the objective mirror.

catastrophic hypothesis (451) Explanation for natural processes that depends on dramatic and unlikely events, such as the collision of two stars to produce our solar system.

celestial equator (20) The imaginary line around the sky directly above Earth's equator.

celestial sphere (20) An imaginary sphere of very large radius surrounding Earth and to which the planets, stars, sun, and moon seem to be attached.

center of mass (95) The balance point of a body or system of bodies.

Cepheid variable star (275) Variable star with a period of 1 to 60 days; the period of variation is related to luminosity.

Chandrasekhar limit (289) The maximum mass of a white dwarf, about 1.4 $M_{\odot}$; a white dwarf of greater mass cannot support itself and will collapse.

charge-coupled device (CCD) (123) An electronic device consisting of a large array of light-sensitive elements used to record very faint images.

chemical evolution (658) The chemical process that led to the growth of complex molecules on the primitive Earth. This did not involve the reproduction of molecules.

chondrite (622) A stony meteorite that contains chondrules.

chondrule (622) Round, glassy body in some stony meteorites; believed to have solidified very quickly from molten drops of silicate material.

chromatic aberration (114) A distortion found in refracting telescopes because lenses focus different colors at slightly different distances. Images are consequently surrounded by color fringes.

chromosome (655) One of the bodies in a cell that contains the DNA carrying genetic information.

chromosphere (48) Bright gases just above the photosphere of the sun.

circular velocity (94) The velocity required to remain in a circular orbit about a body.

circumpolar constellation (21) Any of the constellations so close to the celestial pole that they never set (or never rise) as seen from a given latitude.

closed orbit (95) An orbit that returns to its starting point; a circular or elliptical orbit. (See *open orbit.*)

closed universe (433) A model universe in which the average density is great enough to stop the expansion and make the universe contract.

cluster method (379) The method of determining the masses of galaxies based on the motions of galaxies in a cluster.

CNO cycle (247) See *carbon–nitrogen–oxygen cycle.*

cocoon (240) The cloud of gas and dust around a contracting protostar that conceals it at visible wavelengths.

cold dark matter (435) Invisible matter in the universe composed by heavy, slow-moving particles such as WIMPs.

collapsar (333) See *hypernova.*

collisional broadening (155) The smearing out of a spectral line because of collisions among the atoms of the gas.

coma (636) The glowing head of a comet.

comet (456) One of the small, icy bodies that orbit the sun and produce tails of gas and dust when they near the sun.

compact object (289) A star that has collapsed to form a white dwarf, neutron star, or black hole.

comparative planetology (476) The study of planets by comparing the characteristics of different examples.

comparison spectrum (124) A spectrum of known spectral lines used to identify unknown wavelengths in an object's spectrum.

composite volcano (528) A volcano built up of layers of lava flows and ash falls. These are steep sided and typically associated with subduction zones.

condensation (461) The growth of a particle by addition of material from surrounding gas, one atom or molecule at a time.

condensation hypothesis (510) The hypothesis that the moon and Earth formed by condensing together from the solar nebula.

condensation sequence (461) The sequence in which different materials condense from the solar nebula at increasing distances from the sun.

conservation of energy law (266) One of the basic laws of stellar structure. The amount of energy flowing out of the top of a shell must equal the amount coming in at the bottom plus whatever energy is generated within the shell.

conservation of mass law (266) One of the basic laws of stellar structure. The total mass of the star must equal the sum of the masses of the shells, and the mass must be distributed smoothly throughout the star.

constellation (14) One of the stellar patterns identified by name, usually of mythological gods, people, animals, or objects; also, the region of the sky containing that star pattern.

continuous spectrum (146) A spectrum in which there are no absorption or emission lines.

convection (164) Circulation in a fluid driven by heat; hot material rises, and cool material sinks.

convective zone (171) The region inside a star where energy is carried outward as rising hot gas and sinking cool gas.

corona (54, 530) The faint outer atmosphere of the sun; composed of low-density, very hot, ionized gas. On Venus, round networks of fractures and ridges up to 1000 km in diameter.

coronagraph (165) A telescope designed to photograph the inner corona of the sun.

coronal gas (228) Extremely high-temperature, low-density gas in the interstellar medium.

coronal hole (181) An area of the solar surface that is dark at X-ray wavelengths; thought to be associated with divergent magnetic fields and the source of the solar wind.

coronal mass ejection (CME) (181) Gas trapped in the sun's magnetic field.

cosmic microwave background radiation (425) Radiation from the hot clouds of the big bang explosion. Because of its large redshift, it appears to come from a body whose temperature is only 2.7 K.

cosmic ray (132) A subatomic particle traveling at tremendous velocity that strikes Earth's atmosphere from space.

cosmological constant (Λ) (438) Einstein's constant that represents a repulsion in space to oppose gravity.

cosmological principle (429) The assumption that any observer in any galaxy sees the same general features of the universe.

cosmology (416) The study of the nature, origin, and evolution of the universe.

Coulomb barrier (169) The electrostatic force of repulsion between bodies of like charge; commonly applied to atomic nuclei.

Coulomb force (142) The repulsive force between particles with like electrostatic charge.

critical density (430) The average density of the universe needed to make its curvature flat.

critical point (558) The temperature and pressure at which the vapor and liquid phases of a material have the same density.

dark age (426) The period of a few hundred million years during which the universe expanded in darkness. Extends from soon after the big bang glow faded into the infrared to the formation of the first stars.

dark energy (439) The energy of empty space that drives the acceleration of the expanding universe.

dark-line spectrum (146) See *absorption spectrum.*

dark matter (348) Nonluminous material that is detected only by its gravitational influence.

dark nebula (221) A nonluminous cloud of gas and dust visible because it blocks light from more distant stars and nebulae.

debris disk (468) A disk of dust found by infrared observations around some stars. The dust is debris from collisions among asteroids, comets, and Kuiper belt objects.

decameter radiation (557) Radio signals from Jupiter with wavelengths of about 10 m.

decimeter radiation (557) Radio signals from Jupiter with wavelengths of about 0.1 m.

deferent (65) In the Ptolemaic theory, the large circle around Earth along which the center of the epicycle moved.

degenerate matter (269) Extremely high-density matter in which pressure no longer depends on temperature, due to quantum mechanical effects.

density (156) The amount of matter per unit volume in a material; measured in grams per cubic centimeter, for example.

density wave theory (357) Theory proposed to account for spiral arms as compressions of the interstellar medium in the disk of the galaxy.

deuterium (170) An isotope of hydrogen in which the nucleus contains a proton and a neutron.

diamond ring effect (48) A momentary phenomenon seen during some total solar eclipses when the ring of the corona and a bright spot of photosphere resemble a large diamond set in a silvery ring.

differential rotation (176) The rotation of a body in which different parts of the body have different periods of rotation; this is true of the sun, the Jovian planets, and the disk of the galaxy.

differentiation (462) The separation of planetary material according to density.

diffraction fringe (115) Blurred fringe surrounding any image caused by the wave properties of light. Because of this, no image detail smaller than the fringe can be seen.

disk component (344) All material confined to the plane of the galaxy.

distance indicator (373) Object whose luminosity or diameter is known; used to find the distance to a star cluster or galaxy.

distance modulus ($m_V - M_V$) (192) The difference between the apparent and absolute magnitude of a star; a measure of how far away the star is.

distance scale (377) The combined calibration of distance indicators used by astronomers to find the distances to remote galaxies.

DNA (deoxyribonucleic acid) (654) The long carbon-chain molecule that records information to govern the biological activity of the organism. DNA carries the genetic data passed to offspring.

Doppler broadening (155) The smearing of spectral lines because of the motion of the atoms in the gas.

Doppler effect (153) The change in the wavelength of radiation due to relative radial motion of source and observer.

double-exhaust model (397) The theory that double radio lobes are produced by pairs of jets emitted in opposite directions from the centers of active galaxies.

double-lobed radio source (397) A galaxy that emits radio energy from two regions (lobes) located on opposite sides of the galaxy.

Drake equation (669) A formula for the number of communicative civilizations in our galaxy.

dust (type II) tail (636) The tail of a comet formed of dust blown outward by the pressure of sunlight. (See *gas tail.*)

dynamo effect (177) The process by which a rotating, convecting body of conducting matter, such as Earth's core, can generate a magnetic field.

east point (20) The point on the eastern horizon exactly halfway between the north point and the south point; exactly east.

eccentric (62) In astronomy, an off-center circular path.

eccentricity, *e* (76) A measure of the flattening of an ellipse. An ellipse of $e = 0$ is circular. The closer to 1 e becomes, the more flattened the ellipse.

eclipse season (50) That period when the sun is near a node of the moon's orbit and eclipses are possible.

eclipse year (51) The time the sun takes to circle the sky and return to a node of the moon's orbit; 346.62 days.

eclipsing binary system (204) A binary star system in which the stars eclipse each other.

ecliptic (24) The apparent path of the sun around the sky.

ejecta (518) Pulverized rock scattered by meteorite impacts on a planetary surface.

electromagnetic radiation (110) Changing electric and magnetic fields that travel through space and transfer energy from one place to another—for example, light, radio waves, and the like.

electron (138) Low-mass atomic particle carrying a negative charge.

ellipse (76) A closed curve enclosing two points (foci) such that the total distance from one focus to any point on the curve back to the other focus equals a constant.

elliptical galaxy (371) A galaxy that is round or elliptical in outline; it contains little gas and dust, no disk or spiral arms, and few hot, bright stars.

emission line (146) A bright line in a spectrum caused by the emission of photons from atoms.

emission nebula (220) A cloud of glowing gas excited by ultraviolet radiation from hot stars.

emission spectrum (bright-line spectrum) (146) A spectrum containing emission lines.

energy (96) The capacity of a natural system to perform work—for example, thermal energy.

energy level (144) One of a number of states an electron may occupy in an atom, depending on its binding energy.

enzyme (654) Special protein that controls processes in an organism.

epicycle (65) The small circle followed by a planet in the Ptolemaic theory. The center of the epicycle follows a larger circle (deferent) around Earth.

equant (66) The point off-center in the deferent from which the center of the epicycle appears to move uniformly.

equatorial mounting (121) A telescope mounting that allows motion parallel to and perpendicular to the celestial equator.

ergosphere (324) The region surrounding a rotating black hole within which one could not resist being dragged around the black hole. It is possible for a particle to escape from the ergosphere and extract energy from the black hole.

escape velocity (95) The initial velocity an object needs to escape from the surface of a celestial body.

evening star (28) Any planet visible in the sky just after sunset.

event horizon (323) The boundary of the region of a black hole from which no radiation may escape. No event that occurs within the event horizon is visible to a distant observer.

evolutionary hypothesis (451) Explanation for natural events that involves gradual changes as opposed to sudden catastrophic changes—for example, the formation of the planets in the gas cloud around the forming sun.

excited atom (144) An atom in which an electron has moved from a lower to a higher orbit.

extrasolar planet (469) A planet orbiting a star other than the sun.

eyepiece (114) A short-focal-length lens used to enlarge the image in a telescope; the lens nearest the eye.

fall (621) A meteorite seen to fall. (See *find.*)

false-color image (123) A representation of graphical data in which the colors are altered or added to reveal details.

field (92) A way of explaining action at a distance; a particle produces a field of influence (gravitational, electric, or magnetic) to which another particle in the field responds.

filament (165) On the sun, a prominence seen silhouetted against the solar surface.

filtergram (165) An image (usually of the sun) taken in the light of a specific region of the spectrum—e.g., an H-alpha filtergram.

find (621) A meteorite that is found but was not seen to fall. (See *fall.*)

fission hypothesis (510) The theory that the moon formed by breaking away from Earth.

flare (181) A violent eruption on the sun's surface.

flatness problem (436) In cosmology, the circumstance that the early universe must have contained almost exactly the right amount of matter to close space-time (to make space-time flat).

flat universe (433) A model of the universe in which space-time is not curved.

flocculent (359) Woolly, fluffy; used to refer to certain galaxies that have a woolly appearance.

flux (191) A measure of the flow of energy onto or through a surface. Usually applied to light.

focal length (114) The distance from a lens to the point where it focuses parallel rays of light.

folded mountain range (486) A long range of mountains formed by the compression of a planet's crust—for example, the Andes on Earth.

forbidden line (218) A spectral line that does not occur in the laboratory because it depends on an atomic transition that is highly unlikely.

forward scattering (562) The optical property of finely divided particles to preferentially direct light in the original direction of the light's travel.

free-fall contraction (240) The early contraction of a gas cloud to form a star during which internal pressure is too low to resist contraction.

frequency (111) The number of times a given event occurs in a given time; for a wave, the number of cycles that pass the observer in 1 second.

galactic cannibalism (388) The theory that large galaxies absorb smaller galaxies.

galactic corona (348) The low-density extension of the halo of a galaxy; now suspected to extend many times the visible diameter of the galaxy.

galactic fountain (351) A region of the galaxy's disk in which gas heated by supernova explosions throws gas out of the disk where it can fall back and spread metals through the disk.

galaxy (8) A very large collection of gas, dust, and stars orbiting a common center of mass. The sun and Earth are located in the Milky Way Galaxy.

Galilean satellites (455) The four largest satellites of Jupiter, named after their discoverer, Galileo.

gamma-ray burster (332) An object that produces a sudden burst of gamma rays; thought to be associated with neutron stars and black holes.

gas (type I) tail (636) The tail of a comet produced by gas blown outward by the solar wind. (See *dust tail.*)

gene (655) A unit of DNA containing genetic information that influences a particular inherited trait.

general theory of relativity (102) Einstein's more sophisticated theory of space and time, which describes gravity as a curvature of space-time.

geocentric universe (61) A model universe with Earth at the center, such as the Ptolemaic universe.

geosynchronous satellite (94) An Earth satellite in an eastward orbit whose period is 24 hours. A satellite in such an orbit remains above the same spot on Earth's surface.

giant molecular cloud (227) Very large, cool cloud of dense gas in which stars form.

giants (196) Large, cool, highly luminous stars in the upper right of the H–R diagram; typically 10 to 100 times the diameter of the sun.

glitch (313) A sudden change in the period of a pulsar.

global warming (490) The gradual increase in the surface temperature of Earth caused by human modifications to Earth's atmosphere.

globular cluster (276) A star cluster containing 50,000 to 1 million stars in a sphere about 75 ly in diameter; generally old, metal-poor, and found in the spherical component of the galaxy.

gossamer ring (563) The dimmest part of Jupiter's ring produced by dust particles orbiting near small moons.

grand unified theory (GUT) (436) Theory that attempts to unify (describe in a similar way) the electromagnetic, weak, and strong forces of nature.

granulation (163) The fine structure visible on the solar surface caused by rising currents of hot gas and sinking currents of cool gas below the surface.

grating (124) A piece of material in which numerous microscopic parallel lines are scribed; light encountering a grating is dispersed to form a spectrum.

gravitational collapse (460) The stage in the formation of a massive planet when it grows massive enough to begin capturing gas directly from the nebula around it.

gravitational lensing (382) The effect of the focusing of light from a distant galaxy or quasar by an intervening galaxy to produce multiple images of the distant body.

gravitational radiation (318) As predicted by general relativity, expanding waves in a gravitational field that transport energy through space.

gravitational redshift (326) The lengthening of the wavelength of a photon due to its escape from a gravitational field.

greenhouse effect (490) The process by which a carbon dioxide atmosphere traps heat and raises the temperature of a planetary surface.

grooved terrain (566) Region of the surface of Ganymede consisting of parallel grooves; believed to have formed by repeated fracture of the icy crust.

ground state (145) The lowest permitted electron orbit in an atom.

half-life (457) The time required for half of the atoms in a radioactive sample to decay.

halo (341) The spherical region of a spiral galaxy containing a thin scattering of stars, star clusters, and small amounts of gas.

heat of formation (463) In planetology, the heat released by the infall of matter during the formation of a planetary body.

heavy bombardment (466) The period of intense meteorite impacts early in the formation of the planets when the solar system was filled with debris.

heliocentric universe (63) A model of the universe with the sun at the center, such as the Copernican universe.

helioseismology (167) The study of the interior of the sun by the analysis of its modes of vibration.

helium flash (270) The explosive ignition of helium burning that takes place in some giant stars.

Herbig–Haro object (245) Small nebula that varies irregularly in brightness; believed to be associated with star formation.

Hertzsprung–Russell diagram (194) A plot of the intrinsic brightness versus the surface temperature of stars; it separates the effects of temperature and surface area on stellar luminosity; commonly absolute magnitude versus spectral type, but also luminosity versus surface temperature or color.

HI cloud (233) An interstellar cloud of neutral hydrogen.

HII region (220) A region of ionized hydrogen around a hot star.

high-velocity star (347) A star with a large space velocity. Such stars are halo stars passing through the disk of the galaxy at steep angles.

Hirayama family (627) Family of asteroids with orbits of similar size, shape, and orientation; believed to be fragments of larger bodies.

homogeneity (429) The assumption that, on the large scale, matter is uniformly spread through the universe.

horizon (20) The line that marks the apparent intersection of Earth and the sky.

horizon problem (436) In cosmology, the circumstance that the primordial background radiation seems much more isotropic than could be explained by the standard big bang theory.

horizontal branch (277) In the H–R diagram of a globular cluster, the sequence of stars extending from the red giants toward the blue side of the diagram; includes RR Lyrae stars.

horoscope (28) A chart showing the positions of the sun, moon, planets, and constellations at the time of a person's birth; used in astrology to attempt to read character or foretell the future.

hot dark matter (434) Invisible matter in the universe composed of low-mass, high-velocity particles such as neutrinos.

hot spot (398) In radio astronomy, a bright spot in a radio lobe.

H–R diagram (194) (See *Hertzsprung–Russell diagram.*)

Hubble constant *(H)* (378) A measure of the rate of expansion of the universe; the average value of velocity of recession divided by distance; about 70 km/s/megaparsec.

Hubble law (378) The linear relation between the distance to a galaxy and its apparent radial velocity.

Hubble time (422) An upper limit on the age of the universe derived from the Hubble constant.

hydrostatic equilibrium (249) The balance between the weight of the material pressing downward on a layer in a star and the pressure in that layer.

hypernova (333) The explosion produced as a very massive star collapses into a black hole; thought to be responsible for at least some gamma-ray bursts.

hypothesis (77) A conjecture, subject to further tests, that accounts for a set of facts.

inflationary universe (436) A version of the big bang theory that includes a rapid expansion when the universe was very young.

infrared cirrus (227) A fine network of filaments covering the sky detected in the far infrared by the IRAS satellite; believed to be associated with dust in the interstellar medium.

infrared radiation (111) Electromagnetic radiation with wavelengths intermediate between visible light and radio waves.

instability strip (278) The region of the H–R diagram in which stars are unstable to pulsation; a star passing through this strip becomes a variable star.

intercloud medium (223) The hot, low-density gas between cooler clouds in the interstellar medium.

intercrater plain (515) The relatively smooth terrain on Mercury.

interferometry (122) The observing technique in which separated telescopes combine to produce a virtual telescope with the resolution of a much-larger-diameter telescope.

interstellar absorption lines (222) Dark lines in some stellar spectra that are formed by interstellar gas.

interstellar dust (219) Microscopic solid grains in the interstellar medium.

interstellar extinction (219) The dimming of starlight by gas and dust in the interstellar medium.

interstellar medium (216) The gas and dust distributed between the stars.

interstellar reddening (219) The process in which dust scatters blue light out of starlight and makes the stars look redder.

intrinsic variable (275) A variable star driven to pulsate by processes in its interior.

inverse square law (90) The rule that the strength of an effect (such as gravity) decreases in proportion as the distance squared increases.

Io flux tube (558) A tube of magnetic lines and electric currents connecting Io and Jupiter.

ion (142) An atom that has lost or gained one or more electrons.

ionization (142) The process in which atoms lose or gain electrons.

Io plasma torus (558) The doughnut-shaped cloud of ionized gas that encloses the orbit of Jupiter's moon Io.

iron meteorite (622) A meteorite composed mainly of iron–nickel alloy.

irregular galaxy (375) A galaxy with a chaotic appearance, large clouds of gas and dust, and both population I and population II stars, but without spiral arms.

island universe (371) An older term for a galaxy.

isotopes (142) Atoms that have the same number of protons but a different number of neutrons.

isotropy (429) The assumption that in its general properties the universe looks the same in every direction.

joule (J) (96) A unit of energy equivalent to a force of 1 newton acting over a distance of 1 meter; 1 joule per second equals 1 watt of power.

Jovian planet (454) Jupiterlike planet with large diameter and low density.

jumbled terrain (509) Strangely disturbed regions of the moon opposite the locations of the Imbrium Basin and Mare Orientale.

Kelvin temperature scale (138) The temperature, in Celsius (centigrade) degrees, measured above absolute zero.

Keplerian motion (348) Orbital motion in accord with Kepler's laws of planetary motion.

Kerr black hole (324) A solution to the equations of general relativity that describes the properties of a rotating black hole.

kiloparsec (kpc) (341) A unit of distance equal to 1000 pc, or 3260 ly.

kinetic energy (96) Energy of motion; depends on mass and velocity of a moving body.

Kirchhoff's laws (146) A set of laws that describe the absorption and emission of light by matter.

Kirkwood's gap (630) Region in the asteroid belt in which there are very few asteroids; caused by resonances with Jupiter.

Kuiper belt (456) The collection of icy planetesimals believed to orbit in a region from just beyond Neptune out to about 50 AU.

Lagrangian point (293) Point of stability in the orbital plane of a binary star system, planet, or moon. One is located 60° ahead and one 60° behind the orbiting bodies; another is located between the orbiting bodies.

large-impact hypothesis (511) The theory that the moon formed from debris ejected during a collision between Earth and a large planetesimal.

large-scale structure (440) The distribution of galaxy clusters and superclusters in walls and filaments surrounding voids mostly empty of galaxies.

L dwarf (150) A type of star that is even cooler than the M stars.

life zone (665) A region around a star within which a planet can have temperatures that permit the existence of liquid water.

light curve (204) A graph of brightness versus time commonly used in analyzing variable stars and eclipsing binaries.

light-gathering power (115) The ability of a telescope to collect light; proportional to the area of the telescope objective lens or mirror.

lighthouse model (312) The explanation of a pulsar as a spinning neutron star sweeping beams of radio radiation around the sky.

light pollution (117) The illumination of the night sky by waste light from cities and outdoor lighting, which prevents the observation of faint objects.

light-year (ly) (7) The distance light travels in 1 year.

limb (164, 498) The edge of the apparent disk of a body, as in "the limb of the moon."

limb darkening (164) The decrease in brightness of the sun or other body from its center to its limb.

line of nodes (50) The line across an orbit connecting the nodes; commonly applied to the orbit of the moon.

liquid metallic hydrogen (557) A form of hydrogen under high pressure that is a good electrical conductor.

lobate scarp (514) A curved cliff such as those found on Mercury.

local bubble or void (229) A region of high-temperature, low-density gas in the interstellar medium in which the sun happens to be located.

look-back time (377) The amount by which you look into the past when you look at a distant galaxy; a time equal to the distance to the galaxy in light-years.

luminosity ***(L)*** (193) The total amount of energy a star radiates in 1 second.

luminosity class (198) A category of stars of similar luminosity; determined by the widths of lines in their spectra.

lunar eclipse (42) The darkening of the moon when it moves through Earth's shadow.

Lyman series (147) Spectral lines in the ultraviolet spectrum of hydrogen produced by transitions whose lowest orbit is the ground state.

MACHO (435) Massive Compact Halo Object; a low-luminosity object such as a planet or a brown dwarf that contributes to the mass of the halo.

Magellanic Clouds (340) Small, irregular galaxies that are companions to the Milky Way; visible in the southern sky.

magnetar (317) A class of neutron stars that have exceedingly strong magnetic fields; thought to be responsible for soft gamma-ray repeaters.

magnetic carpet (166) The widely distributed, low-level magnetic field extending up through the sun's visible surface.

magnetosphere (484) The volume of space around a planet within which the motion of charged particles is dominated by the planetary magnetic field rather than the solar wind.

magnifying power (116) The ability of a telescope to make an image larger.

magnitude–distance formula (192) The mathematical formula that relates the apparent magnitude and absolute magnitude of a star to its distance.

magnitude scale (16) The astronomical brightness scale; the larger the number, the fainter the star.

main sequence (196) The region of the H–R diagram running from upper left to lower right, which includes roughly 90 percent of all stars.

mantle (483) The layer of dense rock and metal oxides that lies between the molten core and Earth's surface; also, similar layers in other planets.

mare (498) One of the lunar lowlands filled by successive flows of dark lava; from the Latin word for *sea.*

mass (90) A measure of the amount of matter making up an object.

mass–luminosity relation (207) The more massive a star is, the more luminous it is.

Maunder butterfly diagram (174) A graph showing the latitude of sunspots versus time; first plotted by W. W. Maunder in 1904.

Maunder minimum (175) A period of less numerous sunspots and other solar activity from 1645 to 1715.

megaparsec (Mpc) (373) A unit of distance equal to 1 million pc.

metals (349) In astronomical usage, all atoms heavier than helium.

metastable level (218) An atomic energy level from which an electron takes a long time to decay; responsible for producing forbidden lines.

meteor (457) A small bit of matter heated by friction to incandescent vapor as it falls into Earth's atmosphere.

meteorite (457) A meteor that has survived its passage through the atmosphere and strikes the ground.

meteoroid (457) A meteor in space before it enters Earth's atmosphere.

micrometeorite (501) Meteorite of microscopic size.

midocean rise (486) One of the undersea mountain ranges that push up from the seafloor in the center of the oceans.

Milankovitch hypothesis (30) The hypothesis that small changes in Earth's orbital and rotational motions cause the ice ages.

Milky Way (9) The hazy band of light that circles the sky, produced by the combined light of billions of stars in our Milky Way Galaxy.

Milky Way Galaxy (9) The spiral galaxy containing the sun; visible at night as the Milky Way.

Miller experiment (657) An experiment that reproduced the conditions under which life began on Earth and manufactured amino acids and other organic compounds.

millisecond pulsar (319) A pulsar with a period of approximately a millisecond, a thousandth of a second.

minute of arc (21) An angular measure; each degree is divided into 60 minutes of arc.

molecular cloud (227) An interstellar gas cloud that is dense enough for the formation of molecules; discovered and studied through the radio emissions of such molecules.

molecule (142) Two or more atoms bonded together.

momentum (89) The tendency of a moving object to continue moving; mathematically, the product of mass and velocity.

morning star (28) Any planet visible in the sky just before sunrise.

multiringed basin (501) Very large impact basin in which there are concentric rings of mountains.

mutant (656) Offspring born with altered DNA.

nadir (20) The point on the bottom of the sky directly under your feet.

nanometer (nm) (111) A unit of length equal to 10^{-9} m.

natural law (77) A conjecture about how nature works in which scientists have overwhelming confidence.

natural motion (87) In Aristotelian physics, the motion of objects toward their natural places—fire and air upward and earth and water downward.

natural selection (653) The process by which the best traits are passed on, allowing the most able to survive.

neap tide (98) Ocean tide of low amplitude occurring at first- and third-quarter moon.

nebula (218) A cloud of gas and dust in space.

nebular hypothesis (451) The proposal that the solar system formed from a rotating cloud of gas.

neutrino (170) A neutral, massless atomic particle that travels at or nearly at the speed of light.

neutron (141) An atomic particle with no charge and about the same mass as a proton.

neutron star (310) A small, highly dense star composed almost entirely of tightly packed neutrons; radius about 10 km.

Newtonian focus (120) The focal arrangement of a reflecting telescope in which a diagonal mirror reflects light out the side of the telescope tube for easier access.

node (50) A point where an object's orbit passes through the plane of Earth's orbit.

nonbaryonic matter (434) In cosmology, a suspected component of the dark matter composed of matter that does not contain protons and neutrons.

north celestial pole (20) The point on the celestial sphere directly above Earth's North Pole.

north point (20) The point on the horizon directly below the north celestial pole; exactly north.

nova (284) From the Latin "new," a sudden brightening of a star, making it appear as a "new" star in the sky; believed associated with eruptions on white dwarfs in binary systems.

nuclear bulge (341) The spherical cloud of stars that lies at the center of spiral galaxies.

nuclear fission (169) Reaction that splits nuclei into less massive fragments.

nuclear fusion (169) Reaction that joins the nuclei of atoms to form more massive nuclei.

nucleosynthesis (350) The production of elements heavier than helium by the fusion of atomic nuclei in stars and during supernovae explosions.

nucleus (of an atom) (141) The central core of an atom containing protons and neutrons; carries a net positive charge.

O association (242) A large, loosely bound cluster of very young stars.

objective lens or mirror (114) The main optical element in an astronomical telescope. The large lens at the top of the telescope or large mirror at the bottom.

oblateness (556) The flattening of a spherical body; usually caused by rotation.

observable universe (420) The part of the universe that is visible from Earth's location in space and time.

occultation (597) The passage of a larger body in front of a smaller body.

Olbers's paradox (419) The conflict between observation and theory as to why the night sky should or should not be dark.

Oort cloud (639) The cloud of icy bodies—extending from the outer part of our solar system out to roughly 100,000 AU from the sun—that acts as the source of most comets.

opacity (248) The resistance of a gas to the passage of radiation.

open cluster (276) A cluster of 10 to 10,000 stars with an open, transparent appearance and stars not tightly grouped; usually relatively young and located in the disk of the galaxy.

open orbit (95) An orbit that does not return to its starting point; an escape orbit. (See *closed orbit.*)

open universe (433) A model universe in which the average density is less than the critical density needed to halt the expansion.

oscillating universe theory (430) The theory that the universe begins with a big bang, expands, is slowed by its own gravity, and then falls back to create another big bang.

outflow channel (540) Geological feature on Mars that appears to have been caused by sudden flooding.

outgassing (463) The release of gases from a planet's interior.

ovoid (600) Geological feature on Uranus's moon Miranda thought to be produced by circulation in the solid icy mantle and crust.

ozone layer (489) In Earth's atmosphere, a layer of oxygen ions (O_3) lying 15 to 30 km high that protects the surface by absorbing ultraviolet radiation.

paradigm (68) A commonly accepted set of scientific ideas and assumptions.

parallax (64) The apparent change in the position of an object due to a change in the location of the observer. Astronomical parallax is measured in seconds of arc.

parsec (pc) (189) The distance to a hypothetical star whose parallax is one second of arc; 1 pc = 206,265 AU = 3.26 ly.

partial eclipse (44, 45) A lunar eclipse in which the moon does not completely enter Earth's shadow; a solar eclipse in which the moon does not completely cover the sun.

Paschen series (147) Spectral lines in the infrared spectrum of hydrogen produced by transitions whose lowest orbit is the third.

passing star hypothesis (450) The proposal that our solar system formed when two stars passed near each other and material was pulled out of one to form the planets.

path of totality (46) The track of the moon's umbral shadow over Earth's surface. The sun is totally eclipsed as seen from within this path.

penumbra (42) The portion of a shadow that is only partially shaded.

penumbral eclipse (54) A lunar eclipse in which the moon enters the penumbra of Earth's shadow but does not reach the umbra.

perigee (48) The orbital point of closest approach to Earth.

perihelion (27) The orbital point of closest approach to the sun.

period–luminosity relation (275) The relation between period of pulsation and intrinsic brightness among Cepheid variable stars.

permitted orbit (143) One of the energy levels in an atom that an electron may occupy.

photon (111) A quantum of electromagnetic energy; carries an amount of energy that depends inversely on its wavelength.

photosphere (47) The bright visible surface of the sun.

planet (6) A nonluminous object, larger than a comet or asteroid, that orbits a star.

planetary nebula (288) An expanding shell of gas ejected from a star during the latter stages of its evolution.

planetesimal (461) One of the small bodies that formed from the solar nebula and eventually grew into protoplanets.

plastic (483) A material with the properties of a solid but capable of flowing under pressure.

plate tectonics (486) The constant destruction and renewal of Earth's surface by the motion of sections of crust.

plutino (612) One of the icy Kuiper belt objects that, like Pluto, are caught in a 3:2 orbital resonance with Neptune.

polar axis (121) The axis around which a celestial body rotates.

poor cluster (384) An irregularly shaped cluster that contains fewer than 1000 galaxies, many spiral, and no giant ellipticals.

population I star (349) Star rich in atoms heavier than helium; nearly always a relatively young star found in the disk of the galaxy.

population II star (349) Star poor in atoms heavier than helium; nearly always a relatively old star found in the halo, globular clusters, or the nuclear bulge.

positron (170) The anti-particle of the electron.

potential energy (96) The energy a body has by virtue of its position. A weight on a high shelf has more potential energy than a weight on a low shelf.

precession (22) The slow change in the direction of Earth's axis of rotation; one cycle takes nearly 26,000 years.

pressure (224) A force exerted over a surface. Expressed as force per unit area.

pressure *(P)* wave (482) In geophysics, a mechanical wave of compression and rarefaction that travels through Earth's interior.

primary lens or mirror (114) The main optical element in an astronomical telescope. The large lens at the top of the telescope tube or the large mirror at the bottom.

prime focus (120) The point at which the objective mirror forms an image in a reflecting telescope.

primeval atmosphere (488) Earth's first air, composed of gases from the solar nebula.

primordial soup (657) The rich solution of organic molecules in Earth's first oceans.

prominence (48, 180) Eruption on the solar surface; visible during total solar eclipses.

proper motion (190) The rate at which a star moves across the sky; measured in seconds of arc per year.

protein (654) Complex molecule composed of amino acid units.

proton (141) A positively charged atomic particle contained in the nucleus of atoms; the nucleus of a hydrogen atom.

proton–proton chain (170) A series of three nuclear reactions that build a helium atom by adding together protons; the main energy source in the sun.

protoplanet (462) Massive object resulting from the coalescence of planetesimals in the solar nebula and destined to become a planet.

protostar (240) A collapsing cloud of gas and dust destined to become a star.

protostellar disk (240) A gas cloud around a forming star flattened by its rotation.

pulsar (335) A source of short, precisely timed radio bursts; believed to be a spinning neutron star.

pulsar wind (313) The flow of high-energy particles that carries most of the energy away from a spinning neutron star.

quantum mechanics (143) The study of the behavior of atoms and atomic particles.

quasar (quasi-stellar object, or QSO) (405) Small, powerful source of energy believed to be the active core of a very distant galaxy.

quasi-periodic oscillation (QPO) (329) A high-speed flickering in the radiation from an accretion disk evidently caused by material spiraling inward.

quintessence (439) The proposed energy of empty space that causes the acceleration of the expanding universe.

radial velocity (V_r) (153) That component of an object's velocity directed away from or toward Earth.

radiant (619) The point in the sky from which meteors in a shower seem to come.

radiation pressure (466) The force exerted on the surface of a body by its absorption of light. Small particles floating in the solar system can be blown outward by the pressure of the sunlight.

radiative zone (170) The region inside a star where energy is carried outward as photons.

radio galaxy (394) A galaxy that is a strong source of radio signals.

radio interferometer (127) Two or more radio telescopes that combine their signals to achieve the resolving power of a larger telescope.

ray (500) Ejecta from a meteorite impact, forming white streamers radiating from some lunar craters.

recombination (426) The stage within a million years of the big bang when the gas became transparent to radiation.

reconnection (181) The process in the sun's atmosphere by which opposing magnetic fields combine and release energy to power solar flares.

red dwarf (197) Cool, low-mass star on the lower main sequence.

redshift (153) The lengthening of the wavelengths of light seen when the source and observer are receding from each other.

reflecting telescope (114) A telescope that uses a concave mirror to focus light into an image.

reflection nebula (220) A nebula produced by starlight reflecting off dust particles in the interstellar medium.

refracting telescope (114) A telescope that forms images by bending (refracting) light with a lens.

regolith (506) A soil made up of crushed rock fragments.

reionization (426) The stage in the early history of the universe when ultraviolet photons from the first stars ionized the gas filling space.

relative age (502) The age of a geological feature referred to other features. For example, relative ages reveal that the lunar maria are younger than the highlands.

relativistic Doppler formula (407) Einstein's description of the redshift or blueshift observed when an object moves through space at a very high velocity.

resolving power (115) The ability of a telescope to reveal fine detail; depends on the diameter of the telescope objective.

resonance (512) The coincidental agreement between two periodic phenomena; commonly applied to agreements between orbital periods, which can make orbits more or less stable.

retrograde motion (64) The apparent backward (westward) motion of planets as seen against the background of stars.

revolution (23) The motion of an object in a closed path about a point outside its volume; Earth revolves around the sun.

rich cluster (384) A cluster containing over 1000 galaxies, mostly elliptical, scattered over a volume about 3 Mpc in diameter.

rift valley (486) A long, straight, deep valley produced by the separation of crustal plates.

ring galaxy (389) A galaxy that resembles a ring around a bright nucleus; believed to be the result of a head-on collision of two galaxies.

RNA (ribonucleic acid) (655) A long carbon-chain molecule that uses the information stored in DNA to manufacture complex molecules necessary to the organism.

Roche limit (562) The minimum distance between a planet and a satellite that holds itself together by its own gravity. If a satellite's orbit brings it within its planet's Roche limit, tidal forces will pull the satellite apart.

Roche lobe (293) In a system with two bodies orbiting each other, the volume of space dominated by the gravitation of one of the bodies.

Roche surface (293) In a system with two bodies orbiting each other, the outer boundary of the volume of space dominated by the gravitation of one of the bodies.

rotation (23) The turning of a body about an axis that passes through its volume; Earth rotates on its axis.

rotation curve (348, 379) A graph of orbital velocity versus radius in the disk of a galaxy.

rotation curve method (379) The procedure for finding the mass of a galaxy from its rotation curve.

RR Lyrae variable star (275) Variable star with a period of 12 to 24 hours; common in some globular clusters.

Sagittarius A* (361) The powerful radio source located at the core of the Milky Way Galaxy.

saros cycle (52) An 18-year $11\frac{1}{3}$-day period after which the pattern of lunar and solar eclipses repeats.

Schmidt-Cassegrain focus (120) The optical design of a reflecting telescope in which a thin correcting lens is placed at the top of a Cassegrain telescope.

Schwarzschild radius (R_s) (323) The radius of the event horizon around a black hole.

scientific model (19) An intellectual concept designed to help you think about a natural process without necessarily being a conjecture of truth.

scientific notation (6) The system of recording very large or very small numbers by using powers of 10.

secondary atmosphere (488) The gases outgassed from a planet's interior; rich in carbon dioxide.

secondary crater (500) A crater formed by the impact of debris ejected from a larger crater.

secondary mirror (120) In a reflecting telescope, the mirror that reflects the light to a point of easy observation.

second of arc (21) An angular measure; each minute of arc is divided into 60 seconds of arc.

seeing (115) Atmospheric conditions on a given night. When the atmosphere is unsteady, producing blurred images, the seeing is said to be poor.

seismic wave (482) A mechanical vibration that travels through Earth; usually caused by an earthquake.

seismograph (482) An instrument that records seismic waves.

selection effect (622) An influence on the probability that certain phenomena will be detected or selected, which can alter the outcome of a survey.

self-sustaining star formation (359) The process by which the birth of stars compresses the surrounding gas clouds and triggers the formation of more stars; proposed to explain spiral arms.

semimajor axis, *a* (76) Half of the longest axis of an ellipse.

SETI (668) Search for Extra-Terrestrial Intelligence.

Seyfert galaxy (396) An otherwise normal spiral galaxy with an unusually bright, small core that fluctuates in brightness; believed to indicate the core is erupting.

Shapley–Curtis Debate (371) The 1920 debate between Harlow Shapley and Heber Curtis over the nature of the spiral nebulae.

shear (*S*) wave (482) A mechanical wave that travels through Earth's interior by the vibration of particles perpendicular to the direction of wave travel.

shepherd satellite (577) A satellite that, by its gravitational field, confines particles to a planetary ring.

shield volcano (528) Wide, low-profile volcanic cone produced by highly liquid lava.

shock wave (239) A sudden change in pressure that travels as an intense sound wave.

sidereal drive (121) The motor and gears on a telescope that turn it westward to keep it pointed at a star.

sidereal period (41) The period of rotation or revolution of an astronomical body relative to the stars.

singularity (323) The object of zero radius into which the matter in a black hole is believed to fall.

sinuous rille (499) A narrow, winding valley on the moon caused by ancient lava flows along narrow channels.

small-angle formula (54) The mathematical formula that relates an object's linear diameter and distance to its angular diameter.

smooth plain (515) Apparently young plain on Mercury formed by lava flows at or soon after the formation of the Caloris Basin.

soft gamma-ray repeater (SGR) (332) An object that produces repeated bursts of lower-energy gamma rays; thought to be produced by magnetars.

solar constant (182) A measure of the energy output of the sun; the total solar energy striking 1 m^2 just above Earth's atmosphere in 1 second.

solar eclipse (45) The event that occurs when the moon passes directly between Earth and the sun, blocking your view of the sun.

solar nebula hypothesis (452) The theory that the planets formed from the same cloud of gas and dust that formed the sun.

solar system (6) The sun and the nonluminous objects that orbit it, including the planets, comets, and asteroids.

solar wind (166) Rapidly moving atoms and ions that escape from the solar corona and blow outward through the solar system.

south celestial pole (20) The point of the celestial sphere directly above Earth's South Pole.

south point (20) The point on the horizon directly above the south celestial pole; exactly south.

special relativity (101) The first of Einstein's theories of relativity, which dealt with uniform motion.

spectral class or type (149) A star's position in the temperature classification system O, B, A F, G, K, and M. Based on the appearance of the star's spectrum.

spectral sequence (149) The arrangement of spectral classes (O, B, A, F, G, K, M) ranging from hot to cool.

spectrograph (123) A device that separates light by wavelength to produce a spectrum.

spectroscopic binary system (202) A star system in which the stars are too close together to be visible separately. You see a single point of light, and only by taking a spectrum can you determine that there are two stars.

spectroscopic parallax (199) The method of determining a star's distance by comparing its apparent magnitude with its absolute magnitude, as estimated from its spectrum.

spherical component (344) The part of the galaxy including all matter in a spherical distribution around the center (the halo and nuclear bulge).

spicule (165) Small, flamelike projection in the chromosphere of the sun

spiral arm (9) Long, spiral pattern of bright stars, star clusters, gas, and dust that extends from the center to the edge of the disk of spiral galaxies.

spiral galaxy (374) A galaxy with an obvious disk component containing gas; dust; hot, bright stars; and spiral arms.

spiral nebula (371) Nebulous object with a spiral appearance observed in early telescopes; later recognized as a spiral galaxy.

spiral tracer (355) Object used to map the spiral arms (e.g., O and B associations, open clusters, clouds of ionized hydrogen, and some types of variable stars).

sporadic meteor (620) A meteor not part of a meteor shower.

spring tide (98) Ocean tide of high amplitude that occurs at full and new moon.

standard candle (373) Object of known brightness which astronomers use to find distance—for example, Cepheid variable stars and supernovae.

star (6) A celestial object composed of gas held together by its own gravity and supported by nuclear fusion occurring in its interior.

starburst galaxy (386) A bright blue galaxy in which many new stars are forming, believed to be caused by collisions between galaxies.

star-formation pillar (243) The column of gas produced when a dense core of gas protects the nebula behind it from the energy of a nearby hot star that is evaporating and driving away a star-forming nebula.

steady state theory (425) The theory (now generally abandoned) that the universe does not evolve.

stellar model (261) A table of numbers representing the conditions in various layers within a star.

stellar parallax (*p*) (189) A measure of stellar distance. (*See* parallax.)

stony-iron meteorite (624) A meteorite that is a mixture of stone and iron.

stony meteorite (622) A meteorite composed of silicate (rocky) material.

stromatolite (659) A layered fossil formation caused by ancient mats of algae or bacteria that build up mineral deposits season after season.

strong force (169) One of the four forces of nature; the strong force binds protons and neutrons together in atomic nuclei.

subduction zone (486) A region of a planetary crust where a tectonic plate slides downward.

subsolar point (524) The point on a planet that is directly below the sun.

summer solstice (26) The point on the celestial sphere where the sun is at its most northerly point; also, the time when the sun passes this point, about June 22, and summer begins in the northern hemisphere.

sunspot (162) Relatively dark spot on the sun that contains intense magnetic fields.

supercluster (440) A cluster of galaxy clusters.

supergiant (197) Exceptionally luminous star 10 to 1000 times the sun's diameter.

supergranule (164) A large granule on the sun's surface including many smaller granules.

superluminal expansion (410) The apparent expansion of parts of a quasar at speeds greater than the speed of light.

supernova (284) A new star appearing in Earth's sky and lasting for a year or so before fading. Caused by the violent explosion of a star.

supernova remnant (301) The expanding shell of gas marking the site of a supernova explosion.

supernova (type I) (299) The explosion of a star, believed caused by the transfer of matter to a white dwarf.

supernova (type II) (299) The explosion of a star, believed caused by the collapse of a massive star.

synchrotron radiation (301) Radiation emitted when high-speed electrons move through a magnetic field.

synodic period (41) The period of rotation or revolution of a celestial body with respect to the sun.

T association (242) A large, loosely bound group of T Tauri stars.

T dwarf (151) A very-low-mass star at the bottom end of the main sequence with a cool surface and a low luminosity.

temperature (138) A measure of the velocity of random motions among the atoms or molecules in a material.

terminator (498) The dividing line between daylight and darkness on a planet or moon.

terrestrial planet (454) Earthlike planet—small, dense, rocky.

theory (77) A system of assumptions and principles applicable to a wide range of phenomena that have been repeatedly verified.

thermal energy (139) The energy stored in an object as agitation among its atoms and molecules.

thermal pulse (287) Periodic eruptions in the helium fusion shell in an aging giant star; thought to aid in ejecting the surface layers of the stars to form planetary nebulae.

tidal coupling (498) The locking of the rotation of a body to its revolution around another body.

tidal heating (567) The heating of a planet or satellite because of friction caused by tides.

tidal tail (388) A long strand of gas, dust, and stars drawn out of a galaxy interacting gravitationally with another galaxy.

time dilation (325) The slowing of moving clocks or clocks in strong gravitational fields.

total eclipse (42, 45) A solar eclipse in which the moon completely covers the bright surface of the sun; a lunar eclipse in which the moon completely enters Earth's dark shadow.

transition (147) The movement of an electron from one atomic orbit to another.

transition region (164) The layer in the solar atmosphere between the chromosphere and the corona.

triple alpha process (270) The nuclear fusion process that combines three helium nuclei (alpha particles) to make one carbon nucleus.

Trojan asteroid (632) Small, rocky body caught in Jupiter's orbit at the Lagrangian points, 60° ahead of and behind the planet.

T Tauri star (244) Young star surrounded by gas and dust, believed to be contracting toward the main sequence.

turnoff point (276) The point in an H–R diagram where a cluster's stars turn off the main sequence and move toward the red giant region, revealing the approximate age of the cluster.

21-cm radiation (224) Radio emission produced by cold, low-density hydrogen in interstellar space.

ultraluminous infrared galaxy (386) A highly luminous galaxy so filled with dust that most of its energy escapes as infrared photons emitted by warmed dust.

ultraviolet radiation (112) Electromagnetic radiation with wavelengths shorter than visible light but longer than X rays.

umbra (42) The region of a shadow that is totally shaded.

uncompressed density (460) The density a planet would have if its gravity did not compress it.

unified model (402) The attempt to explain the different kinds of active galaxies and quasars by a single model.

uniform circular motion (60) The classical belief that the perfect heavens could move only by the combination of constant motion along circular orbits.

valley networks (540) Dry drainage channels resembling streambeds found on Mars.

Van Allen belt (485) One of the radiation belts of high-energy particles trapped in Earth's magnetosphere.

variable star (275) A star whose brightness changes periodically.

velocity (89) A rate of travel that specifies both speed and direction.

velocity dispersion method (380) A method of finding a galaxy's mass by observing the range of velocities within the galaxy.

vernal equinox (26) The place on the celestial sphere where the sun crosses the celestial equator moving northward; also, the time of year when the sun crosses this point, about March 21, and spring begins in the northern hemisphere.

vesicular basalt (505) A porous rock formed by solidified lava with trapped bubbles.

violent motion (87) In Aristotelian physics, motion other than natural motion. (See *natural motion.*)

visual binary system (202) A binary star system in which the two stars are separately visible in the telescope.

water hole (668) The interval of the radio spectrum between the 21-cm hydrogen radiation and the 18-cm OH radiation; likely wavelengths to use in the search for extraterrestrial life.

wavelength (110) The distance between successive peaks or troughs of a wave; usually represented by λ.

wavelength of maximum intensity (λ_{max}) (139) The wavelength at which a perfect radiator emits the maximum amount of energy; depends only on the object's temperature.

weak force (169) One of the four forces of nature; the weak force is responsible for some forms of radioactive decay.

west point (20) The point on the western horizon exactly halfway between the north point and the south point; exactly west.

white dwarf (197) The remains of a dying star that has collapsed to the size of Earth and is slowly cooling off; at the lower left of the H–R diagram.

Widmanstätten pattern (622) Bands in iron meteorites due to large crystals of nickel–iron alloys.

winter solstice (26) The point on the celestial sphere where the sun is farthest south; also, the time of year when the sun passes this point, about December 22, and winter begins in the northern hemisphere.

X-ray burster (329) An object that produces occasional X-ray flares; believed to be caused by mass transfer in a close binary star system.

Zeeman effect (175) The splitting of spectral lines into multiple components when the atoms are in a magnetic field.

zenith (20) The point on the sky directly overhead.

zero-age main sequence (ZAMS) (265) The locus in the H–R diagram where stars first reach stability as hydrogen-burning stars.

zodiac (28) The band around the sky centered on the ecliptic within which the planets move.

zone (561) One of the yellow-white regions that circle Jupiter parallel to its equator; believed to be areas of rising gas.

Answers to Even-Numbered Problems

Chapter 1:
2. 3475 km; **4.** 1.05×10^8 km; **6.** about 1.2 seconds; **8.** 75,000 years; **10.** about 27

Chapter 2:
2. 4; **4.** 2800; **6.** A is brighter than B by a factor of 170; **8.** 66.5°; 113.5°

Chapter 3:
2. a) full; b) first quarter; c) waxing gibbous; d) waxing crescent; **4.** 29.5 days later on about March 30; 27.3 days later on about March 24; **6.** 6850 arc seconds or about 1.9°; **8.** a) The moon won't be full until Oct. 17; b) The moon will no longer be near the node of its orbit; **10.** August 12, 2026 [July 10, 1972 + 3 × ($6585\frac{1}{3}$ days)]. In order to get Aug. 12 instead of Aug. 11, you must take into account the number of leap days in the interval.

Chapter 4:
2. Retrograde motion: Jupiter, Saturn, Uranus, Neptune, and Pluto; Never seen as crescents: Jupiter, Saturn, Uranus, Neptune, and Pluto; **4.** Mars, about 18 seconds of arc; the maximum angular diameter of Jupiter is 50 seconds of arc; **6.** $\sqrt{27} = 5.2$ years

Chapter 5:
2. The force of gravity on the moon is about $\frac{1}{6}$ the force of gravity on Earth; **4.** 7350 m/s; **6.** 5070 s (1 hr and 25 min); **8.** The cannonball would move in an elliptical orbit with Earth's center at one focus of the ellipse; **10.** 6320 s (1 hr and 45 min)

Chapter 6:
2. 3m; **4.** Either Keck telescope has a light-gathering power that is 1.56 million times greater than the human eye; **6.** No, his resolving power should have been about 5.8 seconds of arc at best; **8.** 0.013 m (1.3 cm or about 0.5 inches); **10.** about 50 cm (From 400 km above, a human is about 0.25 seconds of arc from shoulder to shoulder.)

Chapter 7:
2. 150 nm; **4.** by a factor of 16; **6.** 250 nm; **8.** a) B; b) F; c) M; d) K; **10.** about 0.58 nm

Chapter 8:
2. 730 km; **4.** 9×10^{16} J; **6.** 0.222 kg; **8.** about 3.5 times; **10.** 400,000 years

Chapter 9:
2. 63 pc; absolute magnitude is 2; **4.** about B7; **6.** about 1580 solar luminosities; **8.** 160 pc; **10.** a, c, c, d (use Figure 9-13 to determine the absolute magnitudes); **12.** 3.69 days or 0.010 years; about 1.2 solar masses; about 0.67 and 0.53 solar masses; **14.** 1.38×10^6 km, about the size of the sun

Chapter 10:
2. 60,000 nm; **4.** 0.0001; **6.** 1.5×10^6 years; **8.** 4.2×10^{35} kg or 210,000 solar masses (*Note:* Each hydrogen molecule contains two H atoms.)

Chapter 11:
2. 24.5 km/s; **4.** 2.98 km/s; **6.** 9.46×10^{10} s (about 3000 years); **8.** There are four ^1H nuclei in the figure. They are used in a series of reactions to build up nuclei from ^{12}C to ^{15}N. In the last reaction (with the ^{15}N) a ^{12}C and a ^4He are produced. Because the ^{12}C can be used in the next cycle, the net is four ^1H nuclei in with one ^4He nucleus out; **10.** 7.9×10^{33} kg

Chapter 12:
2. about 9.8×10^6 years for a 16-solar-mass star; about 5.7×10^5 years for a 50-solar-mass star; **4.** about 1×10^6 times less than present or about 1.4×10^{-6} g/cm^3; **6.** 2.4×10^{-9} or 1/420,000,000; **8.** about 3 pc; **10.** 3.04 minutes early after 1 year; 30.4 minutes early after 10 years

Chapter 13:
2. about 1.8 ly; **4.** about 16,000 years old; **6.** about 940 years ago (approximately 1060 AD); **8.** 2400 pc

Chapter 14:
2. 7.1×10^{25} J/s or about 0.19 solar luminosity; **4.** 820 km/s (assuming mass is one solar mass); **6.** about 11 seconds of arc; **8.** about 490 seconds

Chapter 15:
2. about 16 percent; **4.** 3.8×10^6 years; **6.** overestimate by a factor of 1.58; **8.** about 21 kpc; **10.** 7.8×10^{10} solar masses; **12.** 1500 K

Chapter 16:
2. 2.58 Mpc; **4.** 131 km/s; **6.** 28.6 Mpc; **8.** 1.64×10^8 yr; **10.** 4.49×10^{41} kg

Chapter 17:
2. 7.8×10^6 years; **4.** 0.024 pc; **6.** −28.5; **8.** about 29,800 km/s; **10.** 0.16

Chapter 18:
2. 57; 17.5 billion years; 11.7 billion years; **4.** 1.6×10^{-30} gm/cm^3; **6.** 76 km/Mpc; **8.** 16.6 billion years; 11 billion years; the universe could be older

Chapter 19:
2. It will look $206{,}265^2 = 4.3 \times 10^{10}$ times fainter, which is about 26.6 magnitudes fainter; 22.6 mag; **4.** about 3.3 times the half-life, or 4.3 billion years; **6.** large amounts of methane and water ices; **8.** about 1300

Chapter 20:
2. about 17 percent; **4.** 81×10^6 yr; they have been subducted; **6.** 0.22 percent

Chapter 21:
2. The rate at which an object radiates energy is proportional to its surface area, which is proportional to its radius squared (r^2). However, the energy an object has stored as heat is proportional to its mass and hence to its volume, and that is proportional to its radius cubed (r^3). So the cooling time will be proportional to the amount of heat divided by the rate of cooling, which is the same as the radius cubed divided by the radius squared (r^3/r^2). That shows that the cooling time is proportional to the radius (r); and that means that the bigger an object is, the longer it takes to cool; **4.** No. Their angular diameter would be only 0.5 second of arc. They would be visible in photos taken from orbit around the moon; **6.** 0.5 second of arc (assuming an astronaut seen from above is 0.5 meter in diameter); no; **8.** 10.0016 cm; **10.** Mercury, $V_e = 4250$ m/s; moon, $V_e = 2380$ m/s; Earth, $V_e = 11{,}200$ m/s

Chapter 22:
2. 33,400 km (39,500 km from the center of Venus); **4.** 61 seconds of arc; **6.** 260 km; **8.** 82 seconds of arc

Chapter 23:
2. 35 Earth days; **4.** 4.4°; **6.** about 0.056 nm; **8.** 5.2 m/s

Chapter 24:
2. 42 seconds of arc; **4.** 256 m/s; **6.** 8.8 s; yes; **8.** 12.3 km/s; **10.** 1.04×10^{26} kg (17.2 Earth masses)

Chapter 25:
2. one billion; **4.** 79 km; **6.** 0.18 km/s; **8.** 3.28 AU and 2.5 AU; **10.** 9 million km (0.06 AU); too small; **12.** 33 Earth masses

Chapter 26:
2. 8.9 cm; 0.67 mm; **4.** about 1.3 solar masses; **6.** 380 km; **8.** pessimistic, 2×10^{-5}; optimistic, 10^7

Credits

This page constitutes an extension of the copyright page. We have made every effort to trace the ownership of all copyrighted material and to secure permission from copyright holders. In the event of any question arising as to the use of any material, we will be pleased to make the necessary corrections in future printings. Thanks are due to the following authors, publishers, and agents for permission to use the material indicated.

Chapter 1. 2–3: T. Rector, University of Alaska, and WIYN/NOAO/AURA/NSF **4, left:** M. Seeds **4, right:** USGS **5, left:** NASA infrared photograph **5, right:** NASA **6:** NOAO **6:** NASA **8:** NOAO **9, bottom:** Anglo-Australian Telescope Board **9, top:** Detail of galaxy map from M. Seldner, B. L. Siebers, E. J. Groth, and P. J. E. Peebles, *Astronomical Journal 82* (1977) **10:** Bill Schoening/NOAO/AURA/NSF

Chapter 2. 12–13: Courtesy of Kris Koenig **33:** USPS **15, top:** From Duncan Bradford, *Wonders of the Heavens,* Boston: John B. Russell, 1837 **16:** William Hartman **18:** Roger Ressmeyer/Corbis **20:** AURA/NOAO/NSF **23:** Royalty-Free/Corbis **27:** NASA

Chapter 3. 36–37: NSO/AURA/NSF **38, top:** Yerkes Observatory **38, bottom:** T. Scott Smith **39:** UCO/Lick Observatory **43:** © 1982 by Dr. Jack B. Marling **44:** Celestron International **47:** Daniel Good **48, left:** Daniel Good **48, right:** NOAO **49:** Daniel Good **49, left:** UCO/Lick Observatory **54, top:** Laurence Marschall

Chapter 4. 59, top: Jamie Backman **59, bottom:** Benelux Press/Index Stock Imagery **60:** Courtesy NPS Chaco Culture National Historical Park **64:** Yerkes Observatory **64:** From *Cosmographica* by Peter Apian (1539) **68:** NOAO and Nigel Sharp **69:** Grundy Observatory **71:** Yerkes Observatory **74:** The Granger Collection, New York **77:** From the collection of John Coolidge III

Chapter 5. 86: NASA/JSC **88:** NASA **91:** ESA/STScI and NASA **95:** NASA **99:** Nigel Sharp/NOAO/AURA/NSF **106:** Larry Mulvehill/The Image Works **107:** NASA/JSC

Chapter 6. 108: Robert Frost, "The Star-Splitter," *New Hampshire: A Poem with Notes and Grace Notes* (New York: Henry Holt and Co., 1923), pp. 27–30. **108–109:** NOAO/AURA/NSF **110:** NOAO/AURA/NSF **116:** NASA **117, top:** Courtesy of William Keel **117, center:** NOAA **117, bottom:** ESO **119, bottom:** NOAO/AURA/NSF **119, left:** NOAO/AURA/NSF **119, right:** ESO **120:** AURA/NOAO/NSF **121, left:** W. M. Keck Observatory **121, right:** Paul Kalas **123, left and bottom:** C. Hawk, B. Savage, N. A. Sharp, NOAO/WIYN/NSF **123, right:** NOAO/WIYN/NSF **124:** NASA/CXC/Rutgers/J. Hughes **126, bottom:** NRAO **126, top:** Courtesy Seth Shostak/SETI Institute **127:** NRAO/AUI/NSF **127, bottom:** NRAO/AUI **127, center:** David Parker/SPL/Photo Researchers, Inc. **129, left:** William Keel/IRAF **129, center:** NASA **129, bottom:** Kris Koenig/Coast Learning Systems **130, 131:** NASA **134, left:** ESO **134, right:** NASA/CXC/PSU/S. Park

Chapter 7. 136, 137: ESO **138:** NASA Hubble Heritage Team/STScI/AURA **147:** AURA/NOAO/NSF **150:** AURA/NOAO/NSF **152:** Adapted from Thomas R. Geballe, Gemini Observatory, from a graph that originally appeared in *Sky and Telescope Magazine,* February 2005, p. 37. **158:** T. Rector, University of Alaska and WIYN/NURO/AURA/NSF

Chapter 8. 160–161: NASA/SOHO **163:** Daniel Good **164:** P. N. Brant, G. Scharmer, G. W. Simpson, Swedish Vacuum Solar Telescope, La Palma **165, left:** NOAA/SEL/USAF **165, right:** NOAO/NSO **166:** SOHO/ESA/NASA **167:** Stanford-Lockheed Institute for Space Research, Palo Alto, CA and NASA GSFC **172, left:** Brookhaven National Laboratory **172, right:** Courtesy SNO **174, background:** Royal Swedish Academy of Science **174, right, and 175, top left:** NASA **175, top right:** J. Harvey/NSO and HAO/NCAR **175, center:** M. Seeds **175, bottom left:** SOHO/EIT, ESA and NASA **175, bottom right:** TRACE/NASA **179:** K. Strassmeier, Vienna, AURA/NOAO /NSF **180, left:** TRACE/NASA **180, right:** Sacramento Peak Observatory **180, bottom:** SOHO, EIT, ESA and NASA **181, top left:** NASA **181, center left:** SOHO/MDI, ESA and NASA **181, background:** © Jan Curtis **181, bottom left:** SOHO/LASCO, ESA, and NASA **181, bottom right:** Yohkoh/ISAS/NASA **184, top :** NOAO and Daniel Good **184, bottom:** NASA/SOHO

Chapter 9. 186–187: Mark McCaughrean and Morten Andersen of the Astrophysical Institute, Potsdam, and the ESO **188:** 2MASS Sky Survey and IPAC **194:** Krafft/Photo Researchers, Inc. **196:** M. Seeds **196:** Heidi Schweiker/NAOA/AURA/NSF **201:** USGS **202:** UCO/Lick Observatory **204:** The Observations of the Carnegie Institution of Washington **207:** M. Seeds **211, bottom left, and 213:** NASA

Chapter 10. 216–217: NASA, ESA, and the Hubble Heritage Team, AURA/STScI **218:** Daniel Good **219 and 220, top:** ESO **220, bottom:** Anglo-Australian Telescope Board **221:** Caltech **221, top center:** NASA **221, top right:** Daniel Good **221, center:** Anglo-Australian Telescope Board **221, bottom left:** Hubble Heritage Team, NASA **222:** The Observations of the Carnegie Institution of Washington **223:** Anglo-Australian Telescope Board **226:** J. Dickey, UMn, F. Lockman, NRAO, Sky View **228:** NASA/IPAC, Courtesy Deborah Levine **230:** M. Seeds **231, top:** Caltech **231, center:** Courtesy Richard E. White; image rendered by Duncan Chesley of American Image, Inc. **231, bottom left:** NASA and The Hubble Heritage Team, AURA/STScI **231, bottom right:** R. Chris Smith, CTIO and Y.-H. Chu, UICU **232:** AURA/NOAO/NSF and Nigel Sharp **233:** NASA and The Hubble Heritage Team, AURA/STScI

Chapter 11. 236–237: ESO **238, left:** ESO **238, right:** W. Brander, JPL/IPAC, E. K. Grebel, University of Washington, Y. Chu, University of Illinois, and NASA **240:** NASA/JPL-Caltech/V. Gorjian and NOAO **241:** ESO **243, top:** C. Burrows, STScI & ESA, WFPC 2 Investigation Definition Team, NASA **243, bottom:** N. Walborn and J. Maiz-Apellaniz, STScI, R. Barba, La Plata Observatory, and NASA **244, background, and 244, bottom right:** NASA **245, top left:** Richard Mundt, Calar Alto Observatory **245, top, and 245, bottom left:** NASA **245, bottom right:** AURA/NOAO/NSF **246, left:** NASA, ESA, The Heritage Hubble Team **246, right:** ESO **252, center:** NASA **252, bottom:** NASA/CXC/SAO **253, left:** ESO **253, top right:** Anglo-Australian Telescope Board **253, center:** Dan Gezari, Dana Backman, and Mike Werner **253, bottom:** NASA **256:** STScI/NASA

Chapter 12. 258–259: T. A. Rector and B. A. Wolpa, NRAO/NOAO/AUI/NSF **260:** NOAO/AURA/NSF **262:** The Boeing Company **264, left:** NASA/CXC/SAO/HST **264, bottom:** ISO, and courtesy ESA **264, inset:** Jon Morse and J. Hester **265, left:** Gemini Observatory/University of Hawaii Institute for Astronomy/Michael Liu/NSF **265, right:** STSci and NASA **276, right:** Anglo-Australian Telescope Board **276, background:** AURA/NOAO/NSF **277, top:** Anglo-Australian Telescope Board **277, center:** Caltech **277, bottom:** Nigel Sharp, Mark Hanna/AURA/NOAO/NSF **282:** NASA/Walborn, Maiz-Apellaniz, and Barba

Chapter 13. 284–285: NASA/ESA/HEIC/Hubble Heritage Team **286, left:** 2MASS **286, center:** NASA/CXC/SAO **286, right:** NASA/CXC/Rutgers/J. Warren et al., STScI/University of Illinois **290:** Hubble Heritage Team, STScI/AURA/NASA **291:** NASA **292, left:** Hubble Heritage Team, STScI/AURA/NASA **292, center and right** NASA/IPAC courtesy of Deborah Levine **295, top:** © UC Regents **295, bottom:** M. Shara and R. Williams, STScI **295, bottom:** R. Gilmozzi, ESO, and NASA **297, left:** Gemini Observatory/NSF/C. Aspin **297, right:** NASA **299:** Adam Burrows/University of Arizona, Department of Astronomy **300:** High-Z Supernova Search Team, HST, NASA **301:** ESO **302, top left:** Caltech **302, top right:** NASA/SAO/CXC **302, bottom left:** 2MASS **302, bottom right:** NRAO/AUI/NSF **303, left:** Anglo-Australian Telescope Board **303, bottom:** NASA, Challis, Kirshner and Sugerman **303, right:** Christopher Burrows, ESA/STcI, NASA **306, top:** NASA/Hubble Heritage Team/STScI/AURA **306, bottom:** NASA/CAC/SAO/CSIRO/ATNF/ATCA

Chapter 14. 308–309: NASA/CXC/SAO **310:** NASA/SAO/CXC/P. Slane et al. **311:** M. Seeds **312:** Caltech **313, left:** NASA/CXC/SAO/University of Massachusetts **313, right:** F. Lu/McGill **313, bottom:** V. Kaspi **315:** NASA **316, top:** AURA/NOAO/KPNO **316, bottom:** Figure 2 from F. R. Harnden, Jr., in *The Astrophysical Journal,* 283 (August 1, 1984): 279–285. Copyright 1984 by The American Astronomical Society. Reprinted with permission of F. R. Harnden, Jr., and the AAS.

317: NASA and F. M. Walter **318:** John Rowe Animations **319:** J. Trumper, Max-Planck Institute **320, bottom:** NASA/CXC/ASTRON/B. Stappers et al. **320, bottom:** AAO/J. Bland-Hawthorn & H. Jones **320, top:** SOHO/MDI **323:** John P. Hughes, Rutgers University **330:** Ivan King and NASA/ESA **331, bottom:** NASA **331, top:** NASA/CXC/SAO **332:** NASA/CXC/M. Weiss. Found at: http://www.chandra.harvard.edu/press/04_releases/press_010504.html **333, left:** ESO and NASA **333, right:** NASA/Skyworks Digital **335:** NASA/McGill, V. Kaspi et al. **336:** CXC/M. Weiss

Chapter 15. 338–339: ESO and Nico Housen **340:** AURA/NOAO/NSF **342, center:** AURA/NOAO/NSF **342, bottom:** Doug Williams, N. A. Sharp/NOAO/AURA/NSF **345:** Daniel Good **346, top left:** 2MASS **346, bottom left:** DIRBE courtesy Henry Freudenreich **346, top right:** Todd Boroson/NOAO/AURA/NSF **346, bottom right:** ESO **351:** NASA/CXC/PSU/S. Park et al. **354:** Kris Koenig/Coast Learning Systems **355:** NASA/Heritage Hubble Team **357, center:** Painting by M. Seeds, based on a study by G. De Vaucouleurs and W. D. Pence **357, bottom:** ESO **358:** NASA/Hubble Heritage Team **359:** Anglo-Australian Telescope Board **361:** NASA/Hubble Heritage Team **362, top right:** NRAO/AUI/NSF **362, bottom left:** 2MASS **362, center:** N. Killeen and Kwok-Yung Lo **363, center:** NASA/CXC/MIT/F. K. Baganoff et al. **363, bottom left:** ESO **366:** NASA/Hubble Heritage Team/STScI/AURA

Chapter 16. 368–369: NASA/Hubble Heritage Team/STScI/AURA **370, left:** Hubble Heritage Team/AURA/StScI/NASA **370, top right:** ESO **370, bottom right:** Hubble Heritage Team/Aura/StScI/NASA **371:** R. Williams and the Hubble Deep Field Team /STScI/NASA **372:** M. Seeds **374, bottom:** Anglo-Australian Telescope Board **374, top right:** AURA/NOAO/NSF **374, top right:** Anglo-Australian Telescope Board **375, top:** NASA/Hubble Heritage Team **375, center left:** Rudolph E. Schild **375, center right:** G. J. Jacoby, M. J. Pierce/AURA/NOAO/NSF **375, bottom left:** AURA/NOAO/

NSF **375, bottom right:** © R. J. Dufour, Rice University **376:** J. Trauger, JPL; Wendy Freedman, Observatories of the Carnegie Institution of Washington **376:** NASA **377:** NASA/ESA/GOODS Team and M. Giavgalisco./STScI **379:** Hubble Heritage Team, STScI/AURA/NASA **380:** Courtesy Vera Rubin **381:** Bill Schoening/AURA/NOAO/NSF **382, left:** L. Thompson/NOAO **382, right:** Simon D. M. White, Ulrich G. Briel, and J. Patrick Henry **383:** W. N. Colley and E. Turner, Princeton, and J. A. Tyson Bell Labs, and NASA **384:** NOAO/AURA/NSF **386:** Courtesy Nicolas Martin and Rodrigo Ibata, Observatoire de Strasbourg-France **387, left:** ESA/NASA, P. Anders **387, center and right:** NASA/ESA/Hubble Heritage Team/AURA/STScI **388, top and bottom right:** NOAO **388, bottom left:** NOAO **389, top right:** Brad Whitmore, STScI/NASA **389, center right:** NASA/CXC/SAO/G. Fabbiano et al. **389, center left:** Francois Schweizer, Carnegie Institution of Washington, Brad Whitmore, STScI **389, center left:** NASA/Heritage Hubble Team **389, bottom left:** Michael J. West **389, right center:** Robin Braun, NRAO /AUI/NSF **389, bottom right:** NASA/Hubble Heritage Team/ESA/AURA/STScI **390, left:** NASA, W. Keel, F. Owen, M. Ledlow, and D. Wang **390, right:** NASA, H.-J. Yan, R. Windhorst, and S. Cohen, Arizona State University **392, top:** P. Massey, Lowell, N. King, STScI, S. Holmes, Charleston, G. Jacoby, WIYN/AURA/NSF **392, bottom:** WIYN/NOAO/NSF

Chapter 17. 394–395: Courtesy NRAO/AUI **396, left:** Anglo-Australian Telescope Board **396, center:** Hubble Heritage Team, AURA/STScI, NASA **396, right:** NASA, Andrew S. Wilson; Patrick L. Shopbell, Chris Simpson, Thaisa Storchi-Bergman, F. K. B. Barbosa and Martin J. Ward **397, top:** John W. Mackenty, Institute for Astronomy, University of Hawaii **397, center:** NRAO/AUI/NSF **398, top:** NASA/CXC/NRAO **398, center:** E. Schreier, STScI, NASA **398, center right :** Used with permission, Fosbury R. A. E., Vernet, J., Vilar-Martin, M., Cohen, M. H., Ogle, P. M., Tran, H. D, & Hook, R. N. 1998, Optical continuum structure of Cygnus A. "KNAW colloquium on: The most distant radio galaxies," Amsterdam, 15–17 October 1997, Roettgering H, Best, P., and Lehnert, M., eds, Reidel, astro-ph/9803310 **398, bottom left :** X-ray: NASA/CXC/M. Karovska, et al., Radio: NRAO/VLA/Schiminovich, et al., Radio: NRAO/VLA/J. Condon et al., Optical: Digitized Sky Survey U. K. Schmidt Image/STScI **398, center left:** NRAO/AUI/NSF **399, top right:** NRAO/AUI/NSF and C. O'Dea and F. Owen **399, center right:** NRAO/AUI/NSF and C. O'Dea, F. Owen, M. Inoue, and J. Eilek **399, bottom:** Adapted from a diagram by Ann Field, NASA, STSci **400:** NASA **401, left:** AURA/NOAO/NSF **401, top center:** NASA/STScI **401, center:** NASA/CXC/W. Forman et al. **401, right:** NRAO and J. Biretta **401, left:** L. Ferrarese, Johns Hopkins University, and NASA **401, right:** The Hubble Heritage Team/NASA **402, top:** Roeland, P. van der Marel, STScI, Frank C. van den Bosch, University of Washington, and NASA **402, bottom left:** NASA/CXC/MIT/UCSB/P/Ogle et al. **402, bottom right:** NASA/STScI/A. Capetti et al. **404, left:** AURA/NOAO /NSFA; NASA/CXC/SAO **404, right:** NASA/CXC/SAO **405:** ESA **406, top:** C. Steidel, Caltech, NASA **406, bottom:** Courtesy Maarten Schmidt **408, top left:** George Rhee, NASA, STScI **408, top right:** Chris Impy, University of Arizona and NASA **408, bottom:** Image by W. Keel from data in the NASA/ESA Hubble Space Telescope Archive, originally obtained with J. Westphal as Principal Investigator **409:** AURA/NOAO/NSF **410, top:** NRAO/AUI/NSF **410, bottom:** J. Bahcall, Institute for Advanced Study, Mike Disney, University of Wales, NASA **414, top, center, and bottom:** NRAO/AUI

Chapter 18. 416–417: Courtesy of MPA and Joerg Colberg **418:** R. Williams, STScI HDF-South Team, NASA **419:** Courtesy of Janet Seeds **423:** Ken Lanzetta and Amos Yahil, Stony Brook and NASA **424, left:** AT&T Archives **424, right:** NASA **427:** K. Lanzetta, SUNY, A. Schaller for STScI, and NASA **428:** NASA/WMAP Science Team **431:** NASA, Benites, Broadhurst, Ford, Clampin, Hartig, Illingworth, ACS Science Team and ESA **432:** Hubble Heritage Team/STScI/AURA/NASA **438:** ESO **438:** Adam Riess, STScI/NASA **440:** NASA/CXC/IoA/S. Allen et al. **441:** Sloan Digital Sky Survey **442:** Adapted from a model by Kauffmann, Colberg, Diaferio, & White: Max-Planck Institute für Astronomie **443, top:** NASA/WMAP Science Team **443, bottom:** Courtesy of the BOOMERANG Collaboration **445, top:** NASA, Benites, Broadhurst, Ford, Clampin, Hartig, Illingworth, ACS Science Team and ESA **445, bottom:** Courtesy of the BOOMERANG Collaboration

Chapter 19. 448–449: NASA **451:** Janet Seeds **454:** John Hopkins University, Applied Physics Laboratory, NASA **455:** Celestron International **456, 457:** UCO/Lick Observatory **457, bottom right:** Grundy Observatory **458, left:** Daniel Good **458, right:** R. Kempton, New England Meteoritical Services **462:** NASA **464:** JPL/NASA **467:** NASA **468, center:** C. R. O'Dell, Rice, NASA **468, top right:** M. McCaughrean, Max Plank Inst. für Astronomie, C. R. O'Dell, Rice, NASA **468, bottom left:** J. Bally, H. Throop, C. R. O'Dell, NASA **469:** D. Padgett, IPAC/Caltech, W. Brandner IPAC, K. Stapelfeldt, JPL/NASA **470, top:** C. Burrows and J. Krist, STScI and NASA **470, center left:** A. Weinberger, E. Becklin, UCLA, G. Schneider, U. of Arizona, and NASA **470 center right:** Joint Astronomy Center **470, bottom left:** NASA **470, bottom right:** NASA, ESA, Kalas, Graham, and Clampin **473:** ESO **475:** NASA

Chapter 20. 476–477: Jean-Bernard Carillet/Getty Images **478:** NASA **478, top right:** UCO/Lick Observatory **484:** From *Virus Ultrastructure: Electron Micrograph Images.* Copyright 1995 by Linda M. Stannard. Reprinted with permission from Linda M. Stannard. Found at: http://web.uct.ac.za/depts/mmi/stannard/adeno.html **485:** Jimmy Westlake **486, left:** William K. Hartmann **486 top center:** Janet Seeds **486, top right and top left;** National Geophysical Data Center **487, bottom:** M. Seeds **491:** Ozone Processing Team/NASA/Goddard Space Flight Center **493, top:** William K. Hartmann **493, bottom:** USGS

Chapter 21. 496–497: JSC/NASA **498:** UCO/Lick Observatory **499:** NASA **500, 501:** NASA **502, left:** NASA **502 center and right:** UCO/Lick Observatory **503:** Phyllis Leber **504:** NASA **505, left:** UCO/Lick Observatory **505, right and bottom, and 506:** NASA **507 :** M. Seeds **508:** NASA/Clementine **509, right:** NASA **509, left:** Don Davis **512, 513, 514, 515:** NASA **516:** Martin Slade/JPL/NASA **518, 520, 521:** NASA

Chapter 22. 522: © Calvin J. Hamilton **523 and 525, top and bottom:** NASA **526, top:** M. Seeds **526:** NASA **526, top:** USGS **527:** NASA **528, bottom left:** USGS **529:** NASA **530:** PhotoDisc/Getty Images **531, left:** NASA **531, right:** © 1992, David P. Anderson, Southern Methodist University **532:** NASA **534, left:** Courtesy of Lowell Observatory **534, right:** USGS **535:** NASA **535:** Philip James, University of Toledo, Steven Lee, University of CO, Boulder, and NASA **536, 537:** NASA **538:** NASA/JPL/Cornell **539:** NASA **540:** NASA/JPL/Malin Space Science Systems **541, right:** NASA/USGS **541, left:** © Calvin J. Hamilton **542:** Malin Space Science Systems/NASA **543:** NASA/JPL/Arizona State University **544:** NASA/JPL/Cornell/USGS **545, top:** NASA **545, bottom left:** Malin Space Science Systems/NASA **545, bottom right:** NASA **546:** NASA/JPL/Malin Space Science Systems **547, left:** Damon Simonelli and Joseph Ververka, Cornell University/NASA **547, right, and 550:** NASA

Chapter 23. 552–553: NASA/JPL **555:** NASA/JPL/Space Science Institute/University of Arizona **557:** NASA/JPL/University of Arizona **559, top:** John P. Clarke, University of Michigan, NASA **559, bottom and 560, top left:** NASA **560, top right:** NASA/JPL/University of Arizona **560, left:** Hughes Aircraft Company **560, bottom:** NASA/JPL **561, bottom:** NASA/JPL/Caltech **562:** NASA/JPL/Space Science Institute **563, bottom, and 564, left:** NASA **564, center:** Mike Skrutskie **564, right:** University of Hawaii **565-569:** NASA **570, top:** NASA/JPL/Bruce A. Goldberg **570, center:** NASA/JPL/John Hopkins University Applied Physics Lab **570, bottom:** NASA/JPL **572:** NASA, ESA, J. Clarke, Boston University and X. Levay, STScI **573:** NASA/STScI **574:** NASA and E. Karkoschka **575, 576, 577:** NASA **577:** NASA/JPL/Science Institute **579, 580** NASA/ESA/JPL/University of Arizona **581, 582:** NASA/JPL/Space Science Institute **585:** NASA/JPL

Chapter 24. : 588–589: NASA/JHUAPL/SwRI **590, 592, 593:** NASA **595:** Lawrence Sromovsky, UW-Madison Space Science and Engineering Center **596:** Courtesy Floyd Herbert, LPL **597:** ESO **599, 601:** NASA **601, center and bottom right; 602, bottom left:** USGS **602, bottom right:** P. Bridges **604:** NASA, L. Sromovsky, and P. Fry, UW-Madison **605:** NASA **605, left:** NASA and Erich Karkoschka **605, right:** U.C. Berkeley/W. M. Keck Observatory **606:** NASA **608:** Courtesy of Lowell Observatory **609:** NASA **610, top:** NASA/McGill/V. Kaspi et al. **610, bottom left:** NASA **610, bottom right:** R. Albrecht, ESA/ESO Space Telescope European Coordinating Facility, NASA **614:** NASA and Heidi Hammel

Chapter 25. 616–617: NASA/JPL **618:** Courtesy Tod Lauer **620:** M. Seeds **621:** NASA/JPL-Caltech/M. Kelley, University of Minnesota **622:** USGS **622, 623:** M. Seeds **623:** Courtesy Dr. Monica Kress **624:** Courtesy of Russell Kemton, New England Meteoritical **625, top:** Hubble Heritage Team/STScI/AURA/NASA **625, bottom; 628, 629:** NASA **632:** Reprinted with permission of MIT Lincoln Laboratory, Lexington, Massachusetts **633:** Courtesy of Gareth Williams, Minor Planet Center **634:** T. Rector, University of Alaska Anchorage, Z. Levy, and L. Frattare, Space Telescope Science Institute and WIYN/NOAO /AURA/NSF **635, top left:** © 1986 Max-Planck Institiute **635, center:** Steven Larson **635:** NASA **636:** Caltech **637, top left:** ESO **637, center and bottom:** NASA **637, top right:** Dean Ketelsen **638, top left:** NASA/JPL-Caltech/UMD **639:** NASA **640:** SOHO **641, left:** NASA and G. Bernstein **641, right:** John Hopkins University Applied Physics Lab./Southwest Research Institute, JHUAPL/SwRI **643, center:** Virgil L. Sharpton, University of Alaska, Fairbanks **647:** NASA **647, bottom:** Courtesy Russell Kempton, New England Meteoritical

Chapter 26. 650–651: Kris Koenig/Coast Learning Systems **653, left:** M. Seeds **653, right:** L. D. Simon **654:** M. Seeds **657, left:** Courtesy J. William Schopf **657, right:** Grundy Observatory **658:** Courtesy Stanley Miller **659, left:** Courtesy Chip Clark, National Museum of Natural History **659, right:** Courtesy Sidney Fox and Randall Grubbs **660, inset:** Courtesy Chip Clark, National Museum of Natural History **660:** Peter Sawyer/National Museum of Natural History **663, left:** D. W. Miller **663, right:** NASA/JPL/Cornell **664:** NASA **669:** Isaac Gary/SETI Institute **671, top:** T. A. Rector, B. A. Wolpa, NRAO /NOAO/AURA/NSF **671, bottom:** ESO

Afterword. 675: NASA

Q

R

S